THE THEORY OF TRANSITION-METAL IONS

THE THEORY OF TRANSITION-METAL IONS

BY

J. S. GRIFFITH

Professor of Chemistry, University of Pennsylvania
Berry-Ramsey Fellow in Mathematics,
King's College, Cambridge

CAMBRIDGE
AT THE UNIVERSITY PRESS
1961

PUBLISHED BY
THE SYNDICS OF THE CAMBRIDGE UNIVERSITY PRESS
Bentley House, 200 Euston Road, London, N.W. 1
American Branch: 32 East 57th Street, New York 22, N.Y.

Printed in Great Britain at the University Press, Cambridge
(Brooke Crutchley, University Printer)

CONTENTS

PREFACE

I have tried to present a unified and deductive introduction to that part of theoretical physics which is becoming known as ligand-field theory. The field of application of this theory is rapidly spreading at the present time and, because of this, it appears more suitable and more helpful to concentrate upon the methods of the theory rather than the details of applications. Hence, although a considerable survey of experimental measurements appears in chs. 10–12, it is by no means exhaustive. The exclusion of almost all reference to rare-earth or actinide ions and to chemical applications—for which the reader is referred to L. E. Orgel's book, *Transition-metal Chemistry* (London: Methuen, 1960)—stems from similar considerations.

There are a number of essential prerequisites to a proper understanding of the theory of the physical properties of ions in compounds. Chs. 1–6, together with §§8.4, 8.6 and 8.7, include those things I deem necessary. Three seem to me especially important: a detailed understanding of the selection rules and other numerical restrictions upon matrix elements implied by the classification of the behaviour of the constituent operators and functions under the elements of symmetry groups; the use of Dirac's equation to derive the spin-orbit coupling and nuclear hyperfine energy; the complex of ideas which has as particular manifestations Kramers's theorem on degeneracy, Wigner's operation of time reversal, and Frobenius and Schur's discussion of the relation between an irreducible representation and its complex conjugate. These matters play a central role in my presentation of the theory.

When deciding the contents of the book it soon became apparent to me that there were many important propositions which workers in the field regarded as 'obvious' and used in order to streamline calculations, but which had never been formally proved. A particular example is the relation between 'holes' and 'particles'. Following the work of Shortley and Racah, it is to be expected that the matrix elements of quantities of interest between hole states are simply related to those between particle states in ligand-field configurations as they are in atomic configurations. But to use this relationship with confidence in calculations it is necessary to know and prove its precise form, including the specification of relative phases. In this case and otherwise I have tried to present and prove results in the forms which are actually needed in calculations.

With a book of this size in a fast-expanding field it is inevitable that the contents should represent in the main the author's position at a time past. Most of the book was written in 1958 and it was submitted in the spring of 1959 but I have referred to later work when it has cast a genuinely new light on some topic.

Finally, it is a pleasure to acknowledge my benefit from many discussions on theoretical physics and theoretical chemistry with Professor H. C. Longuet-Higgins and members of his department, especially with L. E. Orgel on the theory of transition-metal ions. I am also indebted to C. K. Jørgensen for his kindness in preparing Table 11.3 for me and to him and C. E. Schäffer for the data in Appendix A 40.

J. S. GRIFFITH

June 1960

CHAPTER 1

INTRODUCTION

1.1. Transition metals and their compounds

The transition metals are those which have partly filled shells of *d*-electrons in some, at least, of their compounds. In a similar way, the rare-earth metals have partly filled shells of *f*-electrons. Among the stable elements there are three series of transition metals and one of rare-earth metals. We are concerned in this book with the three stable series of transition metals. In order of increasing atomic number, they are called the first, second and third transition series. (See Appendix 1 for an enumeration of the elements in the three series.) The theory we shall develop is applicable, with minor modifications, to rare-earth metal compounds and to those compounds of the unstable elements at the end of the periodic table which contain partly filled shells.

Two extreme classes of transition-metal compound are conveniently distinguished—the metallic and the non-metallic. The former class includes the alloys, interstitial hydrides, borides, carbides and, stretching the use of the word compound slightly, the metals themselves. Typical members of the latter class are inorganic salts such as copper sulphate or potassium ferricyanide. From a theoretical standpoint the essential distinction between the two classes is that in the latter the *d*-electrons of the partly filled shells may be assigned individually to particular metal atoms. Each metal atom (or ion) has its own set of *d*-electrons localized near it and having little interaction with the sets belonging to neighbouring metal atoms. In a metallic compound the *d*-electrons are owned collectively by all the metal atoms and they cannot be separated into nearly non-interacting sets.

In this book we are concerned only with the non-metallic type of compound. The implied division of the electrons into localized groups is strictly never more than an approximation to the truth but it turns out to be a good approximation for a wide range of compounds which may, therefore, be called non-metallic.

It is usually possible to decide unambiguously how many *d*-electrons are localized near a particular transition-metal atom in a compound. There are transition-metal compounds, $KMnO_4$ or CrO_2Cl_2 are examples, in which there are no *d*-electrons. Our theory says little about such compounds although it can compare them with similar compounds which do contain *d*-electrons.

Finally, what is a *d*-electron in an atom, an ion or a compound? This will emerge slowly as we pass through the book. The main general discussion of the meaning of a *d*-electron in a compound is in ch. 7 and there the sense in which we may say that $KMnO_4$ contains no *d*-electrons will become apparent.

1.2. Stereochemistry

X-ray crystallographic determinations of the structures of a great number of transition-metal compounds have been made. A survey of all the data available on the class of compounds in which we are interested shows that the most common arrangement of nearest neighbours about the metal ion is that of a more or less distorted octahedron. Thus the anhydrous fluorides of Mn^{++}, Fe^{++}, Co^{++} and Ni^{++} crystallize in the rutile structure, each metal atom being surrounded by an almost regular octahedron of fluoride ions (as shown in Fig. 12.8). The hydrated ions $Mn(H_2O)_6^{++}$, $Fe(H_2O)_6^{++}$, $Co(H_2O)_6^{++}$, $Ni(H_2O)_6^{++}$ in crystals all have almost regular octahedral symmetry. An ion such as $Co(NH_3)_5Cl^{++}$ has the five ammonia molecules at the vertices of an octahedron and the chloride ion at a slightly greater distance from the metal atom along the line joining the latter to the sixth vertex.

Two other important types of stereochemical arrangement must be mentioned. Tetrahedral coordination of the metal ion is fairly common, particularly among the Co^{++} compounds. Some examples are $CoCl_4^{--}$, $ReCl_4^{-}$ and the blue form of $CoCl_2(NH_3)_2$. Planar complexes in which the metal is surrounded by a square of four molecules or ions are very common for metal ions with eight *d*-electrons and occur also for those with seven and nine. The Cu^{++} ion with nine *d*-electrons exhibits the whole range of stereochemistries from slightly distorted octahedral to planar. Sometimes there are four close neighbours of the Cu^{++} ion in a plane and two more distant ones completing a distorted octahedron. Other stereochemical arrangements are found, but not nearly as commonly as the three types already mentioned.

The more distant environment of the metal ion, namely the environment of the central octahedral, tetrahedral or planar group, is extremely variable. Fortunately groups other than nearest neighbours are relatively unimportant in determining at least the coarse chemical and physical properties of the ion. Consequently we can often neglect all but the nearest neighbours of the metal ion. These latter we shall usually refer to as the ligands.

Many of our detailed spectroscopic and magnetic data refer to ions in solution. The few detailed comparisons which can be made between the properties of ions in solids of known structure and those of the corresponding ions in solution show that the inner coordination group is usually maintained almost unchanged in solution. For example, the optical spectra of hydrated transition-metal ions in aqueous solution are almost identical with those of the same hydrated ions in crystalline solids. It will appear in due course that this usually implies that the stereochemistry is the same in the two situations. The physicist, however, must be warned that solutions of transition-metal compounds often contain unexpected molecular species. In general, solutions of transition-metal ions may contain equilibrium mixtures of different complexes, that is, the metal may occur in association with different sets of ligands. Nevertheless, many different types of evidence show that each species maintains its own particular stereochemistry with regard to the orientation of nearest neighbours about the metal ion.

In textbooks dealing with transition-metal chemistry it is usual to distinguish between compounds and complex ions. This particular type of classification, while it may be useful in certain contexts, tends to hide the important features common to both classes. In general, the electronic properties of a metal ion will be determined by its nearest neighbours, and it makes little difference whether these are part of a binary solid or of a discrete complex ion. In fact it is even true that the optical absorption spectrum of the 'compound' MnF_2 is very similar to that of the 'complex ion' $Mn(H_2O)_6^{++}$ in aqueous solution.

1.3. The valencies of the transition metals

The concept of valency as used by the chemist in connexion with transition-metal compounds will be regarded as a purely formal one. The molecular species under consideration is supposed to be made up of metal ions and other molecules or ions. To each of the latter a certain characteristic charge is assigned, usually that necessary to give it a closed shell, e.g. zero for a molecule such as water or ammonia, minus one for a halide or nitrate ion, minus two for an oxide, sulphide or sulphate ion, and so on. (See ch. 7 for a discussion of closed shells.) The charge which must then be assigned to the metal ion in order to give the correct charge for the molecular species is termed its valency. To take an example, we consider CrO_4^{--}, which is said to be a compound of hexavalent chromium, since a charge of $+6$ on the chromium, together with four charges of -2 on the oxygens, add up to give a charge of -2 on the species CrO_4. It must be emphasized that there is no implication that anything like a Cr^{+6} ion is present in the CrO_4^{--} anion. The term 'valency' is also used in many other, usually imprecisely defined, ways. For the purposes of the present book the formal definition given above is almost always unambiguous and then has the advantage that it involves no preconceptions about the electronic structure of particular transition-metal compounds. In other contexts a different definition might be more convenient.

It is next interesting to ask whether there is ever a direct correspondence between the formal valency and the electronic distribution in the region near the metal ion. Since the second and subsequent ionization potentials of the metals are larger than the ionization potential of any other molecule or ion in the environment, it is clear that the consequent electron-attracting tendency of the metal ions in high valencies must be neutralized in one way or another by their nearest neighbours. The manner of this neutralization will appear in ch. 7; for the present we note that as the formal valency of the metal ion increases, the degree of its correspondence with the actual electronic structure usually decreases. It is for this reason that the compounds discussed in this book are nearly all formally of low valency.

The valencies of a metal ion which are stable will of course depend on the nature of the ligands. Manganese, for example, forms hexa- and heptavalent compounds with oxygen, MnO_4^{--}, MnO_4^{-}; di- and trivalent compounds with water; and zerovalent compounds with carbon monoxide. A full discussion of the stabilization of valencies by different ligands would take us too far into general chemistry. The reader is referred to works on inorganic chemistry for further details.

The valencies exhibited by different metals in the presence of the same ligand also depend on a number of factors. One of these is very simple and of great importance. The ease of oxidation from the nth to the $n+1$th valency is closely related to the $n+1$th ionization potential, particularly for the lower valencies of the transition metals. The observed ionization potentials for the three lowest valencies of the transition metals are given in Appendix 1.

In the first transition series only copper can exist as an aqueous monovalent ion. This is presumably to be correlated with the high second ionization potential of copper, which is more than 2 eV larger than that of any other of these elements. In a similar way the trivalent compounds of Sc, Ti, V, Cr, Mn, Fe, Co are obtained more easily than those of Ni, Cu, Zn; that is, the metals with lower third ionization potentials are more easily obtained in the trivalent state. Many other examples of this sort could be given, but it must also be emphasized that a number of other factors are involved which depend on the interaction of the metal with its environment. The chemist describes this situation by saying that the oxidation-reduction potential of a metal ion depends on the nature of the ligands.

1.4. Theories of chemical bonding

A few remarks on the more important contemporary theories of the electronic structure of transition-metal compounds may help to clarify our approach to the subject.

The simplest possible picture of the electronic structure of transition-metal compounds is the purely ionic one. Here the formal valencies of the different ions in the structure are interpreted literally to imply the presence of the corresponding ions. This theory, which was adequate for the description of the stoichiometric properties of most compounds, has been more or less abandoned by chemists, since it is clearly unable to account for many experimental observations and is, in any case, quite unrealistic in the light of our present knowledge of the ionization potentials and electron affinities of atoms, ions and molecules.

The most influential 'chemical' theory of transition-metal structure has undoubtedly been the valence-bond theory, as developed by Pauling. Here an attempt is made to distinguish between 'ionic' compounds which are held together by electrostatic forces, much as in alkali halide crystals, and 'covalent' compounds which are held together by directed bonds. Conceptually this theory has proved so attractive to chemists that it has been the basis of most recent chemical thinking on the subject. Despite its usefulness in this direction it has not proved fruitful in the field of quantitative calculation. It now seems certain that some of the postulates of the scheme require revision, but its simplicity guarantees it a central importance in qualitative chemical discussions of the theory of valency.

While chemists were developing the valence-bond theory, the ionic theory was being adapted for use in a more quantitative way. Largely through the influence of Bethe and Van Vleck, a detailed understanding of the magnetic properties of divalent and trivalent ions in certain types of environment was built up. This theory was restricted in its range of application since failure to include 'chemical' interactions excluded most metal compounds of high valency and also many others. Within its range, by applying perturbation theory to the quantum mechanical description of the free ions, it achieved many notable quantitative successes. For reasons which it is now difficult to understand no chemical applications of the theory were made before 1950. In other fields of theoretical chemistry the molecular-orbital method had, by 1950, achieved a central position. It was natural, therefore, that when certain experimental observations demanded an explanation in terms of electron-sharing between atoms the electrostatic theory was extended in the direction suggested by the molecular-orbital method. The theory which has resulted is a hybrid which still depends heavily on the simple electrostatic theory, but which can be justified as an approximation to a more complete molecular-orbital treatment.

CHAPTER 2

ANGULAR MOMENTUM AND RELATED MATTERS

2.1. The Hamiltonian for an atomic system

In order to understand the behaviour of a metal ion in the environment imposed upon it in a compound, it is necessary first to understand the electronic structure of free atoms and ions. This is obvious for compounds in which one believes that the environment represents a rather small disturbance of this electronic structure. However, it is also very useful to be able to refer to the theory of the structure of free atoms even when the disturbance is large. This is especially true of the sort of compounds that we consider in this book, namely those in which unpaired metal electrons are localized on or at least near to their parent metal ions. In this chapter and the next three, therefore, I describe those features of the theory of atomic structure which are most relevant to our later treatment of metal compounds. I start from the beginning, but it is rather desirable that the reader should possess already a little knowledge of quantum mechanics: §§ 1–22, 27, 42 and 43 of Professor Dirac's book *Quantum Mechanics* probably cover all that is really necessary.[1] I use the bra and ket notation almost exclusively throughout the book because I feel that, as with vectors in classical mechanics, it makes the intuitive significance of equations and results much easier to grasp.[2]

An atom or ion consists of a relatively massive positively charged nucleus together with a number of electrons. The electrons are held near the nucleus by their electrostatic attraction to the latter and to some extent apart from each other by their mutual electrostatic repulsions. Because of its relatively large mass it is a good approximation to regard the nucleus as being at rest. This means, then, that we have a classical Hamiltonian

$$\mathscr{H} = \sum_{\kappa=1}^{n}\left(\frac{1}{2m}\mathbf{p}_\kappa^2 - \frac{Ze^2}{r_\kappa}\right) + \sum_{\kappa<\lambda}^{n}\frac{e^2}{r_{\kappa\lambda}}, \tag{2.1}$$

for the system. In (2.1) $\mathbf{p}_\kappa$ is the momentum vector of the κth electron and r_κ its distance from the nucleus, m the mass of the electron, $-e$ its charge, $+Ze$ the charge on the nucleus and $r_{\kappa\lambda}$ the distance from the κth to the λth electron. There are n electrons and the whole system has a charge $(Z-n)\,e$, so for a neutral atom $n = Z$ and for a positive ion $n < Z$.

[1] Dirac (1947). §§ 42 and 43 are perturbation theory. A summary of parts of this, together with some further developments, is in Appendix 3.

[2] I quite often use single symbols to stand for kets or bras and may write, for example, $\psi = |t\mathfrak{F}\rangle$. I do not apologize for this apparent confusion of types because it does not occur in places where it could lead to difficulty.

Naturally (2.1) is not a complete and accurate Hamiltonian for the system. The effect of regarding the nucleus as being at rest is truly a small one and the whole part of it which is relevant to us can be taken into account by interpreting m in (2.1) as the reduced mass $\mu M/(\mu+M)$, where μ is the true mass of an electron and M the mass of the nucleus. This result is obtained by referring all distances in the more exact theory to the centre of mass of the system whereupon one obtains (2.1) to an approximation quite sufficient for our purposes. There are other ways in which (2.1) is incomplete which will prove important in our theory. The electron has a magnetic moment and so, quite often, does the nucleus. This introduces extra terms into (2.1) which represent magnetic interactions between the various particles of the system. These we consider in ch. 5, but until the end of ch. 4 we derive only the consequences of the Hamiltonian (2.1). The main reason for this particular division of labour is that we find later that the effect of the environment of a metal ion is usually larger, in terms of the energies associated with it, than most or all of the magnetic interactions but usually a little smaller than the electrostatic interactions arising from (2.1). As a consequence of this, the electronic structure of atoms or ions having the Hamiltonian (2.1) is specially important and so we now pass on to consider this.

In quantum theory the classical Hamiltonian $\mathscr{H}$ is taken over directly, but $\mathbf{p}_\kappa$ and $\mathbf{r}_\kappa$ are now regarded as operators, the components of which do not all commute. If we write q and q' for typical, but not necessarily different, coordinates x, y or z then

$$\left.\begin{aligned} p_{\kappa q}p_{\lambda q'} &= p_{\lambda q'}p_{\kappa q} \quad \text{for all} \quad \kappa, \lambda,\\ q_\kappa q'_\lambda &= q'_\lambda q_\kappa \qquad \text{for all} \quad \kappa, \lambda,\\ q_\kappa p_{\lambda q'} - p_{\lambda q'}q_\kappa &= i\hbar\delta_{qq'}\delta_{\kappa\lambda}, \end{aligned}\right\} \tag{2.2}$$

where the Kronecker delta symbol is defined by the equations $\delta_{ab} = 0$, unless $a = b$ when $\delta_{ab} = 1$. In (2.2) $p_{\kappa q}$ represents the q component of $\mathbf{p}_\kappa$ and similarly for the others. Any pair of components of the same or different momenta commute with each other and the corresponding statement is true for coordinates. The only non-commuting pairs of components are those referring to a coordinate of a single particle and its conjugate momentum.

Schrödinger's form for the equations of motion is

$$i\hbar\frac{d}{dt}|X\rangle = \mathscr{H}\,|X\rangle, \tag{2.3}$$

and, if we express the ket $|X\rangle$ in terms of the coordinates of the electrons and the time, the $\mathbf{p}_\kappa$ may be equated to differential operators

$$\mathbf{p}_\kappa = -i\hbar\nabla_\kappa = -i\hbar\left(\frac{\partial}{\partial x_\kappa}, \frac{\partial}{\partial y_\kappa}, \frac{\partial}{\partial z_\kappa}\right). \tag{2.4}$$

It is immediate that these $\mathbf{p}_\kappa$ satisfy (2.2). Equation (2.3) then becomes Schrödinger's wave-equation for the system

$$\mathscr{H}(-i\hbar\nabla_\kappa, \mathbf{r}_\kappa)\,\phi(\mathbf{r}_\kappa, t)\rangle = i\hbar\frac{\partial}{\partial t}\phi(\mathbf{r}_\kappa, t)\rangle. \tag{2.5}$$

The Hamiltonian $\mathscr{H}$ represents the total energy of the system. Since it does not involve the time explicitly we may find stationary states which are eigenfunctions of $\mathscr{H}$. In fact we write

$$\phi(\mathbf{r}_\kappa, t)\rangle = e^{-iEt/\hbar}\psi(\mathbf{r}_\kappa)\rangle, \tag{2.6}$$

and

$$\mathscr{H}\psi\rangle = E\psi\rangle, \tag{2.7}$$

where ψ is independent of time and ϕ then satisfies (2.5). We shall be interested in these stationary states for atomic systems and when discussing them shall sometimes drop the ket symbols in equations such as (2.7). We always retain the symbols to express average values and for matrix elements between different states, e.g. $\langle\bar{\psi}|\,\mathbf{p}_\kappa^2\,|\psi\rangle$ is the average value of the square of the momentum of the κth electron for the state ψ and $\langle\bar{\psi}_1|\,x_\lambda\,|\psi_2\rangle$ is the matrix element of x_λ between the states ψ_1 and ψ_2. We now have the wave equation

$$\left\{\sum_{\kappa=1}^{n}\left(-\frac{\hbar^2}{2m}\nabla_\kappa^2 - \frac{Ze^2}{r_\kappa}\right) + \sum_{\kappa<\lambda}^{n}\frac{e^2}{r_{\kappa\lambda}}\right\}\psi = E\psi, \tag{2.8}$$

corresponding to (2.1).

It is an experimental fact that electrons obey Fermi–Dirac statistics and this appears in the theory as an assertion that electronic systems can only be represented by solutions of (2.8) which are fully antisymmetric with respect to interchange of electrons. We shall see precisely what this means in § 2.6. It is also an experimental fact that an electron has a magnetic moment and that there are two independent internal states for the electron, called states of spin, which are associated with different orientations of the magnetic moment. The extra terms in the Hamiltonian due to the magnetic moment of the electron are usually small and as already remarked are not considered until ch. 5.

Because of the requirement of Fermi–Dirac statistics it turns out that the existence of two independent spin states for the electron actually has a very large influence upon the energy of many-electron systems even though the terms in the Hamiltonian which involve the spin are small. In § 2.4 we shall consider the algebraic techniques for dealing with the spin and then discuss its influence on the electronic energy. First, however, we must discuss the properties of orbital angular momentum.

2.2. Orbital angular momentum

In classical mechanics there are always two important constants of the motion for an isolated system. The first is the total energy and we recognize this in quantum theory by taking it for our Hamiltonian. The second is the total angular momentum. This is also important in quantum theory and, as we shall see shortly, it commutes with the Hamiltonian (2.1) and so can still be taken as a constant of the motion. Because of this it is very useful to examine its properties, eigenstates and eigenvalues in considerable detail. We shall find actually that if we use (2.1) there are two kinds of angular momentum which are constants of the motion in quantum theory. The first corresponds most closely to the total angular momentum in classical mechanics. We now call it the total orbital angular momentum to distinguish it from the second kind, the spin angular momentum. Then we shall find later that if magnetic interactions between the particles are included, only the sum of the orbital and spin momentum is a constant of the motion. This, in quantum mechanics, is called the total angular momentum. The third common constant of the motion for a classical system is the total linear momentum. This is not relevant for us because our system is not really isolated; it is congregated round the fixed nucleus. In the words of Weyl, we have transformed space from a homogeneous space to a centred space. The centre is the nucleus and so long as we regard it as a fixed point charge we have full rotational symmetry about it but no longer translational symmetry. This is why angular but not linear momentum is useful as a tool in the analysis of structure.

Just as in the classical theory, then, the orbital angular momentum of a particle about the origin is defined as the vector $\mathbf{l} = \mathbf{r} \wedge \mathbf{p}$. Using the commutation relations (2.2) we find readily that

$$\mathbf{r} \wedge \mathbf{p} = -\mathbf{p} \wedge \mathbf{r}, \tag{2.9}$$

as in the classical theory, but that

$$l_x l_y - l_y l_x = i\hbar l_z, \tag{2.10}$$

and similarly for cyclic permutations of x, y, z. The three relations of the type (2.10) may be abbreviated in vector notation to the equivalent single equation $\mathbf{l} \wedge \mathbf{l} = i\hbar\mathbf{l}$ showing that in quantum theory, unlike the classical theory, the vector product of a vector with itself is not necessarily zero.

It is convenient to introduce a notation for the commutator of two quantities a and b, of which one may be a vector:

$$[a, b] \equiv ab - ba. \tag{2.11}$$

The reader should note that we are following here the normal mathematical usage rather than the Poisson bracket notation of Dirac. The two notations differ, of course, only by a factor $i\hbar$. A number of useful relations follow immediately from (2.11). We have the anticommutation relations for the commutator:

$$[b, a] = -[a, b] \quad \text{and} \quad [a, a] = 0,$$

and the distributive laws

$$[a, b+c] = [a, b] + [a, c],$$

$$[b+c, a] = [b, a] + [c, a].$$

Then there is also a pair of relationships which are rather like the rule for differentiating a product

$$\left.\begin{aligned} [ab, c] &= a[b, c] + [a, c]b, \\ [c, ab] &= a[c, b] + [c, a]b. \end{aligned}\right\} \tag{2.12}$$

We shall refer to all these relations collectively as (2.12). In the new notation, (2.10) becomes $[l_x, l_y] = i\hbar l_z$.

For a set of n particles, the total orbital angular momentum about the origin is defined as the sum

$$\mathbf{L} = \sum_{\kappa=1}^{n} \mathbf{l}_\kappa = \sum_{\kappa=1}^{n} \mathbf{r}_\kappa \wedge \mathbf{p}_\kappa, \tag{2.13}$$

of the angular momentum vectors for each particle separately. Using the relations (2.12) we readily obtain the commutation relations for the components of $\mathbf{L}$ from those of $\mathbf{l}$. For example

$$\begin{aligned}[L_x, L_y] &= [\sum_\kappa l_{\kappa x}, \sum_\lambda l_{\lambda y}] \\ &= \sum_\kappa \sum_\lambda [l_{\kappa x}, l_{\lambda y}] \\ &= \sum_\kappa [l_{\kappa x}, l_{\kappa y}] = \sum_\kappa i\hbar l_{\kappa z} = i\hbar L_z,\end{aligned} \tag{2.14}$$

where we have used the fact, which follows from the definition of $\mathbf{l}_\kappa$ and the commutation relations for $\mathbf{p}_\kappa$ and $\mathbf{r}_\kappa$, that all the components of $\mathbf{l}_\kappa$ commute with all those of $\mathbf{l}_\lambda$ when $\kappa \neq \lambda$. The only two other independent commutation relations for components of $\mathbf{L}$ are obtained by cyclically permuting x, y and z. Thus the relations for $\mathbf{L}$ are formally identical with those for $\mathbf{l}$ and may be written

$$\mathbf{L} \wedge \mathbf{L} = i\hbar \mathbf{L}.$$

It follows that any result we prove for $\mathbf{L}$, using its commutation relations only, remains true on replacing $\mathbf{L}$ by $\mathbf{l}$ and vice versa.

We have already remarked that the great interest in $\mathbf{L}$ in the theory of atomic structure derives from the fact that each of its components commutes with the Hamiltonian $\mathscr{H}$ of (2.1). This is shown most conveniently by introducing spherical polar coordinates r, θ, ϕ having the z-axis as pole. Then

$$\left.\begin{aligned} x &= r \sin\theta \cos\phi, \\ y &= r \sin\theta \sin\phi, \\ z &= r \cos\theta, \end{aligned}\right\} \tag{2.15}$$

and we have

$$\begin{aligned} L_z = \sum_{\kappa=1}^{n} l_{\kappa z} &= \sum_{\kappa=1}^{n} \left\{ -i\hbar \left(x_\kappa \frac{\partial}{\partial y_\kappa} - y_\kappa \frac{\partial}{\partial x_\kappa} \right) \right\} \\ &= -i\hbar \sum_{\kappa=1}^{n} \frac{\partial}{\partial \phi_\kappa}, \end{aligned} \tag{2.16}$$

on performing the substitution for x_κ and y_κ. In $\mathscr{H}$, ∇_κ^2 involves ϕ_κ only as $\partial^2/\partial\phi_\kappa^2$ and $1/r_\kappa$ does not involve ϕ_κ at all. So the only part of $\mathscr{H}$ with which L_z might not commute is the last term. However

$$\begin{aligned} \left[L_z, \frac{e^2}{r_{\kappa\lambda}}\right] &= -i\hbar e^2 \sum_{\mu=1}^{n} \frac{\partial}{\partial \phi_\mu} (r_{\kappa\lambda}^{-1}) \\ &= -i\hbar e^2 \frac{\partial}{\partial \phi_\kappa} (r_{\kappa\lambda}^{-1}) - i\hbar e^2 \frac{\partial}{\partial \phi_\lambda} (r_{\kappa\lambda}^{-1}) = 0 \end{aligned} \tag{2.17}$$

as $r_{\kappa\lambda}$ depends on $\phi_\kappa - \phi_\lambda$ but not on $\phi_\kappa + \phi_\lambda$ and so $\mathscr{H}$ commutes with L_z and hence, by symmetry, also with L_x and L_y.

The entire set $\mathscr{H}$, L_x, L_y and L_z do not form a commuting set of observables because L_x, L_y and L_z do not commute amongst themselves. However, L_z and $\mathbf{L}^2$ do commute, for

$$\begin{aligned}[\mathbf{L}^2, L_z] &= [L_x^2+L_y^2+L_z^2, L_z]\\ &= L_x[L_x, L_z]+[L_x, L_z]L_x+L_y[L_y, L_z]+[L_y, L_z]L_y\\ &= -i\hbar L_x L_y - i\hbar L_y L_x + i\hbar L_y L_x + i\hbar L_x L_y\\ &= 0.\end{aligned}$$

More trivially, we also have

$$[\mathbf{L}^2, \mathscr{H}] = [L_x^2+L_y^2+L_z^2, \mathscr{H}] = 0,$$

by (2.12) and the commutation of $\mathscr{H}$ with each of L_x, L_y and L_z separately. So $\mathscr{H}$, $\mathbf{L}^2$ and L_z (or, of course, L_x or L_y) form a commuting set of observables.

At this stage we recall the well-known proposition (Dirac, p. 49) that a set of commuting observables possesses a complete set of simultaneous eigenkets. This means that we can expand an arbitrary ket as a sum and perhaps also an integral over kets each of which is simultaneously an eigenket of $\mathscr{H}$, $\mathbf{L}^2$ and L_z. This particular set of simultaneous eigenkets plays a large part in what is to follow.

As I have already indicated, we shall be interested later in more general types of angular momenta. We define a general angular momentum to be any real vector having the commutation relations (2.10). Such a vector does not necessarily possess any expression like (2.16) in terms of a Schrödinger representation. We shall, therefore, deduce various properties of the simultaneous eigenkets of $\mathscr{H}$, $\mathbf{L}^2$ and L_z using only the commutation relations. This means that our results will also be true for general angular momenta which commute with $\mathscr{H}$. We write $|E\beta L_z'\rangle$ for a simultaneous eigenket where E, β and L_z' are the eigenvalues of $\mathscr{H}$, $\mathbf{L}^2$ and L_z, respectively, and define the shift operator

$$L^+ = L_x + iL_y.$$

L^+ is not real; its complex conjugate is

$$L^- = L_x - iL_y.$$

Then, using relations of the type of (2.14),

$$[L_z, L^\pm] = \pm\hbar L^\pm, \tag{2.18}$$

whence

$$L_z L^+ |E\beta L_z'\rangle = (L_z'+\hbar)\, L^+ |E\beta L_z'\rangle,$$

$$L_z L^- |E\beta L_z'\rangle = (L_z'-\hbar)\, L^- |E\beta L_z'\rangle.$$

This justifies the name 'shift operator' because it shows that the shift operators shift one simultaneous eigenket of $\mathbf{L}^2$ and L_z into another having the same eigenvalue for $\mathbf{L}^2$ but a different eigenvalue for L_z. We shall find this operation of shifting to be a very useful technique in the theory. Repeating the shift operations n times we find

$$L_z(L^\pm)^n |E\beta L_z'\rangle = (L_z \pm n\hbar)(L^\pm)^n |E\beta L_z'\rangle. \tag{2.19}$$

Note that an equation such as (2.18) or (2.19) represents two distinct equations in one of which we take the upper and in the other the lower sign throughout. Now L^+ and L^- commute with $\mathscr{H}$ and $\mathbf{L}^2$ and so, unless they are zero, each of the kets $(L^\pm)^n\,|E\beta L_z'\rangle$ is an eigenket of $\mathscr{H}$ and $\mathbf{L}^2$ with eigenvalues E and β, respectively.

However, only a finite number of the kets $(L^\pm)^n\;|E\beta L_z'\rangle$, for a given $|E\beta L_z'\rangle$ and varying n, can be non-zero. We can see this by taking any particular non-zero one having, for example, L_z'' for the eigenvalue of L_z. It is convenient to normalize it and then it is written $|E\beta L_z''\rangle$. We now consider the two kets $L^\pm\,|E\beta L_z''\rangle$ obtained by shifting this ket. Using the formula

$$L^\mp L^\pm = \mathbf{L}^2 - L_z^2 \mp \hbar L_z, \tag{2.20}$$

which follows immediately from (2.14) and the definition of the operators $L^\pm$, and the requirement that the square of the length of any ket vector must be non-negative we obtain the pair of restrictions

$$0 \leqslant \langle E\beta L_z''\,|\,L^\mp L^\pm\,|\,E\beta L_z''\rangle = \beta - (L_z'')^2 \mp \hbar L_z'', \tag{2.21}$$

on the possible values of L_z''. Adding the two forms of (2.21) we obtain

$$0 \leqslant \beta - (L_z'')^2$$

which means that β is positive and $|L_z''| \leqslant \beta^{\frac{1}{2}}$. Thus the range of possible values of L_z'' is limited. This is only possible if there are eigenvalues of L_z, $\alpha^\pm$ say, with $\alpha^+ \geqslant \alpha^-$, such that

$$L^\pm\,|E\beta\alpha^\pm\rangle = 0, \quad |E\beta\alpha^\pm\rangle \neq 0.$$

Multiplying on the left by $\langle E\beta\alpha^\pm|\,L^\mp$ we deduce immediately from (2.20) that

$$\left.\begin{aligned} &\beta - (\alpha^+)^2 - \hbar\alpha^+ = 0 \\ \text{and}\qquad &\beta - (\alpha^-)^2 + \hbar\alpha^- = 0. \end{aligned}\right\} \tag{2.22}$$

Subtracting the first equation from the second and factorizing yields

$$(\alpha^+ + \alpha^-)(\alpha^+ - \alpha^- + \hbar) = 0,$$

which means that $\alpha^+ = -\alpha^-$ (because $\alpha^+ \geqslant \alpha^-$). We now put $\alpha^+ = L\hbar$, where L is a dimensionless number, and have from (2.22),

$$\beta = L(L+1)\hbar^2. \tag{2.23}$$

Equation (2.19) shows that any pair of eigenvalues of L_z belonging to our set must differ by an integral multiple of $\hbar$. So $2L$ is a non-negative integer. It is customary, for simplicity, to label the eigenkets of $\mathbf{L}^2$ and L_z by the dimensionless numbers L and $\hbar^{-1}L_z'$, respectively, rather than by the actual eigenvalues. So a typical one is written $|ELM\rangle$. It may be that there are other observables α which commute amongst themselves and also with $\mathscr{H}$, $\mathbf{L}^2$ and L_z. Here we write just α, for brevity, although in general α would stand for a set $\alpha_1, \ldots, \alpha_n$ of commuting observables. In such a case the simultaneous eigenkets would be written $|\alpha' ELM\rangle$ and these also will be connected, for different values of M, by the shift operators $L^\pm$. However, the proof that the value of α' is not changed by the operation of shifting requires that α commutes with $L^{\pm}$ and hence with L_x and L_y also. For observables α which turn up in practice this is usually the case and we shall understand a set of commuting observables α, $\mathbf{L}^2$ and L_z to have α commuting with L_x and L_y as well as with $\mathbf{L}^2$ and L_z.

We now have the eigenvalue equations

$$\left.\begin{aligned} \alpha\,|\alpha' ELM_L\rangle &= \alpha'\,|\alpha' ELM_L\rangle, \\ \mathscr{H}\,|\alpha' ELM_L\rangle &= E\,|\alpha' ELM_L\rangle, \\ \mathbf{L}^2\,|\alpha' ELM_L\rangle &= L(L+1)\,\hbar^2\,|\alpha' ELM_L\rangle, \\ L_z\,|\alpha' ELM_L\rangle &= M_L\hbar\,|\alpha' ELM_L\rangle, \end{aligned}\right\} \tag{2.24}$$

where the suffix L in M_L is included if there is any possibility of doubt about the angular momentum vector to which it refers. It is unnecessary in (2.24) in the form in which we have written them. However, it is necessary when one has more than one angular momentum and is a notation we use in the next section because there would be a possibility of ambiguity there. We have seen that L is an integer or an integer plus a half. For a given α', E and L, M_L takes all values from $-L$ to $+L$ which differ from L by integers. We suppose all the kets (2.24) to be normalized (I shall always use the word normalized to mean normalized to unity) and to be obtained, for any given α', E and L, by multiplying a set of the kind given in (2.19) by positive normalizing factors. This definition imposes phase relations between those kets having the same α', E and L but different M_L. These phase relations are the ones which are universally adopted and so, whenever we write $|\alpha' ELM\rangle$ or $|\alpha' ELM_L\rangle$ in this book, we always mean kets which have their phases related in the aforementioned way. We almost invariably normalize them too; in the rare cases when this is not so, explicit mention will be made of the fact. Using (2.20) and the value $L(L+1)\,\hbar^2$ for β, we have in detail

$$L^{\pm}\,|\alpha' ELM\rangle = \hbar\{(L\mp M)\,(L\pm M+1)\}^{\frac{1}{2}}\,|\alpha' ELM\pm 1\rangle, \tag{2.25}$$

for all M if we define $|\alpha' ELM\rangle$ to be zero for $|M| > L$.

Equations (2.24) and (2.25) show how various operators affect the kets $|\alpha' ELM\rangle$. They can be expressed equivalently as formulae for matrix elements of the operators and it is instructive to rewrite them in this way by operating with $\langle\alpha'' E' L' M'|$ on the left of each:

$$\left.\begin{aligned} \langle\alpha'' E' L' M'|\,\alpha\,|\alpha' ELM\rangle &= \alpha'\delta_{\alpha'\alpha''}\delta_{EE'}\delta_{LL'}\delta_{MM'}, \\ \langle\alpha'' E' L' M'|\,\mathscr{H}\,|\alpha' ELM\rangle &= E\delta_{\alpha'\alpha''}\delta_{EE'}\delta_{LL'}\delta_{MM'}, \\ \langle\alpha'' E' L' M'|\,\mathbf{L}^2\,|\alpha' ELM\rangle &= L(L+1)\,\hbar^2\delta_{\alpha'\alpha''}\delta_{EE'}\delta_{LL'}\delta_{MM'}, \\ \langle\alpha'' E' L' M'\,|\,L_z\,|\alpha' ELM\rangle &= M\hbar\delta_{\alpha'\alpha''}\delta_{EE'}\delta_{LL'}\delta_{MM'} \end{aligned}\right\} \tag{2.24'}$$

and

$$\left.\begin{aligned} \langle\alpha'' E' L' M'|\,L^{+}\,|\alpha' ELM\rangle &= \hbar\{(L-M)\,(L+M+1)\}^{\frac{1}{2}}\,\delta_{\alpha'\alpha''}\delta_{EE'}\delta_{LL'}\delta_{M'M+1}, \\ \langle\alpha'' E' L' M'|\,L^{-}\,|\alpha' ELM\rangle &= \hbar\{(L+M)\,(L-M+1)\}^{\frac{1}{2}}\,\delta_{\alpha'\alpha''}\delta_{EE'}\delta_{LL'}\delta_{M'M-1}. \end{aligned}\right\} \tag{2.25'}$$

If we have an angular momentum **L** which does not commute with $\mathscr{H}$ our results remain true provided merely that we drop the parameter E and the equation involving $\mathscr{H}$ from (2.24), (2.24′), (2.25) and (2.25′).

Example

If γ commutes with $\mathscr{H}$, L_x, L_y and L_z, and if $\alpha = \gamma + L_z$ then α, $\mathscr{H}$, $\mathbf{L}^2$, L_z forms a commuting set. However, the shift operators $L^{\pm}$ change the eigenvalues α' as well as L_z'.

2.3. Addition of angular momenta

We obtained in the last section the eigenvalues of L_z and $\mathbf{L}^2$ for a single angular momentum vector **L**. If we use the Hamiltonian (2.1) there are two total angular momentum vectors of an atomic system which are constants of the motion. The first is the total orbital angular momentum which we have discussed and the second is the total spin angular momentum which is introduced in the next section. Because of this it is interesting to ask what are the eigenvalues of the components of the sum of two angular momenta. The answer to this question will be useful in a wide variety of contexts and, as we shall see later, often even when neither of the two angular momenta which are added together is a constant of the motion.

Let us take, then, two angular momenta $\mathbf{L}_1$ and $\mathbf{L}_2$ which commute with one another. This means that each of the nine possible pairs of their components commutes separately. Thus

$$[L_{1x}, L_{2x}] = [L_{1x}, L_{2y}] = [L_{1x}, L_{2z}] = 0, \quad \text{etc.}$$

Their sum $\mathbf{J} = \mathbf{L}_1 + \mathbf{L}_2$ is also an angular momentum because, using the properties (2.12) of commutators

$$\begin{aligned} [J_x, J_y] &= [(L_{1x} + L_{2x}), (L_{1y} + L_{2y})] \\ &= i\hbar L_{1z} + i\hbar L_{2z} = i\hbar J_z, \end{aligned} \tag{2.26}$$

and cyclically for the other components. So J_z and $\mathbf{J}^2$ commute and we have already found their possible eigenvalues. We wish to relate them to the eigenvalues of the constituent angular momenta $\mathbf{L}_1$ and $\mathbf{L}_2$. Suppose that the commuting set $\mathbf{L}_1^2$, $\mathbf{L}_2^2$, L_{1z}, L_{2z} is extended by adding further operators γ to form a complete commuting set. This would have simultaneous eigenkets $|\gamma' L_1 L_2 M_1 M_2\rangle$, where M_1 and M_2 refer to L_{1z} and L_{2z}, respectively. It is an easy calculation to show that $\mathbf{L}_1^2$ and $\mathbf{L}_2^2$ both commute with $\mathbf{J}^2$ and J_z and that, therefore, γ, $\mathbf{L}_1^2$, $\mathbf{L}_2^2$, $\mathbf{J}^2$, J_z forms a commuting set. Its eigenkets will be written $|\gamma' L_1 L_2 J M_J\rangle$.

One immediately asks what is the relationship between the sets of eigenkets $|\gamma' L_1 L_2 M_1 M_2\rangle$ and $|\gamma' L_1 L_2 J M_J\rangle$? And if, as we have assumed, γ, $\mathbf{L}_1^2$, $\mathbf{L}_2^2$, L_{1z}, L_{2z} is a complete commuting set, is γ, $\mathbf{L}_1^2$, $\mathbf{L}_2^2$, $\mathbf{J}^2$, J_z one also? We shall find the answer to the second question to be in the affirmative and shall give a partial answer to the first here and the rest of the answer in § 2.5.

It is immediate that for given $\gamma' L_1 L_2$ there are just $(2L_1+1)(2L_2+1)$ states $|\gamma' L_1 L_2 M_1 M_2\rangle$, one for each different possible pair of $M_1 M_2$ values. Each of these is an eigenket of J_z with $M_J\hbar = (M_1+M_2)\hbar$. The largest positive eigenvalue obtained in this way is $(L_1+L_2)\hbar$. On applying the shift operators

$$J^{\pm} = J_x \pm iJ_y = L_{1x}+L_{2x} \pm i(L_{1y}+L_{2y})$$
$$= (L_{1x} \pm iL_{1y})+(L_{2x} \pm iL_{2y}) = L_1^{\pm}+L_2^{\pm}, \qquad (2.27)$$

to this ket all the resultant kets must be linear combinations of the $|\gamma' L_1 L_2 M_1 M_2\rangle$. This is because, as we have just seen, $J^{\pm}$ are expressible as the sum of $L_1^{\pm}$ and $L_2^{\pm}$. Hence we have

$$J^+ |\gamma' L_1 L_2 L_1 L_2\rangle = 0,$$

since otherwise it would be an eigenket of J_z with eigenvalue $(L_1+L_2+1)\hbar$, which is impossible. Multiplying by J^- on the left and using (2.20) we find

$$0 = (\mathbf{J}^2 - J_z^2 - \hbar J_z)\,|\gamma' L_1 L_2 L_1 L_2\rangle$$
$$= (\mathbf{J}^2 - (L_1+L_2)^2\hbar^2 - (L_1+L_2)\hbar^2)\,|\gamma' L_1 L_2 L_1 L_2\rangle,$$

and on rearrangement

$$\mathbf{J}^2\,|\gamma' L_1 L_2 L_1 L_2\rangle = (L_1+L_2)(L_1+L_2+1)\hbar^2\,|\gamma' L_1 L_2 L_1 L_2\rangle.$$

So we have found an eigenket of $\mathbf{J}^2$ as well as of J_z and having $J = M_J = L_1+L_2$. By applying J^- to this we obtain successively $|\gamma' L_1 L_2, L_1+L_2, M_J\rangle$ for all M_J from L_1+L_2 to $-(L_1+L_2)$ as linear combinations of the $|\gamma' L_1 L_2 M_1 M_2\rangle$.

We now consider the next highest value of $M_J\hbar$, namely $(L_1+L_2-1)\hbar$. There are, in general, two $|\gamma' L_1 L_2 M_1 M_2\rangle$ with this eigenvalue for J_z; they are $|\gamma' L_1 L_2, L_1-1, L_2\rangle$ and $|\gamma' L_1 L_2, L_1, L_2-1\rangle$. Clearly we can continue the procedure and classify all the $|\gamma' L_1 L_2 M_1 M_2\rangle$ by their M_J values, as shown in Table 2.1.

Table 2.1. *Classification of* $|\gamma' L_1 L_2 M_1 M_2\rangle$ *according to* M_J *values*

M_J	(M_1, M_2) values
L_1+L_2	(L_1, L_2)
L_1+L_2-1	(L_1, L_2-1), (L_1-1, L_2)
L_1+L_2-2	(L_1, L_2-2), (L_1-1, L_2-1), (L_1-2, L_2)
L_1+L_2-n	(L_1, L_2-n), (L_1-1, L_2-n+1), ..., (L_1-n, L_2)

Each time we decrease M_J by one we get one more ket with the new M_J value so long as $L_2-n \geqslant -L_2$ and $L_1-n \geqslant -L_1$. This means that the last increase in the number of linearly independent kets occurs for $M_J = |L_1-L_2|$. In view of the discussion in the last paragraph it is natural to expect that there will be just enough kets to give the $2J+1$ eigenkets, having $|M_J| \leqslant J$, for $\mathbf{J}^2$ for each J from $J = L_1+L_2$ to $J = |L_1-L_2|$.

This is true and we establish it with the help of the observation that if $|X\rangle$ is an eigenket of $\mathbf{J}^2$ and J_z and $|Y\rangle$ is any ket then if $|X\rangle$ is orthogonal to $|Y\rangle$ so also is $J^+|X\rangle$ to $J^+|Y\rangle$. For

$$\langle X|\,J^-.J^+\,|Y\rangle = \langle X|\,(\mathbf{J}^2 - J_z^2 - \hbar J_z)\,|Y\rangle, \quad \text{from} \quad (2.20)$$
$$= (J(J+1) - M_J^2 - M_J)\hbar^2 \langle X|Y\rangle$$
$$= 0,$$

using our assumed properties of $|X\rangle$ and $|Y\rangle$, and remembering that $\langle X|J^-$ is the bra conjugate to the ket $J^+|X\rangle$. We now use induction on n and suppose that from the $(n+1)$ kets $|\gamma' L_1 L_2, L_1-\epsilon, L_2+\epsilon-n\rangle$, n linear combinations are eigenkets of J_z with $M_J = L_1+L_2-n$ and having $J = L_1+L_2, L_1+L_2-1, \ldots, L_1+L_2-n+1$. Then there is one linear combination of them, $|Y\rangle$ say, which is orthogonal to all those n eigenkets $|JM_J\rangle$. So $J^+|Y\rangle$ is orthogonal to the n shifted kets $J^+|JM_J\rangle$. But this means that $J^+|Y\rangle$ is orthogonal to all the kets $|\gamma' L_1 L_2 M_1 M_2\rangle$ with $M_1+M_2 = L_1+L_2-n+1$, i.e. it is zero. Hence as before $|Y\rangle$ must be an eigenket of $\mathbf{J}^2$ with $J = L_1+L_2-n$. This establishes the induction which terminates, as we have seen, with $J = |L_1-L_2|$.

The total number of the kets $|\gamma' L_1 L_2 JM\rangle$, for fixed $\gamma' L_1 L_2$ and writing M for M_J, must of course be the same as the total number of the $|\gamma' L_1 L_2 M_1 M_2\rangle$. The identity

$$\sum_{J=|L_1-L_2|}^{L_1+L_2} (2J+1) = (2L_1+1)(2L_2+1)$$

assures us that this is so.

As there is only one ket for each set of γ', L_1, L_2, J, M values we have shown that γ, $\mathbf{L}_1^2$, $\mathbf{L}_2^2$, $\mathbf{J}^2$, J_z also form a complete commuting set, and that each $|\gamma' L_1 L_2 JM\rangle$ is expressible as a linear combination of those $|\gamma' L_1 L_2 M_1 M_2\rangle$ which have the same eigenvalues for γ, $\mathbf{L}_1^2$ and $\mathbf{L}_2^2$. As an example we might have taken $\mathbf{L}_1$ and $\mathbf{L}_2$ as $\mathbf{l}_1$ and $\mathbf{l}_2$, the angular momenta of the first and second electrons, respectively. In this case $\mathbf{L}_1$ and $\mathbf{L}_2$ would not commute with the Hamiltonian $\mathscr{H}$ so we could not take $\mathscr{H}$ as one of the complete commuting set. Later, we shall be concerned with pairs of angular momenta which do commute with $\mathscr{H}$ and each other, e.g. the total spin and total orbital angular momenta, and shall then have two alternative modes of representation of the basic eigenkets.

Using the process of vector addition, we can now establish a useful result about the eigenvalues of orbital angular momenta. If we consider a single particle first, the Z-component of its orbital angular momentum is l_z and so, by (2.16), the wave-equation determining an eigenfunction of l_z is

$$l_z \psi = -i\hbar \frac{\partial \psi}{\partial \phi} = l'_z \psi, \tag{2.28}$$

with the solution
$$\psi = f(r,\theta)\, e^{il'_z \phi \hbar^{-1}}, \tag{2.29}$$

and in order for ψ to be a single-valued function of ϕ it is necessary that

$$l'_z = m_l \hbar,$$

where m_l is integral (one uses small letters for the angular momenta of single electrons and capital letters for those pertaining to many-electron systems). This is, of course, consistent with our earlier finding that m_l must be half an integer, but is more restrictive. Not only m_l but also l, defined according to (2.24), must be integral. However, we have just shown that the resultant J arising from the sum $\mathbf{J}$ of two angular momenta $\mathbf{l}_1$ and $\mathbf{l}_2$ only takes values differing by integers from (l_1+l_2). So the sum of two, and hence of any number, of orbital angular momenta can only have eigenvalues for its components which are integral multiples of $\hbar$. This means that the allowed values of L for the total orbital angular momentum of a system are all non-negative integers.

Examples

1. Verify the assertion made in the text that $\mathbf{L}_1^2$ commutes with both $\mathbf{J}^2$ and J_z.
2. λ and μ are two real numbers and $\mathbf{L}_1$ and $\mathbf{L}_2$ are commuting angular momenta. In what circumstances is $\mathbf{K} = \lambda\mathbf{L}_1 + \mu\mathbf{L}_2$ also an angular momentum? Why does your result not conflict with the fact that $\mathbf{L}_1$ and $\mathbf{J} = \mathbf{L}_1 + \mathbf{L}_2$ are angular momenta and so is $\mathbf{L}_2 = \mathbf{J} - \mathbf{L}_1$?

2.4. The spin of the electron

We have seen that, although a consideration of the commutation relations amongst components of an angular momentum $\mathbf{L}$ only requires that the eigenvalues L shall be half-integral, all orbital angular momenta actually have L integral. This was proved by induction from the fact that each component of $\mathbf{l}$ can be expressed as an operator of infinitesimal rotation about the axis to which it refers (e.g. (2.28) for l_z). Alternatively it can be derived directly from (2.16) for a general orbital angular momentum $\mathbf{L}$ using similar arguments (see example 1 at the end of this section).

Angular momenta, however, do exist which take non-integral values and we now describe these. There are good experimental and theoretical reasons for supposing that electrons (and also protons and neutrons) each possess an intrinsic angular momentum $\mathbf{s}$ of magnitude $\frac{1}{2}\hbar$, i.e. $\mathbf{s}^2$ has the eigenvalue $\frac{3}{4}\hbar^2$. This means that there are just two independent internal states for the electron, corresponding to the eigenvalues $\pm\frac{1}{2}\hbar$ of the component of $\mathbf{s}$ along some specified axis. The momentum $\mathbf{s}$ is called a spin angular momentum. There are various ways in which this spin can be introduced into the theory. The first is purely phenomenological. We say that the electron has this intrinsic angular momentum $\mathbf{s}$ of magnitude $\frac{1}{2}\hbar$; that the components of $\mathbf{s}$ commute with the observables $\mathbf{r}$ and $\mathbf{p}$ describing spatial properties of the electron and with all observables referring to other particles of the system; also that the electron has an intrinsic magnetic moment aligned along its axis of spin. We shall adopt this approach until the middle of ch. 5 when an account is given of Dirac's relativistic equation for an electron. The latter resulted from an attempt to provide an equation for the electron which satisfied simultaneously the requirements of special relativity theory and quantum mechanics. The existence of the spin and the magnetic moment are a consequence of this equation. Then finally the spin of the electron is intimately related to the connectivity of the group of rotations of Euclidean space. I give a brief description of this in § 6.10 after we have become familiar with the concept of a rotation group. Now, however, we merely accept that experimentally, and in accord with Dirac's theory, the electron has the spin $\mathbf{s}$ and a magnetic moment $\boldsymbol{\mu}$ tied to the spin according to the equation

$$\boldsymbol{\mu} = -\frac{e}{mc}\mathbf{s}. \tag{2.30}$$

The experimental value of the magnetic moment is in very good agreement with the value, $e\hbar/2mc$, deduced from (2.30) (although the agreement is not quite perfect, see § 5.4).

From our discussion of orbital angular momentum, it is evident that as the magnitude of $\mathbf{s}$ is not an integral multiple of $\hbar$ we cannot interpret the components of $\mathbf{s}$ as differential operators in ordinary three-dimensional space. Fortunately this is unnecessary because, by using (2.24) and (2.25) which hold for any angular momentum, we can write down explicitly the results of the operation of the three components of $\mathbf{s}$ on the eigenkets for a single electron. Writing γ, $\mathbf{s}^2$, s_z for a complete commuting set of observables for an electron, we note that $\mathbf{s}^2$ has always the value $\frac{1}{2}(\frac{1}{2}+1)\hbar^2 = \frac{3}{4}\hbar^2$ for every ket. We then suppose the γ to have the values γ' and can write down immediately (from (2.24) and (2.25)) the effect of $s^{\pm}$ and s_z:

$$\left.\begin{aligned} &s_z\,|\gamma'\tfrac{1}{2}\tfrac{1}{2}\rangle = \tfrac{1}{2}\hbar\,|\gamma'\tfrac{1}{2}\tfrac{1}{2}\rangle, &&s_z\,|\gamma'\tfrac{1}{2}-\tfrac{1}{2}\rangle = -\tfrac{1}{2}\hbar\,|\gamma'\tfrac{1}{2}-\tfrac{1}{2}\rangle,\\ &s^+\,|\gamma'\tfrac{1}{2}\tfrac{1}{2}\rangle = 0, &&s^+\,|\gamma'\tfrac{1}{2}-\tfrac{1}{2}\rangle = \hbar\,|\gamma'\tfrac{1}{2}\tfrac{1}{2}\rangle,\\ &s^-\,|\gamma'\tfrac{1}{2}\tfrac{1}{2}\rangle = \hbar\,|\gamma'\tfrac{1}{2}-\tfrac{1}{2}\rangle, &&s^-\,|\gamma'\tfrac{1}{2}-\tfrac{1}{2}\rangle = 0.\\ &\text{From the definition of } s^{\pm} \text{ it follows that}\\ &s_x\,|\gamma'\tfrac{1}{2}\tfrac{1}{2}\rangle = \tfrac{1}{2}\hbar\,|\gamma'\tfrac{1}{2}-\tfrac{1}{2}\rangle, &&s_x\,|\gamma'\tfrac{1}{2}-\tfrac{1}{2}\rangle = \tfrac{1}{2}\hbar\,|\gamma'\tfrac{1}{2}\tfrac{1}{2}\rangle,\\ &s_y\,|\gamma'\tfrac{1}{2}\tfrac{1}{2}\rangle = \tfrac{1}{2}i\hbar\,|\gamma'\tfrac{1}{2}-\tfrac{1}{2}\rangle, &&s_y\,|\gamma'\tfrac{1}{2}-\tfrac{1}{2}\rangle = -\tfrac{1}{2}i\hbar\,|\gamma'\tfrac{1}{2}\tfrac{1}{2}\rangle. \end{aligned}\right\} \quad (2.31)$$

We call all these equations, collectively, (2.31).

Because of the occurrence of the factor $\frac{1}{2}\hbar$ in the equations for s_x, s_y and s_z, it is convenient to consider with $\mathbf{s}$ the associated vector $\boldsymbol{\sigma}$ defined by

$$\mathbf{s} = \tfrac{1}{2}\hbar\boldsymbol{\sigma}. \qquad (2.32)$$

An angular momentum $\mathbf{s}$ with a square $\mathbf{s}^2$ which can take only the one value $\frac{3}{4}\hbar^2$ has certain algebraic properties which are not shared by general angular momenta. These are exhibited more concisely in terms of $\boldsymbol{\sigma}$. From (2.31) and (2.32) we see that σ_x^2, σ_y^2 and σ_z^2 each have the same effect on the two eigenstates $|\gamma'\frac{1}{2}\frac{1}{2}\rangle$ and $|\gamma'\frac{1}{2}-\frac{1}{2}\rangle$ as the number one. This is therefore also true for their effect on any linear combination of $|\gamma'\frac{1}{2}\frac{1}{2}\rangle$ and $|\gamma'\frac{1}{2}-\frac{1}{2}\rangle$ and so we have the operator equations

$$\sigma_x^2 = \sigma_y^2 = \sigma_z^2 = 1. \qquad (2.33)$$

In a similar way one sees directly from (2.31) and (2.32) that any pair of components of $\boldsymbol{\sigma}$ anticommute, that is

$$\left.\begin{aligned} \sigma_x\sigma_y + \sigma_y\sigma_x &= 0,\\ \sigma_y\sigma_z + \sigma_z\sigma_y &= 0,\\ \sigma_z\sigma_x + \sigma_x\sigma_z &= 0. \end{aligned}\right\} \qquad (2.34)$$

To complete the description of the algebraic scheme used to treat the spin we repeat that a spin vector of a particle commutes with the spin vector of any other particle and with any operator, referring either to the same or different particles, which involves only space or time coordinates. Thus $\mathbf{s}_\kappa$ commutes with $\mathbf{p}_\kappa$, $\mathbf{p}_\lambda$, $\mathbf{r}_\kappa$, $r_{\kappa\lambda}$ and $\mathbf{s}_\lambda$ $(\lambda \neq \kappa)$.

The total spin of a system of n particles, each of spin $\frac{1}{2}\hbar$, is given by

$$\mathbf{S} = \sum_{\kappa=1}^{n} \mathbf{s}_\kappa, \tag{2.35}$$

and the process of vector addition shows that S has only integral values for n even and half-odd integral ones for n odd. Clearly the largest possible value of S_z is $\frac{1}{2}n\hbar$, so the largest possible value of S is $\frac{1}{2}n$ for an n-electron system. The effect of the operation of the components of $\mathbf{S}$ on an arbitrary ket $|X\rangle$ is most readily obtained by expanding $|X\rangle$ in terms of the simultaneous eigenkets of a commuting set of observables which contains $s_{\kappa z}$ for each electron. Such an eigenket could be written $|\gamma' m_{s_1} m_{s_2} \ldots m_{s_n}\rangle$ and then the operation of each component of each $\mathbf{s}_\kappa$ on it is determined by (2.31). Hence so also is the operation of each component of $\mathbf{S}$.

The total angular momentum of a system is defined simply as the sum

$$\mathbf{J} = \mathbf{L} + \mathbf{S} \tag{2.36}$$

of the total orbital and the total spin angular momenta. It is customary to write $\mathbf{L}$ (or $\mathbf{l}$) for an orbital angular momentum, $\mathbf{S}$ (or $\mathbf{s}$) for a spin angular momentum and $\mathbf{J}$ (or $\mathbf{j}$) for either a total angular momentum defined according to (2.36) or for an arbitrary angular momentum whose nature has not yet been specified. In future we, too, shall use the symbols in this way and shall write equations such as (2.24) and (2.25) in terms of $\mathbf{J}$ or $\mathbf{j}$ instead of $\mathbf{L}$. From (2.36), J like S has integral or half-odd integral eigenvalues according as to whether the number of electrons in the system is even or odd. Perhaps I should say at this point that we are assuming at present that the nucleus has no spin. This is not always the case. The effects of the nuclear spin are very small, though, and we defer discussion of them until § 5.3.

Examples

1. If ϕ_λ is the ϕ coordinate for the λth electron, write $\phi^\kappa = \sum_{\lambda=1}^{n} a_{\kappa\lambda}\phi_\lambda$, where $\phi^1 = \frac{1}{\sqrt{n}}\sum_{\lambda=1}^{n}\phi_\lambda$, and show directly from (2.16) that an orbital angular momentum $\mathbf{L}$ can only take integral values.

   (Hint: Let $[a_{\kappa\lambda}]$ be an orthogonal matrix.)

2. Show that $\sigma_x\sigma_y = i\sigma_z$.

2.5. Wigner's formula

It is now time to consider the addition of angular momenta in greater detail. Let us write $\mathbf{j}_1$ and $\mathbf{j}_2$ for two commuting angular momenta and $\mathbf{j}$ for their sum. Now we know that if we take the $(2j_1+1)(2j_2+1)$ products $|j_1 m_1\rangle\,|j_2 m_2\rangle$, which we may write $|j_1 j_2 m_1 m_2\rangle$, then suitable linear combinations of them give each of the basic states for $j = |j_1 - j_2|, \ldots, j_1 + j_2$ just once. Clearly, apart from arbitrary phase factors common to all states with the same j, the coefficients in these linear combinations must be completely determined by j, j_1, j_2, m, m_1, m_2, where m is the quantum number corresponding to j_z. In fact it is possible to give an explicit formula for them. This formula was first given by Wigner and is rather complicated. However, it will be useful to us and so we give a derivation here. We shall see that it can be derived in a straightforward and completely elementary manner.[1]

We wish to find the coefficients in the expansion

$$|j_1j_2jm\rangle = \sum_{m_1, m_2} |j_1j_2m_1m_2\rangle \langle j_1j_2m_1m_2 \,|\, j_1j_2jm\rangle, \tag{2.37}$$

and as j_1, j_2 are common to all kets occurring in the problem we abbreviate (2.37) to

$$|jm\rangle = \sum_{m_1, m_2} |m_1m_2\rangle \langle m_1m_2 \,|\, jm\rangle. \tag{2.38}$$

$|m_1m_2\rangle$ is an eigenstate of j_z with $j_z' = (m_1+m_2)\hbar$ and so $\langle m_1m_2 \,|\, jm\rangle$ is zero unless $m_1+m_2 = m$. Therefore $\delta(m, m_1+m_2)$ is a factor of $\langle m_1m_2 \,|\, jm\rangle$. First, we search for the coefficients in the particular case $m = j$. So

$$|jj\rangle = \sum_{m_1, m_2} \delta(j, m_1+m_2) \langle m_1m_2 \,|\, jj\rangle \,|m_1m_2\rangle, \tag{2.39}$$

and we obtain a recurrence relation for the $\langle m_1m_2 \,|\, jj\rangle$ by operating on (2.39) with the shift operator $j^+ = j_1^+ + j_2^+$. Evidently $j^+|jj\rangle = 0$, so this gives

$$\begin{aligned} 0 = j^+|jj\rangle = \hbar \sum_{m_1m_2} \delta(j, m_1+m_2) \langle m_1m_2 \,|\, jj\rangle \\ \times \{(j_1-m_1)^{\frac{1}{2}}(j_1+m_1+1)^{\frac{1}{2}}\,|m_1+1, m_2\rangle + (j_2-m_2)^{\frac{1}{2}}(j_2+m_2+1)^{\frac{1}{2}}\,|m_1, m_2+1\rangle\}, \end{aligned} \tag{2.40}$$

where we have used (2.25). Take the coefficient of $|m_1, m_2+1\rangle$ in (2.40) and we obtain

$$\langle m_1-1, m_2+1 \,|\, jj\rangle = -\left\{\frac{(j_2-m_2)(j_2+m_2+1)}{(j_1-m_1+1)(j_1+m_1)}\right\}^{\frac{1}{2}} \langle m_1m_2 \,|\, jj\rangle,$$

where $m_1+m_2 = j$. Then we use this recurrence relation λ times to give

$$\langle m_1-\lambda, m_2+\lambda \,|\, jj\rangle = (-1)^\lambda \left\{\frac{(j_2-m_2)!\,(j_2+m_2+\lambda)!\,(j_1-m_1)!\,(j_1+m_1-\lambda)!}{(j_2-m_2-\lambda)!\,(j_2+m_2)!\,(j_1-m_1+\lambda)!\,(j_1+m_1)!}\right\}^{\frac{1}{2}} \times \langle m_1m_2 \,|\, jj\rangle. \tag{2.41}$$

Equation (2.41) gives a general formula for the coefficients with $m = j$, apart from a constant independent of λ. We now suppose $j_1 \leqslant j_2$ and put $m_1 = j_1$, $m_1-\lambda = n_1$, $m_2+\lambda = n_2$ and (2.41) becomes

$$\begin{aligned} \langle n_1n_2 \,|\, jj\rangle &= A\delta(j, n_1+n_2)(-1)^\lambda \left\{\frac{(j_2+n_2)!\,(j_1+n_1)!}{(j_2-n_2)!\,(j_1-n_1)!}\right\}^{\frac{1}{2}} \\ &= B\delta(j, n_1+n_2)(-1)^{j_1-n_1} \binom{j_2+n_2}{j_1-n_1}^{\frac{1}{2}} \binom{j_1+n_1}{j_2-n_2}^{\frac{1}{2}}, \end{aligned} \tag{2.42}$$

where A and B are constants independent of n_1, n_2 and in the last line of (2.42) we have used the usual notation $\binom{x}{y}$ for the binomial coefficient $x!/[y!(x-y)!]$.

[1] I am indebted to Racah (1942*b*) here, although his proof is not the same.

In deriving (2.42) we assumed that $j_2 \geqslant j_1$, but the symmetry of the final result shows it to be true for $j_2 \leqslant j_1$ also (the ratio of $(-1)^{j_1-n_1}$ to $(-1)^{j_2-n_2}$ is independent of n_1 and n_2 for fixed j). In (2.40)–(2.42) and also in subsequent ones, although the numbers j_1, j_2, m_1, m_2, etc., are not necessarily integers they always occur in combinations which are integers. So it is perfectly permissible to use the factorial notation.

Next we determine the normalizing coefficient B. If the $\langle n_1 n_2 | jj \rangle$ are orthonormal we must have

$$1 = \sum_{n_1 n_2} |\langle n_1 n_2 | jj \rangle|^2 = |B|^2 \sum_{n_1+n_2=j} \binom{j_2+n_2}{j_1-n_1}\binom{j_1+n_1}{j_2-n_2}. \tag{2.43}$$

The sum in (2.43) over the products of binomial coefficients can be performed explicitly. This is done by observing that

$$(1+x)^{-a-b} = (1+x)^{-a}(1+x)^{-b}$$

and expanding both sides in infinite series, as we may certainly do for any x such that $|x| < 1$. We get

$$\sum_{n=0}^{\infty} \binom{a+b+n-1}{n} (-x)^n = \sum_{m'=0}^{\infty} \binom{a+m'-1}{m'} (-x)^{m'} \sum_{m''=0}^{\infty} \binom{b+m''-1}{m''} (-x)^{m''}$$

and since, for $|x| < 1$, the series concerned are absolutely convergent, we may multiply them out term by term and equate the coefficients of x^n on the two sides to give the formula

$$\binom{a+b+n-1}{n} = \sum_{m'+m''=n} \binom{a+m'-1}{m'}\binom{b+m''-1}{m''}. \tag{2.44}$$

Putting $m' = j_1 - n_1$, $m'' = j_2 - n_2$, $n = m' + m'' = j_1 + j_2 - j$, $a = j_2 - j_1 + j + 1$, $b = j_1 - j_2 + j + 1$ we sum (2.43) by means of (2.44) to obtain

$$1 = |B|^2 \binom{j_1+j_2+j+1}{j_1+j_2-j}. \tag{2.45}$$

Clearly B is undetermined to the extent of a phase factor. It is the usual convention to define this phase factor by taking the positive square root of (2.45), i.e. we put

$$B = \binom{j_1+j_2+j+1}{j_1+j_2-j}^{-\frac{1}{2}}.$$

Substituting this in (2.42) then gives

$$\langle n_1 n_2 | jj \rangle = \delta(j, n_1+n_2)(-1)^{j_1-n_1} \binom{j_2+n_2}{j_1-n_1}^{\frac{1}{2}} \binom{j_1+n_1}{j_2-n_2}^{\frac{1}{2}} \binom{j_1+j_2+j+1}{j_1+j_2-j}^{-\frac{1}{2}}. \tag{2.46}$$

For the particular case $m = j$ (2.46) gives a simple closed formula for the coefficients. Unfortunately for general m it does not appear to be possible to do this and we have to be content with an expression as a rather complicated sum. Nevertheless, (2.46) is really the crux of the derivation, for it gives us $|jj\rangle$ in terms of $|n_1 n_2\rangle$ for all j. Once given the $|jj\rangle$ we have merely to operate on them with the shift operator j^- a sufficient number of times to give us the $|jm\rangle$ for any m. We proceed to do this next.

If one takes k applications of a shift operator J^- to an eigenket $|JM\rangle$ one gets, from (2.25)

$$(\hbar^{-1}J^-)^k\,|JM\rangle = \left\{\frac{(J+M)!\,(J-M+k)!}{(J-M)!\,(J+M-k)!}\right\}^{\frac{1}{2}} |J, M-k\rangle, \tag{2.47}$$

where we interpret the expression on the right to be identically zero if $M-k < -J$. Using (2.47) on the ket $|jj\rangle$ gives

$$(\hbar^{-1}j^-)^k\,|jj\rangle = \left(\frac{(2j)!\,k!}{(2j-k)!}\right)^{\frac{1}{2}} |j, j-k\rangle,$$

which becomes, on putting $j-k = m$,

$$\left(\frac{(2j)!\,(j-m)!}{(j+m)!}\right)^{\frac{1}{2}} |jm\rangle = (\hbar^{-1}j^-)^{j-m}\,|jj\rangle = (\hbar^{-1}j_1^- + \hbar^{-1}j_2^-)^{j-m}\,|jj\rangle$$
$$= \sum_r \binom{j-m}{r} (\hbar^{-1}j_1^-)^r\,(\hbar^{-1}j_2^-)^{j-m-r}\,|jj\rangle. \tag{2.48}$$

We now substitute (2.38) for $|jj\rangle$, actually in terms of $|n_1 n_2\rangle$ not $|m_1 m_2\rangle$, using the expressions (2.46) for the coefficients $\langle n_1 n_2\,|\,jj\rangle$. The operation of the factors $(\hbar^{-1}j_1^-)^r$ and $(\hbar^{-1}j_2^-)^{j-m-r}$ on $|n_1 n_2\rangle$ is then given immediately by further applications of (2.47). The result is the desired expansion of the $|j_1 j_2 jm\rangle$ in terms of the $|j_1 j_2 n_1 n_2\rangle$. The actual coefficients are obtained (as one sees from (2.38)) by multiplying through by the bra $\langle j_1 j_2 m_1 m_2|$. They are

$$\langle j_1 j_2 m_1 m_2\,|\,j_1 j_2 jm\rangle$$
$$= \left\{\frac{(j+m)!\,(j-m)!\,(j_1-m_1)!\,(j_2-m_2)!\,(j_1+j_2-j)!\,(2j+1)}{(j_1+m_1)!\,(j_2+m_2)!\,(j_1-j_2+j)!\,(j_2-j_1+j)!\,(j_1+j_2+j+1)!}\right\}^{\frac{1}{2}}$$
$$\times\,\delta(m, m_1+m_2) \sum_r (-1)^{j_1+r-m_1} \frac{(j_1+m_1+r)!\,(j_2+j-r-m_1)!}{r!(j-m-r)!\,(j_1-m_1-r)!\,(j_2-j+m_1+r)!}, \tag{2.49}$$

where the summation is over all values of r such that all factorials occurring are of non-negative integers ($0! = 1$). Formula (2.49) is known as Wigner's formula and the coefficients $\langle m_1 m_2\,|\,jm\rangle$ as the Wigner coefficients after Wigner who first derived the formula. They are, of course, real and satisfy

$$\langle j_1 j_2 jm\,|\,j_1 j_2 m_1 m_2\rangle = \langle j_1 j_2 m_1 m_2\,|\,j_1 j_2 jm\rangle. \tag{2.50}$$

Formula (2.49) is very fundamental in the theory of atomic structure. As I have already indicated, the coefficients are not completely determined by the requirement that all individual kets should be correctly connected by (2.25). There is still one arbitrary choice to be made for each j. Naturally it is convenient to make a definite choice once and for all and (2.49) contains implicitly such a choice, in fact the conventional one. If we put $j = j_1+j_2$, $m_1 = j_1$, $m_2 = j_2$, $m = j$, in (2.49) we readily obtain

$$\langle j_1 j_2 j_1 j_2\,|\,j_1 j_2, j_1+j_2, j_1+j_2\rangle = 1,$$

showing that $$|j_1 j_2, j_1+j_2, j_1+j_2\rangle = |j_1 j_1\rangle\,|j_2 j_2\rangle.$$

The apparent lack of symmetry between j_1 and j_2 in (2.49) is only an inevitable matter of sign (see ex. 1, below).

Next we remark that although we assumed that the kets $|j_1 j_2 m_1 m_2\rangle$ were normalized this is not entirely necessary. Suppose they are all correctly connected by the shift operators j_1^+ and j_2^+ (i.e. via (2.25)). Then if

$$\langle j_1 j_2 m_1 m_2 | j_1 j_2 m_1 m_2\rangle = N$$

for one of them it follows that

$$\langle j_1 j_2 m_1 m_2 | j_1 j_2 m_1 m_2\rangle = N,$$

with the same N, for all of them. So $(1/\sqrt{N})\,|j_1 j_2 m_1 m_2\rangle$ form a correctly connected and normalized set and hence so also do

$$\frac{1}{\sqrt{N}}\,|j_1 j_2 jm\rangle = \sum_{m_1, m_2} \frac{1}{\sqrt{N}}\,|j_1 j_2 m_1 m_2\rangle\langle j_1 j_2 m_1 m_2 | j_1 j_2 jm\rangle \tag{2.51}$$

with the $\langle j_1 j_2 m_1 m_2 | j_1 j_2 jm\rangle$ given by (2.49). In other words if we know that $|j_1 j_2 m_1 m_2\rangle$ are correctly connected but not necessarily normalized, then the coefficients (2.49) will still transform them into eigenkets of $\mathbf{j}^2$ and j_z which also are correctly connected by the shift operators but not necessarily normalized. The normalization factor required is then the same for all the $(2j_1+1)(2j_2+1)$ kets $|j_1 j_2 jm\rangle$.

We conclude this discussion of Wigner's formula by asking a question that will be very relevant to some of our later applications of the formula. In order for (2.51) to give us eigenstates of j^2 and j_z, is it really necessary that $\mathbf{j}_1$ and $\mathbf{j}_2$ should commute with one another? Could they even perhaps be taken to be the same vector? The answer to this latter question is in the affirmative, but with some reservations in detail. We suppose then that we have kets

$$|j_1 j_2 m_1 m_2\rangle = |j_1 m_1\rangle\,|j_2 m_2\rangle,$$

where the components of the product are eigenkets of the same angular momentum $\mathbf{j}$ say. We write such kets as products of observables $f_{j_1 m_1}, f_{j_2 m_2}$, say, operating on a standard ket (see Dirac § 20) $\rangle = \rangle_1\rangle_2$ referring to the system of product kets. So

$$|j_1 m_1\rangle\,|j_2 m_2\rangle \rightarrow f_{j_1 m_1} f_{j_2 m_2}\rangle. \tag{2.52}$$

If j_α is a component of $\mathbf{j}$ then from the properties (2.12) of commutators

$$[j_\alpha, f_{j_1 m_1} f_{j_2 m_2}] = f_{j_1 m_1}[j_\alpha, f_{j_2 m_2}] + [j_\alpha, f_{j_1 m_1}] f_{j_2 m_2}.$$

We interpret this equation via (2.52), remembering that $j_\alpha\rangle = 0$, to obtain

$$j_\alpha\,|j_1 j_2 m_1 m_2\rangle = |j_1 m_1\rangle(j_\alpha\,|j_2 m_2\rangle) + (j_\alpha\,|j_1 m_1\rangle)\,|j_2 m_2\rangle. \tag{2.53}$$

If, now, we define $\mathbf{j}^a$ and $\mathbf{j}^b$ by the equations

$$\left.\begin{aligned} \mathbf{j}^a\,|j_1 j_2 m_1 m_2\rangle &= (\mathbf{j}\,|j_1 m_1\rangle)\,|j_2 m_2\rangle, \\ \mathbf{j}^b\,|j_1 j_2 m_1 m_2\rangle &= |j_1 m_1\rangle(\mathbf{j}\,|j_2 m_2\rangle), \end{aligned}\right\} \tag{2.54}$$

then $\mathbf{j} = \mathbf{j}^a + \mathbf{j}^b$ and provided that each of the $|j_1 m_1\rangle$ and each of the $|j_2 m_2\rangle$ are correctly connected in phase by the shift operators we can follow through our previous arguments to conclude that $|j_1 j_2 jm\rangle$ given by (2.37) and (2.49) are eigenkets of $\mathbf{j}^2$ and j_z with the indicated j and m values. We cannot, however, arrive so readily at the correct normalizing factors. For if $|j_1 m_1\rangle$ and $|j_2 m_2\rangle$ are separately normalized it will usually be the case that $|j_1 j_2 m_1 m_2\rangle$ is not normalized (because the two constituent kets are functions of the same variables and therefore to normalize a product ket we only integrate one set, not two sets, of times). The $|j_1 j_2 jm\rangle$ are, for a fixed value of j, correctly connected in phase so they are all normalized by the same factor if we restrict attention to one value of j, but varying m.

Examples

1. Show that $\langle j_2 j_1 m_2 m_1 \,|\, j_2 j_1 jm\rangle = (-1)^{j_1+j_2-j}\langle j_1 j_2 m_1 m_2 \,|\, j_1 j_2 jm\rangle$.

(Hint: Change the parameter of summation in (2.49).)

2. If $j_1 = j_2 = l$, then $|\,ll00\rangle = \dfrac{1}{\sqrt{(2l+1)}} \sum_{m=-l}^{+l} (-1)^{l-m}\,|\,llm\;-m\rangle$.

3. Take $j_1 = 1$ and derive the coefficients shown in Table 2.2.

Table 2.2. *Wigner coefficients for* $j_1 = 1$

	$m_1 = -1$	$m_1 = 0$	$m_1 = +1$
$j_2 = j-1$	$\sqrt{\dfrac{(j-m)(j-m-1)}{2j(2j-1)}}$	$\sqrt{\dfrac{j^2-m^2}{j(2j-1)}}$	$\sqrt{\dfrac{(j+m)(j+m-1)}{2j(2j-1)}}$
$j_2 = j$	$-\sqrt{\dfrac{(j-m)(j+m+1)}{2j(j+1)}}$	$\dfrac{-m}{\sqrt{j(j+1)}}$	$\sqrt{\dfrac{(j+m)(j-m+1)}{2j(j+1)}}$
$j_2 = j+1$	$\sqrt{\dfrac{(j+m+1)(j+m+2)}{(2j+2)(2j+3)}}$	$-\sqrt{\dfrac{(j+1)^2-m^2}{(j+1)(2j+3)}}$	$\sqrt{\dfrac{(j-m+1)(j-m+2)}{(2j+2)(2j+3)}}$

2.6. Permutations and the Pauli Exclusion Principle

We now leave our main chain of development to describe, in this and the next section, two rather special topics which are basic to the theory of atomic structure. The first is an investigation of some consequences of the requirement that electrons should obey Fermi–Dirac statistics. For this we need a few elementary properties of permutations.

A permutation of n labelled objects can be written as

$$P = \begin{bmatrix} 1 & 2 & \dots & n \\ i_1 & i_2 & \dots & i_n \end{bmatrix}. \tag{2.55}$$

P, in (2.55), changes the mth object from being the mth object into being the i_mth object. Thus if the objects have labels $1, 2, \dots, n$ then P changes the label from m to i_m. Alternatively we could have kept the labels unchanged and permuted the objects. We do not do this. There is no need to put the numbers $1, 2, \dots, n$ in their natural order in (2.55) and we could replace them by any permutation of them provided we replace $i_1, i_2, \ldots, i_n$ by the same permutation of them. For example, alternative expressions are

$$P = \begin{bmatrix} 2 & 1 & \ldots & n \\ i_2 & i_1 & \ldots & i_n \end{bmatrix} = \begin{bmatrix} n & n-1 & \ldots & 2 & 1 \\ i_n & i_{n-1} & \ldots & i_2 & i_1 \end{bmatrix}, \quad \text{etc.}$$

The simplest permutation is the identity permutation

$$I = \begin{bmatrix} 1 & 2 & \ldots & n \\ 1 & 2 & \ldots & n \end{bmatrix}$$

which leaves all the labels unaltered. Then corresponding to P of (2.55) there is the inverse permutation

$$P^{-1} = \begin{bmatrix} i_1 & i_2 & \ldots & i_n \\ 1 & 2 & \ldots & n \end{bmatrix}$$

which brings the labels back again to their original positions. The product $P_1 P_2$ of permutations P_1 and P_2 is defined in the natural way as P_2 followed by P_1. So we obviously have

$$P^{-1}P = PP^{-1} = I.$$

When there are n different objects there are $n!$ different permutations corresponding to the $n!$ different possible choices of $i_1, i_2, \ldots, i_n$ in (2.55).

A specially simple kind of permutation is a cycle. This is defined to be a permutation which permutes a certain number, r say, of the labels cyclically and leaves the others unaltered. A typical cycle would be

$$Q = \begin{bmatrix} i_1 & i_2 & i_3 & \ldots & i_{r-1} & i_r & i_{r+1} & \ldots & i_n \\ i_2 & i_3 & i_4 & \ldots & i_r & i_1 & i_{r+1} & \ldots & i_n \end{bmatrix}.$$

We say that Q is a cycle of order r and write it in the abbreviated form

$$Q = (i_1 i_2 \ldots i_r).$$

Any permutation P may be broken up into a product of cycles. This can be seen most easily by following the fate of a given label on indefinite repetition of P. Thus if

$$P = \begin{bmatrix} 1 & 2 & 3 & 4 & 5 & 6 & 7 & 8 \\ 3 & 5 & 7 & 8 & 1 & 6 & 2 & 4 \end{bmatrix}$$

we find that $1 \to 3 \to 7 \to 2 \to 5 \to 1$, $4 \to 8 \to 4$ and $6 \to 6$. So

$$P = (13725)(48)(6).$$

It is unnecessary to write down cycles which contain only one label (except of course for I which would have nothing left at all if we adhered rigidly to this convention). We notice that P is expressed as a product of independent cycles, that is no two cycles have a label in common. This means that the order of the cycles in the expression for P is irrelevant; they commute with one another. So

$$P = (13725)(48) = (48)(13725).$$

A cycle of order two is called a transposition. As the name suggests, it simply transposes, or interchanges, two of the labels and leaves all the others unchanged. Any cycle may be expressed as a product of transpositions. Thus

$$(i_1 i_2 \dots i_r) = (i_1 i_r)(i_1 i_{r-1}) \dots (i_1 i_3)(i_1 i_2).$$

The individual transpositions here do not commute with one another. We have seen already that a general permutation P can be expressed as a product of cycles and now we see that it can be broken up still further into a product of transpositions. The number of independent cycles was uniquely determined by the permutation P but this is no longer true for the number of transpositions. For example, we may always add $(12)(21) = I$ without altering P. The particular permutation P we considered above can be written

$$P = (48)(15)(12)(17)(13) = (15)(48)(12)(17)(12)(12)(13)$$

and in many other ways. However, the parity (i.e. evenness or oddness) of the number of transpositions is uniquely determined by P. This is most easily seen by considering the so-called alternant determinant

$$\Delta \equiv \begin{vmatrix} x_1^{n-1} & x_1^{n-2} & \dots & x_1 & 1 \\ x_2^{n-1} & x_2^{n-2} & \dots & x_2 & 1 \\ \dots & \dots & \dots & \dots & \dots \\ x_n^{n-1} & x_n^{n-2} & \dots & x_n & 1 \end{vmatrix} \equiv \prod_{i<j}^{n} (x_i - x_j). \tag{2.56}$$

Any transposition, (ab) say, applied to the suffixes simply interchanges the ath with the bth row in (2.56) and hence just changes the sign of Δ. This means that the product of m transpositions multiplies Δ by $(-1)^m$. So if P is expressed in two ways as the product of m_1 and m_2 transpositions respectively then P multiplies Δ by $(-1)^{m_1}$ and $(-1)^{m_2}$. But the effect of P on Δ is perfectly well defined and therefore $(-1)^{m_1} = (-1)^{m_2}$. The common value is called the parity of P and P is said to be odd or even according as to whether m_1 is odd or even. Evidently the parity of the product $P_\mu P_\nu$ is given by the product $(-1)^{\mu+\nu}$ of the parities.

The reader will probably know that there is an intimate connexion between permutations and determinants.[1] The determinant $|A|$ of an $n \times n$ matrix $A = [a_{ij}]$ satisfies

$$|A| = \sum_\nu (-1)^\nu P_\nu a_{11} a_{22} \dots a_{nn}, \tag{2.57}$$

where the P_ν act only on the second suffixes (i.e. they permute the columns of the matrix). Equation (2.57) would also be true if the P_ν acted only on the first suffixes. ν takes $n!$ integral values and P_ν runs over all $n!$ permutations. We take ν running from 1 to $n!$ and having the same parity as P_ν. This is always possible (see ex. 4, p. 30). Take now a fixed permutation P_σ. Expressing it as a product of transpositions, it follows from the elementary properties of determinants that

$$P_\sigma |A| = (-1)^\sigma |A|, \tag{2.58}$$

which will be important later. Via (2.57), (2.58) is to be related to the fact that $P_\sigma P_\nu$ ranges over all $n!$ permutations if P_ν does itself.

[1] See, for example, Aitken (1946).

We are now in a position to investigate in more detail the meaning and consequences of the requirement that electrons should obey Fermi–Dirac statistics. The wave-function ψ for an n-electron system, including spin, must satisfy

$$P_\nu \psi = (-1)^\nu \psi, \tag{2.59}$$

for each permutation P_ν of the n arguments. Thus for a two-electron system we might take

$$\psi = f(1)\,g(2) - f(2)\,g(1), \tag{2.60}$$

where $f(1)$ is a function for the first electron alone, and similarly for the other functions in (2.60). The identity permutation $P_1 = I$ obviously leaves ψ unchanged. There is only one other permutation P_2 and it transposes 1 and 2. Therefore

$$P_2 \psi = f(2)\,g(1) - f(1)\,g(2) = -\psi,$$

which shows that ψ satisfies (2.59).

Very often we shall express our wave-functions as sums of functions each of which is a simple product of the form

$$\chi = \phi_1(1)\,\phi_2(2)\ldots\phi_n(n). \tag{2.61}$$

In (2.61) each of the constituent functions is a function of the coordinates (space and spin) of a single electron. In ket notation (2.61) would become

$$|\chi\rangle = |\phi_1(1)\rangle\,|\phi_2(2)\rangle\ldots|\phi_n(n)\rangle. \tag{2.62}$$

It is easier to understand the essentials here by using (2.61), but all the equations of this section can be immediately translated into the ket form, (2.62), if the reader prefers.

The function (2.61) does not satisfy (2.59). The effect of P_ν on (2.61) can be written down immediately because it simply permutes the arguments $1, 2, \ldots, n$. If some of the functions ϕ_i are the same then there will be permutations P_ν, apart from the identity permutation I, which leave χ unchanged but none which change its sign. However, from (2.61) it is easy to construct a function which does satisfy (2.59).

Let us write

$$\Psi = \frac{1}{\sqrt{n!}} \sum_\nu (-1)^\nu P_\nu \phi_1(1)\,\phi_2(2)\ldots\phi_n(n), \tag{2.63}$$

in which the νth term of the sum is obtained by applying the permutation P_ν to the arguments and then multiplying by plus or minus one according as to whether P_ν is even or odd. We see that (2.63) is equal to the determinant

$$\Psi = \frac{1}{\sqrt{n!}} \begin{vmatrix} \phi_1(1) & \phi_1(2) & \ldots & \phi_1(n) \\ \phi_2(1) & \phi_2(2) & \ldots & \phi_2(n) \\ \cdots & \cdots & \cdots & \cdots \\ \phi_n(1) & \phi_n(2) & \ldots & \phi_n(n) \end{vmatrix}. \tag{2.64}$$

Ψ is said to be a determinantal wave-function and could also be written

$$\Psi = \frac{1}{\sqrt{n!}}\,|\phi_i(j)|.$$

We can deduce from (2.63) and (2.64) a number of interesting things about Ψ. Equation (2.58) shows that Ψ does satisfy the requirement of (2.59). Conversely if a function

$$\Psi^1 = \sum_\nu \lambda_\nu P_\nu \phi_1(1)\,\phi_2(2)\ldots\phi_n(n)$$

satisfies (2.59) with the λ_ν numbers, then

$$(-1)^\sigma P_\sigma \Psi^1 = \Psi^1.$$

So
$$\begin{aligned} n!\,\Psi^1 &= \sum_\sigma (-1)^\sigma P_\sigma \Psi^1 = \sum_\sigma \sum_\nu (-1)^\sigma P_\sigma \lambda_\nu P_\nu \phi_1(1)\ldots\phi_n(n), \\ &= \sum_\nu \lambda_\nu \sum_\sigma (-1)^\sigma P_\sigma P_\nu \phi_1(1)\ldots\phi_n(n), \\ &= \sum_\nu \lambda_\nu (-1)^\nu \sum_\tau (-1)^\tau P_\tau \phi_1(1)\ldots\phi_n(n), \end{aligned}$$

where we have written $P_\tau = P_\sigma P_\nu$, and used the fact that P_τ runs over all permutations when P_σ does. $\sum_\nu \lambda_\nu(-1)^\nu$ is just a number, possibly zero, and so we have shown that Ψ is the only independent function satisfying (2.59) which can be derived from (2.61). The determinantal form of Ψ shows that it is not identically zero if and only if the functions $\phi_1, \phi_2, \ldots, \phi_n$ are linearly independent. It is convenient to assume that they form an orthonormal set, that is that

$$\langle \overline{\phi}_i \mid \phi_j \rangle = \int \overline{\phi}_i \phi_j \, d\tau = \delta_{ij}.$$

If they do, then the only non-vanishing terms in

$$\langle \overline{\Psi} \mid \Psi \rangle = \int \overline{\Psi} \Psi \, d\tau_1 \ldots d\tau_n$$

are those in which the arguments are in the same order in both $\overline{\Psi}$ and Ψ. So $\langle \overline{\Psi} \mid \Psi \rangle = 1$ which is the reason for the factor $1/\sqrt{n!}$ in (2.63) and (2.64).

If we take any orthonormal set of functions $\chi_1, \chi_2, \ldots$, not necessarily finite in number and select n of them, $\chi_{i_1}, \chi_{i_2}, \ldots, \chi_{i_n}$ say, we can build a determinantal function (2.64) from them. Then these n are linearly independent if and only if they are all different. Now suppose each χ_i is expressed as a product of a space function and a spin function according to the rules

$$\begin{aligned} \chi_{2j-1} &= f_j \alpha, \\ \chi_{2j} &= f_j \beta \quad (j = 1, 2, \ldots). \end{aligned}$$

Here f_j is the jth space function and α and β are the spin functions having $m_s = +\frac{1}{2}$ and $-\frac{1}{2}$, respectively. Then Ψ is non-zero only when it is composed of just n distinct χ_i which means, in turn, that not more than two can have any given f_j as their spatial parts. Translated into words, this is a familiar form of Pauli's exclusion principle: no more than two electrons may occupy the same orbital.

An interesting application of the general theory is to states of a system containing just two electrons. Let us suppose that ϕ_1 in (2.64) has f_1 as its spatial part, ϕ_2 has f_2 and that f_1, f_2 are different functions of an orthonormal set. Then after placing one electron in f_1 and the other in f_2 we obtain four possible functions corresponding to the two choices of spin for each space function. We write them $\Psi_{\alpha\alpha}, \Psi_{\alpha\beta}, \Psi_{\beta\alpha}, \Psi_{\beta\beta}$ in an obvious notation and can then multiply out the determinants to obtain (α_1 is short for $\alpha(1)$ and means that particle 1 has $m_s = +\frac{1}{2}$, etc.)

$$\left.\begin{aligned}
\Psi_{\alpha\alpha} &= \frac{1}{\sqrt{2}}(f_1(1)\,f_2(2)-f_2(1)\,f_1(2))\,\alpha_1\alpha_2,\\
\Psi_{\alpha\beta} &= \frac{1}{\sqrt{2}}(f_1(1)\,f_2(2)\,\alpha_1\beta_2-f_2(1)\,f_1(2)\,\alpha_2\beta_1),\\
\Psi_{\beta\alpha} &= \frac{1}{\sqrt{2}}(f_1(1)\,f_2(2)\,\alpha_2\beta_1-f_2(1)\,f_1(2)\,\alpha_1\beta_2),\\
\Psi_{\beta\beta} &= \frac{1}{\sqrt{2}}(f_1(1)\,f_2(2)-f_2(1)\,f_1(2))\,\beta_1\beta_2.
\end{aligned}\right\} \qquad (2.65)$$

Each Ψ of (2.65) is an eigenfunction of S_z, M_S being $+1$ for $\Psi_{\alpha\alpha}$, 0 for $\Psi_{\alpha\beta}$ and $\Psi_{\beta\alpha}$ and -1 for $\Psi_{\beta\beta}$. Because of their expression in terms of eigenfunctions of s_{1z} and s_{2z} we can apply to the Ψ the shift operators $S^{\pm}$. $S^+\Psi_{\alpha\alpha} = 0$ which shows $\Psi_{\alpha\alpha}$ to be in an eigenstate of $\mathbf{S}^2$ with eigenvalue $S = 1$ (see p. 15). Put $\Psi(SM_S)$ for a function with eigenvalues S and M_S for the total spin and linearly dependent on the four functions (2.65). Then S^- on (2.65) shows that

$$\left.\begin{aligned}
\Psi(11) &= \Psi_{\alpha\alpha} = \frac{1}{\sqrt{2}}(f_1(1)\,f_2(2)-f_2(1)\,f_1(2))\,\alpha_1\alpha_2,\\
\Psi(10) &= \frac{1}{\sqrt{2}}(\Psi_{\alpha\beta}+\Psi_{\beta\alpha}) = \frac{1}{\sqrt{2}}(f_1(1)\,f_2(2)-f_2(1)\,f_1(2))\frac{1}{\sqrt{2}}(\alpha_1\beta_2+\beta_1\alpha_2),\\
\Psi(1\,-1) &= \Psi_{\beta\beta} = \frac{1}{\sqrt{2}}(f_1(1)\,f_2(2)-f_2(1)\,f_1(2))\,\beta_1\beta_2.
\end{aligned}\right\} \qquad (2.66)$$

The fourth linearly independent function is

$$\begin{aligned}
\Psi(00) &= \frac{1}{\sqrt{2}}(\Psi_{\alpha\beta}-\Psi_{\beta\alpha})\\
&= \frac{1}{\sqrt{2}}(f_1(1)\,f_2(2)+f_2(1)\,f_1(2))\,(\alpha_1\beta_2-\beta_1\alpha_2).
\end{aligned} \qquad (2.67)$$

Both $S^+\Psi(00)$ and $S_z\Psi(00)$ are zero which justifies the notation used in (2.67). The functions (2.66) are called collectively a triplet and (2.67) is called a singlet, the name giving the multiplicity due to the spin. We notice that in this particular case all the eigenfunctions of the spin have achieved antisymmetry in a specially simple way. They are all products of two functions, one a function of space coordinates alone and the other of spin coordinates alone. For those with $S = 1$ the space part is antisymmetric and the spin part symmetric. The opposite is true for the function with $S = 0$.

The main point of the preceding analysis is that it shows a way in which the spin can influence the energy even without entering explicitly into the Hamiltonian. Because of the antisymmetry of the total function for the two electrons, the space part of it must be symmetric when $S = 0$ and antisymmetric when $S = 1$. This is in fact a general result for two electrons and does not depend upon choosing functions of the simple type Ψ of (2.65). The symmetric and antisymmetric solutions of the equation

$$\mathscr{H}\psi = E\psi,$$

where ψ is a function of space coordinates alone, have in general quite different energies. At the moment all we can say is that they could have different energies; later we shall see that this is usually the case in fact. In particular this means that energies depend not only on the choice of orbitals but also on the total spin. If there are more than two electrons it is not always possible to factorize the total eigenfunction for definite states of spin in the way we have done above. However, the assertion that the total spin has a strong indirect effect on the allowed energies remains true.

Examples

1. Relate (2.66) and (2.67) to the vector addition procedure for spin vectors.

2. n space functions, not necessarily distinct, are chosen from an orthonormal set of one-electron functions. 2^n functions Ψ are written down, according to (2.64), one for each choice of the set of n spin functions. What condition must the n space functions satisfy in order for there to be a non-zero linear combination of the Ψ having $S = \frac{1}{2}n$? When the condition is satisfied, write down explicit expressions for these linear combinations.

3. A function Ψ referring to three or more electrons and satisfying (2.59) is expressible as a product of a space part F and a spin part A. The space part is totally symmetric, that is $P_\nu F = F$ for all ν. Show that $\Psi \equiv 0$.

(Hint: There are just two independent spin functions for one electron, hence 2^n for an n-electron system.)

4. If $n > 1$, show that there are $\frac{1}{2}(n!)$ even and $\frac{1}{2}(n!)$ odd permutations.

(Hint: Consider what happens to TP_μ, where T is a fixed transposition, as P_μ ranges over the even permutations.)

2.7. The Russell–Saunders coupling scheme

Using the Hamiltonian $\mathscr{H}$ of (2.1) we saw in § 2 that the total orbital angular momentum $\mathbf{L}$ commutes with $\mathscr{H}$. $\mathscr{H}$ does not involve the spin coordinates at all and so the total spin angular momentum $\mathbf{S}$ commutes with it also. Therefore $\mathscr{H}$, $\mathbf{L}^2$, L_z, $\mathbf{S}^2$ and S_z form a set of five commuting observables. This means that we can classify the states of an atomic system by taking them as simultaneous eigenkets of these five observables. From a purely abstract and conceptual point of view it is very illuminating to do this. However, in practice it is not generally possible because the Schrödinger equation (2.8) can only be solved exactly when there is just a single electron in the system. On the other hand, it is comparatively easy to find eigenkets of the remaining four observables and in practical problems it is often useful to use a complete set of kets which are simultaneous eigenkets of $\mathbf{L}^2$, L_z, $\mathbf{S}^2$ and S_z and which are approximate eigenkets of $\mathscr{H}$. Any simultaneous eigenket of $\mathbf{L}^2$, L_z, $\mathbf{S}^2$, S_z and $\mathscr{H}$ can then be expanded in terms of our complete set, and if all the coefficients except one are very small then the corresponding member of our complete set is an approximate eigenket of $\mathscr{H}$. It is convenient to keep such a definition of 'approximate eigenket of $\mathscr{H}$' at the back of our minds although unfortunately we shall be forced to rely on physical, rather than mathematical, arguments to justify the use of the phrase for the particular approximate eigenkets that we ultimately use. To balance against this, in the later chapters of the book we will find considerable *a posteriori* justification for them in our comparison with experimental data.

We return now to concepts which can be precisely and usefully defined and take a complete set of simultaneous eigenkets of $\mathbf{L}^2$, L_z, $\mathbf{S}^2$ and S_z. Our previous discussion of addition of angular momenta shows that there is a complete set of eigenkets of $\mathbf{S}^2$, $\mathbf{L}^2$, $\mathbf{J}^2$ and J_z, where $\mathbf{J} = \mathbf{S}+\mathbf{L}$, associated with our first set through Wigner's formula (2.49). Any ket from either complete set is a simultaneous eigenket of $\mathbf{S}^2$ and $\mathbf{L}^2$. It is customary to call any such complete set of kets, each of which is a simultaneous eigenket of $\mathbf{S}^2$ and $\mathbf{L}^2$ a Russell–Saunders (or SL) coupling scheme. The name of the coupling scheme is a tribute to the pioneer work by Russell and Saunders in the early classification of atomic spectra. The two particular Russell–Saunders coupling schemes mentioned above are called respectively the $SLM_S M_L$ and the $SLJM$ schemes, the quantum numbers referring to spin being written before those referring to orbital angular momentum. Their eigenkets are written $|\alpha' SLM_S M_L\rangle$ and $|\alpha' SLJM\rangle$ respectively, where α represents additional observables needed to make up a complete set and $\mathbf{S}$ corresponds to $\mathbf{j}_1$ and $\mathbf{L}$ to $\mathbf{j}_2$ in (2.49).

We saw in the last section that the eigenvalue S might have a large effect on the allowed energies because of the antisymmetry of the wave-function and anticipated to say that indeed this is usually so. We anticipate again to say that L also has a large effect on the energy. S_z and L_z, however, have none at all because both $\mathbf{S}$ and $\mathbf{L}$ are symmetrical functions of the electronic coordinates, i.e. any permutation of the electrons leaves them unchanged. As a consequence of this, if we have a ket $|\alpha' SLM_S M_L\rangle$ satisfying the antisymmetry requirement (2.59) then all kets obtained by operating with the shift operators $S^\pm$ and $L^\pm$ will also satisfy (2.59). So the kets $|\alpha' SLM_S M_L\rangle$ can be grouped into sets satisfying (2.59) and each set containing $(2S+1)(2L+1)$ individual states, one for each allowed pair $(S_z' L_z')$. Such a set is called a term and is written ${}^{2S+1}L$. If one needs to distinguish different terms one will put additional information before the term symbol, for example, one might write $a\,{}^{2S+1}L$ and $b\,{}^{2S+1}L$ for two different terms having the same S and L values. When we wish to indicate individual states of a term we write the M_S and M_L values after the term symbol in that order, so we write ${}^{2S+1}L$, M_S, M_L or in ket notation $|{}^{2S+1}L, M_S, M_L\rangle$. It is fairly obvious from what has already been said that all states of a given term have the same first-order energies, i.e. that

$$\langle \alpha' SLM_S M_L | \mathscr{H} | \alpha' SLM_S M_L\rangle$$

is the same for all M_S, M_L when α', S and L are given. I leave the reader to establish this for himself at this stage; he will find a formal proof in § 8 which holds also for the exact energies.

The superscript $(2S+1)$ is called the multiplicity of the term and we say the term is a singlet, doublet, triplet, ... according as $S = 0, 1, 2, \ldots$. When $L \geqslant S$ the multiplicity is equal to the number of different J values occurring for the corresponding kets in the $SLJM$ scheme. It is customary to call it the multiplicity, however, whether $L \geqslant S$ or not. In practice one usually knows the numerical value of L and then writes ${}^{2S+1}X$, where X is a capital letter determined by the rule that

$$X = S, P, D, F, G, H, I, K, L, M, N, O, Q, \dots \quad \text{when} \quad L = 0, 1, 2, 3, \dots.$$

The rather peculiar choice of letters for low L values has its etymology in early spectroscopic usage. The first four letters are the initials of the words Sharp, Principal, Diffuse and Fundamental which were used to describe the appearance of certain spectral lines. After F the choice is alphabetical, J being omitted because of the possibility of confusion with a general angular momentum vector. There is not much danger in practice of confusion between $X = S$ and the spin quantum number S. It is usually possible to determine experimentally with reasonable certainty what are the S and L values of the various eigenstates of $\mathscr{H}$. The term which has the lowest energy is called the ground term. As an example, the ground term of the neutral iron atom is known to be ${}^5D$ which means that it has $S = L = 2$.

It is interesting to ask why we use $\mathbf{L}$ and $\mathbf{S}$ rather than $\mathbf{L}$ and all the $\mathbf{s}_i$, one for each electron. For if we used the latter we should, in general, be able to make up a larger commuting set of observables and so we might imagine that this would give a more detailed and useful classification of the states of the system. This is not so, however, just because of the requirement of antisymmetry. First, we remark that $\mathbf{s}_i^2$ has always the value $\frac{3}{4}\hbar^2$ and so tells us nothing useful. Let us suppose then that we have a set of simultaneous eigenkets of the n components s_{iz} which satisfy the antisymmetry requirement (2.59). A typical one may be written $|\alpha' m_{s_1} m_{s_2} \dots m_{s_n}\rangle$, where α' include L and M_L. Then

$$s_{iz}|\alpha' m_{s_1} m_{s_2} \dots m_{s_n}\rangle = m_{s_i}\hbar\,|\alpha' m_{s_1} m_{s_2} \dots m_{s_n}\rangle.$$

Now we operate on this with the transposition $T_{ij} = (ij)$ and obtain

$$-s_{jz}|\alpha' m_{s_1} m_{s_2} \dots m_{s_n}\rangle = -m_{s_i}\hbar\,|\alpha' m_{s_1} m_{s_2} \dots m_{s_n}\rangle,$$

showing that the antisymmetry forces all the m_{s_i} to be equal. But the ket $|\alpha' m_{s_1} m_{s_2} \dots m_{s_n}\rangle$ is clearly an eigenket of S_z' with eigenvalue $\hbar \sum_{i=1}^{n} m_{s_i} = n\hbar m_{s_1}$ by our argument above. So we have not really any extra information at all. On the other hand, s_{iz} does not commute with $\mathbf{S}^2$ and so we would not have thought of including $\mathbf{S}^2$ in our set of observables. In fact then we have in every way a more useful classification when we use $\mathbf{S}^2$ and S_z rather than the individual s_{iz}.

Until we come to ch. 5 we shall use the approximate Hamiltonian $\mathscr{H}$ of (2.1). The main corrections which it ultimately needs arise from our present neglect of the various magnetic interactions both amongst the electrons and also between the electronic motions and the electric field of the nucleus. We consider these in detail in ch. 5 but merely say here that the extra terms which will be added to the Hamiltonian $\mathscr{H}$ do not in general commute with $\mathbf{S}$ or $\mathbf{L}$ but do commute with **J**.[1] This is why the *SLJM* scheme is sometimes useful: we use it when we wish to consider the effect of these extra terms. The extra terms usually have a small effect on the energy and then their main influence will be to separate the various J values of a given term. It is customary to call the set of $(2J+1)$ states with given J and connected by the shift operators $J^{\pm}$ a level. Then the J value is written as a subscript to the term symbol, thus, ${}^{2S+1}L_J$. The five levels of the ${}^5D$ term are ${}^5D_0$, ${}^5D_1$, ${}^5D_2$, ${}^5D_3$ and ${}^5D_4$ with respective degeneracies 1, 3, 5, 7 and 9. Individual states are written by placing the M_J value of the level after the term symbol as a superscript, thus ${}^{2S+1}L_J^M$.

We conclude this section by remarking that there are cases when the separation between terms is small compared with the effects of the parts left out of $\mathscr{H}$ in (2.1). We shall not come across such cases in this book (though see §§ 5.2 and 7.4) but for them the SL coupling schemes are not even useful first approximations.

Example

The functions (2.66) and (2.67) are all eigenstates of $\mathbf{S}^2$ and S_z but only two of them are eigenstates of s_{1z} and s_{2z}.

2.8. Vectors, their matrix elements and selection rules

The z-component of the orbital angular momentum vector for a single electron is a constant multiple of the infinitesimal rotation operator about the z-axis. This is the meaning of equation (2.28). A similar thing is true for the x- and y-components and the only reason that we chose to demonstrate it for l_z in (2.28) is that our spherical polar coordinates were chosen to refer to the z-axis. So when we classify kets for a single electron by the eigenvalues of $\mathbf{l}^2$ and l_z we are really specifying the behaviour of those kets under rotations of the coordinate space for that electron. If the reader did ex. 1 at the end of § 2.4 he will realize that the components of **L**, the total angular momentum vector for an n-electron system, rotate the coordinate spaces for all those electrons together.

It is now possible to distinguish two rather special kinds of vector with respect to an angular momentum **L**, where **L** refers to one or more electrons. The first kind is a vector lying totally within a space that the components of **L** rotate. A typical one is the coordinate vector $\mathbf{r}_i$ of a single one of the electrons to which **L** refers. Then the second kind lies totally outside the spaces which the components of **L** rotate. A typical example would be the total spin vector **S** for the electrons. I follow Condon and Shortley and call the first special kind of vector a 'vector of type T with respect to the angular momentum **L**'. We notice that this concept is one of mutual but unsymmetrical relationship. Finally, there are vectors which have neither of these simple relationships to **L**, and the total angular momentum **J** is an example of such a vector.

I have introduced the two special categories of vector with respect to **L** via the concept of infinitesimal rotation operators in order to clarify their true significance. It is now desirable to characterize them alternatively in terms of ideas which are more fundamental to our present approach to quantum mechanics.

[1] Assuming the nucleus to be a point charge.

A component of $\mathbf{L}$ is an operation of differentiation so if we have a ket $|X\rangle$ and a component v_i of a vector then the rule for differentiating a product shows us that

$$L_x v_i |X\rangle = (L_x v_i)|X\rangle + v_i L_x |X\rangle,$$

and similarly for the other components of $\mathbf{L}$. This equation may be rewritten in terms of commutators as

$$[L_x, v_i]|X\rangle = (L_x v_i)|X\rangle$$

or since $|X\rangle$ is arbitrary as $\quad [L_x, v_i] = (L_x v_i)$.

$(L_x v_i)$ is an infinitesimal rotation of the component v_i and so must be expressible as a linear combination of the three components of $\mathbf{v}$ (this is true for a finite rotation and hence also for an infinitesimal one). All vectors have the same behaviour under rotations, by definition, so the coefficients of this linear combination must be the same for all vectors in a space rotated by $\mathbf{L}$. Thus a vector of type T with respect to $\mathbf{L}$ is completely characterized by the coefficients representing the linear dependence of its commutators with respect to the components of $\mathbf{L}$ on its own components. Let us now be explicit and take the coordinate vector $\mathbf{r}$ of one of the electrons. Its nine commutators with components of $\mathbf{L}$ are

$$\left.\begin{array}{lll} [L_x, x] = 0, & [L_x, y] = i\hbar z, & [L_x, z] = -i\hbar y, \\ [L_y, x] = -i\hbar z, & [L_y, y] = 0, & [L_y, z] = i\hbar x, \\ [L_z, x] = i\hbar y, & [L_z, y] = -i\hbar x, & [L_z, z] = 0. \end{array}\right\} \qquad (2.68)$$

Having reached this characterization we can extend it to a general angular momentum $\mathbf{j}$. It is also possible to interpret the components of $\mathbf{j}$ as infinitesimal rotations in a suitable space but we do not pursue this point at the moment (see §6.10). Let us take then as our definition of $\mathbf{T}$ as a vector of type T with respect to a general angular momentum $\mathbf{j}$ the requirement that it shall have the following commutators with $\mathbf{j}$

$$\left.\begin{array}{lll} [j_x, T_x] = 0, & [j_x, T_y] = i\hbar T_z, & [j_x, T_z] = -i\hbar T_y, \\ [j_y, T_x] = -i\hbar T_z, & [j_y, T_y] = 0, & [j_y, T_z] = i\hbar T_x, \\ [j_z, T_x] = i\hbar T_y, & [j_z, T_y] = -i\hbar T_x, & [j_z, T_z] = 0. \end{array}\right\} \qquad (2.69)$$

Before discussing the consequences of (2.69) we will characterize our second special kind of vector. If $\mathbf{v}$ lies outside the space rotated by $\mathbf{L}$ then $\mathbf{v}$ has constant components when partially differentiated by the variables referring to the space rotated by $\mathbf{L}$. So $(L_x v_i) = 0$ and hence all the nine commutators of $\mathbf{L}$ with $\mathbf{v}$ are zero. We define our second special kind of vector with respect to a general angular momentum $\mathbf{j}$ to be simply any vector which commutes, i.e. all its components commute, with $\mathbf{j}$.

I have now said enough to make it clear which vectors are of type T with respect to a given $\mathbf{j}$ and which commute with it. However, the preceding discussion is suggestive rather than rigorous and so it is desirable to show how one derives the results directly from the definitions in terms of commutators. We do this by a process of induction or building up. It follows immediately from (2.2) and the definition of $\mathbf{l}$ that $\mathbf{r}$ and $\mathbf{p}$ are of type T with respect to $\mathbf{l}$. Then any angular momentum $\mathbf{j}$ is of type T with respect to itself. This follows immediately from the commutation rules (2.10) which hold for a general angular momentum and is, therefore, true in particular for a spin angular momentum of one electron $\mathbf{s}$ with respect to itself. Equipped with these basic results we merely need a number of 'building up' rules:

(*a*) If $\mathbf{T}$ satisfies (2.69) with respect to $\mathbf{j}$ and commutes with $\mathbf{j}_1$ then $\mathbf{T}$ satisfies (2.69) with respect to $\mathbf{j}+\mathbf{j}_1$.

(*b*) If $\mathbf{T}_1$ and $\mathbf{T}_2$ satisfy (2.69) then $\mathbf{T}_1 . \mathbf{T}_2$ commutes with $\mathbf{j}$, and $\mathbf{T}_1 \wedge \mathbf{T}_2$ and $\mathbf{T}_1+\mathbf{T}_2$ satisfy (2.69).

(*c*) If α commutes with $\mathbf{j}$ and $\mathbf{T}$ satisfies (2.69) then so does $\alpha\mathbf{T}$.

All three rules may be verified immediately by direct substitution using the definition of a vector of type T. The physical significance of rule (*a*) is that if a vector is rotated by the rotation operators of a space it is also rotated by the rotation operators of any larger space containing the first one. For example, $\mathbf{r}_i$ is of type T with respect to $\mathbf{j}_i$, $\mathbf{L}$ and $\mathbf{J}$, where $\mathbf{L}$ and $\mathbf{J}$ refer to the n-electron system. Rules (*b*) and (*c*) show that if we start with any set of vectors of type T then any linear combination of these vectors with scalar coefficients is also of type T, always of course with respect to the same angular momentum $\mathbf{j}$. Also the operation of taking a vector product keeps one within the same space. Lastly, the scalar product commutes with $\mathbf{j}$ as one would expect from the fact that scalar products are invariant under rotations. As an example of the combined effects of (*a*), (*b*) and (*c*) we remark that $\mathbf{r}_i^2$, $\mathbf{p}_i^2$, $\mathbf{L}.\mathbf{S}$ and $f(r_i)\,\mathbf{l}_i.\mathbf{s}_i$ all commute with $\mathbf{J}$, the total angular momentum. $\mathbf{L}.\mathbf{S}$, however, commutes with neither $\mathbf{L}$ nor $\mathbf{S}$ separately.

Knowing just the commutation relations of a scalar or a vector with an arbitrary angular momentum $\mathbf{j}$, it is possible to derive a great deal of information about the matrix elements of that scalar or vector in a scheme in which $\mathbf{j}^2$ and j_z are diagonal simply from those commutation relations. We will demonstrate this now, first for an observable α which commutes with all the components of $\mathbf{j}$. α may be a scalar or a component of a vector which commutes with $\mathbf{j}$. We have

$$\alpha j_z = j_z \alpha, \quad \alpha \mathbf{j}^2 = \mathbf{j}^2 \alpha,$$

and take matrix elements of these two equations to obtain

$$\begin{aligned} 0 &= \langle \eta' jm | \,(\alpha j_z - j_z \alpha)\, | \xi'' j'm' \rangle \\ &= \langle \eta' jm | \,\alpha\, | \xi'' j'm' \rangle \,(m'-m)\,\hbar \end{aligned}$$

and

$$0 = \langle \eta' jm | \,\alpha\, | \xi'' j'm' \rangle \,(j'(j'+1) - j(j+1))\,\hbar^2.$$

This means that $\langle \eta' jm | \,\alpha\, | \xi'' j'm' \rangle$ is certainly zero unless $j = j'$ and $m = m'$. It is customary to say that we have found a selection rule on j and m for α and write it

$$\Delta j = 0, \quad \Delta m = 0. \tag{2.70}$$

In our derivation the labels ξ'', η' may represent either the same or different ways of making up the complete commuting set, i.e. we may have ξ, $\mathbf{j}^2$, j_z or η, $\mathbf{j}^2$, j_z.

We can, however, say more than this. Not only is the matrix of α diagonal in m, it is completely independent of it. This follows from an expansion, using the fact that j^+ and j^- also commute with α

$$\alpha\,|\xi''jm\rangle = \sum_{\eta'} |\eta'jm\rangle\langle\eta'jm|\,\alpha\,|\xi''jm\rangle,$$

whence

$$j^{\pm}\alpha\,|\xi''jm\rangle = \sum_{\eta'} \hbar\{(j\mp m)(j\pm m+1)\}^{\frac{1}{2}}\,|\eta'j,m\pm 1\rangle\langle\eta'jm|\,\alpha\,|\xi''jm\rangle, \quad (2.71)$$

but we also have

$$\begin{aligned}\alpha j^{\pm}\,|\xi''jm\rangle &= \hbar\alpha\{(j\mp m)(j\pm m+1)\}^{\frac{1}{2}}\,|\xi''j,m\pm 1\rangle \\ &= \sum_{\eta'} \hbar\{(j\mp m)(j\pm m+1)\}^{\frac{1}{2}}\,|\eta'j,m\pm 1\rangle\langle\eta'j,m\pm 1|\,\alpha\,|\xi''j,m\pm 1\rangle.\end{aligned} \quad (2.72)$$

In deriving (2.71) and (2.72) we have used (2.25) and (2.70). Then comparing coefficients of $|\eta'j,m\pm 1\rangle$ we obtain

$$\langle\eta'jm|\,\alpha\,|\xi''jm\rangle = \langle\eta'j,m\pm 1|\,\alpha\,|\xi''j,m\pm 1\rangle \quad (2.73)$$

provided the kets involved are non-zero. Usually, of course, we shall be interested in matrix elements between kets for which the ξ are the same as the η. The more general (2.73), however, enables us to put $\alpha = 1$ and deduce that the transformation matrix $\langle\eta'jm\,|\,\xi''jm\rangle$ between the ξjm scheme and the ηjm scheme is completely independent of m. We can write

$$\langle\eta'j\,|\,\xi''j\rangle = \langle\eta'jm\,|\,\xi''jm\rangle \quad (2.74)$$

because of this. As an example of (2.73) we remark that $\mathbf{L}.\mathbf{S}$ has matrix elements in an $SLJM$ scheme which are independent of M and verify this independently by observing that

$$\mathbf{L}.\mathbf{S} = \tfrac{1}{2}(\mathbf{J}^2-\mathbf{L}^2-\mathbf{S}^2).$$

However, we have seen that the proposition would remain true even if $\mathbf{L}^2$ and $\mathbf{S}^2$ did not occur in the complete commuting set.

We now pass to a consideration of the matrix elements of a vector of type T with respect to a general angular momentum $\mathbf{j}$ in a jm scheme and anticipate to say that again we can find the complete dependence upon the m values. It is convenient here to replace T_x, T_y and T_z by the linear combinations

$$\left.\begin{aligned} T_1 &= -\frac{1}{\surd 2}T^+ = -\frac{1}{\surd 2}(T_x+iT_y), \\ T_0 &= T_z, \\ T_{-1} &= \frac{1}{\surd 2}T^- = \frac{1}{\surd 2}(T_x-iT_y). \end{aligned}\right\} \quad (2.75)$$

As the notation suggests, it is possible and useful for many purposes to think of T_1, T_0 and T_{-1} as forming three 'operator eigenstates' T_m of j having $j = 1$ and $m = 1$, 0 and -1, respectively. More precisely we note that it follows immediately by direct substitution from (2.69) and (2.75) that

$$\left.\begin{aligned} &[j^+,T_1] = [j_z,T_0] = [j^-,T_{-1}] = 0, \\ &[j^{\pm},T_0] = \hbar\surd 2\,T_{\pm 1}, \quad [j_z,T_{\pm 1}] = \pm\hbar T_{\pm 1}, \\ &[j^{\pm},T_{\mp 1}] = \hbar\surd 2\,T_0. \end{aligned}\right\} \quad (2.76)$$

Equations (2.76) show that the effect of j_z, j^+ and j^- on the kets defined as

$$|Tjm'm\rangle = T_{m'}\,|jm\rangle \tag{2.77}$$

is the same, considering the kets as functions of m' and m, as their effect on the kets

$$|1jm'm\rangle = |1m'\rangle\,|jm\rangle, \tag{2.78}$$

where each of $|1m'\rangle$ and $|jm\rangle$ separately is an eigenstate of $\mathbf{j}^2$ and j_z; in fact we will regard (2.78) as simply a restatement of (2.77) in a different notation. We already discussed this kind of situation in § 2.5 where we defined in (2.54) two vectors $\mathbf{j}^a$ and $\mathbf{j}^b$ operating, respectively, on the first and second of the two constituent kets of the product. Then we saw that we could couple them using Wigner's formula provided we remembered that the eigenstates of $\mathbf{j}$ would not necessarily be normalized by the same factor for each j value although the normalization factor would be independent of m.

In our case we have $j^a = 1, j^b = j$ so let us write J for the eigenvalue referring to $\mathbf{j} = \mathbf{j}^a+\mathbf{j}^b$. The relevant eigenstates are $|1jJM\rangle$, where $J = j$ or $j\pm 1$. Then

$$|1jm'm\rangle = \sum_{JM} |1jJM\rangle\langle 1jJM\,|\,1jm'm\rangle$$

and the matrix elements of $T_{m'}$ are

$$\begin{aligned}\langle j_1m_1|\,T_{m'}\,|jm\rangle &= \langle j_1m_1\,|\,1jm'm\rangle\\ &= \langle j_1m_1\,|\,1jj_1m_1\rangle\langle 1jj_1m_1\,|\,1jm'm\rangle,\end{aligned} \tag{2.79}$$

where $\langle 1jj_1m_1\,|\,1jm'm\rangle$ are Wigner coefficients and we gave a table of closed formulae for them in ex. 3 of § 2.5. Our discussion earlier in this section shows that $\langle j_1m_1\,|\,1jj_1m_1\rangle$ is independent of m_1 and it is customary to write

$$\begin{aligned}\langle j_1m_1\,|\,1j_1-1j_1m_1\rangle &= (j_1(2j_1-1))^{\frac{1}{2}}\langle j_1\vdots T\vdots j_1-1\rangle,\\ \langle j_1m_1\,|\,1j_1j_1m_1\rangle &= -(j_1(j_1+1))^{\frac{1}{2}}\langle j_1\vdots T\vdots j_1\rangle,\\ \langle j_1m_1\,|\,1j_1+1j_1m_1\rangle &= -((j_1+1)(2j_1+3))^{\frac{1}{2}}\langle j_1\vdots T\vdots j_1+1\rangle.\end{aligned}$$

The matrix elements are zero unless $j_1 = j$ or $j\pm 1$. Using (2.79) and the formulae of ex. 3 of § 2.5 we can give the complete dependence of the matrix of $\mathbf{T}$ on m and m_1. For convenience later we give this in terms of $T^\pm$ and T_z and take the expressions (2.75) and substitute in (2.79) to give

$$\left.\begin{aligned}\langle jm|\,T_z\,|j-1m\rangle &= \langle j\vdots T\vdots j-1\rangle(j^2-m^2)^{\frac{1}{2}},\\ \langle jm|\,T^\pm\,|j-1m\mp 1\rangle &= \mp\langle j\vdots T\vdots j-1\rangle(j\pm m-1)^{\frac{1}{2}}(j\pm m)^{\frac{1}{2}},\\ \langle jm|\,T_z\,|jm\rangle &= \langle j\vdots T\vdots j\rangle m,\\ \langle jm|\,T^\pm\,|jm\mp 1\rangle &= \langle j\vdots T\vdots j\rangle(j\mp m+1)^{\frac{1}{2}}(j\pm m)^{\frac{1}{2}},\\ \langle jm|\,T_z\,|j+1m\rangle &= \langle j\vdots T\vdots j+1\rangle((j+1)^2-m^2)^{\frac{1}{2}},\\ \langle jm|\,T^\pm\,|j+1m\mp 1\rangle &= \pm\langle j\vdots T\vdots j+1\rangle(j\mp m+1)^{\frac{1}{2}}(j\mp m+2)^{\frac{1}{2}}.\end{aligned}\right\} \tag{2.80}$$

As we have already remarked, all matrix elements other than those given by (2.80) are zero. We have a selection rule on j for $\mathbf{T}$:

$$\Delta j = 0 \quad \text{or} \quad \pm 1. \tag{2.81}$$

Equations (2.80) and (2.81) are, of course, consistent with our earlier equations (2.24′) and (2.25′) for the matrix elements of the angular momentum itself and combine with them to give the result

$$\langle j' \vdots\, j \,\vdots j'' \rangle = \hbar \delta_{j'j''}.$$

While it is true that formulae (2.80) give us the complete dependence of the matrix of $\mathbf{T}$ on m, they are a little complicated. For this reason I give a simple deduction from them which is easier to use and often quite sufficient in scope. This deduction is that if we have two vectors $\mathbf{T}_1$ and $\mathbf{T}_2$, both of type T with respect to $\mathbf{j}$, then as their matrix elements are both given by formulae of the type (2.80) they must be proportional to each other. There are three constants of proportionality, one for each value of Δj in (2.80). For convenience in future reference we give this result the status of a theorem which is stated formally as

The replacement theorem. If both $\mathbf{T}_1$ and $\mathbf{T}_2$ are of type T with respect to an angular momentum $\mathbf{j}$ then

$$\langle \alpha' jm |\, \mathbf{T}_1 \,| \alpha'' j' m' \rangle = \gamma \langle \alpha' jm |\, \mathbf{T}_2 \,| \alpha'' j' m' \rangle$$

for some constant $\gamma = \gamma(\alpha', \alpha'', j, j')$ independent of m, m' except possibly in the case that $\langle \alpha' j \vdots\, T_2 \,\vdots \alpha'' j' \rangle = 0$.

The extra quantum numbers α', α'' are included so that the observables α, $\mathbf{j}^2$, j_z should make up a complete commuting set. Although for the sake of simplicity we did not include α in the derivation of (2.80) the derivation would still go through in exactly the same way with α present provided that α commutes with j_x and j_y as well as with j_z. In the replacement theorem there is actually no need to have the same eigenvalues of α appearing on the two sides of the equation and this gives a useful corollary by taking $\mathbf{T}_2 = \mathbf{j}$ and noting that $\langle \alpha' j \vdots\, j \,\vdots \alpha'' j' \rangle = \hbar \delta_{\alpha'\alpha''} \delta_{jj'}$.

Corollary to the replacement theorem. For any vector $\mathbf{T}$ of type T with respect to an angular momentum $\mathbf{j}$, the relationship

$$\langle \alpha' jm |\, \mathbf{T} \,| \alpha'' jm' \rangle = \gamma \langle \alpha' jm |\, \mathbf{j} \,| \alpha' jm' \rangle$$

holds with $\gamma = \gamma(\alpha', \alpha'', j)$ independent of m, m'. The reader should note that we have here an equation with α'' appearing explicitly only on the left.

The point of the replacement theorem and its corollary is that it means that, in many circumstances, when we have to determine the matrix elements of a complicated quantity of type T we can replace that quantity by a simpler one whose matrix elements are easier to evaluate. One then determines the constant of proportionality γ by working out the matrix element of the complicated quantity in one particular case. We shall use the corollary in ch. 5 to prove the Landé interval rule. The general idea of replacement has far-reaching usefulness, however, and we shall meet many other examples of it later when we come to consider atoms and ions in sites which have less than the full spherical symmetry.

We shall find on p. 49 that electric dipole radiation can only take place between levels which have a non-vanishing matrix element of the electric dipole moment

$$\mathbf{D} = \sum_j e\mathbf{r}_j$$

between them. $\mathbf{D}$ is of type T with respect to the total angular momentum and so (2.81) then gives us a selection rule for electric dipole radiation. In the Russell–Saunders coupling scheme, (2.81) also gives a selection rule on L whilst (2.70) gives the rule $\Delta S = 0$. We collect these selection rules for electric dipole radiation and have

$$\left.\begin{aligned} \Delta J &= 0 \quad \text{or} \quad \pm 1, \\ \Delta L &= 0 \quad \text{or} \quad \pm 1, \\ \Delta S &= 0, \end{aligned}\right\} \tag{2.82}$$

where $\mathbf{L}$, $\mathbf{S}$ and $\mathbf{J}$ are the total orbital, total spin and total angular momenta, respectively, for the atom.

We conclude the section by discussing another important operator and its selection rules. This is the parity operator P which represents inversion in the origin. It operates on a wave-function in the Schrödinger representation according to the equation

$$Pf(\mathbf{p}_i, \mathbf{r}_i) = f(-\mathbf{p}_i, -\mathbf{r}_i). \tag{2.83}$$

Then

$$P\mathscr{H}f = \mathscr{H}Pf, \quad P^2 = 1,$$

where $\mathscr{H}$ is the Hamiltonian of (2.1). We see that

$$P\mathbf{l}f = P(\mathbf{r} \wedge \mathbf{p})f = \mathbf{r} \wedge \mathbf{p}Pf = \mathbf{l}Pf \tag{2.84}$$

which exhibits the fact that $\mathbf{l}$ is a pseudo-vector (i.e. unlike an ordinary vector it does not change sign on reflection in the origin but is left unchanged). Similarly we define the operation of P on the spin vector $\mathbf{s}$ by generalizing (2.83) to

$$Pf(\mathbf{p}_i, \mathbf{r}_i, \mathbf{l}_i, \mathbf{s}_i) = f(-\mathbf{p}_i, -\mathbf{r}_i, \mathbf{l}_i, \mathbf{s}_i). \tag{2.85}$$

We show now that P is an observable. The equations

$$P\{\tfrac{1}{2}(P+1)\,|X\rangle\} = \tfrac{1}{2}(P+1)\,|X\rangle,$$

$$P\{\tfrac{1}{2}(1-P)\,|X\rangle\} = -\tfrac{1}{2}(1-P)\,|X\rangle,$$

and

$$|X\rangle = \tfrac{1}{2}(P+1)\,|X\rangle + \tfrac{1}{2}(1-P)\,|X\rangle,$$

show that P possesses a complete set of eigenkets. So we only have to show that P is real. We consider two basic product kets expressed in the Schrödinger representation for their spatial parts, $|\phi(\mathbf{r}_i, \mathbf{p}_i)\,\chi(\mathbf{s}_i)\rangle$ and $|\psi(\mathbf{r}_i, \mathbf{p}_i)\,\rho(\mathbf{s}_i)\rangle$ say.

Then

$$\begin{aligned} &\langle\overline{\phi}(\mathbf{r}_i, \mathbf{p}_i)\,\chi(\mathbf{s}_i)\,P\psi(\mathbf{r}_i, \mathbf{p}_i)\,\rho(\mathbf{s}_i)\rangle \\ &= \langle\overline{\phi}(\mathbf{r}_i, \mathbf{p}_i)\,\chi(\mathbf{s}_i)\,\psi(-\mathbf{r}_i, -\mathbf{p}_i)\,\rho(\mathbf{s}_i)\rangle \\ &= \langle\overline{\phi}(-\mathbf{r}_i, -\mathbf{p}_i)\,\chi(\mathbf{s}_i)\,\psi(\mathbf{r}_i, \mathbf{p}_i)\,\rho(\mathbf{s}_i)\rangle \\ &= \overline{\langle\overline{\psi}(\mathbf{r}_i, \mathbf{p}_i)\,\rho(\mathbf{s}_i)\,P\phi(\mathbf{r}_i, \mathbf{p}_i)\,\chi(\mathbf{s}_i)\rangle}, \end{aligned}$$

So P is real and hence an observable.

We have shown that P commutes with $\mathscr{H}$ and from (2.84) and (2.85) it commutes also with $\mathbf{J}$, $\mathbf{S}$ and $\mathbf{L}$ and so P, $\mathscr{H}$, $\mathbf{S}^2$, $\mathbf{L}^2$, $\mathbf{J}^2$ and J_z or P, $\mathscr{H}$, $\mathbf{S}^2$, $\mathbf{L}^2$, S_z and L_z possess a complete set of simultaneous eigenkets (Dirac, p. 49). Because of this it is convenient and customary to use eigenstates of P both for exact and approximate eigenstates of $\mathscr{H}$. The two eigenvalues of P are evidently ± 1 and those states with $+1$ are called even states and those with -1 odd states. There is a simple and rigorous selection rule on P for electric dipole radiation. This follows from the equation

$$P\mathbf{D} = -\mathbf{D}P,$$

whence

$$\langle P'|\,\mathbf{D}\,|P''\rangle\,(P'+P'') = 0.$$

So $\langle P'|\,\mathbf{D}\,|P''\rangle$ is zero unless $P' = -P''$. In other words even-odd transitions are allowed but even-even and odd-odd transitions are forbidden. Electric dipole radiation is always associated with a change of parity. This is called Laporte's rule after Laporte who discovered it.

Examples

1. Prove the building-up rules (*a*), (*b*) and (*c*) given on p. 35.

2. A linear operator A is said to be idempotent if $A^2 = A$. Show that a real idempotent operator A such that $A\,|X\rangle$ is defined for all $|X\rangle$ is an observable.

3. An observable ξ has eigenkets $|\xi'\rangle$. Let α be a linear operator such that $\alpha\,|\xi'\rangle = 0$ when $\xi' < 1$ and $\alpha\,|\xi'\rangle = |\xi'\rangle$ for $\xi' \geqslant 1$. Show that α is an observable.

4. Show that for a single electron $\mathbf{r}\,.\,\mathbf{l} = 0$. Use (2.80) to deduce that $\langle l\,\vdots\, r\,\vdots\, l\rangle = 0$. This means that in an electric dipole transition l must change by one unit. Symbolically $\Delta l = \pm 1$.

5. The individual functions of an orthonormal set of one-electron functions are each eigenfunctions of the parity operator P. Show that any determinantal function derived from them according to (2.64) is also an eigenfunction of P.

6. If $\mathscr{H}' = \sum_{\kappa=1}^{n} f(r_\kappa)\,\mathbf{r}_\kappa\,.\,\mathbf{s}_\kappa$ is added to the Hamiltonian (2.1) for an atomic system, the new Hamiltonian still commutes with $\mathbf{J}$ but no longer commutes with the parity operator P.

$\mathscr{H}'$ is called a pseudo-scalar and we say that parity is not conserved for the system. There is, as yet, no evidence for parity non-conservation in atomic systems although it is known not to be conserved in some nuclear reactions.

CHAPTER 3

ELECTROMAGNETIC RADIATION[1]

3.1. Electromagnetic fields

3.1.1. Maxwell's equations. So far our atoms have been envisaged to lie in free space, subject to no influences from outside. In reality, however, they both emit and absorb radiation and have their energy-levels perturbed by external electric and magnetic fields. All these influences are electromagnetic, so in order to study them we must first study the electromagnetic field itself. This is described by Maxwell's equations:

$$\left.\begin{aligned} \operatorname{div}\mathbf{E} &= 4\pi\rho, \\ \operatorname{div}\mathbf{H} &= 0, \\ \operatorname{curl}\mathbf{E} &= -\frac{1}{c}\frac{\partial\mathbf{H}}{\partial t}, \\ \operatorname{curl}\mathbf{H} &= \frac{1}{c}\frac{\partial\mathbf{E}}{\partial t}+\frac{4\pi}{c}\rho\mathbf{v}; \end{aligned}\right\} \tag{3.1}$$

where $\mathbf{E}$ and $\mathbf{H}$ are the electric and magnetic field strengths, ρ is the density and $\mathbf{v}$ the velocity of electric charge and c is the velocity of light. The units are e.m.u. for magnetic quantities and e.s.u. for electrical quantities. The force on a particle having a charge $+e$ is

$$\mathbf{F} = e\left(\mathbf{E}+\frac{1}{c}\mathbf{v}\wedge\mathbf{H}\right). \tag{3.2}$$

Equations (3.1) show the necessary relationships which hold between $\mathbf{E}$ and $\mathbf{H}$ on the one hand and ρ and $\mathbf{v}$ on the other. However, ρ and $\mathbf{v}$ do not completely determine $\mathbf{E}$ and $\mathbf{H}$. For if we take any solution of (3.1) and add to it an arbitrary solution of

$$\left.\begin{aligned} \operatorname{div}\mathbf{E}^0 &= \operatorname{div}\mathbf{H}^0 = 0, \\ \operatorname{curl}\mathbf{E}^0 &= -\frac{1}{c}\frac{\partial\mathbf{H}^0}{\partial t}, \\ \operatorname{curl}\mathbf{H}^0 &= \frac{1}{c}\frac{\partial\mathbf{E}^0}{\partial t}, \end{aligned}\right\} \tag{3.3}$$

then $\mathbf{E}+\mathbf{E}^0$ and $\mathbf{H}+\mathbf{H}^0$ also satisfy (3.1) with respect to ρ and $\mathbf{v}$. It is usual to derive second-order differential equations from (3.3) by differentiation

$$-\frac{1}{c^2}\frac{\partial^2\mathbf{H}^0}{\partial t^2} = \frac{1}{c}\frac{\partial}{\partial t}(\operatorname{curl}\mathbf{E}^0) = \operatorname{curl}\left(\frac{1}{c}\frac{\partial\mathbf{E}^0}{\partial t}\right)$$

$$= \operatorname{curl}\operatorname{curl}\mathbf{H}^0 = \operatorname{grad}\operatorname{div}\mathbf{H}^0-\nabla^2\mathbf{H}^0 = -\nabla^2\mathbf{H}^0,$$

[1] This chapter is largely based on Heitler (1954).

where $\nabla^2\mathbf{H}^0 \equiv (\nabla^2 H_x^0, \nabla^2 H_y^0, \nabla^2 H_z^0)$. In precisely the same way,

$$\frac{1}{c^2}\frac{\partial^2 \mathbf{E}^0}{\partial t^2} = \nabla^2\mathbf{E}^0.$$

So the solutions of (3.3) satisfy homogeneous wave-equations and are, of course, the mathematical expression of the familiar electromagnetic waves of which light waves and radio waves are examples.

It is well known that a vector field whose divergence is identically zero can be represented as the curl of another vector field.[1] Then let us write $\mathbf{H} = \operatorname{curl}\mathbf{A}$ and the third equation of (3.1) becomes

$$\operatorname{curl}\left(\mathbf{E}+\frac{1}{c}\frac{\partial \mathbf{A}}{\partial t}\right) = 0,$$

which means that $\mathbf{E}+(1/c)\,(\partial\mathbf{A}/\partial t)$ can be represented as the gradient of a scalar function, $-\phi$ say[1]. Thus

$$\mathbf{E} = -\frac{1}{c}\frac{\partial \mathbf{A}}{\partial t} - \operatorname{grad}\phi.$$

$\mathbf{A}$ and ϕ are called the vector and scalar potentials, respectively, of the electromagnetic field and we see that $\mathbf{H}$ and $\mathbf{E}$ are uniquely determined by them. We have now satisfied two of (3.1) identically and, in terms of $\mathbf{A}$ and ϕ, the remaining two become

$$-\frac{1}{c}\operatorname{div}\frac{\partial \mathbf{A}}{\partial t} - \nabla^2\phi = 4\pi\rho,$$

$$\operatorname{grad}\left(\operatorname{div}\mathbf{A}+\frac{1}{c}\frac{\partial\phi}{\partial t}\right) - \nabla^2\mathbf{A} = -\frac{1}{c^2}\frac{\partial^2\mathbf{A}}{\partial t^2}+\frac{4\pi}{c}\rho\mathbf{v}. \tag{3.4}$$

Now although $\mathbf{H}$ and $\mathbf{E}$ are uniquely determined by $\mathbf{A}$ and ϕ, the converse is not true. Suppose we put $\mathbf{A}' = \mathbf{A}-\operatorname{grad}\chi$ and $\phi' = \phi+(1/c)\,(\partial\chi/\partial t)$, where χ is an arbitrary scalar function. $\mathbf{H}$ and $\mathbf{E}$ calculated from $\mathbf{A}'$ and ϕ' are the same as $\mathbf{H}$ and $\mathbf{E}$ calculated from $\mathbf{A}$ and ϕ. Also direct substitution shows that $\mathbf{A}'$ and ϕ' satisfy (3.4) if $\mathbf{A}$ and ϕ do so. This change of the vector and scalar potentials is called a gauge transformation. The name comes from an attempt by Weyl to geometrize the electromagnetic field in a manner similar to the geometrization of the gravitational field by Einstein in his relativity theory of gravitation. In Weyl's theory a standard of length, or gauge, is assigned to every point of space time and the electromagnetic potentials are related to the changes of this standard for infinitesimal displacements in space time. For us the value of the gauge transformation is simply that it enables us without inconsistency to require $\mathbf{A}$ and ϕ to satisfy a subsidiary relation which simplifies (3.4) considerably. $\operatorname{div}\mathbf{A}$ is a scalar field and we choose χ so that $\nabla^2\chi = \operatorname{div}\mathbf{A}$ everywhere. Then $\operatorname{div}\mathbf{A}' = 0$ and equations (3.4) become

$$\left.\begin{aligned} -\nabla^2\phi' &= 4\pi\rho, \\ \frac{1}{c}\operatorname{grad}\frac{\partial\phi'}{\partial t} - \nabla^2\mathbf{A}' + \frac{1}{c^2}\frac{\partial^2\mathbf{A}'}{\partial t^2} &= \frac{4\pi}{c}\rho\mathbf{v}. \end{aligned}\right\} \tag{3.5}$$

[1] See, for example, Rutherford (1954).

Equations (3.5) show that the scalar potential ϕ' is determined by the charges ρ alone just as if the latter were at rest. For this reason any choice of gauge which has $\operatorname{div} \mathbf{A} = 0$ is called a Coulomb gauge. These are the gauges that we shall use in this book.

After this brief discussion of the classical electromagnetic field we need to know next how the field influences an atom. This influence arises via (3.2). In order to take (3.2) over into the quantum theory we must transform it into a statement about a Hamiltonian for the system. This transformation is absolutely basic to the theory. I will not interrupt the continuity of the book by giving it here, but have put it in Appendix 4 at the end of the book. The result of the transformation is a simple rule: to obtain the Hamiltonian for a charged particle in an electromagnetic field we must replace the momentum vector $\mathbf{p}$ by $\mathbf{p}+(e/c)\mathbf{A}$ in the Hamiltonian for the system without the field, where $-e$ is the charge on the particle, and we must also add $-e\phi$ to the Hamiltonian. Thus for one electron, charge $-e$, in a central field due to a nucleus of charge $+Ze$ the Hamiltonian for no field is

$$\mathscr{H} = \frac{1}{2m}\mathbf{p}^2 - \frac{Ze^2}{r}$$

and with a field is

$$\mathscr{H}' = \frac{1}{2m}\left(\mathbf{p}+\frac{e}{c}\mathbf{A}\right)^2 - \frac{Ze^2}{r} - e\phi. \tag{3.6}$$

We could have regarded the nucleus as an external field, had we wished, and would then have

$$\mathscr{H} = \frac{1}{2m}\mathbf{p}^2,$$

$$\mathscr{H}' = \frac{1}{2m}\left(\mathbf{p}+\frac{e}{c}\mathbf{A}\right)^2 - e\left(\frac{Ze}{r}+\phi\right),$$

where the electrostatic potential is now the sum of the potential ϕ due to the genuinely external field and Ze/r due to the nucleus. Thus the two ways of regarding the nucleus are consistent. For the n-electron Hamiltonian (2.1) modified for the presence of an external field we have

$$\mathscr{H}' = \sum_{j=1}^{n}\left(\frac{1}{2m}\left(\mathbf{p}_j+\frac{e}{c}\mathbf{A}_j\right)^2 - \frac{Ze^2}{r_j} - e\phi_j\right) + \sum_{j<k}^{n}\frac{e^2}{r_{jk}}, \tag{3.7}$$

where $\mathbf{A}_j$ and ϕ_j are the values of the electromagnetic potentials at the position of the jth electron. I should add, perhaps, that (3.7) is still incomplete because it refers to electrons which are just point charges. Actually the electron has an intrinsic magnetic moment which produces some modification in (3.7). This is treated in ch. 5.

The reader may possibly be worried by the fact that (3.6) and (3.7) are not gauge invariant. The solutions of the equation $\mathscr{H}'\psi = i\hbar(\partial/\partial t)\psi$ are, however, changed only by a phase factor of modulus unity and therefore a change of gauge makes only a formal difference and no difference at all to the observable consequences of the equation. To see this, let us put

$$\mathbf{A} = \mathbf{A}' - \operatorname{grad}\chi, \quad \phi = \phi' + \frac{1}{c}\frac{\partial \chi}{\partial t}$$

and then $\mathscr{H}'$ becomes

$$\mathscr{H}'_t = \sum_{j=1}^{n} \left(\frac{1}{2m}\left(\mathbf{p}_j + \frac{e}{c}\mathbf{A}'_j - \frac{e}{c}\operatorname{grad}\chi_j\right)^2 - \frac{Ze^2}{r_j} - e\phi'_j - \frac{e}{c}\frac{\partial \chi_j}{\partial t}\right) + \sum_{j<k}^{n} \frac{e^2}{r_{jk}},$$

where $\chi_j = \chi(j)$, $\operatorname{grad}\chi_j = \operatorname{grad}_j \chi(j)$. Now consider a function $\psi_t = e^{i\lambda}\psi$, where λ is some real function of space and time yet to be determined and ψ satisfies $\mathscr{H}'\psi = i\hbar(\partial/\partial t)\psi$. Then

$$i\hbar\frac{\partial}{\partial t}\psi_t = i\hbar e^{i\lambda}\left(i\frac{\partial\lambda}{\partial t} + \frac{\partial}{\partial t}\right)\psi = e^{i\lambda}\left(-\hbar\frac{\partial\lambda}{\partial t}\psi + \mathscr{H}'\psi\right),$$

$$\left(\mathbf{p}_j + \frac{e}{c}\mathbf{A}'_j - \frac{e}{c}\operatorname{grad}\chi_j\right)e^{i\lambda}\psi = e^{i\lambda}\left(\mathbf{p}_j + \frac{e}{c}\mathbf{A}'_j + \hbar\operatorname{grad}_j\lambda - \frac{e}{c}\operatorname{grad}\chi_j\right)\psi.$$

This suggests putting
$$\lambda = \frac{e}{\hbar c}\sum_{j=1}^{n}\chi_j$$

in which case
$$\left(\mathbf{p}_j + \frac{e}{c}\mathbf{A}'_j - \frac{e}{c}\operatorname{grad}\chi_j\right)e^{i\lambda}\psi = e^{i\lambda}\left(\mathbf{p}_j + \frac{e}{c}\mathbf{A}'_j\right)\psi,$$

$$\rightarrow \left(\mathbf{p}_j + \frac{e}{c}\mathbf{A}'_j - \frac{e}{c}\operatorname{grad}\chi_j\right)^2 e^{i\lambda}\psi = e^{i\lambda}\left(\mathbf{p}_j + \frac{e}{c}\mathbf{A}'_j\right)^2\psi,$$

$$\rightarrow \mathscr{H}'_t e^{i\lambda}\psi = e^{i\lambda}\mathscr{H}'\psi - \frac{e}{c}e^{i\lambda}\psi\sum_{j=1}^{n}\frac{\partial\chi_j}{\partial t},$$

and so $\mathscr{H}'_t\psi_t = i\hbar(\partial/\partial t)\psi_t$. This proves, in general, the gauge invariance of the predictions of (3.7). However, in practice we usually regard the external field as giving rise to a small perturbation on $\mathscr{H}$. Then we will wish to know that the predictions of a certain order in perturbation theory are invariant to gauge transformations too. This is so providing we regard the terms linear in $\mathbf{A}$ and ϕ in (3.7) as being first-order small quantities and the terms quadratic in $\mathbf{A}$ as being second-order small quantities. The gauge invariance of the results of second-order perturbation theory is so fundamental to any discussion of the magnetic properties of molecules that I think it is worth while giving a formal proof of it. This proof is to be found in Appendix 5.

3.1.2. Examples of electromagnetic fields. We can now discuss briefly the three most important special cases of an electromagnetic field. The first is a uniform constant magnetic field $\mathbf{H}$. $\mathbf{H}$, then, is independent of position and time and the vector potential $\mathbf{A} = \frac{1}{2}\mathbf{H}\wedge\mathbf{r}$, where $\mathbf{r}$ is the position vector relative to an arbitrary chosen origin, satisfies both $\operatorname{curl}\mathbf{A} = \mathbf{H}$ and $\operatorname{div}\mathbf{A} = 0$. The reader may verify this by differentiation. We pursue the consequences of introducing this value for $\mathbf{A}$ in (3.7) in ch. 5 after we have discovered how to introduce the spin magnetic moment.

The second special case is a uniform constant electric field $\mathbf{E}$. This will not be useful to us, but we note for completeness that the potential $\phi = -\mathbf{E}.\mathbf{r}$ is adequate to describe it.

The third special case is that of a pure radiation field. This is important when we come to emission and absorption of light and other electromagnetic radiation. There are two ways in which a radiation field can be treated in the theory. The first is to regard it as a classical electromagnetic field satisfying Maxwell's equations for empty space (i.e. (3.3)). Then it affects the atom because of the replacement of $\mathbf{p}$ by $\mathbf{p}+(e/c)\mathbf{A}$ in the Hamiltonian for the atom. However, this is a semi-classical approach, for the atom is a 'quantum' atom but the field is still a 'classical' field. The result is an asymmetry in the theory: the field influences the atom but the atom cannot influence the field. The second way treats both the atom and the field as quantum systems and is much more satisfactory from a conceptual point of view. The whole system of atom + field then has a Hamiltonian which consists of three parts

$$\mathscr{H} = \mathscr{H}_a+\mathscr{H}_i+\mathscr{H}_f, \tag{3.8}$$

where $\mathscr{H}_a$ is the Hamiltonian for the atom without a field, $\mathscr{H}_i$ is the change in $\mathscr{H}_a$ on replacing $\mathbf{p}$ by $\mathbf{p}+(e/c)\mathbf{A}$ and $\mathscr{H}_f$ is the Hamiltonian for the field alone. We are able to distinguish atoms and fields so easily in practice just because $\mathscr{H}_i$, the interaction, is small compared with $\mathscr{H}_a$ and $\mathscr{H}_f$. Hence if $|a\rangle$ and $|f\rangle$ are eigenstates of $\mathscr{H}_a$ and $\mathscr{H}_f$, respectively, then $|a\rangle|f\rangle$ is quite a good approximation to an eigenstate for $\mathscr{H}$. In fact if $|a_i\rangle$ and $|f_j\rangle$ are complete sets for $\mathscr{H}_a$ and $\mathscr{H}_f$ it is usual to suppose that $|a_i\rangle|f_j\rangle$ is a complete set for $\mathscr{H}$. However $|a\rangle|f\rangle$ is not an exact eigenstate of $\mathscr{H}$ and hence it is not a stationary state but changes in time into a linear combination $\sum_{ij}\lambda_{ij}|a_i\rangle|f_j\rangle$. So long as we are wedded to a rigid distinction between atoms and fields we interpret this to say that there is an exchange of energy between the atom and the field; the atom absorbs and emits electromagnetic radiation. From a more general point of view we merely say that the states $|a\rangle|f\rangle$ are not eigenstates of the total energy.

We now pursue the second way and already know $\mathscr{H}_a$ and $\mathscr{H}_i$ in (3.8). We use a Coulomb gauge and because of (3.5) can choose the scalar potential to be zero (ρ is zero because the radiation field alone is the field without the atom being present). The actual expression for $\mathscr{H}_i$ is

$$\mathscr{H}_i = \frac{e}{2mc}\sum_{j=1}^{n}\left(\mathbf{A}_j.\mathbf{p}_j+\mathbf{p}_j.\mathbf{A}_j+\frac{e}{c}\mathbf{A}_j^2\right).$$

In the present context, the second-order term is quite negligible compared to the first-order terms and may be neglected. In accordance with our decision to use Coulomb gauge for the electromagnetic field we find

$$\mathbf{p}_j.\mathbf{A}_j\psi = -i\hbar\nabla_j.(\mathbf{A}_j\psi) = -i\hbar\mathbf{A}_j.\nabla_j\psi-i\hbar\psi\,\mathrm{div}_j\,\mathbf{A}_j = \mathbf{A}_j.\mathbf{p}_j\psi,$$

and so $\mathscr{H}_i$ assumes the simpler form

$$\mathscr{H}_i = \frac{e}{mc}\sum_{j=1}^{n}\mathbf{A}_j.\mathbf{p}_j. \tag{3.9}$$

3.1.3. Quantization of the classical radiation field. We now determine the mathematical scheme which describes $\mathscr{H}_f$. Let us consider a classical field first, with no charges. As usual we use Coulomb gauge and so $\phi = 0$, $\operatorname{div} \mathbf{A} = 0$. Then (3.5) yields the wave-equation

$$\nabla^2 \mathbf{A} = \frac{1}{c^2}\frac{\partial^2 \mathbf{A}}{\partial t^2}, \tag{3.10}$$

for $\mathbf{A}$. Take a rectangular box with sides of length L_x, L_y, L_z parallel to the OX, OY, OZ axes, respectively, and seek solutions of (3.10) of the form $\mathbf{A} = q_\alpha \mathbf{A}_\alpha$, where q_α is a function of time alone and $\mathbf{A}_\alpha$ of space alone. Then

$$\ddot{q}_\alpha + \nu_\alpha^2 q_\alpha = 0,$$

$$\nabla^2 \mathbf{A}_\alpha + \frac{\nu_\alpha^2}{c^2}\mathbf{A}_\alpha = 0, \quad \operatorname{div} \mathbf{A}_\alpha = 0. \tag{3.11}$$

The general solution of (3.10) is a sum $\mathbf{A} = \sum_\alpha q_\alpha \mathbf{A}_\alpha$. Each q_α satisfies an equation for a harmonic oscillator with frequency $\nu_\alpha/2\pi$ and the components of the $\mathbf{A}_\alpha$ satisfy the wave-equation for a quantum particle in a box.

It is convenient for our applications to search for solutions which represent running waves. Therefore we take as our boundary condition that $\mathbf{A}$ and its derivatives should have the same values at corresponding positions of opposite surfaces of the box. We take the solutions in the complex form

$$\mathbf{A}_\alpha = \left(\frac{4\pi c^2}{V}\right)^{\frac{1}{2}} \boldsymbol{\pi}_\alpha e^{i\boldsymbol{\kappa}_\alpha.\mathbf{r}},$$

$$c\,|\boldsymbol{\kappa}_\alpha| = \nu_\alpha, \quad q_\alpha = |q_\alpha|\, e^{-i\nu_\alpha t}, \tag{3.12}$$

where $V = L_x L_y L_z$ and then write

$$\mathbf{A} = \sum_\alpha (q_\alpha \mathbf{A}_\alpha + \bar{q}_\alpha \bar{\mathbf{A}}_\alpha), \tag{3.13}$$

where a bar denotes a complex conjugate. It is convenient to define $\mathbf{A}_{-\alpha} = \bar{\mathbf{A}}_\alpha$. The condition $\operatorname{div} \mathbf{A}_\alpha = 0$ implies that $\boldsymbol{\pi}_\alpha . \boldsymbol{\kappa}_\alpha = 0$. For each α there are two linearly independent possibilities for $\boldsymbol{\pi}_\alpha$ and the sum in (3.13) is over these two possibilities as well as over α. $\boldsymbol{\pi}_\alpha$ is the direction of polarization of the light wave represented by $\mathbf{A}_\alpha$ and is a unit vector.

The total momentum in a field is given by

$$\mathbf{G} = \frac{1}{4\pi c}\int \mathbf{E} \wedge \mathbf{H}\, d\tau,$$

and if we take one value of α and of $\boldsymbol{\pi}_\alpha$, we find

$$\left.\begin{aligned} \mathbf{E} &= \frac{i\nu_\alpha}{c}(q_\alpha \mathbf{A}_\alpha - \bar{q}_\alpha \bar{\mathbf{A}}_\alpha), \\ \mathbf{H} &= i\boldsymbol{\kappa}_\alpha \wedge (q_\alpha \mathbf{A}_\alpha - \bar{q}_\alpha \bar{\mathbf{A}}_\alpha), \end{aligned}\right\} \tag{3.14}$$

from the definition of the vector potential. Using (3.12) we find

$$\left.\begin{aligned} \int \bar{\mathbf{A}}_\alpha . \mathbf{A}_\beta\, d\tau &= \int \mathbf{A}_{-\alpha} . \mathbf{A}_\beta\, d\tau = 4\pi c^2 \delta_{\alpha\beta}, \\ \int \mathbf{A}_\alpha^2\, d\tau &= \int \bar{\mathbf{A}}_\alpha^2\, d\tau = 0, \end{aligned}\right\} \tag{3.15}$$

and so $\mathbf{G}_\alpha = 2\nu_\alpha \bar{q}_\alpha q_\alpha \boldsymbol{\kappa}_\alpha$ and $|\mathbf{G}_\alpha| = 2c^{-1}\nu_\alpha^2 \bar{q}_\alpha q_\alpha$, where we have used the fact that $\boldsymbol{\kappa}_\alpha$ is orthogonal to $\mathbf{A}_\alpha$ and $\bar{\mathbf{A}}_\alpha$. In precisely the same way, the energy content of the field is

$$\mathscr{H}_\alpha = \frac{1}{8\pi}\int (\mathbf{H}^2 + \mathbf{E}^2)\, d\tau = 2\nu_\alpha^2 \bar{q}_\alpha q_\alpha. \tag{3.16}$$

It follows from the orthogonality properties (3.15) that $\mathscr{H} = \Sigma \mathscr{H}_\alpha$ is the energy when more than one of the $q_\alpha \mathbf{A}_\alpha$ is non-zero. This quantity $\mathscr{H}$ is the classical Hamiltonian for the system, for we introduce the real canonical variables

$$Q_\alpha = q_\alpha + \bar{q}_\alpha, \quad P_\alpha = -i\nu_\alpha(q_\alpha - \bar{q}_\alpha)$$

and have $\mathscr{H}_\alpha = \frac{1}{2}(P_\alpha^2 + \nu_\alpha^2 Q_\alpha^2) = 2\nu_\alpha^2 \bar{q}_\alpha q_\alpha$ and the equations

$$\frac{\partial \mathscr{H}_\alpha}{\partial Q_\alpha} = -\dot{P}_\alpha, \quad \frac{\partial \mathscr{H}_\alpha}{\partial P_\alpha} = \dot{Q}_\alpha$$

are satisfied because of (3.12). Each $\mathscr{H}_\alpha$ refers to an independent classical oscillation of the system and so $\mathscr{H} = \Sigma \mathscr{H}_\alpha$ is the Hamiltonian for the whole system. I remark that $\mathscr{H}_\alpha$ and $\mathscr{H}_{-\alpha}$ refer to light waves travelling in opposite directions and are therefore independent. This follows from the expression that we found for $\mathbf{G}_\alpha$. There are two independent polarizations $\boldsymbol{\pi}_\alpha$ for each α.

We have shown that a classical radiation field may be regarded as a set of independent harmonic oscillators, each with its own Hamiltonian $\mathscr{H}_\alpha$. The same is true in the quantum theory and we need merely to replace the time-dependent P_α and Q_α by time-independent operators P_α, Q_α satisfying the usual commutation relations $[Q_\alpha, P_\beta] = i\hbar\delta_{\alpha\beta}, [Q_\alpha, Q_\beta] = [P_\alpha, P_\beta] = 0$. The two expressions $\frac{1}{2}(P_\alpha^2 + \nu_\alpha^2 Q_\alpha^2)$ and $2\nu_\alpha^2 \bar{q}_\alpha q_\alpha$ for $\mathscr{H}_\alpha$ are now no longer equal and it is customary to take the second as the correct quantum-mechanical Hamiltonian. It will become apparent shortly that this choice ensures that the vacuum has energy zero when there are no photons present. For similar reasons $\mathbf{G}_\alpha$ is still taken to be $2\nu_\alpha \bar{q}_\alpha q_\alpha \boldsymbol{\kappa}_\alpha$.

We now drop the suffix α and work out the eigenvalues of the harmonic oscillator in the same way as we did for angular momentum in § 2.2. It is more convenient to work in terms of q and $\bar{q}$, for which $[\bar{q}, q] = -\hbar/2\nu$. Then as $\mathscr{H} = 2\nu^2 \bar{q} q$ we deduce $\mathscr{H} q = q\mathscr{H} - \hbar\nu q$ and $\mathscr{H}\bar{q} = \bar{q}\mathscr{H} + \hbar\nu\bar{q}$. Let $|E\rangle$ be an eigenket of $\mathscr{H}$ with eigenvalue E. Then

$$E = \langle E \,|\, \mathscr{H} \,|\, E\rangle = 2\nu^2 \langle E \,|\, \bar{q} q \,|\, E\rangle$$

which is non-negative because it is a positive multiple of the square of the length of the ket $q\,|E\rangle$. $E = 0$ if and only if $q\,|E\rangle = 0$. Next,

$$\mathscr{H} q\,|E\rangle = q\mathscr{H}\,|E\rangle - \hbar\nu q\,|E\rangle = (E - \hbar\nu)\, q\,|E\rangle,$$

$$\mathscr{H}\bar{q}\,|E\rangle = \bar{q}\mathscr{H}\,|E\rangle + \hbar\nu\bar{q}\,|E\rangle = (E + \hbar\nu)\,\bar{q}\,|E\rangle,$$

so just as in § 2.2 we deduce that the eigenvalues of $\mathscr{H}$ are $n\hbar\nu$ for all non-negative integers n. When $E = n\hbar\nu$ we say that there are n photons present. Because of the formula $\mathbf{G} = 2\nu\bar{q}q\boldsymbol{\kappa}$, $\mathbf{G}$ then has the definite value $n\hbar\boldsymbol{\kappa}$ and $|\mathbf{G}| = (n\hbar\nu/c)$. We say that each of the n photons has momentum $\hbar\nu/c$ along the direction of the vector of propagation $\boldsymbol{\kappa}$.

If $|E'\rangle$ is a second eigenstate of $\mathscr{H}$ with $E' = n'\hbar\nu$, then

$$E'\langle E' \mid q \mid E\rangle = \langle E' \mid \mathscr{H}q \mid E\rangle = (E-\hbar\nu)\langle E' \mid q \mid E\rangle,$$

and so if $\langle E' \mid q \mid E\rangle$ is non-zero we must have $n' = n-1$. Similarly for $\langle E' \mid \bar{q} \mid E\rangle$ to be non-zero we need $n' = n+1$. Let $n' = n-1$, then $\langle E \mid \bar{q}q \mid E\rangle = n\hbar/2\nu$, so $|E'\rangle = (2\nu/n\hbar)^{\frac{1}{2}} q\,|E\rangle$. Therefore

$$\langle E' \mid q \mid E\rangle = \left(\frac{2\nu}{n\hbar}\right)^{\frac{1}{2}} \langle E \mid \bar{q}q \mid E\rangle = \left(\frac{n\hbar}{2\nu}\right)^{\frac{1}{2}}.$$

Similarly when $n' = n+1$, $\langle E' \mid \bar{q} \mid E\rangle = (n'\hbar/2\nu)^{\frac{1}{2}}$. In both cases the square of the matrix element is proportional to the number of photons in the upper of the two states concerned.

Finally, the basic kets of the pure radiation field can be specified by giving the number n for each different radiation oscillator. The q_α, $\bar{q}_\alpha$ then have non-zero matrix elements only between pairs of basic kets which differ by one unit in just one of the occupation numbers n. The total Hamiltonian $\mathscr{H} = \Sigma\mathscr{H}_\alpha$ is the Hamiltonian $\mathscr{H}_f$ which we use for the pure radiation field when discussing atoms in radiation fields.

3.1.4. Transition probabilities. Having armed ourselves with a description of the atom, the pure radiation field and the interaction between them we now can pass on to discuss the exchange of energy between the atom and the field. Let us write $\mathscr{H}_0 = \mathscr{H}_a + \mathscr{H}_f$ and suppose the system to be initially ($t = 0$) in a state $|0\rangle = |a_0\rangle|f_0\rangle$ which is a simple product of eigenstates of $\mathscr{H}_a$ and $\mathscr{H}_f$, respectively. Then at a later time ($t > 0$) the system will be in a state

$$|X\rangle = \Sigma c_j\,|j\rangle,$$

where the $|j\rangle$ with $j \neq 0$ range over all other independent simple product states. As in previous cases one symbol stands for the eigenvalues of a number of observables. The c_j are determined by the equation

$$(\mathscr{H}_0 + \mathscr{H}_i)\,|X\rangle = i\hbar\frac{\partial}{\partial t}\,|X\rangle, \tag{3.17}$$

where $|X\rangle = |0\rangle$ at time $t = 0$. Then for a short time after $t = 0$ all c_j except c_0 will be very small and c_0 will be practically unity. This enables us to determine the initial behaviour of the c_j by using also the fact that each $|j\rangle$ is a stationary state of $\mathscr{H}_0$. Thus we have

$$\mathscr{H}_0\,|j\rangle = i\hbar\frac{\partial}{\partial t}\,|j\rangle$$

and

$$\begin{aligned}(\mathscr{H}_0 + \mathscr{H}_i)\sum_j c_j\,|j\rangle &= \sum_j c_j\,\mathscr{H}_0\,|j\rangle + \sum_j c_j\,\mathscr{H}_i\,|j\rangle \\ &= i\hbar\frac{\partial}{\partial t}\sum_j c_j\,|j\rangle = \sum_j i\hbar c_j\frac{\partial}{\partial t}\,|j\rangle + \sum_j i\hbar\frac{dc_j}{dt}\,|j\rangle\end{aligned}$$

from (3.17), whence

$$\sum_j c_j\,\mathscr{H}_i\,|j\rangle = \sum_j i\hbar\frac{dc_j}{dt}\,|j\rangle,$$

and on multiplying this last equation on the left by the bra corresponding to a particular one of the j, $\langle k|$ say, we obtain

$$\sum_j c_j \langle k \,|\, \mathscr{H}_i \,|\, j\rangle = i\hbar \frac{dc_k}{dt}.$$

Initially, however, all c_j except c_0 are zero, so the initial value of $i\hbar(dc_k/dt)$ is simply $\langle k \,|\, \mathscr{H}_i \,|\, 0\rangle$. Thus a direct transition from the state $|0\rangle$ to the state $|k\rangle$ is only possible if the matrix element $\langle k \,|\, \mathscr{H}_i \,|\, 0\rangle$ is non-zero. $|0\rangle = |a_0\rangle |f_0\rangle$ and let us suppose $|k\rangle = |a\rangle|f\rangle$. Then the form (3.9) for $\mathscr{H}_i$ means that

$$\mathscr{H}_i = \frac{e}{mc}\sum_{j=1}^{n}\sum_{\alpha}(q_\alpha \mathbf{A}_{j\alpha}\cdot\mathbf{p}_j + \bar{q}_\alpha \bar{\mathbf{A}}_{j\alpha}\cdot\mathbf{p}_j)$$

and so $\langle k \,|\, \mathscr{H}_i \,|\, 0\rangle$ breaks up into a sum of products like

$$\langle a| \frac{e}{mc}\sum_{j=1}^{n} \mathbf{A}_{j\alpha}\cdot\mathbf{p}_j \,|a_0\rangle \langle f|\, q_\alpha \,|f_0\rangle.$$

We have already seen that $\langle f \,|\, q_\alpha \,|f_0\rangle$ and $\langle f \,|\, \bar{q}_\alpha \,|f_0\rangle$ are zero unless $|f\rangle$ and $|f_0\rangle$ differ by just one in the quantum number n_α. In other words transitions are only possible if they involve the emission or the absorption of just one photon by the field.

The other matrix element may be simplified by observing that visible light has a wavelength lying roughly in the range 4000–7000 Å, whilst the dimensions of an atom are of the order of 1–2 Å. This means that $\mathbf{A}_\alpha$ is approximately constant over the atom and may be taken outside the matrix element to give

$$\frac{e}{mc}\mathbf{A}_\alpha\cdot\sum_j \langle a \,|\, \mathbf{p}_j \,|\, a_0\rangle$$

This can easily be transformed to a sum over matrix elements of the radius vector $\mathbf{r}$. Let E and E_0 be the energies of the two atomic states $|a\rangle$ and $|a_0\rangle$, respectively. Then

$$\begin{aligned}(E-E_0)\langle a \,|\, \mathbf{r}_j \,|\, a_0\rangle &= \langle a|\,(\mathscr{H}_a \mathbf{r}_j - \mathbf{r}_j \mathscr{H}_a)\,|a_0\rangle \\ &= \langle a| \left(\frac{1}{2m}\mathbf{p}_j^2\mathbf{r}_j - \frac{1}{2m}\mathbf{r}_j\mathbf{p}_j^2\right)|a_0\rangle \\ &= \frac{1}{2m}\langle a|\,[\mathbf{p}_j^2, \mathbf{r}_j]\,|a_0\rangle \\ &= \frac{1}{m}\langle a|\,(-i\hbar\mathbf{p}_j)\,|a_0\rangle,\end{aligned}$$

which is the required transformation. We have used (2.1) for $\mathscr{H}_a$ and the commutation relations (2.2) for $\mathbf{p}_j$ and $\mathbf{r}_j$. Our matrix element, then, is proportional to the matrix element of the total dipole moment of the atom, $\mathbf{D} = \sum_j e\mathbf{r}_j$. For this reason such transitions are said to be due to electric dipole radiation. They are not the only possible type of transition because we have incorrectly assumed $\mathbf{A}_\alpha$ to be completely constant over the atom. Later we shall discuss the way in which the small variations of $\mathbf{A}_\alpha$ over the atom can give rise to observable consequences.

Next we derive an expression for the transition probability per unit time. We will also see that energy is conserved in the transition to a very good approximation. Let us write E^k and E^0 for the energies of $|k\rangle$ and $|0\rangle$, respectively. Each of these energies is the sum of the energy of the atom and of the field. Then the dependence of $\langle k\,|\,\mathscr{H}_i\,|\,0\rangle$ on time is given by

$$\langle k\,|\,\mathscr{H}_i\,|\,0\rangle = \exp\left[i\hbar^{-1}t(E^k - E^0)\right] M_k,$$

where M_k is independent of time. Therefore $c_k(t)$ satisfies the equations,

$$i\hbar\,\frac{dc_k(t)}{dt} = M_k \exp\left[i\hbar^{-1}t(E^k - E^0)\right] \quad (c_k(0) = 0),$$

which can be integrated immediately to give

$$c_k(t) = \frac{M_k\{\exp\left[i\hbar^{-1}t(E^k - E^0)\right] - 1\}}{E^0 - E^k},$$

for times for which $c_0(t)$ is not appreciably different from unity. The relevant unit of time here is therefore the natural lifetime of the initial state $|0\rangle$. Subject to this restriction the probability of finding the system in state $|k\rangle$ at the time t is

$$|c_k(t)|^2 = \frac{2\,|M_k\,|^2\{1 - \cos\left[\hbar^{-1}t(E^k - E^0)\right]\}}{(E^0 - E^k)^2}. \tag{3.18}$$

We now assume that the final energy E^k lies in a continuous spectrum. Then we consider the total probability of transition to any state in this spectrum and must integrate $|c_k(t)|^2$ over the energy variable $E = E^k - E^0$. The transition probability per unit time is

$$\begin{aligned} P &= t^{-1}\int |c_k(t)|^2 \rho\, dE \\ &= \int \frac{2|M_k|^2\left[1 - \cos(\hbar^{-1}tE)\right]}{tE^2}\rho\, dE \\ &= \int \frac{2\rho\,|M_k|^2}{\hbar}\,\frac{1 - \cos y}{y^2}\,dy, \end{aligned}$$

where we have written $\hbar y = Et$ and ρ is the density of states around the energy E.

Now the integral

$$\int_{-\infty}^{+\infty} \frac{1 - \cos y}{y^2}\,dy$$

exists and is easily shown by complex variable methods[1] to have the value π. The integrand is everywhere less than or equal to $2y^{-2}$ and so decreases fairly rapidly as $|y| \to \infty$. In more detail we see that the part of the integral which arises from the range $2n\pi \leqslant |y| \leqslant 2(n+1)\pi$ is less than the integral of $(1 - \cos y)/4n^2\pi^2$ over the corresponding range, i.e. less than $1/n^2\pi$. As we have

$$\sum_{n=1}^{\infty} \frac{1}{n^2\pi} = \frac{\pi}{6}$$

[1] See, for example, Whittaker & Watson (1946), ex. 2, p. 116.

we have shown that at least five-sixths of the integral comes from $|y| \leqslant 2\pi$ and the rest is almost all from the region where y is not large compared with 2π. Provided that $\rho\,|M_k|^2$ does not vary appreciably over this range of y we have the approximate value

$$P = \frac{2\pi\rho\,|M_k|^2}{\hbar} \tag{3.19}$$

for the transition probability per unit time. $\rho\,|M_k|^2$ has an approximate constant value over this range of y, for example we might take it at $y = 0$, i.e. at $E^k = E^0$. Almost all the contribution to (3.19) comes from energies E^k such that $E^k - E^0$ is not large compared with h/t. This result has two familiar interpretations. The first is that it shows that energy is conserved in the transition to a very good approximation. The second is that this is simply an expression of the uncertainty relation between the time of existence of a state and the accuracy with which its energy can be determined. In our applications the lines are invariably very much broader than this necessary minimum breadth. The natural line-breadth will always be masked by other broadening influences.

Before applying (3.19) we must understand a little more clearly what it means. The states $|k\rangle$ which have energies near E^0 form a discrete set. This is a consequence of our assumption that the radiation field is confined to a box of finite size. As the size of the box tends to infinity, so the energies of these states get closer together and in deriving (3.19) we have assumed that we may write a continuous function $\rho(E)$ such that $\rho(E)\,dE$ gives the number of states in the energy range E to $E+dE$ sufficiently accurately for our purposes. This is possible for a combination of two reasons. The first is that the states are very dense as a function of energy. This may be seen by examining the applications in § 3.2. The second is that the natural line breadth makes it unimportant whether there are any states $|k\rangle$ with E^k exactly equal to E^0. It is only the density near E^0 which is relevant.

In general, ρ will be a function of E and also other variables, ξ say. Then (3.19) is replaced by

$$P = \frac{2\pi}{\hbar}\int \rho(\xi, E)\,|M_k|^2\,d\xi. \tag{3.20}$$

Evidently the derivation of this formula depends upon $\rho\,|M_k|^2$ being a continuous function of E for fixed values of ξ.

We now pass to the evaluation of M_k. Writing $|k\rangle = |a\rangle|f\rangle$ and $|0\rangle = |a_0\rangle|f_0\rangle$, where all the kets are now independent of time,

$$M_k = \langle k\,|\,\mathscr{H}_i\,|\,0\rangle = \frac{e}{mc}\sum_{j=1}^{n}\sum_{\alpha}\langle a|\langle f|\,(q_\alpha \mathbf{A}_{j\alpha}\cdot\mathbf{p}_j + \bar{q}_\alpha \bar{\mathbf{A}}_{j\alpha}\cdot\mathbf{p}_j)\,|a_0\rangle|f_0\rangle,$$

where $\mathbf{A}_{j\alpha}$ is the value of $\mathbf{A}_\alpha$ at the position of the jth electron. We already know that $\langle f\,|\,q_\alpha\,|f_0\rangle$ and $\langle f\,|\,\bar{q}_\alpha\,|f_0\rangle$ are zero unless $|f\rangle$ differs from $|f_0\rangle$ by just one photon. Therefore we have shown that photons can only be emitted or absorbed one at a time. The conservation of energy then implies that the energy of the photon is the same as the energy change of the atom. The matrix elements of $q_\alpha \mathbf{A}_{j\alpha}\cdot\mathbf{p}_j$ can be non-zero only if $\langle f\,|\,q_\alpha\,|f_0\rangle$ is non-zero, i.e. if $|0\rangle$ has one more photon than $|k\rangle$. Therefore they correspond to absorption of radiation by the atom. Similarly $\bar{q}_\alpha \bar{\mathbf{A}}_{j\alpha} . \mathbf{p}_j$ correspond to emission. q_α and $\bar{q}_\alpha$ can never both have non-zero matrix elements to $|k\rangle$ at the same time. Now suppose that at most the oscillator $\mathbf{A}_\alpha = (4\pi c^2/V)^{\frac{1}{2}} \boldsymbol{\pi}_\alpha e^{i\boldsymbol{\kappa}_\alpha . \mathbf{r}}$ gives a non-zero contribution to M_k. In absorption

$$M_k = \frac{e}{mc} \langle a| \sum_{j=1}^{n} \mathbf{p}_j . \mathbf{A}_{j\alpha} |a_0\rangle \langle f| q_\alpha |f_0\rangle$$

$$= \frac{e}{mc}\left(\frac{4\pi c^2}{V}\right)^{\frac{1}{2}} \left(\frac{n\hbar}{2\nu}\right)^{\frac{1}{2}} \langle a| \sum_{j=1}^{n} \mathbf{p}_j . \boldsymbol{\pi}_\alpha e^{i\boldsymbol{\kappa}_\alpha . \mathbf{r}_j} |a_0\rangle,$$

and in emission

$$M_k = \frac{e}{mc}\left(\frac{4\pi c^2}{V}\right)^{\frac{1}{2}} \left(\frac{n\hbar}{2\nu}\right)^{\frac{1}{2}} \langle a_0| \sum_{j=1}^{n} \mathbf{p}_j . \boldsymbol{\pi}_\alpha e^{-i\boldsymbol{\kappa}_\alpha . \mathbf{r}_j} |a\rangle.$$

In both cases

$$|M_k|^2 = \frac{2\pi n\hbar e^2}{m^2 \nu V} |\langle a| \sum_{j=1}^{n} \mathbf{p}_j . \boldsymbol{\pi}_\alpha e^{i\boldsymbol{\kappa}_\alpha . \mathbf{r}_j} |a_0\rangle|^2.$$

This equality of emission and absorption probabilities ensures that the equilibrium distribution of energy between states is the Boltzmann distribution.

3.2. Emission and absorption of radiation

3.2.1. Spontaneous emission. Our first application of the preceding analysis is to the determination of the probability for spontaneous emission of radiation by an atom which is in an excited state $|a'\rangle$. Suppose the atom has only one state $|a\rangle$ with energy less than $|a'\rangle$ and write $\hbar\nu$ for the energy difference between $|a'\rangle$ and $|a\rangle$. Then the initial state of the system is $|a'\rangle|f'\rangle$, where $|f'\rangle$ is the vacuum, i.e. no photons are present. The final states are $|a\rangle|f\rangle$, where $|a\rangle$ is the ground state of the atom and $|f\rangle$ has now just one photon with its energy close to $\hbar\nu$.

The first task is to determine the density of photon states, i.e. ρ, as a function of E and possibly other variables. In order for $\mathbf{A}_\alpha$ of (3.12) to satisfy the boundary conditions we must have $\kappa_{\alpha x} L_x = 2\pi n_x$, $\kappa_{\alpha y} L_y = 2\pi n_y$, $\kappa_{\alpha z} L_z = 2\pi n_z$, where n_x, n_y and n_z are integers. For simplicity we take a cubic box ($L_x = L_y = L_z = L$). The result is not actually appreciably dependent on the shape of the box. Also define $n^2 = n_x^2 + n_y^2 + n_z^2$. Then $\boldsymbol{\kappa}^2 = (\nu^2/c^2) = (4\pi^2 n^2/L^2)$ and $\boldsymbol{\kappa}$ is in the same direction as the vector $\mathbf{n}$. For n large the number of oscillators in a direction within the solid angle $d\Omega$ of $\mathbf{n}$, and hence also within $d\Omega$ of $\boldsymbol{\kappa}$, and with n lying between n and $n+dn$ is $n^2 dn d\Omega = (L/2\pi c)^3 \nu^2 d\nu d\Omega = \rho dE$. $E = \hbar\nu$. Therefore the density of states is $\rho = \hbar^{-1}\nu^2(L/2\pi c)^3 d\Omega$ for each polarization. In evaluating the transition probability we must sum over the two polarizations and integrate over Ω. The integral over Ω is taken over all angles because $\boldsymbol{\kappa}$ and $-\boldsymbol{\kappa}$ represent propagation vectors for different photons.

Collecting our results, we have found that the probability I for spontaneous emission of a photon with energy close to $\hbar\nu$ is given by

$$I = \frac{2\pi}{\hbar} \sum_{\text{pol.}} \int_\Omega |M_k|^2 \rho$$

$$= \frac{\nu e^2}{\hbar m^2 c^3} \sum_{\text{pol.}} \int_\Omega |\langle a'| \sum_{j=1}^{n} \mathbf{p}_j . \boldsymbol{\pi} e^{i\boldsymbol{\kappa} . \mathbf{r}_j} |a\rangle|^2 d\Omega. \tag{3.21}$$

To proceed, we investigate the magnitude of $\boldsymbol{\kappa}.\mathbf{r}$. It is $\nu r/c = 2\pi r\mu$, where μ is the frequency of the light in cm^{-1}. μ lies in the range 14,000–25,000 for light in the visible region, whereas r is of the order of 10^{-8} cm. So $2\pi r\mu$ is of the order of 10^{-3} and therefore we expand $e^{i\boldsymbol{\kappa}.\mathbf{r}_j}$ and retain only the first two terms. The remaining ones are of no interest to us at all.

It will clarify the formulae and the derivation if we first consider a one-electron atom and neglect even the second term in the expansion of $e^{i\boldsymbol{\kappa}.\mathbf{r}}$. Then

$$I = \frac{\nu e^2}{hm^2c^3} \sum_{\text{pol.}} \int |\langle a' \,|\, \mathbf{p}.\boldsymbol{\pi} \,|\, a\rangle|^2 \, d\Omega.$$

The variable Ω of course refers to the angles in $\boldsymbol{\kappa}$ and $\boldsymbol{\pi}$, not to those in $\mathbf{p}$ or $|a\rangle$ or $|a'\rangle$. First, we transform the matrix element of $\mathbf{p}.\boldsymbol{\pi}$ by using

$$\hbar\nu \langle a' \,|\, \mathbf{r} \,|\, a\rangle = -\frac{i\hbar}{m} \langle a' \,|\, \mathbf{p} \,|\, a\rangle,$$

and so $|\langle a' \,|\, \mathbf{p}.\boldsymbol{\pi} \,|\, a\rangle|^2 = m^2\nu^2 \,|\langle a' \,|\, \mathbf{r}.\boldsymbol{\pi} \,|\, a\rangle|^2$. Then we sum over the polarizations and must specify two directions of polarization for each vector $\boldsymbol{\kappa}$. For example

$$\boldsymbol{\kappa} = \frac{\nu}{c}(\sin\theta \cos\phi, \sin\theta \sin\phi, \cos\theta),$$

$$\boldsymbol{\pi}^1 = (-\cos\theta \cos\phi, -\cos\theta \sin\phi, \sin\theta),$$

$$\boldsymbol{\pi}^2 = (-\sin\phi, \cos\phi, 0).$$

Introducing this into $|\langle a' \,|\, \mathbf{r}.\boldsymbol{\pi} \,|\, a\rangle|^2$ and integrating over θ and ϕ, the cross-terms come to zero and we obtain

$$\sum_{\text{pol.}} \int |\langle a' \,|\, \mathbf{r}.\boldsymbol{\pi} \,|\, a\rangle|^2 \, d\Omega = \tfrac{8}{3}\pi \,|\langle a' \,|\, \mathbf{r} \,|\, a\rangle|^2,$$

and hence

$$I = \frac{8\pi e^2\nu^3}{3hc^3} |\langle a' \,|\, \mathbf{r} \,|\, a\rangle|^2 = \frac{8\pi\nu^3 R^2}{3hc^3}, \tag{3.22}$$

where $R^2 = |\langle a' \,|\, e\mathbf{r} \,|\, a\rangle|^2$ is the square of the matrix element of the electric dipole moment $e\mathbf{r}$ between the two states. In terms of wave-numbers, μ, we have $I = 64\pi^4\mu^3R^2/3h$ and if we take $\mu = 20{,}000$ and $|\langle a' \,|\, \mathbf{r} \,|\, a\rangle|^2 = 10^{-16}\,\text{cm}^2$, then $I = 0{\cdot}6 \times 10^8\,\text{sec}^{-1}$. Hence the lifetime of an excited state is of the order of 10^{-8} sec if $|\mathbf{r}|$ is 1 Å. We shall say that the accompanying transition is strongly allowed for electric-dipole radiation.

The generalization of the preceding analysis to n electrons is completely trivial and $e\mathbf{r}$ becomes replaced by $e\sum_{j=1}^{n}\mathbf{r}_j$. This is because operators referring to different electrons commute with one another. The other simplification we made was to neglect the higher terms in the expansion of $e^{i\boldsymbol{\kappa}.\mathbf{r}}$. They are negligible unless R^2 is either identically zero or accidentally very small. That may easily occur, however. For example, if both $|a'\rangle$ and $|a\rangle$ have the same parity (i.e. behaviour under inversion, see § 2.8), the matrix element of $e\mathbf{r}$ is identically zero. However, the transition probability is not necessarily zero and we must then consider the term $i(\boldsymbol{\kappa}.\mathbf{r})(\mathbf{p}.\boldsymbol{\pi})$.

3.2.2. Magnetic dipole and electric quadrupole transitions. We again start with a one-electron atom and suppose that $\langle a' \mid \mathbf{r} \mid a \rangle = 0$. Then

$$I = \frac{\nu e^2}{hm^2c^3} \sum_{\text{pol.}} \int_{\Omega} |\langle a' \mid (\boldsymbol{\kappa}.\mathbf{r})(\mathbf{p}.\boldsymbol{\pi}) \mid a \rangle|^2 \, d\Omega.$$

It is convenient to separate $(\boldsymbol{\kappa}.\mathbf{r})(\mathbf{p}.\boldsymbol{\pi})$ into two parts. To do this we use suffixes for the three components of each of the vectors and have

$$\begin{aligned}(\boldsymbol{\kappa}.\mathbf{r})(\mathbf{p}.\boldsymbol{\pi}) &= \sum_{ij} \kappa_i r_i p_j \pi_j = \Sigma \kappa_i \pi_j r_i p_j \\ &= \tfrac{1}{2}\Sigma \kappa_i \pi_j (r_i p_j - r_j p_i) + \tfrac{1}{2}\Sigma \kappa_i \pi_j (r_i p_j + r_j p_i) \\ &= \tfrac{1}{2}\boldsymbol{\kappa} \wedge \boldsymbol{\pi}.\mathbf{l} + \tfrac{1}{2}\Sigma \kappa_i \pi_j (r_i p_j + r_j p_i),\end{aligned}$$

where $\mathbf{l} = \mathbf{r} \wedge \mathbf{p}$ is the orbital angular momentum vector. The matrix element of the second term is transformed in just the way we transformed the matrix element of $\mathbf{p}$ in § 3.1.4 and gives

$$\langle a' | (r_i p_j + r_j p_i) | a \rangle = im\nu \langle a' | r_i r_j | a \rangle$$

and so

$$\begin{aligned}I &= \frac{\nu e^2}{4hm^2c^3} \sum_{\text{pol.}} \int_{\Omega} \{ |\langle a' | \boldsymbol{\kappa} \wedge \boldsymbol{\pi}.\mathbf{l} | a \rangle|^2 + m^2\nu^2 |\langle a' | (\boldsymbol{\kappa}.\mathbf{r})(\boldsymbol{\pi}.\mathbf{r}) | a \rangle|^2 \} \, d\Omega \\ &= I_1 + I_2, \quad \text{say}^1. \end{aligned} \qquad (3.23)$$

We now simplify I_1 by performing the indicated summation and integration. In the notation of the last section,

$$\boldsymbol{\kappa} \wedge \boldsymbol{\pi}^1 = \frac{\nu}{c}(\sin\phi, \ -\cos\phi, \ 0),$$

$$\boldsymbol{\kappa} \wedge \boldsymbol{\pi}^2 = \frac{\nu}{c}(-\cos\theta \cos\phi, \ -\cos\theta \sin\phi, \ \sin\theta)$$

and

$$\begin{aligned} I_1 &= \frac{\nu e^2}{4hm^2c^3} \frac{\nu^2}{c^2} \tfrac{8}{3}\pi \, |\langle a' \mid \mathbf{l} \mid a \rangle|^2 \\ &= \frac{8\pi\nu^3}{3hc^3} \left| \langle a' | \frac{e}{2mc} \mathbf{l} | a \rangle \right|^2. \end{aligned} \qquad (3.24)$$

This is just the same as (3.22) except that the electric-dipole moment $e\mathbf{r}$ has been replaced by $(e/2mc)\,\mathbf{l}$. We shall see in §§ 5.5 and 5.6 that this latter vector is the magnetic dipole moment due to the orbital motion of the electron. The electron also possesses a spin magnetic moment which we have neglected so far. The spin magnetic moment vector is $-(e/mc)\,\mathbf{s}$, where $\mathbf{s}$ is the spin vector of the electron. Its interaction with the magnetic field of the photon is then $\mathscr{H}_s = (e/mc)\,\mathbf{s}\,.\,\text{curl}\,\mathbf{A}$ (see (5.35)). We evaluate curl $\mathbf{A}$ in terms of the vectors $\boldsymbol{\pi}$ and $\boldsymbol{\kappa}$ and then put $e^{i\boldsymbol{\kappa}.\mathbf{r}} = 1$ to obtain

$$\mathscr{H}_s = \frac{e}{mc}\left(\frac{4\pi c^2}{L^3}\right)^{\frac{1}{2}} i(q - \bar{q})\,\boldsymbol{\kappa} \wedge \boldsymbol{\pi}.\mathbf{s},$$

[1] The cross-terms which should formally be present in I vanish on summing and integrating.

which gives a contribution to I identical with I_1, except that **l** is replaced by **2s**. Therefore the correct expression for I_1 is actually

$$I_1 = \frac{8\pi\nu^3}{3hc^3} |\langle a'| \frac{e}{2mc} (\mathbf{l}+2\mathbf{s}) |a\rangle|^2. \tag{3.25}$$

The corresponding transitions are called magnetic dipole transitions.

I_2 is simplified in the same manner and we obtain

$$I_2 = \frac{\nu e^2}{4hm^2c^3} \frac{m^2\nu^4}{c^2} \tfrac{4}{5}\pi \sum_{ij} |\langle a' | \mathcal{N}_{ij} | a\rangle|^2$$

$$= \frac{\pi e^2 \nu^5}{5hc^5} \sum_{ij} |\langle a' | \mathcal{N}_{ij} | a\rangle|^2, \tag{3.26}$$

where $\mathcal{N}_{ij}$ is the quadrupole tensor $r_i r_j - \frac{1}{3}r^2\delta_{ij}$. The corresponding transitions are called electric quadrupole transitions.

As before, the generalization to n electrons is straightforward. $(e/2mc)(\mathbf{l}+2\mathbf{s})$ is replaced by $(e/2mc)(\mathbf{L}+2\mathbf{S})$ and $\mathcal{N}_{ij}$ by the sum of the quadrupole tensors for each of the n electrons.

In Russell–Saunders coupling the vector $\mathbf{L}+2\mathbf{S}$ only has non-vanishing matrix elements within a term and so, to this approximation, magnetic-dipole radiation can only take place between the levels of a term. If the matrix element of $\mathbf{L}+2\mathbf{S}$ between the two states $|a\rangle$ and $|a'\rangle$ has magnitude $\hbar$ we deduce that the relative probability of a magnetic-dipole transition to that for an electric-dipole transition for $e^{-2}R^2 = 10^{-16}\,\text{cm}^2$ is 3×10^{-6}. The relative probability for an electric-quadrupole transition is proportional to the square of the frequency. For $\mu = 20{,}000\,\text{cm}^{-1}$ it is about 10^{-7} of that for an allowed electric-dipole transition if we assume $\sum_{ij} |\langle a' | \mathcal{N}_{ij} | a\rangle|^2 = 10^{-32}\,\text{cm}^4$.

3.2.3. Absorption spectra. The theory of absorption is practically identical with that of emission and we have already seen that the probabilities for emission and absorption between a given pair of states are equal. From an experimental point of view, absorption spectra of metal compounds are easier to obtain than emission spectra simply because we cannot use very high temperatures or the compounds would decompose. Therefore, in thermal equilibrium, the occupied states are all fairly close to the ground state and radiation cannot be emitted in the visible region. However, it can be absorbed. It can then either be re-emitted or the energy can be 'degraded'—that is it can be transformed into energy of thermal motion, mainly lattice vibration, of the solid. On the other hand, it may be partly degraded and then re-emitted at a lower frequency. Then we have the phenomena of fluorescence and phosphorescence.

In this section we discuss the theory of an experiment in which a beam of light with a definite frequency, polarization and direction passes through an aggregate of metal ions. It is assumed that any radiation which is absorbed by the aggregate is completely degraded. For convenience of conception and calculation the experiment is idealized somewhat as illustrated in Fig. 3.1. A beam of n photons is travelling from left to right in a rectangular box of length L along the direction of propagation. This situation is represented in the theory by a value n for the occupation number of the radiation oscillator having

$$\mathbf{A} = \left(\frac{4\pi c^2}{LD}\right)^{\frac{1}{2}} \boldsymbol{\pi}(q\,e^{i\boldsymbol{\kappa}.\mathbf{r}} + \bar{q}\,e^{-i\boldsymbol{\kappa}.\mathbf{r}}),$$

where D is the cross-sectional area of the box, and $\boldsymbol{\kappa}$ is the direction of propagation. $\boldsymbol{\pi}$ is a fixed unit vector and $\boldsymbol{\pi}.\boldsymbol{\kappa} = 0$. The box has periodic walls in the sense described in § 3.1.

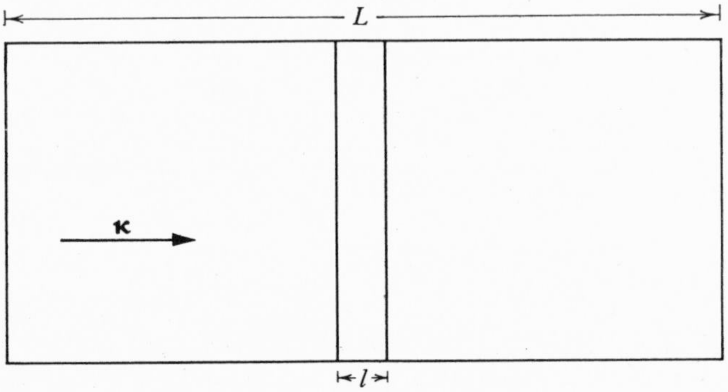

Fig. 3.1. The partial absorption of a beam of light in a thin specimen.

The specimen which absorbs the light is an aggregate of metal ions spread homogeneously (from the macroscopic point of view) as a thin lamina of width l and area D lying normal to the propagation vector $\boldsymbol{\kappa}$ of the light. We are going to calculate the energy E_0 incident on the lamina per second and the energy E_1 absorbed by the lamina per second. The ratio $\alpha = E_1/lE_0$ is then the absorption coefficient of the specimen as it is normally defined.

Let us write $|n\rangle$ for the state vector of the radiation field. Then E_0 is obtained from the integral of the Poynting vector $(c/4\pi)\,\mathbf{E} \wedge \mathbf{H}$ for $|n\rangle$ over a cross-section of the box. But the integral of $(c/4\pi)\,\mathbf{E} \wedge \mathbf{H}$ over the whole box is $c^2\mathbf{G}$ and we have already seen that $\mathbf{G}$ lies along $\boldsymbol{\kappa}$ and has magnitude $2c^{-1}\nu^2\bar{q}q = c^{-1}n\hbar\nu$ for the state $|n\rangle$. Therefore $E_0 = L^{-1}n\hbar\nu c$.

To calculate E_1 we must specify a little more about the metal ions. In solids and liquids various factors, including the thermal motions of adjacent atoms, ensure that a given electronic transition is spread over a range of frequencies—in other words the absorption line is broadened. Therefore we assign a distribution function for the energy of the excited state $|a'\rangle$ of the atom relative to the ground state $|a\rangle$. For simplicity we suppose each atom to have one excited state and that the distribution function is $g(\nu)$. Then $\int g(\nu)\,d\nu = 1$ and a fraction $g(\nu)\,d\nu$ of the ions have their excited states in the energy range $\hbar\nu$, $\hbar(\nu+d\nu)$ above the ground state. If the concentration of metal ions is c_0 gram ion per litre and N is Avogadro's number, the transition probability per second is

$$\frac{4\pi^2 c_0 N l e^2 n g}{10^3 \hbar L m^2 \nu} |\langle a' \mid \mathbf{p}.\boldsymbol{\pi} e^{i\boldsymbol{\kappa}.\mathbf{r}} \mid a\rangle|^2 = \frac{4\pi^2 c_0 N l e^2 n \nu g}{10^3 \hbar L} |M|^2, \quad \text{say,}$$

where $M = \langle a' \mid \boldsymbol{\pi}.\mathbf{r} \mid a\rangle$ for electric dipole transitions.

It follows immediately that

$$\alpha = \frac{4\pi^2 c_0 N e^2 \nu g}{10^3 \hbar c} |M|^2.$$

It is customary to express absorption spectra in terms of the molecular extinction coefficient ϵ which is defined by the relation $\alpha = \epsilon c_0 \log_{10} e$. A convenient measure of the total absorption due to the electronic transition is the oscillator strength. This dimensionless quantity f is defined by

$$f = \frac{10^3 mc \log_{10} e}{2\pi^2 e^2 N} \int \epsilon(\nu)\, d\nu.$$

Our calculation gives
$$f = \frac{4\pi m}{h} \int |M|^2 \nu g(\nu)\, d\nu.$$

For a narrow band this may be replaced by the approximate expression

$$f = \frac{4\pi m \nu}{h} |M|^2.$$

In general, the matrix element

$$M = \langle a' \mid \{\boldsymbol{\pi}.\mathbf{r} + \frac{1}{2m\nu} \boldsymbol{\kappa} \wedge \boldsymbol{\pi}.(\mathbf{l} + 2\mathbf{s}) + \tfrac{1}{2} i(\boldsymbol{\kappa}.\mathbf{r})(\boldsymbol{\pi}.\mathbf{r})\} \mid a\rangle,$$

for one electron, to the first order in $\boldsymbol{\kappa}.\mathbf{r}$.

The formulae we have just derived are appropriate for polarized light passing through an anisotropic medium. In many experiments, however, unpolarized light is passed through an isotropic medium. This is always true for spectra of ions in solution. In such cases we need the mean of the preceding expressions over angle and polarization. The cross-terms between the three different elements then integrate out. We have evaluated the other integrals earlier in the section, whence

$$|M|^2 = \frac{1}{3}\left\{ |\langle a' \mid \mathbf{r} \mid a\rangle|^2 + |\langle a' \mid \frac{1}{2mc}(\mathbf{l} + 2\mathbf{s}) \mid a\rangle|^2 + \frac{3\nu^2}{40c^2} \sum_{ij} |\langle a' \mid \mathcal{N}_{ij} \mid a\rangle|^2 \right\}. \quad (3.27)$$

CHAPTER 4

THE STRUCTURE OF FREE ATOMS AND IONS[1]

4.1. One-electron atoms

After our somewhat long introduction to the various basic techniques we can now pass on to use them to describe the electronic structure of atoms. The simplest type of atom is one in which there is just one electron moving in the field of a single nucleus of charge Ze. When $Z = 1$ we have atomic hydrogen, and when $Z = 2$ singly ionized helium, etc. This is the only type of atom for which it has proved possible to obtain explicit closed expressions for the energies and eigenfunctions of the stationary states. For this reason and also because the theory forms a useful starting point for treating more complicated systems we discuss the theory of this atom first. In accordance with our general plan we neglect relativistic effects in this chapter and energies associated with the spin magnetic moment of the electron (or of the nucleus if it has a spin).

We suppose at first that the single electron moves in a general field possessing spherical symmetry. The theory will then also apply to those other atoms which may be considered, to a good approximation, to have all their electrons except one coupled together to form an inert core. Then the last electron moves in the field of the nucleus and this inert core. The core is spherically symmetrical and so the Hamiltonian for the outer electron is

$$\mathscr{H} = \frac{1}{2m}\mathbf{p}^2 + V, \tag{4.1}$$

with V a function of the radius r alone. Typical examples of such atoms are the alkali metal atoms. We shall see later why we can separate off part of the atomic system in this way: here we merely investigate the consequences of (4.1).

It is convenient to use spherical polar coordinates

$$\left.\begin{aligned} x &= r\sin\theta\cos\phi, \\ y &= r\sin\theta\sin\phi, \\ z &= r\cos\theta, \end{aligned}\right\} \tag{4.2}$$

and remembering that $\mathbf{p} = -i\hbar\nabla$ the Schrödinger equation corresponding to (4.1) is

$$-\frac{\hbar^2}{2mr^2}\left(r\frac{\partial^2}{\partial r^2}r + \frac{1}{\sin\theta}\frac{\partial}{\partial\theta}\sin\theta\frac{\partial}{\partial\theta} + \frac{1}{\sin^2\theta}\frac{\partial^2}{\partial\phi^2}\right)\psi + V\psi = E\psi. \tag{4.3}$$

[1] I would like to express my debt to Condon & Shortley's book, *The Theory of Atomic Spectra.* My knowledge of the theory is founded on this book.

We seek solutions of the form $\psi = R(r)\,\Theta(\theta)\,\Phi(\phi)$ and have, on substituting for ψ in (4.3) and multiplying through by $r^2\psi^{-1}$

$$-\frac{\hbar^2 r}{2mR}\frac{\partial^2(rR)}{\partial r^2}-\frac{\hbar^2}{2m\Theta\sin\theta}\frac{\partial}{\partial\theta}\left(\sin\theta\frac{\partial\Theta}{\partial\theta}\right)-\frac{\hbar^2}{2m\Phi\sin^2\theta}\frac{\partial^2\Phi}{\partial\phi^2}+(Vr^2-Er^2)=0. \tag{4.4}$$

In (4.4), the first and fourth terms depend on r but not θ or ϕ, whilst the second and third depend on θ and ϕ but not r. It follows that

$$-\frac{\hbar^2 r}{2mR}\frac{d^2(rR)}{dr^2}+(Vr^2-Er^2)=\lambda, \tag{4.5}$$

where λ is independent of r, θ and ϕ. Similarly we may separate the equation for θ and ϕ into two:

$$\left.\begin{aligned}-\frac{\hbar^2}{2m\Theta\sin\theta}\frac{d}{d\theta}\left(\sin\theta\frac{d\Theta}{d\theta}\right)-\frac{\hbar^2\mu}{2m\sin^2\theta}&=-\lambda,\\ \frac{1}{\Phi}\frac{d^2\Phi}{d\phi^2}&=\mu,\end{aligned}\right\} \tag{4.6}$$

where μ also is constant. The equation for Φ integrates immediately to give

$$\Phi = e^{ip\phi} \quad (p = 0, \pm 1, \pm 2, \ldots), \tag{4.7}$$

where $\mu = -p^2$. Comparing this with (2.29) we see that ψ obtained in this way is an eigenfunction of l_z with eigenvalues $p\hbar$ and, as before, p must be integral in order for ψ to be a single-valued function of ϕ.

Putting $\mu = -p^2$ in the first of (4.6) we find that $w = \cos\theta$ satisfies the associated Legendre equation

$$(1-w^2)\frac{d^2\Theta}{dw^2}-2w\frac{d\Theta}{dw}-\frac{p^2\Theta}{1-w^2}-\frac{2m\lambda\Theta}{\hbar^2}=0. \tag{4.8}$$

In order for (4.8) to have single-valued solutions without singularities we require that

$$\lambda = -\frac{\hbar^2}{2m}l(l+1)$$

with l a non-negative integer, and $l \geqslant |p|$. The solutions are called the associated Legendre polynomials and written $P_l^p(w)$, where p is now positive. Explicitly they are

$$P_l^p(w) = \frac{(1-w^2)^{\frac{1}{2}p}}{2^l l!}\frac{d^{p+l}(w^2-1)^l}{dw^{p+l}}. \tag{4.9}$$

One easily sees that the P_l^p satisfy (4.8) by direct substitution. The P_l^p are in fact the only single-valued functions without singularities which are solutions of (4.8). By expressing the condition

$$\mathbf{L}^2\,|L\rangle = L(L+1)\,\hbar^2\,|L\rangle$$

in terms of spherical polar coordinates one also obtains equations (4.6) with $\lambda = -(\hbar^2/2m)\,L(L+1)$ so the functions $P_l^p(\cos\theta)\,e^{\pm ip\phi}$ are eigenfunctions of $\mathbf{L}^2$ and L_z with eigenvalues $l(l+1)\,\hbar^2$ and $\pm p\hbar$, respectively. The condition $l \geqslant p$ also reflects the findings of § 2.2.

The functions P_l^p and Φ are not normalized and we normalize each separately to unity, defining new functions:

$$\left.\begin{aligned} \Phi_p &= \frac{1}{\sqrt{2\pi}} e^{ip\phi} \quad (p = 0, \pm 1, \pm 2, \ldots), \\ \Theta_{lp} &= (-1)^p \sqrt{\left(\frac{2l+1}{2}\frac{(l-p)!}{(l+p)!}\right)} P_l^p(\cos\theta) \quad (p \geqslant 0), \\ \Theta_{lp} &= \sqrt{\left(\frac{2l+1}{2}\frac{(l-|p|)!}{(l+|p|)!}\right)} P_l^{|p|}(\cos\theta) \quad (p \leqslant 0), \\ Y_{lp} &= \Theta_{lp}\Phi_p. \end{aligned}\right\} \tag{4.10}$$

We use these precise definitions of Φ_p, Θ_{lp} and Y_{lp} throughout this book. The Y_{lp}, thus defined, have their phases connected by the shift operators $L^{\pm}$ as discussed already in connexion with (2.24) and (2.25). This is the reason for the factor $(-1)^p$ for $p \geqslant 0$ in (4.10), and means that $\Theta_{lp} = (-1)^p \Theta_{l-p}$. The explicit forms of the Θ_{lp} are shown in Table 4.1 for all $l \leqslant 4$.

Table 4.1. *The functions Θ_{lp} for $l \leqslant 4$*

$l=0$	$l=1$	$l=2$
$\Theta_{00} = \frac{1}{2}\sqrt{2}$	$\Theta_{10} = \frac{\sqrt{3}}{\sqrt{2}}\cos\theta$	$\Theta_{20} = \frac{\sqrt{5}}{2\sqrt{2}}(3\cos^2\theta - 1)$
	$\Theta_{1\pm1} = \mp\frac{1}{2}\sqrt{3}\sin\theta$	$\Theta_{2\pm1} = \mp\frac{1}{2}\sqrt{15}\sin\theta\cos\theta$
		$\Theta_{2\pm2} = \frac{1}{4}\sqrt{15}\sin^2\theta$

$l=3$	$l=4$
$\Theta_{30} = \frac{\sqrt{7}}{2\sqrt{2}}(2\cos^3\theta - 3\sin^2\theta\cos\theta)$	$\Theta_{40} = \frac{3}{8\sqrt{2}}(35\cos^4\theta - 30\cos^2\theta + 3)$
$\Theta_{3\pm1} = \mp\frac{\sqrt{21}}{4\sqrt{2}}\sin\theta(5\cos^2\theta - 1)$	$\Theta_{4\pm1} = \mp\frac{3\sqrt{5}}{4\sqrt{2}}\sin\theta\cos\theta(7\cos^2\theta - 3)$
$\Theta_{3\pm2} = \frac{1}{4}\sqrt{105}\sin^2\theta\cos\theta$	$\Theta_{4\pm2} = \frac{3\sqrt{5}}{8}\sin^2\theta(7\cos^2\theta - 1)$
$\Theta_{3\pm3} = \mp\frac{\sqrt{35}}{4\sqrt{2}}\sin^3\theta$	$\Theta_{4\pm3} = \mp\frac{3\sqrt{35}}{4\sqrt{2}}\sin^3\theta\cos\theta$
	$\Theta_{4\pm4} = \frac{3\sqrt{35}}{16}\sin^4\theta$

We now return to (4.5). We cannot solve it in general, but for the special case of one electron in a Coulomb field ($V = -Ze^2r^{-1}$) one can obtain closed solutions. Using our value for λ we have

$$-\frac{\hbar^2 r}{2m}\frac{d^2(rR)}{dr^2} - Ze^2rR - Er^2R = -\frac{\hbar^2}{2m}l(l+1)R,$$

which becomes, on putting $\chi = rR$,

$$-\frac{\hbar^2}{2m}\left(\frac{d^2\chi}{dr^2}-\frac{l(l+1)\chi}{r^2}\right)-\frac{Ze^2\chi}{r}-E\chi = 0. \tag{4.11}$$

The bound states must have $\chi \to 0$ as $r \to \infty$ and because R is finite at the origin must also have $\chi \to 0$ as $r \to 0$. When r is large, (4.11) becomes approximately

$$-\frac{\hbar^2}{2m}\frac{d^2\chi}{dr^2} = E\chi$$

with solutions $\chi = Ae^{-r\sqrt{(-2mE\hbar^{-2})}}$. We have chosen the positive square root and must have $E < 0$ in order to satisfy $\chi \to 0$ at infinity. This result suggests putting $\chi = e^{-ar}f$, with $a = \sqrt{(-2mE\hbar^{-2})}$. Doing this we readily obtain the equation

$$f''-2f'a-\frac{l(l+1)f}{r^2}+\frac{2Ze^2mf}{\hbar^2 r} = 0 \tag{4.12}$$

for f, with dashes denoting differentiation. Let us now seek a solution as a power series in r and write

$$f = r^s \sum_{t=0}^{\infty} a_t r^t,$$

with $a_0 = 1$. As $r \to 0$, f and also χ behave in the same way as r^s and so in order for χ to tend to zero we must have $s > 0$. If we substitute for f in (4.12) we find that the coefficient of r^{s-2} is $s(s-1)-l(l+1)$. As this must be zero for f to satisfy (4.12) and as also $s > 0$ we have $s = l+1$. This shows that, in a state having orbital angular momentum l, R and hence also ψ behaves as r^l at the origin. In particular, all states except S states vanish at the origin.

Next, the vanishing of the coefficient of a general power, r^{s+t-2} say, of r leads to the recurrence formula

$$a_t\{t^2+(2l+1)t\} = a_{t-1}\left\{2a(l+t)-\frac{2Ze^2m}{\hbar^2}\right\}$$

for the a_t. As $t \to \infty$, so $(a_t/a_{t-1}) \to (2a/t)$ and hence for large r, $f \sim e^{2ar}$. As $a > 0$ this means that χ does not tend to zero at infinite distance from the nucleus unless the a_t are all zero from some point on. If $a_t = 0$ but $a_{t-1} \neq 0$ we find

$$a(l+t) = \frac{Ze^2m}{\hbar^2}$$

and hence
$$E = -\frac{\hbar^2}{2m}a^2 = -\frac{me^4Z^2}{2\hbar^2(l+t)^2} \quad (t = 1, 2, 3, \ldots). \tag{4.13}$$

It is customary to write $n = l+t$. n is called the principal quantum number and l the azimuthal quantum number. Then the energies of the bound states are given as a series of discrete levels by formula (4.13). For a given value of n, l can take any integral value from 0 up to $n-1$. Then for each l there are $(2l+1)$ possible values for l_z and two possible values for s_z. In all we have a degeneracy of $2n^2$ for the energy level corresponding to the principal quantum number n.

It is customary to write small letters for the angular momentum operators and eigenvalues for single electrons. Thus the space eigenfunctions[1] of the hydrogen atom may be written in the sequence

$$1s;\quad 2s\,2p;\quad 3s\,3p\,3d;\quad \ldots;\quad nl;\quad \ldots, \tag{4.14}$$

of increasing energy. We will often write a general one-electron eigenket $|nlm_s m_l\rangle$. For hydrogen, all states with the same principal quantum number have the same energy. As an example, the normalized spatial eigenfunction of lowest energy for hydrogen itself is

$$\psi_{1s} = \frac{1}{\pi^{\frac{1}{2}}}\left(\frac{me^2}{\hbar^2}\right)^{\frac{1}{2}} e^{-me^2r/\hbar^2}.$$

The probability per unit range of r of finding the electron at distance r from the nucleus is

$$P(r) = 4\pi r^2\psi_{1s}^2,$$

and the most probable distance is given by $dP/dr = 0$, i.e. when

$$2r\psi_{1s}^2 + r^2\frac{d(\psi_{1s}^2)}{dr} = 0,$$

and is $r = \hbar^2/me^2$ which is also the radius of the first Bohr orbit in the old quantum theory.

In the alkali metal atoms there is one electron outside an inert inner core. A statement of this kind is, of course, only an approximation to the truth but it enables us to treat these atoms as one-electron atoms. Then V is no longer the Coulomb field and states with the same n but different l do not now necessarily have the same energy (n may be given a meaning for general V either by enumerating the states, as in (4.14), or by letting $-Ze^2r^{-1}$ change continuously into V). As we shall see in the next section, because of the Pauli Principle, there is a sense in which one can say that a certain number of the lowest energy sets of states for the single electron are filled in the alkali metal atoms. Because of this the odd electron can only occupy the unfilled states of (4.14). Thus in potassium, the $1s$, $2s$, $2p$, $3s$ and $3p$ states are entirely filled and so the odd electron can only go into $3d$ or nl with $n > 3$.

Examples

1. The mean value of r for the $1s$ state of hydrogen is $3\hbar^2/2me^2$.

2. Take $E > 0$ in (4.11) and show that there is a solution which is finite everywhere whatever the value of E. Such solutions do not vanish at infinity and so do not represent bound states for the atom.

3. By considering (4.10) show that the states of the hydrogen atom have the same parity as l. This gives an alternative derivation of the selection rule $\Delta l \neq 0$ for electric-dipole radiation.

4. Show that the $2p$ functions have the same angular variation as z, $(1/\sqrt{2})\,(x-iy)$ and $(1/\sqrt{2})\,(-x-iy)$.

5. Show that $\Theta = \sin^l\theta$ satisfies (4.6) when $\mu = -l^2$, $\lambda = -(\hbar^2/2m)\,l(l+1)$ and deduce that the mean value of $\cos^2\theta$ for the state $|nlm_s l\rangle$ is $(2l+3)^{-1}$.

[1] One-electron functions are also called orbitals; we usually, but not invariably, use the word to mean spatial functions without specification of spin.

4.2. Electron configurations

We now pass on to the main object of our present interest, the many-electron atom. As elsewhere in this chapter we use the approximate Hamiltonian $\mathscr{H}$ given in (2.1). The most important term which yet remains to be added to $\mathscr{H}$ is the so-called spin-orbit coupling energy. The spin-orbit coupling is a magnetic interaction between the magnetic moment of the electron and the electric field of the nucleus. I describe its effects in the next chapter. However, for almost all the atoms and ions which concern us in this book the main effect of the spin-orbit coupling is to remove the degeneracy of the various levels belonging to a term. It gives rise to energies which are usually small compared with the separation between terms, and so although in an exact theory the concept of term as applied to eigenstates of the energy has no precise meaning, nevertheless, in practice our eigenstates normally do belong to levels of a term to a rather good approximation.

As I remarked in the last section it is not possible to solve the wave equation associated with $\mathscr{H}$ of (2.1) exactly if $n > 1$. So we are forced to use an approximation in order to get any results. In our later applications we will find that the broad qualitative behaviour of ions in compounds depends very markedly upon the nature of the ground term and of the lowest-lying excited terms—that is just on the L and S values—and not so much on other features of the eigenfunctions of the ion. We now develop an approximate theory of the electrostatic energies of many-electron atoms. Compared with the problem of the exact solution of the Schrödinger equation it is a very simple theory. However, it has the great merit of generality and manageability and is sufficiently exact for most of our purposes. I indicate later in the chapter how to refine the theory when this is necessary.

We have already met with one method of writing down functions for an n-electron system which satisfy the antisymmetry requirement. We take an orthonormal set of one-electron functions and then build determinantal functions from them according to (2.64). This is the method that we adopt now in a particular form which is called the central-field approximation. From a physical point of view one supposes that each electron moves in an average field due to the nucleus and all the other electrons and, further, that this average field is spherically symmetric. Then its eigenfunctions would be of the one-electron type considered in the last section, namely, products of a radial function and an angular function which may be conveniently taken to be one of the Y_{lp} of (4.10). Before developing this method it is desirable to emphasize that the nature of our assumption is that we are taking the sort of basic one-electron functions which would be appropriate if each electron moved independently in a spherically symmetric field. This is all we actually assume. Therefore neither is it relevant to ask what this field actually is, nor is it necessary to ask for any detailed justification for this idea of an average field. The field does not appear in our calculations and we calculate the approximate energies using the exact Hamiltonian (2.1), not any sort of averaged-field Hamiltonian. Average central fields do appear in certain specific methods for obtaining the radial parts of our one-electron functions but are not an essential part of the central-field method.

First, therefore, we obtain approximate electrostatic energies and eigenkets for the terms and consider the effect of the spin-orbit coupling as a perturbation on these later (in ch. 5). In the central-field approximation we search for eigenfunctions which are finite sums of determinants of one-electron functions. Further we require that the one-electron functions should be solutions of a central field problem. This means that the wave-function is written:

$$\Psi = \sum_a c_a \sum_\mu (n!)^{-\frac{1}{2}} (-1)^\mu P_\mu\{\phi_1^a(1)\,\phi_2^a(2)\ldots\phi_n^a(n)\}, \tag{4.15}$$

where each ϕ is of the form

$$\phi_i^a(j) = R_{n^il^i}(j)\,Y_{l^ip^i}(j)\,\tau_{s^i}(j), \tag{4.16}$$

with τ_{s^i} spin kets and $s^i = \pm\frac{1}{2}$ corresponding to a spin of $\pm\frac{1}{2}$ along OZ. Without any real loss of generality we take R real. We remember that P_μ permutes the electrons, i.e. the arguments, in (4.15), not the functions. Clearly in a determinant only $2(2l+1)$ one-electron functions with a given n and l can occur, otherwise the determinant is identically zero. For a fixed n and l, R_{nl} is, of course, the same for all p and s in (4.16). We discuss its computation in a later section, but in the meantime assume it to be known for all relevant nl values.

The determinants in (4.15) may be partially specified by giving the nl values occurring in each term. If we specify the number of times each nl value occurs and then take all determinants for these numbers we say we have a configuration of electrons and write it $(n_1l_1)^{a_1}(n_2l_2)^{a_2}\ldots(n_rl_r)^{a_r}$, where a_i is an integer giving the number of electrons with quantum numbers n_il_i. In particular cases we often omit the brackets, and the index if it is unity. Thus we write $1s^2\,2s^2\,2p$ for $(1s)^2\,(2s)^2\,(2p)^1$. As another example the configuration $3d\,4s$ means that one electron is in a $3d$ orbital (ten in all) and one in a $4s$ orbital (two in all), so the configuration consists of 20 independent determinantal functions. Any sum (4.15) over such determinants is said to belong to the configuration. Since all the operators **S**, **L** and **J** clearly turn a function of a configuration into another of the same configuration we can break up any configuration into terms and levels just as with general functions. (This, of course, depends on the fact that the operators commute with each of the P_μ.)

We next consider how to determine which terms occur for a given configuration. With two electrons nl, $n'l'$ not having both $n = n'$ and $l = l'$ (inequivalent electrons; in the contrary case they are said to be equivalent) basic functions of the configurations $nl\,n'l'$ are $|nlsp, n'l's'p'\rangle$ (the basic functions are not simple products of the constituent one-electron functions, but 2×2 determinants as in (4.15)). For fixed s, s' we apply the procedure of § 2.3 to find that there are terms with all values of L from $l+l'$ to $|l-l'|$. As in the discussion of § 2.6 (see (2.66) and (2.67)) there is clearly a singlet and a triplet for each value of L. For example, the configuration $3d\,4d$ contains the terms ^1S, ^3S, ^1P, ^3P, ^1D, ^3D, ^1F, ^3F, ^1G and ^3G. This procedure is quite general and we can always obtain the terms of any configuration consisting of two or more inequivalent electrons (i.e. pairwise inequivalent, for all pairs) by applying this so-called vector-coupling procedure to both orbital and spin functions. In fact, one finds the allowed terms of a general configuration inductively by adding one electron at a time. The occurrence of singlets and triplets for two electrons represents the coupling of the two spins of $\frac{1}{2}$ to give a resultant of 1 or 0.

With equivalent electrons, matters are not quite so simple. For the exclusion principle eliminates certain terms. For example, the function $|nlsp, n'lsp\rangle$ is identically zero when $n = n'$. Because the shift operators connect the states of a term, it follows that only whole terms are removed from the set of states allowed for inequivalent electrons.

A convenient method of obtaining the allowed terms is to classify the determinantal functions by their M_S, M_L values. We take $3d^2$ as an example. The highest possible M_L value is $M_L = 4$ and this is achieved by $|3\,2\,\tfrac{1}{2}\,2, 3\,2\,-\tfrac{1}{2}\,2\rangle$. There are no others, because

$$|3\,2\,\tfrac{1}{2}\,2, 3\,2\,\tfrac{1}{2}\,2\rangle \equiv |3\,2\,-\tfrac{1}{2}\,2, 3\,2\,-\tfrac{1}{2}\,2\rangle \equiv 0$$

and

$$|3\,2\,-\tfrac{1}{2}\,2, 3\,2\,\tfrac{1}{2}\,2\rangle = -|3\,2\,\tfrac{1}{2}\,2, 3\,2\,-\tfrac{1}{2}\,2\rangle.$$

The M_S value is $M_S = 0$. Thus already we have shown that ^3G cannot occur in $3d^2$ but ^1G must.

One then considers $M_L = 3$. There are four independent functions: those obtained by giving the four different possible pairs of values to s, s' in $|3\,2\,s\,2, 3\,2\,s'\,1\rangle$. The four M_S values are $1, 0, 0, -1$ so we have one triplet and one singlet having an $M_L = 3$ component. The singlet is the ^1G found above and the triplet must be ^3F.

We can clearly continue this procedure to give all the allowed terms of $3d^2$. The numbers of determinantal functions with given M_S, M_L values are conveniently put in a table (Table 4.2a) from which the allowed terms are easily

Table 4.2a. *M_S, M_L table for $3d^2$*

M_L \ M_S	$+1$	0	-1	Implied terms
4	0	1	0	^1G
3	1	2	1	^3F
2	1	3	1	^1D
1	2	4	2	^3P
0	2	5	2	^1S
-1	2	4	2	—
-2	1	3	1	—
-3	1	2	1	—
-4	0	1	0	—

Table 4.2b. *Part of the M_S, M_L table for d^3*

M_L \ M_S	$\frac{3}{2}$	$\frac{1}{2}$	Implied terms
5	0	1	^2H
4	0	2	^2G
3	1	4	^4F, ^2F
2	1	6	^2D, ^2D
1	2	8	^4P, ^2P
0	2	8	—

read off. So the allowed terms of $3d^2$ are ^1G, ^3F, ^1D, ^3P and ^1S. One notices that one need only write down that part of the M_S, M_L table for which $M_S \geqslant 0$, $M_L \geqslant 0$.

As a second example we give, in Table 4.2*b*, that part of the M_S, M_L table for which $M_S \geqslant 0$ and $M_L \geqslant 0$, together with the allowed terms deduced from the table, for d^3 (i.e. for nd^3, for any $n \geqslant 3$). The reader will have no difficulty in reconstructing this table for himself. We see that there are two ^2D terms in a d^3 configuration.

The allowed terms for all configurations of equivalent p or d electrons can be easily obtained by the same method and are shown in table 4.3. The allowed terms for any configuration involving only s, p and d electrons can now be obtained by the vector-coupling procedure. When we have a configuration which has the maximum possible number of electrons with given nl (i.e. $4l+2$) we say that the

Table 4.3. *Allowed terms for p and d configurations*

	Singlets	Doublets	Triplets	Quartets	Quintets	Sextets
p, p^5	—	P	—	—	—	—
p^2, p^4	SD	—	P	—	—	—
p^3	—	PD	—	S	—	—
d, d^9	—	D	—	—	—	—
d^2, d^8	SDG	—	PF	—	—	—
d^3, d^7	—	$PDDFGH$	—	PF	—	—
d^4, d^6	$SSDDFGGI$	—	$PPDFFGH$	—	D	—
d^5	—	$SPDDD$ $FFGGHI$	—	$PDFG$	—	S

configuration contains a closed shell of nl electrons (nl is the type, e.g. $1s$ or $3d$, not the number). The M_S, M_L table for a configuration consisting only of closed shells evidently has just the one entry under $M_S = M_L = 0$ and so contains only the one term ^1S. As the reader will readily verify, the addition of any number of closed shells to a configuration makes no difference to the M_S, M_L table. Thus the allowed terms of a configuration consisting of closed shells and partly filled shells are the same as those of the corresponding configuration obtained by omitting the closed shells. For example, the ground configuration of the chromic ion is $1s^2\,2s^2\,2p^6\,3s^2\,3p^6\,3d^3$ and therefore its allowed terms are the same as those for $3d^3$, that is ^2H, ^2G, ^4F, ^2F, ^2D, ^2D, ^4P, ^2P.

Having discussed the question of allowed terms we now pass on to the problem of determining the first-order electrostatic energies of those terms. To do this we must diagonalize the matrix of the Hamiltonian $\mathscr{H}$ of (2.1) within the configuration. Since $\mathscr{H}$ commutes with $\mathbf{L}^2$ and $\mathbf{S}^2$, there will only be non-diagonal elements between terms of the same type. Thus when d^n is expressed in the form of SL terms, $\mathscr{H}$ is already diagonal for d, d^2, d^8, d^9 and d^{10} but is not completely so for the other d^n configurations. For some of their terms we will have secular equations to solve, for example to find the approximate ^2D eigenfunctions one must solve a quadratic for d^3 and a cubic for d^5.

In section 4.6, p. 87, we will see how to determine the form of the eigenfunctions belonging to the various terms. At the moment, however, we merely have the determinantal functions Ψ' of (2.64) and these are not necessarily eigenfunctions of $\mathbf{S}^2$ or $\mathbf{L}^2$ (though they are eigenfunctions of S_z and L_z). They form a particularly simple basic set of functions for working out the matrix elements of $\mathscr{H}$, and in working out term-energies it is usually more convenient to use them than to use eigenfunctions of $\mathbf{S}^2$ and $\mathbf{L}^2$. For a given configuration we write them ψ_i, say, and our problem then falls into two parts. The first is to determine the energies of the terms as functions of the matrix elements

$$\mathscr{H}_{ij} = \langle \overline{\psi}_i \,|\, \mathscr{H} \,|\, \psi_j \rangle,$$

where ψ_j in detail is

$$\psi_j = \sum_{\mu} (n!)^{-\frac{1}{2}} (-1)^{\mu} P_{\mu}\{\phi_1^j(1)\, \phi_2^j(2) \ldots \phi_n^j(n)\} \tag{4.17}$$

with P_μ operating on the electrons. The second part is then to work out the $\mathscr{H}_{ij}$ in terms of more fundamental quantities. We discuss the first part here and then pass on to the second part in the next two sections.

Our immediate problem is to find the roots of the equation

$$|\mathscr{H}_{ij} - E\delta_{ij}| = 0,$$

as simply as possible in terms of the $\mathscr{H}_{ij}$. However, if we tried solving this equation as it stands we should have high-order secular equations to solve in order to obtain the energies of the terms. Fortunately there is a way of avoiding many of these equations. $\mathscr{H}$ commutes with S_z and L_z, so the secular equation for the whole configuration breaks up into a number of small equations, one for each possible pair of values M_S, M_L. On multiplying out one of these equations we find that its leading terms are

$$E^m - E^{m-1} \sum_{i=1}^{m} \mathscr{H}_{ii} + \ldots,$$

and hence the sum of the roots is just $\sum_{i=1}^{m} \mathscr{H}_{ii}$. This is called the diagonal-sum rule.

Armed with the diagonal-sum rule we write down the sum of the $\mathscr{H}_{ii}$ for the functions ψ_i occurring in a given square of the M_S, M_L table for the configuration (for example, Tables 4.2*a* or *b*). This sum then gives the sum of the energies of all the terms which have a state with the M_S, M_L values corresponding to the square. By subtracting the sums for suitable adjacent squares we can obtain the energy of any term which occurs only once in the configuration.

As an example we consider d^3. Write $\Sigma(M_S, M_L)$ for the sum $\Sigma\mathscr{H}_{ii}$ for all ψ_i occurring in the square M_S, M_L and $E(^{2S+1}L)$ for the energy of the term ^{2S+1}L. Then

$$\begin{aligned}
E(^2H) &= \Sigma(\tfrac{1}{2}, 5),\\
E(^2G) &= \Sigma(\tfrac{1}{2}, 4) - \Sigma(\tfrac{1}{2}, 5),\\
E(^4F) &= \Sigma(\tfrac{3}{2}, 3),\\
E(^2F) &= \Sigma(\tfrac{1}{2}, 3) - \Sigma(\tfrac{3}{2}, 3) - \Sigma(\tfrac{1}{2}, 4),\\
E(^4P) &= \Sigma(\tfrac{3}{2}, 1) - \Sigma(\tfrac{3}{2}, 2),\\
E(^2P) &= \Sigma(\tfrac{1}{2}, 1) - \Sigma(\tfrac{3}{2}, 1) - \Sigma(\tfrac{1}{2}, 2) + \Sigma(\tfrac{3}{2}, 2).
\end{aligned}$$

The same procedure applied to ^2D gives us only the sum of the energies of the two ^2D terms as $\Sigma(\frac{1}{2}, 2) - \Sigma(\frac{1}{2}, 3)$. We see later (§ 4.6) how to find the separate energies of terms which occur more than once in a configuration.

We have found that whenever a term appears only once in a configuration, the diagonal-sum rule gives its first-order energy as a definite linear combination of diagonal elements $\mathscr{H}_{ii}$ of $\mathscr{H}$. If a term appears more than once, however, we must evaluate some of the off-diagonal $\mathscr{H}_{ij}$ and solve a secular equation. In § 4.6 we shall separate the two ^3F terms of d^4 in this way (the corresponding separation for the two ^2D of d^3 is given in Condon & Shortley, p. 233, by the same method) and give the energies of all the terms of the d^n configurations. First, however, we calculate the $\mathscr{H}_{ii}$ in terms of more fundamental quantities.

Examples

1. One electron is placed in each of four distinct functions of an orthonormal set of spatial functions. Show that the resulting system consists of one quintet, three triplets and two singlets.
2. Show that f^3 contains one ^2P but no ^2S.
3. The allowed terms of l^2 are $^1(2a)$, $a = 0, 1, \ldots, l$ and $^3(2a+1)$, $a = 0, 1, \ldots, l-1$.
4. The allowed terms of l^{2l+1} having $S = l - \frac{1}{2}$ have all values of L from $L = 1$ to $L = 2l$, each just once.
5. All states of the same configuration have the same parity.
6. The allowed terms of l^ϵ are the same as those of $l^{4l+2-\epsilon}$.

4.3. Evaluation of the matrix elements $\mathscr{H}_{ij}$

We now turn to the second part of the evaluation of the energies and consider the individual $\mathscr{H}_{ij}$. We may write $\mathscr{H}$ in the form

$$\mathscr{H} = \sum_{\kappa=1}^{n} U(\kappa) + \sum_{\kappa>\lambda}^{n} V(\kappa\lambda) \tag{4.18}$$

to illustrate that it is a sum of one- and two-electron operators. Then

$$U(\kappa) = \frac{1}{2m}\mathbf{p}_\kappa^2 - \frac{Ze^2}{r_\kappa} \quad \text{and} \quad V(\kappa\lambda) = \frac{e^2}{r_{\kappa\lambda}},$$

from (2.1). Then corresponding to (4.18) we have

$$\mathscr{H}_{ij} = U_{ij} + V_{ij},$$

where the matrix elements are those of $\mathscr{H}$ between the functions ψ_j defined in (4.17). On substituting (4.17) for ψ_i and ψ_j each matrix element becomes a sum over $(n!)^2$ terms as well as the sum already in equation (4.18). Fortunately we can simplify these sums considerably.

We start with U_{ij}. Using (4.17) we find

$$U_{ij} = (n!)^{-1} \sum_\mu \sum_\nu (-1)^\mu (-1)^\nu P_\mu^i P_\nu^j \times \langle \bar{\phi}_1^i(1)\, \bar{\phi}_2^i(2) \ldots \bar{\phi}_n^i(n) \,|\, \sum_{\kappa=1}^{n} U(\kappa) \,|\, \phi_1^j(1)\, \phi_2^j(2) \ldots \phi_n^j(n) \rangle. \tag{4.19}$$

In (4.19) P^i_μ operates only on the electrons of ψ_i and P^j_ν only on those of ψ_j. Then each term in (4.19) is just a number and so we may permute *all* the electrons in any particular term without altering that number. Further we may use different permutations for different terms. This enables us to effect a considerable simplification. With the notation of (4.19) we use P_ν^{-1} to permute the $\mu\nu$th term. This means we multiply each term by $(P^i_\nu P^j_\nu)^{-1}$ and then the electrons of ψ_j are in their natural order for every term. Thus $P^i_\mu P^j_\nu$ at the beginning of (4.19) is replaced by $(P^i_\nu)^{-1} P^i_\mu$. But, for fixed ν, as P^i_μ runs through all the $n!$ permutations so also does $(P^i_\nu)^{-1} P^i_\mu$ (see § 2.6). Applying these results to (4.19), we find

$$\begin{aligned} U_{ij} &= (n!)^{-1} \sum_{\mu\nu} (-1)^{\mu+\nu} (P^i_\nu)^{-1} P^i_\mu \langle \bar{\phi}^i_1(1) \ldots \bar{\phi}^i_n(n) \mid \sum_{\kappa=1}^n U(\kappa) \mid \phi^j_1(1) \ldots \phi^j_n(n) \rangle \\ &= (n!)^{-1} \sum_{\sigma\nu} (-1)^\sigma P^i_\sigma \langle \bar{\phi}^i_1(1) \ldots \bar{\phi}^i_n(n) \mid \sum_{\kappa=1}^n U(\kappa) \mid \phi^j_1(1) \ldots \phi^j_n(n) \rangle \\ &= \sum_{\sigma} (-1)^\sigma P^i_\sigma \langle \bar{\phi}^i_1(1) \ldots \bar{\phi}^i_n(n) \mid \sum_{\kappa=1}^n U(\kappa) \mid \phi^j_1(1) \ldots \phi^j_n(n) \rangle \\ &= \sum_{\kappa=1}^n \sum_{\sigma} (-1)^\sigma P^i_\sigma \langle \bar{\phi}^i_1(1) \ldots \bar{\phi}^i_n(n) \mid U(\kappa) \mid \phi^j_1(1) \ldots \phi^j_n(n) \rangle. \end{aligned} \tag{4.20}$$

Each matrix element in (4.20) breaks up into a product

$$\begin{aligned} &P^i_\sigma \langle \bar{\phi}^i_1(1) \ldots \bar{\phi}^i_n(n) \mid U(\kappa) \mid \phi^j_1(1) \ldots \phi^j_n(n) \rangle \\ &\quad = \langle \bar{\phi}^{i'}_1(1) \mid \phi^j_1(1) \rangle \ldots \langle \bar{\phi}^{i'}_\kappa(\kappa) \mid U(\kappa) \mid \phi^j_\kappa(\kappa) \rangle \ldots \langle \bar{\phi}^{i'}_n(n) \mid \phi^j_n(n) \rangle, \text{ say,} \end{aligned}$$

and since the ϕ_a form an orthonormal set, unless $\phi^{i'}_a \equiv \phi^j_a$ for all a except possibly for $a = \kappa$, the matrix element is zero. This means that U_{ij} itself is necessarily zero unless ψ_i and ψ_j differ in at most one pair of functions $\phi^i_{\kappa'}$, ϕ^j_κ.

Thus we have two cases. If $\psi_i \equiv \psi_j$ then, because of the orthonormality, only those terms of (4.20) with the ϕ^i in the same order as the ϕ^j can be non-zero. So each sum over σ reduces to one term only and

$$U_{ii} = \sum_{\kappa=1}^n \langle \bar{\phi}_\kappa(\kappa) \mid U(\kappa) \mid \phi_\kappa(\kappa) \rangle. \tag{4.21}$$

The κth term of (4.21) is the contribution to U_{ii} from the κth orbital, not from the κth electron. So we can rewrite it

$$U_{ii} = \sum_{\kappa=1}^n \langle \bar{\phi}_\kappa(1) \mid U(1) \mid \phi_\kappa(1) \rangle$$

if we wish, or more concisely

$$U_{ii} = \sum_{\kappa=1}^n \langle \bar{\phi}_\kappa \mid U \mid \phi_\kappa \rangle. \tag{4.22}$$

Another way of seeing this is to observe that κ appears in two ways, one labelling the orbitals and the other as a parameter labelling the variable of integration. The matrix elements depend on the particular orbitals, or functions, which appear in the integral but not of course on the name of the parameter of integration. Thus is the indistinguishability of electrons maintained. It is often convenient, however, to talk loosely of electrons as if they were in particular orbitals. For example, one might say that a $4s$ electron had such and such an energy. Such a statement should always be interpreted in terms of an equation such as (4.22) to mean that a $4s$ orbital is occupied and gives a certain contribution via (4.22) to the total energy.

If ψ_i differs from ψ_j in just one function ϕ^i_λ, $\phi^j_{\lambda'}$ then we have again only one term in the sum over σ but this time have also only one term in the sum over κ, namely that for which $\kappa = \lambda'$. So U_{ij} has at most one non-vanishing term, namely

$$U_{ij} = (-1)^\tau \langle \bar{\phi}^i_\lambda \mid U \mid \phi^j_{\lambda'} \rangle, \tag{4.23}$$

where P_τ is the permutation needed to put the functions of ψ_i in the same order as those of ψ_j and again we omit the parameter of integration. As we have already remarked, U_{ij} is zero if ψ_i differs from ψ_j in more than one constituent one-electron function.

Now $U = (1/2m)\,\mathbf{p}^2 - (Ze^2/r)$ which commutes with both the spin and the orbital angular momentum and therefore we can simplify (4.22) and (4.23) still further. We substitute (4.16) for ϕ_κ and have

$$U_{ii} = \sum_{\kappa=1}^{n} \langle \bar{R}_{n^\kappa l^\kappa} \mid U \mid R_{n^\kappa l^\kappa} \rangle, \tag{4.24}$$

$$U_{ij} = (-1)^\tau \, \delta(l^i, l^j)\, \delta(p^i, p^j)\, \delta(s^i, s^j) \langle \bar{R}_{n^i l^i} \mid U \mid R_{n^j l^j} \rangle, \tag{4.25}$$

where in (4.25) the symbols l^i, l^j, etc., refer to the pair of functions ϕ^i_λ, $\phi^j_{\lambda'}$ which differ in ψ_i and ψ_j.

The sum in (4.24) depends only on the number of electrons having each nl value. But, by definition, this is the same for all states ψ_i of a configuration. Hence U_{ii} is the same for all states of a given configuration. If ψ_i and ψ_j belong to the same configuration then $n^i l^i$ must be the same as $n^j l^j$ in (4.25) because all the pairs of functions which were integrated out to obtain (4.25) had pairwise the same nl values as each other. But since ϕ^i_λ and $\phi^j_{\lambda'}$ do differ, they must differ then in either $p^i \neq p^j$ or $s^i \neq s^j$.

In either case U_{ij} is zero because of the delta functions in (4.25). We have shown, then, that U_{ij} is diagonal within any configuration and has its diagonal elements the same for all states of the same configuration. In other words U_{ij} just adds the same constant quantity to the energies of every term in a configuration. It is only the two-electron operator part $\Sigma V(\kappa\lambda)$ of the electrostatic energy which separates the terms.

We now pass on to the evaluation of the matrix elements V_{ij} of the two-electron operators. This is precisely analogous to the evaluation of U_{ij} but is slightly more complicated just because each $V(\kappa\lambda)$ contains the coordinates of two different electrons. Corresponding to (4.20) we now have

$$V_{ij} = \sum_{\kappa>\lambda}^{n} \sum_{\sigma} (-1)^\sigma P^i_\sigma \langle \bar{\phi}^i_1(1) \dots \bar{\phi}^i_n(n) \mid V(\kappa\lambda) \mid \phi^j_1(1) \dots \phi^j_n(n) \rangle. \tag{4.26}$$

From (4.26) we can only deduce immediately that V_{ij} is identically zero if ψ_i differs from ψ_j in at least three constituent one-electron functions. Further, if a matrix element of $V(\kappa\lambda)$ is non-zero, then in general so is the matrix element in which ϕ^i_κ and ϕ^i_λ have changed places. If $\psi_i \equiv \psi_j$ this gives us

$$V_{ii} = \sum_{\kappa>\lambda}^{n} \{\langle\overline{\phi}_\kappa(1)\,\overline{\phi}_\lambda(2)\,|\,V(12)\,|\,\phi_\kappa(1)\,\phi_\lambda(2)\rangle \\ -\langle\overline{\phi}_\lambda(1)\,\overline{\phi}_\kappa(2)\,|\,V(12)\,|\,\phi_\kappa(1)\,\phi_\lambda(2)\rangle\}.$$

We now adopt a definite convention throughout this book about matrix elements involving four one-electron functions. We write

$$\langle ab\,|\,X\,|\,cd\rangle \equiv \langle a(1)\,b(2)\,|\,X(12)\,|\,c(1)\,d(2)\rangle \tag{4.27}$$

and can then write V_{ii} more succinctly as

$$V_{ii} = \sum_{\kappa>\lambda}^{n} (\langle\overline{\phi}_\kappa\overline{\phi}_\lambda\,|\,V\,|\,\phi_\kappa\phi_\lambda\rangle - \langle\overline{\phi}_\lambda\overline{\phi}_\kappa\,|\,V\,|\,\phi_\kappa\phi_\lambda\rangle). \tag{4.28}$$

It follows from our discussion of the diagonal-sum rule method of obtaining term-energies that these diagonal V_{ii} have a special importance in the theory. It is customary to introduce the notation

$$\left.\begin{aligned} J(\kappa,\lambda) &\equiv \langle\overline{\phi}_\kappa\overline{\phi}_\lambda\,|\,V\,|\,\phi_\kappa\phi_\lambda\rangle, \\ K(\kappa,\lambda) &\equiv \langle\overline{\phi}_\lambda\overline{\phi}_\kappa\,|\,V\,|\,\phi_\kappa\phi_\lambda\rangle. \end{aligned}\right\} \tag{4.29}$$

Simplifying (4.16) to $\phi_i = f_i \tau_{s^i}$, where f_i is a function of spatial coordinates alone, we have

$$J(\kappa,\lambda) = \int \frac{e^2\,|f_\kappa(1)|^2\,|f_\lambda(2)|^2}{r_{12}}\,d\tau_1 d\tau_2 \tag{4.30}$$

which is the classical expression for the Coulomb interaction between two charge clouds of density $e\,|f_\kappa|^2$ and $e\,|f_\lambda|^2$. For this reason $J(\kappa,\lambda)$ is called a Coulomb integral. $K(\kappa,\lambda)$ has no classical analogue, but because of the exchange of ϕ_κ with ϕ_λ is called an exchange integral. Then V_{ii} may be expressed even more briefly as

$$V_{ii} = \sum_{\kappa>\lambda}^{n} (J(\kappa,\lambda) - K(\kappa,\lambda)). \tag{4.31}$$

Next we consider ψ_i differing from ψ_j in just one function ϕ^i_κ, ϕ^j_κ, say. If P_τ is the permutation which puts the functions of ψ_i in the same order as those of ψ_j we then have

$$V_{ij} = (-1)^\tau \sum_{\lambda(\neq\kappa)} \{\langle\overline{\phi}^i_\kappa\overline{\phi}_\lambda\,|\,V\,|\,\phi^j_\kappa\phi_\lambda\rangle - \langle\overline{\phi}_\lambda\overline{\phi}^i_\kappa\,|\,V\,|\,\phi^j_\kappa\phi_\lambda\rangle\}, \tag{4.32}$$

where we omit the superscripts i and j on ϕ_λ because $\phi^i_\lambda \equiv \phi^j_\lambda$ for $\lambda \neq \kappa$ (after applying P_τ). We may now argue as we did with U_{ij} and say that when ψ_i and ψ_j belong to the same configuration then ϕ^i_κ and ϕ^j_κ must have the same nl values. So they differ either in p^i or s^i (see (4.16)). In the matrix element

$$\langle\overline{\phi}^i_\kappa\overline{\phi}_\lambda\,|\,V\,|\,\phi^j_\kappa\phi_\lambda\rangle$$

V commutes with $L_z = l_{1z} + l_{2z}$ and $S_z = s_{1z} + s_{2z}$. It follows immediately from our discussion of ϕ^i_κ and ϕ^j_κ that $\phi^j_\kappa\phi_\lambda$ and $\phi^i_\kappa\phi_\lambda$, considered as two-electron functions, are each simultaneous eigenfunctions of L_z and S_z but their eigenvalues differ for at least one of those operators. Therefore the matrix element is zero and the same is obviously true for $\langle\bar{\phi}_\lambda\bar{\phi}^i_\kappa \mid V \mid \phi^j_\kappa\phi_\lambda\rangle$ for precisely the same reasons. So V_{ij} is zero for any pair ψ_i, ψ_j from the same configuration and differing by just one function.

Finally, when ψ_i, ψ_j differ in two functions ϕ^i_κ, ϕ^i_λ as against ϕ^j_κ, ϕ^j_λ say, we have

$$V_{ij} = (-1)^\tau (\langle\bar{\phi}^i_\kappa\bar{\phi}^i_\lambda \mid V \mid \phi^j_\kappa\phi^j_\lambda\rangle - \langle\bar{\phi}^i_\lambda\bar{\phi}^i_\kappa \mid V \mid \phi^j_\kappa\phi^j_\lambda\rangle), \tag{4.33}$$

where the permutation P_τ is defined as before. V_{ij} in (4.33) is not necessarily zero when ψ_i and ψ_j belong to the same configuration.

Just as we simplified the U_{ii} further by introducing the explicit expressions (4.16) for the one-electron functions so we can simplify the V_{ii} and V_{ij} further. In this latter case, though, the results are not so simple and we defer the investigation to the next section. However, we can already at this stage show that the term energies of a configuration which contains completely-filled shells of electrons are the same, apart from a constant energy common to all terms of the configuration, as those of the corresponding configuration in which the filled shells have been omitted. This is obvious for off-diagonal matrix elements because the only non-vanishing ones are those V_{ij} between functions ψ_i and ψ_j differing in just two pairs of the one-electron functions. Then these pairs are the same in the two correlated configurations and since, from (4.33), V_{ij} depends only on the functions which actually differ, V_{ij} must be identically the same in the two configurations. The reader may be worried here about the possibility of a change of sign; however, provided the one-electron functions of any filled shell are kept together in (4.17) the sign cannot change.

We now discuss the diagonal elements and break them up into three parts

$$\begin{aligned} \mathscr{H}_{ii} &= \mathscr{H}_{ii}\,(\text{filled shells}) + \mathscr{H}_{ii}\,(\text{interaction}) + \mathscr{H}_{ii}\,(\text{partly-filled shells}) \\ &= \mathscr{H}^{(f)}_{ii} + \mathscr{H}^{(i)}_{ii} + \mathscr{H}^{(p)}_{ii}, \quad \text{say.} \end{aligned} \tag{4.34}$$

This is possible because of the expressions (4.22) and (4.31). $\mathscr{H}^{(f)}_{ii}$ and $\mathscr{H}^{(p)}_{ii}$ are, respectively, the energies of atoms containing the filled shells or the partly-filled shells alone. Then the difference in energy between corresponding diagonal elements of the two configurations is $\mathscr{H}_{ii} - \mathscr{H}^{(p)}_{ii}$ which, from (4.34), is $\mathscr{H}^{(f)}_{ii} + \mathscr{H}^{(i)}_{ii}$. We have to prove that this latter sum is the same for all i. $\mathscr{H}^{(f)}_{ii}$ is trivially always the same so we need consider only $\mathscr{H}^{(i)}_{ii}$ and split it up as follows

$$\mathscr{H}^{(i)}_{ii} = \sum_f \sum_p (J(\phi_f, \phi_p) - K(\phi_f, \phi_p)), \tag{4.35}$$

where f and p refer, as before, to the filled shells and to the partly-filled shells, respectively. U makes no contribution to $\mathscr{H}^{(i)}_{ii}$. We now show that

$$E_p = \sum_f (J(\phi_f, \phi_p) - K(\phi_f, \phi_p)), \tag{4.36}$$

depends on the nl values but not on the $m_s m_l$ values of ϕ_p. When we have done this, it will follow immediately that (4.35) gives the same value for all ψ_i belonging to the same configuration. E_p simply gives the energy of a ^2L state consisting of a number of filled shells and one electron outside with $l = L$, less the energy of the filled shells. This is the same for all components of that ^2L term and hence E_p is independent of m_s and m_l. We have established our result using the fact that $\mathscr{H}$ commutes with **L** and **S**.

The virtual elimination of any effect of closed shells that we have just made is very fundamental to the theory. It reduces very considerably the amount of work necessary in giving an approximate description of the states and energies of many-electron atoms. For we may completely disregard the filled electron shells of atoms and discuss them in terms of their outer electrons only. We often call the filled shells 'inner shells' and the partly-filled shells 'valence shells', because for the low-lying states in which we are interested, an inner shell is usually nearer to the nucleus and a partly-filled shell usually consists of orbitals which can take part in bonding in a chemical sense. The electrons are called, correspondingly, inner electrons and valence electrons, although we always remember that they are really inner orbitals and valence orbitals in accordance with our discussion of (4.22). Lastly, one should recognize clearly that it is because we are working out the energies only to a first order in perturbation theory that we have found it possible to 'eliminate' the closed shells. In a more complete theory it is no longer possible. However, as we see in more detail later, the experimental data suggest strongly that a classification of atomic spectra based on this 'irrelevance of closed shells' is a sensible and useful one and so afford a certain *a posteriori* justification for the first-order perturbation theory.

We comment here also that we now have a justification for our earlier description of an alkali metal atom as one valence electron moving in a central field: all the other electrons were in filled shells. For example, the potassium atom is $1s^2\,2s^2\,2p^6\,3s^2\,3p^6\,nl^1$ with $n \geqslant 4$ or $n = 3$, $l = 2$. Core excitations of potassium can be observed but they have very high energies—much higher than the energy required to ionize the valence electron completely.

Example

By considering one term from a suitable configuration, show that the sum (4.36) is still independent of the $m_s m_l$ values of ϕ_p when the sum over f is restricted to a sum over those electrons of an inner shell which have α-spin.

4.4. Slater–Condon parameters

In order to proceed, we must now evaluate the two-electron matrix elements occurring in (4.28–31) and (4.33). They are all of the form

$$\langle \bar{\phi}_i \bar{\phi}_j \,|\, V \,|\, \phi_m \phi_t \rangle. \tag{4.37}$$

It is natural, therefore, to try to expand V as a series of products of one-electron operators referring to each argument separately. This is in fact possible and we start by writing

$$V = \frac{e^2}{r_{12}} = \frac{e^2}{(r_1^2 + r_2^2 - 2r_1 r_2 \cos\omega)^{\frac{1}{2}}},$$

where ω is the angle between the vectors $\mathbf{r}_1$ and $\mathbf{r}_2$. Then for $r_1 \neq r_2$ we put $r_<$, $r_>$ for the smaller and larger of r_1, r_2, respectively, and can expand V as a power series in $r_</r_>$ giving

$$V = \frac{e^2}{r_>}\left(1+\frac{r_<^2}{r_>^2}-2\frac{r_<}{r_>}\cos\omega\right)^{-\frac{1}{2}}$$

$$= \frac{e^2}{r_>}\sum_{k=0}^{\infty}\frac{r_<^k}{r_>^k}P_k(\cos\omega), \tag{4.38}$$

where the $P_k(\cos\omega)$ are Legendre polynomials. They are sometimes defined to be the coefficients in the expansion (4.38). They are the associated Legendre polynomials of (4.9) for $w = \cos\omega$ and $p = 0$ (see Whittaker & Watson, 1946, ch. 15).

We now wish to express $P_k(\cos\omega)$ in terms of products of one-electron functions. This is done by means of the spherical harmonic addition theorem. We prove this theorem by using ex. 2, p. 24. This immediate consequence of Wigner's formula is

$$|ll00\rangle = \frac{1}{\sqrt{(2l+1)}}\sum_{m=-l}^{+l}(-1)^{l-m}\,|llm-m\rangle.$$

We interpret this formula for our present purpose by taking $\mathbf{j}_1 = \mathbf{l}_1$ to refer to the first argument and $\mathbf{j}_2 = \mathbf{l}_2$ to refer to the second. Then the right-hand side becomes, in the Schrödinger representation

$$\frac{1}{\sqrt{(2l+1)}}\sum_{m=-l}^{+l}(-1)^{l-m}Y_{lm}(1)\,Y_{l-m}(2) = \frac{(-1)^l}{\sqrt{(2l+1)}}\sum_{m=-l}^{+l}Y_{lm}(1)\,\overline{Y}_{lm}(2).$$

The left-hand side is an eigenstate of $\mathbf{j}^2 = (\mathbf{l}_1+\mathbf{l}_2)^2$ and of $j_z = l_{1z}+l_{2z}$ and we now show its relationship to $P_l(\cos\omega)$:

$$(l_{1z}+l_{2z})\,P_l(\cos\omega) = -i\hbar\left(\frac{\partial}{\partial\phi_1}+\frac{\partial}{\partial\phi_2}\right)P_l(\cos\omega)$$

which is zero because $\cos\omega = \cos\theta_1\cos\theta_2+\sin\theta_1\sin\theta_2\cos(\phi_1-\phi_2)$ and so $P_l(\cos\omega)$ involves ϕ_1 and ϕ_2 only in the combination $(\phi_1-\phi_2)$. By symmetry, the other components of $\mathbf{j}$ give zero when operating on $P_l(\cos\omega)$. So $P_l(\cos\omega)$ is an eigenstate of $\mathbf{j}^2$ and j_z with the eigenvalues zero. However, if we regard it alternatively as a Legendre function of $\cos\omega$ and let θ_2, ϕ_2 be fixed it is then an eigenstate of $\mathbf{l}_1^2$ and l_{1z} having $l_1 = l$, $l_{1z} = 0$ but referred to the θ_2-, ϕ_2-axis rather than the z-axis. Therefore its dependence on θ_1, ϕ_1 is of the general form

$$P_l(\cos\omega) = \sum_m a_m Y_{lm}(1)$$

and similarly for θ_2, ϕ_2. The only independent function satisfying these conditions is $|ll00\rangle$. We must normalize $P_l(\cos\omega)$ and then have

$$\gamma\frac{1}{\sqrt{4\pi}}\left(\frac{2l+1}{4\pi}\right)^{\frac{1}{2}}P_l(\cos\omega) = \frac{(-1)^l}{\sqrt{(2l+1)}}\sum_{m=-l}^{+l}Y_{lm}(1)\,\overline{Y}_{lm}(2).$$

The two normalizing factors for $P_l(\cos\omega)$ arise from the need to integrate over the variables for both arguments. $|\gamma| = 1$. γ is obtained by putting $\theta_2 = 0$ in the equation to give us finally

$$P_l(\cos\omega) = \frac{4\pi}{2l+1}\sum_{m=-l}^{+l} Y_{lm}(1)\,\overline{Y}_{lm}(2). \tag{4.39}$$

This is called the spherical harmonic addition theorem.

We now use (4.16) for the ϕ and the results (4.38) and (4.39) to evaluate (4.37):

$$\begin{aligned}\langle\overline{\phi}_i\overline{\phi}_j\,|\,V\,|\,\phi_m\phi_t\rangle &= \langle\overline{\phi}_i\overline{\phi}_j\,|\sum_{k=0}^{\infty}\frac{e^2r_<^k}{r_>^{k+1}}P_k(\cos\omega)\,|\,\phi_m\phi_t\rangle\\ &= \sum_{k=0}^{\infty}\langle R_{n^il^i}R_{n^jl^j}\,|\,\frac{e^2r_<^k}{r_>^{k+1}}\,|\,R_{n^ml^m}R_{n^tl^t}\rangle\\ &\qquad\times\langle\overline{Y}_{l^ip^i}\overline{Y}_{l^jp^j}\,|\,P_k(\cos\omega)\,|\,Y_{l^mp^m}Y_{l^tp^t}\rangle\,\delta(s^i,s^m)\,\delta(s^j,s^t)\\ &= \sum_{k=0}^{\infty}R^k(n^il^i,n^jl^j,n^ml^m,n^tl^t)\,A_k\,\delta(s^i,s^m)\,\delta(s^j,s^t),\quad\text{say.}\end{aligned} \tag{4.40}$$

A_k involves functions of angle only and it can be simplified with the help of (4.39) to give

$$\begin{aligned}A_k &= \frac{4\pi}{2k+1}\sum_{p=-k}^{+k}\langle\overline{Y}_{l^ip^i}\overline{Y}_{l^jp^j}\,|\,Y_{kp}(1)\,\overline{Y}_{kp}(2)\,|\,Y_{l^mp^m}Y_{l^tp^t}\rangle\\ &= \frac{4\pi}{2k+1}\sum_{p=-k}^{+k}\langle\overline{Y}_{l^ip^i}Y_{kp}Y_{l^mp^m}\rangle\langle\overline{Y}_{l^jp^j}\overline{Y}_{kp}Y_{l^tp^t}\rangle.\end{aligned} \tag{4.41}$$

$|Y_{kp}Y_{l^mp^m}\rangle$ is an eigenket of l_{1z} with eigenvalue $(p+p^m)\hbar$ which means that $\langle\overline{Y}_{l^ip^i}Y_{kp}Y_{l^mp^m}\rangle$ is zero unless $p^i = p+p^m$. The other term is zero unless $p^j+p = p^t$. So A_k is zero unless $p^i+p^j = p^m+p^t$, a fact that also follows directly from (4.37) by observing that $V(12)$ commutes with $l_{1z}+l_{2z}$. When this condition is satisfied, the sum (4.41) reduces to a single term which we write using the notation

$$\begin{aligned}c^k(lp,l'p') &= \left(\frac{4\pi}{2k+1}\right)^{\frac{1}{2}}\langle\overline{Y}_{lp}Y_{k\,p-p'}Y_{l'p'}\rangle,\\ &= \left(\frac{2}{2k+1}\right)^{\frac{1}{2}}\langle\Theta_{lp}\Theta_{k\,p-p'}\Theta_{l'p'}\rangle,\end{aligned} \tag{4.42}$$

which yields $$A_k = c^k(l^ip^i,l^mp^m)\,c^k(l^tp^t,l^jp^j)\,\delta(p^i+p^j,p^m+p^t). \tag{4.43}$$

It follows immediately from the definition that

$$c^k(lp,l'p') = (-1)^{p-p'}c^k(l'p',lp).$$

The c^k are of course just numbers. It is surprisingly difficult to find general methods for integrals of this kind. Even the case of the integral of three Legendre polynomials is not easy and was guessed from a consideration of special cases by Adams and subsequently proved by him. A general formula for the c^k has been worked out by Gaunt (1929), but it is very complicated and hence in practice one works out the c^k once and for all and uses tables for them. Gaunt's formula can also be deduced from Wigner's and Adams's. The reason why it is not a trivial deduction from Wigner's formula alone is the difficulty of normalization referred to at the end of § 2.5. We give the values of the c^k for s, p and d electrons and for pure f electron configurations in Table 4.4. They are zero unless k, l^i and l^m satisfy the conditions

$$k+l^i+l^m \quad \text{even}$$

and

$$|l^i-l^m| \leqslant k \leqslant l^i+l^m. \tag{4.44}$$

From (4.40) we have

$$\langle\bar{\phi}_i\bar{\phi}_j \mid V \mid \phi_m\phi_t\rangle = \delta(s^i, s^m)\,\delta(s^j, s^t)\,\delta(p^i+p^j, p^m+p^t)$$
$$\times \sum_{k=0}^{\infty} R^k(n^il^i, n^jl^j, n^ml^m, n^tl^t)\, c^k(l^ip^i, l^mp^m)\, c^k(l^tp^t, l^jp^j), \tag{4.45}$$

where the sum is in fact finite because of (4.44).

Corresponding to the notation J and K introduced in (4.29) for matrix elements arising from diagonal V_{ii} one defines the Slater–Condon parameters

$$\left.\begin{aligned} F^k(n^il^i, n^jl^j) &\equiv R^k(n^il^i, n^jl^j, n^il^i, n^jl^j),\\ G^k(n^il^i, n^jl^j) &\equiv R^k(n^jl^j, n^il^i, n^il^i, n^jl^j).\end{aligned}\right\} \tag{4.46}$$

When dealing with a configuration of equivalent electrons $n^i = n^j$ and $l^i = l^j$ so $F^k(n^il^i, n^il^i) \equiv G^k(n^il^i, n^il^i)$ and we need use only the F^k. Also, as at the end of the last sentence, we may then often drop the arguments. One also defines

$$\left.\begin{aligned} a^k(l^ip^i, l^jp^j) &= c^k(l^ip^i, l^ip^i)\, c^k(l^jp^j, l^jp^j),\\ b^k(l^ip^i, l^jp^j) &= \{c^k(l^ip^i, l^jp^j)\}^2.\end{aligned}\right\} \tag{4.47}$$

From (4.45), (4.46) and (4.47) follows

$$\left.\begin{aligned} J(\phi_i, \phi_j) &= \sum_{k=0}^{\infty} a^k(l^ip^i, l^jp^j)\, F^k(n^il^i, n^jl^j),\\ K(\phi_i, \phi_j) &= \delta(s^i, s^j) \sum_{k=0}^{\infty} b^k(l^ip^i, l^jp^j)\, G^k(n^il^i, n^jl^j).\end{aligned}\right\} \tag{4.48}$$

For a configuration of equivalent nl electrons one may rewrite (4.48) as

$$J(\phi_i, \phi_j) = \sum_{k=0}^{2l} a^kF^k, \quad K(\phi_i, \phi_j) = \delta(s^i, s^j) \sum_{k=0}^{2l} b^kF^k \tag{4.49}$$

with the sums over even values of k only. Thus the energies V_{ii} (and actually V_{ij} also) are expressed in terms of $l+1$ undetermined parameters F^k. a^k and b^k are fractions and it is usual to define new parameters $F_k = (1/d_k^2)\,F^k$, where d_k is the common denominator occurring in Table 4.4 for a given set of c^k. Thus for p electrons one writes

$$\left.\begin{aligned} F_0 &= F^0,\\ F_2 &= \tfrac{1}{25}F^2.\end{aligned}\right\} \tag{4.50}$$

We can now express the energies of the terms of any configuration as functions (usually linear) of the parameters F^k (or F_k) and G^k. For most configurations there are usually considerably more terms than parameters. Hence one can either calculate the R^k with some assumed central field or can obtain them empirically by fitting observed spectral data. We shall indicate in a later section

Table 4.4 *Values of* $c^k(l^i p^i, l^m p^m)$. *See Condon & Shortley* (1953) *and Shortley & Fried* (1938) *for further values*

$l^i l^m$	p^i	p^m	$c^1\sqrt{3}$	c^3		
$s\,p$	0	± 1	-1	0		
	0	0	$+1$	0		
			$c^1\sqrt{15}$	$c^3\sqrt{245}$		
$p\,d$	± 1	± 2	$-\sqrt{6}$	$+\sqrt{3}$		
	± 1	± 1	$+\sqrt{3}$	-3		
	± 1	0	-1	$+\sqrt{18}$		
	0	± 2	0	$+\sqrt{15}$		
	0	± 1	$-\sqrt{3}$	$-\sqrt{24}$		
	0	0	$+2$	$+\sqrt{27}$		
	± 1	∓ 2	0	$+\sqrt{45}$		
	± 1	∓ 1	0	$-\sqrt{30}$		
			c^0	$5c^2$		
$s\,s$	0	0	1	0		
$p\,p$	± 1	± 1	1	-1		
	± 1	0	0	$+\sqrt{3}$		
	0	0	1	$+2$		
	± 1	∓ 1	0	$-\sqrt{6}$		
			c^0	$c^2\sqrt{5}$		
$s\,d$	0	± 2	0	$+1$		
	0	± 1	0	-1		
	0	0	0	$+1$		
			c^0	$7c^2$	$21c^4$	
$d\,d$	± 2	± 2	1	-2	$+1$	
	± 2	± 1	0	$+\sqrt{6}$	$-\sqrt{5}$	
	± 2	0	0	-2	$+\sqrt{15}$	
	± 1	± 1	1	$+1$	-4	
	± 1	0	0	$+1$	$+\sqrt{30}$	
	0	0	1	$+2$	$+6$	
	± 2	∓ 2	0	0	$+\sqrt{70}$	
	± 2	∓ 1	0	0	$-\sqrt{35}$	
	± 1	∓ 1	0	$-\sqrt{6}$	$-\sqrt{40}$	
			c^0	$15c^2$	$33c^4$	$\frac{429}{5}c^6$
$f\,f$	± 3	± 3	1	-5	$+3$	-1
	± 3	± 2	0	$+5$	$-\sqrt{30}$	$+\sqrt{7}$
	± 3	± 1	0	$-\sqrt{10}$	$+\sqrt{54}$	$-\sqrt{28}$
	± 3	0	0	0	$-\sqrt{63}$	$+\sqrt{84}$
	± 2	± 2	1	0	-7	$+6$
	± 2	± 1	0	$+\sqrt{15}$	$+4\sqrt{2}$	$-\sqrt{105}$
	± 2	0	0	$-\sqrt{20}$	$-\sqrt{3}$	$+4\sqrt{14}$
	± 1	± 1	1	$+3$	$+1$	-15
	± 1	0	0	$+\sqrt{2}$	$+\sqrt{15}$	$+5\sqrt{14}$
	0	0	1	$+4$	$+6$	$+20$
	± 3	∓ 3	0	0	0	$-\sqrt{924}$
	± 3	∓ 2	0	0	0	$+\sqrt{462}$
	± 3	∓ 1	0	0	$+\sqrt{42}$	$-\sqrt{210}$
	± 2	∓ 2	0	0	$+\sqrt{70}$	$+\sqrt{504}$
	± 2	∓ 1	0	0	$-\sqrt{14}$	$-\sqrt{378}$
	± 1	∓ 1	0	$-\sqrt{24}$	$-\sqrt{40}$	$-\sqrt{420}$

how one can calculate the R^k, but as we are only interested in using the R^k in this book, we shall take the empirical values, these by definition agreeing with experiment as well or better, at least in the field of atomic spectra.

There are certain properties of the F_k and G_k which follow from their definitions without making any specific assumption about the actual form of the radial functions. Equation (4.30) shows that a Coulomb integral is always positive. From (4.46), we have

$$F^k(nl, n'l') = e^2 \int \frac{r_<^k}{r_>^{k+1}} (R_{nl}(1))^2 (R_{n'l'}(2))^2 r_1^2 r_2^2 dr_1 dr_2$$

and so each individual F^k is also positive. But since $r_<^k/r_>^{k+1}$ is a decreasing function of k we may also deduce that $F^k > F^{k'}$ whenever $k > k'$. In terms of F_k this reads

$$d_k^2 F_k > d_{k'}^2 F_{k'} \quad \text{for} \quad k > k'. \tag{4.51}$$

It is not so obvious that the G^k must be non-negative. This was shown by Racah (1942*b*) to be true also and I give his proof here.

Equation (4.46) yields

$$G^k = e^2 \int \frac{r_<^k}{r_>^{k+1}} R_{nl}(1)\, R_{n'l'}(1)\, R_{nl}(2)\, R_{n'l'}(2)\, r_1^2 r_2^2 dr_1 dr_2.$$

Then we put $f(r) \equiv er^2 R_{nl} R_{n'l'}$ and have

$$G^k = \int_0^\infty \int_0^\infty \frac{r_<^k}{r_>^{k+1}} f(r_1)\, f(r_2)\, dr_1 dr_2 = \int_0^\infty f(r_1)\, \phi(r_1)\, dr_1,$$

where
$$\phi(r_1) = r_1^{-k-1} \int_0^{r_1} r_2^k f(r_2)\, dr_2 + r_1^k \int_{r_1}^\infty r_2^{-k-1} f(r_2)\, dr_2.$$

We multiply both sides by r_1^{k+1} and differentiate, whereupon

$$[r_1^{k+1} \phi(r_1)]' = (2k+1)\, r_1^{2k} \int_{r_1}^\infty r_2^{-k-1} f(r_2)\, dr_2,$$

and therefore
$$\{r_1^{-2k}[r_1^{k+1}\phi(r_1)]'\}' = -(2k+1)\, r_1^{-k-1} f(r_1).$$

So
$$G^k = -(2k+1)^{-1} \int_0^\infty r_1^{k+1} \phi(r_1)\, \{r_1^{-2k}\, [r_1^{k+1}\phi(r_1)]'\}'\, dr_1,$$

which on integration by parts becomes

$$G^k = (2k+1)^{-1} \int_0^\infty r_1^{-2k} \{[r_1^{k+1}\phi(r_1)]'\}^2\, dr_1 \geqslant 0$$

and is the desired result.

Examples

1. Derive the following properties of the c^k directly from their definition.

(*a*) $c^k(l-p, l'-p') = c^k(lp, l'p')$.

(*b*) $c^k(lp, l'p') = 0$ unless $|l-l'| \leqslant k \leqslant l+l'$.

(*c*) $c^k(lp, l'p') = 0$ for $k < |p-p'|$.

(*d*) $c^k(lp, l'p') = 0$ unless $l+l'+k$ is even.

(*e*) $c^0(lp, l'p') = \delta(l, l')\, \delta(p, p')$.

(*f*) $c^k(lp, 00) = (2l+1)^{-\frac{1}{2}} \delta_{kl}$.

2. (Racah.) Write $$(2k+1)^{-1}G^k-(2k+3)^{-1}G^{k+1}=\int_0^\infty f(r_1)\,\psi(r_1)\,dr_1$$

and show that $$f(r_1)=(2k+2)^{-1}r_1^{k+2}(r_1^{-2k}(r_1^{k+2}\psi(r_1))'')''.$$

Deduce that $(2k+1)^{-1}G^k$ is a non-increasing function of k.

4.5. p^n configurations

In this section we illustrate the theory by obtaining the electrostatic energies of all terms of all p^n configurations as functions of the Slater–Condon parameters F_k. For p electrons there are just two, F_0 and F_2, and they are given by (4.46) and (4.50). The p^2 configuration has the three terms ^3P, ^1S, ^1D. A single p function would be written $|n1m_sm_l\rangle$ but we abbreviate it here to $|m_l^+\rangle$ or $|m_l^-\rangle$ according to whether m_s is $+\frac{1}{2}$ or $-\frac{1}{2}$, respectively. Then there is only one triplet and so any state having $M_S=1$ must belong to it. A particular one is the single determinant function

$$|1^+0^+\rangle\in{}^3P$$

($\in$ means logical inclusion), whence by (4.31), (4.47) and (4.48)

$$\begin{aligned}E(^3P)&=J(1,0)-K(1,0)\\&=(F_0-2F_2)-3F_2=F_0-5F_2.\end{aligned}\tag{4.52}$$

In (4.52) we have written only the electrostatic energy of interaction between the two electrons in the ^3P state. The reader will remember that the contributions from the U_{ii} are the same for all states of a configuration. Also if there are closed shells as well as our two p electrons they give rise to a contribution to the energy which again is the same for all states. So in discussing a single configuration we may leave these out by referring all the states to a suitable zero of energy. We must, however, include them when we compare different configurations.

We now continue and observe that any state with $M_L=2$ must belong to ^1D so we have

$$\left.\begin{aligned}&|1^+1^-\rangle\in{}^1D,\\ E(^1D)&=J(1,1)=F_0+F_2.\end{aligned}\right\}\tag{4.53}$$

Finally, each of ^1S, ^1D and ^3P has just one state for which $M_S=M_L=0$ and so we use the diagonal-sum rule to derive

$$\begin{aligned}&E(^1S)+E(^1D)+E(^3P)\\&\quad=E(|1^+-1^-\rangle)+E(|0^+0^-\rangle)+E(|-1^+1^-\rangle)\\&\quad=2J(1,-1)+J(0,0)=3F_0+6F_2,\end{aligned}$$

whence $$E(^1S)=F_0+10F_2,\tag{4.54}$$

completing our determination of the relative first-order energies of the p^2 configuration.

Now if we write, as usual, $\mathbf{s}_i$ and $\mathbf{l}_i$ for the spin and orbital angular momentum vectors for the ith electron, it follows from our general discussion of vectors of type **T** that both $\mathbf{l}_i.\mathbf{l}_j$ and $\mathbf{s}_i.\mathbf{s}_j$ commute with both **L** and **S**. Therefore they will have zero matrix elements between any pair of terms which differ in their L or S values, but each have constant diagonal matrix elements for any given term. For the p^2 configuration their matrix elements are

$$\left.\begin{aligned} \langle {}^3P \,|\, \mathbf{l}_1.\mathbf{l}_2 \,|\, {}^3P\rangle &= -1, & \langle {}^3P \,|\, \mathbf{s}_1.\mathbf{s}_2 \,|\, {}^3P\rangle &= +\tfrac{1}{4},\\ \langle {}^1D \,|\, \mathbf{l}_1.\mathbf{l}_2 \,|\, {}^1D\rangle &= +1, & \langle {}^1D \,|\, \mathbf{s}_1.\mathbf{s}_2 \,|\, {}^1D\rangle &= -\tfrac{3}{4},\\ \langle {}^1S \,|\, \mathbf{l}_1.\mathbf{l}_2 \,|\, {}^1S\rangle &= -2, & \langle {}^1S \,|\, \mathbf{s}_1.\mathbf{s}_2 \,|\, {}^1S\rangle &= -\tfrac{3}{4}, \end{aligned}\right\} \tag{4.55}$$

where we have used $\mathbf{L}^2 = \mathbf{l}_1^2+\mathbf{l}_2^2+2\mathbf{l}_1.\mathbf{l}_2$ and $\mathbf{S}^2 = \mathbf{s}_1^2+\mathbf{s}_2^2+2\mathbf{s}_1.\mathbf{s}_2$ and have written the results in units of $\hbar^2$. It follows from (4.52–55) that the electrostatic energy e^2/r_{12} has identically the same matrix elements within p^2 as the quantity

$$w_{12} = F_0-(5+3\mathbf{l}_1.\mathbf{l}_2+12\mathbf{s}_1.\mathbf{s}_2)\,F_2. \tag{4.56}$$

In (4.56) the integrals F_0 and F_2 are, of course, just numbers.

It now follows immediately from (4.31) that the matrix elements of the electrostatic energy $\sum_{i<j} e^2/r_{ij}$ are identically the same within any p^n configuration as are those of $W = \sum_{i<j} w_{ij}$. But

$$\begin{aligned} W &= \sum_{i<j}^{n} (F_0-(5+3\mathbf{l}_i.\mathbf{l}_j+12\mathbf{s}_i.\mathbf{s}_j)\,F_2)\\ &= \tfrac{1}{2}n(n-1)\,F_0-\left\{\tfrac{5}{2}n(n-1)+\tfrac{3}{2}\mathbf{L}^2-\tfrac{3}{2}\sum_{i=1}^{n}\mathbf{l}_i^2+6\mathbf{S}^2-6\sum_{i=1}^{n}\mathbf{s}_i^2\right\}F_2, \end{aligned}$$

whence

$$\langle {}^{2S+1}L \,|\, W \,|\, {}^{2S+1}L\rangle = \tfrac{1}{2}n(n-1)\,F_0-\tfrac{1}{2}(5n^2-20n+3L(L+1)+12S(S+1))\,F_2. \tag{4.57}$$

Formula (4.57), with its method of derivation, is due to Van Vleck (1934). It gives the electrostatic energies of all terms of all p^n configurations as a polynomial function of n, L and S. It is, incidentally, also true for $n = 0$ or 1 and gives zero in both those cases. One sees from (4.57) that the relative energies of the terms of a given configuration depend only on L and S not on n. But we have already seen (ex. 6, p. 68) that the allowed terms of l^ϵ are the same as those of $l^{4l+2-\epsilon}$. So (4.57) shows that for p^ϵ configurations the same is true for the first-order energies. This can also be shown by a more general method due to Shortley. We write $\mathscr{H}_{ii}^{\epsilon}$ and $\mathscr{H}_{ii}^{4l+2-\epsilon}$ for diagonal elements corresponding in the sense that they both refer to determinantal functions, but the $(4l+2-\epsilon)$ individual one-electron functions (including spin) for $\mathscr{H}_{ii}^{4l+2-\epsilon}$ are just those which did not occur for $\mathscr{H}_{ii}^{\epsilon}$. Let us also put $f(jk) = J(jk)-K(jk)$. Clearly $f(jj) \equiv 0$. Now l^{4l+2} has only a ${}^1S$ term and l^{4l+1} only a ${}^2L$ term and therefore

$$\sum_{j>k=1}^{4l+2} f(jk)-\sum_{j>k=1}^{4l+1} f(jk) = \sum_{k=1}^{4l+1} f(4l+2,k) = \sum_{k=1}^{4l+2} f(4l+2,k) = a,$$

where a is independent of which function has been chosen to be number $4l+2$, because a is essentially just the difference in energy of the ^1S and ^2L terms. Then

$$\mathscr{H}_{ii}^{4l+2-\epsilon} = \sum_{j>k=\epsilon+1}^{4l+2} f(jk) = \frac{1}{2}\sum_{j=\epsilon+1}^{4l+2}\sum_{k=\epsilon+1}^{4l+2} f(jk)$$

$$= \frac{1}{2}\sum_{j=1}^{4l+2}\sum_{k=1}^{4l+2} f(jk) + \frac{1}{2}\sum_{j=1}^{\epsilon}\sum_{k=1}^{\epsilon} f(jk) - \frac{1}{2}\sum_{j=1}^{\epsilon}\sum_{k=1}^{4l+2} f(jk) - \frac{1}{2}\sum_{j=1}^{4l+2}\sum_{k=1}^{\epsilon} f(jk)$$

$$= \mathscr{H}_{ii}^{\epsilon} + \sum_{j>k=1}^{4l+2} f(jk) - \epsilon a, \tag{4.58}$$

which establishes the result (see also § 9.7).

We have found that the relative energies of p^2 and p^4 are

$$\left.\begin{aligned} E(^1S) &= F_0 + 10F_2, \\ E(^1D) &= F_0 + F_2, \\ E(^3P) &= F_0 - 5F_2, \end{aligned}\right\} \tag{4.59}$$

and, from (4.57), those of p^3 are

$$\left.\begin{aligned} E(^2P) &= 3F_0, \\ E(^2D) &= 3F_0 - 6F_2, \\ E(^4S) &= 3F_0 - 15F_2. \end{aligned}\right\} \tag{4.60}$$

These formulae imply that the intervals between the terms of a configuration should stand in a definite numerical ratio to one another. In fact we should have

$$\left.\begin{aligned} r_2 = r_4 &= \frac{E(^1S)-E(^1D)}{E(^1D)-E(^3P)} = \frac{3}{2}, \\ r_3 &= \frac{E(^2P)-E(^2D)}{E(^2D)-E(^4S)} = \frac{2}{3}. \end{aligned}\right\} \tag{4.61}$$

The experimental ratios are smaller than the theoretical ones by about 25 % for the first short-period elements (e.g. C, N, O). For the second short period, probably fortuitously, the ratios are much closer to the theoretical values. The present theory is not entirely relevant for later periods because we have neglected the spin-orbit coupling entirely and this is a worse approximation the further one goes down the periodic table. In spite of this expectation, the actual agreement is usually not too bad even far down the periodic table although there are several cases of very bad disagreement.

Another thing that one notices immediately about (4.59) and (4.60) is that in both cases the lowest-lying term is the one with the highest spin. This is a theoretical justification of a part of Hund's rule for determining the ground term of a configuration. Hund's rule asserts that the ground term of a configuration is a term with the highest allowed value of the total spin and the highest value of the total orbital angular momentum consistent with the first requirement. In the case that the configuration is the ground configuration (we use this to mean that it contains the ground term of the atom or ion) there are no known exceptions to this rule. Thus in the doubly ionized ions of the first long period lying in the range Ca^{++} to Zn^{++} the ground term in the gas phase is a term from the configuration $1s^2\,2s^2\,2p^6\,3s^2\,3p^6\,3d^n$, where n runs from 0 to 10. In each case the ground term is the one suggested by Hund's rule and is, for example, ${}^3F$ for Ti^{++} ($n = 2$) and ${}^5D$ for Fe^{++} ($n = 6$); see Table 4.3 and also Appendix 1.

A more general form of Hund's rule would be the assertion that in a given configuration the terms had energies ranged according to their spins, and within each group with the same spin according to their orbital angular momenta. This is true for p^n. However, it is not true, in general, even for the configuration containing the ground term. As an example, the known terms of V III (we follow the spectroscopist's usage in writing An for the $(n-1)$th ionized atom A rather than the physical chemist's in which An is the nth state of ionization; so V III $= V^{2+}$) belonging to the configuration $1s^2\,2s^2\,2p^6\,3s^2\,3p^6\,3d^3$ are in the order

$$ {}^4F < {}^2P < {}^4P < {}^2G < {}^2D < {}^2H. $$

Actually the suggestion that the terms are ordered roughly according to the total spin is not too bad an approximation to the truth although there is not usually much correlation with the total orbital angular momentum.

I conclude this section by showing a rather interesting method of improving the agreement with experiment whilst retaining the simplicity of the method. If we work out our electrostatic energies as far as the expression (4.31) we notice that each term $J(\kappa\lambda)-K(\kappa\lambda)$ in the sum is actually the energy of a two-electron function. Now we know that the terms of the configuration p^2 do not satisfy their ratio rule and we ask whether the deviation from the ratio rule for the other p^n configurations could be explained solely in terms of the deviation for p^2. To be specific, let us introduce two parameters giving the separations for p^2 between ${}^3P$ and ${}^1D$ and between ${}^1D$ and ${}^1S$, respectively, and use these to derive the ratio of the separations in p^3 and p^4. This can be done very simply by Van Vleck's method which we used earlier in this section.

We write

$$ E({}^3P) = c, \quad E({}^1D) = c+a, \quad E({}^1S) = a+b+c $$

for the energies of p^2. Then it is easy to see from (4.55) that these energies are produced by the operator

$$ E_2 = -(a+\tfrac{2}{3}b)\,\mathbf{s}_1.\mathbf{s}_2-\tfrac{1}{3}b\mathbf{l}_1.\mathbf{l}_2, $$

apart from a constant energy added to all three terms. The relative energies of p^n are then given by

$$ E_n = -(\tfrac{1}{2}a+\tfrac{1}{3}b)\,S(S+1)-\tfrac{1}{6}bL(L+1), \tag{4.62} $$

and we see that

$$ r_2 = \frac{b}{a}, \quad r_3 = \frac{4b}{9a} = \tfrac{4}{9}r_2, \quad r_4 = r_2. \tag{4.63} $$

The theoretical ratios r_2/r_3 and r_2/r_4 are therefore 2·25 and 1, respectively. Experimentally we have for the neutral atoms of the first two short periods, omitting the inner shells:

$$\left.\begin{array}{lll} \text{C I} & 2p^2 & r_2 = 1{\cdot}13, \\ \text{N I} & 2p^3 & r_3 = 0{\cdot}500, \\ \text{O I} & 2p^4 & r_4 = 1{\cdot}14, \end{array}\right\} \quad \frac{r_2}{r_3} = 2{\cdot}26, \quad \frac{r_2}{r_4} = 0{\cdot}99,$$

$$\left.\begin{array}{lll} \text{Si I} & 3p^2 & r_2 = 1{\cdot}48, \\ \text{P I} & 3p^3 & r_3 = 0{\cdot}648, \\ \text{S I} & 3p^4 & r_4 = 1{\cdot}43, \end{array}\right\} \quad \frac{r_2}{r_3} = 2{\cdot}28, \quad \frac{r_2}{r_4} = 1{\cdot}03. \tag{4.64}$$

The agreement is striking and suggests that this way of looking at the deviations from the theoretical ratios for r_2, r_3 and r_4 separately must be in some sense a meaningful one. In terms of perturbation theory one would say that there must be some sort of approximate additivity of the second-order perturbation corrections. This would not be surprising.

Historically the treatment I have just given arose from a discovery by Trees (1951*b*) that the iron group spectra can be made to agree very much better with the theoretical first-order formulae if a term equal to $\alpha L(L+1)$ is added to those first-order formulae. Racah (1952) then observed that the correction could be introduced via the two-electron functions as we have done above. A detailed analysis of the first short-period spectra based on these ideas has been given by Rohrlich (1956). This type of refinement of the theory is generally called the polarization effect.

Examples

1. Show that the polarization effect in p^n as we treated it above is mathematically equivalent to Trees's $L(L+1)$ correction. (Remember that we regard the F_k as empirically determined parameters.)

2. By choosing suitable orders for the one-electron functions in the basic determinantal functions show that the off-diagonal matrix elements $\mathscr{H}_{ij}$ in $l^{4l+2-\epsilon}$ can be taken to be the same as the corresponding ones in l^ϵ. Deduce that the energies of corresponding terms of l^ϵ and $l^{4l+2-\epsilon}$ are the same functions of the basic parameters, apart from a constant diagonal energy, even when the electrostatic energy is diagonalized within a configuration containing more than one term of the same kind (see also §9.7).

4.6. d^n and other configurations

The d^n configurations are the most interesting from the point of view of our applications. This is because the important transition group ions have d^n for their ground configuration. It is convenient to follow Racah and introduce three linear combinations of the F_k as the basic parameters for d electrons. We define

$$\left.\begin{aligned} A &= F_0 - 49F_4, \\ B &= F_2 - 5F_4, \\ C &= 35F_4 \end{aligned}\right\} \tag{4.65}$$

and see from (4.51) and Table 4.4 that A, B and C are necessarily positive. As we shall use Racah's parameters a great deal in this book I give in Table 4.5 the Coulomb and exchange integrals for d electrons expressed in terms of them.

Table 4.5. *Coulomb and exchange integrals between d electrons expressed in Racah parameters*

m_l values		J	K
± 2	± 2	$A+4B+2C$	$(A+4B+2C)$
± 2	∓ 2	$A+4B+2C$	$2C$
± 2	± 1	$A-2B+C$	$6B+C$
± 2	∓ 1	$A-2B+C$	C
± 2	0	$A-4B+C$	$4B+C$
± 1	± 1	$A+B+2C$	$(A+B+2C)$
± 1	∓ 1	$A+B+2C$	$6B+2C$
± 1	0	$A+2B+C$	$B+C$
0	0	$A+4B+3C$	$(A+4B+3C)$

We now proceed as before and work out the energies of the terms of the d^2 configuration first. These terms are ^1G, ^3F, ^1D, ^3P and ^1S and we calculate their energies in that order, using the diagonal-sum rule. Thus

$$\begin{aligned} E(^1G) &= J(2,2) = A+4B+2C, \\ E(^3F)+E(^1G) &= 2J(2,1) = 2A-4B+2C, \\ E(^1D)+E(^3F)+E(^1G) &= 2J(2,0)+J(1,1) \\ &= 3A-7B+4C, \\ E(^3P)+E(^1D)+E(^3F)+E(^1G) &= 2J(2,-1)+2J(1,0) = 4A+4C, \\ E(^1S)+E(^3P)+E(^1D)+E(^3F)+E(^1G) &= 2J(2,-2)+2J(1,-1)+J(0,0) \\ &= 5A+14B+11C. \end{aligned}$$

From these equations we deduce immediately that the first-order electrostatic energies for d^2 are

$$\left.\begin{aligned} E(^1S) &= A+14B+7C, \\ E(^1D) &= A-\ \ 3B+2C, \\ E(^1G) &= A+\ \ 4B+2C, \\ E(^3P) &= A+\ \ 7B, \\ E(^3F) &= A-\ \ 8B. \end{aligned}\right\} \qquad (4.66)$$

A, B and C are positive and we predict that ^3F should lie lowest, in accordance with Hund's rule.

Unfortunately it is not possible to proceed with the same simplicity and generality for d^n configurations as we did for p^n configurations. Nevertheless, we can obtain a general formula for the energies of the terms which have spin $S = \frac{1}{2}n$. We do this by taking as our basic determinantal functions only those which have also $M_S = \frac{1}{2}n$. Then every one-electron function has α spin and so (4.31) becomes a sum over $J(\kappa,\lambda)-K(\kappa,\lambda)$ for pairs of electrons each having α spin. This means that it is a sum over quantities referring only to the triplets in (4.66) and therefore if we can find an expression like (4.56) for the triplets we can deduce a general formula for d^n when $S = \frac{1}{2}n$. Naturally this is very easy, the expression being

$$w_{12} = A-(8+3\mathbf{l}_1.\mathbf{l}_2)\,B,$$

whence the formula for d^n is

$$\langle {}^{n+1}L\,|\,W\,|\,{}^{n+1}L\rangle = \tfrac{1}{2}n(n-1)\,(A-8B)+\tfrac{3}{2}[6n-L(L+1)]\,B. \qquad (4.67)$$

This result is due to Racah (1942*a*). It only holds for $n \leqslant 5$ although we can pass to $n > 5$ via the relationship (4.58).

The states of highest spin are the most interesting from our point of view because they include the ground term and as the separation between terms is usually very large compared with kT, the unit of thermal energy, the ground term is normally the only one which is appreciably occupied. The quartets of d^5 are interesting in applications and we derive their energies next. Let us take the $M_S = \frac{3}{2}$ states and consider the 25 basic determinantal functions $\phi(m,m')$, where m is the m_l value for the single one-electron function with β spin and m' is the negative of the m_l value of the missing one-electron function with α spin. Thus

$$\phi(2,1) = |2^+2^-1^+0^+-2^+\rangle,$$

where the one-electron functions are enumerated in order of m_l and then in order of m_s. We now obtain the diagonal elements for $\phi(m,m')$ by relating them to corresponding diagonal elements of the ${}^5D$ terms of d^4 and d^6. We have, from (4.31), by simply breaking up the sum appropriately,

$$E(d^6;\,{}^5D) = E[\phi(m,m')]+J(m,m')+\sum_{m_1 \neq -m'} f(m_1,\,-m'),$$

$$E(d^5;\,{}^6S) = E(d^4;\,{}^5D)+\sum_{m_1 \neq -m'} f(m_1,\,-m'),$$

where $f = J-K$ as before. Actually $f(-m',\,-m') = 0$ so there is no need to restrict the sums over m_1. Subtracting these two equations we obtain

$$E[\phi(m,m')] = E(d^6;\,{}^5D)+E(d^4;\,{}^5D)-E(d^5;\,{}^6S)-J(m,m'),$$

and so the only dependence of $E[\phi(m,m')]$ on m and m' is through the one quantity $J(m,m')$. The M_L value of $\phi(m,m')$ is $m+m'$ and so, apart from the reversal of sign and an additive constant, the energies of the five terms ${}^6S$, ${}^4P$, ${}^4D$, ${}^4F$ and ${}^4G$ of d^5 are given by the diagonal-sum rule to be the same functions of A, B and C as the terms of d^2 with the same L value. This is because the $J(m,m')$ were the only integrals which appeared in the determination of the latter energies and then in precisely corresponding places. Now (4.67) gives us $E(d^5;\,{}^6S) = 10A-35B$ and therefore the quartet energies are

$$E({}^4L) = 11A-21B+7C-E(d^2;\,{}^{2S+1}L).$$

Equations (4.66) then yield the four energies

$$\left.\begin{aligned} E(^4P) &= 10A-28B+7C,\\ E(^4D) &= 10A-18B+5C,\\ E(^4F) &= 10A-13B+7C,\\ E(^4G) &= 10A-25B+5C.\end{aligned}\right\} \tag{4.68}$$

This relation of the energies of the terms with $S = l-\frac{1}{2}$ and $S = l+\frac{1}{2}$ in the configuration l^{2l+1} to the energies of l^2 is completely general (Racah, 1942*b*), as the argument given above can be immediately extended from $l = 2$ to general l. Another example of this is to be found in (4.59) and (4.60) and gives a deeper significance to the fact that $r_2 r_3 = 1$ in (4.61).

Table 4.6. *Theoretical energies of terms of d^n configurations for $n \leqslant 5$. The relative energies of the terms of d^{10-n} are the same as those of d^n*

d^2	d^3
$^3F = A-8B$	$^4F = 3A-15B$
$^3P = A+7B$	$^4P = 3A$
$^1G = A+4B+2C$	$^2H = {}^2P = 3A-6B+3C$
$^1D = A-3B+2C$	$^2G = 3A-11B+3C$
$^1S = A+14B+7C$	$^2F = 3A+9B+3C$
	$^2D = 3A+5B+5C \pm (193B^2+8BC+4C^2)^{\frac{1}{2}}$
d^4	**d^5**
$^5D = 6A-21B$	$^6S = 10A-35B$
$^3H = 6A-17B+4C$	$^4G = 10A-25B+5C$
$^3G = 6A-12B+4C$	$^4F = 10A-13B+7C$
$^3F = 6A-5B+5\frac{1}{2}C \pm \frac{3}{2}(68B^2+4BC+C^2)^{\frac{1}{2}}$	$^4D = 10A-18B+5C$
$^3D = 6A-5B+4C$	$^4P = 10A-28B+7C$
$^3P = 6A-5B+5\frac{1}{2}C \pm \frac{1}{2}(912B^2-24BC+9C^2)^{\frac{1}{2}}$	$^2I = 10A-24B+8C$
$^1I = 6A-15B+6C$	$^2H = 10A-22B+10C$
$^1G = 6A-5B+7\frac{1}{2}C \pm \frac{1}{2}(708B^2-12BC+9C^2)^{\frac{1}{2}}$	$^2G = 10A-13B+8C$
$^1F = 6A+6C$	$^2G' = 10A+3B+10C$
$^1D = 6A+9B+7\frac{1}{2}C \pm \frac{3}{2}(144B^2+8BC+C^2)^{\frac{1}{2}}$	$^2F = 10A-9B+8C$
$^1S = 6A+10B+10C \pm 2(193B^2+8BC+4C^2)^{\frac{1}{2}}$	$^2F' = 10A-25B+10C$
	$^2D' = 10A-4B+10C$
	$^2D = 10A-3B+11C \pm 3(57B^2+2BC+C^2)^{\frac{1}{2}}$
	$^2P = 10A+20B+10C$
	$^2S = 10A-3B+8C$

It is possible by somewhat similar methods to determine the first-order energies of the remaining terms of the d^n configurations. They are not so useful to us but I have put them in Table 4.6, which was taken from one of Racah's papers,[1] for reference if required, and plotted them as functions of the ratio B/C in Figs. 4.1–4. The relative energies of d^6–d^{10} are the same as those of d^{10-n} (ex. 2, p. 83, see also, the notes to § 9.7).

[1] Racah (1942*b*). For earlier work see Ufford & Shortley (1932); Ostrofsky (1934); Condon & Shortley (1953); Laporte (1942) and Laporte & Platt (1942). The degeneracies for $B = 0$ are not entirely accidental (see Racah, 1942*b*, 1949, and note, p. 245).

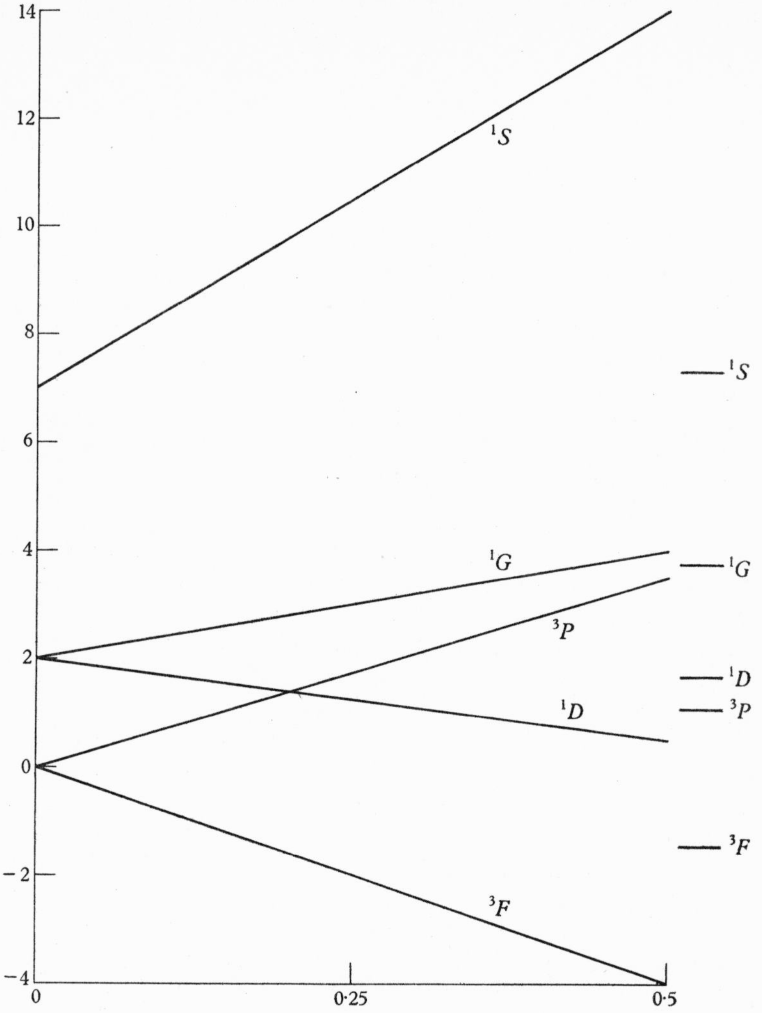

Fig. 4.1. Energy-level diagram for d^2. Energies of terms in Figs. 4.2–5 are given in units of C and plotted against B/C. The observed positions of the terms of $4d^2 5s^2$ for Zr I, relative to the ^3F term, are shown at the right of Fig. 4.2 assuming $B/C = 0{\cdot}1785$.

Apart from general methods of obtaining energies it is always possible to obtain them in a straightforward manner by using the shift operators which connect the various $M_S M_L$ values of a given term to determine actual eigenfunctions belonging to each of the terms. As this is a method which is of rather general use I will demonstrate it by obtaining the energies of the ^3F terms of the configuration d^4 as functions of A, B and C. We take only states having $M_S = 1$ and, as in the use of the diagonal-sum rule, start with the highest M_L value and work downwards. Let us use the abbreviations

$$|\phi_1, \phi_2, \ldots, \phi_n\rangle = (n!)^{-\frac{1}{2}} \sum_{\mu} (-1)^{\mu} P_{\mu} |\phi_1(1) \ldots \phi_n(n)\rangle$$

as usual and also

$$p^+ = R_{n2} Y_{2p} \alpha, \quad p^- = R_{n2} Y_{2p} \beta, \quad p^2 = p^+ p^-,$$

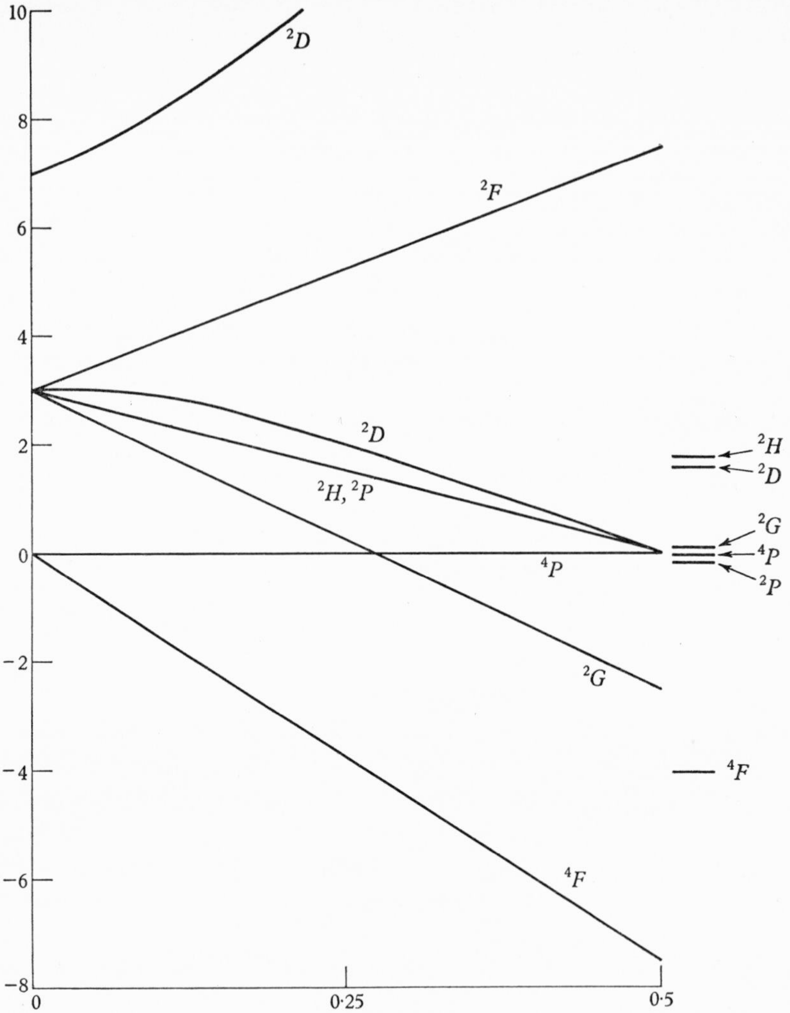

Fig. 4.2. Energy-level diagram for d^3. The observed positions of the terms of $3d^3$ for V III, relative to the ^4F term, are shown at the right assuming $B/C = 0{\cdot}2683$.

and write $|^{2S+1}L, M_L\rangle$ for the state of ^{2S+1}L having $M_L\hbar$ for L_z and $M_S = 1$. Then

$$|2^2, 1^+, 0^+\rangle \in {}^3H$$

with
$$E(^3H) = 6A - 17B + 4C.$$

There are two of our states having $M_L = 4$, namely

$$|2^2, 1^+, -1^+\rangle \quad \text{and} \quad |2^+, 1^2, 0^+\rangle$$

with
$$\mathscr{H}_{ii}(|2^2, 1^+, -1^+\rangle) = 6A - 15B + 4C,$$

$$\mathscr{H}_{ii}(|2^+, 1^2, 0^+\rangle) = 6A - 14B + 4C,$$

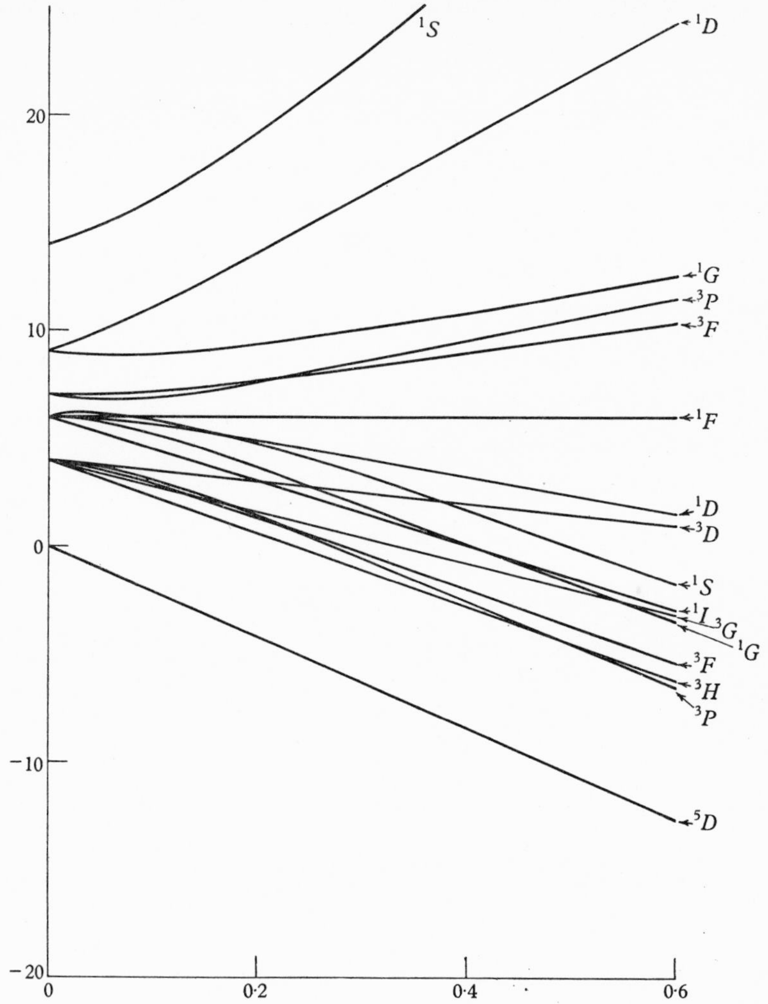

Fig. 4.3. Energy-level diagram for d^4.

so by the diagonal sum rule

$$E(^3G) = 6A - 12B + 4C.$$

The actual eigenstates are given by operating on $|2^2, 1^+, 0^+\rangle$ with L^- to obtain

$$|^3H, 4\rangle = (\tfrac{3}{5})^{\frac{1}{2}}\,|2^2, 1^+, -1^+\rangle - (\tfrac{2}{5})^{\frac{1}{2}}\,|2^+, 1^2, 0^+\rangle$$

and then $|^3G, 4\rangle$ is orthogonal to it and so we may write

$$|^3G, 4\rangle = (\tfrac{2}{5})^{\frac{1}{2}}\,|2^2, 1^+, -1^+\rangle + (\tfrac{3}{5})^{\frac{1}{2}}\,|2^+, 1^2, 0^+\rangle.$$

Four of our states have $M_L = 3$ and we abbreviate them as

$$a = |2^2, 1^+, -2^+\rangle, \quad b = |2^2, 0^+, -1^+\rangle,$$

$$c = |2^+, 1^2, -1^+\rangle, \quad d = |2^+, 1^+, 0^2\rangle,$$

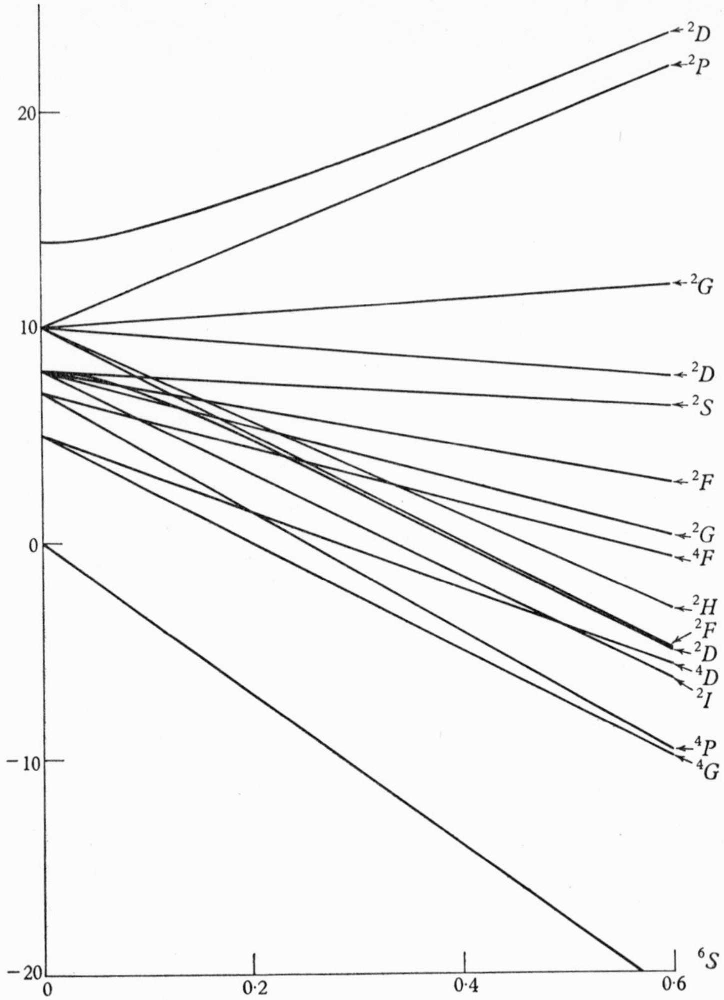

Fig. 4.4. Energy-level diagram for d^5.

and operate on $|^3H, 4\rangle$ and $|^3G, 4\rangle$ to obtain

$$\left.\begin{aligned} |^3H, 3\rangle &= \frac{1}{\sqrt{15}}(\sqrt{2}\,a+\sqrt{3}\,b-2\sqrt{2}\,c+\sqrt{2}\,d), \\ |^3G, 3\rangle &= \frac{1}{2\sqrt{5}}(2a+\sqrt{6}\,b+c-3d), \\ |\psi_1\rangle &= \frac{1}{\sqrt{5}}(\sqrt{3}\,a-\sqrt{2}\,b), \\ |\psi_2\rangle &= \frac{1}{\sqrt{15}}(a+\tfrac{1}{2}\sqrt{6}\,b+\tfrac{5}{2}c+\tfrac{5}{2}d). \end{aligned}\right\} \quad (4.69)$$

Then our two $|^3F, 3\rangle$ are linear combinations of $|\psi_1\rangle$ and $|\psi_2\rangle$. Using (4.31) and (4.33) we easily obtain the matrix elements of electrostatic interaction between a, b, c and d. These are shown in Table 4.7 from which one derives the matrix of electrostatic interaction for the two states $|\psi_1\rangle$ and $|\psi_2\rangle$. This is given in Table 4.8. The solution of the corresponding secular equation then gives the energies of the two states

$$E(^3F) = 6A-5B+5\tfrac{1}{2}C \pm \tfrac{3}{2}(68B^2+4BC+C^2)^{\frac{1}{2}}.$$

Table 4.7. *Electrostatic matrix elements for the $M_S = 1$, $M_L = 3$ states of d^4*

V_{ij}	a	b	c	d
a	$6A+5C$	$-3\sqrt{6}B$	$6B+C$	$4B+C$
b	$-3\sqrt{6}B$	$6A-11B+4C$	$-\sqrt{6}B$	$-2\sqrt{6}B$
c	$6B+C$	$-\sqrt{6}B$	$6A-15B+5C$	$B+C$
d	$4B+C$	$-2\sqrt{6}B$	$B+C$	$6A-13B+5C$

Table 4.8. *Electrostatic matrix elements for the two ^3F terms of d^4*

V_{ij}	ψ_1	ψ_2
ψ_1	$6A+\frac{14}{5}B+\frac{23}{5}C$	$\frac{48}{5}B+\frac{6}{5}C$
ψ_2	$\frac{48}{5}B+\frac{6}{5}C$	$6A-\frac{64}{5}B+\frac{32}{5}C$

As a further illustration I mention the f^n configurations. These occur in the rare-earth ions. The general theory of these is rather more difficult than for p^n or d^n and is outside the scope of this book.[1] However, we can simply assume Hund's rule to obtain the ground terms and then calculate their energies by elementary methods. We follow Racah (1949) and write

$$\left.\begin{aligned} E^0 &= F_0-10F_2-33F_4-286F_6,\\ E^1 &= \tfrac{7}{9}(10F_2+33F_4+286F_6),\\ E^2 &= \tfrac{1}{9}(F_2-3F_4+7F_6),\\ E^3 &= \tfrac{1}{3}(5F_2+6F_4-91F_6).\end{aligned}\right\} \tag{4.70}$$

Then for $n \leqslant 7$ the first-order electrostatic energies of the ground terms are

$$\left.\begin{aligned} E(f^1;\,^2F) &= 0,\\ E(f^2;\,^3H) &= E^0-9E^3,\\ E(f^3;\,^4I) &= 3E^0-21E^3,\\ E(f^4;\,^5I) &= 6E^0-21E^3,\\ E(f^5;\,^6H) &= 10E^0-9E^3,\\ E(f^6;\,^7F) &= 15E^0,\\ E(f^7;\,^8S) &= 21E^0.\end{aligned}\right\} \tag{4.71}$$

Each of the results in (4.71) can be obtained from the energy of a single determinantal function.

[1] See Racah (1949) and Elliott, Judd & Runciman (1957).

We have now derived practically all our basic results in the theory of the electrostatic energies of atoms and it is probably desirable to emphasize the approximate nature of our calculations. We have really done little more than determine these energies by first-order perturbation theory and have carried our calculations just far enough to lift the accidental degeneracy which is present in a configuration if one neglects the electrostatic interaction altogether. It is true that for some terms (e.g. ^3F of d^4) we solve a secular equation of low degree exactly, but we never go outside the configuration. One method of further refinement was outlined at the end of the last section, but although the polarization method leads to considerably better agreement with experiment in many cases it is not immediately clear what is its precise relationship to the approximate theory which it 'refines'.

Another method of refinement which is of long standing is that of configuration interaction. In this method one simply takes account of the matrix elements between corresponding terms of a finite number of neighbouring configurations and diagonalizes $\mathscr{H}$ within more than one configuration. From a formal point of view it is obvious what is going on when one does this and so it might seem that this is a very good method. Further, if we make the usual assumption that our determinantal functions form a complete set then by taking enough configurations we can approximate as closely as we wish to the exact eigenfunctions of $\mathscr{H}$. So far so good. But there are serious disadvantages. In our theory so far we have had a small number of parameters—the F_k—which we are going to determine by making a best fit with experimental energies of atomic transitions. There are usually many more transitions than F_k and so we can not only obtain the F_k but can also get some idea of the internal consistency of the method by calculating the root-mean-square or some other measure of the deviation between the 'best-fitted theory' and experiment. More configurations mean more than proportionately more parameters. This in turn means that one must either calculate some or all of the parameters or restrict oneself to interaction between only a very few configurations. Also one must include configurations from the continuous spectrum to get an accurate function. Calculation is certainly possible and we consider this way briefly in § 8. However, it tends to be rather hard work and not always particularly rewarding, because the results are often not in a form which is very easy to use in applications to ions in compounds. Also, of course, empirical fitting is especially satisfactory for us because we are not really interested in atoms or ions as such but rather in their behaviour in compounds. Therefore it is more important to get parameters which make our theory agree with experiment than to be able to derive those parameters completely theoretically. On the other hand, one must always bear in mind the possibility that empirically fitted parameters may change their values radically from one problem to the next. There is no mathematical reason which prevents them from doing this; we hope, however, that our physical intuition will warn us of the situations when they are likely to do so.

We must certainly be thankful that the first-order theory gives us such a good qualitative understanding of atomic structure and spectra. It is this that makes it possible to develop reasonably useful general procedures; otherwise we should merely have a collection of detailed calculations of eigenvalues and eigenfunctions connected by little more than a formal similarity between the basic eigenvalue equations (2.8).

Before leaving the matter of refinements to the theory it would be well to contemplate what may reasonably be called the classic example of configuration interaction. This will at least invoke in us a feeling of surprise that the first-order theory does seem to work quite well and also remove any feeling that it may well be that the matrix elements we are neglecting are small anyway even though we cannot see quite why they should be. They simply are not, usually, and Bacher's example of Mg I is an excellent one to show this. The ground configuration is $1s^2\,2s^2\,2p^6\,3s^2$ and we omit the inner shells to write it $3s^2$. It has just the one term ${}^1S_0$. One of the low-lying excited configurations is $3s\,3d$ having the two terms ${}^3D$ and ${}^1D$. The energies may be calculated using the diagonal-sum rule to be

$$\left.\begin{aligned} E({}^3D) &= F_0 - G_2, \\ E({}^1D) &= F_0 + G_2, \end{aligned}\right\} \tag{4.72}$$

where $G_2 = \frac{1}{5}G^2(3s,3d)$. Now we already saw at the end of § 4 that G_k is non-negative and it is actually obviously positive in this case. So we have the firm prediction that ${}^3D$ must lie lower in energy than ${}^1D$. The reverse is true in fact, not only for the configuration $3s\,3d$ but for $3s\,nd$ configurations for all n up to $n = 12$. One cannot explain away all these disagreements within the framework of definite configurations on the assumption that one or both of the component terms has been incorrectly assigned to the configuration.

Bacher suggested that the explanation for $3s\,3d$ lay in configuration interaction between this configuration and the configuration $3p^2$. There is little doubt that this is a correct explanation (the use of the indefinite article is intentional) and a related but more complicated discussion could be given for the whole series. In calculating the interaction we must include explicitly the contribution from $\sum_{\kappa=1}^{n} U(\kappa)$ in (4.18), because although this one-electron part of the Hamiltonian is diagonal within any configuration it may have different diagonal values for different configurations. Also it may have non-vanishing matrix elements between configurations. We can still neglect the inner shells. Then

$$E(3s\,3d) = I(3s) + I(3d) + F_0(3s,3d)\begin{cases} +G_2(3s,3d) & a\,{}^1D, \\ -G_2(3s,3d) & {}^3D, \end{cases}$$

$$E(3p^2) = 2I(3p) + F_0(3p,3p)\begin{cases} +10F_2(3p,3p) & {}^1S, \\ +\;\;\;F_2(3p,3p) & b\,{}^1D, \\ -\;\;5F_2(3p,3p) & {}^3P, \end{cases}$$

where $I(\phi_\kappa)$ is $\langle \overline{\phi}_\kappa \mid U \mid \phi_\kappa \rangle$ and a, b distinguish the two ${}^1D$ terms. Because $\mathscr{H}$ commutes with $\mathbf{L}$ and $\mathbf{S}$ the only non-vanishing off-diagonal matrix elements of $\mathscr{H}$ are between the two ${}^1D$ states. For the same reason $\langle a\,{}^1D\,M_L \mid \mathscr{H} \mid b\,{}^1D\,M_L' \rangle$

is diagonal with the diagonal element independent of M_L. Also any pair of states, one from each of the two configurations, differ in two one-electron functions and so $\sum_{\kappa=1}^{n} U(\kappa)$ has a zero matrix element between them. There is therefore essentially only one new matrix element to evaluate. Let us take $M_L = 2$ and have

$$\psi_1 = \frac{1}{\sqrt{2}}(|3s^+, 3d2^-\rangle - |3s^-, 3d2^+\rangle) \in a\,^1D,$$

$$\psi_2 = |3p1^+, 3p1^-\rangle \in b\,^1D.$$

Equation (4.33) then shows that

$$\begin{aligned}\mathscr{H}_{12} = \langle\bar{\psi}_1|\mathscr{H}|\psi_2\rangle &= \sqrt{2}\,\langle 3s0, 3d2\,|\,V\,|\,3p1, 3p1\rangle \\ &= \sqrt{2}\sum_k c^k(11,00)\,c^k(22,11)\,R^k(3p,3p,3s,3d) \\ &= \sqrt{2}\,c^1(11,00)\,c^1(22,11)\,R^1(3p,3p,3s,3d), \\ &= \frac{2}{\sqrt{15}}R^1(3p,3p,3s,3d). \qquad (4.73)\end{aligned}$$

The reader may find it instructive to evaluate $\mathscr{H}_{12}$ for himself using (4.45), Table 4.4 and ex. 1, p. 78. We write $R^1 = R^1(3p, 3p, 3s, 3d)$ for short. Then the energies of the two ^1D terms are the solutions of the secular equation

$$\begin{vmatrix} E(3s3d;\,^1D) - W & \frac{2}{\sqrt{15}}R^1 \\ \frac{2}{\sqrt{15}}R^1 & E(3p^2;\,^1D) - W \end{vmatrix} = 0. \qquad (4.74)$$

To proceed we need the numerical values of the quantities in (4.74). In the configurations $3p^2$ only ^3P is known and so Bacher used this ^3P and the ^3D of $3s3d$ to give an empirical measure of the separation between the configurations and calculated F_2, G_2 and R^1 for a reasonable choice of one-electron $3s$, $3p$ and $3d$ functions. He obtained

$$6F_2 = 7186\,\text{cm}^{-1}, \quad G_2 = 2038\,\text{cm}^{-1}, \quad R^1 = 21{,}387\,\text{cm}^{-1}$$

which gives for the relevant $\mathscr{H}_{ij}$:

$$\mathscr{H}_{11} = E(^3D) + 2G_2 = 52{,}033\,\text{cm}^{-1},$$

$$\mathscr{H}_{22} = E(^3P) + 6F_2 = 65{,}040\,\text{cm}^{-1},$$

$$\mathscr{H}_{12} = \mathscr{H}_{21} = \frac{2}{\sqrt{15}}R^1 = 11{,}044\,\text{cm}^{-1},$$

and

$$E(^1D) = 71{,}355 \quad \text{or} \quad 45{,}718\,\text{cm}^{-1},$$

$$E(^3D) = 47{,}957\,\text{cm}^{-1},$$

where all energies are relative to the ground $3s^2$ as zero. The agreement with the experimental ^1D, which lies at $46{,}403\,\text{cm}^{-1}$, is very good.[1] Without configuration

[1] In Condon & Shortley (1953), p. 367, the agreement is stated to be not so good. This is because Bacher made a numerical error in his calculation (Bacher, 1933) which he subsequently corrected (Bacher, 1939). Condon & Shortley use the earlier and I the later value.

interaction it would have been at 52,033 cm^{-1}. This example speaks for itself. Perhaps one might underline the fact, however, that not only the energies but also the eigenfunctions are altered by this interaction. So when we later discuss the details of functions belonging to particular configurations we must always remember that the actual eigenstate of $\mathscr{H}$ will be, in this language, a mixture of states from many configurations, and the most that we can hope for is that there is one state from one configuration which has a much larger coefficient in the mixture than any other state. In particular, we must never be misled by the possibility of assigning terms to configurations. Configuration interaction does not alter the number of terms arising from the configurations and so we could still assign the ^1D of Mg I lying at 46,403 cm^{-1} to $3s\,3d$ if we wished, and this is done in Moore's *Atomic Energy Levels*. However, it tends to obscure the fact that it is apparently actually three-quarters $3s\,3d$ and one-quarter $3p^2$.

We conclude this section by remarking that the assumption of definite configurations for terms implies certain selection rules for optical emission and absorption. For electric dipole radiation the matrix element of $\mathbf{D} = \Sigma e\mathbf{r}_j$ between the two states must be non-zero and as this is a sum of one-electron operators we deduce that it must vanish if the states belong to configurations which differ in the nl values of more than one electron. Two-electron jumps are forbidden. Then if the two configurations differ in the nl values of just one electron, $n_1 l_1$ and $n_2 l_2$ say for the two configurations, the matrix element of $\mathbf{D}$ reduces according to (4.23) to $(-1)^r \langle n_1 l_1 m_{s_1} m_{l_1} | e\mathbf{r} | n_2 l_2 m_{s_2} m_{l_2} \rangle$ and so this pair of one-electron functions must satisfy the selection rules for a single electron. We can always choose the $m_s m_l$ values so that they match by selecting suitable states from the configurations. So the configuration selection rules are the ones on n and l, and we have just the rule for l that $\Delta l = \pm 1$ (ex. 4, p. 40). When $\Delta l = \pm 1$ Laporte's selection rule for parity is automatically satisfied. Finally, electric-dipole transitions are forbidden between states of the same configuration, because such states all have the same parity (ex. 5, p. 68). Except for the selection rule on parity, which is rigorous, these configuration selection rules are only approximate and have precise meaning only within the framework of configuration assignment.

Examples

1. Determine the first-order formulae for the energies of the terms of sp^2 and derive a ratio rule. The terms of C II $2s2p^2$ referred to the ^4P as zero are

$$^2D = 31{,}898\ \mathrm{cm}^{-1}, \quad ^2P = 67{,}646\ \mathrm{cm}^{-1}, \quad ^2S = 53{,}460\ \mathrm{cm}^{-1}.$$

Show that these values do not fit your ratio rule at all well.

2. The term energies for $nl\,n's$ are $F_0(nl, n's) \pm G_l(nl, n's)$.

3. The energy of any state of f^n which has $S = \frac{1}{2}n$ is $\frac{1}{2}n(n-1)E^0$ if $E^3 = 0$. Does this latter condition conflict with (4.51)?

4. Al II is isoelectronic with Mg I. The five lowest-lying states are $3s^2\,{}^1S$, $3s\,3p\,{}^3P$, $3s\,3p\,{}^1P$, $3p^2\,{}^1D$, $3s\,4s\,{}^3S$. Unlike the case of Mg I the $3s\,3d\,{}^3D$ lies below $3s\,3d\,{}^1D$. Show that these facts are consistent with our discussion of Mg I.

5. Considering in each case the ground term of the configuration show that

$$E(l^{2l+1+\epsilon}) = E(l^{2l+1}) + E(l^{\epsilon}) + (2l+1)\,\epsilon F_0, \quad \text{where} \quad 0 \leqslant \epsilon \leqslant 2l+1.$$

(Hint: Use (4.39) to sum over the half-filled shell.)

6. To which of d^3f, spd^2, s^2d^2, f^4, pd^3 is electric-dipole radiation from sd^3 allowed?

7. (Van Vleck.) Show that the difference in energy between the two terms of $l^\epsilon s$ obtained by coupling a given term ${}^{2S+1}L$ of l^ϵ in the two possible ways with the ${}^2S$ term of s is $(2S+1)\,G_l(l,s)$. We note that ex. 2 is a special case of this and that the result is independent of L.

4.7. Fitting parameters to the observed spectrum

It is useful to try fitting theoretical formulae to the observed spectra both in order to see how well those formulae agree with experiment and also to obtain values for the parameters occurring in the formulae which can then be used in further calculations. We shall discuss only the parameters B and C relating to the d^n configurations, but other parameters can be treated in a similar manner.

Let us start with the configuration d^2. We let the lowest term have an energy a and then have

$$\left.\begin{aligned} E({}^3F) &= a, & E({}^1D) &= a+\ 5B+2C,\\ E({}^3P) &= a+15B, & E({}^1G) &= a+12B+2C,\\ & & E({}^1S) &= a+22B+7C. \end{aligned}\right\} \tag{4.75}$$

As we shall see in the next chapter the spin-orbit coupling separates the levels of a term so the experimental energies corresponding to a term comprise one for each J value. However, the theory shows (§ 5.2) that the centre of gravity of a term is unaffected by the spin-orbit coupling to first order in perturbation theory. This means that if we take the arithmetic mean of the energies for the different levels of a term, weighting each level according to its multiplicity, we usually get a good estimate for the purely electrostatic energy of the term. Thus for d^2 we write

$$E({}^3F) = \tfrac{5}{21}E({}^3F_2)+\tfrac{1}{3}E({}^3F_3)+\tfrac{3}{7}E({}^3F_4),$$
$$E({}^3P) = \tfrac{1}{9}E({}^3P_0)+\tfrac{1}{3}E({}^3P_1)+\tfrac{5}{9}E({}^3P_2).$$

This procedure is more reliable for the first transition series than for the later ones because the spin-orbit splitting is smaller (see ch. 5, especially Fig. 1).

Given the experimental term energies it is convenient to fit them to (4.75) by the method of least squares. Formulae (4.75) have the form

$$z_i = a+x_iB+y_iC,$$

where z_i are the energies of the terms and i runs from 1 to 5. If we give each level a weight f_i we have to minimize

$$g = \sum_i f_i(z_i-a-x_iB-y_iC)^2$$

with respect to a, B and C. If g is a minimum then

$$\frac{\partial g}{\partial a} = \frac{\partial g}{\partial B} = \frac{\partial g}{\partial C} = 0,$$

which leads to the equations

$$\left.\begin{aligned} \bar{z} &= a+B\bar{x}_1+C\bar{x}_2,\\ \overline{(zx_1)} &= a\bar{x}_1+B\overline{x_1^2}+C\overline{(x_1x_2)},\\ \overline{(zx_2)} &= a\bar{x}_2+B\overline{(x_1x_2)}+C\overline{x_2^2}. \end{aligned}\right\} \tag{4.76}$$

It is customary to make the weight $f_i = 1$ for each term. However, this has the disadvantage that if, later, one is interested in fitting formulae which do include the effects of spin-orbit coupling one will then naturally give a weight of 1 to each level. And when one considers ions in complexes the degeneracy is lifted still further, often completely. It would therefore seem preferable always to take a weight f_i equal to the multiplicity of the energy level concerned, for then there is no discontinuous change, from one situation to another, in the parameters which fit best. This weight also has the advantage that it gives less weight to terms of low multiplicity and those terms tend to be the ones that fit the theoretical formulae worst. Then we take $f_i = (2L+1)(2S+1)$ for the term ${}^{2S+1}L$. It is unlikely that this makes much difference in practice because the parameters do not lead to very good agreement with experiment anyway, but it seems a more consistent thing to do.

Having chosen the weights f_i it is easy to solve (4.76) for a, B and C in terms of the z_i. If we number the z_i from 1 to 5 in the order ${}^3F$, ${}^3P$, ${}^1D$, ${}^1G$, ${}^1S$, and using $f_i = (2L+1)(2S+1)$, we obtain

$$\left.\begin{aligned} B &= \tfrac{3}{50}(z_2-z_1)+\tfrac{1}{70}(z_4-z_3),\\ C &= \tfrac{1}{350}(21z_1-84z_2+40z_3+37z_4+28z_5).\end{aligned}\right\} \tag{4.77}$$

It is often the case that the ${}^1S$ term is missing from the observed terms; we then have to fit our a, B and C to the four remaining terms. In that case (4.76) gives

$$\left.\begin{aligned} B &= \tfrac{3}{50}(z_2-z_1)+\tfrac{1}{70}(z_4-z_3),\\ C &= -\tfrac{1}{10}(2z_1+3z_2)+\tfrac{1}{14}(3z_3+4z_4).\end{aligned}\right\} \tag{4.78}$$

We could of course also give a formula for a, but B and C are the parameters which determine the separations of the terms, a only giving the position of the configuration relative to other ones (in this method we cannot put $a = 0$). The most interesting thing about (4.77) and (4.78) is that we get the same estimate of B whether we fit all five terms or miss ${}^1S$ out. So when we have only the four terms ${}^3F$, ${}^3P$, ${}^1D$, ${}^1G$ we get the 'best' value for B but must be prepared for C to change if we find and include ${}^1S$.

A fairly typical example is furnished by the neutral zirconium atom Zr I. The ground term is $1s^2\,2s^2\,2p^6\,3s^2\,3p^6\,3d^{10}\,4s^2\,4p^6\,4d^2\,5s^2$; ${}^3F$, and all five terms of this $4d^2$ configuration are known. Using (4.78) we obtain

$$B = 254\,\text{cm}^{-1}, \quad C = 1423\,\text{cm}^{-1},$$

but it we omit the ${}^1S$ term then

$$B = 254\,\text{cm}^{-1}, \quad C = 1975\,\text{cm}^{-1}.$$

This gives Table 4.9 which gives some idea of the extent of fit (see also, Fig. 4.1). It also shows that the ${}^1S$ fits worst. The parameter C obtained without using ${}^1S$ is probably more useful for calculations on transition-metal ions in compounds, because then we are usually only interested in fairly low-lying terms. The bad fit of ${}^1S$ is probably due to the fact that the theoretical formulae predict ${}^1S$ to be high because the electrons are close together, while in the real eigenfunction the electrons adjust their positions relative to each other by including r_{12} explicitly in the eigenfunction. In the theory we can replace the function $\psi(d^2; {}^1S)$ by the functions $\psi(d^2; {}^1S)f(r_{12})$ to give the formally simplest function which allows for this effect.

Table 4.9. *Fitting of theoretical formulae to the observed energies for the terms of* $4d^2$ *of* Zr I *without* ($C = 1975$) *and with* ($C = 1423$) *the* ${}^1S$ *term. Energies in* cm^{-1}

	Experiment	$B = 254$, $C = 1,975$	Difference	$B = 254$, $C = 1,423$	Difference
${}^3F$	0	−82	+ 82	192	− 192
${}^3P$	3,529	3,728	− 199	4,002	− 473
${}^1D$	4,380	5,138	− 758	4,308	+ 72
${}^1G$	7,335	6,916	+ 419	6,086	+1,247
${}^1S$	12,420	(19,331)	(−6,911)	15,741	−3,321

Our second example is Zr III. This has the ground configuration obtained by omitting the two $5s$ electrons from Zr I. We leave the ${}^1S$ out of discussion (in any case its position seems a little doubtful)[1] and this time determine the parameters a, B and C first with the weighting factor 1 and then $(2S+1)(2L+1)$. This gives the parameters in Table 4.10 and we see they are not very different in the two cases.

Table 4.10. *Parameters obtained by fitting* $4d^2$ *of* Zr III *using the two alternative statistical weights*

	$f = 1$	$f = (2L+1)(2S+1)$
a	−326	−110
B	558	539
C	1557	1640

When we try fitting $d^3, \ldots, d^7$ some of the theoretical formulae involve square roots so we do not have simple expressions for B and C in terms of the observed energies. As an example we take V III whose ground term is $1s^2\, 2s^2\, 2p^6\, 3s^2\, 3p^6\, 3d^3$. This was fitted by Many (1946) and he found $B = 766\,\text{cm}^{-1}$, $C = 2855\,\text{cm}^{-1}$. The comparison with experiment is shown in Table 4.11 and Fig. 4.2 using the energies calculated by Many. We see that the fit is fairly good except for ${}^2P$. ${}^2P$ is so bad that Many did not include it in his calculation. ${}^2P$ is also in bad disagreement with the theoretical formulae in the isoelectronic Cr IV and is also lower than ${}^2H$ (with which it 'should' be degenerate) in Mn V and Fe VI. It is not known in the higher members of this isoelectronic sequence. These facts are not easy to explain although we may make the same remark as we did about ${}^1S$ of d^2. Also an inclusion of a polarization correction[2] does lead to a separation of ${}^2P$ and ${}^2H$.

[1] Compare Moore (1952) with Kiess (1956).
[2] For some discussion of this problem, see Racah (1955).

Our three examples of Zr I, Zr III and V III, which are typical, show the kind of agreement with experiment that we ought to expect when we use a theory of a similar kind to calculate energies of ions in compounds. One can obtain better agreement by taking configuration interaction or polarization into account. The polarization correction is most simply introduced by adding $\alpha L(L+1)$ to the theoretical formulae. α is then determined by fitting. For example, it comes to 69 cm^{-1} in Mn II (Trees, 1951) and this is a fairly typical value. Other values are given in the references to Appendix 6.

Table 4.11. *Comparison of theoretical formulae with experiment for $3d^3$ of* V III *with $B = 766$ cm^{-1}, $C = 2855$ cm^{-1}*

	Experiment	Calculated	Difference
^4F	336	387	− 51
^4P	11,668	11,873	− 205
^2G	12,089	12,015	+ 74
^2D	16,317	17,197	− 880
^2H	16,907	15,844	+1,063
^2P	11,327	15,844	−4,517
^2F	—	27,329	—
^2D	—	42,756	—

For the first transition series a considerable collection of values of B and C have been obtained by fitting to experimental data. I give in Appendix 6 the values which appear to me most reliable for atoms from titanium to copper together with their ions up to the fourth ionization. There is much less data for the second and third series (see Appendix 6 for what is available). It appears that B is approximately half as large in the second and third series as it is in the first. The values of C in the later series are a little discordant but, at least at the beginning of the series, do not seem to decrease quite as much as those of B relative to the first series.

It is interesting to plot the value of B for the first series against the atomic number of the element. This is done in Fig. 4.5. B is much more sensitive to degree of ionization than C and a plot of C is not very illuminating. If we write n for the position of the element in its period (i.e. $n = Z-18$) and m for its degree of ionization (i.e. $m = 0$ for a neutral atom) then the formula

$$B = 145+80n+95m, \tag{4.79}$$

gives a good estimate.[1] C is roughly given by

$$\left.\begin{aligned} C &= 520+370n, \quad m = 0, 1, \\ C &= 1400+300n, \quad m = 2. \end{aligned}\right\} \tag{4.79a}$$

Our results can be used to explain the form of the variation of ionization potential with atomic number in ions. We do this for the divalent ions of the first transition series because they all have d^n as their ground configuration and

[1] See also Racah (1954) and Racah & Shadmi (1959), *Rull. Res. Counc. Israel*, *8F*, 15.

adequate experimental data is available. Similar discussions would apply to other ions. The energy of the d^n shell of one of these ions is given by the sum of its own electrostatic energy and of its energy in the field of the core. Suppose a single d electron has energy $-U$ in the field of the core. Then the total energy of the d^n shell is easily written down (half of the relevant energies are in Table 4.6) and is given for each value of n in Table 4.12, column 2. The ionization potentials are then just the differences in energy between successive ions. The theoretical formulae are given in column 3 of Table 4.12 and we have set $\phi = U-(n-1)A$.

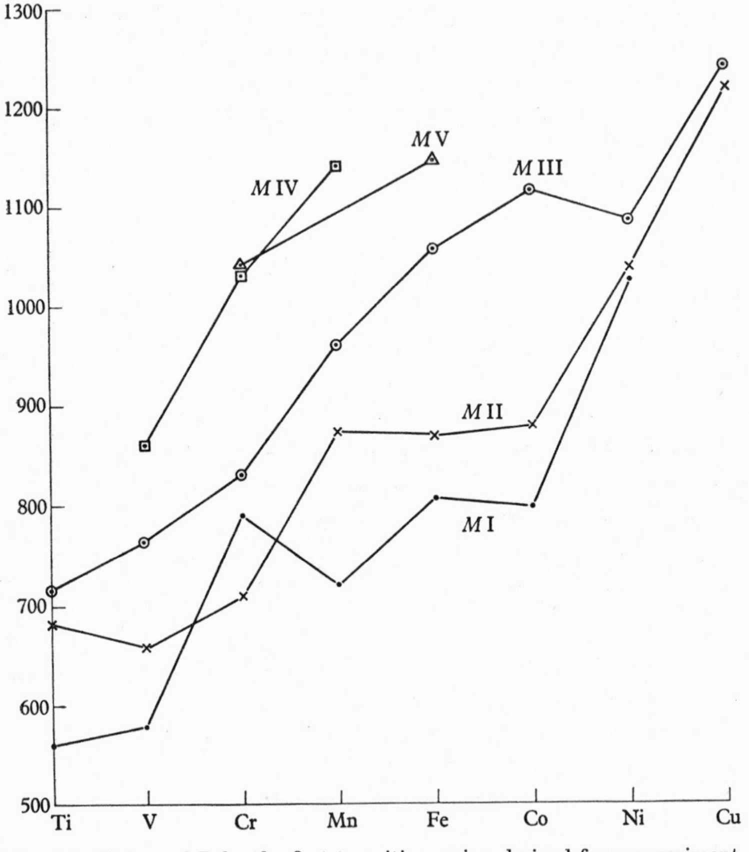

Fig. 4.5. Values of B for the first transition series, derived from experiment.

The clearest way to compare our theoretical formulae with experiment is to note that ϕ ought to change smoothly with increasing n (A and B, as well as n itself, change as we change n). Then we use the experimental values of the ionization potentials together with the values of B and C given in Appendix 6 to deduce ϕ ((4.79) and (4.79a) were used for Zn^{++}). The ionization potentials and ϕ derived from them are plotted in Fig. 4.6 and we see how well our hope is borne out. The main irregularity in the ionization potentials is to be correlated algebraically with the sudden appearance of $7C$ in the theoretical formulae at the same place as the disappearance of B. Physically we say that the ions up to d^5 are being progressively stabilized by exchange energy because the electrons all

Table 4.12. *Theoretical energies and ionization potentials for d^n ions*

$$\phi = U-(n-1)A$$

n	$E(d^n)$	$E(d^{n-1})-E(d^n)$
0	0	—
1	$-U$	ϕ
2	$-2U+A-8B$	$\phi+8B$
3	$-3U+3A-15B$	$\phi+7B$
4	$-4U+6A-21B$	$\phi+6B$
5	$-5U+10A-35B$	$\phi+14B$
6	$-6U+15A-35B+7C$	$\phi-7C$
7	$-7U+21A-43B+14C$	$\phi+8B-7C$
8	$-8U+28A-50B+21C$	$\phi+7B-7C$
9	$-9U+36A-56B+28C$	$\phi+6B-7C$
10	$-10U+45A-70B+35C$	$\phi+14B-7C$

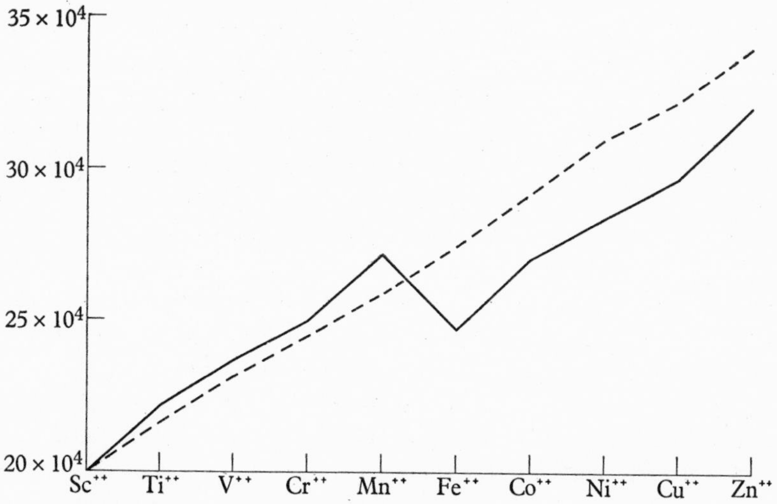

Fig. 4.6. The observed ionization potentials (——) for divalent ions of the first transition series and the corrected ionization potentials ϕ (----) derived from them.

go in with their spins parallel. But at d^5 the d shell is half full so that the next electron goes in with its spin antiparallel to those which are already there and hence the ion is less stable relative to d^5 than one might otherwise have expected. One should emphasize that the parameters U, A, B and C vary from element to element and the overall increase of ionization potential along the series is due mainly to the influence of the increased nuclear charge on U. Finally, the varying contribution of B is also clearly reflected in the experimental data.

Examples

1. Assuming, as in § 6, that formulae for d^n can be worked out correctly in terms of the separations for d^2, show that the $L(L+1)$ correction holds exactly for all d^n configurations if and only if

$$35E(^3F)+14E(^1S) = 40E(^1D)+9E(^1G)$$

holds for d^2.

2. Using the polarization correction, fit the four lowest terms of Zr I ($4d^2$) and show that the ^1S then deviates by 4808 cm^{-1} from its theoretical value. ($\alpha = 70$ for Zr I and 91 for Zr III by this method.)

4.8. Radial functions and the Hartree–Fock method

An alternative method of obtaining the parameters describing electrostatic interactions in atoms is by taking suitable radial functions R_{nl} for the one-electron orbitals and working the R^k out directly from their definitions (through (4.40)). Naturally this method requires a previous knowledge of the radial functions. The most important non-empirical way of obtaining the latter is the Hartree–Fock method. This involves determining, by an iterative procedure, the best approximate solution of a particularly simple type for the Schrödinger equation.

I will outline how this is done for a special case which in fact includes most of the cases which interest us. We search for that determinantal function which is the best approximate solution. Many of the functions we have discussed so far were not themselves determinantal functions but merely finite sums of them. However, providing Hund's rule is true, the ground term always contains at least one state which is itself a determinantal function to the approximation we have used so far (neglecting configuration interaction). Therefore we can always obtain an approximate eigenfunction of the ground term in this way. Similarly for any other configuration we can obtain an approximate eigenfunction of at least one term in it. If, then, we use the same set of radial functions for each term of a configuration we have obtained the radial functions for all terms of the configuration. This does involve an extra assumption, however. We relied on it in the last section when we sought the best B and C to fit the formulae for a configuration: we kept B and C constant within a configuration. So the particular case is not so restrictive after all.

To obtain the best single-determinant solutions of Schrödinger's equation

$$\mathscr{H}\psi = \left\{\sum_{i=1}^{n} U(i) + \sum_{i>j=1}^{n} V(ij)\right\}\psi = E\psi, \tag{4.80}$$

we put

$$\psi = (n!)^{-\frac{1}{2}}\Sigma(-1)^{\mu} P_{\mu}\{\phi_1(1)\ldots\phi_n(n)\}. \tag{4.81}$$

Then it is possible to show[1] that we obtain them by minimizing $E = \langle\bar{\psi}\mathscr{H}\psi\rangle$ subject to the conditions $\langle\bar{\phi}_i\phi_j\rangle = \delta_{ij}$. Then, because one can vary the real or imaginary parts of ϕ_i separately, we obtain, using the methods of reduction shown in § 3,

$$\left.\begin{aligned} 0 = \langle\delta\bar{\psi}\,|\,\mathscr{H}\,|\,\psi\rangle &= \sum_i\langle\delta\bar{\phi}_i U\phi_i\rangle + \sum_{i\neq j}\langle\delta\bar{\phi}_i\bar{\phi}_j\,|\,V\,|\,\phi_i\phi_j\rangle \\ &\qquad - \sum_{i\neq j}\langle\delta\bar{\phi}_i\bar{\phi}_j\,|\,V\,|\,\phi_j\phi_i\rangle, \\ \langle\delta\bar{\phi}_i\phi_j\rangle &= 0, \end{aligned}\right\} \tag{4.82}$$

[1] See Hartree (1957) for this and for a complete account of the Hartree–Fock method.

whence by Lagrange's method of undetermined multipliers

$$U\phi_i\rangle + \sum_{j \neq i} \langle \overline{\phi}_j | V | \phi_j \rangle \phi_i \rangle - \sum_{j \neq i} \langle \overline{\phi}_j | V | \phi_i \rangle \phi_j \rangle = \sum_j \epsilon_{ij} \phi_j \rangle. \qquad (4.83)$$

Equations (4.83) are called the Hartree–Fock equations and could be solved by an iterative procedure to give the best possible solutions of (4.80) of the form (4.81). However, in practice one makes two simplifications. The coefficients of $|\phi_i\rangle$ and $|\phi_j\rangle$ in (4.83) are not necessarily spherically symmetrical. This lack of spherical symmetry is unlikely to make a great deal of difference to the accuracy of the solutions and so at each stage of iteration one averages these coefficients over the angular space (the individual functions ϕ_j of course are not taken to be spherically symmetric in general). The second simplification is made less frequently and is to neglect the exchange term in (4.83). Then $\epsilon_{ij} = 0$ when $i \neq j$ and we have Hartree's equations

$$U\phi_i\rangle + \sum_{j \neq i} \langle \overline{\phi}_j | \frac{e^2}{r_{12}} | \phi_j \rangle \phi_i \rangle = \epsilon_{ii} \phi_i \rangle. \qquad (4.84)$$

In (4.84) we simply solve the one-electron problem for ϕ_i in the field of the nucleus and the average field due to all the other electrons. Energies and eigen-

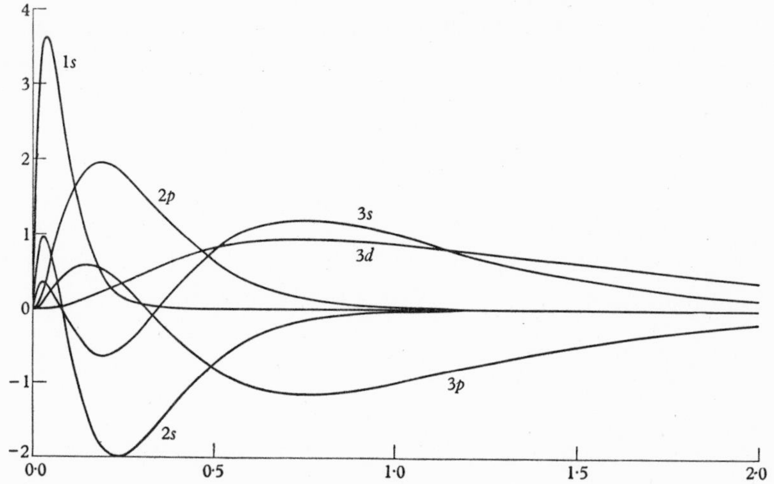

Fig. 4.7. Hartree–Fock radial functions χ for Mn^{++}. r is in units of $a = \hbar^2/me^2 = 0{\cdot}528$ Å.

functions determined by (4.84) do not appear to differ greatly from those determined by (4.83). Both methods give 'best possible' absolute energies for atoms which differ from the experimental ones (and from accurate solutions of (4.80) when those exist) by about $\frac{1}{2}$ e.v. per electron. Relative energies within and between configurations are usually considerably better, however.

In Fig. 4.7 the radial functions χ (of (4.11)) obtained by Hartree (1955) for Mn^{++} by the Hartree–Fock method are plotted. We see the separation into electron shells; as we shall see in a moment, neutron diffraction measurements suggest a rather more diffuse $3d$ orbital.

In very approximate work it is a nuisance to have to use the Hartree-type orbitals because they only come in the form of numerical tables. Slater, therefore, proposed using functions of the simple form[1]

$$\phi = N r^{n^*-1} \exp\left[-(Z-\sigma)\frac{r}{n^* a}\right] Y_{lp} \tau_s \tag{4.85}$$

which are solutions of the one-electron problem with a potential energy

$$V(r) = -\frac{(Z-\sigma)e^2}{r} + \frac{n^*(n^*-1)\hbar^2}{2mr^2} \tag{4.86}$$

and $a = \hbar^2/me^2$. Equation (4.84) justifies the consideration of such an approximation. Slater gave the following rules to determine n^* and σ. n^* is equal to n for $n = 1, 2, 3$ but for $n = 4, 5, 6$ is 3·7, 4·0 and 4·2, respectively. Z is the nuclear charge and σ effectively a shielding constant determined by dividing the electrons into groups: $1s$; $2s, 2p$; $3s, 3p$; $3d$; $4s, 4p$; $4d$; $4f$; $5s, 5p$; $5d$; and taking contributions to σ for a given orbital:

(i) None from an orbital in a group outside the one considered.

(ii) 0·35 from each other occupied orbital in the group (0·30 for $1s$).

(iii) 1·00 from each orbital inside a d or f one. For s or p ones 0·85 for each with n less by 1 and 1·00 for each orbital further in.

As an example we may consider the Mn^{++} ion which has $1s^2\,2s^2\,2p^6\,3s^2\,3p^6\,3d^5$; ^6S as its ground state. The (unnormalized) orbitals determined by Slater's rules are

$$\left.\begin{aligned} \phi_{1s} &= e^{-24\cdot7/a}, & \phi_{2s} &= re^{-10\cdot43r/a}, \\ \phi_{2p} &= re^{-10\cdot43r/a}Y_{1q}, & \phi_{3s} &= r^2e^{-4\cdot58r/a}, \\ \phi_{3p} &= r^2e^{-4\cdot58r/a}Y_{1q}, & \phi_{3d} &= r^2e^{-1\cdot87r/a}Y_{2q}. \end{aligned}\right\} \tag{4.87}$$

The maximum probability for a Slater orbital considered as a function of the radius is obtained by differentiating $f(r) = \int \phi^2 r^2 \sin\theta\, d\theta\, d\phi$ and is

$$r_{\text{max.}} = \frac{an^{*2}}{Z-\sigma}. \tag{4.88}$$

Table 4.13. *Comparison of Hartree–Fock and Slater* $r_{\text{max.}}$ *for* Mn^{++}

(The units are $a = \hbar^2/me^2 = 0\cdot528$ Å)

Orbital	$1s$	$2s$	$2p$	$3s$	$3p$	$3d$
Hartree–Fock	0·04	0·23	0·19	0·78	0·78	0·78
Slater	0·04	0·19	0·19	0·65	0·65	1·64

It is interesting to compare the values for $r_{\text{max.}}$ given by (4.88) with those obtained from Hartree's data for Mn^{++} (see Table 4.13). Slater's functions do not contain radial nodes, yet there is fair agreement on the whole. The $3d$ function differs most. For $3d$ in Mn^{++} an experimental value of 1·25 has been obtained from neutron diffraction experiments (Shull, Straussen & Wollan (1951)).

[1] For a recent discussion of Slater functions for transition-metal ions see Brown (1958).

Table 4.14. *The approximate constancy of the products of calculated mean radii, $\bar{r}$, with the electrostatic parameter B or C. Interpolated values of $\bar{r}$ are in brackets (after Hartree* (1956))

	$\bar{r}$	$\bar{r}B$	$\bar{r}C$		$\bar{r}$	$\bar{r}B$	$\bar{r}C$
Ti^+	1·645	1122	4081	Ti^{++}	(1·44)	1034	3786
V^+	(1·46)	962	3529	V^{++}	1·28	980	3654
Cr^+	(1·33)	944	3711	Cr^{++}	(1·17)	971	4013
Mn^+	1·248	1090	3906	Mn^{++}	1·082	1039	3598
Fe^+	(1·17)	1017	4256	Fe^{++}	(1·02)	1079	3979
Co^+	(1·095)	961	4192	Co^{++}	(0·97)	1082	4235
Ni^+	(1·035)	1073	4465	Ni^{++}	(0·925)	1003	4469
Cu^+	0·979	1190	4645	Cu^{++}	(0·89)	1102	4147

Hartree finds that calculated atomic orbitals with the same n and l in different atoms and ions tend to be very nearly the same except for a scale factor. It follows from this that the Slater–Condon parameters between orbitals with the same n and l should be inversely proportional to the mean radii of an nl orbital. This follows immediately on substituting $R'_{nl}(r) = R_{nl}(\lambda r)$ in the integrals defining the F_k and $\bar{r}$. So $\bar{r}F_k$ would be constant. Hartree (1956) gave a table showing that this was roughly true for $\bar{r}B = \bar{r}(F_2 - 5F_4)$ when calculated $\bar{r}$ and empirical B are used. We show this for B and C in Table 4.14 using Hartree's $\bar{r}$ (the figures in brackets were interpolated) and the B and C from Appendix 6. The agreement is a little better for B than for C.

CHAPTER 5

MAGNETIC EFFECTS IN ATOMIC STRUCTURE

5.1. Spin-orbit coupling

In the previous chapter we met and used the concept of spin for the electron. However, although the spin influenced the energies and allowed eigenstates by introducing an internal degree of freedom for the electron, there were no terms in the Hamiltonian showing an explicit dependence upon it. In a more accurate theory such terms appear as corrections to our previous Hamiltonian and are often small. In the present chapter we give the theory of these corrections and shall see that they arise physically because the electron possesses a magnetic moment. There are two rather different ways of introducing the spin and magnetic moment of the electron into the theory.[1] One is phenomenological: an appeal is made to experiment to show first that the electron has two independent internal states and secondly that it has a magnetic moment. The other proceeds via Dirac's linear relativistic wave-equation for the electron and deduces both these properties from that equation. The latter approach is more satisfactory from a fundamental point of view, but is mathematically a little more difficult. I shall describe it later in this chapter, but for the convenience of the reader who does not wish to work through that theory I use the former method first and describe the specific magnetic effects in atomic structure from a rather empirical basis.

Experimentally the electron has a magnetic moment of magnitude $e\hbar/2mc$. Because the electron has a negative charge it would be natural to expect, from classical physics, that if this moment arises from a rotation of the negative charge then the magnetic moment would lie in a direction directly opposed to the vector representing the rotation. In other words the magnetic moment $\boldsymbol{\mu}$ would be given by $-e/(mc)\,\mathbf{s}$. This is in agreement with experiment and also with the Dirac theory of the electron (although see p. 121).

Now let us discuss the energy of interaction of this magnetic moment with the electric field of a nucleus. Suppose the electron to be moving around the nucleus with an angular velocity $\boldsymbol{\omega}$. An observer standing on the electron would see a positively charged nucleus moving round him also with angular velocity $\boldsymbol{\omega}$. This motion of the nucleus would set up a magnetic field at the electron in the direction of $\boldsymbol{\omega}$ and hence there is a different energy for different orientations of the magnetic moment of the electron relative to $\boldsymbol{\omega}$. Clearly the system has its lowest energy when $\boldsymbol{\mu}$, the magnetic moment, is parallel to $\boldsymbol{\omega}$ and hence also to $\mathbf{l}$ the orbital angular momentum of the electron. Since, as we have just seen, the

[1] A third way of introducing the spin appears in § 6.9.

spin angular momentum vector is directly opposed to the magnetic moment vector this means that there is a coupling between the spin and orbital moments of an electron in an atomic system, the sign of the coupling constant being positive.

Strangely enough, the actual value of this coupling constant cannot be easily obtained by purely classical arguments. As we see later, however, one can obtain it from Dirac's relativistic equation and it is actually, for a central field,

$$\delta\mathscr{H} = -\frac{e}{2m^2c^2r}\frac{dA_0}{dr}\mathbf{l}.\mathbf{s} = \xi(r)\,\mathbf{l}.\mathbf{s}, \quad \text{say,}$$

where A_0 is the electrostatic potential due to the central field. A_0 is equal to Zer^{-1} for a nucleus of charge Ze when there are no other electrons present. It then follows that dA_0/dr is negative everywhere and this shows that the sign of the coupling constant is indeed positive. For any reasonable choice of a central field for an atom one again finds that $\delta\mathscr{H}$ is positive.

Our discussion so far has referred only to one electron in a central field. The generalization to n electrons is obtained by simply using the same formula for each one individually and summing over the electrons. This gives us

$$\delta\mathscr{H} = \sum_{\kappa=1}^{n} \xi(r_\kappa)\,\mathbf{l}_\kappa.\mathbf{s}_\kappa. \tag{5.1}$$

This, then, is the desired addition to the Hamiltonian (2.1) when we include the magnetic interaction between the electron moment and the electric field of the nucleus. First, let us determine certain selection rules for the quantity $\delta\mathscr{H}$. It is a sum of products, a particular product being $\xi(r_\kappa)\mathbf{l}_\kappa.\mathbf{s}_\kappa$. As $\xi(r_\kappa)\mathbf{l}_\kappa$ and $\mathbf{s}_\kappa$ are each of type T with respect to the angular momentum vector $\mathbf{J}$, it follows from the results of § 2.8 that $\delta\mathscr{H}$ commutes with all the components of $\mathbf{J}$. This means that $\mathbf{J}^2$ and J_z remain good quantum numbers for the atomic system even when we include the spin-orbit coupling. This is really only to be expected because the system still retains its spherical symmetry and the components of $\mathbf{J}$ can be interpreted as infinitesimal operators of rotation for the system (see § 6.9). Another way of putting this is to say that we have the selection rules $\Delta J = \Delta M_J = 0$ for J.

$\mathbf{J}$ is the only one of our classifying vectors which commutes rigorously with the Hamiltonian for an n-electron system with $n > 1$ when we include the spin-orbit coupling. However, as we anticipated to remark in the last chapter, L and S are usually 'quite good quantum numbers'. So we also ask what selection rules are satisfied by $\delta\mathscr{H}$ in a Russell–Saunders coupling scheme. The best way to do this is to take an SLM_SM_L scheme and expand an arbitrary matrix element of $\delta\mathscr{H}$ as a sum of products of matrix elements of $\xi(r_\kappa)\mathbf{l}_\kappa$ and $\mathbf{s}_\kappa$ separately. This gives us

$$\langle\alpha''SLM_SM_L\,|\,\delta\mathscr{H}\,|\,\alpha'S'L'M'_SM'_L\rangle = \sum_{\kappa=1}^{n}\sum_{\alpha'''}\langle\alpha''SLM_SM_L\,|\,\xi(r_\kappa)\mathbf{l}_\kappa\,|\,\alpha'''SL'M_SM'_L\rangle. \\ \times\langle\alpha'''SL'M_SM'_L\,|\,\mathbf{s}_\kappa\,|\,\alpha'S'L'M'_SM'_L\rangle. \tag{5.2}$$

Using the selection rules for vectors of type T with respect to angular momenta that we found in § 2.8 and combining our results with the findings of the previous paragraph, we obtain the list $\Delta L = 0, \pm 1$, $\Delta S = 0, \pm 1$, $\Delta J = 0$, $\Delta M = 0$ for the spin-orbit coupling matrix elements in a Russell–Saunders coupling scheme.

One should not think that $\delta\mathscr{H}$ represents the only modification to the Hamiltonian which is necessary to give a completely exact description of an atomic system. We have at the present stage omitted a number of other effects. We have neglected to include the mutual magnetic interactions between the spin magnetic moments of different electrons and also those arising from the magnetic moments associated with the orbital motions of the electrons. Such interactions would give rise to terms in the Hamiltonian depending on $\mathbf{l}_\kappa . \mathbf{l}_\lambda$, $\mathbf{l}_\kappa . \mathbf{s}_\lambda$ and on $\mathbf{s}_\kappa . \mathbf{s}_\lambda$. They are always, or at least almost always, much smaller than the spin-orbit coupling energy in the atoms in which we are interested. This is because our atoms have a Z which is considerably larger than 1 and each of these effects is proportional to the product of the charges of the two particles taking part. Hence the interaction with the nucleus, i.e. the spin-orbit coupling energy, is larger than these other interactions.[1] Another kind of term which we have neglected completely so far arises from the possibility that the nucleus itself may possess a magnetic moment. In fact it is also the case that nuclei are not strictly point charges but have an internal structure which gives rise for many nuclei to an electrostatic potential which is not spherically symmetrical. Both these nuclear effects, however, always give rise to very small energies. In spite of this they do sometimes produce important experimental consequences. For this reason we consider them again in § 5.5 and in ch. 12, although we may safely neglect them at the moment in our discussion of the gross features of atomic structure.

We conclude this section by showing that the matrix elements of $\delta\mathscr{H}$ assume a specially simple form when they are evaluated within a single Russell–Saunders term. In this case (5.2) gives

$$\langle \alpha' SLM_S M_L \,|\, \delta\mathscr{H} \,|\, \alpha' SLM'_S M'_L \rangle$$

$$= \sum_{\kappa=1}^{n} \sum_{\alpha''} \langle \alpha' SLM_S M_L \,|\, \xi(r_\kappa)\,\mathbf{l}_\kappa \,|\, \alpha'' SLM_S M'_L \rangle . \langle \alpha'' SLM_S M'_L \,|\, \mathbf{s}_\kappa \,|\, \alpha' SLM'_S M'_L \rangle. \quad (5.3)$$

We simplify (5.3) by using the corollary to the replacement theorem derived in § 2.8, combined with (2.73) which shows that the first of the two matrix elements in any one of the terms of (5.3) is independent of M_S and the second of M_L. Our matrix elements of $\delta\mathscr{H}$ then become

$$\sum_{\kappa=1}^{n} \sum_{\alpha''} \gamma_L(\alpha', \alpha'', S, L)\, \gamma_S(\alpha'', \alpha', S, L) \langle \alpha' SLM_S M_L \,|\, \mathbf{L} \,|\, \alpha' SLM_S M'_L \rangle$$
$$\times \langle \alpha' SLM_S M'_L \,|\, \mathbf{S} \,|\, \alpha' SLM'_S M'_L \rangle$$
$$= \lambda \langle \alpha' SLM_S M_L \,|\, \mathbf{L}.\mathbf{S} \,|\, \alpha' SLM'_S M'_L \rangle, \quad (5.4)$$

where $$\lambda = \lambda(\alpha', \alpha'', S, L) = \sum_{\kappa=1}^{n} \sum_{\alpha''} \gamma_L(\alpha', \alpha'', S, L)\, \gamma_S(\alpha'', \alpha', S, L)$$

[1] This is not strictly true. Actually by taking a general central field for each electron to move in rather than Zer^{-1} we include the main part of the influence of the other electrons. The shielding effect of the inner electrons reduces the spin-orbit coupling constant calculated for the bare nucleus.

is just a number independent of M_S, M_L, M'_S and M'_L. Having obtained our relationship it is obviously more convenient to transform back to the $SLJM$ scheme. We do this by expanding a typical ket $|\alpha' SLJM\rangle$ in that scheme in terms of the kets of the $SLM_S M_L$ scheme. Then we have

$$|\alpha' SLJM\rangle = \Sigma |\alpha' SLM_S M_L\rangle \langle \alpha' SLM_S M_L \,|\, \alpha' SLJM\rangle$$

and use the relationship of (5.3) and (5.4) and then transform back to the $SLJM$ scheme to obtain

$$\begin{aligned}\langle \alpha' SLJM \,|\, \delta\mathscr{H} \,|\, \alpha' SLJM\rangle &= \lambda \langle \alpha' SLJM \,|\, \mathbf{L}.\mathbf{S} \,|\, \alpha' SLJM\rangle \\ &= \tfrac{1}{2}\lambda[J(J+1) - L(L+1) - S(S+1)]. \qquad (5.5)\end{aligned}$$

Apart from the constant λ which depends only on the particular term we have chosen, (5.5) gives us the complete matrix of $\delta\mathscr{H}$ within a Russell–Saunders term (we remember that $\delta\mathscr{H}$ is diagonal in J and M). The difference in energy between the level with J and the level with $J-1$ is seen immediately to be λJ. This is the mathematical expression of what is known as the Landé interval rule, which states that the separation between two adjacent levels of a term is proportional to the higher J value. It is named after Landé who discovered the rule from an extensive analysis of experimental data before modern quantum theory arrived on the scene to give the justification for his rule.

Example

Show that the parity operator P still commutes with the Hamiltonian after the spin-orbit coupling has been added. The classification into even and odd states still holds rigorously, therefore.

5.2. Examples of calculations of spin-orbit coupling energies

We have obtained the general form of the spin-orbit coupling energy when the latter is regarded as a small perturbation on the Russell–Saunders terms. Its matrix elements are proportional to those of $\lambda \mathbf{L}.\mathbf{S}$, where λ is a constant for any given term but may vary from one term to another.

Now the spin-orbit energy (5.1) is a sum of one-electron operators and so (by the discussion of § 4.3) we deduce that for any determinantal function

$$\left.\begin{aligned}\psi &= |\phi_1(1) \dots \phi_n(n)\rangle, \\ \langle \overline{\psi} \,|\, \delta\mathscr{H} \,|\, \psi\rangle &= \sum_{\kappa=1}^{n} \langle \overline{\phi}_\kappa \,|\, \xi(r)\mathbf{l}.\mathbf{s} \,|\, \phi_\kappa\rangle.\end{aligned}\right\} \qquad (5.6)$$

If, as usual, each ϕ_κ is expressed as a simple product $R_{nl} Y_{lp} \tau_s$ of a radial part, an angular part and a spin part then the integrals in (5.6) over angle and spin can be performed in any particular case and we are left with the radial integral

$$\int R_{nl}^2 \xi(r)\, r^2 dr$$

which cannot be evaluated without a knowledge of R_{nl} and ξ. It is customary to set this latter integral equal to $\hbar^{-2}\zeta_{nl}$, where ζ_{nl} is called the spin-orbit coupling constant for an nl orbital. These parameters ζ_{nl} play the part in the theory of spin-orbit coupling that the Slater–Condon parameters play in the theory of electrostatic energies. ζ_{nl} depends on n and l but not on p and s and is positive.

Our general problem is to determine λ in terms of these more basic quantities ζ_{nl}. However, let us first consider a simple example. If we have one p electron then $l = 1$ and $s = \frac{1}{2}$ so $j = \frac{1}{2}$ or $\frac{3}{2}$ and there are the two levels $^2P_{\frac{1}{2}}$ and $^2P_{\frac{3}{2}}$ of the ^2P term. The six basic states of ^2P may be taken as

$$\psi_1 = |1^+\rangle, \quad \psi_3 = |0^+\rangle, \quad \psi_5 = |-1^+\rangle,$$
$$\psi_2 = |1^-\rangle, \quad \psi_4 = |0^-\rangle, \quad \psi_6 = |-1^-\rangle,$$

where the numbers give the m_l values and + and − refer to the spin. ψ_1 is the only state having $m_j = 1\frac{1}{2}$ and so it belongs to $^2P_{\frac{3}{2}}$. Its energy is

$$\begin{aligned} E(^2P_{\frac{3}{2}}) &= \langle 1^+ \,|\, \xi(r)\,\mathbf{l}.\mathbf{s} \,|\, 1^+\rangle \\ &= \langle 1^+ \,|\, \xi(r)\{\tfrac{1}{2}l^+s^- + \tfrac{1}{2}l^-s^+ + l_z s_z\} \,|\, 1^+\rangle \\ &= \tfrac{1}{2}\hbar^2 \langle 1^+ \,|\, \xi(r) \,|\, 1^+\rangle = \tfrac{1}{2}\zeta_p, \end{aligned} \tag{5.7}$$

where ζ_p is the spin-orbit parameter for the p electron. Both ψ_2 and ψ_3 have $m_j = \frac{1}{2}$ and so, by the diagonal-sum rule, the sum of their energies must equal $E(^2P_{\frac{3}{2}}) + E(^2P_{\frac{1}{2}})$. In fact

$$E(\psi_2) = -\tfrac{1}{2}\zeta_p, \quad E(\psi_3) = 0,$$

whence using (5.7)

$$E(^2P_{\frac{1}{2}}) = -\zeta_p. \tag{5.8}$$

A number of points arise out of this calculation. First, we notice that the weighted mean of the energies $E(^2P_{\frac{1}{2}})$ and $E(^2P_{\frac{3}{2}})$ is zero and confirm from (5.5) that this is true for the spin-orbit energy of any term in the present approximation. Next, as ζ_p is positive, the lower j value lies lower.

It is clear that, just as in the electrostatic case, closed shells may be disregarded so long as we consider only relative energies and stay within a configuration. As an example of this and of our calculation for a single p electron we remark that in B III ($1s^2\,2p$) the $^2P_{\frac{3}{2}}$ lies 34·1 cm^{-1} above the $^2P_{\frac{1}{2}}$ which makes $\zeta_p = 22{\cdot}7$ cm^{-1} for B III. This is said to be the normal order for levels in a term. It corresponds to a positive λ in (5.5) and usually occurs for terms from a shell which is less than half full.[1] λ is usually negative for shells which are more than half full and then the term is said to have its levels inverted.

Formula (5.5) for ^2P gives

$$E(^2P_{\frac{1}{2}}) = -\lambda, \quad E(^2P_{\frac{3}{2}}) = \tfrac{1}{2}\lambda$$

so (5.7) and (5.8) show that $\lambda = \zeta$. This relation always holds when there is only one electron as one immediately verifies directly.

As a second example we take the ^3P ground term of $2p^2$. Making use of (5.5) we have only to determine the energy of one suitable state in terms both of ζ_p and of λ. This then gives λ in terms of ζ_p and (5.5) gives the energies of all the levels. As before we select the highest M_J value first. This has $M_J = 2$ and the corre-

[1] There are exceptions to this rule. Glaring examples occur in some terms of the alkali and alkaline earth spectra and are almost certainly connected with the inadequacy of neglecting configuration interactions involving core excitations. See Phillips (1933).

sponding function is $\psi = |1^+0^+\rangle$. Its energy, worked out in the two ways, is $\frac{1}{2}\zeta_p = \lambda$ so the levels have energies $E(^3P_0) = -2\zeta_p$, $E(^3P_1) = -\zeta_p$, $E(^3P_2) = \zeta_p$. In C III they are -52, -29 and $28\,\text{cm}^{-1}$, respectively, referred to the centre of gravity.

Passing now to the ground term, ^{2S+1}L say, of a general l^n configuration, with $n \leqslant 2l+1$, the function $|l^+, (l-1)^+, \ldots, (l-n+1)^+\rangle$ has the highest M_J value. Dropping the suffix on ζ, its energy is $\frac{1}{2}L\zeta = \lambda LS$ and so $\lambda = \zeta/2S$. A similar argument for a more-than-half-filled shell yields $\lambda = -\zeta/2S$, and explains why the levels in these latter terms are in inverted order.

A slightly more complicated example is furnished by the ^3P term of d^2. Again we take the highest M_J, which is 2, and arises on the one hand from $^3F_2$, $^3F_3$, $^3F_4$, $^3P_2$, ^1D and ^1G and on the other from $|1^+0^+\rangle$, $|2^+-1^+\rangle$, $|2^+0^-\rangle$, $|2^-0^+\rangle$, $|1^+1^-\rangle$, $|2^-1^-\rangle$. Then we find the equation

$$E(^3F_2)+E(^3F_3)+E(^3F_4)+E(^3P_2) = -\tfrac{1}{2}\zeta$$

which, combined with (5.5) and $\lambda(^3F) = \frac{1}{4}\zeta$ which we found above, yields $\lambda(^3P) = \frac{1}{2}\zeta$. We expect, therefore, a ratio of 7 : 3 for the total spreads of the ^3F and ^3P terms of d^2. In V IV it is actually 730 : 332.

The reader will have noticed that in calculating the energies of levels in a configuration we diagonalize the electrostatic energy first and then apply the spin-orbit energy as a perturbation. This is because in cases of interest to us the spin-orbit energy is quite a bit smaller than the electrostatic energy. This is not the case for all configurations, however, and so the alternative procedure of diagonalizing the spin-orbit energy first and then considering the electrostatic energy by applying it through first-order perturbation theory is sometimes useful. This latter procedure is called using a jj-coupling scheme and configurations for which one can obtain reasonable agreement with experiment by this procedure are said to obey jj-coupling. It will often be the case that, for a given atom, some configurations obey SL-coupling and others jj-coupling. It is clear, however, that if we diagonalize both the electrostatic energy and the spin-orbit energy within a configuration we obtain the same result whichever first-order functions we start from. In this latter case we say that we are making calculations in intermediate coupling. We use the expression 'coupling scheme' to indicate the kind of basic functions we use, not to indicate the suitability of using those functions from the point of view of obtaining good first-order agreement with experiment. Thus C III which we mentioned earlier has a very small ζ compared with F_2 for the configuration $1s^2\,2p^2$ so we say it obeys SL-coupling. We would be sensible, therefore, to use an SL-coupling scheme but we could if we wished use a jj-coupling scheme with basic states from the three jj-coupling configurations $(2p_{\frac{1}{2}})^2$, $(2p_{\frac{1}{2}})^1\,(2p_{\frac{3}{2}})^1$ and $(2p_{\frac{3}{2}})^2$. The jj-coupling scheme would then give a very bad first-order agreement with experiment and we would need to take into account the off-diagonal electrostatic matrix elements. This logical distinction between an atomic system obeying a coupling scheme and the mere use of that coupling scheme for the system will recur when we come to contrast the weak-field, strong-field and rare-earth coupling schemes for ions in compounds in § 7.4.

Of course, in practice it is usually too much labour to diagonalize both the electrostatic and the spin-orbit energy even within one configuration and so we are often interested in second-order corrections to the energy due to spin-orbit coupling. We take an example which has some relevance to the theory of electron resonance in ferric compounds (§ 12.4.12). The ground term of d^5 is ^6S and this cannot be split by spin-orbit coupling to any order. However, it is possible for off-diagonal matrix elements of the spin-orbit coupling energy to mix into the ground state one or more of the excited levels of d^5. So the ground state will not be exactly ^6S but will be $\psi = c_0\psi(^6S)+\Sigma c_i X_i$, where c_0, c_i are numbers and X_i are excited levels. We use perturbation theory to determine the c_0, c_i and X_i to first order.

First, we use the selection rules to determine, in the Russell–Saunders coupling scheme, which levels of d^5 can have non-vanishing matrix elements with ^6S. $\Delta S = 0, \pm 1$ so the levels must come from quartets. Then $\Delta L = 0, \pm 1$ so the only possible quartet is ^4P. Finally, $\Delta J = 0$ which means that $^4P_{\frac{5}{2}}$ is the only level of d^5 which has a non-vanishing matrix element with ^6S. Then we have

$$\langle ^6S, M_S 0 \,|\, \delta\mathscr{H} \,|\, ^4P^{M_J}_{\frac{5}{2}}\rangle = \mu\delta_{M_S M_J}, \tag{5.9}$$

where μ is to be determined. The only other quantity we need to know is $F = E(^4P_{\frac{5}{2}})-E(^6S)$. One readily verifies that $\lambda(^4P) = 0$ and so $F = 7B+7C$. In order to find μ we first find the function $X = {}^4P^{\frac{5}{2}}_{\frac{5}{2}}$. This is straightforward and very similar to our determination of functions of ^3F of d^4 in § 4.6. As before we write $p^2 = p^+, p^-$, where p is the m_l value of a d orbital and also define

$$a = |2^2, 0^+, -1^+, -2^+\rangle,$$

$$b = |2^+, 1^2, -1^+, -2^+\rangle,$$

$$c = |2^+, 1^+, 0^2, -2^+\rangle,$$

$$d = |2^+, 1^+, 0^+, -1^2\rangle,$$

$$e = |2^+, 1^+, 0^+, -1^+, -2^+\rangle.$$

Then
$$X = \frac{1}{\sqrt{10}}\{\sqrt{2}\,a+\sqrt{3}\,b+\sqrt{3}\,c+\sqrt{2}\,d\}. \tag{5.10}$$

It follows from the form of a, b, c and d that their matrix elements with e are

$$\langle a \,|\, \delta\mathscr{H} \,|\, e\rangle = \langle d \,|\, \delta\mathscr{H} \,|\, e\rangle = \zeta,$$

$$\langle b \,|\, \delta\mathscr{H} \,|\, e\rangle = \langle c \,|\, \delta\mathscr{H} \,|\, e\rangle = \tfrac{1}{2}\sqrt{6}\,\zeta,$$

whence $\mu = \langle X \,|\, \delta\mathscr{H} \,|\, e\rangle = \zeta\sqrt{5}$. Therefore the lowest level of d^5 is given approximately by

$$\psi = \psi(^6S, M0) - \frac{\zeta\sqrt{5}}{7(B+C)}\psi(^4P^M_{\frac{5}{2}}), \tag{5.11}$$

where $\psi(^4P^{\frac{5}{2}}_{\frac{5}{2}}) = X$ of (5.10).

Finally, I remark that the parameters ζ can be obtained by fitting observed spectra to the theoretical formulae, just as we saw for the Slater–Condon parameters in § 4.7. I give tables of them in Appendix 6 where some indication of the methods of estimation is given, together with references. It is instructive to plot their values against atomic number and I have done this for the neutral atoms in Fig. 5.1. We see that the parameters for the second series are $2\frac{1}{2}$–3 times as large as those for the first and that those for the third are between 7 and 10 times as large. I have also scaled down the available parameters for the third series by a factor of four and plotted their scaled-down values in Fig. 5.1 to show that the

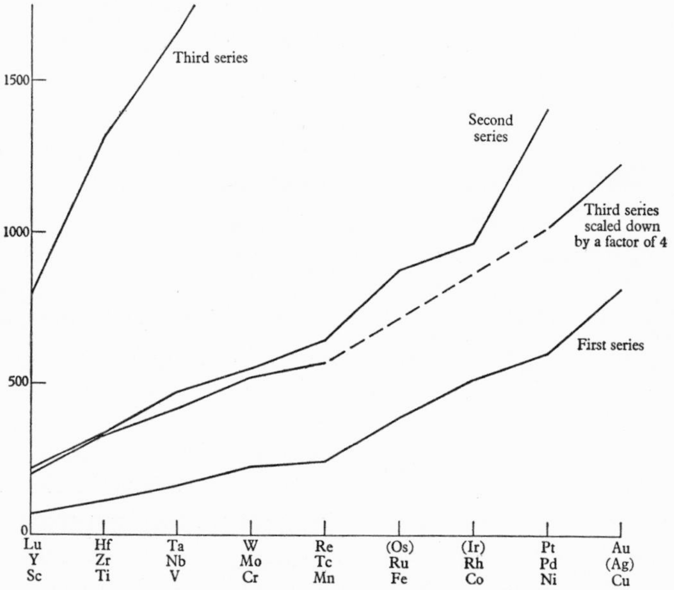

Fig. 5.1. The spin-orbit coupling constant ζ for d electrons in neutral atoms.

general variation in all three series is very similar. It is clear that one ought to take the spin-orbit coupling into account very carefully in the third series, and fairly carefully for later elements of the second, in fitting theoretical formulae for energies to experimental data.

Examples

1. Show that $\lambda(d^3; {}^4F) = \frac{1}{3}\zeta$.

2. Considering only the interaction of ${}^6S$ and ${}^4P$ in d^5 show that the states of the lowest level are given by $\psi = \psi({}^6S, M0)\cos\theta + \psi({}^4P^M_{\frac{5}{2}})\sin\theta$, where $\tan\theta = \frac{1}{2}\gamma - (\frac{1}{4}\gamma^2 + 1)^{\frac{1}{2}}$ and $\gamma\zeta\sqrt{5} = 7(B+C)$.

3. Show that the states of the lowest level of p^3 are approximately

$$\psi({}^4S, M0) - \frac{\zeta}{15F_2}\psi({}^2P^M_{\frac{3}{2}}).$$

4. Show that the total spread, χ, of the ground term of an l^c configuration is given by the formula $\chi = (L+\frac{1}{2})\zeta$, in the usual notation, except when $L = 0$. (Assume Hund's rule.)

5. In an l^{2l+1} configuration with basic states determinantal functions made up of one-electron functions $|nlm_s m_l\rangle$, take those functions which have a given M_S, M_L for some chosen M_S, M_L. Number them from 1 to a, call them ψ_i and regard two as identical whenever they differ only in phase. Define an operator P on one of them by first replacing the $(2l+1)$ one-electron functions by the $(2l+1)$ which do not occur in the determinant and after that changing every m_s, m_l to $-m_s$, $-m_l$. Show that if ψ_i is a fixed one of the functions:

(i) $P\psi_i = \psi_j$ for some ψ_j of the set;

(ii) $P^2 = 1$;

(iii) The spin-orbit energy of $P\psi_i$ is minus the spin-orbit energy of ψ_i.

Deduce that the first-order spin-orbit energy of any term which occurs only once in l^{2l+1} is zero.

5.3. The nucleus in atomic structure

So far the nucleus has been treated as a fixed positive point charge. This, of course, is not strictly true. The nucleus is a dynamical system of finite mass and non-zero mean diameter and its associated electric field is not rigorously the field expected from a point charge. Finally, and most important for us, the nucleus often has an angular momentum and an associated magnetic moment. From a purely energetic point of view all these effects are always extremely small. Nevertheless, they often show up in accurate measurements, especially magnetic ones. They then give us useful information, not only about the nucleus itself, but also about the electronic structure of the system.

The nucleus consists of nucleons (neutrons and protons) held together by strongly attractive forces, whose detailed nature is unknown. It appears, however, that these forces act approximately equally between any pair of nucleons. There is also the normal Coulombic repulsive force between protons but this is rather smaller in magnitude. However, the specifically nuclear forces have a small range and so the total energy associated with them does not add up so fast as the total Coulomb energy. Very large nuclei are therefore liable to break up. This is called fission. Both neutrons and protons have intrinsic spins of $\frac{1}{2}\hbar$ and magnetic moments about a thousandth of the electron moment. Finally, there is a very large spin-orbit coupling energy of unexplained origin[1] which tends to align the spin of a nucleon parallel to its orbital moment.

Just as with an electronic system a nucleus has a series of energy levels. Fortunately, however, they are so far apart that only the lowest is occupied at temperatures which are at all relevant to us. Also the mixing of nuclear levels under the effect of the perturbation caused by the electrons is completely negligible. The only properties of a nucleus which are relevant, then, are those of its ground level. The total angular momentum of the ground level is called the spin of the nucleus and is written I corresponding to the vector $\mathbf{I}$. The energy of the nucleus in a magnetic field is $-\boldsymbol{\mu}.\mathbf{H}$ and $\boldsymbol{\mu}$ is generally parallel to $\mathbf{I}$, otherwise it is antiparallel. It is convenient to write $\boldsymbol{\mu} = \hbar^{-1}\gamma\beta_N\mathbf{I}$, where $\beta_N = e\hbar/2Mc$ is the nuclear magneton, M is the proton mass and γ is a pure number of order of magnitude unity. Then if $\mathbf{H}$ is along OZ we have a series of equally spaced levels with energies $-\gamma\beta_N M_I$.

Rather more nuclei than one might perhaps expect have $I = 0$. This is for much the same reason that states of high S lie lowest in the atom. For the specifically nuclear forces, unlike electrostatic forces between like particles, are attractive which tends to make low I lie low[2] for many nuclei. The spin-orbit coupling, however, works in the opposite direction and I is quite large for some nuclei. For example, $I = 4\frac{1}{2}$ for ^{93}Nb and ^{209}Bi. The number of nucleons is written as a left superscript.

[1] It has been suggested that this is not due to a term of the form $\Sigma\xi(r_\kappa)\mathbf{l}_\kappa.\mathbf{s}_\kappa$. (Feingold & Wigner, quoted in Mayer & Jensen (1955), p. 60.) It does not seem to me that such a suggestion could, in itself, even if true, be regarded as an improvement (i.e. simplification) of the theory.

[2] Of course it is not quite so simple as this because the nuclear spin is the total angular momentum of the nucleus, not the spin in the ordinary sense.

The magnetic interaction with the nucleus may usually be regarded as a small perturbation on the level system of the electrons as well as on that of the nucleus. In that case, corresponding to the angular momentum **I** of the nucleus and **J** of the electrons, the usual set of states corresponding to vector coupling of **I** with **J** are eigenstates for the whole system to a good approximation. The total angular momentum vector of the nucleus together with electrons is written $\mathbf{F} = \mathbf{I}+\mathbf{J}$ and so for a given electronic level F ranges from $|I-J|$ to $I+J$. Each level, then, is split up into a series of sublevels. This extra structure to atomic energy levels is said to be hyperfine structure. As the name suggests, the splittings concerned are always very small and are usually in the range 10^{-3} to $10^{-1}\,\text{cm}^{-1}$. Formulae for this magnetic coupling with the nucleus are derived in §§ 5.5 and 12.2, but I anticipate to say that the extra terms in the Hamiltonian commute with **F** and give rise to energies which can be represented in the general form

$$E(F) = \tfrac{1}{2}A[F(F+1)-I(I+1)-J(J+1)].$$

In other words the theory predicts that the hyperfine structure components satisfy the Landé interval rule.

As an example we consider the ground term $3d^5\,4s^2$; ${}^6S_{\frac{5}{2}}$ of the neutral manganese atom. There is only one stable manganese nucleus, ^{55}Mn. It has a spin of $\frac{5}{2}$ and so the values of F range from 0 to 5, with energies

$$E(F) = \tfrac{1}{2}A[F(F+1)-17\tfrac{1}{2}],$$

and the intervals are $\quad I_F = E(F)-E(F-1) = AF.$

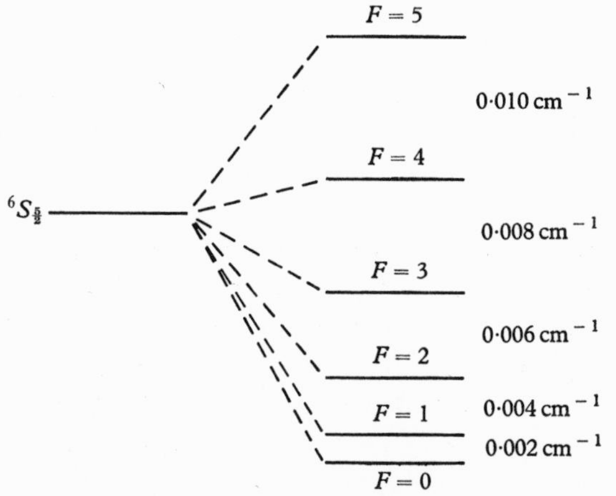

Fig. 5.2. Hyperfine structure in the ground term of Mn I. Data from White & Ritschl (1930).

Hyperfine structure has been observed in the optical spectrum by White & Ritschl (1930) and fits the theoretical formula nicely (see Fig. 5.2). The parameter $A = 0{\cdot}002\,\text{cm}^{-1}$. The first excited level of Mn I lies at $17{,}052\,\text{cm}^{-1}$ above the ground term which gives some indication of how relatively small hyperfine structure is. In other cases A ranges up to about $10^{-1}\,\text{cm}^{-1}$.

The nucleus has no electric dipole moment[1] but can have a quadrupole moment. A table of properties of nuclei of interest to us is given in Appendix 1.

[1] Experimentally it does not. Theoretically it cannot, but only if one assumes that parity is conserved.

5.4. Dirac's linear wave-equation for the electron

Thus far we have treated the spin of the electron on a purely phenomenological basis. In this section and the next we see how it arises from a very natural marriage of two physical principles, one from quantum theory and the other from special relativity theory. This marriage has as its offspring Dirac's linear wave-equation for the electron and we follow Dirac's own treatment closely in order to obtain it. Then in the next section we give a number of useful and interesting deductions which follow in a rather straightforward manner from the equation.

We turn first to special relativity theory. As the reader will know,[1] the analytic form of this theory derives from the interpretation by Einstein of the set of transformation equations for the electromagnetic field discovered by Lorentz. Einstein interpreted these as being orthogonal transformations (i.e. proper and improper rotations) of a four-dimensional continuum, the points of this continuum being events. This means that each point is specified by four coordinates, three of space and one of time. We write a point (x_0, x_1, x_2, x_3), where x_1, x_2, x_3 refer to space and $x_0 = ct$ refers to time. c is the velocity of light and so all the x_μ have dimensions of length. Common sense asserts that time is a very different sort of thing from space and this is reflected in the theory by the fact that the four-dimensional continuum is not a Euclidean space but a slightly different one. Distance in it is defined to be given by the formula

$$s^2 = x_0^2 - x_1^2 - x_2^2 - x_3^2, \tag{5.12}$$

rather than by the sum of four squares which would be appropriate to a Euclidean space. The transformations which leave (5.12) invariant are called Lorentz transformations and consist of a six-parameter set which leave some arbitrarily chosen point of the continuum fixed (these are sometimes called orthogonal transformations relative to the quadratic form (5.12), when the fixed point is the origin, but this is not the sense in which the word orthogonal is usually applied to matrices) and a four-parameter set of translations. These Lorentz transformations have the property that they leave Maxwell's equations for the electromagnetic field unchanged in form[2] provided that we interpret the scalar and vector potential as forming the time and space components of a vector in the four-dimensional continuum. Let us write A_0 for the scalar potential ϕ, and $\mathbf{A}$ for the vector potential, and then the four-vector is (A_0, A_x, A_y, A_z).

Special relativity theory asserts that not merely the electromagnetic equations but also the correct mechanical equations must be invariant in form under Lorentz transformations. This means in practice that the entities occurring in the

[1] Introductions to the theory are found in Synge, *Relativity: the Special Theory* (North-Holland Publishing Company, 1956); Tolman, *Relativity, Thermodynamics and Cosmology* (Oxford, 1946). See also, Eddington, *The Mathematical Theory of Relativity* (Cambridge, 1952). Alternatively, the reader may omit §§ 5.4 and 5.5 without encountering appreciable difficulty later.

[2] Actually these two sets are not the only transformations having this property, but the remaining ones have been found, as yet, to have no real significance for physics. See Cunningham, *Proc. Lond. Math. Soc.* (1909), **8**, 77; and Bateman, *ibid.* (1910), **8**, 223.

mathematical theory, as representatives of physical quantities, must be scalars, vectors or tensors in the four-dimensional space.[1] The mechanical momentum **p** does not satisfy this requirement. Actually $\mathbf{p} = m(d\mathbf{r}/dt)$ and so not merely does **p** have only three components but also it involves differentiation with respect to a particular coordinate t rather than with respect to a scalar. The appropriate generalization of **p** to four dimensions is achieved by taking the time component p_0 to be c^{-1} times the energy E of the system so we have the four vector (Ec^{-1}, p_x, p_y, p_z). The difficulty about differentiation is dealt with by writing

$$p_\mu = m_0 c \frac{dx_\mu}{ds} \quad (\mu = 0, 1, 2, 3), \tag{5.13}$$

where m_0 is the rest mass (i.e. the mass of the particle measured in a frame of reference in which it is at rest), and s is defined as in (5.12). Equation (5.13) is consistent with the original definition of **p** provided we interpret m as $m_0 c(dt/ds)$. Equation (5.12) gives us

$$ds^2 = c^2 dt^2 - dx^2 - dy^2 - dz^2,$$

which implies that $(ds/dt)^2 = c^2 - v^2$, where v is the velocity of the particle. So $m = m_0/\sqrt{(1-v^2/c^2)}$. The time component gives the well-known relationship between mass and energy

$$E = m_0 c^3 \frac{dt}{ds} = mc^2. \tag{5.14}$$

After these preliminary remarks, which are not intended to introduce the reader to relativity theory but rather to remind him of a few of the basic definitions and properties, we pass on to discuss the wave-equation for a free particle. It is natural to take the energy of the particle as our quantum Hamiltonian and the easiest way to obtain an equation for this is to consider the scalar connected with p_μ by (5.12) and (5.13)

$$\begin{aligned} p_0^2 - p_1^2 - p_2^2 - p_3^2 &= m_0^2 c^2 \left[\left(\frac{dx_0}{ds}\right)^2 - \left(\frac{dx_1}{ds}\right)^2 - \left(\frac{dx_2}{ds}\right)^2 - \left(\frac{dx_3}{ds}\right)^2\right] \\ &= m_0^2 c^2. \end{aligned} \tag{5.15}$$

We already know the operators corresponding to the components of p_μ, they are

$$\left.\begin{aligned} p_0 &= Ec^{-1} = c^{-1} i\hbar \frac{\partial}{\partial t} = i\hbar \frac{\partial}{\partial x_0}, \\ p_1 &= -i\hbar \frac{\partial}{\partial x_1}, \\ p_2 &= -i\hbar \frac{\partial}{\partial x_2}, \\ p_3 &= -i\hbar \frac{\partial}{\partial x_3}. \end{aligned}\right\} \tag{5.16}$$

[1] Or spinors. See §§ 6.9 and 6.10.

The change of sign between space and time components in (5.16) is connected with the fact that (5.12) is not the metric for a Euclidean space. This leads to the introduction of contravariant vectors which transform like x_μ or p_μ and covariant ones which transform like $\partial/\partial x_\mu$. It is because we have equated these two different types of vector that we have the sign change in (5.16). On allowing (5.15) to operate on a function ψ in accordance with (5.16) we obtain the relativistic equation

$$(p_0^2-p_1^2-p_2^2-p_3^2-m_0^2c^2)\,\psi = 0, \tag{5.17}$$

for a free particle. Equation (5.17) is known as the Klein–Gordon equation.

The Klein–Gordon equation has the disadvantage that it involves a second differential coefficient with respect to the time. This means that in order to specify a system at a particular time we must give both ψ and $\partial\psi/\partial t$. While it is not impossible that this should be necessary it would be out of line with the general spirit and formulation of non-relativistic quantum mechanics. Dirac, therefore, proposed factorizing the operator in (5.17) into two linear factors, thus

$$\begin{aligned} &p_0^2-p_1^2-p_2^2-p_3^2-m_0^2c^2 \\ &\quad = (p_0-\alpha_1 p_1-\alpha_2 p_2-\alpha_3 p_3-\alpha_4 m_0 c)(p_0+\alpha_1 p_1+\alpha_2 p_2+\alpha_3 p_3+\alpha_4 m_0 c). \end{aligned}$$

By comparing coefficients of the various terms we obtain

$$\left.\begin{aligned} \alpha_1^2 = \alpha_2^2 = \alpha_3^2 = \alpha_4^2 = 1, \\ \alpha_1\alpha_2+\alpha_2\alpha_1 = \alpha_1\alpha_3+\alpha_3\alpha_1 = \alpha_2\alpha_3+\alpha_3\alpha_2 = 0, \\ \alpha_1\alpha_4+\alpha_4\alpha_1 = \alpha_2\alpha_4+\alpha_4\alpha_2 = \alpha_3\alpha_4+\alpha_4\alpha_3 = 0. \end{aligned}\right\} \tag{5.18}$$

If the α_i satisfy (5.18) and commute with the p_μ, then any solution of the linear equation

$$(p_0+\alpha_1 p_1+\alpha_2 p_2+\alpha_3 p_3+\alpha_4 m_0 c)\,|\psi\rangle = 0 \tag{5.19}$$

is also a solution of (5.17) as we see immediately by multiplying on the left by $(p_0-\alpha_1 p_1-\alpha_2 p_2-\alpha_3 p_3-\alpha_4 m_0 c)$.

It is of course obvious from (5.18) that the α_i cannot be numbers. They are algebraic quantities. We now represent them in terms of matrices. First, we consider the four quantities $1, \sigma_x, \sigma_y, \sigma_z$, where the σ_i are the components of the vector $\boldsymbol{\sigma}$ associated with the spin vector **s** (see (2.32–34)). They can be represented as matrices by the correspondence

$$1 = \begin{bmatrix} 1 & 0 \\ 0 & 1 \end{bmatrix}, \quad \sigma_x = \begin{bmatrix} 0 & 1 \\ 1 & 0 \end{bmatrix}, \quad \sigma_y = \begin{bmatrix} 0 & -i \\ i & 0 \end{bmatrix}, \quad \sigma_z = \begin{bmatrix} 1 & 0 \\ 0 & -1 \end{bmatrix}. \tag{5.20}$$

One verifies immediately that these matrices satisfy

$$\left.\begin{aligned} \sigma_x^2 = \sigma_y^2 = \sigma_z^2 = 1, \\ \sigma_x\sigma_y+\sigma_y\sigma_x = \sigma_x\sigma_z+\sigma_z\sigma_x = \sigma_y\sigma_z+\sigma_z\sigma_y = 0, \end{aligned}\right\} \tag{5.21}$$

which are (2.33) and (2.34). Further, any 2×2 matrix can be expressed as a linear combination of the four matrices (5.20) with complex coefficients and determines uniquely a linear combination of $1, \sigma_x, \sigma_y, \sigma_z$.

Equations (5.21) are very like (5.18) except that we have not enough anti-commuting quantities. Therefore we consider a second set of quantities 1, ρ_x, ρ_y, ρ_z commuting with the σ_i and with the p_μ. If we then let

$$\alpha_1 = \rho_x \sigma_x, \quad \alpha_2 = \rho_x \sigma_y, \quad \alpha_3 = \rho_x \sigma_z, \quad \alpha_4 = \rho_z,$$

these α_i satisfy (5.18). Although we have used x, y and z as suffixes the α_i do not actually involve x, y or z as variables which is witnessed by the fact that α_i commutes with p_μ. We are now in a position to write Dirac's equation in the form which is most useful for our purpose.

We write $\boldsymbol{\sigma}$ for the vector $(\sigma_x, \sigma_y, \sigma_z)$ and take the 2×2 matrix representation (5.20) only for the ρ_i. We have

$$\rho_x = \begin{bmatrix} 0 & 1 \\ 1 & 0 \end{bmatrix}, \quad \rho_z = \begin{bmatrix} 1 & 0 \\ 0 & -1 \end{bmatrix},$$

and Dirac's equation becomes first

$$(p_0 + \rho_x \boldsymbol{\sigma}.\mathbf{p} + \rho_z m_0 c)\,|\psi\rangle = 0, \tag{5.22}$$

and then

$$\left(p_0 \begin{bmatrix} 1 & 0 \\ 0 & 1 \end{bmatrix} + \boldsymbol{\sigma}.\mathbf{p} \begin{bmatrix} 0 & 1 \\ 1 & 0 \end{bmatrix} + m_0 c \begin{bmatrix} 1 & 0 \\ 0 & -1 \end{bmatrix}\right) |\psi\rangle = 0. \tag{5.23}$$

This means that we should take $|\psi\rangle$ as a vector

$$|\psi\rangle = \begin{bmatrix} \psi_1 \\ \psi_2 \end{bmatrix}$$

and let us also drop the suffix on m_0, to obtain the simultaneous equations

$$\left.\begin{aligned} (p_0 + mc)\,\psi_1 + \boldsymbol{\sigma}.\mathbf{p}\psi_2 &= 0, \\ (p_0 - mc)\,\psi_2 + \boldsymbol{\sigma}.\mathbf{p}\psi_1 &= 0. \end{aligned}\right\} \tag{5.24}$$

Equations (5.24) are the form in which we shall apply Dirac's equation to obtain various corrections to the non-relativistic Hamiltonian for a single electron. It needs merely to be supplemented by the remark that we have so far been neglecting the fact that the electron has a charge $-e$ and may be in an electromagnetic field with potentials (A_0, A_x, A_y, A_z). We introduce the influence of this field into (5.19) or (5.24) by the usual rule of replacing p_μ by $p_\mu + (e/c)\,A_\mu$. Equations (5.24) then become

$$\left.\begin{aligned} \left(p_0 + mc + \frac{e}{c}A_0\right)\psi_1 + \boldsymbol{\sigma}.\left(\mathbf{p} + \frac{e}{c}\mathbf{A}\right)\psi_2 &= 0, \\ \left(p_0 - mc + \frac{e}{c}A_0\right)\psi_2 + \boldsymbol{\sigma}.\left(\mathbf{p} + \frac{e}{c}\mathbf{A}\right)\psi_1 &= 0. \end{aligned}\right\} \tag{5.25}$$

Equations (5.25), or rather the linear equation corresponding to them, are not simply the linearized form of the Klein–Gordon equation for an electron in a field and it is primarily because of this that Dirac's equation predicts observable effects which do not arise out of the Klein–Gordon equation.

We conclude this section with various general remarks. First, we note that although we have represented the ρ_i as matrices (and can represent α_μ, see ex. 2,

p. 122) this is not quite the same thing as saying that they are matrices. They are just algebraic quantities over the complex field which satisfy (5.21). We introduce a matrix representation because matrices make it easier for us to study the consequences of (5.19). It is perfectly possible to study them without introducing representations and we should come to the same conclusions.

Next we remark that it is not obvious that (5.19) is relativistically invariant, i.e. that the consequences of it remain unchanged when we use it in a new coordinate system. One can show that this is actually so[1] but we shall not do so here. Gauge invariance is easier to establish, for the truth of the equation

$$\left[p_0+\frac{e}{c}A_0+\sum_{i=1}^{3}\alpha_i\left(p_i+\frac{e}{c}A_i\right)+\alpha_4 mc\right]|\psi\rangle = 0$$

assures the truth of the equation

$$\left[p_0+\frac{e}{c}A_0^1+\sum_{i=1}^{3}\alpha_i\left(p_i+\frac{e}{c}A_i^1\right)+\alpha_4 mc\right]|\psi\rangle^1 = 0,$$

and conversely, where

$$A_0 = A_0^1+\frac{1}{c}\frac{\partial\chi}{\partial t},\quad \mathbf{A} = \mathbf{A}^1-\operatorname{grad}\chi,\quad |\psi\rangle = e^{ie\chi/(\hbar c)}|\psi\rangle^1.$$

This follows immediately on substitution.

If we set $A_0 = (e/r)$ and $\mathbf{A} = 0$ we have the case of an electron moving under the influence of a fixed stationary point positive charge of magnitude $+e$. The stationary states of this system are given by taking the energy, i.e. cp_0, as a constant, W say. It is possible to solve this problem exactly[2] and one obtains the formula

$$W = mc^2\left\{1+\frac{\alpha^2}{\{n-j-\frac{1}{2}+\sqrt{[(j+\frac{1}{2})^2-\alpha^2]}\}^2}\right\}^{-\frac{1}{2}}, \tag{5.26}$$

where $\alpha = e^2/\hbar c$ is a small pure number (approximately $1/137$), n is the principal quantum number and j the quantum number total of angular momentum. If we expand (5.26) and neglect α^2 compared with $(j+\frac{1}{2})^2$ the first two terms are

$$W = mc^2\left(1-\frac{\frac{1}{2}\alpha^2}{n^2}\right)$$

$$= mc^2-\frac{me^4}{2\hbar^2 n^2},$$

which corresponds to (4.13), apart from the addition of the rest mass mc^2, and confirms the identification of n as the principal quantum number. For $j \neq n-\frac{1}{2}$ there are two levels for each j corresponding to $l = j\pm\frac{1}{2}$ and (5.26) predicts that they should still be rigorously degenerate. The four lowest levels of the system then lie in the order

$$1s\,{}^2S_{\frac{1}{2}};\quad (2s\,{}^2S_{\frac{1}{2}},\ 2p\,{}^2P_{\frac{1}{2}});\quad 2p\,{}^2P_{\frac{3}{2}};\quad \ldots$$

according to (5.26). The separation between ${}^2P_{\frac{1}{2}}$ and ${}^2P_{\frac{3}{2}}$ would be ascribed to spin-orbit coupling.

[1] Dirac (1947), p. 257.

[2] Dirac (1947), p. 268.

The passage from these results for a fixed nucleus to the corresponding results for a hydrogen atom is by no means an entirely trivial matter. For our linear equation refers to the motion of an electron in an externally defined electromagnetic field.[1] The equation for the hydrogen atom involves the motion of both nucleus and electron and when we perform a separation of variables it would seem strange to call the equation of relative motion of the two particles an equation for the electron alone. The large ratio of the masses of the nucleus and electron is not relevant here; we cannot get out of the difficulty by saying that the nucleus is 'practically fixed'. Therefore when we use Dirac's equation for a hydrogen atom we must remember that it really refers to an internal coordinate and, therefore, to a mutual property of the nucleus and electron and

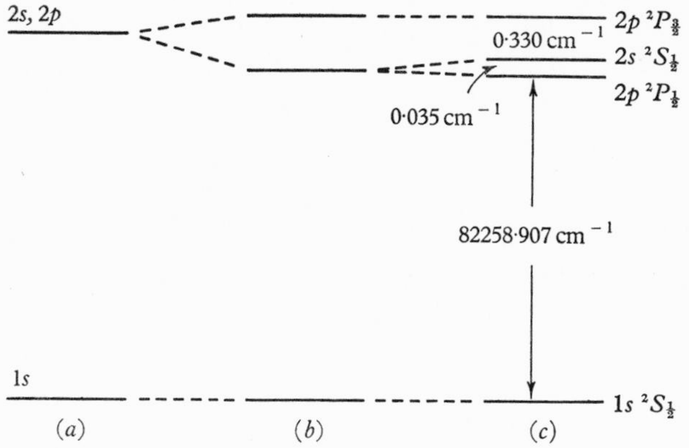

Fig. 5.3. The lowest levels of hydrogen in various approximations, neglecting nuclear hyperfine structure. (*a*) Non-relativistic theory; (*b*) Dirac theory; (*c*) experiment, in agreement with quantum electrodynamic refinement of Dirac theory. The separations are not to scale!

not to the electron alone. The reader may think this change of attitude to Dirac's equation when discussing bound electrons is perfectly unexceptionable or he may feel there is a real difficulty here.[2]

From a practical point of view, however, it is abundantly clear that Dirac's equation works for bound as well as for free electrons providing we make the natural interpretation of m in it as the reduced mass of the electron. Then (5.26) gives the energy levels of the hydrogen atom to the approximation in which we neglect the small effects due to the magnetic moment of the proton.

Our final general point about Dirac's equation is that it is now known that the energy levels predicted by (5.26) are not in perfect agreement with experiment. This had been suspected for a long time but direct proof came finally from the work of Lamb and Retherford, who obtained transitions between the $2s\,^2S_{\frac{1}{2}}$ and $2p\,^2P_{\frac{1}{2}}$ levels which should be strictly degenerate according to the Dirac theory.

[1] Inevitably, therefore, a classical field.

[2] For two opposing points of view on this matter see Dirac, Peierls & Pryce, *Proc. Camb. Phil. Soc.* (1942), **38**, 193; and Eddington, *ibid.* p. 201. The logical problem has, I think, disappeared now: one should derive properties of bound electrons from a relativistic formulation adequate for many particles and then obtain the Dirac equation as a very good first approximation for 'one-particle' systems.

These transitions corresponded to a separation of 0·035 cm^{-1} between the two levels and it was also shown that the $^2S_{\frac{1}{2}}$ lay above the $^2P_{\frac{1}{2}}$. The discussion of ch. 3 lays open to us the origin of this effect, namely that we have rigidly separated the electron from the electromagnetic field and have given a quantum mechanical treatment of the first only. It is, in fact, possible to account for the Lamb–Retherford shift quantitatively by using modern quantum electrodynamics but we shall not discuss this here.

The level scheme for $n = 1$ and 2 is shown schematically in Fig. 5.3 according to the three main approximations. The Lamb–Retherford shift is too small to matter to us, but there is another related deviation from Dirac's equation that is important in the interpretation of some electron resonance experiments. This is that the spin magnetic moment of the electron, which we derive in the next section to be $e\hbar/2mc$, is not exactly that but rather $1{\cdot}001145(e\hbar/2mc)$. This is a very small difference but it can be detected in accurate measurements in certain circumstances.

Examples

1. Taking the Hamiltonian cp_0 from (5.19) show that it commutes with $\mathbf{j} = \mathbf{l}+\frac{1}{2}\hbar\boldsymbol{\sigma}$.
2. Show that the set of all linear combinations of the sixteen quantities $1, \rho_i, \sigma_j, \rho_i\sigma_j$ can be represented as the set of all 4×4 matrices with complex elements.
3. Show that $\rho_z(\boldsymbol{\sigma}\,.\,\mathbf{l}+\hbar)$ commutes with the Hamiltonian cp_0.
4. If $\boldsymbol{\sigma}$ commutes with the vectors (three-component vectors) $\mathbf{A}$ and $\mathbf{B}$, then

$$(\boldsymbol{\sigma}\,.\,\mathbf{A})(\boldsymbol{\sigma}\,.\,\mathbf{B}) = \mathbf{A}\,.\,\mathbf{B}+i\boldsymbol{\sigma}\,.\,\mathbf{A}\wedge\mathbf{B}.$$

5.5. Deductions from Dirac's linear wave-equation[1]

In the last section we obtained Dirac's equation in the presence of an electromagnetic field in the form

$$\left.\begin{aligned}\left(p_0+mc+\frac{e}{c}A_0\right)\psi_1+\boldsymbol{\sigma}\,.\left(\mathbf{p}+\frac{e}{c}\mathbf{A}\right)\psi_2 &= 0,\\ \left(p_0-mc+\frac{e}{c}A_0\right)\psi_2+\boldsymbol{\sigma}\,.\left(\mathbf{p}+\frac{e}{c}\mathbf{A}\right)\psi_1 &= 0.\end{aligned}\right\} \tag{5.27}$$

We now derive a number of consequences and shall be primarily interested in estimates for the energies of stationary states. So we replace cp_0 in (5.27) by the eigenvalue W of the energy. Now as we saw in connexion with (5.26), W usually lies close to mc^2 and as we shall only be interested in this case we write

$$W = E+mc^2,$$

where E is the energy of the system if we omit the rest mass. For our purposes E is small compared with mc^2 (with a ratio of the order of 1 in 10^4). Equations (5.27) now become

$$\left.\begin{aligned}(E+2mc^2+eA_0)\psi_1+c\boldsymbol{\sigma}\,.\left(\mathbf{p}+\frac{e}{c}\mathbf{A}\right)\psi_2 &= 0,\\ (E+eA_0)\psi_2+c\boldsymbol{\sigma}\,.\left(\mathbf{p}+\frac{e}{c}\mathbf{A}\right)\psi_1 &= 0.\end{aligned}\right\} \tag{5.28}$$

[1] For the early literature on this subject see Fermi, *Z. Phys.* (1930), **60**, 320; Hargreaves, *Proc. Roy. Soc.* A (1929), **124**, 568; (1930), **127**, 141, 407; Breit, *Phys. Rev.* (1930), **35**, 1447; (1931), **37**, 51.

Although this may not seem quite obvious, the presence of $2mc^2$ in the first line, but not the second, of (5.28) has actually the effect of making ψ_1 small compared with ψ_2. For this reason it is convenient to eliminate ψ_1 (ψ_1 and ψ_2 are often called the small and large components, respectively). We first introduce the abbreviation

$$f = \frac{2mc^2}{2mc^2+E+eA_0}, \tag{5.29}$$

where f is evidently approximately unity. Then the first line of (5.28) becomes

$$\psi_1 = -\frac{f}{2mc}\boldsymbol{\sigma}.\left(\mathbf{p}+\frac{e}{c}\mathbf{A}\right)\psi_2, \tag{5.30}$$

which on substitution into the second line gives

$$\left(\frac{1}{2m}\boldsymbol{\sigma}.\left(\mathbf{p}+\frac{e}{c}\mathbf{A}\right)f\boldsymbol{\sigma}.\left(\mathbf{p}+\frac{e}{c}\mathbf{A}\right)-eA_0-E\right)\psi_2 = 0. \tag{5.31}$$

Equation (5.31) is our general equation for E and ψ_2. $\boldsymbol{\sigma}.[\mathbf{p}+(e/c)\mathbf{A}]$ is of course just the scalar $\boldsymbol{\sigma}.\mathbf{p}+(e/c)\boldsymbol{\sigma}.\mathbf{A}$. We have made no assumption so far about A_0 or $\mathbf{A}$, except that they are independent of the time, in particular there is no need for A_0 and f in (5.29) and (5.31) to represent a spherically symmetrical electrostatic field.

Except very near a source or a sink of the field, f will be very nearly unity and so we may expand it to give

$$f = 1-\frac{E+eA_0}{2mc^2}; \quad \frac{\partial f}{\partial x_i} = -\frac{e}{2mc^2}\frac{\partial A_0}{\partial x_i}, \tag{5.32}$$

to the first order in small quantities, where x_i is a spatial coordinate. In discussing (5.31) we shall use the formula

$$(\boldsymbol{\sigma}.\mathbf{B})(\boldsymbol{\sigma}.\mathbf{C}) = \mathbf{B}.\mathbf{C}+i\boldsymbol{\sigma}.\mathbf{B}\wedge\mathbf{C}, \tag{5.33}$$

where $\mathbf{B}$ and $\mathbf{C}$ are any vectors which each commute with $\boldsymbol{\sigma}$ but not necessarily with each other. Equation (5.33) follows easily from the relations (5.21) for the vector $\boldsymbol{\sigma}$. If $\mathbf{B} = \mathbf{C}$ we have $(\boldsymbol{\sigma}.\mathbf{B})^2 = B^2+i\boldsymbol{\sigma}.\ \mathbf{B}\wedge\mathbf{B}$, but $\mathbf{B}\wedge\mathbf{B}$ need not be zero.

We now pass on to simplify (5.31) further in a number of special cases.

5.5.1. The non-relativistic equation. The lowest-order approximation is the one in which we suppose $f = 1$ everywhere. Then (5.31) becomes

$$\left\{\frac{1}{2m}\left[\boldsymbol{\sigma}.\left(\mathbf{p}+\frac{e}{c}\mathbf{A}\right)\right]^2-eA_0\right\}\psi_2 = E\psi_2. \tag{5.34}$$

Using (5.33) we find

$$\begin{aligned}\left[\boldsymbol{\sigma}.\left(\mathbf{p}+\frac{e}{c}\mathbf{A}\right)\right]^2 &= \left(\mathbf{p}+\frac{e}{c}\mathbf{A}\right)^2+i\boldsymbol{\sigma}.\left(\mathbf{p}+\frac{e}{c}\mathbf{A}\right)\wedge\left(\mathbf{p}+\frac{e}{c}\mathbf{A}\right)\\ &= \left(\mathbf{p}+\frac{e}{c}\mathbf{A}\right)^2+i\boldsymbol{\sigma}.\left(\mathbf{p}\wedge\frac{e}{c}\mathbf{A}+\frac{e}{c}\mathbf{A}\wedge\mathbf{p}\right)\\ &= \left(\mathbf{p}+\frac{e}{c}\mathbf{A}\right)^2+\frac{e\hbar}{c}\boldsymbol{\sigma}.\operatorname{curl}\mathbf{A},\end{aligned}$$

where we have used $\mathbf{p} = -i\hbar\nabla$. Equation (5.34) now becomes

$$\left\{\frac{1}{2m}\left(\mathbf{p}+\frac{e}{c}\mathbf{A}\right)^2+\frac{e\hbar}{2mc}\boldsymbol{\sigma}.\mathbf{H}-eA_0\right\}\psi_2 = E\psi_2. \tag{5.35}$$

Equation (5.35) shows that ψ_2 satisfies the ordinary non-relativistic equation, to our approximation, but with the additional term $(e\hbar/2mc)\,\boldsymbol{\sigma}.\mathbf{H}$ which would be interpreted to mean that the electron had a magnetic moment $-(e\hbar/2mc)\,\boldsymbol{\sigma}$. The observed magnetic moment is very close indeed to this value (see the end of § 5.4). Combining this new result with our previous finding (§ 5.4, ex. 1) that $\mathbf{l}+\frac{1}{2}\hbar\boldsymbol{\sigma}$ commutes with the relativistic Hamiltonian we naturally identify the vector $\mathbf{s} = \frac{1}{2}\hbar\boldsymbol{\sigma}$ with the spin vector introduced phenomenologically in the non-relativistic theory.

We must not be deceived by the close analogy between (5.35) and the non-relativistic equation into thinking that (5.35) actually is the non-relativistic equation. The state vector for the electron is not ψ_2 alone but the column vector $\begin{bmatrix}\psi_1\\ \psi_2\end{bmatrix}$. So if we take $\mathbf{H} = 0$ in (5.35) and solve the equation explicitly for a Coulomb field, $A_0 = Ze/r$ the most reasonable unnormalized state vector to take, in this approximation, is

$$\begin{bmatrix}-\dfrac{1}{2mc}\boldsymbol{\sigma}.\mathbf{p}\psi_2\\[2ex] \psi_2\end{bmatrix}.$$

ψ_2 involves a spin function and our expression for ψ_1 can be worked out immediately. The probability density is given by $\bar{\psi}_1\psi_1+\bar{\psi}_2\psi_2$ and mean values of observables must be taken over ψ_1 and ψ_2.

5.5.2. The spin-orbit coupling energy. Here we set $\mathbf{A} = 0$ but retain the first-order terms arising from f. In our present application A_0 is a function of r alone. However, later in the book we discuss spin-orbit coupling for ions in crystals and then the electrostatic field of their environment is not spherically symmetrical. So for the time being we do not impose any restriction on the form of A_0 and specialize it to be a function of r as late in the calculation as possible.

Equation (5.31) now becomes

$$\left\{\frac{1}{2m}(\boldsymbol{\sigma}.\mathbf{p})\,f(\boldsymbol{\sigma}.\mathbf{p})-eA_0-E\right\}\psi_2 = 0. \tag{5.36}$$

By the rule for differentiating a product

$$(\boldsymbol{\sigma}.\mathbf{p})\,f = \boldsymbol{\sigma}.(\mathbf{p}f)+f\boldsymbol{\sigma}.\mathbf{p},$$

where $(\mathbf{p}f)$ means that $\mathbf{p}$ operates only on f. Then

$$\begin{aligned}(\boldsymbol{\sigma}.\mathbf{p})\,f(\boldsymbol{\sigma}.\mathbf{p}) &= (\boldsymbol{\sigma}.\mathbf{p})^2 f-(\boldsymbol{\sigma}.\mathbf{p})\,\boldsymbol{\sigma}.(\mathbf{p}f)\\ &= X_1-X_2, \quad \text{say.}\end{aligned}$$

Now, from (5.32) and (5.33),

$$X_1 = \mathbf{p}^2 f = \mathbf{p}^2\left(1-\frac{E+eA_0}{2mc^2}\right).$$

But as we are searching for ψ_2 which satisfy (5.31) it follows that to the zero order ψ_2 satisfies

$$\left(\frac{1}{2m}\mathbf{p}^2 - eA_0 - E\right)\psi_2 = 0,$$

as we have seen. This implies that, to the first order,

$$X_1 = \mathbf{p}^2\left(1 - \frac{\mathbf{p}^2}{4m^2c^2}\right) = \mathbf{p}^2 - \frac{1}{4m^2c^2}\mathbf{p}^4, \tag{5.37}$$

where we write $\mathbf{p}^4$ for $(\mathbf{p}^2)^2$.

It is a little more complicated to treat X_2. We have, by (5.33)

$$\begin{aligned} X_2 &= \mathbf{p}.(\mathbf{p}f) + i\boldsymbol{\sigma}.\mathbf{p}\wedge(\mathbf{p}f) \\ &= (-i\hbar)^2\nabla.(\nabla f) + i(-i\hbar)^2\boldsymbol{\sigma}.\nabla\wedge(\nabla f) \\ &= -\hbar^2\nabla^2 f - \hbar^2(\nabla f).\nabla + i(-i\hbar)^2\boldsymbol{\sigma}.[-(\nabla f)\wedge\nabla + \operatorname{curl}\nabla f]. \end{aligned}$$

The first term is zero to first order because A_0 satisfies Laplace's equation $\nabla^2 A_0 = 0$ and the last term is also because curl grad $\equiv 0$. So we are left with two terms only

$$X_2 = -\hbar^2(\nabla f).\nabla - h\boldsymbol{\sigma}.(\nabla f)\wedge\mathbf{p}. \tag{5.38}$$

Equation (5.38) is useful for discussing spin-orbit coupling in a crystal field. Now we let A_0 and hence f be a function of r alone whereupon

and

$$\left.\begin{aligned} \nabla f &= r^{-1}\mathbf{r}\frac{df}{dr} \\ X_2 &= -\frac{\hbar^2}{r}\frac{df}{dr}\mathbf{r}.\nabla - \frac{\hbar}{r}\frac{df}{dr}\boldsymbol{\sigma}.\mathbf{r}\wedge\mathbf{p} \\ &= -\hbar^2\frac{df}{dr}\frac{\partial}{\partial r} - \frac{\hbar}{r}\frac{df}{dr}\boldsymbol{\sigma}.\mathbf{l}. \end{aligned}\right\} \tag{5.39}$$

We know that

$$\frac{df}{dr} = -\frac{e}{2mc^2}\frac{dA_0}{dr}$$

and collect our results to give the equation for ψ_2 correct to the first order

$$\left\{\frac{1}{2m}\mathbf{p}^2 - \frac{1}{8m^3c^2}\mathbf{p}^4 - eA_0 - \frac{e\hbar}{4m^2c^2r}\frac{dA_0}{dr}\boldsymbol{\sigma}.\mathbf{l} - \frac{e\hbar^2}{4m^2c^2}\frac{dA_0}{dr}\frac{\partial}{\partial r}\right\}\psi_2 = E\psi_2. \tag{5.40}$$

Our method of derivation is substantially that given in Condon & Shortley (1953), p. 130. That latter derivation is, however, incorrect and leads to an equation differing from (5.40) in the sign of the last term on the left-hand side. The error arose from a false assumption that $(1/2m)\,\mathbf{p}^2 = eA_0 + E$ to zero order. This is only true when the operators in the equation operate on ψ_2. In Condon & Shortley a compensating error occurs on p. 119 so that agreement with experiment is not affected.

The second of the five terms on the left-hand side gives the relativistic variation of mass with velocity. The fourth term is the spin-orbit coupling energy and on writing $\mathbf{s} = \frac{1}{2}\hbar\boldsymbol{\sigma}$ it becomes

$$\delta\mathscr{H} = -\frac{e}{2m^2c^2r}\frac{dA_0}{dr}\mathbf{l}.\mathbf{s}, \tag{5.41}$$

which is the equation we assumed in § 5.1. If $A_0 = Ze/r$ this becomes

$$\delta\mathscr{H} = \frac{Ze^2}{2m^2c^2r^3}\mathbf{l}.\mathbf{s}. \tag{5.42}$$

Equation (5.40) was derived for those parts of space which are not too close to $r = 0$. Now if we use (5.42) and work out a mean value for a function ψ_2 with $l = 0$, this function will not vanish at the origin and so the integral of r^{-3} will be infinite. On the other hand, $\mathbf{l}.\mathbf{s}$ is zero for such a state. It is easy to settle the correct value for the mean of $\delta\mathscr{H}$ by taking f itself in the formula for the spin-orbit coupling to obtain

$$\delta\mathscr{H} = \frac{1}{mr}\frac{df}{dr}\mathbf{l}.\mathbf{s}.$$

Writing $\psi_2 = R(r)\,\tau_s$ as a function of r and spin alone we find

$$\langle\bar{\psi}_2\,|\,\delta\mathscr{H}\,|\,\psi_2\rangle = \langle\tau_s|\int\frac{R^2}{mr}\frac{df}{dr}\mathbf{l}.\mathbf{s}\,r^2\sin\theta\,dr\,d\theta\,d\phi\,|\tau_s\rangle.$$

But
$$\frac{df}{dr} = \left(1+\frac{E+eA_0}{2mc^2}\right)^{-2}\frac{Ze^2}{2mc^2r^2}$$

which is finite at the origin and so $\langle\bar{\psi}_2\,|\,\delta\mathscr{H}\,|\,\psi_2\rangle = 0$ because ψ_2 has all the components of $\mathbf{l}$ zero.

We notice, finally, that we do not take a contribution to $\delta\mathscr{H}$ from ψ_1. This is not because it would be of the second order of small quantities but because we are merely making a first-order perturbation calculation on the equation for ψ_2, and so ψ_1 does not come into it at all.

5.5.3. The interaction with the nuclear magnetic moment. Here we take $A_0 = Ze/r$ and $\mathbf{A} = \gamma\beta_N\hbar^{-1}r^{-3}\mathbf{I}\wedge\mathbf{r}$, where $\mathbf{I}$ is the angular momentum vector for the nucleus, $\beta_N = e\hbar/2Mc$ is the nuclear magneton and γ is a pure number. $\gamma = 2{\cdot}793$ for the proton and would be unity if Dirac's equation applied to the proton. The reason for the deviation is not known. The vector $\mathbf{I}$ commutes with the other observables occurring in the problem. It is convenient to write $\boldsymbol{\mu} = \gamma\beta_N\hbar^{-1}\mathbf{I}$ for the magnetic moment of the nucleus. We then have the classical formula $\mathbf{A} = r^{-3}\boldsymbol{\mu}\wedge\mathbf{r}$ for the vector potential due to a dipole situated at the origin, but must not forget that the components of $\boldsymbol{\mu}$ do not commute with one another.

The only difference between our present calculation and the preceding one is that we now have a non-zero vector potential $\mathbf{A}$. If we write $\Delta\psi_2$ for the additional terms in the equation we are now deriving, then

$$\Delta = \frac{1}{2m}\left\{\boldsymbol{\sigma}.\left(\mathbf{p}+\frac{e}{c}\mathbf{A}\right)f\boldsymbol{\sigma}.\left(\mathbf{p}+\frac{e}{c}\mathbf{A}\right)-\boldsymbol{\sigma}.\mathbf{p}f\boldsymbol{\sigma}.\mathbf{p}\right\},$$

and so $$2m\Delta = \frac{e}{c}(\boldsymbol{\sigma}.\mathbf{A})f(\boldsymbol{\sigma}.\mathbf{p})+\frac{e}{c}(\boldsymbol{\sigma}.\mathbf{p})f(\boldsymbol{\sigma}.\mathbf{A})+\frac{e^2}{c^2}(\boldsymbol{\sigma}.\mathbf{A})f(\boldsymbol{\sigma}.\mathbf{A}).$$

Let us transform $2m\Delta$ in the same way that we have transformed comparable expressions earlier in the section:

$$\begin{aligned}\frac{2mc\Delta}{e} &= f(\mathbf{A}.\mathbf{p}+i\boldsymbol{\sigma}.\mathbf{A}\wedge\mathbf{p})+\boldsymbol{\sigma}.(\mathbf{p}f)(\boldsymbol{\sigma}.\mathbf{A})\\ &\quad +f(\mathbf{p}.\mathbf{A}+i\boldsymbol{\sigma}.\mathbf{p}\wedge\mathbf{A})+\frac{e}{c}f\mathbf{A}^2\\ &= 2f\mathbf{A}.\mathbf{p}+f\hbar\boldsymbol{\sigma}.\operatorname{curl}\mathbf{A}+\boldsymbol{\sigma}.(\mathbf{p}f)(\boldsymbol{\sigma}.\mathbf{A})+\frac{e}{c}f\mathbf{A}^2.\end{aligned} \tag{5.43}$$

We have used $\operatorname{div}\mathbf{A} = 0$. $\operatorname{curl}\mathbf{A} = -\boldsymbol{\mu}/r^3+[3(\boldsymbol{\mu}.\mathbf{r})\,\mathbf{r}]/r^5$ and we now transform the first three terms of (5.43) further:

$$2f\mathbf{A}.\mathbf{p} = \frac{2f}{r^3}\boldsymbol{\mu}\wedge\mathbf{r}.\mathbf{p} = \frac{2f}{r^3}\boldsymbol{\mu}.\mathbf{r}\wedge\mathbf{p} = \frac{2f}{r^3}\boldsymbol{\mu}.\mathbf{l},$$

$$f\hbar\boldsymbol{\sigma}.\operatorname{curl}\mathbf{A} = f\hbar\left(-\frac{\boldsymbol{\mu}.\boldsymbol{\sigma}}{r^3}+\frac{3(\boldsymbol{\mu}.\mathbf{r})(\boldsymbol{\sigma}.\mathbf{r})}{r^5}\right),$$

$$\begin{aligned}\boldsymbol{\sigma}.(\mathbf{p}f)(\boldsymbol{\sigma}.\mathbf{A}) &= -\frac{i\hbar}{r}\frac{df}{dr}(\boldsymbol{\sigma}.\mathbf{r})(\boldsymbol{\sigma}.\mathbf{A})\\ &= -\frac{i\hbar}{r}\frac{df}{dr}(\mathbf{r}.\mathbf{A}+i\boldsymbol{\sigma}.\mathbf{r}\wedge\mathbf{A}) = \frac{\hbar}{r^4}\frac{df}{dr}\boldsymbol{\sigma}.(\mathbf{r}\wedge(\boldsymbol{\mu}\wedge\mathbf{r}))\\ &= \frac{\hbar}{r^2}\frac{df}{dr}(\boldsymbol{\mu}.\boldsymbol{\sigma}-(\boldsymbol{\mu}.\mathbf{r})(\boldsymbol{\sigma}.\mathbf{r})r^{-2}).\end{aligned}$$

On substituting these findings in (5.43) we obtain

$$\Delta = \mathbf{B}.\boldsymbol{\mu}+\frac{e^2}{2mc^2}f\frac{(\boldsymbol{\mu}\wedge\mathbf{r})^2}{r^6}, \tag{5.44}$$

where $$\mathbf{B} = \frac{e}{mc}\left\{f(r^{-3}\mathbf{l}-r^{-3}\mathbf{s}+3r^{-5}(\mathbf{s}.\mathbf{r})\,\mathbf{r})+r^{-2}\frac{df}{dr}(\mathbf{s}-r^{-2}(\mathbf{s}.\mathbf{r})\,\mathbf{r})\right\}. \tag{5.45}$$

The second term in (5.44) belongs to the next order in a perturbation expansion in powers of μ and we therefore reject it.[1] We must now evaluate

$$\Delta = \gamma\beta_N\hbar^{-1}\mathbf{B}.\mathbf{I}$$

for an atomic state. It follows immediately from (5.45) that **B** commutes with **I** and is of type T with respect to the electronic angular momentum vector **j**. The second factor **I** commutes with **j** and is of type T with respect to **I**. Writing $\mathbf{f} = \mathbf{j}+\mathbf{I}$ for the total angular momentum we deduce that Δ commutes with **f**. We wish to know the matrix elements of Δ within the set of states $|jifm_f\rangle$ which arise from a given level of the atom and, of course, from the ground level of the

[1] However, we cannot easily dismiss the term in μ^2 on the ground that it is small, because if we evaluate it for an ns electron we get a divergent integral. If we wish to investigate it we use a two-particle relativistic equation (see Bethe & Salpeter, 1957, p. 196).

nucleus. f ranges from $i+j$ to $|i-j|$ and i is the spin of the nucleus. The commutation properties of Δ and its factors which we listed above show that we can simplify the required matrix elements of Δ in precisely the same way that we simplified those of the spin-orbit coupling in deriving the Landé interval rule in § 5.1. We have

$$E(f) = \langle jifm_f | \Delta | jifm_f \rangle = \tfrac{1}{2}A[f(f+1)-i(i+1)-j(j+1)], \tag{5.46}$$

where A, not to be confused with the electromagnetic potential, is independent of f and m_f. The actual value of A is most easily obtained by methods which we meet later (§ 12.2). It is given in ex. 2, p. 328.

Let us now pass to an n electron system and write

$$\Delta_\kappa = \gamma\beta_N \hbar^{-1}\mathbf{I}.\mathbf{B}_\kappa$$

for Δ for the κth electron. Then it is natural to assume that the correct modification to the n electron Hamiltonian is obtained by adding

$$\delta\mathscr{H} = \sum_{\kappa=1}^{n} \Delta_\kappa = \gamma\beta_N \hbar^{-1}\mathbf{I}.\sum_{\kappa=1}^{n} \mathbf{B}_\kappa. \tag{5.47}$$

This commutes with $\mathbf{F}$, the total angular momentum for the whole system, and in a $JIFM_F$ scheme gives rise to energies

$$E(F) = \tfrac{1}{2}A\{F(F+1)-I(I+1)-J(J+1)\}, \tag{5.48}$$

where we now write I for the nuclear spin. This is proved in the same way as (5.46) and A is a number independent of F and M_F. The value of A is derived using the results for one-electron atoms.

Examples

1. Obtain the terms of Δ, linear in $\mathbf{A}$, in the theory of the interaction with the nuclear magnetic moment directly from Dirac's equation (5.23) by first-order perturbation theory. (Hint: Use (5.30), $W = cp_0$ and remember that the perturbation is $-e\rho_x \boldsymbol{\sigma}.\mathbf{A}$.)

2. Writing $X_1 = \tfrac{1}{2}(\rho_x - i\rho_y)|\psi\rangle$, $X_2 = \tfrac{1}{2}(1-\rho_z)|\psi\rangle$, where $|\psi\rangle$ satisfies (5.22), show that X_2 satisfies (5.31) without using a matrix representation for ρ_x, ρ_y or ρ_z.

5.6. Atoms in external magnetic fields

To pass from the Dirac theory for a single electron to several electrons is not at all a simple matter. We do not possess a relativistically invariant equation, of the same type as Dirac's equation, for more than one particle. This is mainly because of the difficulty of incorporating the electrostatic interaction e^2/r_{12} in the equation. For r_{12} is not a scalar in the space-time continuum. One can, however, in principle still treat the problem fully relativistically but the theory is much more difficult. We therefore adopt an intermediate point of view and, roughly speaking obtain corrections by using Dirac's equation for each electron separately in an average field of nucleus and remaining electrons and work out the other interactions classically.[1] The reader may have omitted the last two sections so I shall now drop explicit reference to Dirac's equation. Other readers will observe that our treatment of the n-electron system is at least partially justified by § 5.5.

We now discuss an atom in a constant external magnetic field **H** described by the vector potential $\mathbf{A} = \frac{1}{2}\mathbf{H}\wedge\mathbf{r}$. We neglect small effects, such as nuclear hyperfine structure, completely in this section. Then the main modifications to our non-relativistic Hamiltonian (2.1) are the inclusion of the spin-orbit coupling energy and the energy of orientation of the spin magnetic moment in the external field. Next in order of importance would usually be the magnetic interactions between the orbital and spin magnetic moments of pairs of electrons. Neglecting these latter, and in the absence of the external field, the modified Hamiltonian is

$$\mathscr{H}_0 = \sum_{\kappa=1}^{n}\left\{\frac{1}{2m}\mathbf{p}_\kappa^2 - \frac{Ze^2}{r_\kappa} + \xi(r_\kappa)\mathbf{l}_\kappa.\mathbf{s}_\kappa\right\} + \sum_{\kappa<\lambda}\frac{e^2}{r_{\kappa\lambda}}. \tag{5.49}$$

In the presence of a field (5.49) becomes

$$\mathscr{H}_0+\mathscr{H}_1 = \mathscr{H}_0 + \sum_{\kappa=1}^{n}\left\{\frac{1}{2m}\left(\mathbf{p}_\kappa+\frac{e}{c}\mathbf{A}_\kappa\right)^2 - \frac{1}{2m}\mathbf{p}_\kappa^2 + \frac{e}{mc}\mathbf{s}_\kappa.\mathbf{H} + \frac{e}{c}\xi(r_\kappa)\,\mathbf{r}_\kappa\wedge\mathbf{A}_\kappa.\mathbf{s}_\kappa\right\}. \tag{5.50}$$

The last term of (5.50) is quite negligible compared with the other terms depending on **H**, the ratio between them being about $2{\cdot}5\times10^{-5}n^{-2}$ for an electron in an nl orbital of hydrogen.

With this observation, and remembering that $\operatorname{div}\mathbf{A} = 0$ we obtain our expression for $\mathscr{H}_1$

$$\mathscr{H}_1 = \sum_{\kappa=1}^{n}\left\{\frac{e}{mc}\mathbf{A}_\kappa.\mathbf{p}_\kappa + \frac{e^2}{2mc^2}\mathbf{A}_\kappa^2 + \frac{e}{mc}\mathbf{H}.\mathbf{s}_\kappa\right\}. \tag{5.51}$$

We have not yet used the particular form $\frac{1}{2}\mathbf{H}\wedge\mathbf{r}$ for **A**. Using the Bohr magneton $\beta = e\hbar/2mc$ we find

$$\begin{aligned}\mathscr{H}_1 &= \sum_{\kappa=1}^{n}\left\{\frac{e}{2mc}\mathbf{H}\wedge\mathbf{r}_\kappa.\mathbf{p}_\kappa + \frac{e^2}{8mc^2}|\mathbf{H}\wedge\mathbf{r}_\kappa|^2 + \frac{e}{mc}\mathbf{H}.\mathbf{s}_\kappa\right\}\\ &= \hbar^{-1}\beta\mathbf{H}.\sum_{\kappa=1}^{n}(\mathbf{l}_\kappa+2\mathbf{s}_\kappa) + \frac{e^2}{8mc^2}\sum_{\kappa=1}^{n}|\mathbf{H}\wedge\mathbf{r}_\kappa|^2\\ &= \hbar^{-1}\beta\mathbf{H}.(\mathbf{L}+2\mathbf{S}) + \frac{e^2}{8mc^2}\sum_{\kappa=1}^{n}|\mathbf{H}\wedge\mathbf{r}_\kappa|^2.\end{aligned} \tag{5.52}$$

The first term of (5.52) is called the paramagnetic part of $\mathscr{H}_1$ and is zero for atoms in ^1S states. The second term is very small and for atoms not in ^1S states is quite negligible compared with the paramagnetic part. It is called the diamagnetic part. In weak fields we regard $\mathscr{H}_1$ as a perturbation small (energies of the order

[1] One can do better than this in problems which merit more detailed treatment. Generalizations of Dirac's equation have been given by Breit, *Phys. Rev.* (1929), **34**, 553; (1932), **39**, 616; Eddington, *Proc. Roy. Soc.* A (1929), **122**, 358; Gaunt, *Proc. Roy. Soc.* A (1929), **122**, 513. For further discussion see Breit, *Phys. Rev.* (1938), **53**, 153; Brown, *Phil. Mag.* (1952), **43**, 467; Bethe & Salpeter, *Quantum Mechanics of One- and Two-Electron Atoms* (Springer-Verlag, 1957).

of 1 cm^{-1}) compared with the separation between levels of a term and have the main contribution to the energy as the first-order perturbation energy due to $\mathscr{H}_1$:

$$E_1 = \langle \overline{\psi} \mid \hbar^{-1}\beta\mathbf{H}.(2\mathbf{J}-\mathbf{L}) \mid \psi \rangle. \tag{5.53}$$

Without loss of generality we take $\mathbf{H}$ parallel to OZ and have

$$E_1 = 2\beta HM - \hbar^{-1}\beta H \langle \overline{\psi} \mid L_z \mid \psi \rangle, \tag{5.54}$$

for a state $\psi = |\alpha' SLJM\rangle$. L_z commutes with J_z and hence $\langle \overline{\psi} \mid L_z \mid \psi \rangle$ is diagonal with respect to J_z (but not with respect to J). By the replacement theorem the matrix element $\langle \overline{\psi} \mid L_z \mid \psi \rangle$ is proportional to M. Its actual value can be derived from Wigner's formula (see Appendix 7) and is

$$\langle \overline{\psi} \mid L_z \mid \psi \rangle = \frac{\hbar M[J(J+1)+L(L+1)-S(S+1)]}{2J(J+1)}, \tag{5.55}$$

whence
$$E = \beta HM\left\{2-\frac{J(J+1)-S(S+1)+L(L+1)}{2J(J+1)}\right\}. \tag{5.56}$$

It is usual to write $E = g\beta HM$, and call g the Landé factor after Landé who discovered the formula empirically before quantum mechanics was able to predict it. Then we have

$$g = 1+\frac{J(J+1)+S(S+1)-L(L+1)}{2J(J+1)}. \tag{5.57}$$

In deriving (5.56) we have neglected the matrix elements of L_z between states of different J. In other words we have supposed those matrix elements small compared with the multiplet splitting between levels. This condition is satisfied in practice for most atoms even for macroscopically strong magnetic fields.

In this book we are not concerned primarily with free atoms and ions but rather with chemical compounds. As we remarked in ch. 1, many compounds can be represented formally as an aggregate of metal ions and other (usually diamagnetic) units. It is therefore instructive to ask what magnetic susceptibility would be expected theoretically from such an aggregate of ions. We shall assume that each ion is in a spherically symmetric electrostatic field due to the rest of the crystal. This, of course, could only be a rough approximation to the truth. However, it is still useful to have the formulae because they often give one a good first approximation even when the energies associated with departure from spherical symmetry of the environment are larger than the energies of interaction with the external magnetic field. We shall see why this is in ch. 10. But there is another reason for obtaining formulae for the interaction of a static magnetic field with ions in spherically symmetric electrostatic fields. We shall show in § 9.5 that there is a detailed correspondence between the theory for d^n in a strong electrostatic field having the symmetry of a regular octahedron and that for the configuration p^{6-n} for the free ions.

In anticipation of this application we now replace $\mathbf{L}$ in (5.52) by $\gamma\mathbf{L}$, where γ

is a real number which is unity for the case of genuine free ions. Then (5.54) becomes

$$E_1 = 2\beta HM + \hbar^{-1}\beta(\gamma-2)\,H\,\langle\overline{\psi}\,|\,L_z\,|\,\psi\rangle, \tag{5.58}$$

and (5.57) is replaced by

$$g = \gamma+(2-\gamma)\frac{J(J+1)+S(S+1)-L(L+1)}{2J(J+1)}. \tag{5.59}$$

If we now assume that E can be expanded as a power series

$$E = W_0+HW_1+H^2W_2+\ldots, \tag{5.60}$$

giving the energy of the atom or ion in the magnetic field H then the magnetic moment in the field direction is

$$\mu = -\frac{\partial E}{\partial H} = -W_1-2W_2H-\ldots. \tag{5.61}$$

The moment per gram ion is then obtained by averaging over all states, weighted with the Boltzmann factor, giving

$$M = N\frac{\Sigma\mu\, e^{-E/kT}}{\Sigma e^{-E/kT}},$$

whence χ, the susceptibility, is

$$\chi = \frac{M}{H} = -\frac{N}{H}\frac{\Sigma\frac{\partial E}{\partial H}e^{-E/kT}}{\Sigma e^{-E/kT}}, \tag{5.62}$$

where N is Avogadro's number.

Now we substitute (5.60) and (5.61) in (5.62) and obtain χ as a power series in H. We also require that the material possesses no mean residual moment in the absence of the field. This is expressed analytically by the equation

$$\Sigma[\mu]_{H=0}\,e^{-W_0/kT} = -\Sigma W_1\,e^{-W_0/kT} = 0,$$

which, from (5.56), is obviously true for free ions. With these assumptions it follows that

$$\chi = NH^{-1}\frac{\Sigma(-W_1-2HW_2-\ldots)\,e^{-W_0/kT}\left(1-\frac{HW_1}{kT}-\ldots\right)}{\Sigma e^{-W_0/kT}\left(1-\frac{HW_1}{kT}-\ldots\right)}$$

$$= N\frac{\Sigma\left(\frac{W_1^2}{kT}-2W_2\right)e^{-W_0/kT}}{\Sigma e^{-W_0/kT}}. \tag{5.63}$$

In (5.63) we have retained only that part of the expansion of χ which is independent of H. This is usually a good approximation except at very low temperatures and high fields. When the approximation breaks down χ depends on the field H and we say that saturation effects are appearing.

It is convenient, before continuing, to rearrange (5.63) slightly. W_0 is the same for all states of a given level so if only one level, ${}^{2S+1}L_J$ say, is occupied thermally then we have

$$\chi(SLJ) = N(2J+1)^{-1}\sum_M\left(\frac{W_1^2}{kT}-2W_2\right), \tag{5.64}$$

and in general
$$\chi = \frac{\Sigma(2J+1)\,\chi(SLJ)\,e^{-W_0/kT}}{\Sigma(2J+1)\,e^{-W_0/kT}}, \tag{5.65}$$

where in (5.65) the sum is over levels not over individual states. Equation (5.65) expresses the fact that the susceptibility is the mean of the susceptibilities of the individual levels weighted by their degeneracy $(2J+1)$ and by the Boltzmann factor $e^{-W_0/kT}$. Because of this, when we know the W_0 and the $\chi(SLJ)$ we can easily deduce the susceptibility χ.

The only states which are appreciably thermally occupied at normal temperatures are those arising from the lowest term. We will assume Russell–Saunders coupling and calculate χ from (5.63). To do this we need to know W_0, W_1 and W_2. It follows from § 5.1 that the relative energies of the levels ${}^{2S+1}L_J$ of a term are given by
$$W_0(JM) = \tfrac{1}{2}\gamma\lambda J(J+1), \tag{5.66}$$

where λ is the spin-orbit parameter for the term (see (5.5)) and γ is the coefficient of **L** introduced earlier in the section. Also we have
$$W_1(JM) = g\beta M. \tag{5.67}$$

We have not yet obtained W_2. It arises because the field **H** gives rise to matrix elements between different levels of a term. The perturbation due to **H** satisfies the same commutation relations with respect to **J** as the electric-dipole moment **D**. It therefore follows from the analysis of § 2.8 that its only non-vanishing elements between different levels of the ground term are from $|JM\rangle$ to $|J\pm 1\,M\rangle$. Since it involves only **L** and **S** there are no non-vanishing elements, in Russell–Saunders coupling, to other terms. J_z itself is diagonal within the ground term in JM quantization so the relevant matrix elements are
$$(\gamma-2)\,\hbar^{-1}\beta H\,\langle J\pm 1\,M\,|\,L_z\,|\,JM\rangle.$$

The actual evaluation of these from Wigner's formula is a little complicated but straightforward. It is given in Appendix 7, where it is shown that
$$\langle J-1\,M\,|\,L_z\,|\,JM\rangle = \langle JM\,|\,L_z\,|\,J-1\,M\rangle$$
$$= -\hbar f(SLJ)\sqrt{(J^2-M^2)},$$
where
$$f(SLJ) = \left(\frac{\{J^2-(L-S)^2\}\{(L+S+1)^2-J^2\}}{4J^2(4J^2-1)}\right)^{\frac{1}{2}}. \tag{5.68}$$

In (5.68) the sign of $f(SLJ)$ depends on the relative phase for the different J values of a term. We have chosen it in accordance with the usual convention, i.e. in accord with (2.49), but it would not make any difference to the results if it were chosen differently. The energy of a level correct to the second order in the magnetic field is now given by second-order perturbation theory as
$$E(JM) = W_0(JM)+HW_1(JM)-(\gamma-2)^2\hbar^{-2}\beta^2H^2\sum_{J'=J\pm 1}\frac{|\langle J'M|\,L_z\,|\,JM\rangle|^2}{E(J')-E(J)},$$

whence using (5.68)

$$W_2 = -(\gamma-2)^2\beta^2\left\{\frac{[f(SL\,J+1)]^2[(J+1)^2-M^2]}{\lambda\gamma(J+1)} - \frac{[f(SLJ)]^2(J^2-M^2)}{\lambda\gamma J}\right\}. \quad (5.69)$$

W_2 depends on M only through the explicit occurrence of M^2 so we can perform the sum over M indicated in (5.63) by using the relations

$$\sum_M 1 = 2J+1, \quad \sum_M M^2 = J(J+1)(2J+1). \quad (5.70)$$

Then by elementary algebraic manipulation we find that

$$\sum_M W_2(JM) = -\frac{1}{3\lambda\gamma}\beta^2(\gamma-2)^2(4J^2-1)\{[f(SL\,J+1)]^2-[f(SLJ)]^2\}, \quad (5.71)$$

where we understand that $f(SLJ) = 0$ when $J = 0$. It is possible to simplify (5.71) further by noticing the algebraic identity

$$(2J-1)(2-\gamma)^2\{[f(SL\,J+1)]^2-[f(SLJ)]^2\} = (g-\gamma)(g-2). \quad (5.72)$$

Combining (5.62), (5.67) and (5.70–2) we finally obtain

$$\chi(SLJ) = \frac{N\beta^2 g^2 J(J+1)}{3kT} + \frac{2N\beta^2(g-\gamma)(g-2)}{3\lambda\gamma}. \quad (5.73)$$

Equation (5.73) is strictly meaningless for $J = 0$ for then g is undefined. If, however, we set $J = S-L$ in (5.59), cancel out the factor $(S-L)$, and then put $J = 0$ we obtain $g = \gamma+(2-\gamma)(L+1)$. This value of g satisfies (5.72) and gives the right value for (5.73) when $J = 0$.

The first part of (5.73) is the ordinary Curie law part of the susceptibility, while the second part is a temperature independent contribution. There should also be a diamagnetic contribution to $\chi(SLJ)$ from the last part of (5.52), but this is relatively very small for an aggregate of paramagnetic ions. It becomes more important when we deal with molecules (see ch. 10). The susceptibility due to all the levels of the ground term is now given by (from (5.65) and (5.66))

$$\chi = \frac{\sum_{J=|L-S|}^{L+S}(2J+1)\chi(SLJ)\exp[-\lambda\gamma J(J+1)/2kT]}{\sum_{J=|L-S|}^{L+S}(2J+1)\exp[-\lambda\gamma J(J+1)/2kT]}. \quad (5.74)$$

The main application in the past of (5.73) and (5.74) (with $\gamma = 1$) has been to the interpretation of the paramagnetic susceptibilities of rare-earth compounds. In those compounds the unpaired electrons are in $4f$ orbitals and it is supposed that they are sufficiently well shielded from the unsymmetrical parts of their environment for the theory appropriate to spherical symmetry about the nucleus to apply (actually it is the smallness of the asymmetry, together with Van Vleck's theorem on the irrelevance of small perturbations, which is responsible here; see § 10.1). Hund originally calculated the susceptibilities of the rare earths using only the first part of (5.73) for the one level which is expected to be the ground level. $J = |L-S|$ for the first half and $J = L+S$ for the second half of the rare-earth series. This is a good approximation to χ at room temperature except for f^5 and f^6 and is equivalent to assuming that kT is small compared with the energy separation between the lowest two levels.

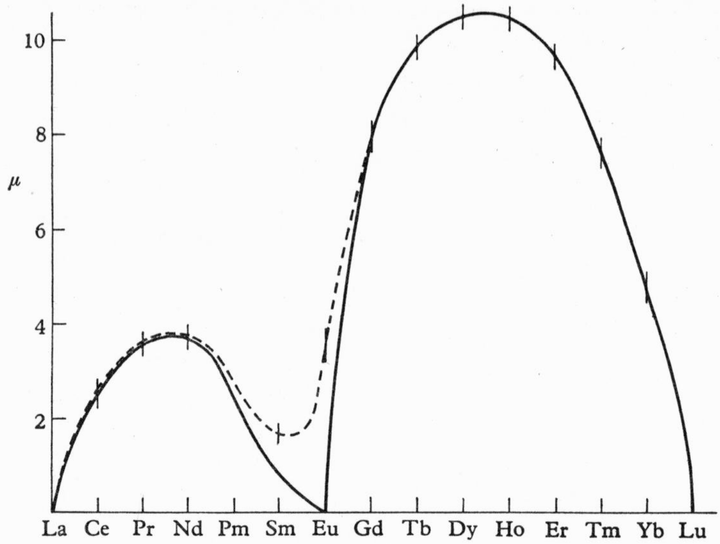

Fig. 5.4. Paramagnetic susceptibilities of rare-earth compounds (after Van Vleck, 1932). $\mu = \sqrt{(3kT\chi/N\beta^2)}$ is actually plotted (see (10.15)). ——, Ground level only; ----, including excited levels; |, experiment.

Van Vleck and Frank calculated the susceptibilities using (5.74) and estimated $\zeta = 1330\,\text{cm}^{-1}$ for Samarium (f^5) and $\zeta = 1530\,\text{cm}^{-1}$ for Europium (f^6). Their comparison with experiment is shown in Fig. 5.4 and is seen to be very good. ζ is now known experimentally for Samarium and Europium and is 1180 and $1360\,\text{cm}^{-1}$, respectively. This would lower the broken curve in Fig. 5.4 slightly.

CHAPTER 6[1]

GROUPS AND THEIR MATRIX REPRESENTATIONS

6.1. Rotations and the concept of a symmetry group

In classifying the states of free atoms we found the parity operator P and the angular momenta $\mathbf{S}$, $\mathbf{L}$ and $\mathbf{J}$ very useful. We saw from the form of L_z in the Schrödinger representation (2.16) that the components of $\mathbf{L}$ are multiples of infinitesimal rotations. Also the eigenfunctions of $\mathbf{L}^2$ and L_z for a single particle (4.10) show that the classification by means of angular momenta reflects the angular behaviour of the wave-functions. This is also true of eigenstates of S_z, $\mathbf{S}^2$, J_z and $\mathbf{J}^2$ and means that we might equivalently have classified the states directly from their behaviour under rotations.

In a free atom or ion it is possible to use angular momenta to classify states because of the spherical symmetry of the system. However, for an ion in a compound we no longer have spherical symmetry nor, in the compounds in which we shall be interested, has one even axial symmetry. Because of this, none of the angular momentum operators commute with the Hamiltonian. If we neglect the spin-orbit coupling energies then $\mathbf{S}$ still commutes with the Hamiltonian and may be used to classify states (we assume this until § 6.9). $\mathbf{L}$, however, does not and so loses a great deal of its usefulness. In spite of this we can still classify states by means of their behaviour under rotations but we must now restrict attention to finite, not infinitesimal, rotations. In this chapter we study this classification for ions in compounds and also see how it can be connected with the classification of states of the free ions by means of $\mathbf{L}^2$ and L_z.

We consider first the crystal field approximation, that is we suppose the ion subject to a static electrostatic field due to its neighbours. Its neighbours will be other atoms or ions which in many cases have a symmetrical arrangement. For example, there may be just six identical neighbours arranged at the vertices of a regular octahedron, i.e. they may be six fluorine ions, with one at $\pm a$ along

[1] The first and most important general discussion of the application of the theory of finite groups and their representations to the theory of ions in crystals, was given by Bethe, *Ann. Phys., Lpz.* (1929), **3**, 133. The concept of a spinor group is due to Cartan. The characters of the irreducible representations of the finite spinor groups T^*, O^* and K^* were first given by Frobenius, *S.B. preuss. Akad. Wiss.* (1899*a*), p. 339.

Following Murnaghan, I do not attempt to give a complete bibliography. My personal knowledge of group theory stems much more from lectures at Cambridge, especially those on group representations by Professor Philip Hall, and from discussions with innumerable people here at Cambridge than from books or journals. The books which are on my shelves and to which I naturally turn for information about finite groups are: Burnside, *Theory of Groups of Finite Order* (Dover, 1955); Ledermann, *The Theory of Finite Groups* (Oliver & Boyd, 1949); Littlewood, *A University Algebra* (Heinemann, 1950), chs. 14 and 15; Murnaghan, *The Theory of Group Representations* (Johns Hopkins, 1938); Rutherford, *Substitutional Analysis* (Edinburgh, 1948); Weyl, *Classical Groups* (Princeton, 1946).

each of the X-, Y- and Z-axes. Then it is clear that there are a number of rotations which leave the overall aspect of the six fluorine ions unchanged and, as we shall say, leave the regular octahedron invariant.

One such rotation would be C_4^z, a rotation of 90° about the OZ-axis. All such rotations must leave V, the electrostatic field due to the neighbours, unchanged. Therefore they commute with V. Any rotation about the nucleus as origin commutes with the Hamiltonian (2.1). The Hamiltonian in the crystal field is

$$\mathscr{H}_c = \sum_{\kappa=1}^{n} \left(\frac{1}{2m}\mathbf{p}_\kappa^2 - \frac{Ze^2}{r_\kappa}\right) + \sum_{\kappa>\lambda}^{n} \frac{e^2}{r_{\kappa\lambda}} + \sum_{\kappa=1}^{n} V(\kappa), \tag{6.1}$$

and we have shown that
$$\mathscr{H}_c R = R\mathscr{H}_c, \tag{6.2}$$
for any rotation which leaves the regular octahedron invariant. We could evidently discuss in the same way any environment of an ion in terms of the rotations which leave the environment invariant. The identity rotation I which leaves everything unchanged trivially always has this property and there may or may not be any other such rotations. However, even if there are not there are very often ones which leave the environment at least approximately invariant and a study of such R is still very useful. Before proceeding, we remark that the restriction to an ion in a static field is not an essential one: to remove it, we merely include some or all of the electrons and nuclei of the environment in the Hamiltonian (6.1) and the problem is formally unchanged. We suppose, however, at the moment that the central nucleus is fixed and that it may be regarded as a point charge.[1]

With the rotations R_i we can also consider the inversion in the origin, P, and reflexions in planes through the origin, σ_i say. We call any of R_i, P, σ_i (or products of them, see next paragraph) which leave the environment of an ion invariant a symmetry element for the system. Any symmetry element is either a rotation or is expressible as the product of P with a rotation. The latter kind is called an improper rotation and, in cases where we need to distinguish them, the former a proper rotation. The symmetry elements all commute with $\mathscr{H}_c$. The aggregate of all symmetry elements is called the symmetry group of the site. We note that we are only considering symmetry groups, each element of which leaves the nucleus of the ion invariant; such groups are called point groups. At the moment we are also only considering the effect of symmetry elements on spatial functions: spin operators and kets are left unchanged by all symmetry elements until § 6.9.

If we apply two symmetry elements in succession, R_1 and R_2 say with R_2 preceding R_1 in order of operation, the net result clearly satisfies the definition of a symmetry element. We write it R_1R_2 and have

$$(R_1R_2)f = R_1(R_2f)$$

for any function f. Evidently R_1 and R_2 do not necessarily commute. We can apply three symmetry elements in succession and obtain the associative law for symmetry elements
$$R_1(R_2R_3) = (R_1R_2)R_3,$$

[1] Logically we are anticipating by talking about crystal fields; the concept is partially justified and discussed in more detail in chs. 7 and 8. The nucleus is allowed to move in § 11.3 and to have some internal structure in § 12.2.

simply because one must operate with R_3, R_2, R_1 in order. Then if we write R^{-1} for the opposite rotation or operation to R and, as above, write I for the identity rotation we have the identities

$$R^{-1}R = RR^{-1} = I,$$

$$RI = IR = R,$$

for all R.

Any set of elements which can be multiplied together and satisfy the identities we have proved for the R_i is called in mathematics a group. This is why we have called the set of R_i a symmetry group.

We give now a formal definition of an abstract group, which is any set G of elements satisfying:

(1) For any $g \in G$, $h \in G$, there is a product gh defined, which also belongs to G.

(2) The associative law $(gh)j = g(hj)$ for any $g, h, j \in G$.

(3) There is a unit element which we write 1 in G such that $1g = g1 = g$ for all g in G.

(4) Every element $g \in G$ has an inverse $g^{-1} \in G$ such that $gg^{-1} = g^{-1}g = 1$.

In defining the group we have used the symbol 'ϵ' of logical inclusion. We always use the multiplicative notation for the group operation.

It is not necessary that the elements of G should be represented as rotations, reflexions, inversions or products of these. An abstract group is simply any set of elements satisfying (1)–(4) above. The importance of showing that the set of all symmetry elements of a site forms a group lies in the fact that there exists a very extensive mathematical theory of groups and we shall shortly use some of the results of that theory in order to study symmetry groups.

Let us now illustrate the concept of a group with a few examples. A group with only two elements is furnished by the pair of numbers 1 and -1. 1 is the unit element and each number is its own inverse. We remark that it is not only true that each element of any group has at least one inverse, it is also true that it has only one inverse. For suppose that h_1 and h_2 are both inverses of the element g. Then

$$gh_1 = h_1g = gh_2 = h_2g = 1$$

and so

$$h_1 = h_1 1 = h_1(gh_2) = (h_1g)h_2 = 1h_2 = h_2.$$

Just as with quantum mechanical operators, the possible non-commutativity of group elements means that we must be careful of their order in products. For gh is not necessarily equal to hg. As a consequence we can divide an equation through by g in two different ways: we multiply it by g^{-1} either on the left or on the right. So $a = b$ implies both $g^{-1}a = g^{-1}b$ and $ag^{-1} = bg^{-1}$. In certain groups, however, all pairs of elements commute. Such groups are called commutative groups (or sometimes they are called Abelian groups). The group containing just 1 and -1 is a commutative group and so also is any group containing only numbers.

Our next example is a permutation group. We met permutations in § 2.6. Their multiplication is clearly associative. The set of all permutations on n symbols contains $n!$ elements and is written S_n. It is called the symmetric group on n symbols. This is normal usage. Be careful not to confuse 'symmetric group' with 'symmetry group'. It is easy to verify that S_n satisfies the definition of a group. The simplest one of any interest is S_3 and this contains six elements. If we write $A = (123)$ and $C = (12)$ we find that

$$A^2 = (132), \quad AC = CA^2 = (13),$$
$$A^3 = C^2 = 1, \quad CA = A^2C = (23),$$

and the multiplication table for the six elements is given in Table 6.1. Evidently S_3 is not a commutative group. The number of elements in a group is called its order. S_3 has order 6 and is the non-commutative group of lowest order.

Table 6.1. *Multiplication table for the group S_3*

S_3	1	A	A^2	C	AC	CA
1	1	A	A^2	C	AC	CA
A	A	A^2	1	AC	CA	C
A^2	A^2	1	A	CA	C	AC
C	C	CA	AC	1	A^2	A
AC	AC	C	CA	A	1	A^2
CA	CA	AC	C	A^2	A	1

Our third example is a rotation group. Take a plane lamina with the shape of a regular n-sided polygon. A rotation of $2\pi m/n$, with m an integer, about the high-symmetry axis leaves the aspect of this lamina unchanged. So also does a rotation of 180° about any axis passing through a vertex and the centre of the polygon. Select the rotation which is $2\pi/n$ in a counterclockwise direction about the high-symmetry axis and call it A. Call any one of the second kind of rotation C. It is not hard to see that A and C satisfy the relations

$$A^n = C^2 = 1, \quad A^mC = CA^{-m}, \quad \text{for all } m.$$

There are $2n$ distinct rotations which leave the lamina invariant and they may be conveniently written A^mC^p, where $m = 0, 1, \ldots, n-1$ and $p = 0$ or 1. The reader may verify that this set of rotations forms a group; it is called the nth dihedral group and is written D_n. The order of D_n is $2n$. D_n is non-commutative for $n > 2$. When $n = 2$ one uses the symbol D_2 to mean the rotation group of a lamina which is rectangular but not square.

The dihedral group D_3 has six elements, just as S_3 did. However, the resemblance goes deeper than this. If we write the multiplication table for D_3 in the same way as for S_3 in Table 6.1 we find that the two tables are identical. So S_3 and D_3 are identical as far as their group structure is concerned. One says they are isomorphic. To be precise one says that two groups G and G' are isomorphic[1] if there exists a $1:1$ correspondence between their elements

$$g \rightleftarrows g',$$

[1] This is the usage of the word isomorphism in modern mathematics and replaces the phrase 'simple isomorphism' which occurs in the older literature.

such that g_1g_2 corresponds to $g_1'g_2'$ or in symbols $(g_1g_2)' = g_1'g_2'$. This definition implies in particular that the unit elements correspond and that inverses correspond. We write $G \cong G'$. In the case of D_3 and S_3 we exhibited the isomorphism by choosing the same symbols for corresponding elements. This is always possible.

We see now the sense in which S_3 is the smallest non-commutative group. There are others of order 6 but they are all isomorphic with it. However, there is another, genuinely different, group of order 6 but it is a commutative one. Suppose we take those rotations which leave a regular hexagonal lamina invariant but do not turn the lamina over. There are six of these, namely A^m for $m = 0, 1, \ldots, 5$ in our previous notation ($A^6 = A^0 = 1$). Each element of this group is expressible as a power of a chosen one of its elements, although not any one as A^2 would not do. Such a group is called a cyclic group. Our cyclic group of order 6 is written C_6 and the general one C_n.

All our examples have been finite groups and finite groups are our main interest. However the concepts of group, commutativity and isomorphism apply equally to infinite groups (i.e. groups which do not contain a finite number of elements). Examples of infinite groups are the set of all non-zero real numbers and the set of all rotations about a fixed point of space.

Examples

1. The four numbers $1, -1, i, -i$ form a group.

2. The set of all real numbers can be made into a group by defining the group product to be the ordinary sum. This group is isomorphic with the multiplicative group of all positive real numbers.

(Hint: Take logarithms.)

3. Any two isomorphic groups have the same order. If two groups are each isomorphic with a third, then they are isomorphic with each other.

4. S_4 is not isomorphic with D_4.

5. The set of all distinct products of the components $\sigma_x, \sigma_y, \sigma_z$ of the vector $\boldsymbol{\sigma}$ of (2.32) forms a group of order 16.

6.2. Elementary properties of groups

The group C_n is, for each n, contained in the group D_n. When such a thing happens we say that the first group is a subgroup of the second. In order for a subset of a finite group to be a subgroup it is both necessary and sufficient for the product of any two elements of the subset to be also in the subset. For if we take an element g of the subset there must then be a number n such that $g^n = 1$. The reader should see why this is so for a finite group.[1] Therefore the unit element is in the subset and so also is the inverse g^{n-1}.

The group S_m is isomorphic with a subgroup of the group S_{m+n} for any $n \geqslant 0$. For we just take those permutations of S_{m+n} which leave the first n symbols unchanged. Any group is a subgroup of itself so S_n is a subgroup of S_n. Any other subgroup is called a proper subgroup.

Sometimes one can find two subgroups H, K of G having only the element 1

[1] It is not necessarily true for an infinite group that there must be such an n for each element.

in common and such that every element of H commutes with every element of K. We then consider the set HK of all products

$$hk, \quad h \in H, \quad k \in K.$$

If two such products are equal, we have

$$h_1 k_1 = h_2 k_2 \to h_2^{-1} h_1 = k_2 k_1^{-1}$$

and since the only element in both H and K is 1 we deduce that

$$h_1 = h_2, \quad k_1 = k_2.$$

Thus all the different products hk are distinct elements of the group and we say that HK is the direct product of H and K and write

$$HK = H \times K.$$

If $HK = G$ we say that we have decomposed G into the direct product of two of its subgroups. For most purposes we may then study G more easily by studying H and K separately. As an example, $C_6 = C_2 \times C_3$ as we see by writing

$$C_2 = (1, A^3) \quad \text{and} \quad C_3 = (1, A^2, A^4).$$

We now take a fixed element h of G and consider the set hG of hg, where g ranges over the whole group G. Then

$$hg = hg' \to h^{-1}hg = h^{-1}hg' \to g = g',$$

so all the different hg are distinct elements of the group. For any $g_1 \in G$ we have

$$g_1 = h(h^{-1}g_1),$$

so the hg comprise all the elements of G once and once only. We could also consider the set Gh and hence

$$hG = Gh = G \quad \text{for any } h \in G,$$

and note that we have already had an example of this result in § 2.6. Table 4.1 illustrates it for S_3 because each element occurs just once in each row and in each column.

We have met the concept of isomorphism. Homomorphism is a related concept. We say the group G' is a homomorphic image of the group G if there is a relation $g \to g'$ between their elements such that $(g_1 g_2)' = g_1' g_2'$. The difference from isomorphism lies in the fact that a given g' may correspond to more than one g in G. But each g of G still corresponds to just one g' in G'. An example is furnished by C_6 and its subgroup $C_2 = (1, A^3)$. We let $g \in C_6$ correspond to g^3 and have C_2 as a homomorphic image of C_6. Each element of C_2 arises from three elements of C_6. We say that there is a homomorphism of C_6 onto C_2 and that C_6 gets mapped onto C_2 under this homomorphism.

Examples

1. Any finite rotation group about a point is a subgroup of the group of all rotations about that point.
2. The group of ex. 1, p. 139 is not the direct product of a pair of its proper subgroups.
3. G is a finite group and H a subgroup. g_1 and g_2 are elements of G but not necessarily of H. Prove that the sets Hg_1 and Hg_2 are identical if they have an element in common. Show that G may be decomposed into a sum of sets Hg_i and that the order of H divides the order of G.
4. The set of even permutations on n objects forms a subgroup of S_n with order $\frac{1}{2}(n!)$.
5. Show that $D_{4n+2} \cong D_{2n+1} \times C_2$.

6.3. The symmetry groups of physics

Before continuing, it is probably useful to give a general survey of the groups which are of interest to us. We discuss pure rotation groups first and have already met two infinite sequences of them, C_n and D_n. It is possible to prove that there are only three more distinct finite rotation groups.[1] These three are called the tetrahedral, octahedral (or cubic) and icosahedral (or dodecahedral) groups and we write them respectively T, O and K. As the names suggest, the groups and their existence are connected with the five Platonic solids. They are in fact the rotation groups of the Platonic solids. A cube may be inscribed in a regular octahedron by placing a vertex in the centre of each face of the latter. Similarly a regular octahedron may be inscribed in a cube. This exhibits the fact that the octahedral group is isomorphic with the cubic group and essentially identical with it. The same holds for the relation between the icosahedral and dodecahedral groups. In Figs. 6.1 and 6.2 the icosahedron and dodecahedron are drawn as a hypothetical observer at infinity might be expected to see them.

We already know the group structures of C_n and D_n. The remaining three groups are conveniently related to permutation groups. First we take T. Here it is useful to extend T to include improper rotations. This extended group is written T_d. A typical reflexion interchanges two vertices of the tetrahedron and leaves the other two unchanged. This naturally suggests that we consider the elements of T_d as permutations of the vertices. Then T_d is related to the symmetric group S_4. Is this relation an isomorphism? Yes, it is. First, we note that every element of the tetrahedral group gets represented as an element of S_4. Then if any $g \in T_d$ is represented as $1 \in S_4$ we must have $g = 1$ because no other symmetry element leaves all four vertices fixed. It follows from the definition that the correspondence of T_d to S_4 is homomorphic and so if g_1 and g_2 both correspond to the same element $h \in S_4$ we find $g_1^{-1}g_2$ corresponds to $1 \in S_4$. But then $g_1^{-1}g_2 = 1$ and so $g_1 = g_2$. Thus the mapping is $1:1$ into S_4 and we have merely to show that every element of S_4 appears. As we saw, every transposition does, and therefore every element does. So we have found that $T_d \cong S_4$.

Every symmetric group S_n contains as a subgroup the set of all even permutations (ex. 4, above). This subgroup is called the alternating group on n

[1] For an elegant and simple proof of this, see Weyl (1952), Appendix A.

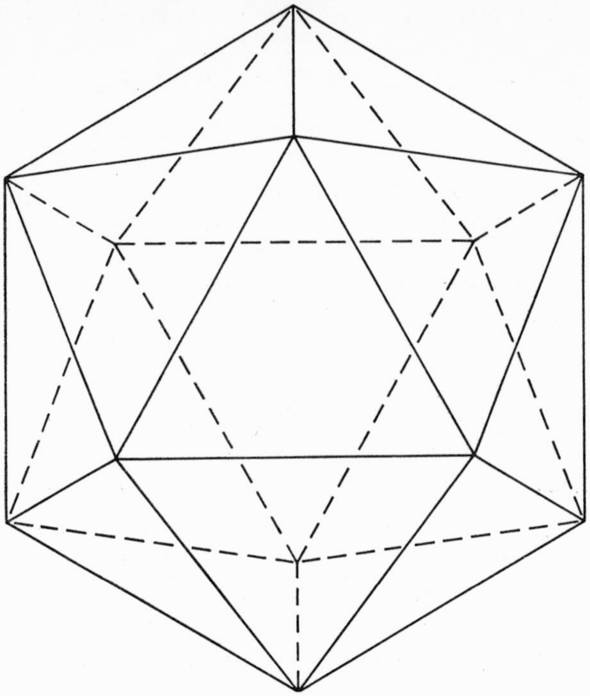

Fig. 6.1. The icosahedron.

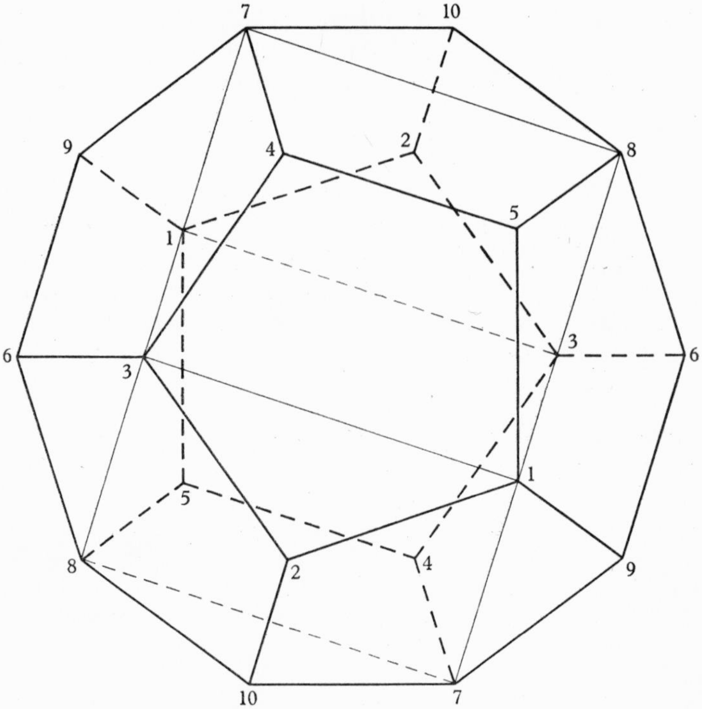

Fig. 6.2. The dodecahedron. The faint line shows a cube drawn in the manner suggested by Euclid.

symbols, is written A_n and has order $\frac{1}{2}(n!)$. Under the correspondence that we have just established, T corresponds to A_4. So T has 12 elements and T_d has 24.

The group O is also isomorphic with S_4. This may be seen by considering O as the rotation group of a cube. A cube has four diagonals which we number from 1 to 4. Let us suppose a diagonal to be unchanged by a rotation if that rotation merely reverses its direction. Then O is exhibited as the symmetric group on the four diagonals. This follows by precisely the same sequence of arguments used for T_d (the transpositions arise from rotations of 180° about axes bisecting the edges of the cube).

The last group K is a little more complicated. Here we use an argument due mainly to Euclid. A cube can be drawn in such a way that each vertex of the cube is a vertex of the dodecahedron and the twelve edges of the cube lie with one in each of the twelve faces of the dodecahedron. This is illustrated in Fig. 6.2 where the cube is seen parallel to two of its faces. Consider now a particular face of the dodecahedron. There are five diagonals and therefore five different cubes can be drawn. It is not possible to draw any more in this manner. On applying a rotation of K these five cubes are permuted amongst each other. We wish to show that K is isomorphic with A_5. This is most easily done by numbering the diagonals of the dodecahedron from 1 to 10 as in Fig. 6.2. Then Euler's five cubes have their diagonals as follows:

cube one	1, 3, 7, 8;
cube two	2, 4, 8, 9;
cube three	3, 5, 9, 10;
cube four	4, 1, 10, 6;
cube five	5, 2, 6, 7.

Any rotation which leaves cube one invariant must permute 1, 3, 7, 8 among themselves and also 2, 4, 5, 6, 9, 10 among themselves. Combining this with the corresponding assertions for cubes two and three we deduce that any rotation which leaves the first three cubes each invariant leaves each of the four diagonals 3, 6, 8 and 9 invariant. Therefore it is the identical rotation. The only essentially different choice of three cubes is one, two and four and in that case diagonals 1, 4, 5 and 8 are left invariant. So any rotation of K which leaves three cubes invariant is the identical rotation and leaves all cubes invariant. This shows that K can be represented as a permutation group on the cubes and then contains no transposition. On the other hand, it contains all the twenty cycles (abc). This is because any pair of cubes d, e say, have a diagonal in common and the two rotations of $\pm\frac{2}{3}\pi$ about this diagonal correspond to just the two cycles (abc) and (acb) which leave both d and e invariant. Apart from the transpositions, there are only two types of odd permutation. These are the thirty cycles $(abcd)$ and the twenty permutations of the type $(ab)(cde)$. As $(abcd) = (ab)(bcd)$, the presence of either type in K would imply the presence of a transposition. Therefore K contains no odd permutation and is isomorphic with A_5 or one of its subgroups. On the other hand, one can count up the number of distinct rotations in

K by considering each kind of possible axis in turn. It is equal to 60, the order of A_5. So $K = A_5$.

Both the cube and the dodecahedron possess a centre of symmetry. We write their groups including improper rotations as O_h and K_i and the group which contains only the inversion P and the identical rotation as C_i. Then $O_h = O \times C_i$ and $K_i = K \times C_i$. The group K_i is not isomorphic with S_5 although it has the same number of elements. T_d does not possess a centre of symmetry and is not isomorphic with $T \times C_i$. The group having the latter symmetry elements is written T_h. A way in which the symmetry T_h could arise is furnished by a cation surrounded by six water molecules. Let us suppose that the cation and all the atoms of any opposite pair of water molecules are coplanar, thus

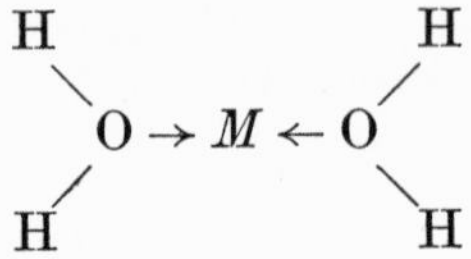

The system has therefore a centre of symmetry and it is possible to arrange the three pairs of H_2O molecules so that the symmetry is T_h. The easiest way to see this is to let each oxygen nucleus lie on an OX-, OY- or OZ-axis and project the

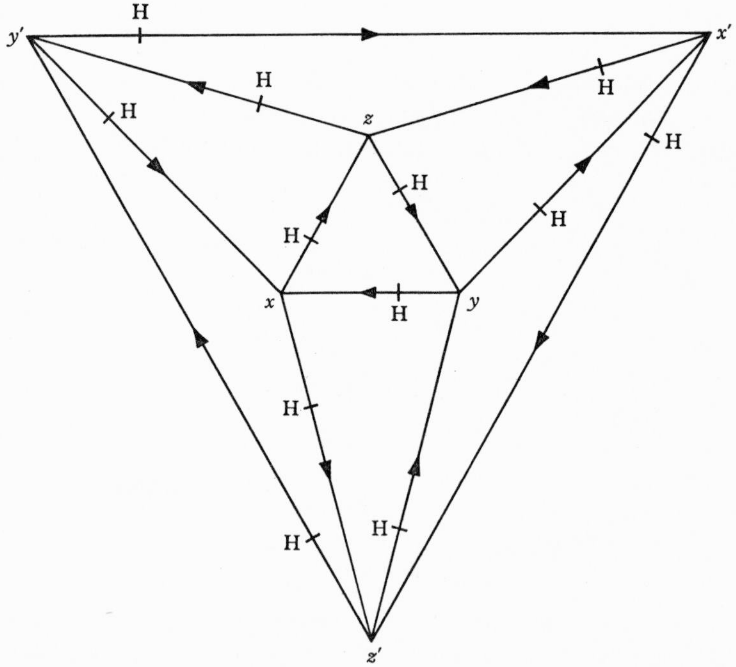

Fig. 6.3. Projection of a regular octahedron showing how six water molecules can achieve T_h symmetry. The oxygen atoms lie at the vertices and are not marked.

octahedron on to a plane from a point just outside one face. This gives Fig. 6.3 where the projections of the oxygen nuclei are marked x, x', y, y', z, z'. The edges have directions marked, and if we arrange the OH_2 so that the OH bonds lie in the directions determined by the outward pair of arrows at each oxygen nucleus then the original system has symmetry T_h. This is quite a reasonable configuration for it on general steric grounds and it would not be surprising if it gave the minimum energy for an isolated complex consisting of just a metal ion and six H_2O molecules. However, in a solid or in solution packing considerations may force the assumption of a lower symmetry.

The groups C_n and D_n have groups, containing improper rotations, which may be said to correspond to them. We note that if a group contains an improper rotation R, then the correspondence $g \to Rg$ within the group is a 1:1 correspondence associating proper rotations with improper rotations. So half the elements of a group are improper if a single one is. Also, if we replace the improper elements h by Ph (P is inversion and so P and Ph may or may not belong to the original group) we obtain either an isomorphic proper rotation group or the proper rotations of the group each repeated twice. So from the point of view of group structure, symmetry groups containing improper rotations add very little. However, they give additional knowledge from the geometrical point of view. I give them all in Table 6.2. A group contains σ_h, the reflexion in a plane at right angles to the high-symmetry axis if and only if it has the suffix h. If it has the suffix v it contains one or more reflexions in a plane containing the high-symmetry axis[1] (D_{nh} does also). S_{2n} is the cyclic group generated by the product $p = r\sigma_h$ of a rotation r through π/n about OZ with σ_h (reflexion in the XY-plane). One is not likely to confuse it with the symmetric group S_{2n}. S_2 has two elements, one being the inversion. C_{1h} has two elements, one being the reflexion σ_h. D_{nh} and D_{nv} are respectively the groups of the regular n-prism and n-antiprism. If we choose the height of the 3-antiprism right its symmetry group changes from D_{3v} to O_h.

Table 6.2. *The distinct improper cyclic and dihedral groups.* $C_i = S_2$ *contains the inversion and* C_{1h} *contains a reflexion*

Group		n even	n odd
C_{nh} S_{2n}	$n \geqslant 1$	$C_n \times C_i$ $\cong C_{2n}$	$C_n \times C_{1h}$ $\cong C_{2n}$
C_{nv} D_{nh} D_{nv}	$n \geqslant 2$	$\cong D_n$ $D_n \times C_i$ $\cong D_{2n}$	$\cong D_n$ $D_n \times C_{1h}$ $D_n \times C_i \cong D_{2n}$

The rotation groups C_n, D_n, T, O and K together with T_d, T_h, O_h, K_i and the groups of Table 6.2 are all the distinct finite point symmetry groups.[2] The most useful in the theory of transition-metal ions are those which in spite of having a low enough symmetry to occur fairly often in practice have a high enough symmetry to contain a fair amount of information. This singles out especially O_h, T_d, D_{4h} and D_{3v}.[3]

[1] D_{nv} is sometimes written D_{nd}.

[2] See, for example, Weyl (1952), p. 80 and Appendix B.

[3] The interesting suggestion has been made (Judd, 1957; Judd & Wong, 1958) that the field around rare-earth ions often possesses, or mimics, K symmetry.

I conclude this section by remarking that the reason that we have arrived at an infinite sequence of groups rather than merely at the 32 crystallographic point groups is that we are interested in local symmetry only and not in space-filling properties. It is only the near environment which concerns the ions to which our theories easily apply. In other words they apply to ions for which cooperative phenomena are not of great importance.

6.4. Matrix representations of groups

The problems involved in the general study and classification of groups are often very difficult and many of them are unsolved. Fortunately, however, they do not concern us at all. For, as we have seen, the kinds of possible symmetry group for the environment of an ion are very limited, each has a simple structure and that structure is in each case completely known. In fact, so long as we restrict ourselves to finite pure rotation groups, there are only three groups other than the two infinite sets of groups C_n and D_n.

The environmental symmetry groups of interest to us are all finite. We assume this property to hold from now on unless we specifically state otherwise. Also the degeneracy of any eigenvalue of $\mathscr{H}_c$ is finite. Then if we operate with a symmetry element R on an eigenstate $|H'\rangle$ of $\mathscr{H}_c$ we have

$$\mathscr{H}_c R\,|H'\rangle = R\mathscr{H}_c\,|H'\rangle = RH'\,|H'\rangle = H'R\,|H'\rangle$$

and so $R\,|H'\rangle$ is also an eigenstate of $\mathscr{H}_c$. If the degeneracy of the eigenvalue H' is n and $|H'i\rangle$ $(i = 1, 2, \ldots, n)$ is an orthonormal set of eigenstates corresponding to this eigenvalue then we must have

$$R\,|H'i\rangle = \sum_{j=1}^{n} a_{ji}\,|H'j\rangle, \tag{6.3}$$

for suitable numbers a_{ji}. We may write (6.3) in matrix form by putting $A = [a_{ji}]$ and taking $|\mathbf{H}'\rangle$ to stand for the row vector with its ith component $|H'i\rangle$, obtaining

$$R\,|\mathbf{H}'\rangle = |\mathbf{H}'\rangle\,A(R), \tag{6.4}$$

where we have written $A(R)$ to show the dependence of A on R. The $|H'i\rangle$ satisfy the orthonormality relations

$$\langle H'i\,|\,H'j\rangle = \delta_{ij}. \tag{6.5}$$

(6.5) is unchanged by a rotation, so substituting in it from (6.3) we get

$$R\,\langle H'i\,|\,H'j\rangle = \sum_{l=1}^{n} \bar{a}_{li}a_{lj} = \delta_{ij}. \tag{6.6}$$

(6.6) shows that A is a unitary matrix whence we deduce

$$\bar{A}'A = A\bar{A}' = I; \quad \sum_{l=1}^{n} \bar{a}_{il}a_{jl} = \delta_{ij}. \tag{6.7}$$

We now operate with R and S successively and obtain

$$SR\,|\mathbf{H}'\rangle = S\,|\mathbf{H}'\rangle\,A(R) = |\mathbf{H}'\rangle\,A(S)\,A(R), \tag{6.8}$$

i.e.

$$A(SR) = A(S)\,A(R), \tag{6.9}$$

and so the group multiplication of S and R goes over into matrix multiplication of $A(S)$ and $A(R)$. In other words we have represented our symmetry group by a group of matrices. Any such representation in terms of unitary matrices, satisfying (6.9) we call a matrix representation and say that the set of kets $|H'i\rangle$ is a basis for it. There is no need for $A(R)$ and $A(S)$ to be distinct if R and S are; for example, we may associate the $n \times n$ unit matrix with every element of the group. In fact we have a homomorphic (sometimes isomorphic) image of our group as a matrix group. From (6.3) we find $A(I)$ is the unit matrix and this result for the unit element of the group is always true for unitary matrix representations.

At this stage it is desirable to be a little more specific about what we mean by the application of a symmetry element to a ket, i.e. in effect to a spatial function. Suppose we take the OXY-plane and use polar coordinates r, ϕ. Consider a function $f(r, \phi)$ and rotate it by an angle α in a counterclockwise direction. Writing R_α for the rotation we find

$$R_\alpha f(r, \phi) = f(r, \phi - \alpha).$$

However, we could alternatively have rotated the coordinate axes and then we would have found

$$f(r, \phi) \to f(r, \phi + \alpha).$$

It is obviously merely a matter of convenience which definition we use. In fact we use the first throughout the book. A particular and very important case is when $f(r,\phi) = h(r)\, e^{ia\phi}$. Then

$$R_\alpha f = h(r)\, e^{ia(\phi-\alpha)} = e^{-ia\alpha} f.$$

We have seen how to derive a matrix representation from a set of eigenstates and evidently the nature of that representation will, conversely, tell us something about the symmetry properties of the eigenstates. However we could, in general, have derived other matrix representations from the same eigenstates. If we take an arbitrary constant $n \times n$ unitary matrix U then

$$|H'_U i\rangle = \sum_{j=1}^{n} U_{ji} \, |H'j\rangle,$$

or, equivalently,

$$|\mathbf{H}'_U\rangle = |\mathbf{H}'\rangle\, U, \tag{6.10}$$

gives an equally acceptable orthonormal set. Conversely every other orthonormal set is related to $|\mathbf{H}'\rangle$ by some unitary matrix U. We now have

$$\begin{aligned} R\,|\mathbf{H}'_U\rangle &= R\,|\mathbf{H}'\rangle\, U = |\mathbf{H}'\rangle\, A(R)\, U \\ &= |\mathbf{H}'\rangle\, U U^{-1} A(R)\, U = |\mathbf{H}'_U\rangle\, U^{-1} A(R)\, U. \end{aligned}$$

In other words the matrices of the representation have been replaced by their transforms by a fixed unitary matrix U. Equation (6.9) is left unchanged in form because

$$\begin{aligned} U^{-1} A(SR)\, U &= U^{-1} A(S)\, A(R)\, U \\ &= U^{-1} A(S)\, U \,.\, U^{-1} A(R)\, U. \end{aligned}$$

Representations related in this way are said to be equivalent and we write $A \sim U^{-1} A U$.

The equivalence relation has the important properties (6.11), (6.12) and (6.13)

$$A \sim A. \tag{6.11}$$

This is because $A = I^{-1}AI$

$$A \sim B \quad \text{implies} \quad B \sim A. \tag{6.12}$$

For $B = U^{-1}AU$ means that

$$A = UBU^{-1} = (U^{-1})^{-1}B(U^{-1}),$$

$$A \sim B \quad \text{and} \quad B \sim C \quad \text{implies} \quad A \sim C. \tag{6.13}$$

For $B = U^{-1}AU$ and $C = V^{-1}BV$, whence

$$C = V^{-1}BV = V^{-1}U^{-1}AUV = (UV)^{-1}A(UV).$$

Hence we can divide the set of all representations into classes of equivalent ones such that any pair in the same class are equivalent and no pair from different classes are equivalent. Pairs from different classes are called inequivalent. Two sets of states $|\alpha' i\rangle$, $|\alpha'' j\rangle$ evidently have essentially the same behaviour under the symmetry group if and only if they form bases for equivalent representations.

If a representation is equivalent to one, A say, which is such that all the matrices can be expressed in the particular form

$$A(R) = \begin{bmatrix} A'(R) & O_{i,n-i} \\ O_{n-i,i} & A''(R) \end{bmatrix}, \tag{6.14}$$

where O_{jk} is the j rowed, k columned zero matrix and i is the same for all R, we say that the representation is reducible. Otherwise it is irreducible. When a representation is reducible and has a set of states $|H'i\rangle$ for a basis this means that we can find U such that the states $|H'_U i\rangle$ separate into two quite independent sets. Then the operation of any symmetry element R on a state of one set yields a linear combination of states of that set only. This means that it is possible to find a perturbation δV to the electrostatic potential having the full symmetry of the site which will separate the energy of the two sets. True cases of accidental degeneracy are rather rare; in practice they usually arise from considering a rough approximation to the Hamiltonian and are then removed by introducing corrections to the approximate Hamiltonian. If a representation A is irreducible and generated by $|H'i\rangle$ then no perturbation with the full symmetry of the site can lift the degeneracy (see § 6.5).

If A is reducible, then if either A' or A'' of (6.14) are reducible we reduce them and so on. Since A has only n rows and columns, the process must terminate (as any 1×1 matrix representation is irreducible) and we have A expressed in terms of irreducible components $A_1, A_2, \ldots, A_m$ ($m \leqslant n$). We write

$$A = A_1 + A_2 + \ldots + A_m, \tag{6.15}$$

and say that A is expressed as the direct sum of its irreducible components. The number of rows of one of the square matrices is called the degree of its representation.

We can define a relationship of equivalence for group elements also but we give it a different name. Two elements g and h of a group G are said to be conjugate[1] if there exists $p \in G$ such that

$$h = p^{-1}gp.$$

In this case the equivalence classes are simply called the classes of the group.

There are a number of remarkable and important properties of irreducible representations which we need to use. The proofs are rather lengthy and it is not necessary to know them in order to use the results. Accordingly, I merely state the properties as propositions without proof and refer the interested reader to standard works on group theory for the proofs.[2]

Proposition 1. The number of inequivalent irreducible representations of a group G is the same as the number of classes.

Writing $A_i(R)_{\mu\nu}$ for the $\mu\nu$th element of the matrix representing R in the ith representation, we state also

Proposition 2.
$$\sum_{R \in G} \bar{A}_i(R)_{\mu\nu} A_j(R)_{\rho\tau} = \frac{\pi}{\lambda_i} \delta_{ij}\delta_{\mu\rho}\delta_{\nu\tau}, \tag{6.16}$$

where π is the order of G and λ_i the degree of A_i. $\bar{A}_i$ is the conjugate complex matrix to A_i. Equation (6.16) involves just products of individual elements of the matrices; there is no matrix multiplication involved.

We now define the character[3] $\chi(R)$ of a group element R represented by the matrix $A(R)$ to be the trace of the matrix, i.e.

$$\chi(R) = \sum_{i=1}^{\lambda} A(R)_{ii}. \tag{6.17}$$

We shall say that the character of a representation is the set of all characters $\chi(R)$ for all the elements of the group. If $A \sim B$, then $B = U^{-1}AU$ and[4]

$$\chi(B) = \chi(U^{-1}AU) = \sum_{i,j,k=1}^{\lambda} U_{ij}^{-1} A_{jk} U_{ki}$$

$$= \sum_{i,j,k=1}^{\lambda} U_{ki} U_{ij}^{-1} A_{jk} = \sum_{j,k=1}^{\lambda} \delta_{kj} A_{jk} = \sum_{j=1}^{\lambda} A_{jj} = \chi(A). \tag{6.18a}$$

Also
$$\chi(A+C) = \chi(A) + \chi(C). \tag{6.18b}$$

Thus equivalent representations have the same set of characters and, similarly, equivalent elements of the group have the same character. So the character

[1] Also sometimes called similar or equivalent.

[2] See, for example: Littlewood, *loc. cit.*, ch. 15; Weyl, ch. 4, §1; Murnaghan, ch. 3, §2. For our purposes we could, of course, by-pass the general theorems and establish propositions 1 and 2 just for the groups we need by actual consideration of the representations. This would be tedious but possible (one would then use proposition 2 to limit the number of representations).

One may comment that to physicists propositions 1 and 2 often seem the most important theorems in group theory. They are still important in the mathematical theory but others probably transcend them there (for example, the Jordan–Hölder theorem and the Sylow theorems about maximal prime-power subgroups).

[3] Sometimes called characteristic in mathematics.

[4] U_{ij}^{-1} stands for the ijth element of U^{-1} here.

is a property of classes rather than individual elements, and if the group has p classes and hence also p representations we may arrange the characters in a table, called the character table of the group, of p rows and p columns.

If we put $\mu = \nu$ and $\rho = \tau$ in (6.16) we obtain

$$\sum_{R \in G} \bar{A}_i(R)_{\mu\mu} A_j(R)_{\rho\rho} = \frac{\pi}{\lambda_i} \delta_{ij} \delta_{\mu\rho}$$

whence on summing over μ and ρ

$$\sum_{R \in G} \bar{\chi}_i(R) \chi_j(R) = \pi \delta_{ij}. \tag{6.19}$$

If we write the classes as c_k $(k = 1, 2, \ldots)$, $\chi_i(c_k)$ to represent the character of each member of the class and π_k for the number of elements in c_k we have

$$\sum_{k=1}^{p} \pi_k \bar{\chi}_i(c_k) \chi_j(c_k) = \pi \delta_{ij},$$

or

$$\sum_{k=1}^{p} \left(\frac{\pi_k}{\pi}\right)^{\frac{1}{2}} \bar{\chi}_i(c_k) \left(\frac{\pi_k}{\pi}\right)^{\frac{1}{2}} \chi_j(c_k) = \delta_{ij}. \tag{6.20}$$

$(\pi_k/\pi)^{\frac{1}{2}} \chi_i(c_k)$ thus forms a unitary matrix and so we have also

$$\sum_{k=1}^{p} \left(\frac{\pi_i}{\pi}\right)^{\frac{1}{2}} \bar{\chi}_k(c_i) \left(\frac{\pi_j}{\pi}\right)^{\frac{1}{2}} \chi_k(c_j) = \delta_{ij}, \tag{6.21}$$

which for $i = j$ gives

$$\sum_{k=1}^{p} |\chi_k(c_i)|^2 = \frac{\pi}{\pi_i}. \tag{6.22}$$

The class containing the unit element of the group has just the one element and for that class $\chi_k(c_1) = \lambda_k$ the degree of the kth representation, so from (6.22)

$$\sum_{k=1}^{p} \lambda_k^2 = \pi. \tag{6.23}$$

We illustrate the preceding results by discussing the simplest possible non-trivial symmetry group, namely C_2. The elements are 1 and A with $A^2 = 1$. Equation (6.23) becomes

$$\sum_{k=1}^{p} \lambda_k^2 = 2,$$

which means that $p = 2$, $\lambda_1 = \lambda_2 = 1$. There are also two classes, for each of 1 and A by itself forms a class. This is proposition 1 for this group. A representation of degree 1 means that we can represent each group element by a number. Let 1 and A be represented respectively by ϵ and γ. Then ϵ and γ satisfy:

(i) $\epsilon^2 = \epsilon$ which means $\epsilon = 1$;

(ii) $\gamma^2 = \epsilon = 1$, so $\gamma = \pm 1$.

This gives us our two representations, A_1 with $\gamma = +1$ and A_2 with $\gamma = -1$. They clearly satisfy (6.16) and (6.19). Equation (6.16) is always equivalent to (6.19) for representations of degree 1. The character table of C_2 is written as shown in Table 6.3. Every group has a representation, like A_1, in which each group element is represented by the number 1. It is called the unit representation.

Table 6.3. *Character table of C_2*

C_2	1	A
A_1	1	1
A_2	1	-1

Examples

1. Derive the representations and character table for C_6.
2. A reducible representation of degree 2 is commutative.
3. Derive the character table of D_3.

(Hint: Consider the pair of functions $(\cos\phi, \sin\phi)$ and use ex. 2.)

4. A representation of G with character χ is irreducible if and only if $\sum_{R\in G}|\chi(R)|^2 = \pi$, where π is the order of G.

6.5. The direct product of two representations

If a function belongs to a basis for an irreducible representation of a group this clearly tells us something about the nature of the function. This information is the analogue in the case of a site with a finite symmetry group of the classification in terms of L, M_L and parity for the free atom. The statement that a function forms part of a basis for a representation A_i is the analogue of the statement that it is an eigenfunction of $\mathbf{L}^2$ and of parity with a particular eigenvalue for each. The M_L value corresponds in the present situation with a statement that the function is the jth component of the basis. In more detail, the operators L_z and $L^{\pm}$ are replaced collectively by all the elements of the finite group. We include spin later (§ 6.9). For the moment we suppose that the elements of the symmetry group have no effect on spin functions. This is the analogue of the SLM_SM_L scheme, for the environment only directly influences spatial functions. Then the behaviour of functions under the group elements tells one both the representation to which they belong and also how to turn one function of a basis for an irreducible representation into any other one of the same basis. The reason for the existence of these analogies is that atomic terms do actually form representations of the group consisting of all rotations about the nucleus. And although L_z and $L^{\pm}$ do not belong to this group they can be obtained from elements of the group by limiting processes (they are multiples of infinitesimal rotations).

What corresponds to the procedure of § 2.3 for vector addition of two angular momenta $\mathbf{L}_1$ and $\mathbf{L}_2$? The answer is the direct product and I describe this now. If we have two sets $|\gamma' i\rangle$ and $|\gamma'' j\rangle$ say, each forming a basis for a representation of G then we can construct a new representation by writing, for $R \in G$,

$$R\,|\gamma' i\rangle|\gamma'' j\rangle = \sum_{\mu\nu} A_1(R)_{\mu i}\,A_2(R)_{\nu j}\,|\gamma'\mu\rangle|\gamma''\nu\rangle. \tag{6.24}$$

We abbreviate (6.24) by writing $|ij\rangle$ for $|\gamma' i\rangle|\gamma'' j\rangle$ and have

$$R\,|ij\rangle = \sum_{\mu\nu} A_1(R)_{\mu i}\,A_2(R)_{\nu j}\,|\mu\nu\rangle. \tag{6.25}$$

Equation (6.25) gives a new representation of G, written A_1A_2,

$$A_1A_2(R)_{\mu\nu;ij} = A_1(R)_{\mu i}\,A_2(R)_{\nu j}$$

and is called the direct product of the representations A_1 and A_2. For the group of order 2 that we discussed in § 6.4 we have

$$A_1^2 = A_2^2 = A_1, \quad A_1 A_2 = A_2.$$

However, in general, the product of two irreducible representations will be reducible and then we can only express it as a direct sum of a number of irreducible components.

The character of $A_1 A_2(R)$ is given by

$$\chi\{A_1 A_2(R)\} = \sum_{\mu\nu} A_1 A_2(R)_{\mu\nu;\mu\nu} = \sum_{\mu\nu} A_1(R)_{\mu\mu} A_2(R)_{\nu\nu}$$

$$= \chi(A_1)\,\chi(A_2), \tag{6.26}$$

So we can work out the character of the direct product of two irreducible representations very easily once we know the characters of each of those representations. This in fact suffices to determine the decomposition of $A_1 A_2$ as a direct sum of irreducible components. For if

$$B = \sum_i n_i A_i$$

then, by (6.18)

$$\chi_B(R) = \sum_i n_i \chi_i(R)$$

whence, multiplying by $\bar{\chi}_j(R)$, summing over R and using (6.19)

$$\sum_R \chi_B(R)\,\bar{\chi}_j(R) = \sum_i n_i \sum_R \chi_i(R)\,\bar{\chi}_j(R)$$

$$= \sum_i n_i \pi \delta_{ij} = \pi n_j. \tag{6.27}$$

We see also that

$$\sum_R \bar{\chi}_B(R)\,\chi_B(R) = \pi \sum_i n_i^2,$$

so, comparing this with (6.19), a representation χ is irreducible if and only if

$$\sum_R |\chi(R)|^2 = \pi. \tag{6.28}$$

Equation (6.27) is of immense importance. It gives a method of determining the irreducible components of a representation from a knowledge of the character alone. In particular it shows that representations are equivalent if and only if they have the same character. This justifies the use of the word character, for it shows that the character really does characterize a representation, and indicates that for many purposes we can discuss characters rather than the actual matrices of representations. This effects a great simplification.

An important practical example of a direct product occurs when we have a two-electron system and wish to express states for the system in terms of ones in which one electron is placed in one set of orbitals $|\gamma' i\rangle$ and the other in a different set $|\gamma'' j\rangle$ say. Then the antisymmetrized kets are

$$|ij\rangle = \frac{1}{\sqrt{2}}[|\gamma' i(1)\rangle|\gamma'' j(2)\rangle - |\gamma' i(2)\rangle|\gamma'' j(1)\rangle],$$

and the set of $|ij\rangle$ forms a basis for the product representation as before. Because of the exclusion principle one has to be more careful when the two electrons are in the same set and then the antisymmetrized $|\gamma' i\rangle|\gamma' j\rangle$ do not form a basis for the direct product.

A very important use of the direct product is that it enables us to deduce some very useful orthogonality relations for integrals. They are the counterpart in the finite group case of the selection rules on $\mathbf{L}^2$ and L_z. Let $|\gamma_1 i\rangle$ be a basis for an irreducible representation A_1 and $|\gamma_2 i\rangle$ for an irreducible representation A_2. We do not assume here that the kets are normalized. If $A_1 = A_2$ then suppose that $|\gamma_1 i\rangle$ spans not merely a representation equivalent to A_1, but identically the same one. The possibility of arranging this follows from the definition of equivalence. Then $\langle\gamma_1 i\,|\,\gamma_2 j\rangle$ is just a number and is therefore unaffected by any rotation R applied to it. In symbols

$$\langle\gamma_1 i\,|\,\gamma_2 j\rangle = R\,\langle\gamma_1 i\,|\,\gamma_2 j\rangle = \sum_{\mu,\nu} \bar{A}_1(R)_{\mu i}\,A_2(R)_{\nu j}\,\langle\gamma_1\mu\,|\,\gamma_2\nu\rangle, \tag{6.29}$$

where we have used (6.3). We now sum over all R in G and use orthogonality relations (6.16) to obtain

$$\begin{aligned}\pi\,\langle\gamma_1 i\,|\,\gamma_2 j\rangle &= \sum_{R\in G} R\,\langle\gamma_1 i\,|\,\gamma_2 j\rangle \\ &= \sum_{\mu,\nu,R} \bar{A}_1(R)_{\mu i}\,A_2(R)_{\nu j}\,\langle\gamma_1\mu\,|\,\gamma_2\nu\rangle \\ &= \frac{\pi}{\lambda_1}\,\delta_{A_1A_2}\sum_{\mu,\nu}\delta_{\mu\nu}\delta_{ij}\,\langle\gamma_1\mu\,|\,\gamma_2\nu\rangle = \frac{\pi}{\lambda_1}\,\delta_{A_1A_2}\delta_{ij}\sum_{\mu}\langle\gamma_1\mu\,|\,\gamma_2\mu\rangle. \end{aligned} \tag{6.30}$$

Equation (6.30) shows a number of things. It shows that $\langle\gamma_1 i\,|\,\gamma_2 j\rangle$ is zero unless the two functions belong to the same irreducible representation. Then the matrix element is zero even when $A_1 = A_2$ unless the two functions correspond, i.e. $i = j$. Finally, when $A_1 = A_2$ and $i = j$ we have

$$\langle\gamma i\,|\,\gamma i\rangle = \lambda^{-1}\sum_{\mu}\langle\gamma\mu\,|\,\gamma\mu\rangle.$$

The right-hand side is independent of i and so $\langle\gamma i\,|\,\gamma i\rangle$ is the same for all i (λ, the degree of the representation, is equal to the number of terms in the sum so everything is internally consistent). This result is the exact analogue of our earlier result that $\langle LM_L\,|\,LM_L\rangle$ is independent of M_L when the kets are properly connected by shift operators. It suggests to us that in the case of finite groups 'spanning an identical representation' should take the place in our analysis previously occupied by 'being properly connected by shift operators'. This proves to be so.

Suppose now we want the matrix elements of an operator ξ which commutes with all rotations of the group. $\mathscr{H}_c$ is a typical example, but any function which is invariant under all elements of the group comes into this category. Then $\xi\,|\gamma_2 j\rangle$ for varying j span identically the same representation as the $|\gamma_2 j\rangle$. So we deduce immediately that $\langle\gamma_1 i\,|\,\xi\,|\,\gamma_2 j\rangle$ is zero unless $A_1 = A_2$ and $i = j$ and that it is independent of j when these conditions are satisfied. This result is a very important one and demonstrates formally the usefulness of classification according to irreducible representations. For we have only to take $\xi = \mathscr{H}_c$ to be able to say that different functions of a basis for an irreducible representation have the same energy.

In the case that A_2 is possibly reducible and ξ does not necessarily commute with all elements of G we write A_3 for the representation for which ξ forms part of a basis. For a completely general ξ, A_3 is a sum over all representations, but this case does not often occur. Then $A_3 A_2 = \Sigma n_i A_i$ and we deduce that $\langle \gamma_1 i \,|\, \xi \,|\, \gamma_2 j\rangle$ is necessarily zero if $n_1 = 0$.

We conclude this section by answering a rather practical question. It is evident from the preceding discussion that we should choose for each kind of irreducible representation of each group of interest a particular representative once and for all. This means choosing a set of matrices, not a particular basis. We call this a choice of standard basis relations between the components of bases of irreducible representations. Then whenever we come across a set of functions forming a basis for an irreducible representation Γ we choose that set of linear combinations which forms a basis for the standard representation which is equivalent to Γ. There is essentially only one because if

$$|a\gamma i\rangle \quad \text{and} \quad |b\gamma i\rangle = \sum_j \alpha_{ij}\,|a\gamma j\rangle$$

are normalized kets spanning identically the same representation it follows from (6.30) and its consequences that for some β independent of i:

$$\beta\delta_{ik} = \langle a\gamma i \,|\, b\gamma k\rangle = \sum_j \alpha_{kj}\langle a\gamma i \,|\, a\gamma j\rangle$$

$$= \sum_j \alpha_{kj}\delta_{ij}\langle a\gamma i \,|\, a\gamma i\rangle = \alpha_{ki}\langle a\gamma i \,|\, a\gamma i\rangle,$$

$\langle a\gamma i \,|\, a\gamma i\rangle$ is independent of i and cannot be zero unless $|a\gamma i\rangle$ itself is zero. So α_{ki} is a multiple of δ_{ik} and since it is a unitary matrix it is equal to $e^{i\phi}\,\delta_{ik}$ for some phase factor $e^{i\phi}$. Thus the $|b\gamma i\rangle$ are determined up to an arbitrary phase factor common to all of them.

Our practical problem is the following. Given a set of functions which forms a basis for an irreducible representation we wish to take linear combinations in such a way that they satisfy the standard basis relations. In order for this to be so, each element of the group must get its correct matrix representation. But we surely do not need to check this for every element of the group. What, then, is the minimum number of group elements for which we must make this check? The answer, for our groups, is never more than two. To see this we introduce the concept of a set of generators for a group. A set of elements g_i is a set of generators for a group G if every element can be expressed as a product of the elements g_i, the product possibly containing many repetitions of any particular g_i. There will usually be many possible sets of generators; one could even take the whole group G but this would not be a very helpful choice. Then if each of the generators gets the right matrix so does every element of G. Given a set of generators, a set of generating relations for G is a set of formulae giving relations between the generators and sufficient of them to ensure that any group with those generators and satisfying the generating relations must be isomorphic with G. This latter concept is less useful to us but I give it for interest and completeness.

Sets of generators and generating relations for our groups are easily written down and are:

For C_n:	generator A	$A^n = 1$;	
D_n:	generators A, C	$A^n = C^2 = 1$,	$A^mC = CA^{-m}$;
$O \cong T_d$:	generators A, C, D	$A^4 = C^2 = D^3 = 1$,	$CDA = 1$;
K:	generators A, C, D	$A^5 = C^2 = D^3 = 1$,	$CDA = 1$;
T:	generators C, D	$C^2 = D^3 = 1$,	$(CD)^3 = 1$.

Suitable choices for A, C and D are illustrated in Fig. 6.4. For both O and K we could take any two of the three generators and this would be sufficient because of $CDA = 1$. It is convenient for many purposes to take A and $B = CD^2$ for the generators of O. It is easy to verify that they are generators and that B is drawn correctly in Fig. 6.4.

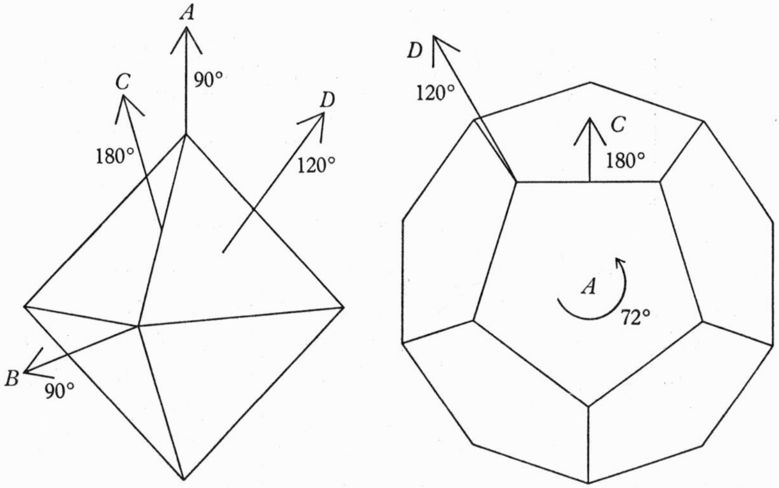

Fig. 6.4. Generators for the octahedral and icosahedral groups. The rotations are counterclockwise by the amount indicated. D for the group K is offset slightly to make its point of origin visible.

Examples

1. Two groups G_1 and G_2, each of order 8, are generated by a and b. $a^4 = 1$ and $aba = b$. $b^2 = 1$ in G_1 and $b^2 = a^2$ in G_2. Show that $G_1 \cong D_4$, that G_1 and G_2 have the same classes and character table and that G_1 is not isomorphic with G_2.

2. (Schur's lemma: Schur, 1906.) Show that any matrix which commutes with all matrices of an irreducible representation must be a multiple of the unit matrix. (In a slightly more general form this lemma is equivalent to the orthogonality relations (6.16).)

3. An orthonormal set of functions f_i ($i = 1, 2, \ldots, n$) form a basis for a matrix representation of a group G. Show that $\sum_{i=1}^{n} |f_i|^2$ forms a basis for the unit representation of G.

6.6. The representations of C_n, D_n, T_d, O and K

6.6.1. Preliminary observations. The actual determination of the representations of the symmetry groups is facilitated by some preliminary observations. First, we note that if $A(R)_{\mu\nu}$, for R ranging over a group G, is a representation then two representations of the direct product $G \times C_2$ of G with a group C_2 containing just two elements can be derived immediately. $C_2 = (1, g)$ and we have A' and A'' defined by

$$A'(R)_{\mu\nu} = A''(R)_{\mu\nu} \quad = A(R)_{\mu\nu} \quad \text{for all } R \in G,$$
$$A'(Rg)_{\mu\nu} = -A''(Rg)_{\mu\nu} = A(R)_{\mu\nu} \quad \text{for all } R \in G.$$

A' and A'' are irreducible if A is (use (6.28)) so if we possess a complete set of irreducible representations of G we derive a complete set for $G \times C_2$. The completeness is assured by (6.23) for $G \times C_2$. Therefore once we know the representations of C_n, D_n, T_d, O and K the representations of all the improper rotation groups follow immediately. In fact, as far as character tables are concerned, if M is the character table of G, then

Table 6.4. *Character table of* $G \times C_2$

$G \times C_2$	$(R_1) - (R_n)$	$(gR_1) \ldots (gR_n)$
A'_1 ⋮ A'_n	M	M
A''_1 ⋮ A''_n	M	$-M$

is the character table of $G \times C_2$. The (R_i), (gR_i) are the classes.

A second point to notice is that useful properties of the representations of T_d, O and K can be derived from the expression of those groups as permutation groups. The symmetric group S_n always has two representations of degree 1. The first Γ_1 associates the number 1 with every element of S_n. It is the unit representation of S_n. The second Γ_2 associates the parity $(-1)^\mu$ with the permutation P_μ. It is called the alternating representation.

Γ_1 and Γ_2 enable us to demonstrate a very important relationship among the representations of the symmetric groups which we can establish easily and then take over as a property of the isomorphic symmetry groups. If Γ_i is any matrix representation, the representation $\Gamma_i^* = \Gamma_2 \Gamma_i$ is also a matrix representation. Γ_i and Γ_i^* are called associated representations. Their relationship is mutual because, clearly, $(\Gamma_i^*)^* = \Gamma_i$.

If the representations Γ_i and Γ_i^* are equivalent then Γ_i is said to be self-associated. We may obtain a necessary and sufficient condition for Γ_i to be self-associated by considering its character $\chi(\Gamma_i)$. If P is a permutation of the group, then

$$\Gamma_i^*(P) = \Gamma_2(P)\, \Gamma_i(P)$$

and we deduce from (6.26) that

$$\chi_i^*(P) = \chi_2(P)\, \chi_i(P) = (-1)^\mu \chi_i(P), \tag{6.31}$$

where μ is the parity of the permutation P. Now Γ_i and Γ_i^* are equivalent if and only if $\chi_i^*(P) = \chi_i(P)$ for all P in the group. From (6.31), this means that $\chi_i(P) = 0$ for all the odd permutations P and is our necessary and sufficient condition for Γ_i to be self-associated.

If we have found a representation Γ_i which is not self-associated then the representation Γ_i^* gives us a completely new one, i.e. a non-equivalent one. We have seen how to recognize if Γ_i is self-associated or not and shall use the process of association to assist in determining the matrix representations of T_d and O.

We next show that the classes of a permutation group are immediately recognizable. Suppose two permutations P and Q are expressed as products of independent cycles. Then if each has the same number of cycles of each order they are said to be conjugate.[1] Thus (123) (45) (6) is conjugate to (135) (46) (2) but is not conjugate to (12) (34) (56). Then we shall show that P and Q are conjugate if and only if there is an R in S_n such that $P = R^{-1}QR$. For let us write

$$Q = \begin{pmatrix} 1 & 2 & \dots & n \\ i_1 & i_2 & \dots & i_n \end{pmatrix}, \quad R = \begin{pmatrix} j_1 & j_2 & \dots & j_n \\ 1 & 2 & \dots & n \end{pmatrix} = \begin{pmatrix} k_1 & k_2 & \dots & k_n \\ i_1 & i_2 & \dots & i_n \end{pmatrix},$$

whence

$$R^{-1}QR = \begin{pmatrix} i_1 & i_2 & \dots & i_n \\ k_1 & k_2 & \dots & k_n \end{pmatrix}\begin{pmatrix} 1 & 2 & \dots & n \\ i_1 & i_2 & \dots & i_n \end{pmatrix}\begin{pmatrix} j_1 & j_2 & \dots & j_n \\ 1 & 2 & \dots & n \end{pmatrix} = \begin{pmatrix} j_1 & j_2 & \dots & j_n \\ k_1 & k_2 & \dots & k_n \end{pmatrix}.$$

In other words $R^{-1}QR$ is obtained from Q by applying the permutation R^{-1} to *all* the numbers in the bracket representation of Q. But this means that $R^{-1}QR$ has the same cycle structure as Q, i.e. they are conjugate. Conversely, if P and Q are conjugate there exists a permutation R which, when applied to the numbers of P, turns it into Q (because R cannot affect the cycle structure). Hence $P = R^{-1}QR$ as above.

The foregoing argument shows that P and Q are conjugate as permutations when and only when they are conjugate as group elements of the group S_n. This means that if we can show that a group G is isomorphic with one of the S_n we can immediately recognize its classes.

Equipped with these results, we discuss S_3. There are three classes:

$$1; \quad (123), (132); \quad (13), (23), (12).$$

Therefore there are three representations of which Γ_1 and Γ_2 are two. Then

$$\lambda_1^2 + \lambda_2^2 + \lambda_3^2 = 6 \quad \text{implies} \quad \lambda_3 = 2.$$

So the last representation has degree 2. It must be self-associated, because it is the only one of degree 2, and so the character of the third class is zero. Our character table (Table 6.5) now reads

Table 6.5

S_3	1	(123), (132)	(12), (23), (31)
$\chi_1 = \chi(\Gamma_1)$	1	1	1
$\chi_2 = \chi(\Gamma_2)$	1	1	−1
$\chi_3 = \chi(\Gamma_3)$	2	x	0

[1] Sometimes called similar.

In order for χ_3 to be orthogonal to χ_1 and χ_2 in the sense of (6.19) we must have $x = -1$. This gives the character table for S_3. The matrices of the representation Γ_3 are most easily obtained from $D_3 \cong S_3$. For clearly the pair ($\cos\phi$, $\sin\phi$) form a basis for Γ_3 of D_3.

6.6.2. The groups C_n, D_n, T_d, O and K. Let us now return to the groups C_n, D_n, T_d, O and K. First, we ask what representations they have of degree 1. Take C_n first and let A be represented by x. Then $x^n = 1$ and $x = e^{2\pi i m/n}$ with m an integer. Letting m be successively $0, 1, \ldots, n-1$ we have n distinct representations of C_n and hence all the irreducible representations. Next we take D_n and let A and C be represented by x and y, respectively. Then $x^n = y^2 = 1$ and, from the generating relations, $x^m y = y x^{-m}$ for each integer m, Therefore $x^2 = y^2 = 1, x = \pm 1, y = \pm 1$. If n is odd, $x^n = 1$ then implies $x = 1$. So for n odd, D_n has two representations of degree 1 and for n even it has four. The reader may like to use the same arguments to show that T_d and O have two representations of degree 1 and K has only one. The former representations correspond to Γ_1 and Γ_2 of S_4.

The remaining representations of D_n are found most easily by considering the pair of functions ($\cos m\phi$, $\sin m\phi$) for a general integer m. We let $m = 1, 2, \ldots, \lambda$ and then for both $D_{2\lambda+1}$ and $D_{2\lambda+2}$ the representations spanned by ($\cos m\phi$, $\sin m\phi$) are all inequivalent and non-commutative. These assertions may be verified immediately. Ex. 2, p. 151, then shows that they are irreducible. Together with the representations of degree 1, there are just enough in each case to satisfy (6.23), and so we have all the irreducible representations for all D_n. An interesting example is the character table of D_6 which is given in Table A 2. It illustrates the relationship of the table for a group $G \times C_2$ to the table for G which we discussed earlier. For $D_6 \cong D_3 \times C_2$.

Next we have $T_d \cong O \cong S_4$. There are five classes of S_4 so, with our knowledge that there are just two representations of degree 1, we deduce

$$\lambda_3^2 + \lambda_4^2 + \lambda_5^2 = 22; \quad \lambda_3 > 1, \quad \lambda_4 > 1, \quad \lambda_5 > 1,$$

which has only one distinct kind of solution, namely $\lambda_3 = 2$, $\lambda_4 = \lambda_5 = 3$. The set of functions (x, y, z) are the basis for a representation of O which is irreducible because there is no vector left invariant, or multiplied only by a phase factor, by all elements of O. We write it Γ_4. It is not self-associated, so $\Gamma_5 = \Gamma_4^*$. Γ_3 must be self-associated, so has character zero for the two classes of odd permutations. Its character for the two remaining classes is then determined immediately by the orthogonality relations (6.19) for characters. The complete table is given in Table A 5. The classes are given in terms of the kind of operation to be found in them; for example, $8C_3$ means the class consisting of the eight rotations of $\frac{2}{3}\pi$ or $\frac{4}{3}\pi$ about a threefold axis. $6\sigma_d$ is the class containing the six reflexions of the regular tetrahedron. There are two forms of notation in common usage for these groups. The first, Bethe's notation, is that of the Γ_i. The second, Mulliken's notation, takes the English letters A, B for representations of degree 1, E for degree 2 and T for degree 3. Mulliken's notation is given at the left of the Γ_i in

Table A 5. The Γ notation is more useful in the general theory but in particular cases such as the present many people, including myself, prefer to use the Mulliken notation. I use it in this book.

While we are on the subject of notation it is suitable to say that it is customary, as in the theory of free atoms, to use small letters to refer to one-electron functions and capital letters to refer to n-electron or to general functions. Thus a one-electron system under the cubic group can span the representations a_1, a_2, e, t_1 or t_2, but the states of an n-electron system can span A_1, A_2, E, T_1 or T_2. This is analogous to the relation between $s, p, d, \ldots$ orbitals and $S, P, D, \ldots$ terms.

The last group of interest is K. It is isomorphic with A_5. The best ways of obtaining the characters use more powerful methods than I have described here. However, it may interest the reader to see that the character table can be determined by elementary methods without great difficulty. I do this only in outline. If the reader fills in the details he will find it to be good practice.

Two elements which are conjugate in A_5 are also conjugate in S_5, but the converse is not always true. So each class of A_5 is contained in a class of even permutations of S_5. There are four of the latter: the class of the unit element; the cycles of order 3; the cycles of order five; and the class of products of two commuting transpositions. Typical representatives of the last three classes are (123), (12345), (12)(34). We write the four classes C^1, C^2, C^3, C^4, respectively. C^1 is a class of A_5. If $x, y \in C^i$ then there is a $g \in S_5$ such that $g^{-1}xg = y$. Then if $h \in G$ is odd and commutes with x we have $(hg)^{-1}x(hg) = g^{-1}h^{-1}xhg = g^{-1}xg = y$. So either g or hg is even and then x is conjugate to y in A_5. For C^2 with $x = (123)$ a suitable h is (45) and for C^4 with $x = (12)(34)$, $h = (12)$. Therefore C^1, C^2 and C^4 are classes of A_5. C^3 splits into two classes of A_5 because ($abcde$) can never be turned into an expression in which the $abcde$ undergo an odd permutation. On the other hand, we can obtain any even permutation (see the discussion of P, Q, R earlier in the section).

Then K has 5 classes and just the one representation Γ_1 of degree 1. So $\Sigma\lambda_k^2 = 60$ implies that the remaining representations have degrees (3, 3, 4, 5). The set of three p-functions and the set of five d-functions form bases for irreducible representations of K because their characters satisfy (6.28). See Table A 13 for the information necessary to write down their characters. The character of the representation arising from the seven f-functions satisfies $\Sigma\chi^2 = 120$ whence (see (6.27) and (6.28)) the representation breaks up into two irreducible representations of degrees 3 and 4. The nine g-functions have a character χ which is orthogonal to χ_1 and satisfies $\Sigma\chi^2 = 120$. So χ is the sum of two irreducible characters one of degree 4, the other 5. This last result gives us the character of degree 4 and the f-functions give us a new character of degree 3 which is different from our earlier one.[1] This completes the derivation. The character table is given in Table A 7.

The character tables of those groups which are of most use to us are given in Tables A 1–7. I have omitted most of the groups, such as O_h, which are merely direct products of a pure rotation group with the group C_i which contains the inversion. As we saw at the beginning of this section, such character tables can

[1] The two representations of degree three are, of course, not associated, unlike the case of O.

be written down immediately once one knows the character table of the corresponding pure rotation group.

The decomposition of the direct product of a pair of representations occurring in one of Tables A 1–7 can be easily obtained. As an example let us take the representation E of the group O and consider its direct product E^2 with itself. It is

	E	$8C_3$	$3C_2$	$6C_2'$	$6C_4$
E^2	4	1	4	0	0

Using (6.27) we work out $\sum_R \chi_{E^2}(R)\overline{\chi}_j(R)$ for the five representations of O. It is 24, 24, 24, 0, 0 for A_1, A_2, E, T_1, T_2, respectively. Equation (6.27) shows that these numbers are $24n_j$ and so we have found that

$$E^2 = A_1 + A_2 + E.$$

Naturally in a simple case like this one could just as quickly perform the decomposition by inspection, but in more complicated cases it is easiest to use (6.27). I have put the decomposition of all products of irreducible representations of the three groups D_4, O and K in Table A 9 for convenient reference. The product of any pair of reducible representations B and B' is easily deduced, because if

$$B = \sum_i n_i A_i, \quad B' = \sum_i n_i' A_i$$

then

$$BB' = \sum_i n_i n_i' A_i^2 + \sum_{i<j} (n_i n_j' + n_j n_i') A_i A_j$$

and A_i^2 and $A_i A_j$ are in Table A 9.

6.6.3. The Pauli exclusion principle re-examined. The Pauli principle can be regarded as a proposition about representations of permutation groups. It is evidently equivalent to the assertion that the wave-function for an n-electron system must form a basis for the alternating representation Γ_2 of the symmetric group on the n arguments occurring in the function. As the Hamiltonian is a basis for the unit representation Γ_1 it has no non-zero matrix elements between a state belonging to Γ_2 and any state belonging to any other representation of S_n. Therefore if we use perturbation theory and take our basic states as components of bases for representations of S_n, the wave-function will be expressed as a sum only over basic states belonging to Γ_2 (see Appendix 3).

Sometimes, however, we break up a matrix element of a perturbing energy

$$V = \sum_{\kappa=1}^{n} a(\kappa)\, b(\kappa)$$

as

$$\langle \overline{\psi} \,|\, V \,|\, \psi \rangle = \sum_{\kappa=1}^{n} \sum_{\phi} \langle \overline{\psi} \,|\, a(\kappa) \,|\, \phi \rangle \langle \overline{\phi} \,|\, b(\kappa) \,|\, \psi \rangle.$$

We did this in (5.3) when discussing the spin-orbit coupling energy. In such a case $a(\kappa)$ and $b(\kappa)$ do not form bases for Γ_1 and hence the sum is not only over ϕ

belonging to Γ_2 but also over ϕ which belong to other representations of S_n. It is important to remember this because the equation

$$n\langle\overline{\psi} \mid a(\kappa) \mid \phi\rangle = \langle\overline{\psi} \mid \sum_{\kappa=1}^{n} a(\kappa) \mid \phi\rangle$$

is true if ψ and ϕ both belong to Γ_2, but not necessarily if only ψ does.

Examples

1. A wave-function ψ for an n-electron system belongs to Γ_2 if and only if $(1i)\,\psi = -\psi$ for each of the $n-1$ transpositions $(1i)$, $i = 2, 3, \ldots, n$.

2. Take a regular tetrahedron with centre at the origin. The trio x, y, z form a basis for a representation of T_d. So do the sets $x^2, y^2, z^2, xy, yz, zx$ and $x_1y_2 - x_2y_1, y_1z_2 - y_2z_1, z_1x_2 - z_2x_1$, letting x_1, y_1, z_1 and x_2, y_2, z_2 each behave separately as x, y, z under T_d. Determine their characters. Use these facts together with $T_d \cong S_4$ to determine the character table of the group S_4.

3. Take a regular solid in four (Euclidean) dimensions with five vertices and centre at the origin. Show that its full symmetry group is isomorphic with S_5. Use the three types of representation described in ex. 2 to determine the character table of S_5. (This is a little hard but can be done by thinking carefully what type of symmetry operations correspond to each class of S_5 and choosing axes sensibly. Because of the orthogonality relations one need not find to which operations a permutation such as (12345) corresponds.)

It is interesting to notice that, just as in ordinary space, the 'd-functions' break up into two irreducible representations, one the same as that spanned by the 'p-functions'.

6.7. Relations between representations

If a site has a symmetry group G then a small distortion will in general remove some of its symmetry elements. The new site has lower symmetry and its symmetry group, H say, will be a subgroup of G. Therefore, if we enumerate all the subgroups of G we shall have a list of the possible site symmetries which can be obtained by a small distortion of the original site. Of course, there is no logical necessity for a small distortion to lower the site symmetry. If we apply the reverse distortion to the one mentioned above, we can increase the site symmetry group from H to G. However, in practice we usually exclude this possibility by choosing G to be the largest symmetry group which is attainable by a small distortion of the site. Thus we even think of the group T_h of Fig. 6.3 as being related to O_h because the elecrostatic field at the ion will be predominantly octahedral.

It is not difficult to determine the subgroups of a finite symmetry group[1] and I give a list of them for the more important symmetry groups in Table A 10. The table gives the number of essentially different subgroups of each kind. For example, in O_h all the subgroups represented by the symbol C_3 are essentially the same[2] because they all refer to a threefold axis of the octahedron and all threefold axes are equivalent. However, there are two types of twofold axis

[1] The improper groups are dealt with by the method described in Murnaghan, *loc. cit.* ch. 11 and Weyl (1952), Appendix B.

[2] Technically one says that for any pair of them there exists an inner automorphism of O turning one into the other.

and this is reflected in the occurrence of $2C_2$ in the table. We also recall that C_n, D_n, T, O, K, T_d, T_h, O_h, K_i and the groups of Table 7.2 give all the distinct finite subgroups of the infinite group R_{3i} consisting of all rotations, reflexions and inversion at a point.

It follows from the definition of a representation that if Γ is a representation of a group G it is also a representation of any subgroup H of G. It would be tedious but straightforward to pursue this relationship for all the subgroups of all the important symmetry groups. We therefore determine the relationship for the most useful subgroups only. Our task is lessened by observing that we need only do it for irreducible representations of G. Further, if G and H are pure rotation groups and Γ of G breaks up into the direct sum of Γ_1 and Γ_2 for H, then the representation Γ_g of $G \times C_i$ breaks up into $\Gamma_{1g} + \Gamma_{2g}$ of $H \times C_i$ and similarly for Γ_u. A representation symbol with the suffix g or u means that the functions have even or odd parity respectively.

As an example let us take O and D_4 and consider the irreducible representation E of O. Every element of D_4 belongs also to O and so we find the character of E as a representation of D_4 immediately from Table A 5. In fact we have

$$\chi(1) = \chi(C_2^x) = \chi(C_2^y) = \chi(C_2^z) = 2,$$

$$\chi(C_4^z) = \chi(C_2') = 0,$$

and so E breaks up into A_1 and B_1. We write

$$E(O) = A_1(D_4) + B_1(D_4).$$

Tables A 11 and 12 give these relationships for the most useful subgroups of O_h and of K. Because O is isomorphic with T_d it might seem natural to choose the correspondence between the representations which arises from this isomorphism. It is, however, the geometrical relation between the groups which is physically significant in the theory rather than this purely mathematical one.

It is clear that a connected set of the $(2L+1)$ atomic states having a given value L for the total orbital angular momentum and the $(2L+1)$ possible values for M_L forms a basis for a representation of any finite symmetry group about the origin. In fact we used this result in determining the character table of K. One naturally asks how such a representation breaks up when it is expressed as a sum of irreducible components. For this we need to know the character of an element of the finite symmetry group in this representation. First, take a rotation C_ϕ through an angle ϕ in a counterclockwise direction about the Z-axis. Then

$$C_\phi |LM\rangle = e^{-iM\phi} |LM\rangle. \tag{6.32}$$

Equation (6.32) follows for a one-electron system from (2.28) and for a many-electron system from ex. 1, p. 19. The minus sign in the exponent in (6.32) arises because we are rotating the ket (i.e. the function) not the coordinate system. The character of C_ϕ is therefore given by

$$\chi(C_\phi) = \sum_{M=-L}^{+L} e^{-iM\phi} = \frac{\sin(L+\frac{1}{2})\phi}{\sin\frac{1}{2}\phi}. \tag{6.33}$$

If we now carry out the rotation ϕ about any other axis we still get the same value for $\chi(C_\phi)$. One may see this by observing that the new rotation is conjugate[1] in the group of all rotations to the rotation about OZ and using (6.18a). Alternatively, one quantizes the kets with respect to the component of $\mathbf{L}$ along the new axis and remarks that the new kets are an orthonormal transform of the old ones which again implies that they have the same character. Equation (6.33), then, gives us the characters for all elements of all pure rotation groups and I give a table of them for small L and relevant ϕ in Table A 13. C_n is a rotation of $\phi = 2\pi/n$ and I have written J instead of L. It probably seems fairly obvious to the reader that the characters for integral J are the same as those for the corresponding L and this can be proved quite easily.[2]

We now possess sufficient information to determine how the kets with given L break up into irreducible representations of any finite symmetry group. This is a completely straightforward matter for any particular L. For a general L we note that $\chi(C_n)$ is the same for L as it is for $L+mn$ for any integer m, except for the identity element C_1 for which is differs by $2mn$. So if N is the lowest common multiple of the n for all C_n of the finite group we have $\chi(C_n)$ for $L+mN$ the same as $\chi(C_n)$ for L for all elements of the group except the identity. If follows from this that if we write χ^L for the character for L then

$$\chi^{L+mN} - \chi^L = 2mN, 0, 0, \ldots, 0.$$

Put $b(L) = \Sigma n_i \Gamma_i$ for the break-up into irreducible representations of the representation spanned by the $|LM\rangle$. Then it follows from (6.27) that

$$b(L+mN) - b(L) = \frac{2mN}{\pi} \Sigma \lambda_i \Gamma_i, \tag{6.34}$$

where π is the order of the group and λ_i the degree of the ith representation. Therefore in order to give a complete solution to our problem we need only enumerate the $b(L)$ for $L < N$ and then use (6.34) for $L \geqslant N$. These are given for O ($N = 12$) in Table A 14 and for D_4 ($N = 4$) and D_3 ($N = 6$) in Table A 15. For K ($N = 30$) they are given for $L \leqslant 12$ in Table A 14.

If one is interested in a symmetry group containing improper elements one must first specify the behaviour of the kets $|LM\rangle$ under the inversion. Then if the symmetry group is the direct product of a pure rotation group with C_i we simply add the suffix g or u to all representation symbols in Tables A 14 and 15 according as to whether the $|LM\rangle$ all have even or all have odd parity. This suffices to deal with O_h, K_i, D_{4h} and D_{3v}. For T_d with $|LM\rangle$ of even parity the behaviour of a ket under an improper rotation is the same as its behaviour under the associated proper one. So Table A 14 for O applies also to T_d in this case (see Table A 11). When $|LM\rangle$ has odd parity we must interchange A_1 and A_2 and also T_1 and T_2 in Table A 14. Thus a p-function for a single electron behaves as T_1 under O and as T_2 under T_d. However, a d-function behaves as $E+T_2$ under both.

[1] The transforming rotation U of (6.18a) is the rotation about a line at right-angles to both axes which takes the new axis into OZ.

[2] See p. 173 for the relationship between **S**, **J** and **L**.

6.8. Relations between functions belonging to representations

We now pursue, for O, D_4 and D_3, the decision we made at the end of § 6.5 to choose our sets of functions forming bases for irreducible representations in such a way that when a pair form bases for the same representation they both give rise to that representation in identically the same form, not merely in equivalent ones. For representations of degree 1 there is no choice to be made. For each of those of higher degree we choose a suitable basis and this then defines, by implication, the standard form for the representation. For T_1 of the group O we naturally choose x, y and z. Then in T_1 the elements of O are just represented by the matrices which transform the vector $\mathbf{r}$. Thus

$$C_4^z = \begin{bmatrix} 0 & -1 & 0 \\ 1 & 0 & 0 \\ 0 & 0 & 1 \end{bmatrix} \tag{6.35}$$

in accordance with (6.3). We recall that in the matrix representation x, y, z form a row vector which multiplies the matrix for C_4^z on the left. I also remark again that our rotations rotate the functions not the coordinate system. So

$$[C_4^z x, C_4^z y, C_4^z z] = C_4^z[x, y, z]$$

$$= [x, y, z]\begin{bmatrix} 0 & -1 & 0 \\ 1 & 0 & 0 \\ 0 & 0 & 1 \end{bmatrix} = [y, -x, z],$$

whence $C_4^z x = y$, $C_4^z y = -x$, $C_4^z z = z$, which may help to clarify this point. I have used non-normalized functions for simplicity. If this worries the reader he may replace x, y, z by $xf(r)$, $yf(r)$, $zf(r)$ for some suitable function $f(r)$.

For E and T_2 of O we choose the functions $(2z^2-x^2-y^2, \sqrt{3}\,(x^2-y^2))$ for E and (yz, zx, xy) for T_2. Having made these choices we also introduce a definite notation for a function which is a particular component of a representation. We take θ and ϵ for the two components of E and would then write

$$E\theta = 2z^2-x^2-y^2,$$

$$E\epsilon = \sqrt{3}\,(x^2-y^2),$$

for the particular pair of functions given above. This is the precise analogue of giving LM_L values. We say that we are giving ΓM_Γ values and remark that this pair of values is the most complete description possible of the behaviour of a function under the group O. For when we know Γ and M_Γ we know how every element of O operates on the function. We write ξ, η, ζ for the three components yz, zx, xy of T_2, x, y, z for T_1, a_1 for A_1 and a_2 for A_2. The behaviour of all these functions under typical elements of O is given in Table A 16, together with corresponding definitions for D_4 and D_3. C_4^z and C_4^x by themselves form a pair of generators. We note that the set of products a_2T_1x, a_2T_1y, a_2T_1z are the ξ, η, ζ

components, respectively, of T_2. Similarly $a_2 T_2 \xi$, etc., are the x, etc., components of T_1.

The three functions x, y, z belonging to T_1 may be regarded as p-functions and we remember that in the free atom it was more useful to take the linear combinations $-2^{-\frac{1}{2}}(x+iy)$, z, $2^{-\frac{1}{2}}(x-iy)$ quantized with respect to L_z. This is still useful in the group O and so I have added definitions of $T_1 i$, $T_2 i$ for $i = 1, 0, -1$ in the middle of Table A 16 for reference and we shall make use of this alternative scheme later.

Having agreed upon standard basis relations one can ask how a basis for a representation Γ of O breaks up into bases for representations Γ' of a subgroup of O. As an example let us take the representation T_2 of O and the subgroup D_4. Then

$$\begin{aligned} C_4^z \xi &= -\eta, & (C_4^x)^2 \xi &= \xi, \\ C_4^z \eta &= \xi, & (C_4^x)^2 \eta &= -\eta, \\ C_4^z \zeta &= -\zeta, & (C_4^x)^2 \zeta &= -\zeta, \end{aligned}$$

whence we see that ζ is a basis for the representation B_2 of D_4 and ξ, $-\eta$ form the x-, y-components, respectively, of the representation E of D_4. We remark that as our basis relations are defined only up to a phase factor common to all functions forming a basis for an irreducible representation there is inevitably a certain arbitrariness in the relations between bases for groups and their subgroups. We could replace ξ, $-\eta$ by $e^{i\theta}\xi$, $-e^{i\theta}\eta$ for any real number θ. The relationship between the functions belonging to representations of O and those of its subgroups is shown in Table A 17.

In just the same way we can ask how sets of functions with given L and varying M_L span representations of finite symmetry groups. Tables A 14 and 15 show which representations they span for O, K, D_4 and D_3 and we must find out how the individual components behave. Let us take D_4 first and see how $|LM\rangle$ behaves under the two generators C_4^z and C_2^x of D_4. Equation (6.32) shows that $C_4^z|LM\rangle = e^{-\frac{1}{2}i\pi M}|LM\rangle$. If a is an integer then

$$\left.\begin{aligned} C_4^z|LM\rangle &= |LM\rangle \quad (M = 4a), \\ C_4^z|LM\rangle &= -|LM\rangle \quad (M = 4a+2), \\ C_4^z|LM\rangle &= \pm i\,|LM\rangle \quad (M = 4a\mp 1). \end{aligned}\right\} \tag{6.36}$$

The effect of C_2^x is most easily found by considering the spherical harmonics (4.10) and is obtained by replacing (θ, ϕ) in them by $(\pi-\theta, -\phi)$. So

$$C_2^x|LM\rangle = (-1)^L|L-M\rangle. \tag{6.36a}$$

It follows from this that the pair $|LM\rangle$, $|L-M\rangle$ always forms a representation of D_4. This holds for C_n, D_n for all n. When $M = 0$ there is just the one function $|L0\rangle$. The character of C_2^z for this representation with $M \neq 0$ follows from (6.36) and is 2 for M even and -2 for M odd. The character table for D_4 (Table A 3) then shows us that when M is odd the pair $|LM\rangle$, $|L-M\rangle$ form a basis for E. When M is even the representation is reducible and (6.36a) enables us to achieve the reduction.

The results can be given much more concisely if we introduce a new notation and write

$$\left.\begin{aligned} Z_{L0} &= Z^c_{L0} = Y_{L0}, \\ Z^c_{LM} &= \frac{1}{\sqrt{2}}(Y_{L-M}+\overline{Y}_{L-M}), \\ Z^s_{LM} &= \frac{i}{\sqrt{2}}(Y_{L-M}-\overline{Y}_{L-M}). \end{aligned}\right\} \tag{6.37}$$

All these new functions are real, Z^c_{LM} involves $\cos M\phi$, Z^s_{LM} involves $\sin M\phi$ and the negative signs which occur in some of the Y_{LM} of (4.10) have disappeared. They are often more satisfactory functions to use with finite symmetry groups. When I wish to speak of them collectively I shall refer to the $Z_{L\alpha}$. The way in which the $Z_{L\alpha}$ form bases for representations of D_4 is shown in Table A 18. Naturally it is a straightforward matter to perform the same reduction for C_n and D_n for any n. I give the reduction for D_3 in Table A 18 without separating the components of E.

Unlike that relative to the cyclic and dihedral groups the decomposition of the $Z_{L\alpha}$ relative to the octahedral and icosahedral groups cannot be completely accomplished in a finite closed form.[1] This is because the effect of the group generator C^x_4 is not expressible in closed form (an alternative choice of generators does not improve matters). However, we can obtain a considerable amount of information by comparing Table A 17 with Table A 18. Let us take T_{1z} as an example. Table A 17 shows that this only gives rise to A_2 in D_4 and hence whenever A_2 of D_4 turns up it must be the z-component of a T_1 representation of O. But we know all the A_2 states of D_4. From Table A 18 they are Z^s_{LM} when L is even and Z^c_{LM} when L is odd, for those M which are multiples of four. So we have obtained the z-component of all T_1 states. By the same argument, using B_2, the ζ-component of all T_2 states is given by Z^s_{LM} for L even and Z^c_{LM} for L odd, and M twice an odd integer.

So far so good. However, unfortunately, a consideration of the other representations of D_4 only gives us partial information for O. Consider B_1 of D_4 and take L even. B_1 arises from both A_2a_2 and $E\epsilon$ of O and so all we can say about the functions A_2a_2 of O is that they are linear combinations of functions Z^c_{LM} with M twice an odd integer. Of course if, for a given L, only one representation of symmetry A_2 or E appears then we can go further. Thus for $L = 2$ we deduce that $E\epsilon = Z^c_{22}$ and for $L = 3$, $A_2a_2 = Z^s_{32}$. But in general we cannot.

d^n configurations contain terms with L up to 6 and therefore it is desirable to possess the decomposition relative to O for $L \leqslant 6$. One way to do this is to take the spherical harmonics and express them in terms of x, y and z. It is then a straightforward matter to determine the effect of C^x_4 on them. The complete decomposition is given in Table A 19. The bases are connected in accordance with the definitions of Table A 16.

From a mathematical point of view it is interesting to remark that a purely

[1] See Lage & Bethe, *Phys. Rev.* (1947), **71**, 612; Meyer, *Canad. J. Math.* (1954), **6**, 135; Altmann, *Proc. Camb. Phil. Soc.* (1957), **53**, 343.

formal solution of our problem can be written down. The reader may omit this paragraph and the next without loss of continuity. Let R be represented, according to (6.3), by $a^p_{ji}(R)$ in the pth irreducible representation. Define the quantities

$$e^p_{ij} = \frac{\lambda_p}{\pi} \sum_R a^p_{ji}(R^{-1})\, R, \tag{6.38}$$

where λ_p is the degree of the representation and π the order of the group. Rotations are added by the rule $(c_1 R_1 + c_2 R_2) f = c_1 R_1 f + c_2 R_2 f$, where c_1, c_2 are numbers. Then if S is any fixed element of the group we have

$$\begin{aligned} S e^p_{ij} &= \frac{\lambda_p}{\pi} \sum_R a^p_{ji}(R^{-1})\, SR \\ &= \frac{\lambda_p}{\pi} \sum_T a^p_{ji}(T^{-1}S)\, T \\ &= \frac{\lambda_p}{\pi} \sum_{T,k} a^p_{jk}(T^{-1})\, a^p_{ki}(S)\, T \\ &= \sum_k e^p_{kj}\, a^p_{ki}(S), \end{aligned} \tag{6.39}$$

where we wrote $T = SR$ and used the fact that $a^p_{ji}(R)$ is a matrix representation (6.9). If we now take an arbitrary function f and put $f_i = e^p_{ij} f$, (6.39) yields

$$S f_i = \sum_k f_k\, a^p_{ki}(S),$$

showing that the functions f_i form a basis for the pth irreducible representation or, alternatively, are all identically zero. Taking $|LM\rangle$ for f the set $e^p_{ij}\,|LM\rangle$ for fixed j, p, L, M are either all identically zero or give a basis for the pth irreducible representation. This basis will not in general be normalized but, if a^p_{ji} were chosen in accordance with our standard choice of bases, will be correctly connected. All irreducible representations are obtainable by this method, with considerable repetition if one uses all possible values of j and M. In the case of repeated representations one would get various linear combinations among the repetitions.

However, although this method looks promising at first sight it is not really very useful because its use requires a knowledge of the effect of all rotations of the group on a given $|LM\rangle$. And this, for general M, we do not possess in a simple form for all rotations of O or K. We do for C_n and D_n, but for them the problem is trivial anyway. The e^p_{ii} have the property $e^p_{ii} e^p_{ii} = e^p_{ii}$, which can be proved using the orthogonality relations (6.16) and remembering that for unitary matrices $[a_{ij}]^{-1} = [\bar{a}_{ji}]$. They are called primitive idempotents (or projection operators) and have the property of selecting the ith component of the pth irreducible representation. They satisfy

$$\sum_{i,p} e^p_{ii} = 1,$$

so they break an arbitrary function into components of the various possible symmetries.[1] They assume quite a simple form for D_4 and O if we write $A = C^z_4$,

[1] For more about the primitive idempotents see Weyl, *Classical Groups*, especially chs. 3 and 4.

$C = C_2^x$, $D = C_3^{(111)}$ (i.e. about the $x = y = z > 0$ axis). See Table 6.6. I have just numbered the primitive idempotents for D_4 as $\epsilon_1, \epsilon_2, \epsilon_3, \epsilon_4, \epsilon_5^{\pm}$ and expressed those for O, in an obvious notation, in terms of those for D_4. The reader will see that the expressions for the primitive idempotents for O confirm some of our earlier findings about the relationship of components of representations of O to those of D_4.

Table 6.6. *Primitive idempotents for the groups D_4 and O*

Representation of D_4	Idempotents	Representation of O	Idempotents
A_1	$\epsilon_1 = (1+C)(1+A)(1+A^2)$	A_1	$a_1 = \frac{1}{3}(1+D+D^2)\,\epsilon_1$
A_2	$\epsilon_2 = (1-C)(1+A)(1+A^2)$	A_2	$a_2 = \frac{1}{3}(1+D+D^2)\,\epsilon_3$
B_1	$\epsilon_3 = (1+C)(1-A)(1+A^2)$	E	$e_\theta = \frac{2}{3}(1-\frac{1}{2}D-\frac{1}{2}D^2)\,\epsilon_1$
B_2	$\epsilon_4 = (1-C)(1-A)(1+A^2)$		$e_\epsilon = \frac{2}{3}(1-\frac{1}{2}D-\frac{1}{2}D^2)\,\epsilon_3$
$E\begin{cases}x\\y\end{cases}$	$e_5^+ = (1+C)(1-A^2)$ $\epsilon_5^- = (1-C)(1-A^2)$	T_1	$t_{1x} = \frac{1}{2}(1+D^2A)\,\epsilon_5^+$ $t_{1y} = \frac{1}{2}(1-DA)\,\epsilon_5^-$ $t_{1z} = \epsilon_2$
		T_2	$t_{2\xi} = \frac{1}{2}(1-D^2A)\,\epsilon_5^+$ $t_{2\eta} = \frac{1}{2}(1+DA)\,\epsilon_5^-$ $t_{2\zeta} = \epsilon_4$

We have not yet met the analogue of Wigner's formula. This involves giving the values of the coefficients in the expansion

$$|\Gamma_1\Gamma_2\Gamma c\rangle = \sum_{a,b} |\Gamma_1\Gamma_2 ab\rangle\langle\Gamma_1\Gamma_2 ab\,|\,\Gamma_1\Gamma_2\Gamma c\rangle \tag{6.40}$$

expressing kets which form a basis for an irreducible representation Γ occurring in a direct product $\Gamma_1\Gamma_2$ in terms of the simple product kets

$$|\Gamma_1\Gamma_2 ab\rangle = |\Gamma_1 a\rangle|\Gamma_2 b\rangle.$$

We shall determine the $\langle\Gamma_1\Gamma_2 ab\,|\,\Gamma_1\Gamma_2\Gamma c\rangle$ for the octahedral group O with, as usual, regard for the standard basis relationships between components.[1] As a simple example consider $\Gamma = E$, $\Gamma_1 = A_2$, $\Gamma_2 = E$. There are just two functions $|\Gamma_1\Gamma_2 ab\rangle$, namely $|A_2 a_2\rangle|E\theta\rangle$ and $|A_2 a_2\rangle|E\epsilon\rangle$. Then

$$\left.\begin{aligned}
C_4^z\,|A_2Ea_2\theta\rangle &= -|A_2Ea_2\theta\rangle,\\
C_4^z\,|A_2Ea_2\epsilon\rangle &= |A_2Ea_2\epsilon\rangle,\\
C_4^x\,|A_2Ea_2\theta\rangle &= \tfrac{1}{2}\,|A_2Ea_2\theta\rangle+\tfrac{1}{2}\sqrt{3}\,|A_2Ea_2\epsilon\rangle,\\
C_4^x\,|A_2Ea_2\epsilon\rangle &= \tfrac{1}{2}\sqrt{3}\,|A_2Ea_2\theta\rangle-\tfrac{1}{2}\,|A_2Ea_2\epsilon\rangle,
\end{aligned}\right\} \tag{6.41}$$

where we have substituted directly from Table A 16. Formulae (6.41) show that

$$\left.\begin{aligned}
|A_2EE\theta\rangle &= |A_2Ea_2\epsilon\rangle,\\
|A_2EE\epsilon\rangle &= -|A_2Ea_2\theta\rangle,
\end{aligned}\right\} \tag{6.42}$$

[1] $\langle\Gamma_1\Gamma_2 ab\,|\,\Gamma_1\Gamma_2\Gamma c\rangle$ were given for O by Tanabe & Sugano (1954) and differ from those of Table A 20 in some of their phases but not in any essential way.

where, as usual, we have an arbitrary phase factor common to both kets at our disposal. All other non-vanishing coupling coefficients can be derived in a similar way and they are given in Table A 20.

We have now reached the end of our introduction to the necessary basic group theory with one very important exception. The classification of spin has not appeared and the rest of the chapter is devoted to repairing this omission. From the point of view of group theory, and of topology, spin has a rather deep significance. I give some description of this in the next two sections. To give an adequate description would take a great deal of space and would not be directly relevant to our subject. However, it may be that some readers will find a rather general account of a small part of the theory of interest and so I have included it. I then repeat a number of the calculations we have already made for groups not referring to spin for those referring to spin. We call these latter groups spinor groups. However, the reader who wishes may omit the rest of the chapter without meeting much difficulty later on. For his benefit I will remark that the spinor groups are each just twice as large as the ordinary groups to which they are related. They do not contain the latter groups; the latter groups are homomorphic images of them, two elements of the larger group corresponding to each one of the smaller. Their representations include the representations we have already met and also new ones—the so-called two-valued representations. With these preliminary remarks it will be possible for the reader to use the parts of the tables which refer to two-valued representations in just the same way that he uses those which refer to ordinary representations without any great fear of error. One point to note is that I always write G^* for a spinor group which corresponds to G not referring to spin. Thus O^* corresponds to O.

Example

Prove that the e^p_{ij} of (6.38) satisfy $e^p_{ij}e^q_{kl} = \delta_{pq}\delta_{jk}e^q_{il}$.

6.9. Spinors and spinor groups

The existence of spinors is connected with a rather remarkable algebraic relationship between the group R_3 of proper rotations of three-dimensional Euclidean space about a fixed origin and the group U_2 of 2×2 unitary unimodular matrices. A unimodular matrix has determinant one. Let A belong to U_2, then it satisfies

$$\bar{A}'A = I \quad \text{and} \quad |A| = 1. \tag{6.43}$$

The first condition means that $A^{-1} = \bar{A}'$ which in terms of the elements a_{ij} of A reads

$$a_{11} = \bar{a}_{22}, \quad a_{12} = -\bar{a}_{21}$$

and so the most general 2×2 unitary unimodular matrix is given by the expression

$$A = \begin{bmatrix} \kappa+i\lambda & \mu+i\nu \\ -\mu+i\nu & \kappa-i\lambda \end{bmatrix}, \tag{6.44}$$

where κ, λ, μ, ν are real numbers satisfying $\kappa^2+\lambda^2+\mu^2+\nu^2 = 1$.

We now show that R_3 is a homomorphic image of U_2 and that this homomorphism is such that each element of the group R_3 arises from just two distinct elements of U_2. The homomorphism is most conveniently exhibited by considering the general 2×2 Hermitian matrix with trace zero

$$H = \begin{bmatrix} z & x+iy \\ x-iy & -z \end{bmatrix}.$$

We associate with each unitary unimodular matrix A the transformation

$$H \to A'H\bar{A}, \tag{6.45}$$

which effects a linear transformation on the elements of H.

$\overline{(A'H\bar{A})'} = (\bar{A}'\bar{H}A)' = A'\bar{H}'\bar{A} = A'H\bar{A}$ so the transformed matrix is still Hermitian. Because of (6.18a) it also has trace zero and so can be written

$$\begin{bmatrix} z' & x'+iy' \\ x'-iy' & -z' \end{bmatrix}.$$

Because the determinant of a product of matrices is the product of the determinants and A is unimodular, the transformation (6.45) requires that

$$-|H| = x^2+y^2+z^2$$

is the same as $$-|A'H\bar{A}| = (x')^2+(y')^2+(z')^2.$$

Therefore we have associated an orthogonal transformation of three-dimensional Euclidean space with the matrix A. Let us write it $O(A)$. It is uniquely determined by A.

If A is the unit 2×2 matrix, then (6.45) leaves H unchanged, and so $O(A)$ is the unit 3×3 matrix. Further, $A_1 A_2$ gives rise to

$$\begin{aligned} H &\to (A_1 A_2)' H(\overline{A_1 A_2}) \\ &= A_2' A_1' H \bar{A}_1 \bar{A}_2 = A_2'(A_1' H \bar{A}_1)\bar{A}_2 \end{aligned}$$

which is the transformation of H induced by A_1 followed by that induced by A_2. Hence $O(A_1 A_2) = O(A_1)\, O(A_2)$ and we have a 3×3 matrix representation of the group U_2 by orthogonal matrices.

To proceed we determine the matrices $O(A)$ for two important special cases. First, suppose

$$B_1 = \begin{bmatrix} e^{-\frac{1}{2}i\phi} & 0 \\ 0 & e^{\frac{1}{2}i\phi} \end{bmatrix}. \tag{6.46}$$

Then $$B_1 H \bar{B}_1' = \begin{bmatrix} z & e^{-i\phi}(x+iy) \\ e^{i\phi}(x-iy) & -z \end{bmatrix}$$

and so $$\begin{aligned} x &\to x\cos\phi + y\sin\phi, \\ y &\to -x\sin\phi + y\cos\phi, \\ z &\to z \end{aligned}$$

which shows that $O(B_1)$ is a rotation through an angle ϕ in a counterclockwise direction about the OZ-axis. As usual, we interpret matrices as rotations of functions, rather than coordinate systems. Equation (6.3) then gives

$$O(B_1) = \begin{bmatrix} \cos\phi & -\sin\phi & 0 \\ \sin\phi & \cos\phi & 0 \\ 0 & 0 & 1 \end{bmatrix}.$$

The second special unitary unimodular matrix is

$$B_2 = \begin{bmatrix} \cos\frac{1}{2}\psi & -i\sin\frac{1}{2}\psi \\ -i\sin\frac{1}{2}\psi & \cos\frac{1}{2}\psi \end{bmatrix}. \tag{6.47}$$

A similar calculation shows that B_2 induces the transformation

$$\left.\begin{aligned} x &\to x, \\ y &\to y\cos\psi + z\sin\psi, \\ z &\to -y\sin\psi + z\cos\psi, \end{aligned}\right\} \quad \text{with matrix} \quad O(B_2) = \begin{bmatrix} 1 & 0 & 0 \\ 0 & \cos\psi & -\sin\psi \\ 0 & \sin\psi & \cos\psi \end{bmatrix}.$$

This is a rotation through an angle ψ in a counterclockwise direction about the OX-axis. Now any rotation may be built up as a product of rotations of these two special types. Actually one never needs a product of more than three. Therefore the collection $O(A)$ includes all the three-dimensional rotations. If it includes an improper rotation then, because we have just shown that it must contain the corresponding proper rotation, it includes the inversion matrix

$$-I = \begin{bmatrix} -1 & 0 & 0 \\ 0 & -1 & 0 \\ 0 & 0 & -1 \end{bmatrix}.$$

An easy calculation shows that if the matrix A of (6.44) turns z into $-z$ then $\kappa = \lambda = 0$. It is then impossible to choose μ and ν so that x and y simply change sign. So $O(A)$ contains no improper rotation.

Finally, can two distinct matrices A_1 and A_2 give the same rotation? If so, then

$$O(A_1) = O(A_2)$$

which implies $O(A_1^{-1}A_2) = I$. Hence we seek matrices A such that $O(A) = I$. Just as we found $\kappa = \lambda = 0$ if $O(A) = -I$ so we find here that $\mu = \nu = 0$. This means that A is a matrix of the form (6.46) which tells us immediately that $O(A) = I$ if and only if

$$\mu = \nu = 0, \quad \kappa + i\lambda = e^{-\frac{1}{2}i\phi}, \quad \phi = 2n\pi, \quad n \text{ integral.}$$

So there are just two distinct matrices A, the unit matrix and

$$R = \begin{bmatrix} -1 & 0 \\ 0 & -1 \end{bmatrix}. \tag{6.48}$$

I always denote this matrix by the symbol R. Similarly $O(A_1) = O(A_2)$ implies $A_1 = A_2$ or $A_1 = RA_2$. R commutes with all 2×2 matrices.

It is, I think, useful to sum up what we have found.

Theorem. The group R_3 of proper rotations of three-dimensional Euclidean space is a homomorphic image of the group U_2 of 2×2 unitary unimodular matrices. Under this homomorphism each element g of R_3 arises from just two distinct elements, g' and g'' say, of U_2 where

$$g' = -g''.$$

The group $(1, R)$ is a subgroup of U_2 but it is not true that U_2 is the direct product of $(1, R)$ with R_3. Nor is R_3 a subgroup of U_2.

We can now introduce the concept of a spinor. Just as a vector at the origin is an entity with three components which form a basis for the standard (defining) 3×3 matrix representation of R_3, so a spinor at the origin has two components which form a basis for the 2×2 matrix representation of U_2 which is connected with R_3 via (6.45). Let us suppose the two components are α_1, α_2, then, with $[A]_{ji} = a_{ji}$

$$\alpha_i \to \sum_j \alpha_j[A]_{ji} \quad \text{when} \quad x_i \to \sum_j x_j[O(A)]_{ji},$$

where (x_1, x_2, x_3) has been written for (x, y, z). In accordance with this we call U_2 the spinor group.

An apparent difficulty now arises because to each rotation $O(A)$ corresponds two matrices A. How does one choose which to take? Actually this is easily settled by considerations of continuity. Consider rotations about the z-axis through an angle ϕ. Then

$$A = \pm\begin{bmatrix} e^{-\frac{1}{2}i\phi} & 0 \\ 0 & e^{\frac{1}{2}i\phi} \end{bmatrix}, \quad O(A) = \begin{bmatrix} \cos\phi & -\sin\phi & 0 \\ \sin\phi & \cos\phi & 0 \\ 0 & 0 & 1 \end{bmatrix},$$

as we saw before. First, take the $+$ sign for A and let $\phi = 0$. Then the two unit matrices correspond. Now let ϕ increase continuously from 0 to 2π. $O(A)$ passes continuously through all angles of rotation and returns finally to the unit matrix, i.e. zero rotation, again. A, however, passes continuously to $R = -I$. On increasing ϕ further continuously from 2π to 4π, $O(A)$ repeats its previous performance while A passes continuously from R to I. The matrices A are clearly all distinct for $0 \leqslant \phi < 4\pi$. This shows that if we regard a physical rotation as specified not merely by the initial and final positions of a body it rotates, but also by the sequence of intermediate positions, we obtain a physical way of distinguishing between A and $-A$. The same argument may be applied to any other axis of rotation. In particular, a rotation of 2π about any axis is always the operation R of U_2. It is not immediately obvious that this scheme of association of the A with physical rotations is internally consistent. It is in fact, however (see § 6.10).

Having defined a spinor at a point, and determined its law of transformation under a physical rotation, we should naturally pass to a spinor field by assigning a spinor in a continuous manner to every point of space. We shall not use the spinor formulation as such, and so we do this in a way which relates spinors rather directly with our earlier way of treating electron spin. We define a standard spinor with components α_1, α_2 by requiring that α_1, α_2 transform as the two components of a spinor at any point of space and that they are totally unaffected by translations. I leave it to the reader to prove that this is a consistent definition (see also the examples at the end of the section). These components α_1 and α_2 are to be identified with the spin functions α and β introduced in § 2.4. We define our spinor field ψ_i by putting $\psi_1 = f_1(\mathbf{r})\,\alpha$, $\psi_2 = f_2(\mathbf{r})\,\beta$. Then a rotation $O(A)$ operates on f_1 and f_2 as a rotation in the normal way, while the associated matrix A operates on α and β. The great advantage of this new way of introducing the spin functions is that it shows us immediately how, and to some extent why, physical rotations operate on them.

The next step is to discover what the spin vector $\mathbf{s}$ corresponds to in terms of spinors. We do this by a process of analogy. The operator l_z on ordinary functions is an operator of differentiation with respect to ϕ and so

$$l_z = i\hbar \lim_{\phi\to 0}\frac{R_\phi - I}{\phi}, \tag{6.49}$$

where R_ϕ is a rotation through an angle of ϕ about the z-axis. So it is natural to expect that the quantity

$$j_z = i\hbar \lim_{\phi\to 0}\frac{T_\phi - I}{\phi}, \tag{6.50}$$

where T_ϕ is the corresponding transformation for a spinor field, will play a similar role for the group U_2. Now a spinor field has components $\psi_i = f_i\alpha_i$ and so

$$T_\phi\psi_i = T_\phi(f_i\alpha_i) = (T_\phi f_i)\,(T_\phi\alpha_i), \tag{6.51}$$

$T_\phi f_i$ and $T_\phi\alpha_i$ are determined respectively by the matrices $O(A)$ and A corresponding to the rotation. Then we use the simple identity

$$(T_\phi f_i)\,(T_\phi\alpha_i) - f_i\alpha_i = \{(T_\phi - I)f_i\}\{(T_\phi - I)\,\alpha_i\} + \{(T_\phi - I)f_i\}\,\alpha_i + f_i\{(T_\phi - I)\,\alpha_i\},$$

in which the first term on the right-hand side is a second-order small quantity when $\phi \to 0$, to substitute in (6.50) via (6.51) and obtain

$$\begin{aligned} j_z(f_i\alpha_i) &= \left\{i\hbar \lim_{\phi\to 0}\frac{T_\phi - I}{\phi} f_i\right\}\alpha_i + f_i\left\{i\hbar \lim_{\phi\to 0}\frac{T_\phi - I}{\phi}\,\alpha_i\right\} \\ &= (l_z f_i)\,\alpha_i + f_i\left\{i\hbar \lim_{\phi\to 0}\frac{T_\phi - I}{\phi}\,\alpha_i\right\}. \end{aligned} \tag{6.52}$$

Let us now define s_z to be $i\hbar \lim\limits_{\phi\to 0}(T_\phi - I)/\phi$ restricted to operating only on α_i and l_z may be equivalently defined to be the restriction to f_i. Equation (6.52) then becomes

$$j_z f_i\alpha_i = l_z f_i\alpha_i + s_z f_i\alpha_i$$

for every spinor field $f_i\alpha_i$. In other words we have the operator equation

$$j_z = l_z + s_z.$$

Before we interpret s_z and j_z as the z-components of the spin and total angular momentum vectors we must just confirm that s_z, like l_z, operates in accordance with our previous independent definition. The matrix A_ϕ corresponding to T_ϕ is

$$A_\phi = \begin{bmatrix} e^{-\frac{1}{2}i\phi} & 0 \\ 0 & e^{\frac{1}{2}i\phi} \end{bmatrix},$$

and so $$i\hbar \lim_{\phi\to 0} \frac{A_\phi - I}{\phi} = i\hbar \lim_{\phi\to 0} \begin{bmatrix} \dfrac{e^{-\frac{1}{2}i\phi}-1}{\phi} & 0 \\ 0 & \dfrac{e^{\frac{1}{2}i\phi}-1}{\phi} \end{bmatrix} = \begin{bmatrix} \frac{1}{2}\hbar & 0 \\ 0 & -\frac{1}{2}\hbar \end{bmatrix}.$$

This justifies the interpretation. We define s_x and s_y in a precisely analogous way by considering rotations about the OX- and OY-axes, respectively.

So far we have omitted to mention the role of inversion. This is because it has no very natural place and has to be somewhat artificially tacked on. This is done by simply noting that the group of orthogonal 3×3 matrices is the direct product of the group R_3 of rotations with the group containing just two elements, the unit matrix and minus the unit matrix. There is no 2×2 matrix which can be taken to be the inversion for U_2. Therefore we take an abstract group S containing two elements $(1, \iota)$ say. 1 leaves any spinor field invariant and ι has the effect

$$\iota f_i(\mathbf{r})\,\alpha_i = f_i(-\mathbf{r})\,\alpha_i. \tag{6.53}$$

This corresponds to the electron having positive intrinsic parity. For a spinor particle with negative intrinsic parity, (6.53) is replaced by

$$\iota f_i(\mathbf{r})\,\alpha_i = -f_i(-\mathbf{r})\,\alpha_i.$$

ι commutes with all elements of U_2 so we have extended U_2 to a new group which is the direct product of U_2 with the group S. We call this new group the augmented spinor group. It is convenient to write $O(\iota A) = -O(A)$.

We now have our complete mathematical apparatus for describing the effect of rotations, inversions and hence of finite symmetry groups on systems with spin. The reader will have noticed that we have restricted attention to spatial dimensions and may wonder how spinors would fit into a relativistic theory. This may be accomplished very simply for special relativity theory by replacing the Hermitian matrix of trace zero which we used previously by the general Hermitian 2×2 matrix

$$H = \begin{bmatrix} t+z & x+iy \\ x-iy & t-z \end{bmatrix}, \tag{6.54}$$

and allowing the matrices A to be arbitrary unimodular 2×2 matrices. The transformation (6.45) now induces a Lorentz transformation on the variables x, y, z, t (with $c = 1$). It is not difficult to show that we get all matrices of the proper Lorentz group, each twice, in this way.[1] However, when we consider an ion in a

[1] For details see Murnaghan, ch. 12, or Naĭmark, *Uspehi Mat. Nauk* (1954) (N.S.), **9**, no. 4 (62), p. 19 = *Amer. Math. Soc. Transl.* (2), **6**, 379. For a rather different approach to spinors see Raševskiĭ, *Uspehi Mat. Nauk* (1955) (N.S.), **10**, no. 2 (64), p. 3 = *Amer. Math. Soc. Transl.* (2), **6**, 1. For yet another see Kramers, *Quantum Mechanics* and Brinkman, *Applications of Spinor Invariants in Atomic Physics* (both North-Holland Publishing Company).

crystal we already have a particular time axis singled out for us, namely, that axis with respect to which the crystal appears stationary on the large scale. Then the only transformations of interest are those which leave this axis unchanged, i.e. just the elements of U_2 and ι. Therefore we are little concerned with the genuinely relativistic aspects of the theory.

We now determine the finite subgroups of U_2. Knowing those of R_3 this is rather easy. Start with a finite subgroup G^* of U_2 and replace each matrix $A \in G^*$ by $O(A)$. If $R \in G^*$ each matrix $O(A)$ must appear twice. This is because if $A \in G^*$ so also must AR. Conversely, if any matrix $O(A)$ appears twice then $R \in G^*$ and so they all do. When $R \in G^*$ we have a group of rotations associated with G^*. We call this group G. G^* has twice as many elements as G. There will be one group G^* for each of the finite subgroups of R_3; i.e. there will be two infinite sequences C_n^* and D_n^* and three high-symmetry groups T^*, O^* and K^*.

Next suppose R is not in G^*. G^* is now isomorphic with the group of matrices $O(A)$. If G^* contains an element of even order g, say, then we have $g^{2m} = 1$ for some smallest positive integer m. Then g^m is an element of G^* satisfying $h^2 = 1$, $h \neq 1$. But if one seeks such an element h in U_2 by squaring A of (6.44) one finds the only possible h is R in contradiction to our hypothesis. So the group of matrices $O(A)$ can only contain elements of odd order and must therefore be one of the infinite sequences C_n, with n odd. The corresponding subgroup of U_2, C_n' say, can be generated by

$$A = \begin{bmatrix} e^{-2\pi i/n} & 0 \\ 0 & e^{2\pi i/n} \end{bmatrix}.$$

The finite subgroups of U_2 are shown in Table 6.7 together with a certain amount of other information. I shall call a finite subgroup of U_2 which contains the element R the spinor group associated with the corresponding rotation group. If the latter group is G, I shall always write the former group as G^*. G is not a subgroup of G^* nor, in general, is it isomorphic with the direct product of G with $(1, R)$.

Table 6.7. *The finite subgroups of the spinor group U_2 and their generators*

(1) $C_n^* . A^n = R \quad R^2 = 1$
(2) $C_n' . A^n = 1 \quad n$ odd $\quad (C_{2n}' \equiv C_n^*)$
(3) $D_n^* . A^n = C^2 = R \quad R^2 = 1 \quad AC = CA^{-1}$
(4) $O^* . A^4 = C^2 = D^3 = CDA = R \quad R^2 = 1$
(5) $K^* . A^5 = C^2 = D^3 = CDA = R \quad R^2 = 1$
(6) $T^* . A^3 = C^2 = (AC)^3 = R \quad R^2 = 1$

Remarks about improper groups: T_d^* as for O^* but A is now replaced by its product in the augmented spinor group with the inversion element ι. $O^* \times S$ contains $D_4^* \times S$, $D_3^* \times S$, O^*, T_d^*, $T^* \times S$, where $S = (1, \iota)$. We write this O_h^* contains D_{4h}^*, D_{3v}^*, O^*, T_d^*, T_h^*.

The augmented unitary group is the direct product $U_2 \times S$. So to every spinor group G^* there is an augmented spinor group $G^* \times S$. We write these according to the rules used for rotation groups but with a star as a superscript. Thus the octahedral spinor group including inversion is written O_h^* and $O_h^* = O^* \times S$.

As before there are many improper groups which do not contain the inversion element but the only really interesting one is T_d^*. This has 48 elements and is isomorphic with O^*.

Examples

R_3 and U_2 are about the origin. Elements $r, r' \in R_3$; $A, A' \in U_2$; t, t', t_1 are translations. E_3 is three-dimensional Euclidean space.

1. If $tr = r't'$ then $r = r'$.
2. Let G be the group of all translations and rotations of E_3. Then the set of all products of the type rt contains each element of G just once.
3. Let G^* be the collection of ordered pairs (A, t). Define multiplication by the rule $(A,t)(A',t') = (AA', t_1 t')$, where $tO(A') = O(A')\,t_1$. Show that G^* is a group.
4. Take a fixed point P of E_3. The set of elements of G which leave P fixed may be expressed in the form $t_2^{-1} r t_2$ for some fixed t_2. Deduce that this set forms a group, $R_3(P)$ say, isomorphic with R_3.
5. Take the subset U_{2P} of all $(A,t) \in G^*$ such that $O(A)\,t \in R_3(P)$. Prove that U_{2P} is a subgroup of G^* isomorphic with U_2.

(These examples show how we may free U_2 from its dependence on a particular origin. It is easiest to do them in numerical order!)

6.10. Some mathematical aspects of the groups R_3 and U_2

I know many physicists are interested in the pure mathematical aspects of their subject. For this reason I include here a brief description of a part of the significance of the relationship of R_3 to U_2 from the topological point of view. Intentionally I eschew rigour and try merely to get the ideas across.[1] The reader who is not interested may pass straight on to the next section.

So far we have really been regarding R_3 and U_2 just as groups. That is, we have recognized them as concrete realizations of abstract groups, but have paid little explicit attention to the fact that they possess other structure as well. We have, however, used the other structure in our proofs and it is time to state clearly that it was not merely a useful device, but rather enters into our theory in an essential manner.

First, then, R_3 and U_2 are topological spaces. This means that the limit points of a set of elements are known when that set is known. Instead of giving a general definition of a topological space,[2] however, it is easier simply to demonstrate the topology for these two special cases. A general element of U_2 is described, according to (6.44), by a set of four real numbers $(\kappa, \lambda, \mu, \nu)$ satisfying $\kappa^2 + \lambda^2 + \mu^2 + \nu^2 = 1$, i.e. the unit sphere in a hypothetical four-dimensional real Euclidean space. It is obvious what continuity, convergence and limit points mean on this unit sphere and the associated topology is the topology which is normally introduced into U_2. The unit element of U_2 is $(1, 0, 0, 0)$ and R is the antipodal point $(-1, 0, 0, 0)$. We call these the North Pole N and the South Pole S, respectively. AR is always antipodal to A. So if we consider the set of all straight lines through the origin of the four-dimensional space each line is associated with just a pair, A, AR, say,

[1] For an excellent and easily read, though somewhat long, introduction to most of the relevant mathematics see Pontrjagin, *Topological Groups* (Princeton, 1946).

[2] Pontrjagin, p. 26. Different authors use slightly different definitions.

of U_2. The set of lines may be assigned a topology in the obvious and natural way[1] and this is the topology which is normally assigned to R_3. Equivalently we may stick to the surface of the sphere and merely identify antipodal points.

Secondly, R_3 and U_2 are topological groups. This means that the group operations are continuous functions of the elements. More precisely, if a sequence $a_n \to a$ then $a_n^{-1} \to a^{-1}$ and if also $b_n \to b$ then $a_n b_n \to ab$. This can be proved immediately.

U_2 is simply connected. In other words if one draws any closed curve on its surface it can be contracted over the sphere in a continuous manner to a single point. For example, a great circle may be contracted through a series of ever smaller circles until it finally becomes a point. On the other hand, R_3 is not. Take half a great circle drawn starting at N and finishing at S. This is closed because N and S are identified in R_3. But it is not contractible via a series of curves all lying in R_3. All other paths starting and finishing at N are contractible to this path or to a point. The only obstacle to contraction is the passage 'through to the antipodes' made possible by the identification of antipodal points. So if we can show that the path from N to S can be continuously deformed into the path from S to N we shall always be able to eliminate these passages in pairs. A method of deforming the path from N to S to one from S to N is shown in Fig. 6.5. I have labelled the three successive parts of a path by the letters a, b, c. One performs a first, then b, then c. So we can divide all paths starting and ending at N into two classes, those with an even and those with an odd number of passages.[2] We will call them even and odd paths, respectively.

We multiply classes of closed paths starting at N by the following rule. If a path p_1 belongs to a class C_1 and p_2 to C_2 then C_1C_2 is the class containing p_1 followed by p_2. The latter path is well defined because p_1 and p_2 both start and end at N. The resulting class C_1C_2 is independent of the particular choices of p_1, p_2. The multiplication is associative and the inverse of C_1 is the class containing the path in the reverse direction to p_1. So the classes form a group. This is called the fundamental group of the topological space. It contains only the unit element when and only when the space is simply-connected, as, for example, for U_2. It contains two elements for R_3 and Fig. 6.5 shows that the second element is equal to its inverse.

But whilst on the large scale R_3 and U_2 are different, on the small scale they are structurally identical, considering them as topological groups. I shall not prove this here but merely state it as an interesting fact which the reader may pursue if he wishes.[3] We say then that R_3 and U_2 are locally isomorphic but globally distinct. U_2 is in fact the unique simply-connected topological group which is locally isomorphic with R_3. It is called the universal covering group of R_3. This, from the topological point of view, is the significant relation[4] between U_2 and R_3.

[1] This shows that the underlying topological space of R_3 is the same as that of three-dimensional real projective space.

[2] The possibility of an infinite number of passages causes little extra difficulty because a path must have a finite total length.

[3] Examples 4 and 5 at the end of the section actually provide the essentially preliminary calculations. Then use Pontrjagin, theorem 73, p. 257.

[4] Pontrjagin, ch. 8.

I can now explain the reason for the particular procedure we adopted for deciding which element of U_2 should correspond to a physical rotation which produced a given element of R_3. As we go continuously through our physical rotation the associated matrix A has a representative point which starts at N on the sphere and finishes up at one of two possible points antipodal to each other. Assuming A describes a continuous path on the sphere it is clear that the particular point it finishes at is uniquely defined by the sequence of elements of R_3. Our procedure was merely a method of deciding which point the path finished at. Corresponding to the two points there are two classes of paths in R_3, a path of one class being not deformable continuously into a path of the other.

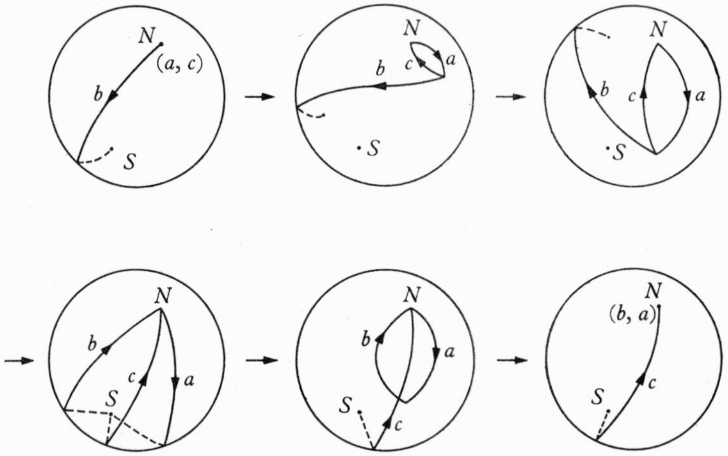

Fig. 6.5. Deformation of the path from N to S into the path from S to N for the group R_3. S lies behind the sphere. The paths a, b, c are always traversed in that order.

The last, and in many ways the most important, extra structure that R_3 and U_2 possess is that they are Lie groups (after S. Lie, Norwegian mathematician). This means that a coordinate system can be introduced around the identity element in terms of which the group operations are analytic functions. For example, an arbitrary element of R_3 near the identity is

$$M = \begin{bmatrix} 1 & -\epsilon_z & \epsilon_y \\ \epsilon_z & 1 & -\epsilon_x \\ -\epsilon_y & \epsilon_x & 1 \end{bmatrix} \tag{6.55}$$

as the reader can verify directly from the definition of an orthogonal matrix. Both R_3 and U_2 are three-dimensional Lie groups because the coordinate system has three real coordinates. Every Lie group has associated with it its Lie algebra (or infinitesimal group) of infinitesimal operations.[1] However, we have already met this in disguise—it is the set of all components of the angular momentum $\mathbf{L}$ for R_3 and of $\mathbf{J}$ for U_2.

R_3 and U_2 possess matrix representations.[2] The set of $|LM\rangle$ for given L and

[1] Pontrjagin, ch. 9.

[2] There are also counterparts of the orthogonality relations. See Pontrjagin, ch. 4.

varying M forms a basis for an irreducible representation of R_3 of degree $2L+1$. Thus there is a set of inequivalent irreducible representations, one for each odd positive integral degree. There are no others. Similarly the $|JM\rangle$ yield one for each positive integral degree for U_2. Again there are no others.

Any representation of R_3 immediately induces a representation of U_2. For we associate with A and AR of U_2 the matrix which has been assigned to $O(A)$. The $|LM\rangle$ induce the representations of U_2 derived from $|JM\rangle$ with integral $J = L$. When J is half an odd integer there are no corresponding representations of R_3. This is because, then, the element R of U_2 gets assigned minus the unit matrix of the appropriate degree. So A and AR always get different matrices and therefore we have two matrices associated with each $O(A) = O(AR)$ of R_3. It is, however, customary to call these representations of U_2 'two-valued representations' of R_3. This is not an entirely satisfactory terminology but should not muddle us. We just remember that, in spite of their name, they are actually just those representations of U_2 which do not give rise in any natural way to representations of R_3. We also see immediately why there are not three-valued, etc., representations of R_3.[1]

If the reader thinks of the meaning of the elements 1 and R of U_2 in terms of physical rotations he will see that any closed even path of R_3 starting at N on the sphere finishes up with 1 and any closed odd path with R. So we may reasonably identify 1 and R with the two elements of the fundamental group of R_3. This group has two irreducible representations, Γ_1 and Γ_2 say, according to whether R is assigned $+1$ or -1. Any state $|JM\rangle$ is a basis for an irreducible representation of the fundamental group, Γ_1 when J is integral and Γ_2 when J is half an odd integer. Therefore we have Γ_1 for an even electron system and Γ_2 for an odd electron system. We have neglected nuclear particles, mesons, neutrinos and hyperons. If we include them we should talk of an even or odd number of particles with spin half (or even more generally of particles with half-odd integral spin). Taking this into account and also the possibility of the creation and annihilation of fundamental particles, there is no experimental evidence suggesting that it is possible for a system to make a transition from a state which is a basis for one irreducible representation of the fundamental group of R_3 to a state which is a basis for the other.

Examples

1. Show that the group R_2 of rotations about a fixed point of two-dimensional Euclidean space has the topology of the circumference of a circle.

2. Show that the fundamental group of R_2 is an infinite commutative group and that it can be generated by one of its elements.

3. Any rotation of R_3 may be represented by a vector $\boldsymbol{\xi}$ from the origin with $|\boldsymbol{\xi}| \leqslant \pi$ in such a way that: (i) the line defined by $\boldsymbol{\xi}$ is its axis, (ii) it is a rotation of $|\boldsymbol{\xi}|$ in a counter-clockwise direction about $\boldsymbol{\xi}$. Introduce a topology naturally into R_3 via this set of vectors and prove that it is the same as the topology we defined earlier in the section. (You will need to identify certain pairs of vectors.)

[1] See also, Littlewood, *The Theory of Group Characters and Matrix Representations of Groups* (Oxford, 1950), p. 248.

4. Write the matrix M of equation (6.55) as

$$M = I+\epsilon_x m_x+\epsilon_y m_y+\epsilon_z m_z = I+\boldsymbol{\epsilon}\,.\,\mathbf{m}.$$

Prove that the vector $\boldsymbol{\mu} = i\hbar\mathbf{m}$ has the same commutation relations as an angular momentum vector.

5. Show that a matrix near the identity of U_2 may be written

$$A = \begin{bmatrix} 1+i\epsilon_z & \epsilon_y+i\epsilon_x \\ -\epsilon_y+i\epsilon_x & 1-i\epsilon_z \end{bmatrix} = I+\boldsymbol{\epsilon}\,.\,\boldsymbol{\eta}$$

and that $-\frac{1}{2}i\hbar\boldsymbol{\eta}$ has the commutation relations of an angular momentum vector.

6.11. The finite spinor groups and their representations

When spin is taken into account, the symmetry group of a site is that subgroup of the augmented spinor group $U_2 \times S$ which leaves the aspect of the site unchanged. This simply means that if, before, we had G as the symmetry group we realize now that we really have G^*. G^* always has twice as many elements as G and is the set of all unitary unimodular matrices A such that $O(A) \in G$.

Given a representation of G, under which $O \in G$ is assigned the matrix $M(O)$ we immediately deduce a representation of G^* by the rule $A \in G^* \rightarrow M[O(A)]$. It is irreducible when the representation of G is and conversely (use (6.28)). Because of the necessary relationship (6.23) there must also be others which cannot be so derived from representations of G. In these the element R always gets assigned minus the unit matrix. This is because R commutes with all elements of G^* and therefore by Schur's lemma (ex. 2, p. 155) must be represented by a multiple of the unit matrix, λI say. $R^2 = I$ and so $\lambda = \pm 1$. If $\lambda = +1$, then A and AR always get the same matrix and hence we derive an irreducible representation of G which by the previous rule gives the representation of G^*. $\lambda = -1$ for all the other irreducible representations. Therefore if Γ is one of the latter representations with $A \rightarrow M(A)$ and we assign matrices to the elements of G by the rule $O(A) \rightarrow M(A)$ then each element of G get assigned two distinct matrices. For this reason the representation Γ of G^* is called a 'two-valued' representation of G. It is an undesirable expression because Γ is not a representation of G, but only of G^*. However, it is widely used so. reluctantly, I do likewise.

We already know the representations of G for all relevant symmetry groups. Their bases are functions referring to even electron systems or to systems without spin. The odd electron systems give rise to the 'two-valued' representations. Let us see how we supplement the character tables with these extra representations.

The octahedral group O^* is a typical example. The pair of spin functions α, β must form a basis for a representation of any finite symmetry group. Write C_n for a counterclockwise rotation of $2\pi/n$. Then $\chi(C_n)$ is independent of the axis of the rotation. So we take the Z-axis and can write down $\chi(C_n)$ from (6.46)

$$\chi(C_n) = 2\cos\frac{\pi}{n}.$$

This gives $\chi(C_2) = 0$, $\chi(C_3) = 1$, $\chi(C_4) = \sqrt{2}$, $\chi(R) = -2$. Because $R = -I$ we also have $\chi(RC_2) = 0$, $\chi(RC_3) = -1$, $\chi(RC_4) = -\sqrt{2}$ (and $\chi(I) = 2$). χ for the powers C_n^m, using $\chi(C_n^m) = 2\cos(m\pi/n)$, shows all other elements to have one or other of these χ values and (6.28) shows that the representation is irreducible. We call it E', see Table A 8. O^* therefore has at least eight classes, and at least three representations more than O. Now multiply each matrix of E' by the corresponding number of the representation A_2. This gives a second irreducible representation E'', say. Lastly, the four functions $|jm\rangle$ for $j = \frac{3}{2}$ give a representation U' of O^*. Equation (6.28) shows this also to be irreducible (use Table A 13). The degrees λ_k of the eight representations A_1, A_2, E, T_1, T_2, E', E'', U' now satisfy $\Sigma\lambda_k^2 = 48$ so we have all the irreducible representations and have, incidentally, proved that O^* has the eight classes shown in Table A 8. The 'two-valued' representations of the other symmetry groups can be obtained in a precisely analogous manner and are given for the important groups also in Table A 8.

After this introduction to the detailed properties of the finite spinor groups it is apparent that we have before us merely a rather tedious repetition of the types of calculation we made earlier for the finite symmetry groups which neglected spin. There is no point in describing again the details of the procedures adopted for this new, but trivially different, situation. The results are given in the tables which are largely self-explanatory. Standard basis relations are defined in Table A 16 after which Table A 19 shows in detail how the kets $|JM\rangle$ give rise to irreducible representations of O^*. J is never larger than $\frac{9}{2}$ in the ground term of d^n. I have chosen as a defining basis for the standard relations in O^* the spin functions α, β for E', $a_2\alpha$, $a_2\beta$ for E'' and the four functions $|\frac{3}{2}m\rangle$, correctly connected by shift operators, for U'. Finally, Table A 20 contains the most useful coupling coefficients.

Example

(The reader may also confirm any or all of the entries in the tables.) A finite symmetry group G has c classes C_j and irreducible representations Γ_i. Regard each χ_i as a vector $\mathbf{v}^i$ in a c-dimensional complex space S with components $\chi_i(C_j)$. The scalar product of $\mathbf{u}$ and $\mathbf{w}$ is $\sum_j \bar{u}_j w_j$. Show that if $\mathbf{w}$ is orthogonal to all the $\mathbf{v}^i$ then $\mathbf{w} = 0$. Deduce directly that in a 'two-valued' representation of G, R is assigned minus the unit matrix.

(Hint: Determine $\chi(R)$ and use (6.6).)

CHAPTER 7

COMPLEX IONS

7.1. The concept of a complex ion

In aqueous solution a metal ion becomes surrounded by a number of other ions or small neutral molecules, for example, water molecules, in a fairly permanent association. The aggregate consisting of the metal ion together with its collection of followers is called a complex ion.

Let us take the example of the nickel ion. In solution this is $Ni(H_2O)_6{}^{++}$, where six water molecules are actually linked to the nickel ion in an approximately octahedral arrangement. We may perhaps have produced these aqueous nickel ions by dissolving nickel sulphate in water and then there will be an equal number of sulphate ions $SO_4{}^{--}$ giving overall electrical neutrality to the solution. If we now concentrate the solution by evaporation, solid nickel sulphate crystallizes out. At room temperature this has the composition $NiSO_4 . 7H_2O$ and X-ray crystallographic studies show that six water molecules surround each nickel ion, the sulphate ions and the remaining water molecules lying in between the $Ni(H_2O)_6{}^{++}$ complexes.

The structure of an isolated water molecule is

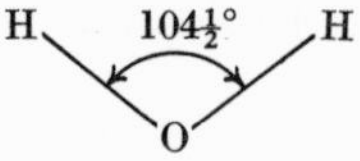

with each OH bond equal and of length 0·96 Å. It has the fairly high electric dipole moment of $1{\cdot}84 \times 10^{-18}$ e.s.u. pointing along the axis of symmetry and with its positive end on the side of the hydrogen atoms. However, the detailed electron distribution is such that there are maxima of electron density in approximately regular tetrahedral directions from the oxygen nucleus—in the directions of the hydrogen nuclei and also in the plane of symmetry at right-angles to the HOH plane, thus

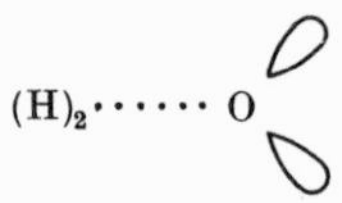

where $(H)_2 \ldots$ represents the projection of OH bonds onto this plane of symmetry. We will see the reason for this distribution of electron density in § 7.2. Here we observe that it means that there are two rather distinct ways in which a water molecule may arrange itself relative to the positive metal ion. These are

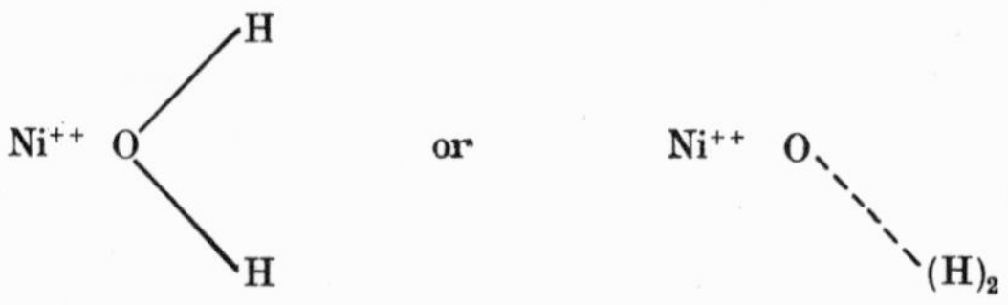

(a) (b)

where in (a) all the atoms are coplanar and in (b) the nickel ion and the two hydrogen atoms occupy three vertices of a tetrahedron about the oxygen atom. The tetrahedron is not regular—the HOH angle may change slightly in forming these hydrates but will still deviate from that appropriate to a regular tetrahedron ($109°\,28'$). Also, packing considerations in the crystal may dictate arrangements intermediate between (a) and (b). In $NiSO_4.7H_2O$, four of the H_2O molecules attached to each nickel ion have arrangement (a) and the remaining two have arrangement (b).

The groups which get attached to the metal ions are called ligands. Other important examples of ligands are NH_3 (ammonia), CN^- (the cyanide ion), and the halide ions F^-, Cl^-, Br^- and I^-. The ammonia molecule is linked so that the metal ion and the three hydrogen atoms are approximately regularly tetrahedrally arranged about the nitrogen atom. The cyanide ion is linked collinearly with the metal ion, and with the carbon atom nearest to it

$$M^{n+}\quad N\begin{array}{l} \diagup H \\ \cdots (H)_2 \end{array} \qquad\qquad M^{n+}\quad C\!-\!N^-$$

An octahedral arrangement of six ligands is the most common arrangement (one often says the ligands are octahedrally coordinated) and a tetrahedral with four less so. Other possibilities include square planar

$$\begin{array}{ccc} & X & \\ X & M & X \\ & X & \end{array}$$

with four ligands and tetragonal pyramidal or trigonal bi-pyramidal with five.

7.2. Electron configurations

Just as with atoms we first discuss the electronic structure of a complex ion with the assumption that the nuclei are fixed. This is not such a good approximation here, but serves for many purposes. I describe its removal in § 11.3.

Our treatment of the electronic structure of a complex ion will parallel very closely our earlier treatment of atoms and ions. The important difference is that we no longer have spherical symmetry but merely the symmetry group G of the arrangement of nuclei (regarded as fixed point charges). Therefore, instead of the classification of one-electron functions in terms of l, m_s and m_l that we had in § 4.1 we must use γ, m_s and m_γ, where γ is an irreducible representation of G and m_γ a quantum number classifying the components of γ. We will now have electron configurations $(n_1\gamma_1)^{a_1}(n_2\gamma_2)^{a_2}\ldots(n_r\gamma_r)^{a_r}$ with the a_i integers. n_i is introduced as an extra classifying parameter because there will usually be several sets of one-electron functions forming bases for each representation γ_i. It is not necessarily the same as n_i of § 4.2, but plays a similar role. The one-electron functions form an orthonormal set.

In free ions we remember that so long as we remain within a configuration all matrix elements are the same as they would be with the omission of any or all of the filled shells of electrons, apart from a constant energy common to all states of the configuration. Empirically all low-lying states of a free di- or tri-positive transition-metal ion belong to one configuration (with a partly-filled d shell), and therefore the simplification of rejecting the filled shells when determining relative energies within a configuration is a very useful one. In complex ions formed from transition-metal ions it appears empirically that all the low-lying states derive from just a few configurations

$$(n_1\gamma_1)^{a_1}\dots(n_r\gamma_r)^{a_r}(n_{r+1}\gamma_{r+1})^{a_{r+1}}\dots(n_{r+s}\gamma_{r+s})^{a_{r+s}},$$

where the part $(n_1\gamma_1)^{a_1}\dots(n_r\gamma_r)^{a_r}$ consists of filled shells and is the same for each configuration and the total degeneracy of $\gamma_{r+1}\dots\gamma_{r+s}$ is five. A typical example of the second part of the configuration assignment (we call this the partly-filled shells even though some may actually be full), for G the octahedral group O, would be $t_2^3e^2$. Here t_2 and e have a total degeneracy of five and we have dropped the symbols n_{r+1}, etc. We see why the degeneracy is five, it is of course the same as the degeneracy of the d orbitals in the free ion.

Let us now 'eliminate the filled shells'. For one-electron operators this is immediate. A diagonal element is a sum $\sum_{\kappa=1}^{n}\langle\overline{\phi}_\kappa\mid U\mid\phi_\kappa\rangle$ (see (4.21)) which can be broken up into two parts, the first arising from the filled shells and the second from the partly-filled shells. The first is always the same and the second is the same as it would be without the filled shells. Next, an off-diagonal element can only be non-zero when the occupancy of the orbitals differs by one. This difference of necessity takes place in the partly-filled shells and as the matrix element is then equal to the matrix element of the one-electron operator between the pair of orbitals which differ it follows that this is the same as it would be in the absence of the filled shells.

For simplicity we suppose no repeated one-electron representation occurs in the partly-filled shells. This supposition is usually true.[1] Then there will be three types of two-electron matrix elements which turn up in evaluating the matrix elements of the electrostatic energy depending on whether they contain 0, 2 or 4 one-electron functions from partly-filled shells. They cannot contain an odd number because of the orthogonality of the one-electron functions, combined with the assumed constancy of $a_1, \dots, a_r$. For the same reason if they contain 2, the two functions from the filled shells must be the same functions and have the same spin. We take a pair of determinantal functions ψ_i, ψ_j and write

$$V_{ij} = V_{ij}^{(f)} + V_{ij}^{(i)} + V_{ij}^{(p)}$$

to distinguish the three parts (f = filled shell, i = interaction, p = partly-filled shell) and have for an off-diagonal element

$$V_{ij}^{(i)} = (-1)^{\tau}\sum_{\kappa=1}^{r}\sum_{m_s=-\frac{1}{2}}^{+\frac{1}{2}}\sum_{m_{\gamma_\kappa}}\{\langle n_\kappa\gamma_\kappa m_s m_{\gamma_\kappa}, n_\lambda\gamma_\lambda m_s' m_{\gamma_\lambda}\mid V\mid n_\kappa\gamma_\kappa m_s m_{\gamma_\kappa}, n_\mu\gamma_\mu m_s'' m_{\gamma_\mu}\rangle$$
$$-\langle n_\kappa\gamma_\kappa m_s m_{\gamma_\kappa}, n_\lambda\gamma_\lambda m_s' m_{\gamma_\lambda}\mid V\mid n_\mu\gamma_\mu m_s'' m_{\gamma_\mu}, n_\kappa\gamma_\kappa m_s m_{\gamma_\kappa}\rangle\}, \quad (7.1)$$

[1] Really only for the high-symmetry components of the field (see §7.4).

where P_τ gets ψ_i in the same order as ψ_j and $\lambda, \mu > r$ and the kets $|n_\lambda \gamma_\lambda m'_s m_{\gamma\lambda}\rangle$, $|n_\mu \gamma_\mu m''_s m_{\gamma_\mu}\rangle$ are the (only) ones in which ψ_i differs from ψ_j. We now show $V_{ij}^{(i)}$ is zero. This is in fact very easy and we simplify our notation to show it. After integrating out the spins in (7.1) we are left with sums of the kind $\sum_a \langle \bar{a}\bar{b} \mid V \mid ac\rangle$ and $\sum_a \langle \bar{a}\bar{b} \mid V \mid ca\rangle$, where a belongs to a particular filled shell and we are summing over m_γ. b and c belong to partly-filled shells. Using Schrödinger's representation the sums become

$$\sum_a \langle \bar{a}\bar{b} \mid V \mid ac\rangle = \int (\sum_a \bar{a}(1)\, a(1))\, \bar{b}(2)\, c(2)\, V(12)\, d\tau_1 d\tau_2,$$

$$\sum_a \langle \bar{a}\bar{b} \mid V \mid ca\rangle = \int (\sum_a \bar{a}(1)\, a(2))\, \bar{b}(2)\, c(1)\, V(12)\, d\tau_1 d\tau_2. \tag{7.2}$$

By ex. 3, p. 155, $\sum_a \bar{a}(1)\, a(1)$ belongs to the unit representation of G and by an obvious extension of that example so also does $\sum_a \bar{a}(1)\, a(2)$. $V(12)$ also is totally symmetric. As b and c are different functions (if they were the same the matrix elements would have vanished already through the necessary orthogonality of the spin functions), it follows from our assumption of no repeated one-electron representation, together with the orthogonality properties for functions belonging to irreducible representations that both the sums are zero. Therefore $V^{(i)}$ is zero. For off-diagonal V_{ij} the orthogonality of the one-electron functions ensures that $V^{(f)}$ is zero too. Therefore $V_{ij} = V_{ij}^{(p)}$ which proves the result for off-diagonal elements.

Precisely the same kind of argument shows us that the diagonal elements are composed of the three parts $V_{ii}^{(f)}$, which is the same for all states in the configurations being considered, $V_{ii}^{(i)}$, which depends on the representations γ from which the partly-filled shell functions come but is independent of the m_s, m_γ values, and finally $V_{ii}^{(p)}$, which is the same as it would be in the absence of the filled shells. We now examine the various parts of the diagonal elements in more detail, writing as usual $\mathscr{H}_{ii}$ for a diagonal element of the whole electronic Hamiltonian including the potential energy in the field of the nuclei and the spin-orbit coupling energy and, as for V_{ij}, putting

$$\mathscr{H}_{ij} = \mathscr{H}_{ij}^{(f)} + \mathscr{H}_{ij}^{(i)} + \mathscr{H}_{ij}^{(p)}.$$

The one-electron operators do not contribute to $\mathscr{H}_{ij}^{(i)}$ at all.

$\mathscr{H}_{ij}^{(f)}$ is just the energy which we would calculate for the filled shells if the partly-filled shells were absent altogether. It is in fact the energy of the singlet, totally symmetric, determinantal function ψ which belongs to the filled-shell configuration $(n_1\gamma_1)^{a_1} \ldots (n_r\gamma_r)^{a_r}$. Let us now take our $Ni(H_2O)_6^{++}$ ion and ask what we expect those filled shells to be. In order for us to satisfy our assumption about not repeating representations we may suppose the ion has symmetry group $G = T_h$ (see note 1, p. 192).

Before doing this we ask what are the filled shells of the Ni^{++} ion and the H_2O molecules taken on their own. For the Ni^{++} ion we know already that they are $1s^2\, 2s^2\, 2p^6\, 3s^2\, 3p^6$ leaving $3d^8$ as a partly-filled shell. The water molecule is a singlet ($S = 0$) and has ten electrons. Two go into an orbital centred on the oxygen nucleus and not very different from the oxygen $1s^2$ inner shell. The remaining eight go two each into orbitals not too unlike oxygen $2s$ and $2p$ orbitals. This gives something rather like an O^{--} ion with configuration $1s^2\,2s^2\,2p^6$ and two protons in the outer reaches of the ion, 0·96 Å from the oxygen nucleus. The wave-function is now a single determinant function ϕ and can be got into a physically more acceptable form by writing

$$\left.\begin{aligned} t_1 &= \tfrac{1}{2}\psi_{2s}+\tfrac{1}{2}\psi_{2p_x}+\tfrac{1}{2}\psi_{2p_y}+\tfrac{1}{2}\psi_{2p_z},\\ t_2 &= \tfrac{1}{2}\psi_{2s}+\tfrac{1}{2}\psi_{2p_x}-\tfrac{1}{2}\psi_{2p_y}-\tfrac{1}{2}\psi_{2p_z},\\ t_3 &= \tfrac{1}{2}\psi_{2s}-\tfrac{1}{2}\psi_{2p_x}+\tfrac{1}{2}\psi_{2p_y}-\tfrac{1}{2}\psi_{2p_z},\\ t_4 &= \tfrac{1}{2}\psi_{2s}-\tfrac{1}{2}\psi_{2p_x}-\tfrac{1}{2}\psi_{2p_y}+\tfrac{1}{2}\psi_{2p_z}.\end{aligned}\right\} \quad (7.3)$$

Then ϕ is unaltered by replacing the $2s$ and $2p$ functions by the t_i. The t_i are now directed along, respectively, the vectors $(1, 1, 1)$, $(1, -1, -1)$, $(-1, 1, -1)$, $(-1, -1, 1)$ and if we put the protons along the $(1, 1, 1)$ and $(1, -1, -1)$ directions we have a picture of a water molecule with a $1s^2$ inner shell, two electron pair bonds (t_1^2 and t_2^2) with hydrogen atoms (two of the electrons originally came from the hydrogen atoms) and two lone pairs (t_3^2 and t_4^2) of electrons. The lone pairs and the bonds are directed tetrahedrally. (See also Appendix 8.)

The process of replacing the $2s$ and $2p$ functions by the t_i is called hybridization, or sometimes 'the formation of directed bond orbitals' and has, of course, no observable consequences in itself. However, it makes it easier to appreciate the structure of the water molecule, which differs from the O^{--} ion mainly in that t_1 and t_2 become elongated in the direction of the hydrogen atoms and are better approximated by expressions $c_1t_1+c_2\psi_{1s}(H_1)$, $c_1t_2+c_2\psi_{1s}(H_2)$, where $c_1^2+c_2^2 = 1$ and $\psi_{1s}(H_1)$, $\psi_{1s}(H_2)$ are the $1s$ orbitals centred on the first and the second hydrogen atom, respectively. The result of this is that the electron density is no longer spherically symmetric but has maxima in the tetrahedral directions. Then, also, the directions deviate from the regular tetrahedral ones.

Before returning to the complex ions we remark that ammonia is, approximately, N^{3-} with three protons in tetrahedral directions (the HNH angle is 107°). CN^- is also filled shell and a singlet.[1]

It appears experimentally to be the case that both the Ni^{++} ion and the six H_2O molecules retain their identity and individuality to a considerable extent in the complex ion. Hence it is natural to try to identify the filled shells of the complex ion with the inner shells of the free Ni^{++} ion together with the electrons of the H_2O molecules. There is a slight difficulty here, for the orbitals localized on a single water molecule clearly cannot form a basis for a representation of G. However, the totality of the occupied orbitals on all the water molecules can

[1] Taking the OZ-axis to contain the CN-axis we take for each atom the digonal hybrids $(1/\sqrt{2})\,(\psi_{2s}+\psi_{2p_z})$ and $(1/\sqrt{2})\,(\psi_{2s}-\psi_{2p_z})$ one of which, t_1 say, points towards the other atom and the other, t_2 say, away. Then in a simple molecular orbital approximation the 14 electrons of CN^- go 4 into the inner $1s$ orbitals, 4 into the lone-pair orbitals t_2 and 6 into one σ- and two π-bond orbitals formed, as in Appendix 8, from t_1, $2p_x$ and $2p_y$ orbitals, respectively. CN^- is a filled shell ion.

and does, because any operation of G turns an orbital on one water molecule into an orbital on the same or on a different water molecule. And as we naturally assume that the same orbitals are occupied on all the water molecules this means the totality does form a representation of G. Therefore we form the filled-shell function ψ in the following way: write down the determinantal function composed of all inner-shell orbitals of Ni^{++} and all occupied orbitals of the H_2O molecules; replace the H_2O orbitals by an orthonormal transform which forms a set of bases for irreducible representations of G. This does not alter the value of the determinantal function and gives us ψ in the form we assumed for it earlier.

We have given a certain physical justification for taking the filled shells as the inner shells of the Ni^{++} ion and the occupied orbitals of the water molecules. Then the partly-filled shells would naturally be taken as the $3d$ orbitals of the Ni^{++} ion. This gives the simplest possible model for the $Ni(H_2O)_6{}^{++}$ complex ion which we hastily replace with a more realistic one.

Clearly there is no reason at all why the filled-shell orbitals in the $Ni(H_2O)_6{}^{++}$ ion should be exactly the same as orbitals of the free Ni^{++} ion and of the free H_2O molecules. We expect, however, that when we take suitable linear combinations of them that 18 will be approximately localized around the Ni^{++} ion and have much the same distribution of density, maxima and nodes as the inner shells of a free Ni^{++} ion and that 10 will be approximately localized around each trio of 2H and an O nucleus in much the same way as those of free water molecules. We expect especially that if the symmetry group of the ion is G then the number of times a particular irreducible representation γ occurs among the filled-shell one-electron orbitals is the same number that one would calculate assuming that the orbitals were those of the Ni^{++} inner shells and of the free water molecules. These assumptions about closed shells are unnecessary in a sense—in all our calculations (except those of § 11.7) we only assume that these shells are filled —but are necessary in order to make our whole treatment of complex ions seem plausible.

The important assumption concerns the partly-filled shells—that they span the same set of irreducible representations of G as the d orbitals. This is supported by a vast amount of *a posteriori* evidence, often indirect, and is plausible but not at all obviously true *a priori*. At the time of writing there is no definite evidence that it is ever untrue for the ground state of any compound of any element of the first, second or third transition series.[1] It is upon this assumption that the usefulness of the formal machinery of ligand-field theory largely rests. It is quite usual for all states within 30,000 cm^{-1} of the ground state to satisfy the assumption and very unusual for a state below 10,000 cm^{-1} not to do so. In case the ligands have filled shells the number of electrons in the partly-filled shells is the same as the number of electrons in the partly-filled d shell of the free ion. In all the compounds we treat in this book this condition for the ligands is satisfied also.

[1] It may perhaps be untrue for some octahedrally coordinated nickel compounds (Nyholm, 1956). Formally, at least, it would be an easy matter to extend the theory to cover such cases if they are confirmed.

As we saw $\mathscr{H}_{ij}^{(i)}$ contains only contributions from the electronic electrostatic interaction and is diagonal. It is the sum of quantities of the kind shown in (7.2) with $b \equiv c$. It is convenient to lump all the contributions involving a particular one-electron ket $|\gamma m_s m_\gamma\rangle$ from the partly-filled shells together as $g(\gamma)$ so that we have

$$\mathscr{H}_{ii}^{(i)} = \Sigma g(\gamma), \tag{7.4}$$

where the sum runs over the one-electron kets in the partly-filled shells which turn up in ψ_i. $g(\gamma)$ is independent of m_s and m_γ and

$$g(\gamma) = 2\sum_a J(a, \gamma m_\gamma) - \sum_a K(a, \gamma m_\gamma), \tag{7.5}$$

where a runs over all orbitals (without spin, for example the $1s^2$ shell of Ni^{++} is counted as the one nickel $1s$ orbital) in the filled shells. $g(\gamma)$ is behaving as the matrix element of a one-electron operator for the partly-filled shells would do. It is therefore useful to add to it the interaction between the nuclei and the electrons of the partly-filled shells. This interaction, which is part of $\mathscr{H}_{ii}^{(p)}$, is a sum over $h(\gamma)$ for each one-electron ket, where

$$h(\gamma) = \langle \gamma m_\gamma | \sum_\lambda \frac{-Z_\lambda e^2}{r_\lambda} | \gamma m_\gamma \rangle \tag{7.6}$$

and the λth nucleus has charge Z_λ and distance r_λ from the electron. $h(\gamma)$ is independent of m_s and m_γ because, by definition, the elements of the group G all leave the arrangement of nuclei unchanged.

$h(\gamma)$ together with the Coulomb part of $g(\gamma)$ now give the classical Coulomb energy of interaction of the electron in $|\gamma m_s m_\gamma\rangle$ with the filled shells and the nuclei. The exchange part of $g(\gamma)$ gives an extra non-classical electrostatic energy. If we suppose for a moment that the filled shells are the inner shells of the metal ion together with all the filled orbitals of the ligands (presumed to possess filled shells), in each case unchanged from their form in the free ion or free ligands, and suppose the γm_γ are metal d orbitals we may break $g(\gamma)+h(\gamma)$ into three parts—the interaction with the inner shells and nucleus of the metal, which is unchanged from the free metal ion; the classical electrostatic interaction with the ligands, and the exchange interaction with the ligands. Neglecting the last and remembering that those parts of $\mathscr{H}_{ij}^{(p)}$ which we have not included in $h(\gamma)$ are the same as for the partly-filled d-shell of the free metal ion, our problem then becomes identical with that of the free metal ion placed in the electrostatic field of the ligands, assumed rigid and unpolarizable. Historically this latter model was very important and it was in terms of it that ligand-field theory was first developed. It was called crystal-field theory. However, the essential features of the crystal-field theory are carried over into our more general approach while certain inessential features, which are in disagreement with experiment and derive mainly from the assumption that the partly-filled shells are actually composed of d orbitals, disappear in ligand-field theory.

Returning to our main theme we replace the electrostatic interaction between the partly-filled shells and the rest of the complex ion by one-electron operators, in a purely formal manner, by writing

$$\mathscr{V} = \sum_{\kappa=1}^{n} V(\kappa), \tag{7.7}$$

where the sum runs over the partly-filled shells and the matrix elements of V satisfy

$$\langle \gamma m_s m_\gamma \mid V \mid \gamma' m_s' m_\gamma' \rangle = [g(\gamma)+h(\gamma)]\,\delta(\gamma,\gamma')\,\delta(m_s,m_s')\,\delta(m_\gamma,m_\gamma'). \tag{7.8}$$

$\mathscr{V}$ is only defined within the partly-filled shells and depends upon the filled shells implicitly via (7.5) and (7.6). It behaves as an operator acting only on spatial functions and having the full symmetry of the group. We call it the operator of the electrostatic ligand field. Note that, as we have defined it, it represents an interaction with all nuclei (including that of the central ion) and all the electrons of the filled shells. We must define it in this way so long as we do not use a model which distinguishes rigidly between central ion filled shells and ligand orbitals, but as $\mathscr{V}$ is only useful if we have no changes in the occupancy of the partly-filled shells it is only relative energies which are important. Therefore, if the crystal field approach were valid all the relative energies would arise from the electrostatic field of the ligands, and so for practical purposes our electrostatic ligand field is a straight generalization of the crystal field. If we neglect the exchange part of $g(\gamma)$ we may regard V as an electrostatic field and write it $V(\mathbf{r})$ but they are not pure d orbitals that find themselves in this field but rather more general ones having, however, the same behaviour under the symmetry group G.

The end of this long discussion, through almost all of which we were guided by physical, not mathematical, considerations, is that we have the Hamiltonian

$$\mathscr{H} = \mathscr{H}_0 + \mathscr{H}_e + \mathscr{H}_s + \mathscr{V} \tag{7.9}$$

for the n electrons in the partly-filled shells where $\mathscr{V}$ is given by (7.7) and (7.8) and

$$\mathscr{H}_0 = \sum_{\kappa=1}^{n}\left(\frac{1}{2m}\mathbf{p}_\kappa^2\right); \quad \mathscr{H}_e = \sum_{\kappa<\lambda}\frac{e^2}{r_{\kappa\lambda}}; \quad \mathscr{H}_s = \sum_{\kappa=1}^{n}\mathbf{t}_\kappa\,.\,\mathbf{s}_\kappa. \tag{7.10}$$

$\mathbf{t}_\kappa$ has the same behaviour under G as a pseudo-vector. This follows from (5.38) by noting that any element of G acts on f in the same way that it acts on a scalar, i.e. it leaves it invariant, so $\nabla f \wedge \mathbf{p}$ has the same behaviour under G as a pseudo-vector (i.e. a vector for proper rotations but unchanged by inversion; in general $\nabla f \wedge \mathbf{p}$ is, of course, not actually a pseudo-vector). Just as in the treatment of spin-orbit coupling in atoms we must include the field of the filled shells in f. The main contribution comes from near the nuclei.

7.3. Evaluation of parameters

We must now calculate energy levels using the Hamiltonian (7.9). It has not proved possible so far to make *a priori* calculations of the one- and two-electron matrix elements required and therefore we can only proceed semi-empirically.[1]

[1] Although a start has been made by Tanabe & Sugano (1956). See also, Kleiner (1952).

Let us first consider what we need. $\mathscr{H}_0$ has the full symmetry of the group and so we may conveniently combine it with $\mathscr{V}$. It is convenient to call $\mathscr{H}_0+\mathscr{V}$ the operator of the ligand field.[1] In order to parametrize $\mathscr{H}_0+\mathscr{V}$ we need the relative energies of one-electron functions belonging to each of the irreducible representations occurring in the partly-filled shells. Taking the octahedral group O_h as an example, the d functions span e_g+t_{2g} and it follows from our fundamental assumption that the partly-filled shells are $t_{2g}^m e_g^n$. To parametrize $\mathscr{H}_0+\mathscr{V}$ we therefore need one parameter—the energy separation between t_{2g} and e_g. This is written $\Delta = E(e_g)-E(t_{2g})$ and, empirically, is positive for compounds with an octahedral arrangement of ligands.

The matrix elements of $\mathscr{H}_e$ and $\mathscr{H}_s$ are not quite so easy because one needs a large number of parameters (apart from an additive constant, 9 for $\mathscr{H}_e$ and 2 for $\mathscr{H}_s$ with O_h and more for lower symmetries, see ch. 9) and there is not enough experimental data to determine them all empirically. Had our orbitals been d orbitals we would only have needed B and C for $\mathscr{H}_e$ and ζ for $\mathscr{H}_s$. We now assume that the parameters representing $\mathscr{H}_e$ and $\mathscr{H}_s$ are expressible in terms of just three—B', C' and ζ', say—and that this dependence on B', C', ζ' is just the same as the dependence on B, C, ζ, respectively, would be if the orbitals were d orbitals. B', C', ζ' need not be the same as B, C, ζ determined from atomic spectra and actually turn out always to be reduced but still positive. In the future we will drop the dashes and just talk about B, C, ζ for the compound (or complex ion). This assumption for $\mathscr{H}_e$ and $\mathscr{H}_s$ is far from satisfactory and derives historically from the excessive use of the crystal-field formalism and from the belief that the partly-filled orbitals in the compound are 'not very different' from free ion d orbitals.

The reader may now reasonably object that he had been led to expect a more refined theory than the electrostatic crystal-field theory but, although he has been taken by devious routes rather close to such a theory, when it comes to making actual calculations he is being asked to use something which is not essentially different from the crystal-field theory. Here I think he would be missing the point. It is true that we might have hoped for better—the *a priori* calculation of the various parameters, or at least an empirical assessment of them without any assumption about their relationship to d-orbital parameters. However, in other ways we really have gained something. It is just that our theory is *essentially* different from the electrostatic crystal-field theory; it is the purely formal resemblance that is so close.

We have not only obtained a considerable understanding of why the crystal-field theory works so well; we have been led to expect a change, possibly substantial, in the parameters B, C and ζ; we realize that the partly-filled shell orbitals are not only on the metal ion but also on the ligands, and when we come to discuss magnetic interactions with the nuclear moment will expect to find reasonable contributions from ligand nuclei as well as metal nucleus; we will be always on the look-out for effects arising from the inadequacy of our simplifying assumption about the parameters B, C and ζ; we expect the matrix elements

[1] For d orbitals $\mathscr{H}_0$ gives a zero contribution to the relative energies.

of the orbital angular-momentum vector **L** to be different from those calculated for d functions.

It rarely makes for clarity of exposition if one is over-sophisticated. The changes of B, C, ζ and $\langle \mathbf{L} \rangle$ from the free ion values are often small (10–20 %) and so we shall give many calculations assuming we have the values appropriate to the free-ion d orbitals. It will then be clear that our calculations should be corrected by using lower values of B, C and ζ and also of $\langle \mathbf{L} \rangle$ (see § 10.44).

Lastly, we remark that in rare-earth and actinide ions f electrons take the place of d-electrons, the orbital degeneracy is seven and B, C are replaced by E^1, E^2, E^3. The simple crystal-field model is a better approximation for f than for d electrons.

7.4. The coupling schemes

Let us now write the Hamiltonian (7.9) in the form

$$\mathscr{H} = \mathscr{H}_0 + \mathscr{H}_e + \mathscr{H}_s + \mathscr{V},$$

where $$\mathscr{H}_e = \sum_{\kappa<\lambda} \frac{e^2}{r_{\kappa\lambda}}; \quad \mathscr{H}_s = \sum_{\kappa=1}^{n} \mathbf{t}_\kappa . \mathbf{s}_\kappa; \quad \mathscr{V} = \sum_{\kappa=1}^{n} V(\kappa),$$

and consider diagonalizing $\mathscr{H}$ within a given l^ϵ configuration. We incorporate $\mathscr{H}_0$ with $\mathscr{V}$ and still write it $\mathscr{V}$ and diagonalize $\mathscr{H}_e$, $\mathscr{H}_s$ and $\mathscr{V}$ in some order. Although the final results will be the same in each case, the actual detailed course of the calculation will depend on the order in which we take $\mathscr{H}_e$, $\mathscr{H}_s$ and $\mathscr{V}$. We have already met this point for the free ion when $\mathscr{V} = 0$ (rejecting, of course, a large additive constant) because, then, if $\mathscr{H}_e$ precedes $\mathscr{H}_s$ we have the SL coupling scheme and if $\mathscr{H}_s$ precedes $\mathscr{H}_e$ we have the jj-coupling scheme. When $\mathscr{V} \neq 0$ there are six possible coupling schemes because there are six possible distinct orders for three objects. Three of these, however, correspond to jj-coupling in the sense that we diagonalize $\mathscr{H}_s$ before $\mathscr{H}_e$. All our ions, when free, belong to the SL coupling or, at worst, the intermediate coupling cases and so these three schemes are not very useful. All the remaining three schemes are important in practice and we discuss each in turn.

7.4.1. The rare-earth coupling scheme, $\mathscr{H}_e, \mathscr{H}_s, \mathscr{V}$. In this scheme both the mutual electrostatic energy of the electrons $\mathscr{H}_e$ and the spin-orbit energy $\mathscr{H}_s$ are diagonalized before any account is taken of the influence of the ligand field. In a crystal-field theory $\mathscr{V}$ is now regarded as a perturbation on the levels of the free ion. Often we only consider the matrix elements of $\mathscr{H}$ within the ground term but we still call it a rare-earth coupling scheme if we take $\mathscr{H}_e$, $\mathscr{H}_s$, $\mathscr{V}$ in order. The rare-earth ions satisfy rare-earth coupling and the scheme is really only useful for them and sometimes for f-electron configurations of actinide ions.

7.4.2. The weak-field coupling scheme $\mathscr{H}_e$, $\mathscr{V}$, $\mathscr{H}_s$. In a crystal-field theory this scheme considers $\mathscr{V}$ as a perturbation on the terms of the free ion and introduces the spin-orbit energy after that. A new point arises here. It is generally true for ions which belong to this coupling case that $\mathscr{V} = \mathscr{V}_1 + \mathscr{V}_2$, where $\mathscr{V}_1$ has higher symmetry than $\mathscr{V}_2$, and $\mathscr{V}_2$ has matrix elements comparable in magnitude with $\mathscr{H}_s$. So in practice it is usual to diagonalize $\mathscr{H}_e$, then $\mathscr{V}_1$ and finally $\mathscr{H}_s$ and $\mathscr{V}_2$ together.[1]

The ions of the first transition series belong to the weak-field case in a large number of their compounds. $\mathscr{V}_1$ usually has symmetry O_h or T_d and sometimes D_{4h}.

7.4.3. The strong-field coupling scheme $\mathscr{V}, \mathscr{H}_e, \mathscr{H}_s$. In this scheme in the crystal-field theory the one-electron orbitals are no longer the $|nlm_s m_l\rangle$, but are linear combinations of them which form bases for irreducible representations of the symmetry group of $\mathscr{V}$, with analogous assertions for the ligand-field theory. As in the weak-field scheme we usually diagonalize a large high-symmetry component $\mathscr{V}_1$ of $\mathscr{V}$ first, then $\mathscr{H}_e$ and finally $\mathscr{H}_s + \mathscr{V}_2$.

The ions of the first transition series in some of their compounds and the ions of the second and third series belong to this coupling case.

As in our discussion of jj-coupling in § 5.2, we observe that we may always use any coupling scheme for any ion, if we wish, provided we are prepared to take enough basic states into account. However, only if the ion belongs to a particular coupling case can we expect to obtain reasonable agreement with experiment by using a simple version of the corresponding coupling scheme. Naturally, in many compounds the metal ion will not belong to any coupling case but will be intermediate between cases. Also a particular metal ion may belong to different cases in different compounds.

The classification of coupling schemes applies equally to the crystal-field theory and the ligand-field theory. However, there is a reason why the strong-field scheme is easier to use than the others when we consider ligand-field theory in its full generality. For in the others we are naturally considering the ligand field to be a perturbation on the terms or levels of the free ion. Now changing the values of B, C and ζ alters the energies, but not the degeneracies, of the states of the free ion. But if we use a more general form for the electrostatic interaction and spin-orbit coupling energy, with more independent parameters, the degeneracies may be altered even when the ligand field $\mathscr{V}$ is numerically zero. As an example, the free-ion ground term of Ni^{++} is ^3F with the levels $^3F_4$, $^3F_3$, $^3F_2$ with degeneracies 9, 7, 5, respectively. Using the most general form of the ligand-field theory with, for example, site symmetry O_h, these levels may be split even when $\mathscr{V} = 0$ because the electrostatic and spin-orbit matrix elements no longer satisfy the relations expected for d electrons. Then $^3F_4$ breaks up, in general, into A_1, E, T_1 and T_2 components, $^3F_3$ into A_2, T_1 and T_2 and $^3F_2$ into E and T_2 (see Table A 14). It is easier to take these effects into account in the strong field than in the other schemes.

[1] It is really to the high-symmetry component (and part of $\mathscr{V}_2$) that the treatment of § 7.2 strictly applies in the form in which I give it there.

7.5. Survey of ligand-field theory[1]

I think it will be useful for the reader to have at this stage a brief survey of some parts of ligand-field theory so that he will have some overall physical picture to guide him through the wilderness of the mathematics that is yet to come.

Take, then, regular octahedral complexes of transition metals of the first transition series described by the crystal-field theory. For an ion with configuration d^n we have four parameters, $\Delta = E(e_g) - E(t_{2g})$ for the crystal field, B and C for the mutual electrostatic interaction of the electrons and ζ for the spin-orbit coupling constant. We neglect ζ at first. The free d^1 and d^9 ions each have just the one term ^2D and this splits in the crystal field into $^2T_2$ and ^2E. See Fig. 7.1, in which Δ increases from zero on the left. In Fig. 7.2 the effect of the crystal field on the nickel ion, Ni^{++}, is shown as a function of Δ (calculated by Orgel). We call such pictures energy level diagrams.

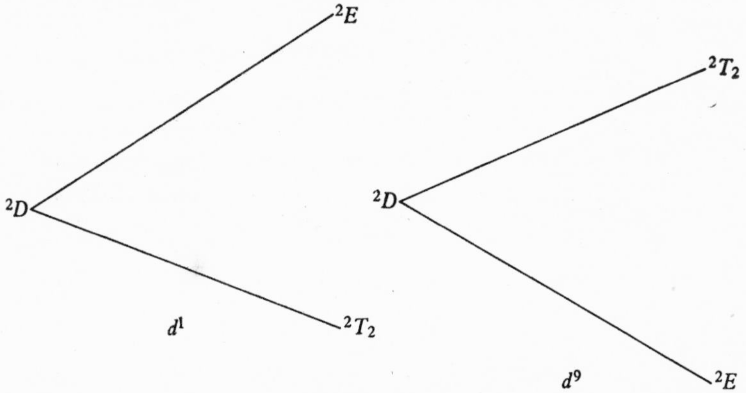

Fig. 7.1. Energy as a function of crystal-field parameter Δ for d^1 and d^9 systems.

When $\Delta = 0$ we have the free gaseous ion terms. Most of these terms (all those having $L > 1$) have spatial functions which form reducible representations of O_h (see Table A 14). As Δ increases from 0 the terms then separate into one or more of the five possibilities

$$^{2S+1}A_{1g},\quad ^{2S+1}A_{2g},\quad ^{2S+1}E_g,\quad ^{2S+1}T_{1g}\quad \text{and}\quad ^{2S+1}T_{2g}.$$

For d^1 the single term ^2D breaks up into $^2T_2$ below and ^2E above as shown in Fig. 7.1. In chs. 8 and 9 we learn how to calculate these energy-level diagrams for all d^n configurations.

In ch. 11 the optical spectra of the visible and near-visible region are interpreted as transitions between states of these diagrams. The intensities are low because the transitions are g–g and hence forbidden as electric-dipole transitions. They become weakly allowed, however, when we take into account the hitherto neglected motion of the nuclei. Also in ch. 11, various regularities in the thermo-

[1] The electrostatic energy level diagrams for d^n in an octahedral field have been calculated for all n by Tanabe & Sugano (1954). They are not the only calculations, but they are the most complete. Important early papers include Finkelstein & Van Vleck (1940); Ilse & Hartmann (1951) and Orgel (1952, 1955*b*).

dynamic properties of d^n compounds are interpreted in terms of the differing effects of the ligand field on ground terms having different values of n.

Inclusion of the spin-orbit coupling reduces the degeneracies further. Having included it, mainly in ch. 9, we calculate the paramagnetic susceptibilities of compounds in ch. 10. Then in ch. 12 we describe the theory of electron resonance, in which electromagnetic radiation of radio frequencies is used to measure the energy separations between nearly degenerate components of the ground state which have been separated by placing the specimen in a strong magnetic field.

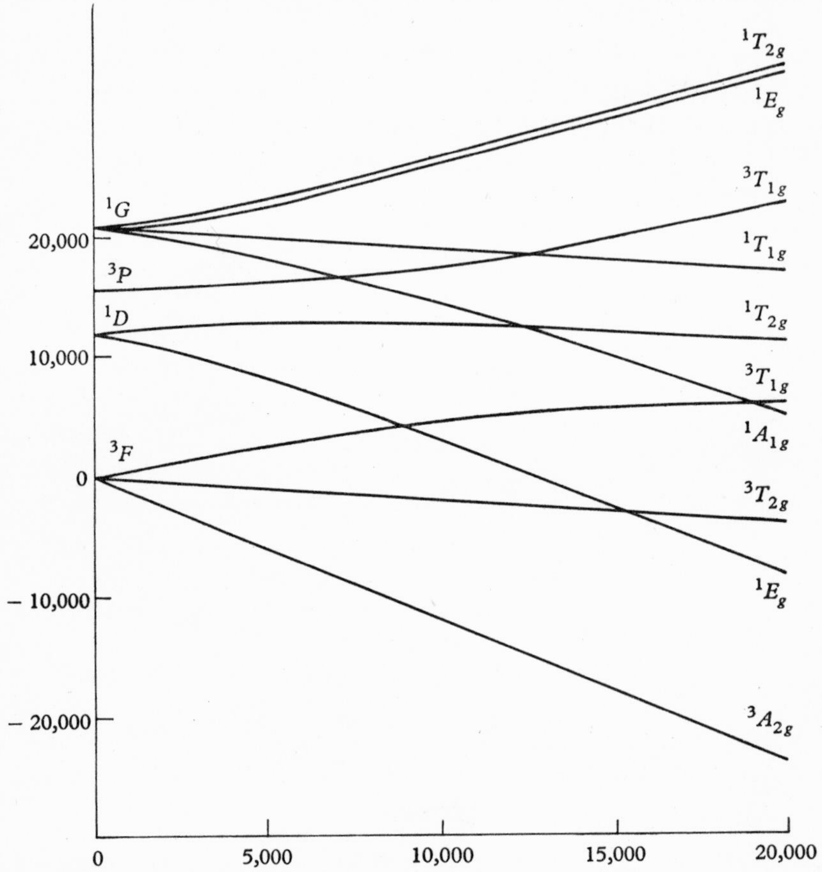

Fig. 7.2. Energy as a function of crystal-field parameter Δ for the d^8 system Ni^{++}. Energies are in cm^{-1} and the ^1S of the free ion is not shown (from Orgel, 1955*a*).

The energy-level diagrams for complex ions for which we do not assume the truth of the crystal-field theory but rather use the ligand-field theory in its most general form would be qualitatively very similar. As we have remarked, however, the electrostatic parameters are no longer just B and C but actually become 9 nearly independent ones for an octahedral field so it is now possible to have the terms of the free ion split up even though $\Delta = 0$. This would in fact happen for suitable choices of the parameters.

CHAPTER 8

CRYSTAL-FIELD THEORY AND THE WEAK-FIELD COUPLING SCHEME

8.1. The influence of a static environment

We approached the electrostatic crystal-field theory from the point of view of electron configurations in a complex ion and saw that it was the same as a theory in which a metal ion is placed in an electrostatic field due to its ligands, the latter being assumed fixed and unpolarizable. This second form is the way in which the crystal-field theory arose historically and is a useful one for giving an intuitive grasp of some features of ligand-field theory. We now examine it in more detail bearing in mind always the many limitations which we have already seen that it must possess.

The crystal-field theory assumes that we may calculate the properties of a transition metal, rare-earth or actinide compound satisfactorily by supposing it to be an aggregate of non-interacting metal ions arranged in an unpolarizable and perfectly insulating medium. Taking an origin at a metal nucleus, the medium surrounding it is not isotropic but has the same point symmetry group as the actual crystal. The main terms in the Hamiltonian for the ion are therefore now

$$\mathscr{H} = \sum_{\kappa=1}^{n}\left\{\frac{1}{2m}\mathbf{p}_\kappa^2 - \frac{Ze^2}{r_\kappa} + \xi(r_\kappa)\mathbf{l}_\kappa\,.\,\mathbf{s}_\kappa + V(\mathbf{r}_\kappa)\right\} + \sum_{\kappa<\lambda}^{n}\frac{e^2}{r_{\kappa\lambda}}, \tag{8.1}$$

where $V(\mathbf{r}_\kappa)$ is the potential energy of the κth electron in the (electrostatic) field of the environment. V, and hence $\mathscr{H}$, commutes with all elements of the site symmetry group. If we say that the environment of the site is approximately octahedral we mean that we can write

$$V = V_1 + V_2, \tag{8.2}$$

where V_1 commutes with all elements of O_h and V_2 is small compared with V_1. We assume that (8.1) applies not merely to the ground but also the excited states of the metal ion, with the same V. Then if our crystal contains N metal ions we work out its properties, according to Maxwell–Boltzmann statistics, as if it were an aggregate of N independent systems each having the Hamiltonian (8.1). We defer till §§ 8.2 and 8.3 the question of how to determine V in greater detail.

I hope the nature of the approximation is clear to the reader. It is difficult to be very precise in such matters without being merely tedious, so I have compromised. A few clarifying remarks may be helpful. First, notice that we have divided the electrons into those belonging to the ion, which are treated quantum mechanically, and those of the medium which are, in effect, treated classically. There is no question of antisymmetrizing between the two kinds. Excited states of the medium are not contemplated and although non-isotropic it possesses no permanent magnetic polarization. As far as a given metal ion is concerned, the other metal ions form part of the rigid environment.

In this chapter we develop mainly the crystal-field method. It will become apparent ultimately that it gives, in general, a good description of the experimental data. However, let us always bear in mind that when we find disagreement between theory and experiment this must be a consequence of a defect in the theoretical model adopted. The form of the disagreement then gives us a hint about the kind of refinement to introduce into the theory in order to improve the agreement.

We use the same description for the metal ion that we did in chs. 2–5 and have approximate eigenstates for it which are finite sums of determinantal functions, using the $|nlm_s m_l\rangle$ scheme for the one-electron orbitals. Therefore we need the matrix elements of the quantity

$$\mathscr{V} = \sum_{\kappa=1}^{n} V(\mathbf{r}_\kappa), \tag{8.3}$$

between determinantal functions. $\mathscr{V}$ is a sum of one-electron operators and so we use the results of § 4.3, especially (4.22) and (4.23), to deduce that if ψ_i and ψ_j are two determinantal functions then

$$\left.\begin{aligned} \langle\bar{\psi}_i|\mathscr{V}|\psi_i\rangle &= \sum_{\kappa=1}^{n} \langle\bar{\phi}_\kappa|V|\phi_\kappa\rangle, \\ \langle\bar{\psi}_i|\mathscr{V}|\psi_j\rangle &= (-1)^\tau \langle\bar{\phi}^i_\lambda|V|\phi^j_{\lambda'}\rangle, \end{aligned}\right\} \tag{8.4}$$

for ψ_i, ψ_j differing by one constituent one-electron function and

$$\langle\bar{\psi}_i|\mathscr{V}|\psi_j\rangle = 0,$$

if they differ by more. P_τ is the permutation which puts ψ_i in the same order as ψ_j and the ϕ are the one-electron functions. V contains no spin-dependent term and so the spin may be integrated straight out of (8.4).

At first we shall diagonalize $\mathscr{V}$ only within the ground term of a free ion configuration and even later shall rarely consider matrix elements between configurations. In case we take all our functions from the same configuration and the configuration contains closed shells the matrix elements of $\mathscr{V}$ are the same, apart from a constant energy common to all diagonal elements, as they would be in the absence of the closed shells. This is for the reasons already explained in §§ 4.3 and 7.2, and means that we shall mainly be interested in the matrix elements of $\mathscr{V}$ within d^ϵ configurations.

Let us be more general for a moment and say that we need the matrix elements of $\mathscr{V}$ within an l^ϵ configuration. These in turn depend, through (8.4), on the $(2l+1)^2$ different one-electron integrals $\langle nlm|V|nlm'\rangle$. Not all these are independent. To find the maximum number of parameters which can be necessary to specify them all it is easiest to replace the $|nlm\rangle$ by an orthonormal transform which is a set of real functions in the Schrödinger representation. This is done by taking R_{nl} real and using the $Z_{l\alpha}$ as defined by (6.37) instead of $|lm\rangle$. All matrix elements of V are now real and the number of distinct ones is the number of ways of selecting two non-ordered objects from $2l+1$, i.e. $(l+1)(2l+1)$. Just as with the F_k, so long as we are only interested in relative energies within a configuration, one of these parameters may be regarded as merely an arbitrary energy added to all states. Then we have $l(2l+3)$ effective parameters. In the next section I prove that we cannot do with fewer in order to be able to parametrize the most general environment.

To describe the effect of the environment of a d^ϵ ion we may need as many as fourteen parameters. This number can be reduced a little if we choose our axes to suit the environment,[1] but is still alarmingly large especially as we envisage usually determining the parameters from experiment. Fortunately, however, in practice the number is almost always reduced very considerably. Let the site have symmetry group G. Then any element g of G commutes with V. So if we apply g to a matrix element $\langle \phi_\kappa | V | \phi_\lambda \rangle$ it turns it into another matrix element of V. This second matrix element must have the same numerical value but, in general, will involve different functions ϕ'_κ and ϕ'_λ, say. So there will be necessary relations between some of the matrix elements and the number of independent parameters will be reduced.

The use of the site symmetry group is rather simpler if we take a different, but equivalent, description of the influence of the environment. Let us denote the $(2l+1)$ real functions considered earlier by f_i, $i = 1, 2, \ldots, (2l+1)$ and write

$$a_{ij} = \langle f_i | V | f_j \rangle; \quad A = [a_{ij}].$$

The matrix A is real and symmetric and it is a known theorem in matrix theory[2] that there exists an orthogonal matrix O such that $O'AO$ is diagonal. Therefore V is diagonal within the set of $(2l+1)$ real functions $g_i = \sum_{j=1}^{2l+1} O_{ji} f_j$. g_i is an orthonormal set. Let the diagonal elements of V within the g_i be written E_i. Conversely given any orthonormal set g_i linearly dependent on the f_j and any $(2l+1)$ real numbers E_i we can deduce the matrix elements of V within the f_j. We call our new description 'specifying the orbital pattern'. Only the relative values of the E_i are effective parameters when we are not interested in absolute energies.

The consequence of a site having a symmetry group G, assumed finite, can now be seen as a restriction on the possible orbital patterns. Choose an orthonormal combination, h_i say, of the f_j which is such that each function h_i is a component of a basis for an irreducible representation of G. If the whole set of h_i spans a particular irreducible representation of degree p more than once, say q times, then choose the h_i in such a way that there are q subsets of the h_i each containing p functions and spanning this representation in identically the same form each time. Then we use our orthogonality relations for integrals (see the discussion of (6.30)) to show that

$$\langle \bar{h}_i | V | h_j \rangle = 0, \tag{8.5}$$

[1] The reduction is really only apparent, rather than real, because the crystal structure will always suggest natural sets of axes to us and these will not *necessarily* be suitable in the other sense.

[2] See, for example, Littlewood, *A University Algebra* (Heinemann, 1950), p. 48.

unless h_i and h_j are corresponding components of bases for the same irreducible representation. If they are, then the integral is unchanged if we replace h_i and h_j by any other pair of corresponding components from the same bases for that representation. Now for all pure rotation groups, except C_n ($n > 2$) and T, and for all the other groups occurring in Tables A 1–7 it is always possible to choose the h_i real. This is proved by selecting for each representation of degree greater than one a particular real basis. Then the matrices of the representation are real and so if a set of the h_i are taken to form a basis for identically the same representation the real parts of the h_i and also $\sqrt{-1}$ times the imaginary parts are real and each separately span the representation. So we can take the h_i real and V is diagonal within the h_i unless there is a repeated representation. If there are repeated representations then we have little blocks within which V is not necessarily diagonal. We may, however, diagonalize it within these blocks with a suitable orthogonal matrix.

We have obtained the restrictions on the orbital pattern implied by the symmetry group G, when the irreducible representations can all be chosen real, and it is clear that there are no further ones. It is easy to give a formula for the number of effective parameters remaining. If n_λ is the number of times the λth irreducible representation occurs, then the number of effective parameters is

$$N = -1 + \tfrac{1}{2}\Sigma n_\lambda(n_\lambda + 1), \tag{8.6}$$

the sum being over the irreducible representations.

The representations which are awkward from our point of view are all of degree one and all of them assign complex numbers to some at least of the classes of the group. Therefore a real function cannot possibly be a basis for any of them. Let a function

$$f = (1/\sqrt{2})(f_0 + if_1)$$

be a basis for one, Γ say. Then

$$\bar{f} = (1/\sqrt{2})(f_0 - if_1)$$

must be a basis for a different one, $\bar{\Gamma}$ say, whose character is the conjugate complex of that of Γ. So they can be grouped into sets of conjugate pairs, for example, $\Gamma_3 = \bar{\Gamma}_2$ in Table A 6. If we write the real expressions $\langle \bar{f} | V | f \rangle$ and $\langle f | V | \bar{f} \rangle$ as integrals they are both the same and so if we also use the orthogonality relation (8.5) between f and $\bar{f}$ we have the relations

$$\langle \bar{f} | V | f \rangle = \langle f | V | \bar{f} \rangle = a, \quad \text{say,}$$

$$\langle f | V | f \rangle = \langle \bar{f} | V | \bar{f} \rangle = 0,$$

which in terms of f_0 and f_1 read

$$\left.\begin{aligned} \langle f_0 | V | f_0 \rangle = \langle f_1 | V | f_1 \rangle = a,\\ \langle f_0 | V | f_1 \rangle = 0. \end{aligned}\right\} \tag{8.7}$$

so we have only one parameter associated with the two irreducible representations Γ and $\bar{\Gamma}$. f_0 and f_1 are normalized when f is and (8.7) shows that if we use them instead of f and $\bar{f}$ they satisfy the relations which would be expected if they formed a basis for an irreducible representation of degree 2. It is not possible to obtain all the corresponding results when Γ is repeated.[1] If

$$f = (1/\sqrt{2})(f_0 + if_1) \quad \text{and} \quad g = (1/\sqrt{2})(g_0 + ig_1)$$

both form bases for Γ, it is not necessarily true that $\langle f_0 | V | g_1 \rangle = 0$.

Table 8.1. *Effective number of parameters needed to parametrize the most general environment for various important symmetry groups and atomic orbitals*

Orbital	K	O, T_d	T, T_h	D_3, D_{3v}	D_4, D_{4h}, C_{4v}	D_5, D_{5v}	C_3	C_4	C_1
p	0	0	0	1	1	1	1	1	5
d	0	1	1	3	3	2	4	4	14
f	1	2	3	6	5	4	9	7	27

Table 8.1 shows the effective number of parameters needed for l^ϵ configurations for $l = 1, 2, 3$ and various site symmetry groups. When no parameters are needed this means the configuration is not split. We shall most often be interested in site symmetry groups for d^ϵ configurations which are, at least approximately, either O or T_d and for either of these groups we need only one parameter. This parameter measures the separation in energy between the three d orbitals belonging to t_2 and the two d orbitals belonging to e. It is important, however, to bear in mind that the reason that we can describe the influence of an arbitrary environment by a finite number of parameters is that we are diagonalizing V only within a configuration. If we include all possible inter-configurational interactions we need an infinite number of parameters.

8.2. The environment as a classical distribution of charge

Let us suppose now that the environment is represented by a charge distribution ρ which is a general function of position but is independent of time. Then the potential energy $V(\mathbf{r})$ to be put into (8.1) is

$$V(\mathbf{r}) = -\int \frac{e\rho(\mathbf{R})}{|\mathbf{R}-\mathbf{r}|}\, d\tau, \tag{8.8}$$

where $\mathbf{r}$ is the position of the electron, $\mathbf{R}$ a general point of the environment, and the integral is over the variables in $\mathbf{R}$, R, Θ and Φ, say. We may expand the potential by using

$$\frac{1}{|\mathbf{R}-\mathbf{r}|} = \sum_{k=0}^{\infty} \frac{r_<^k}{r_>^{k+1}} P_k(\cos\omega),$$

where ω is the angle between $\mathbf{r}$ and $\mathbf{R}$. The environment is envisaged to be mainly

[1] The mathematician will appreciate the difficulty here. We want the h_i to be bases for irreducible representations over the real field, which is not algebraically closed. Therefore the representations may not be absolutely irreducible in which case they do not satisfy the orthogonality relations.

external to the metal ion. For this reason and for mathematical convenience we assume[1] $r < R$, and then

$$\frac{1}{|\mathbf{R}-\mathbf{r}|} = \frac{1}{R}\sum_{k=0}^{\infty}\left(\frac{r}{R}\right)^k P_k(\cos\omega). \tag{8.9}$$

To proceed, we expand $P_k(\cos\omega)$ by the spherical harmonic addition theorem (4.39). However, it is inconvenient to have complex quantities occurring in the potential and so we replace the Y_{lm} by the set of real quantities (6.37)

$$\left.\begin{aligned} Z_{l0} &= Y_{l0}, \\ Z_{lm}^c &= \tfrac{1}{2}\sqrt{2}\,(Y_{l-m}+\overline{Y}_{l-m}), \\ Z_{lm}^s &= \tfrac{1}{2}i\sqrt{2}\,(Y_{l-m}-\overline{Y}_{l-m}), \end{aligned}\right\} \tag{8.10}$$

where $m > 0$. We write $Z_{l\alpha}$ to stand for any of the expressions in (8.10). In terms of these new spherical harmonics (4.39) becomes

$$P_k(\cos\omega) = \frac{4\pi}{2k+1}\sum_{\alpha} Z_{k\alpha}(1)\,Z_{k\alpha}(2). \tag{8.11}$$

Combining (8.8), (8.9) and (8.11) we obtain

$$V(r,\theta,\phi) = \sum_{k=0}^{\infty}\sum_{\alpha} r^k \gamma_{k\alpha} Z_{k\alpha}(\theta,\phi), \tag{8.12}$$

where
$$\gamma_{k\alpha} = -\frac{4\pi e}{2k+1}\int\frac{\rho(\mathbf{R})\,Z_{k\alpha}(\Theta,\Phi)}{R^{k+1}}\,d\tau. \tag{8.13}$$

The $\gamma_{k\alpha}$ are just a set of numbers characteristic of the environment. Evidently $\nabla^2 V = 0$ but this is merely a consequence of our assumption that $r < R$.

If, on the other hand, we wish to find an environmental distribution ρ which will give rise to a pre-assigned potential V satisfying $\nabla^2 V = 0$ in some region surrounding the origin, i.e. to a pre-assigned set of numbers $\gamma_{k\alpha}$, we can write down at once a formal solution

$$\rho(R,\Theta,\Phi) = \sum_{k,\,\alpha}\beta_{k\alpha}(R)\,Z_{k\alpha}(\Theta,\Phi),$$

where $\beta_{k\alpha}(R)$ are any suitable functions of R satisfying

$$\gamma_{k\alpha} = -\frac{4\pi e}{2k+1}\int\frac{\beta_{k\alpha}}{R^{k-1}}\,dR.$$

In the cases in which we use this result we only assign a finite number of the $\gamma_{k\alpha}$ and then there is no difficulty about convergence because we simply put the remaining $\gamma_{k'\alpha'}$ and $\beta_{k'\alpha'}$ equal to zero.

Let us now introduce our findings into (8.4) for the matrix elements of $\sum_j V(\mathbf{r}_j)$ between determinantal functions belonging to the same l^ϵ configuration. We

[1] More precisely, that there exists a such that $\rho(\mathbf{R}) = 0$ for $R < a$, while the electronic wave-function of the ion vanishes for $r > a$. This assumption is, of course, only acceptable in a crystal field approach in which we leave out the supposedly spherically symmetric charge distribution of the central metal ion inner shells.

suppose the spins to have been integrated out and have to determine matrix elements of the general form

$$M = \langle lm' \mid Z_{k\alpha''} \mid lm \rangle.$$

The $Z_{k\alpha''}$ (via the $Y_{lm''}$) may be vector coupled with the $|lm\rangle$ to give states with L varying from $|k-l|$ to $k+l$. Therefore M is zero unless l itself lies in that range, i.e. unless $k \leqslant 2l$. Next we apply the parity operator P to M and deduce that M is zero unless k is even. So the only terms of the sum (8.12) which can contribute to any of the M are the $(l+1)(2l+1)$ ones for which k is even and not greater than $2l$.

If we describe the electronic functions in terms of the real functions $Z_{l\alpha}$ a general matrix element of V between two of them may be written

$$Q_{\alpha\alpha'} = \int Z_{l\alpha} Z_{l\alpha'} R_{nl}^2 V \, d\tau.$$

Now by the same vector coupling arguments that we have just used, the set of distinct products $Z_{l\alpha} Z_{l\alpha'}$ may be expressed as linear combinations of $Z_{p\alpha''}$, where $p = 0, 2, \ldots, 2l$ and all α'' consistent with these p. Conversely the $Z_{p\alpha''}$ may be expressed in terms of the $Z_{l\alpha} Z_{l\alpha'}$. But

$$\int V Z_{p\alpha''} R_{nl}^2 d\tau = \gamma_{p\alpha''} \int r^{p+2} R_{nl}^2 dr,$$

which may be assigned arbitrarily. So, therefore, may the entire set of $Q_{\alpha\alpha'}$ but for the obvious restriction $Q_{\alpha'\alpha} = Q_{\alpha\alpha'}$. In other words we have shown that no matter what set of $(l+1)(2l+1)$ real numbers we choose for the matrix elements of V between the real one-electron functions, it is possible to construct many different charge distributions for the environment which will give rise to exactly this set of numbers for the matrix elements of V.

In § 8.1 we treated the matrix elements of V as if they could be assigned arbitrarily, apart from the consequences of the requirement that V should commute with the elements of some symmetry group, and it may seem merely as if we have finally justified our assumption. This would be slightly to misjudge the situation, however. Although it appeals to our physical intuition, it is not correct to represent the environment by a classical charge distribution, however general. Therefore it is far from clear why a deduction of a rather straightforward and axiomatic scheme from such a premiss is in any reasonable sense a justification. We have shown that two simple ways of looking at the influence of the environment are mathematically equivalent. The relationship of either way to a rigorous quantum-theoretic description of the entire solid is still rather obscure. Some discussion of this matter was given in ch. 7 and we will not repeat it now. Here we content ourselves with the discovery that we have two alternative and equivalent ways of describing an environment and use on any occasion that description which suits us best. It is true, of course, that there are sets of matrix elements for V which correspond to no reasonable charge distribution and this is a point which one will bear in mind in specific applications.

We conclude the section by seeing what form V takes when the site has a symmetry group G. V of (8.12) forms a basis for the unit representation of G. Therefore the same must be true for each term in the sum over k, in fact for

$$X_k = \sum_\alpha \gamma_{k\alpha} Z_{k\alpha}(\theta, \phi)$$

for each k. X_k, then, is a basis for the unit representation of G.

Applying this result to d functions and retaining only those parts of V that give a non-zero contribution to the matrix elements of V within a d^e configuration we find: For D_{3v} with the X-axis intersecting an edge of the prism

$$V = \gamma_{00} + \gamma_{20} r^2 Z_{20} + \gamma_{40} r^4 Z_{40} + \gamma_{43}^c r^4 Z_{43}^c. \tag{8.14}$$

For D_{4h} with the X-axis a twofold axis

$$V = \gamma_{00} + \gamma_{20} r^2 Z_{20} + \gamma_{40} r^4 Z_{40} + \gamma_{44}^c r^4 Z_{44}^c. \tag{8.15}$$

For O (or for T_d with OX, OY, OZ as twofold axes)

$$V = \gamma_{00} + \gamma' r^4 \left(\frac{\sqrt{7}}{2\sqrt{3}} Z_{40} + \frac{\sqrt{5}}{2\sqrt{3}} Z_{44}^c \right). \tag{8.16}$$

The detailed form of V in (8.14)–(8.16) follows immediately from the information in Tables A 18 and 19. We realize, of course, that even in the crystal-field theory it is only as long as we restrict our calculations to the single configuration d^e that such simple expressions for V completely describe the influence of general environments ρ of the above symmetries.

Examples

1. Prove that the effective number of parameters needed to describe a site with symmetry group G determined from the implied restriction on the possible values for the γ_{km} is the same as the number obtained by the methods used in § 8.1. Use this to confirm the numbers given in the Table 8.1. (G may be *any* symmetry group.)

2. All the characters of a group G are real. Show that the direct product $\Gamma_1 \Gamma_2$ of a pair of irreducible representations contains the unit representation just once if $\Gamma_1 = \Gamma_2$ and not at all if $\Gamma_1 \neq \Gamma_2$. Use this to deduce (8.6).

3. Show that if we do not assume $r < R$, then

$$V = \sum_{k, \alpha} (\delta_{k\alpha} r^k + \delta'_{k\alpha} r^{-k-1}) Z_{k\alpha},$$

where $\delta_{k\alpha}$ and $\delta'_{k\alpha}$ are, in general, functions of r.

4. Show that all terms in V of (8.12) except that for which $k = \alpha = 0$ leave the centre of gravity of any term of any l^e configuration unchanged.
(Hint: Use Wigner's formula (2.49) for $j = 0$.)

8.3. Relationships between potentials for different environments

I now show that if we introduce a further assumption about the nature of the environment it is possible to deduce some rather interesting relationships between the magnitudes of the terms in the corresponding potentials. The atoms of a typical environment can usually be clearly divided into two classes, nearest neighbours and those which are not nearest neighbours. We assume that a nearest neighbour atom gives a contribution to ρ which is axially symmetric about the line joining its nucleus to the nucleus of the metal ion and that the remaining atoms of the environment give no contribution. Of course if, as in $K_3Fe(CN)_6$, some of these remaining ions lie on one of the lines defined by the nearest neighbours we may include axially symmetric fields due to them by merely altering the values of some of the parameters occurring in the theory.

If an adjacent atom lies with its nucleus on the Z-axis then the potential due to it (from (8.12)) is

$$V = \sum_{k=0}^{\infty} \gamma_{k0} r^k Z_{k0}(\theta, \phi), \tag{8.17}$$

while if it lies on the axis Θ, Φ then

$$V = \sum_{k,\alpha} \frac{\gamma_{k0}\sqrt{4\pi}}{\sqrt{(2k+1)}} Z_{k\alpha}(\Theta, \Phi)\, r^k Z_{k\alpha}(\theta, \phi), \tag{8.18}$$

where we have used the spherical harmonic addition theorem and the fact that $Z_{k0} = P_k\sqrt{(2k+1)}/\sqrt{4\pi}$. When we have a set of adjacent atoms at Θ_i, Φ_i, then

$$V = \sum_{k,\alpha} \beta_{k\alpha} r^k Z_{k\alpha}(\theta, \phi), \tag{8.19}$$

where

$$\beta_{k\alpha} = \left(\frac{4\pi}{2k+1}\right)^{\frac{1}{2}} \sum_i \gamma^i_{k0} Z_{k\alpha}(\Theta_i, \Phi_i). \tag{8.20}$$

If all the adjacent atoms have identical fields then γ^i_{k0} is independent of i and we just write it γ_{k0} as before.

We now restrict attention to that part of V which is relevant for calculations within a d^ϵ configuration and assume that all the nearest neighbours give identical axial fields (8.17) when referred to suitable axes. The corresponding expression (8.19) is shown in Table 8.2 for various common arrangements. We see that the actual splittings of a d^ϵ configuration for identical atoms at the vertices of a cube, tetrahedron, octahedron, or finally in the centres of the edges of the latter stand in the ratios $-\frac{8}{9}:-\frac{4}{9}:1:-\frac{1}{2}$, respectively. We should, of course, get the same ratios if we assumed further that the adjacent atoms could be adequately represented by point charges or point dipoles and in the literature ratios, such as the above, have often been derived from these rather special models. However, we have derived them from a much more general premiss. One may deduce the potentials for point charges η at distances R from the origin from Table 8.2 by writing

$$\gamma_{k0} = \frac{-e\eta\sqrt{4\pi}}{R^{k+1}\sqrt{(2k+1)}},$$

and for point dipoles of strength d pointing towards the origin by

$$\gamma_{k0} = \frac{-ed(k+1)\sqrt{4\pi}}{R^{k+2}\sqrt{(2k+1)}}.$$

The matrix elements of V between the normalized real one-electron functions $R_{nl}Z_{l\alpha}$ are, in an obvious notation

$$\begin{aligned}\langle \alpha' | V | \alpha'' \rangle &= \sum_{k,\alpha} \beta_{k\alpha} \langle R_{nl} | r^k | R_{nl} \rangle \langle Z_{l\alpha'} | Z_{k\alpha} | Z_{l\alpha''} \rangle \\ &= \sum_{k,\alpha} \beta_{k\alpha} \overline{r^k} \langle Z_{k\alpha} | Z_{l\alpha'} Z_{l\alpha''} \rangle. \end{aligned} \tag{8.21}$$

The products $Z_{l\alpha'}Z_{l\alpha''}$ expressed as linear combinations of the $Z_{k\alpha}$ are given for d functions in Table A 21 with the use of which the matrix elements of V can be written down immediately in terms of the $\beta_{k\alpha}$ and $\overline{r^k}$.

As an example let us take the potential due to six identical neighbours at the vertices of a regular octahedron, which is V_3 of Table 8.2. We readily find that

$$\langle 2^c \mid V \mid 2^c \rangle = \langle 0 \mid V \mid 0 \rangle = \frac{1}{\sqrt{4\pi}}(6\gamma_{00} + 3\gamma_{40}\overline{r^4}),$$

$$\langle 2^s \mid V \mid 2^s \rangle = \langle 1^c \mid V \mid 1^c \rangle = \langle 1^s \mid V \mid 1^s \rangle = \frac{1}{\sqrt{4\pi}}(6\gamma_{00} - 2\gamma_{40}\overline{r^4}) \qquad (8.22)$$

Table 8.2. *The potential for identical nearest neighbours for various arrangements (for d^ϵ configurations)*

Arrangement	Potential
8, at the vertices of a cube in directions $(\pm 1, \pm 1, \pm 1)$	$V_1 = 8\gamma_{00} - \frac{4\sqrt{7}}{9}\gamma_{40}r^4(\sqrt{7}\,Z_{40} + \sqrt{5}\,Z^c_{44})$
4, at the vertices of a regular tetrahedron in directions $(1, 1, 1)$, etc.	$V_2 = 4\gamma_{00} - \frac{2\sqrt{7}}{9}\gamma_{40}r^4(\sqrt{7}\,Z_{40} + \sqrt{5}\,Z^c_{44})$
6, at the vertices of a regular octahedron and lying on the axes	$V_3 = 6\gamma_{00} + \frac{1}{2}\sqrt{7}\,\gamma_{40}r^4(\sqrt{7}\,Z_{40} + \sqrt{5}\,Z^c_{44})$
12, at the centres of the edges of the octahedron of the preceding arrangement	$V_4 = 12\gamma_{00} - \frac{1}{4}\sqrt{7}\,\gamma_{40}r^4(\sqrt{7}\,Z_{40} + \sqrt{5}\,Z^c_{44})$
2, on the $\pm OZ$-axis	$V_5 = 2\gamma_{00} + 2\gamma_{20}r^2Z_{20} + 2\gamma_{40}r^4Z_{40}$
4, on the $\pm OX$-, $\pm OY$-axes	$V_6 = 4\gamma_{00} - 2\gamma_{20}r^2Z_{20} + \frac{3}{2}\gamma_{40}r^4Z_{40} + \frac{\sqrt{35}}{2}\gamma_{40}r^4Z^c_{44}$
3, in a trigonal configuration in the OXY-plane	$V_7 = 3\gamma_{00} - \frac{3}{2}\gamma_{20}r^2Z_{20} + \frac{9}{8}\gamma_{40}r^4Z_{40}$
6, with D_{3v} symmetry about OZ with one lying on the axis $\Theta = \beta$, $\Phi = 0$	$V_8 = 6\gamma_{00} + 3(3\cos^2\beta - 1)\,\gamma_{20}r^2Z_{20}$ $+ \frac{3}{4}\gamma_{40}r^4\{(35\cos^4\beta - 30\cos^2\beta + 3)\,Z_{40} + 2\sqrt{70}\sin^3\beta\cos\beta Z$

and that the remaining $\langle\alpha' \mid V \mid \alpha''\rangle$ are zero. The states $|0\rangle$ and $|2^c\rangle$ belong to the representation e_g of the group 0_h and $|2^s\rangle$, $|1^c\rangle$, $|1^s\rangle$ belong to t_{2g}. The difference of energy between the two representations is

$$\Delta \equiv E(e_g) - E(t_{2g}) = \frac{5}{\sqrt{4\pi}}\gamma_{40}\overline{r^4}. \qquad (8.23)$$

This energy separation Δ is very fundamental in our theory. In the literature $10Dq$ is often used for this quantity. I prefer the single symbol Δ, partly for simplicity, but mainly because the constituent symbols D and q like γ_{40} and $\overline{r^4}$ only have meaning within a particular approximation. Δ, defined as the difference in energy between e_g and t_{2g}, is not nearly so restricted in meaning. If the neighbouring atoms are regarded as being predominantly negatively charged near to the ion then Δ is expected to be positive. It appears to be an empirical fact that Δ is always positive.[1]

The reader who is particularly interested in the point-charge approximation to the environment will naturally ask if any restrictions on the expression for V follow from merely assuming that the environment consists of a finite number of point charges. The answer is that there are none, in the sense that the following proposition is true: if we assign arbitrarily a set of real numbers for a chosen finite number of the $\gamma_{k\alpha}$ of (8.12), then there exists a finite set of positive and negative charges which when arranged suitably on a sphere centred at the nucleus of the ion gives a potential energy V with the chosen $\gamma_{k\alpha}$. The sphere must have its surface outside the ion. I leave the proof as an exercise for the reader.[2]

Example

Take the potential V_8 of Table 8.2 for the case $\cos\Theta = \frac{1}{3}\sqrt{3}$ and show that the orbital pattern consists of one triply degenerate and one doubly degenerate set of orbitals. Why is this?

8.4. Kramers degeneracy[3]

We now interrupt our main line of development for a moment to discuss in this and the next two sections, two very useful general propositions about possibilities of degeneracy. For an ion in free space a non-degenerate state is only possible for an even electron system, because only for such systems can $J = 0$. We are neglecting nuclear degeneracy entirely here. All odd electron states have even degeneracy. In the presence of a magnetic field this degeneracy is, in general, lifted. However, Kramers showed that in the presence of an arbitrary electrostatic field, in zero magnetic field, all states of an odd electron system must still have even degeneracy.

Before proving it in general, let us consider a one-electron system in order to understand what the result really means. Take first the non-relativistic Schrödinger Hamiltonian

$$\mathscr{H}_0 = \frac{1}{2m}\mathbf{p}^2 - e\phi = -\frac{\hbar^2}{2m}\nabla^2 - e\phi,$$

where ϕ is the electrostatic potential due both to nucleus and environment. $\mathscr{H}_0$ is real in two quite different senses. It is real as a linear operator and also real in the elementary complex-number sense of reality. These two definitions of reality do not necessarily coincide even when they are both defined. For example, $\mathbf{p} = -i\hbar\nabla$ is real as a linear operator but pure imaginary in the elemen-

[1] For a simple-minded treatment of the origin of part of Δ as a molecular orbital quantity, see Appendix 8.

[2] Hint. Solve the problem for $\gamma_{k\alpha} = \delta_{kk_1}\delta_{\alpha\alpha_1}$ with k_1, α_1 chosen but arbitrary. First, use a continuous distribution ρ as a solution and then approximate to it by a finite set of points. Having approximated closely enough choose a suitable linear combination $\sum_{k_1,\alpha_1} c_{k_1\alpha_1} S_{k_1\alpha_1}$ of your approximate solutions $S_{k_1\alpha_1}$.

[3] I follow Kramers, *K. Akad. van Wetenschappen, Amsterdam* (1930), **33**, 959. These matters are also connected with time reversal. See Wigner, *Nachr. Ges. Wiss. Göttingen* (1932), p. 546.

tary sense. Thus **p** has two kinds of conjugate operator. We write the conjugate linear operator $\bar{\mathbf{p}}$ and have $\bar{\mathbf{p}} = \mathbf{p}$, and the conjugate in the other sense $\mathbf{p}^*$ and have $\mathbf{p}^* = -\mathbf{p}$. As we saw, $\bar{\mathscr{H}}_0 = \mathscr{H}_0^* = \mathscr{H}_0$.

The star operation applied to a spatial wave-function ψ gives its conjugate complex ψ^*. So if

$$\mathscr{H}_0\psi = E\psi$$

we can deduce

$$\mathscr{H}_0\psi^* = (\mathscr{H}_0\psi)^* = (E\psi)^* = E\psi^*,$$

because E is real. So ψ and ψ^* have both the same energy and therefore if ψ is non-degenerate we must have $\psi^* = e^{ia}\psi$ for some real number a. In that case $e^{\frac{1}{2}ia}\psi$ is real.

The Hamiltonian including spin-orbit coupling is

$$\mathscr{H} = \frac{1}{2m}\mathbf{p}^2 - e\phi + \xi(r)\,\mathbf{l}\,.\,\mathbf{s}.$$

$\mathbf{l}^* = -\mathbf{l}$, but as there is no Schrödinger representation for **s** the star operation on it is not yet defined. It is natural to require **s** to behave in the same way as **l** and to define $\mathbf{s}^* = -\mathbf{s}$. We do this and then have $\mathscr{H}^* = \mathscr{H}$ as before. Finally, we need α^* and β^* for the two spin functions α and β. They must be defined consistently with the definition of $\mathbf{s}^*$. The equation $s_z\alpha = \frac{1}{2}\hbar\alpha$ becomes

$$s_z^*\alpha^* = \tfrac{1}{2}\hbar\alpha^*,$$

whence $s_z\alpha^* = -\frac{1}{2}\hbar\alpha^*$. So $\alpha^* = \lambda\beta$ for some number λ and similarly $\beta^* = \mu\alpha$. Also $s^-\alpha = \hbar\beta$ becomes $-s^+\alpha^* = -s^+\lambda\beta = -\hbar\lambda\alpha = \hbar\beta^*$, so $\lambda = -\mu$. The remaining requirements are then satisfied. λ is any complex number of modulus unity and we choose $\lambda = i$. Thus $\alpha^* = i\beta$ and $\beta^* = -i\alpha$. The reason for this choice will appear in a moment.

If $\psi = f_1\alpha + f_2\beta$ is an eigenfunction of $\mathscr{H}$, then $\psi^* = i(f_1^*\beta - f_2^*\alpha)$ is also an eigenfunction of $\mathscr{H}$ with the same energy. The proof is as previously. We now show that ψ and ψ^* are independent functions. Suppose that $\psi^* = e^{ia}\psi$ for some real number a. Then

$$\left.\begin{aligned} if_1^* &= e^{ia}f_2 \\ -if_2^* &= e^{ia}f_1 \end{aligned}\right\} \rightarrow f_1 = -ie^{-ia}f_2^* = -e^{-ia}e^{ia}f_1 = -f_1;$$

and similarly $f_2 = -f_2$ so $\psi \equiv 0$, which is a contradiction. So ψ^* and ψ are independent states with the same energy. Clearly ψ^* is orthogonal to ψ. Also $(\psi^*)^* = -\psi$ and so we have arranged the set of all eigenstates of $\mathscr{H}$ into subsets, each subset containing just one degenerate pair. This is what is known as Kramers degeneracy.

As an example consider the effect of the star operation, when the electrostatic potential ϕ is a function of r alone, on kets in the $|nlm_s m_l\rangle$ scheme. These correspond to Schrödinger functions $R_{nl}Y_{lm}\tau_{m_s}$. Let R_{nl} be real. It follows from the form of the Y_{lm} that $Y_{lm}^* = (-1)^m Y_{l-m}$. Therefore

$$|nlm_s m_l\rangle^* = (-1)^{m_s+m_l}|nl-m_s-m_l\rangle.$$

The sign of both m_s and m_l is reversed and so the spin-orbit coupling energy is unchanged. This gives some physical insight into the significance of the star operation. The phase of the ket is multiplied by $(-1)^{m_s+m_l} = (-1)^{m_j}$ which is the reason for our particular choice for λ.

When a magnetic field is present $\mathscr{H}^*$ is no longer equal to $\mathscr{H}$. For

$$\mathscr{H} = \frac{1}{2m}\left(\mathbf{p}+\frac{e}{c}\mathbf{A}\right)^2 - e\phi + \xi(r)\mathbf{l}\,.\,\mathbf{s},$$

$$\mathscr{H}^* = \frac{1}{2m}\left(\mathbf{p}-\frac{e}{c}\mathbf{A}\right)^2 - e\phi + \xi(r)\mathbf{l}\,.\,\mathbf{s}.$$

Let us now define a new operation, a dagger operation. This is identical with the star except that it also reverses the sign of $\mathbf{A}$ wherever it occurs. It is the same as star for wave-functions. Then even in a magnetic field $\mathscr{H}^\dagger = \mathscr{H}$. From a physical point of view the extra feature of the dagger is that it reverses the direction of the external magnetic field at all points. It changes both the wave-function and the external field in such a way that the energy of the system remains unchanged. For clearly $\psi^\dagger = \psi^*$ is an eigenstate of $\mathscr{H}^\dagger$ with eigenvalue E whenever ψ is an eigenstate of $\mathscr{H}$ with eigenvalue E.

A general state of an n-electron system may be written

$$\psi = \Sigma f_{m_1 \ldots m_n} \tau_{m_1} \ldots \tau_{m_n},$$

where each τ_{m_i} is a spin function α or β for the ith electron, the $f_{m_1 \ldots m_n}$ are spatial functions and the sum extends over 2^n terms. We apply the dagger by using our previous definition of star for each of the f and τ separately. Then one verifies immediately that

$$\psi^\dagger = \Sigma(-1)^{\Sigma m_i} f^*_{m_1 \ldots m_n} \tau_{-m_1} \ldots \tau_{-m_n}.$$

Evidently $(\psi^\dagger)^\dagger = (-1)^n \psi$. The non-relativistic Hamiltonian for the n-electron system still satisfies $\mathscr{H}^\dagger = \mathscr{H}$. Now put $\psi^\dagger = e^{ia}\psi$. We deduce that

$$(-1)^n \psi = (\psi^\dagger)^\dagger = (e^{ia}\psi)^\dagger = e^{-ia}\psi^\dagger = \psi,$$

which is a contradiction unless n is even. Letting $\mathbf{A} = 0$ we have Kramers's result for general n. When n is odd, the pair $(\psi^\dagger, \psi)$ is often referred to as a Kramers doublet and again $\psi^\dagger$ is orthogonal to ψ and $\psi^\dagger$ is normalized if ψ is.

The preceding results remain true when we add the usual corrections arising from relativistic equations. For one electron this follows immediately from (5.31).

Let us now consider a system in a general electrostatic field but a constant external magnetic field given by the vector potential $\mathbf{A} = \frac{1}{2}\mathbf{H} \wedge \mathbf{r}$. Take an eigenfunction ψ of $\mathscr{H}$ and write its energy $E(H)$ to show its dependence on the numerical parameter $H = |\mathbf{H}|$. $\psi^\dagger$ also has energy $E(H)$ but only if we first reverse the direction of $\mathbf{H}$. Now change the sign of the number H throughout and $\psi^\dagger$ will have energy $E(-H)$, but in the same external magnetic field that ψ was in originally. So in the field $\mathbf{A} = \frac{1}{2}\mathbf{H} \wedge \mathbf{r}$, $\psi^\dagger$ has energy $E(-H)$ when ψ has energy $E(H)$. If the state is non-degenerate this implies that $E(H)$ is an even function of H. In case $E(H)$ can be expanded in a power series in H

$$E(H) = E_0 + E_1 H + E_2 H^2 + \ldots,$$

E_1 must be zero for a non-degenerate state. If $\psi^\dagger$ is not a numerical multiple of ψ then the state, which is doubly degenerate for $H = 0$, breaks up into two which diverge linearly as functions of H with equal and opposite gradients $\pm E_1$. This, of course, is only true for H sufficiently small.

We are now in a position to pursue in a little more detail the problem of those irreducible representations which have complex characters. Define $\langle X|^{\dagger} = \overline{(|X\rangle^{\dagger})}$ and then if V is real we have the equations

$$\langle \bar{f}|\, V\,|f\rangle = (\langle \bar{f}|\, V\,|f\rangle)^{\dagger} = \langle \bar{f}^{\dagger}|\, V\,|f^{\dagger}\rangle = \langle f|\, V\,|\bar{f}\rangle.$$

This is just a translation of part of the argument we used at the end of § 8.1 to show that functions belonging to conjugate representations could be grouped in degenerate pairs. The new formulation is no simpler than the old, but has the advantage that it may be extended immediately to discuss systems with spin. Then let V be any real function which commutes with all elements of the site spinor symmetry group G^*. Suppose Γ, $\overline{\Gamma}$ are conjugate representations of degree one of G^* with ψ, a state of an odd-electron system, forming a basis for Γ. Then $\psi^{\dagger}$ forms a basis for Γ and V is diagonal within the pair of states ψ, $\psi^{\dagger}$ with the same diagonal element for each. This argument can be extended immediately to discuss E'' and E''' of the group T^*.

In a sense the considerations of the preceding paragraph are implicit in the treatment earlier in the section, but I inserted them in order to throw further light on why it is that the group character tables of D_3^*, D_5^* and T^* appear to suggest that certain degeneracies could be lifted by a suitably chosen electrostatic environment while in fact they never can.

Examples

1. Show that the star operation on kets is not a linear operator.

2. When the term $\delta\mathscr{H} = \sum_i f(r_i)\,\mathbf{r}_i\,.\,\mathbf{s}_i$ is added to the Hamiltonian, odd electron systems are no longer doubly degenerate, in general. Why does this not conflict with the analysis of the present section? Show that if there were such a term in the Hamiltonian it would be possible for the ground state of an ion in a crystal to be non-degenerate yet have a permanent magnetic moment.

3. We have defined $\alpha^* = i\beta$, $\beta^* = -i\alpha$. Deduce that the spin functions α_1, β_1 referred to any other set of axes satisfy $\alpha_1^* = i\beta_1$, $\beta_1^* = -i\alpha_1$.

(Hint: Use § 6.9 and find the effect of a coordinate transformation by noting that a rotation applied both to a function and its coordinate system must leave it unchanged.)

4. An orthonormal set of m states ϕ_i of an n-electron system has the property that

$$\phi_i^* = \sum_{j=1}^{m} \phi_j a_{ji}.$$

Show that the matrix $A = [a_{ji}]$ satisfies $A' = (-1)^n A$.

5. Under the hypotheses of ex. 4 and with n even show that there exists an orthonormal set of m states ψ_j each linearly dependent on the ϕ_i and each satisfying $\psi_j^* = \psi_j$.

6. Suppose we have an orthonormal set of $m(2S+1)$ kets $|Mi\rangle$, $i = 1, 2, \ldots, m$, each an eigenstate of $\mathbf{S}^2$ and S_z with eigenvalues S, M and, for fixed i, correctly connected in phase. Suppose, further, that $|Mi\rangle^* = \sum_{j=1}^{m} a_{ji}(M)\,|-Mj\rangle$. Prove that there exists an orthonormal set of m kets $|\alpha Mj\rangle = \sum_{i=1}^{m} y_{ji}\,|Mi\rangle$, with y_{ji} independent of M, satisfying

$$|\alpha Mj\rangle^* = (-1)^M\,|\alpha - Mj\rangle.$$

The point of this example is that it shows that we may always take the basic kets for a system as eigenstates of $\mathbf{S}^2$ and S_z and 'spatially real' in a certain well-defined sense. Note that the function $\alpha(1)\,\beta(2) - \alpha(2)\,\beta(1)$ is not spatially real in this sense.

7. Use the star operation to show that

$$\langle j_1 j_2 -m_1 -m_2 \mid j_1 j_2 j -m\rangle = (-1)^{j_1+j_2-j}\langle j_1 j_2 m_1 m_2 \mid j_1 j_2 j m\rangle.$$

(Hint: It is helpful to use (2.80) with $T_z = j_{1z}$.)

8.5. The Jahn–Teller effect

Kramers's results show is that there is one kind of degeneracy which can never be removed by a purely electrostatic field. Jahn and Teller have established a result of the opposite kind.[1] They showed that, in general, a system in a static environment which possesses any degeneracy other than Kramers degeneracy will achieve a lower energy and a lower degeneracy by distorting the environment. There is one exception to this, when the environment consists just of a set of collinear atoms, but this case is not very important in discussing ions in crystals.

What do we mean by 'in general'? From a purely logical point of view we mean that the effect occurs, unless it does not occur, and have therefore said nothing useful. In practice we do mean something useful, for we establish not only that there is no group-theoretic reason why the environment should not distort in the way mentioned but also that it will do so unless, entirely accidentally, a matrix element happens to be exactly zero. Experience in calculation tells us that these accidents do not happen.[2]

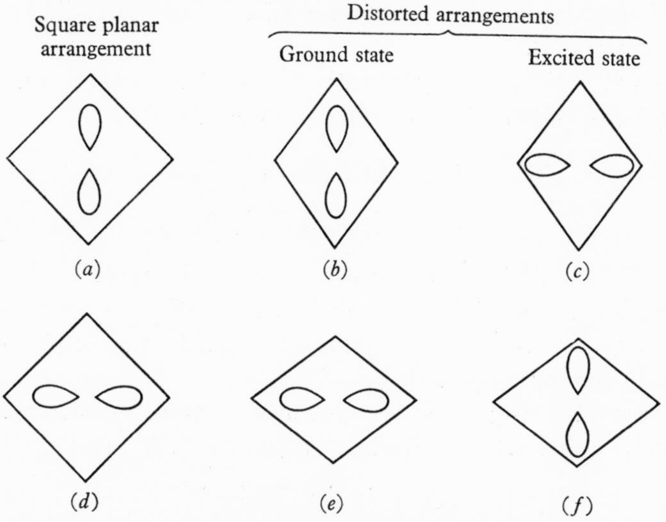

Fig. 8.1. The Jahn–Teller effect in a planar compound for p_x and p_y orbitals. The Y-axis is horizontal.

For the rest of this section we neglect spin and spin-orbit coupling entirely. First, we take an example considered by Jahn and Teller. Consider four identical closed shell ions arranged at the vertices of a square and an ion at the centre having a configuration consisting of closed shells with a single p electron outside. This is illustrated schematically in Fig. 8.1. p_z is an a_{2u} orbital under the sym-

[1] Jahn & Teller, *Proc. Roy. Soc.* A (1937), **161**, 220.

[2] Except for reasons which can be called group-theoretic. This is true throughout the whole of that part of physics which is understood.

metry group D_{4h} and will not interest us further. (p_x, p_y) however are e_u and are necessarily degenerate. The part of the electrostatic potential due to the environment which is relevant for p orbitals is

$$V = \beta_{00} + \beta_{20} Z_{20}.$$

Now let us distort the environment in the manner shown in Fig. 8.1 by pushing two neighbours in and pulling the other two out by the same amount. We may now write

$$V(\eta) = \beta_{00} + \beta_{20} Z_{20} + \beta_{22}^c(\eta) Z_{22}^c,$$

where η is a parameter describing the extent of the distortion. η is chosen so that $\beta_{22}^c = 0$ when $\eta = 0$ and so that if $\eta = \eta_1$ for situation (*b*) and (*c*) in Fig. 8.1, then $\eta = -\eta_1$ for the equal and opposite distortion shown in (*e*) and (*f*). Now the environment for (*e*) and (*f*) is also obtained from that for (*b*) and (*c*) by rotating through 90° about OZ, i.e.

$$V(-\eta) = C_4^z V(\eta),$$

hence

$$\beta_{22}^c(-\eta) = -\beta_{22}^c(\eta).$$

So β_{22}^c is an odd function of η and, unless there is an accident, $V(\eta)$ contains a term linear in η. If the reader is doubtful about the improbability of an 'accident', he has merely to write down $V(\eta)$ explicitly for four point charges as 'adjacent ions' to see that the linear term is in fact non-zero. That the rest of the environment should exactly cancel the influence of the charges could be reasonably regarded as accidental.

The energy of a distorted configuration associated with the value η_1 for the parameter η is now different for p_x and p_y. For write

$$E_{xx} = \langle p_x | \beta_{22}^c Z_{22}^c | p_x \rangle,$$

$$E_{yy} = \langle p_y | \beta_{22}^c Z_{22}^c | p_y \rangle,$$

and we have $E_{xx} = C_4^z E_{xx} = -E_{yy}$ and, for small η, E_{xx} and E_{yy} diverge linearly. We have considered only the crystal-field contribution to the energy, but the linear divergence for this contribution shows that the square planar configuration cannot, in general, be stable.

In our example the reason that the system distorts may be described in a more group-theoretic manner. The three integrands of E_{xx}, E_{yy} and of

$$E_{xy} = \langle p_x | \beta_{22}^c Z_{22}^c | p_y \rangle$$

form a basis for a representation of D_{4h}. Z_{22}^c spans B_{1g} and $p_x^2, p_x p_y, p_y^2$ span part of E_u^2. They span what may be called the symmetric part of E_u^2. We write this $[E_u^2]$ and call it the symmetrized direct product. In general, the symmetrized direct product of Γ with itself has as basis the $\frac{1}{2}n(n+1)$ products $\gamma_1^2 \dots \gamma_n^2$, $\gamma_1\gamma_2, \dots, \gamma_{n-1}\gamma_n$ when Γ has basis $\gamma_1 \dots \gamma_n$. $[E_u^2] = A_{1g} + B_{1g} + B_{2g}$ so the integrands of E_{xx}, E_{yy}, E_{xy} span $B_{1g} + A_{1g} + A_{2g}$. As this contains the unit representation, E_{xx}, E_{yy}, E_{xy} will not in general all be zero. A condition for a distortion is, therefore, that if the degenerate level belongs to Γ_a and the distortion to Γ_b then the unit representation Γ_1 must belong to $[\Gamma_a^2]\Gamma_b$. On the other hand, if

Γ_b is itself Γ_1, the distorted environment will have the same symmetry as the undistorted one. Such a distortion lowers neither the symmetry nor the degeneracy. So what we really require is that for any degenerate Γ_a there should exist a distortion $\Gamma_b \neq \Gamma_1$ such that Γ_1 is contained in $[\Gamma_a^2]\,\Gamma_b$.

We now establish the existence of the Jahn–Teller effect in general and take Γ_a to be a representation of degree greater than one and irreducible in terms of real matrices. Then Γ_a is either irreducible in the ordinary sense (i.e. in terms of complex matrices) or is a sum of two irreducible representations having conjugate complex characters. It then follows at once from the orthogonality properties of characters (6.19) that Γ_1 occurs at most twice in $[\Gamma_a^2]$. However, $[\Gamma_a^2]$ has degree greater than or equal to three and so contains at least one irreducible representation not equal to Γ_1. Call it Γ_c. If we can show that for any Γ_c there is always a distortion having symmetry Γ_c (or $\Gamma_c + \bar{\Gamma}_c$ for Γ_c having a complex character) we shall have established the result.

In any macroscopic crystal there is always an atom in general position with respect to a particular point-symmetry group G. By this we mean that the atom is transformed into a different atom by every element g of G. The set of atoms S thus obtained has the order of the group and, clearly, most atoms in a macroscopic crystal are in general position. Now consider the set of all distortions obtained by moving an atom of S along the line joining that atom to the central ion. This set gives rise to a representation Γ in which every element of the group except the unit element has character zero. The unit element has character π, where π is the order of the group, so from (6.27),

$$\Gamma = \Sigma \lambda_i \Gamma_i,$$

where λ_i is the degree of the ith irreducible representation. This completes the derivation.

Two comments are appropriate. First, we have established the result for a static environment which can undergo classical distortions. We defer consideration of the situation when both ion and environment are regarded as quantum dynamical systems until § 11.3.

Next we see that we have here one reason why it will generally be the case that the correct Hamiltonian to use for a static environment will include low-symmetry components. This was mentioned in § 7.4 when we discussed coupling schemes.

Thirdly, from a quantitative point of view, one expects the nearest neighbours to be the most important causes of Jahn–Teller distortions. Therefore it is desirable to know if the result is still true if we restrict attention to the nearest neighbours. One does this for finite symmetry groups by working through the various possibilities and finds that it is true providing the nearest neighbours give rise to a finite symmetry group. The one exception is when there are just one or two nearest neighbours lying on a line passing through the nucleus. This case is rarely of interest in crystals or in solution although one might regard the uranyl ion UO_2^{++} as an example of it.

8.6. Real and complex representations[1]

In this section we consider the slightly complicated question of groups which are not of the first kind—that is, whose irreducible representations are not all equivalent to representations by means of real matrices—and also discuss the Jahn–Teller effect for systems with spin.

Following Frobenius and Schur, it is convenient to classify irreducible representations into three categories, namely those which are:

(1) equivalent to a real representation;

(2) equivalent to their conjugate complex representation, but to no real representation;

(3) not equivalent to their conjugate complex representation.

The conjugate complex representation to $A(R)_{ij}$ is $\bar{A}(R)_{ij}$. As representations are equivalent if and only if they have the same character, categories 1 and 2 have real characters and category 3 have characters which are not wholly real. A group of the first kind has representations only of the first category.

It is not always easy to prove directly that a representation is not equivalent to any real one, but fortunately there is a simple test which shows to which category an irreducible representation belongs. Let Γ be an irreducible representation of degree n of a group G and choose two bases x_i, y_i each spanning Γ in identically the same form. Then all operations of G turn the $\frac{1}{2}n(n+1)$ symmetrized products $x_1y_1, \ldots, x_ny_n, x_1y_2+x_2y_1, \ldots$ into linear combinations of themselves and similarly for the $\frac{1}{2}n(n-1)$ antisymmetrized products $x_1y_2-x_2y_1, \ldots$. Thus the direct product has been decomposed into a symmetrized direct product, written $[\Gamma^2]$, and an antisymmetrized direct product, written (Γ^2). We shall prove that the unit representation belongs to $[\Gamma^2]$ but not (Γ^2) for category 1, to (Γ^2) but not $[\Gamma^2]$ for category 2, and to neither for category 3.

First, we derive an expression for the character $\chi^{\Gamma^2}(R)$ of Γ^2. Let us write $[R_{ij}]$ for the matrix corresponding to R in the representation Γ. Then

$$Rx_i = \sum_k x_k R_{ki},$$

$$Rx_iy_j = \sum_{k,l} x_k y_l R_{ki} R_{lj},$$

whence

$$\chi^{\Gamma^2}(R) = \sum_{i,j} R_{ii} R_{jj} = [\chi^{\Gamma}(R)]^2. \tag{8.24}$$

The number of times the unit representation occurs in Γ^2 is given, from (6.27), as

$$n_1 = \frac{1}{\pi}\sum_R \chi^{\Gamma^2}(R)\,.\,1 = \frac{1}{\pi}\sum_R \chi^{\Gamma}(R)\,\chi^{\Gamma}(R).$$

π is the order of the group. If Γ belongs to category 1 or 2 then χ^{Γ} is real and the orthogonality relation (6.19) for characters shows that $n_1 = 1$. If Γ is in category 3, then because Γ and $\bar{\Gamma}$ are inequivalent, $n_1 = 0$. This proves our result for category 3.

[1] Mainly adapted from: Frobenius & Schur, *S.B. preuss. Akad. Wiss.* (1906 I), p. 186; Jahn, *Proc. Roy. Soc.* A (1938), **164**, 117.

Next we discuss the character $\chi^{(\Gamma^2)}(R)$ of (Γ^2). As

$$R(x_i y_j - x_j y_i) = \sum_{kl} x_k y_l (R_{ki} R_{lj} - R_{kj} R_{li})$$

$$= \sum_{kl} \tfrac{1}{2}(x_k y_l - x_l y_k)(R_{ki} R_{lj} - R_{kj} R_{li}),$$

then
$$\chi^{(\Gamma^2)}(R) = \tfrac{1}{2}\sum_{i,j}(R_{ii} R_{jj} - R_{ij} R_{ji}) = \tfrac{1}{2}[\chi^\Gamma(R)]^2 - \tfrac{1}{2}\chi^\Gamma(R^2). \tag{8.25}$$

Combining this with (8.24) we have

$$\chi^{[\Gamma^2]}(R) = \tfrac{1}{2}[\chi^\Gamma(R)]^2 + \tfrac{1}{2}\chi^\Gamma(R^2), \tag{8.26}$$

and for representations of categories 1 or 2 can deduce

$$n_1^s = \tfrac{1}{2} + \frac{1}{2\pi}\sum_R \chi^\Gamma(R^2), \tag{8.27}$$

where n_1^s is the number of times the unit representation occurs in $[\Gamma^2]$. From our earlier result $n_1^s = 0$ or 1. If all the matrices are real, then using the orthogonality relations

$$\sum_R \chi^\Gamma(R^2) = \sum_{R,i,j} R_{ij} R_{ji} = \sum_{R,i} R_{ii}^2 = \pi, \quad n_1^s = 1.$$

This must also be true for any equivalent representation, so we have established our result for category 1.

To prove the result for category 2 it is clearly sufficient to show that the converse of the previous result is true, namely, that if the unit representation Γ_1 belongs to $[\Gamma^2]$ then Γ is equivalent to a representation by means of real matrices. Then suppose Γ_1 belongs to $[\Gamma^2]$. This means that there exists a symmetrical bilinear form, not identically zero,

$$\mathbf{x}'B\mathbf{y} = \sum_{i,j} b_{ij} x_i y_j \quad (b_{ij} = b_{ji}),$$

which is left invariant by all elements of G. The conjugate complex matrix $\bar{B}$ gives another symmetrical bilinear form which must be a multiple $\mathbf{x}'B\mathbf{y}\, e^{i\theta}$ of $\mathbf{x}'B\mathbf{y}$ because Γ_1 occurs only once in $[\Gamma^2]$. So the matrix $B\, e^{\frac{1}{2}i\theta}$ is real. Therefore, without loss of generality, we take B to be real.

Because B is real and symmetric, there exists a real orthogonal matrix Q such that QBQ' is diagonal. Now define another diagonal matrix E by the rule that a diagonal element of E is $\sqrt{-1}$ or 1 according as the corresponding diagonal element of QBQ' is negative or not. E is unitary. Then we change the basis of Γ by writing

$$\mathbf{x}' = \boldsymbol{\xi}'EQ, \quad \mathbf{y} = Q'E'\boldsymbol{\eta} = Q'E\boldsymbol{\eta},$$

$$\mathbf{x}'B\mathbf{y} = \boldsymbol{\xi}'EQBQ'E\boldsymbol{\eta} = \boldsymbol{\xi}'\Lambda\boldsymbol{\eta},$$

where Λ is a diagonal matrix with non-negative elements, λ_i say, and $\boldsymbol{\xi}$, $\boldsymbol{\eta}$ are both obtained from $\mathbf{x}$, $\mathbf{y}$, respectively by the same unitary transformation.

Writing R for a typical one of the matrices of the representation Γ, referred to the new basis, the condition of invariance reads $R'\Lambda R = \Lambda$. This is equivalent to $\Lambda R = \bar{R}\Lambda$ or to $\Lambda\bar{R} = R\Lambda$ because R is unitary and Λ real. Therefore

$$\Lambda^2 R = \Lambda\bar{R}\Lambda = R\Lambda^2.$$

Thus Λ^2 commutes with all the matrices of the irreducible representation and is therefore a multiple of the unit matrix (Schur's lemma, ex. 2, p. 155). Λ^2 cannot be identically zero and we deduce that Λ itself is a non-zero multiple of the unit matrix. $\Lambda R = \bar{R}\Lambda$ then implies that all the matrices R are real orthogonal matrices, so Γ is of the first category. It is, of course, merely a matter of substitution to show that $\boldsymbol{\xi}'\boldsymbol{\eta}$ is left invariant by any real orthogonal matrix, so our results are consistent.

We can use (8.25) and (8.26) to give an alternative form of the test. Write

$$\sum_R \chi^{\Gamma}(R^2) = c\pi. \tag{8.28}$$

Then $c = 1, -1$ or 0 according as Γ belongs to category 1, 2 or 3, respectively.

As a second straightforward corollary to our findings we may say that if Γ is of the second category then there exists an antisymmetric bilinear form, not identically zero,

$$\mathbf{x}'B\mathbf{y} = \sum_{i,j} b_{ij}x_i y_j \quad (b_{ij} = -b_{ji}),$$

which is left invariant by all elements of G. Apart from multiplication by a constant number, $\mathbf{x}'B\mathbf{y}$ is the only invariant bilinear form. We remember that Γ is assumed irreducible. As before, B^2 commutes with all matrices R of Γ and is therefore a multiple of the unit matrix. So B is non-singular. However, $B = -B'$ implies $|B| = (-1)^n |B|$, where n is the degree of Γ which means that n must be even.

Let us now discuss degeneracy in systems having spin and placed in an external electrostatic field with spinor symmetry group G^*. There are two kinds of degeneracy which are forced upon the system, one being Kramers degeneracy and the other implied by the nature of the representation to which the functions belong. For even electron systems all irreducible representations are either category 1 or 3. If Γ is category 1 take a basis ψ_i on which Γ is represented by real matrices. Then the two sets $\psi_i + \psi_i^*$ and $\sqrt{-1}\,(\psi_i - \psi_i^*)$ each separately form a basis for Γ and each constituent function is invariant under the star operation. So the star introduces no new degeneracy. If Γ is of category 3 with basis ψ_i then $\bar{\Gamma}$ has basis ψ_i^* and the same energy. Therefore, from the point of view of degeneracy, spin introduces nothing new in even electron systems. In particular the Jahn–Teller theorem applies just as in the absence of spin because the group representations involved are precisely the same.

For an odd electron system, take an eigenvalue of the Hamiltonian with degeneracy n and an orthonormal set of n eigenfunctions ψ_i belonging to this eigenvalue. The star operation enables us to arrange these in pairs and so n is even. The n functions ψ_i form a basis for a representation $\Gamma = \Sigma c_j \Gamma_j$ of the site spinor symmetry group. Then if $\phi_1 \dots \phi_a$ are linear combinations of the ψ_i and form a basis for Γ_j it follows that $\phi_1^* \dots \phi_a^*$ are also linear combinations of the ψ_i and form a basis for $\bar{\Gamma}_j$. So if there is no accidental degeneracy Γ has at most two components, Γ_j and $\bar{\Gamma}_j$. If Γ_j is of category 1 then $\bar{\Gamma}_j = \Gamma_j$. From Table A 22 we see that Γ_j has degree 1 and so $n = 2$ and we just have the two components of a Kramers doublet unconnected by any operation of the symmetry group. If

Γ_j is of category 3 then $\bar{\Gamma}_j \neq \Gamma_j$ and the pair Γ_j, $\bar{\Gamma}_j$ appear together with the same energy. The Kramers degeneracy has doubled the necessary group-theoretic degeneracy.

When Γ_j is of the second category, $\bar{\Gamma}_j = \Gamma_j$, but it does not follow immediately that $\phi_1^* \dots \phi_a^*$ are linearly dependent on $\phi_1 \dots \phi_a$. I now show that except in case of accidental degeneracy this linear dependence must hold or, more precisely, that if the linear dependence does not hold we may construct two sets of linear combinations of the ϕ_i, ϕ_i^* each of which separately forms a basis for Γ_j and which do not get mixed up by the star operation.

First, we remark that as $\Gamma_j = \bar{\Gamma}_j$ there exists a unitary matrix Q such that $\sum_j \phi_j^* Q_{ji}$ form a basis spanning Γ_j in identically the same form as ϕ_i do. In matrix notation this reads $\boldsymbol{\phi}^{*\prime}Q$. If R is the matrix representing a group element in the basis $\boldsymbol{\phi}$ then $\bar{R}$ represents it in $\boldsymbol{\phi}^*$. Therefore from the definition of Q

$$\boldsymbol{\phi}^{*\prime}QR = \boldsymbol{\phi}^{*\prime}\bar{R}Q$$

which, as R is unitary, means that Q is the matrix of an invariant form for Γ_j. As Γ_j is of category 2 we therefore have $Q' = -Q$. Now define

$$\boldsymbol{\phi}'^{\pm} = \boldsymbol{\phi}' \pm \boldsymbol{\phi}^{*\prime}Q.$$

Then $\quad (\boldsymbol{\phi}'^{\pm})^* = \boldsymbol{\phi}^{*\prime} \mp \boldsymbol{\phi}'\bar{Q} = (\boldsymbol{\phi}^{*\prime}Q' \mp \boldsymbol{\phi}')\bar{Q} = \mp(\pm\boldsymbol{\phi}^{*\prime}Q + \boldsymbol{\phi}')\bar{Q} = \mp\boldsymbol{\phi}'^{\pm}\bar{Q}.$

On the other hand ϕ_i^+ and ϕ_i^- separately span Γ_j so we have achieved our object. Thus for category 2 representations the Kramers star operation introduces no extra necessary degeneracy.

Having seen the forms of degeneracy which appear for odd electron systems with spin, we pass on to discuss the Jahn–Teller theorem for this case. First, we must discover what necessary relations among the matrix elements of a real operator V hold because of the star operation. Let us write the eigenvectors for the unperturbed configuration in ket notation $|X_\mu\rangle$, where $\mu = \pm 1, \pm 2, \dots, \pm n$ and

$$|X_\mu^*\rangle = |X_\mu\rangle^* = \pm|X_{-\mu}\rangle,$$

where the sign is the sign of μ. Then $|X_\mu^{**}\rangle = -|X_\mu\rangle$ always. By definition, $\langle X_\mu^*| = \overline{|X_\mu^*\rangle}$. Then if R is an operation of the symmetry group,

$$R|X_\mu\rangle = \sum_\nu |X_\nu\rangle R_{\nu\mu},$$

$$R|X_\mu^*\rangle = \sum_\nu |X_\nu^*\rangle \bar{R}_{\nu\mu},$$

$$R\langle X_\mu^*| = \sum_\nu \langle X_\nu^*| R_{\nu\mu},$$

so the $\langle X_\mu^*|$ span the same representation as the $|X_\mu\rangle$ and in identically the same form.

We have already discussed the way in which the reality of an operator V introduces restrictions on its possible matrix elements. Repeating our earlier calculation in the new notation we find

$$\begin{aligned}\langle X_\mu^*|V|X_\nu\rangle &= -(\langle X_\mu|V|X_\nu^*\rangle)^* = -\overline{\langle X_\mu|V|X_\nu^*\rangle} \\ &= -\langle X_\nu^*|V|X_\mu\rangle \\ &= \tfrac{1}{2}\{\langle X_\mu^*|V|X_\nu\rangle - \langle X_\nu^*|V|X_\mu\rangle\}.\end{aligned} \tag{8.29}$$

If V is a component of a basis V_i of an irreducible representation Γ_1 and $|X_\mu\rangle$ is a basis of Γ_2 then, because $\langle X_\mu^*|$ transforms as $|X_\mu\rangle$, we deduce that the matrix elements of V_i which are not forced to be zero by the star operation transform as $\Gamma_1(\Gamma_2^2)$. In order for it to be possible on group-theoretic grounds for there to be non-vanishing matrix elements of V_i, the unit representation must occur in the product $\Gamma_1(\Gamma_2^2)$. For there to be a Jahn–Teller distortion one such occurrence must arise from Γ_a contained in Γ_1 and Γ_b contained in (Γ_2^2) with neither Γ_a nor Γ_b itself the unit representation.

For future reference we remark here that the derivation (8.29) depended on the fact that $\overline{V} = V^*$. If $\overline{V} = -V^*$, as for example for **L** or **S**, we would finish up with the symmetrized sum $\frac{1}{2}\{\langle X_\mu^* \mid V \mid X_\nu\rangle + \langle X_\nu^* \mid V \mid X_\mu\rangle\}$ and would then deduce that there could only be non-vanishing elements of V if the unit representation occurred in $\Gamma_1[\Gamma_2^2]$.

When Γ_2 has degree 2 we know this cannot be lifted for an odd electron system and therefore (Γ_2^2), which is now of degree 1, must be the unit representation. This is easily proved directly. If $|X_1\rangle$, $|X_{-1}\rangle$ is a basis for Γ_2 and if $\Gamma_2 = \Gamma_2' + \Gamma_2''$, where Γ_2' is of category 1 or 3, then $\Gamma_2'\Gamma_2''$ is equal to the unit representation because $\Gamma_2' = \Gamma_2''$ or $\Gamma_2' = \overline{\Gamma_2''}$ for the two categories, respectively. On the other hand, if Γ_2 is of category 2 then we proved earlier that (Γ_2^2) contains the unit representation.

For Γ_2 of degree greater than 2 there are only four possibilities for pure rotation groups:

For T^*: $$(U'^2) = A + E + T,$$

For O^*: $$(U'^2) = A_1 + E + T_2,$$

For K^*: $$(U'^2) = A + V,$$

$$(W'^2) = A + U + 2V.$$

Each contains a representation other than the unit one which establishes the theorem for a crystal site. It is a straightforward matter to verify it for molecules containing a small number of atoms. Groups containing inversion are dealt with by observing that if $(\Gamma_2^2) = \Sigma c_i \Gamma_i$ then $(\Gamma_{2g}^2) = (\Gamma_{2u}^2) = \Sigma c_i \Gamma_{ig}$.

A point which may worry the reader is the following. The ground state of Mn^{++} is $3d^5$; ^6S which is spherically symmetric. How can a purely electrostatic field lift any of the sixfold spin degeneracy? The answer in general is that the spin-orbit coupling produces a coupling between spin direction and the electrical structure of the environment. A particular way in which this can occur, though not the only one, is that the ground state is not accurately ^6S but contains an admixture of $3d^5$; $^4P_{\frac{5}{2}}$ as we saw in § 5.2.

8.7. Various group-theoretic propositions

It is convenient to collect here various propositions about matrix elements, some proved earlier, so that the reader who does not wish to work through the proofs may have them in one place to refer to them.

First, we recall that we make a standard choice of basis relations for irreducible representations and this is given for O, O^*, D_4, D_4^*, D_3 and D_3^* in Table A 16. Any set of functions $|\Gamma M\rangle$ belonging to an irreducible representation is arranged so as to satisfy this standard choice, which, apart from one arbitrary phase factor common to all the $|\Gamma M\rangle$, can only be done in one way.

Then our first proposition (derived in § 6.5) is:

Theorem 1. (The group orthogonality relations for integrals.) Let Γ_1 and Γ_2 be two irreducible representations of a group G and suppose ξ commutes with all elements of G. Then
$$\langle\Gamma_1 M_1 | \xi | \Gamma_2 M_2\rangle = a\delta_{\Gamma_1\Gamma_2}\delta_{M_1M_2},$$
where a is independent of M_1 and M_2.

This proposition is easily extended to cover the case when we have a set of operators ξ_M which form a basis for an irreducible representation Γ of G. Let us suppose first that we have a group, such as O, D_4, D_4^*, D_3 or D_3^*, in which the product of two irreducible representations never contains any particular irreducible representation more than once. Then the linear combinations
$$|\alpha\Gamma' M'\rangle = \sum_{M, M_2} \langle\Gamma\Gamma_2 M M_2 | \Gamma\Gamma_2\Gamma' M'\rangle \xi_M |\Gamma_2 M_2\rangle$$
form a basis for the irreducible representation Γ'. α is just to distinguish these kets from the $|\Gamma_1 M_1\rangle$ and $|\Gamma_2 M_2\rangle$ with which we started. We deduce from Theorem 1 that
$$\langle\Gamma_1 M_1 | \alpha\Gamma' M'\rangle = a\delta_{\Gamma_1\Gamma'}\delta_{M_1M'}, \tag{8.30}$$
where a is independent of M_1, M' and (trivially, by so defining it) of Γ'. We work now in the reverse direction and have
$$\xi_M |\Gamma_2 M_2\rangle = \sum_{\Gamma', M'} \langle\Gamma\Gamma_2\Gamma' M' | \Gamma\Gamma_2 M M_2\rangle |\alpha\Gamma' M'\rangle,$$
and hence
$$\langle\Gamma_1 M_1 | \xi_M | \Gamma_2 M_2\rangle = a\langle\Gamma\Gamma_2\Gamma_1 M_1 | \Gamma\Gamma_2 M M_2\rangle, \tag{8.31}$$
where a is independent of M_1, M and M_2.

In the above derivation it was necessary that the representation Γ_1 should only occur once in the decomposition of the direct product $\Gamma\Gamma_2$. If it occurs n times we then have n separate sets $|\alpha_i\Gamma' M'\rangle$, where i runs from 1 to n, and n numbers a_i. In this case
$$\langle\Gamma_1 M_1 | \xi_M | \Gamma_2 M_2\rangle = \sum_{i=1}^{n} a_i \langle\Gamma\Gamma_2 i\Gamma_1 M_1 | \Gamma\Gamma_2 M M_2\rangle, \tag{8.32}$$
the extra parameter i in the coupling coefficients telling one that they refer to the ith set $|\alpha_i\Gamma_1 M_1\rangle$. We sum up our findings in three more Theorems. In the statement of each it is assumed that Γ_1, Γ_2 and Γ are irreducible representations of a group G and that ξ_M forms a basis for Γ.

Theorem 2. The matrix element $\langle\Gamma_1 M_1 | \xi_M | \Gamma_2 M_2\rangle$ is zero unless Γ_1 is contained in $\Gamma\Gamma_2$ or, equivalently, unless Γ is contained in $\Gamma_1\bar{\Gamma}_2$.

Theorem 3. Suppose Γ_1 is contained once in $\Gamma\Gamma_2$. Then the array of matrix elements $\langle\Gamma_1 M_1 | \xi_M | \Gamma_2 M_2\rangle$ for the various possible choices of M_1, M and M_2 shows its dependence on the detailed properties of the kets $|\Gamma_1 M_1\rangle$, $|\Gamma_2 M_2\rangle$ and the operators ξ_M only through a single parameter a common to all the matrix elements. In fact $\langle\Gamma_1 M_1 | \xi_M | \Gamma_2 M_2\rangle = a\langle\Gamma\Gamma_2\Gamma_1 M_1 | \Gamma\Gamma_2 M M_2\rangle$.

Theorem 3 shows that the matrix elements are completely determined by the behaviour of the kets and linear operators under the operations of the group G apart from a single multiplying number common to all the matrix elements of the array. Equation (8.32) shows how this is generalized when Γ_1 is contained n times in $\Gamma\Gamma_2$. n, of course, is also the number of times Γ is contained in $\Gamma_1\overline{\Gamma}_2$ and the number of times the unit representation occurs in $\overline{\Gamma}_1\Gamma\Gamma_2$.

Theorem 3 may be written in the useful form:

THEOREM 4. (The group replacement theorem.) Suppose Γ_1 is contained once in $\Gamma\Gamma_2$. Let $|\alpha\Gamma_1 M_1\rangle$ and $|\alpha'\Gamma_1 M_1\rangle$ each form a basis for Γ_1, and similarly $|\beta\Gamma_2 M_2\rangle$, $|\beta'\Gamma_2 M_2\rangle$ for Γ_2. Let, also, ξ_M and η_M both form bases for Γ. Then

$$\langle\alpha\Gamma_1 M_1|\,\xi_M\,|\beta\Gamma_2 M_2\rangle = c\,\langle\alpha'\Gamma_1 M_1|\,\eta_M\,|\beta'\Gamma_2 M_2\rangle,$$

where c is a constant independent of M_1, M and M_2, except possibly when all the matrix elements $\langle\alpha'\Gamma_1 M_1|\,\eta_M\,|\beta'\Gamma_2 M_2\rangle$ are zero.

We will often wish to know matrix elements in which the kets $|\Gamma_1 M_1\rangle$ are identically the same as $|\Gamma_2 M_2\rangle$. This, of course, occurs whenever one tries to diagonalize operators within a set of states which belong to a known irreducible representation and in this case there are further restrictions on the array of matrix elements $\langle\Gamma_1 M_1|\,\xi_M\,|\Gamma_1 M_1'\rangle$ as well as those implied by Theorems 1–4. It is easiest to illustrate the nature of these by taking an irreducible representation Γ_1 of category 1 and taking a set of real functions f_{M_1} forming a basis for Γ_1. Then the set of $f_{M_1}\xi_M f_{M_1'}$ form a basis for $\Gamma_1^2\Gamma$, while the subset

$$g_{M_1MM_1'} = f_{M_1}\xi_M f_{M_1'} + f_{M_1'}\xi_M f_{M_1}$$

form a basis for $[\Gamma_1^2]\,\Gamma$ and

$$h_{M_1MM_1'} = f_{M_1}\xi_M f_{M_1'} - f_{M_1'}\xi_M f_{M_1}$$

form a basis for $(\Gamma_1^2)\,\Gamma$. We now take linear combinations of the g and h which are components of bases of irreducible representations and then integrate over the space of the f_{M_1}. The orthogonality relations for integrals show that the result is zero except for those linear combinations which form a basis for the unit representation. If the unit representation occurs n times in $\Gamma_1^2\Gamma$ we have n parameters so we have merely rephrased our earlier results. Suppose now, however, that the ξ_M are real in both senses, that is $\overline{\xi}_M = \xi_M^* = \xi_M$. Then

$$\langle f_{M_1}\xi_M f_{M_1'}\rangle = \overline{\langle f_{M_1'}\xi_M f_{M_1}\rangle} = \langle f_{M_1'}\xi_M f_{M_1}\rangle^* = \langle f_{M_1'}\xi_M f_{M_1}\rangle$$

and so all the integrals arising from the $h_{M_1MM_1'}$ are zero. Similarly if

$$\overline{\xi}_M = \xi_M = -\xi_M^*$$

all the integrals arising from the $g_{M_1MM'}$ are zero. These results are clearly true in general for category 1 representations and we rewrite them in ket notation as:

THEOREM 5. If Γ_1 is of category 1 and ξ_M satisfy $\overline{\xi}_M = \xi_M^*$ then all the matrix elements $\langle\Gamma_1 M_1|\,\xi_M\,|\Gamma_1 M_1'\rangle$ are zero unless Γ is contained in $[\Gamma_1^2]$. If $\overline{\xi}_M = -\xi_M^*$, then they are zero unless Γ is contained in (Γ_1^2).

THEOREM 6. In the notation of Theorem 5, the number of non-group-theoretically determined parameters necessary to specify the array of matrix elements is

m, where m is the number of times Γ is in $[\Gamma_1^2]$ when $\bar{\xi}_M = \xi_M^*$ and m is the number of times Γ is in (Γ_1^2) when $\bar{\xi}_M = -\xi_M^*$.

The corresponding theorems for representations of category 2 are obtained from the discussion of (8.29). They are:

THEOREM 7. If Γ_1 is of category 2 and ξ_M satisfy $\bar{\xi}_M = \xi_M^*$ then all the matrix elements $\langle \Gamma_1 M_1 \mid \xi_M \mid \Gamma_1 M_1' \rangle$ are zero unless Γ is contained in (Γ_1^2). If $\bar{\xi}_M = -\xi_M^*$, then they are zero unless Γ is contained in $[\Gamma_1^2]$.

THEOREM 8 is the same as Theorem 6 but for an interchange of $[\Gamma_1^2]$ and (Γ_1^2).

This completes our set of theorems. The categories of the representations are given in Table A 22 and the squares $[\Gamma_1^2]$ and (Γ_1^2) are given for the groups O^*, D_4^* and D_3^* in Table A 23.

8.8. Examples of calculations in the weak-field scheme

In this section we consider a few straightforward but very important examples in the weak-field scheme. We neglect the spin-orbit coupling completely and have then to consider diagonalizing $\mathscr{V}$ within a d^n configuration. $\mathscr{V}$ has no non-zero matrix elements between states of different spin and so the diagonalization is a completely separate problem for each multiplicity. We know that in the free ions the ground term always has the highest possible multiplicity and so for a small electrostatic field from the environment a state of highest multiplicity will still lie lowest. Therefore we examine in this section the matrix elements of $\mathscr{V}$ within and between the terms of highest multiplicity.

We perform our calculations for the case in which $\mathscr{V}$ represents an external field of octahedral symmetry. This is very often a good approximation in practice. We suppose the centre of gravity of the d shell to be unaltered by $\mathscr{V}$ and write $+\frac{3}{5}\Delta$ and $-\frac{2}{5}\Delta$, respectively, for the energies of the e and t_2 representations for a single d electron. Ex. 4, p. 202, then shows that the centre of gravity is unaltered for all terms of all d^n configurations to first order in perturbation theory. This assumption for $\mathscr{V}$ is convenient when we are only interested in relative energies; if we are interested in absolute ones we must add a constant energy to each d orbital.

We now pass to examples.

(1) d^1. Here the ^2D term is split into two, ^2E and $^2T_2$ with energies $E(^2E) = \frac{3}{5}\Delta$, $E(^2T_2) = -\frac{2}{5}\Delta$, and separation Δ.

(2) d^2. The terms of highest multiplicity are ^3F and ^3P. We consider only the ground term ^3F for the moment. Take $M_S = 1$ and the states of ^3F are:

$$\left.\begin{aligned}
|33\rangle &= |2^+1^+\rangle, & |3-3\rangle &= |-1^+-2^+\rangle,\\
|32\rangle &= |2^+0^+\rangle, & |3-2\rangle &= |0^+-2^+\rangle,\\
|31\rangle &= \frac{\sqrt{2}}{\sqrt{5}}|1^+0^+\rangle+\frac{\sqrt{3}}{\sqrt{5}}|2^+-1^+\rangle, & |3-1\rangle &= \frac{\sqrt{2}}{\sqrt{5}}|0^+-1^+\rangle+\frac{\sqrt{3}}{\sqrt{5}}|1^+-2^+\rangle,\\
|30\rangle &= \frac{2}{\sqrt{5}}|1^+-1^+\rangle+\frac{1}{\sqrt{5}}|2^+-2^+\rangle, & &
\end{aligned}\right\} \quad (8.33)$$

where the two numbers in the kets on the left represent LM_L values. The states of (8.33) are correctly connected by shift operators.

^3F breaks up into $^3A_2$, $^3T_1$ and $^3T_2$ (see Table A 14), and we determine their energies in a manner strictly analogous to that used in the case of a free ion. That is we see if there are any components whose energies can be written straight down and then use the diagonal sum rule to determine the others. Table A 19 shows that $|30\rangle$ is the z-component of $^3T_1$. So

$$E(^3T_1) = E(|30\rangle) = \tfrac{4}{5}E(1)+\tfrac{4}{5}E(-1)+\tfrac{1}{5}E(2)+\tfrac{1}{5}E(-2).$$

The quantities $E(1)$, etc., are just the diagonal elements of $\mathscr{V}$ for a single d electron. The matrix of $\mathscr{V}$ for one electron is written down immediately and is given in Table 8.3. Therefore $E(^3T_1) = -\tfrac{3}{5}\Delta$.

Table 8.3. *The matrix of $\mathscr{V}$ for a single d electron in an octahedral field*

m_l value	2	1	0	-1	-2
2	$\frac{1}{10}\Delta$	0	0	0	$\frac{1}{2}\Delta$
1	0	$-\frac{2}{5}\Delta$	0	0	0
0	0	0	$\frac{3}{5}\Delta$	0	0
-1	0	0	0	$-\frac{2}{5}\Delta$	0
-2	$\frac{1}{2}\Delta$	0	0	0	$\frac{1}{10}\Delta$

Next, by the diagonal sum rule, $|32\rangle$ and $|3-2\rangle$ have the same total energy as $^3A_2$ and $^3T_2\zeta$. Hence

$$E(^3A_2)+E(^3T_2) = E(|32\rangle)+E(|3-2\rangle) = \tfrac{7}{5}\Delta.$$

Lastly the centre of gravity is unchanged, so

$$E(^3A_2)+3E(^3T_1)+3E(^3T_2) = 0,$$

whence $$E(^3A_2) = \tfrac{6}{5}\Delta, \quad E(^3T_1) = -\tfrac{3}{5}\Delta, \quad E(^3T_2) = \tfrac{1}{5}\Delta.$$

If we wish to know the actual functions, we have only to combine (8.33) with Tables A 16 and 19. For example

$$i|^3T_2 1\zeta\rangle = \frac{1}{\sqrt{2}}|2^+0^+\rangle+\frac{1}{\sqrt{2}}|0^+-2^+\rangle,$$

where we have inserted the M_S values before ζ for completeness.

An interesting point is that although we evaluated the energy of $^3T_1$ by using the exact expression for $|30\rangle$, we could have obtained it without ever finding $|30\rangle$. For the two functions $|1^+-1^+\rangle$ and $|2^+-2^+\rangle$ between them make up the $M_L = 0$ states of both ^3F and ^3P. But the latter behaves as $^3T_1$ under the group O and therefore has zero diagonal matrix elements of $\mathscr{V}$. Accordingly

$$\begin{aligned} E(^3F;\, ^3T_1) &= E(^3F;\, ^3T_1)+E(^3P;\, ^3T_1) \\ &= E(|1^+-1^+\rangle)+E(|2^+-2^+\rangle) \\ &= -\tfrac{3}{5}\Delta. \end{aligned}$$

We notice also that the energy of $|30\rangle$ as calculated from (8.33) is simply the weighted mean of the energies of $|1^+ -1^+\rangle$ and $|2^+ -2^+\rangle$; there is no matrix element between the latter two states. This is a general result for matrix elements between determinants built up from the $|nlm_s m_l\rangle$ so long as we keep within a set of states all having the same M_S, M_L values. It follows because any two such states must differ, if they differ at all, in at least two constituent one-electron functions.

(3) d^3. The terms of highest multiplicity are ^4F and ^4P. We use the device explained in the last paragraph but one to determine the energy of $^4T_1$ of ^4F. Take $M_S = 1\frac{1}{2}$ and then the $M_L = 0$ states here are made up from $|2^+0^+-2^+\rangle$ and $|1^+0^+-1^+\rangle$, whence

$$E(^4F;\, ^4T_1) = E(|2^+0^+-2^+\rangle) + E(|1^+0^+-1^+\rangle)$$
$$= \tfrac{3}{5}\Delta.$$

$^4A_2$ and $^4T_2\zeta$ are composed of $|2^+1^+-1^+\rangle$ and $|1^+-1^+-2^+\rangle$, so

$$E(^4A_2) + E(^4T_2) = -\tfrac{7}{5}\Delta.$$

Lastly $$E(^4A_2) + 3E(^4T_1) + 3E(^4T_2) = 0,$$

whence $$E(^4A_2) = -\tfrac{6}{5}\Delta, \quad E(^4T_1) = \tfrac{3}{5}\Delta, \quad E(^4T_2) = -\tfrac{1}{5}\Delta.$$

(4) d^4. The term of highest multiplicity is ^5D which splits into ^5E and $^5T_2$ with $E(^5E) = -\frac{3}{5}\Delta$, $E(^5T_2) = \frac{2}{5}\Delta$. This follows at once from the fact that $|2^+1^+-1^+-2^+\rangle$ is the $M_S = 2$ state of $^5E\theta$.

(5) d^5. ^6S is not split.

(6) Some general comments.

We could repeat the above arguments to obtain the splitting of the ground terms of d^ϵ for $\epsilon > 5$. However, this is unnecessary because the matrix of $\mathscr{V}$ for the states of highest multiplicity of d^ϵ and $d^{5+\epsilon}$ are the same. This may be seen by taking determinantal functions for $d^{5+\epsilon}$ with minimum M_S, for example, $M_S = -2$ for ^5D of d^6. Then we associate with each function that function of d^ϵ which is obtained by removing the five orbitals having β spin.

When we have done this we notice that the splitting of the ground terms of d^ϵ and $d^{10-\epsilon}$ are also the same except for a change of the sign of Δ. This result is true with much greater generality. The general theorem states that it is possible to set up a correlation between the states of d^ϵ and $d^{10-\epsilon}$ in such a way that the matrix of the crystal field changes sign on passing from d^ϵ to $d^{10-\epsilon}$. The complete matrix of the electrostatic energy $\Sigma(e^2/r_{\kappa\lambda})$ is the same for d^ϵ as for the correlated states of $d^{10-\epsilon}$. In each case these assertions are only true apart, possibly, from a common constant diagonal energy difference. This result is proved for individual states belonging to the strong-field scheme in § 9.7. In the weak-field scheme we require our individual states to belong to free-ion terms. We shall not prove it here, but one can introduce additional parameters to classify the repeated terms of a configuration in such a way that the theorem remains true. Having introduced this classification, the so-called seniority classification, the proof is the same as that of § 9.7 (see note, p. 245).

(7) The T_1 states of highest multiplicity.

We have already diagonalized the octahedral field $\mathscr{V}$ completely within the states of highest multiplicity except for the T_1 states of d^2, d^3, d^7 and d^8. For each of those configurations T_1 occurs twice and so we have to solve a secular equation of degree 2. We already know the diagonal elements of the corresponding secular matrix and the considerations of subsection 6 show that, apart from sign, the off-diagonal element is the same for all four configurations. We therefore work it out for the simplest one, namely d^2.

The states of ^3F are given in (8.33) and those of ^3P are obtained by orthogonality. They are

$$\left.\begin{aligned} |11\rangle &= \frac{\sqrt{2}}{\sqrt{5}}|2^+-1^+\rangle - \frac{\sqrt{3}}{\sqrt{5}}|1^+0^+\rangle, \\ |10\rangle &= \frac{2}{\sqrt{5}}|2^+-2^+\rangle - \frac{1}{\sqrt{5}}|1^+-1^+\rangle, \\ |1-1\rangle &= \frac{\sqrt{2}}{\sqrt{5}}|1^+-2^+\rangle - \frac{\sqrt{3}}{\sqrt{5}}|0^+-1^+\rangle. \end{aligned}\right\} \tag{8.34}$$

The relevant matrix element is

$$\begin{aligned} \langle 30\,|\,\mathscr{V}\,|\,10\rangle &= \tfrac{2}{5}E(|2^+-2^+\rangle) - \tfrac{2}{5}E(|1^+-1^+\rangle) \\ &= \tfrac{2}{5}\Delta, \end{aligned}$$

with the resulting secular equation

$$f(E,\Delta) \equiv \begin{vmatrix} -\frac{3}{5}\Delta - E & \frac{2}{5}\Delta \\ \frac{2}{5}\Delta & 15B - E \end{vmatrix} = 0. \tag{8.35}$$

The same equation holds for d^7, while for d^3 and d^8 we have $f(E, -\Delta) = 0$.

(8) Tetrahedral symmetry. As we saw in § 8.3, from a purely formal point of view, the theory of the influence of an environment possessing tetrahedral symmetry is identical with the theory for an octahedral environment. The parameter Δ entering the theory has now the opposite sign and so the most obvious correlation is between d^ϵ in an octahedral and $d^{10-\epsilon}$ in a tetrahedral environment. For the model discussed in § 8.3, the absolute magnitude of Δ for four neighbours arranged tetrahedrally is four-ninths of the value for six of the same neighbours arranged octahedrally.

Once we include interactions with configurations outside d^ϵ, however, this simple relationship vanishes; for a tetrahedral environment has matrix elements between d^ϵ and $d^{\epsilon-1}p$, while an octahedral one does not.

Example

Establish the selection rule $\Delta M_L = 0, \pm 4$ for matrix elements of an octahedral environment within a d^ϵ configuration.

8.9. Spin-orbit coupling

We shall not discuss the spin-orbit coupling in any detail at this stage. When it is included it will break down our classification into states of definite spin and, just as for the free ion only **J** and not **L** or **S** could be used to classify the exact eigenfunctions, so here it is only the representation of the spinor group for the whole system which can be used rigorously. These latter representations are obtained by taking the direct product of the space representation with the appropriate spin representation. Thus for d^1 we have

$$^2E \rightarrow U',$$

$$^2T_2 \rightarrow E'' + U'$$

from Table A 9. Then the two representations U' have non-zero matrix elements of spin-orbit coupling between them. The same is true for d^9 and will become important when we discuss the magnetic properties of the cupric ion.

Examples

1. Show that there are five excited states with which the ground state, $^3A_2$, of the Ni^{++} ion could mix by spin-orbit coupling under the octahedral group.

2. The spin-orbit coupling has the free-ion form $\sum_\kappa \xi(r_\kappa)\,\mathbf{l}_\kappa \cdot \mathbf{s}_\kappa$ with $\xi(r_\kappa)$ a function only of the scalar distance r_κ from the nucleus. Calculate its matrix elements as a function of ζ within $d^2\,{}^3F$ and $d^3\,{}^4F$. (The answers are in Table A 32. This example is good practice at this stage but it will be profitable to re-examine it after reading § 9.6.)

8.10. The method of operator equivalents

In § 2.8 we proved that the matrix elements of any vector of type T with respect to an angular momentum **J** are proportional, within a given level, to those of **J** itself. This was the corollary to the replacement theorem and we have already found it very useful in discussing spin-orbit coupling energies and the magnetic interaction with the nuclear magnetic moment. We now consider how to generalize this idea.

A vector **T** and the vector **J** had their matrix elements proportional because their components each formed a set of operator states $|JM\rangle$ with $J = 1$. We now remove the restriction that $J = 1$ and consider the example $J = 2$ first. Suppose we want to find the matrix of a set of quantities $V_m = \sum_{i=1}^{n} f(r_i)\,Y_{2m}(i)$, where the sum is over the electrons in the system, within a set of states having a given J. We may couple the V_m to the kets $|JM\rangle$ of the set and then use Wigner's formula. Just as we saw in § 2.8 for $J = 1$, this shows that the matrix elements are determined apart from one arbitrary constant which depends on the detailed nature of the V_m and the $|JM\rangle$. Therefore we may replace V_m by any set of operators which are eigenkets of $\mathbf{J}^2$ and J_z with the values $J = 2$ and $J_z = m\hbar$. The obvious one to use is the set obtained by vector coupling **J** with itself. Thus if we write J_m for the components of **J**, quantized with respect to J_z (2.75), then the appropriate set of operators is

$$W_m = \sum_{m_1, m_2} \langle 11 m_1 m_2 \,|\, 112m\rangle J_{m_1} J_{m_2}.$$

It follows from ex. 1, p. 24, that $\langle 11 m_1 m_2 \,|\, 112m\rangle = \langle 11 m_2 m_1 \,|\, 112m\rangle$ and therefore the coefficients of $J_{m_1} J_{m_2}$ and $J_{m_2} J_{m_1}$ are equal. W_m is a symmetrical quadratic form in the components of **J**. This remains true when we replace the J_{m_i} by linear combinations (the same for both vectors **J**). So if we have to determine matrix elements of the set of quantities $2z^2-x^2-y^2$, x^2-y^2, xy, yz, zx we first express them as symmetrical quadratic forms, for example, $xy = \frac{1}{2}(xy+yx)$, and then replace each of x, y, z by the corresponding component of **J**. Then we have

$$\left.\begin{aligned} 2z^2-x^2-y^2 &= \alpha(2J_z^2-J_x^2-J_y^2), \quad & yz &= \tfrac{1}{2}\alpha(J_y J_z+J_z J_y),\\ x^2-y^2 &= \alpha(J_x^2-J_y^2), & zx &= \tfrac{1}{2}\alpha(J_z J_x+J_x J_z),\\ xy &= \tfrac{1}{2}\alpha(J_x J_y+J_y J_x). \end{aligned}\right\} \tag{8.36}$$

α must be determined by actually evaluating one of the matrix elements. This method is called the method of operator equivalents.[1] It cannot be used between states differing in J because then all products of components of **J** have zero matrix elements.

We now pass to general J for the V and write x_i for the three components of **r**. We take a homogeneous polynomial V of degree n in these components and suppose it has the eigenvalue n for L. In (8.36) we had $n = 2$ and $L = 2$. What is the correct operator equivalent? Just as before, each component x_i is replaced by the corresponding J_i but this time we must completely symmetrize V first. So V becomes $V_s = (1/n!)\sum_{\mu} P_\mu V$, where μ runs over all $n!$ permutations of the positions of the x_i in V. Of course $V_s \equiv V$. Now replace x_i by J_i in V_s and the result is the appropriate operator equivalent, W_s say. The simplest way to see that this really is the correct procedure is to generalize (2.54) slightly. If $x_{i_1} x_{i_2} \dots x_{i_n}$ is a term of V then define J_ϵ^j by the rule

$$J_\epsilon^j x_{i_1} \dots x_{i_n} = x_{i_1} \dots x_{i_{j-1}} (J_\epsilon^j x_{i_j}) x_{i_{j+1}} \dots x_{i_n},$$

where J_ϵ^j only acts on x_{i_j}. Define also J_ϵ^{n+1} to act on none of the x_i but only on the kets $|J'M'\rangle$ between which we are determining the matrix elements of V. Then

$$J_\epsilon = \sum_{j=1}^{n+1} J_\epsilon^j$$

is a symmetrical operator and commutes with all the P_μ. As a consequence $(J_\epsilon V)_s = J_\epsilon V_s$ which shows that if we take a set of $(2L+1)$ V, V^a say, forming a complete orthonormal set for the value L then the matrix of **J** operating on V^a is the same as that for V_s^a and hence for W_s^a, which justifies the procedure.

Let us now consider a particularly interesting example. It is evident that $x^4+y^4+z^4$ has cubic symmetry so we consider a potential of the form

$$\sum_{i=1}^{n} f(r_i)\,(x_i^4+y_i^4+z_i^4),$$

where i now numbers the electrons. Amongst the spherical harmonics for $L \leqslant 4$ the representation A_1 only occurs twice. One of the times is the trivial case of Z_{00} and the other is $(\sqrt{7}/2\sqrt{3})\,Z_{40}+(\sqrt{5}/2\sqrt{3})\,Z_{44}^c$. When we write out this latter function it is a multiple of $r^{-4}(x^4+y^4+z^4-\frac{3}{5}r^4)$. So let us take

$$\mathscr{V} = \sum_{i=1}^{n} f(r_i)\,(x_i^4+y_i^4+z_i^4-\tfrac{3}{5}r_i^4)$$

[1] See Abragam & Pryce (1951a); Stevens (1952); Bleaney & Stevens (1953).

which will leave the centre of gravity of all terms unchanged. The operator equivalent to $\mathscr{V}$ within a set of states with given L (we neglect spin-orbit coupling here so replace **J** by **L**) is then, after some algebraic manipulation,[1]

$$W = \alpha\{14L_z^4 - 12L(L+1)L_z^2 + 10L_z^2 + \tfrac{6}{5}L(L^2-1)(L+2) + (L^+)^4 + (L^-)^4\}, \quad (8.37)$$

where **L** is measured in units of $\hbar$. If we want W to represent a crystal field with separation Δ between e and t_2 for a single d electron we must choose α appropriately. Consider the state $|20\rangle \in e$. From Table 8.3

$$\tfrac{3}{5}\Delta = \langle 20|W|20\rangle = \tfrac{144}{5}\alpha \quad (\alpha = \tfrac{1}{48}\Delta).$$

Similarly for ^3F of d^2, using § 8.8,

$$-\tfrac{3}{5}\Delta = \langle 30|W|30\rangle = 144\alpha \quad (\alpha = -\tfrac{1}{240}\Delta).$$

This gives W for the ^2D of d^1 as

$$W(d^1; {}^2D) = \tfrac{1}{48}\Delta\{14L_z^4 - 62L_z^2 + \tfrac{144}{5} + (L^+)^4 + (L^-)^4\} \quad (8.38)$$

and

$$W(d^2; {}^3F) = -\tfrac{1}{240}\Delta\{14L_z^4 - 134L_z^2 + 144 + (L^+)^4 + (L^-)^4\}. \quad (8.39)$$

Of course $W(d^2; {}^3P) = 0$. Also $W(d^{5+\epsilon}) = -W(d^{5-\epsilon}) = W(d^\epsilon)$.

Examples

1. Show that the operator equivalent to r^4 is $\mathbf{J}^2(\mathbf{J}^2 - \frac{1}{3})$. Hence, or otherwise, prove (8.37).

2. Work out the non-vanishing matrix elements of W within ^3F of d^2, using (8.39), and derive the energies of $^3A_2$ and $^3T_2$.

[1] There is a misprint in Bleaney & Stevens's expression for the cubic potential discussed here $\left(\text{their Table 1: } \dfrac{\beta\overline{r^4}}{5} \text{ should read } \dfrac{\beta\overline{r^4}}{8}\right)$.

CHAPTER 9

THE STRONG-FIELD COUPLING SCHEME[1]

9.1. Strong-field configurations and terms

In the strong-field coupling scheme we diagonalize the high-symmetry component of the ligand field first, then the electrostatic repulsion between the electrons and finally the spin-orbit coupling and the low-symmetry components of the ligand field. The most important example occurs when the high-symmetry component represents an octahedral field and we consider this case exclusively in this chapter.

An octahedral crystal field separates the d orbitals into two sets, e and t_2, which we take to have energies $+\frac{3}{5}\Delta$ and $-\frac{2}{5}\Delta$, respectively. In real form they are

$$\left.\begin{aligned} e\theta &= |20\rangle \sim \tfrac{1}{2}(2z^2-x^2-y^2), \\ e\epsilon &= \frac{1}{\sqrt{2}}(|22\rangle+|2-2\rangle) \sim \frac{\sqrt{3}}{2}(x^2-y^2), \\ t_2\xi &= \frac{i}{\sqrt{2}}(|21\rangle+|2-1\rangle) \sim \sqrt{3}\,yz, \\ t_2\eta &= -\frac{1}{\sqrt{2}}(|21\rangle-|2-1\rangle) \sim \sqrt{3}\,zx, \\ t_2\zeta &= \frac{1}{i\sqrt{2}}(|22\rangle-|2-2\rangle) \sim \sqrt{3}\,xy. \end{aligned}\right\} \quad (9.1)$$

If we take as basic states determinantal functions having m electrons in t_2 and n electrons in e then the crystal field will be diagonal within these basic states. For fixed m, n we call the entire set of such states a strong-field configuration (or often just a configuration) $t_2^m e^n$. Next we diagonalize the electrostatic energy within each strong-field configuration but neglect the matrix elements between them. Each configuration then breaks up into subsets which may be written ${}^{2S+1}\Gamma$, where S is the spin and Γ is a representation of O. In general Γ is irreducible and when it is we call ${}^{2S+1}\Gamma$ a strong-field term (or often just a term). Finally, when we include the spin-orbit coupling each term breaks up into a number of strong-field levels which are classified according to the representations of O^*. We use O, O^* rather than O_h, O_h^* because all functions are g under O_h, O_h^* so long as we stay within a d^ϵ atomic configuration.

Just as in the theory of atomic structure, our first task is to determine what terms belong to each configuration. A general configuration $t_2^m e^n$ has two con-

[1] I have learnt a great deal from Stevens (1953), Tanabe & Sugano (1954) and Racah (1942*a*,*b*, 1943, 1949). The paper by Tanabe & Kamimura (1958) (and that by Koster, 1958, which is related to § 8.7 especially) came too late to influence the chapter and contains a development parallel, but not identical, with § 9.6.

stituent partly-filled shells and the terms are obtained by coupling the allowed terms of t_2^m with those of e^n in all possible ways. This is easy because the coupling of spins is by the vector-coupling procedure and the coupling of the space representations is given in Table A 9. The only problem left is to find the allowed terms of t_2^m and e^n.

Let us take e^n first. When $n = 1$ we just have ^2E. For $n = 2$ there is one triplet state and three singlet ones. The triplet state with $M_S = 1$ is

$$\psi = |\theta^+\epsilon^+\rangle,$$

where we write $|a_1 \dots a_n\rangle$ to denote the determinantal function based on the one-electron functions according to (2.64). As there is only one such ψ it must be a basis for A_1 or A_2. From Table A 16 we see that $C_4^z\psi = -\psi$ so ψ belongs to A_2 and the triplet of e^2 is $^3A_2$. In deriving our results we are supposing rotations to act only on space functions. Thus $C_4^z\alpha = \alpha$ for a spin function α. We assume this until we introduce the spin-orbit coupling energies in § 9.6.

The singlet states are

$$\psi_1 = |\theta^+\theta^-\rangle, \quad \psi_2 = |\epsilon^+\epsilon^-\rangle,$$

$$\psi_3 = \frac{1}{\sqrt{2}}(|\theta^+\epsilon^-\rangle - |\theta^-\epsilon^+\rangle),$$

and satisfy

$$\left.\begin{aligned} C_4^z\psi_1 &= \psi_1, & C_4^x\psi_1 &= \tfrac{1}{4}\psi_1 + \tfrac{3}{4}\psi_2 + \frac{\sqrt{3}}{2\sqrt{2}}\psi_3, \\ C_4^z\psi_2 &= \psi_2, & C_4^x\psi_2 &= \tfrac{3}{4}\psi_1 + \tfrac{1}{4}\psi_2 - \frac{\sqrt{3}}{2\sqrt{2}}\psi_3, \\ C_4^z\psi_3 &= -\psi_3, & C_4^x\psi_3 &= \frac{\sqrt{3}}{2\sqrt{2}}\psi_1 - \frac{\sqrt{3}}{2\sqrt{2}}\psi_2 + \tfrac{1}{2}\psi_3. \end{aligned}\right\} \tag{9.2}$$

It follows at once from (9.2) that $(1/\sqrt{2})(\psi_1 + \psi_2)$ is left invariant by C_4^z and C_4^x and hence by all elements of O. It therefore forms a $^1A_1$ term. The functions $\psi' = (1/\sqrt{2})(\psi_2 - \psi_1)$ and ψ_3 then satisfy

$$\left.\begin{aligned} C_4^z\psi' &= \psi', & C_4^x\psi' &= -\tfrac{1}{2}\psi' - \tfrac{1}{2}\sqrt{3}\,\psi_3, \\ C_4^z\psi_3 &= -\psi_3, & C_4^x\psi_3 &= -\tfrac{1}{2}\sqrt{3}\,\psi' + \tfrac{1}{2}\psi_3, \end{aligned}\right\} \tag{9.3}$$

which, on comparison with Table A 16 shows that ψ' and ψ_3 form the θ- and ϵ-components, respectively, of an E representation and are correctly connected in phase. Lastly, e^3 gives one ^2E term and the closed shell e^4 is $^1A_1$.

We have seen that e^2 comprises $^3A_2$, $^1A_1$ and ^1E. Two inequivalent e-electrons would give all the terms of $^2E \times {}^2E$, i.e. $^3A_2$, $^1A_2$, $^3A_1$, $^1A_1$, ^3E and ^1E. The Pauli exclusion principle has eliminated half these terms. If we write each of the six states of $^3A_2$, $^1A_1$, ^1E as a product of a space function and a spin function α or β, we find the spin factor is symmetric and the space factor antisymmetric for the $^3A_2$ functions and conversely for the singlets. Thus

$$|^3A_2\,1a_2\rangle = (1/\sqrt{2})\,[\theta(1)\,\epsilon(2) - \epsilon(1)\,\theta(2)]\,\alpha(1)\,\alpha(2),$$

$$|^1E\epsilon\rangle = \tfrac{1}{2}[\theta(1)\,\epsilon(2) + \epsilon(1)\,\theta(2)]\,[\alpha(1)\,\beta(2) - \beta(1)\,\alpha(2)]$$

which reminds us of the discussion in § 2.6. It also shows us that we could have used the coupling coefficients of Table A 20 to derive the allowed terms and eigenfunctions of e^2. For we determine the spin of the allowed terms by observing the behaviour of the spatial functions under interchange of the two electrons.

The allowed terms of t_2^2 follow from Table A 20 in the manner just indicated and are ${}^1A_1$, ${}^1E$, ${}^3T_1$ and ${}^1T_2$. The allowed terms of t_2^4 are the same and those of t_2^3 are derived without great difficulty. There is one quartet which is ${}^4A_2$. The nature of the doublets can be found from the character of the representation they span, but to obtain the actual functions we must follow through the same procedure as we used earlier to derive the functions for e^2. Table A 24 gives the functions of t_2^m and e^n. An abbreviation such as ξ^2 stands for $\xi^+\xi^-$. The functions are all correctly connected in phase and the phases of t_2^2 and e^2 are consistent with those of Table A 20 together with the convention implicit in Wigner's formula for the addition of the spins.

Armed with the list of allowed terms for t_2^m and e^n we deduce the allowed terms for a general $t_2^m e^n$ configuration. They are given in Table A 25. Table A 24 together with the rules for coupling representations gives us a definition of all states of $t_2^m e^n$ which includes a choice of phase. We write a general function determined in this way as $|t_2^m(S'\Gamma')\,e^n(S''\Gamma'')\,S\Gamma M_S M_\Gamma\rangle$, where Γ', Γ'' are coupled to form Γ and S', S'' to form S and antisymmetrization between the electrons of t_2^m and e^n is implied. It has already been proved (2.73 and 6.30) that the matrix of the Hamiltonian $\mathscr{H}$ (without spin-orbit coupling) is diagonal in S, Γ, M_S and M_Γ and is independent of M_S and M_Γ for a given term. So we may speak of the Hamiltonian as having a matrix element between the terms

$$|t_2^m(S'\Gamma')\,e^n(S''\Gamma'')\,S\Gamma\rangle \quad \text{and} \quad |t_2^{m'}(S_1\Gamma_1)\,e^{n'}(S_2\Gamma_2)\,S\Gamma\rangle.$$

This matrix element is equal to

$$\langle t_2^m(S'\Gamma')\,e^n(S''\Gamma'')\,S\Gamma M_S M_\Gamma\,|\,\mathscr{H}\,|\,t_2^{m'}(S_1\Gamma_1)\,e^{n'}(S_2\Gamma_2)\,S\Gamma M_S M_\Gamma\rangle,$$

where, of course, $m+n = m'+n'$.

As a simple example of this let us consider the three ${}^4T_2\,\tfrac{3}{2}\zeta$ states of d^5. These are derived from $t_2^4({}^3T_1)\times e^1({}^2E)$, $t_2^3({}^2T_1)\times e^2({}^3A_2)$ and $t_2^2({}^3T_1)\times e^3({}^2E)$. Using Table A 20, they are

$$\begin{aligned}
\psi_1 &= |t_2^4({}^3T_1)\,e^1({}^2E)\,{}^4T_2\,\tfrac{3}{2}\zeta\rangle \\
&= |t_2^4({}^3T_1)\,1z, e^1({}^2E)\,\tfrac{1}{2}\epsilon\rangle = |\xi^+\eta^+\zeta^2\epsilon^+\rangle, \\
\psi_2 &= |t_2^3({}^2T_1)\,e^2({}^3A_2)\,{}^4T_2\,\tfrac{3}{2}\zeta\rangle = |t_2^3({}^2T_1)\,\tfrac{1}{2}z, e^2({}^3A_2)\,1a_2\rangle \\
&= \frac{1}{\sqrt{2}}\,|\xi^2\zeta^+\theta^+\epsilon^+\rangle - \frac{1}{\sqrt{2}}\,|\eta^2\zeta^+\theta^+\epsilon^+\rangle, \\
\psi_3 &= |t_2^2({}^3T_1)\,e^3({}^2E)\,{}^4T_2\,\tfrac{3}{2}\zeta\rangle \\
&= |t_2^2({}^3T_1)\,1z, e^3({}^2E)\,\tfrac{1}{2}\epsilon\rangle = |\eta^+\xi^+\theta^2\epsilon^+\rangle.
\end{aligned} \tag{9.4}$$

In (9.4), all pairs of electrons are antisymmetrized, so an expression such as $|t_2^4({}^3T_1)\,1z, e^1({}^2E)\,\tfrac{1}{2}\epsilon\rangle$ is not just a simple product. When there is only one possible term which can arise from t_2^m or from e^n we shall often omit the term symbol. Thus ψ_1 could be written $|t_2^4({}^3T_1)\,e\,{}^4T_2\,\tfrac{3}{2}\zeta\rangle$ for short.

Finally, let us note that the whole discussion of this section has been based upon the behaviour of the t_2 and e orbitals under the group O. Nowhere have we used (9.1) in terms of d orbitals. So the conclusions are applicable not only to d orbitals but also to any which transform according to the representations T_2 and E of the octahedral group.

9.2. Electrostatic matrix elements

The evaluation of matrix elements of the electrostatic energy between two determinantal functions belonging to strong-field configurations requires a knowledge of the two-electron integrals

$$\langle ac \mid V \mid bd \rangle = \int \bar{a}(1)\, b(1) \frac{e^2}{r_{12}} \bar{c}(2)\, d(2)\, d\tau_1 d\tau_2. \tag{9.5}$$

Our orbitals a, b, c, d are no longer, in general, eigenstates of l_z and so we cannot use the $c^k(lp, l'p')$ to obtain a general expression for (9.5). In spite of this, however, we can easily determine all the integrals.

As we have seen in ch. 7, it is unlikely that the t_2 and e orbitals approximate very closely to d orbitals in compounds. With this in mind we ask first how much information about the integrals can be obtained by assuming only that the basic orbitals form bases for t_2 and e and not that they are also d orbitals. It is always possible (see p. 198) to take these bases real and we do so. Then it is convenient to introduce a slightly different notation and write

$$(ab; cd) = \langle ac \mid V \mid bd \rangle. \tag{9.6}$$

a and b refer to the first electron and c and d to the second. It is immediate that

$$(ab; cd) = (ba; cd) = (ab; dc) = (cd; ab)$$

and

$$[(\rho_1 a_1 + \rho_2 a_2)(\nu_1 b_1 + \nu_2 b_2); cd]$$
$$= \rho_1 \nu_1 (a_1 b_1; cd) + \rho_2 \nu_1 (a_2 b_1; cd) + \rho_1 \nu_2 (a_1 b_2; cd) + \rho_2 \nu_2 (a_2 b_2; cd), \tag{9.7}$$

etc., where ρ_i, ν_i are real numbers. We can choose a non-ordered pair ab in 15 distinct ways and so there are 120 non-trivially different integrals $(ab; cd)$. However, a large proportion are zero as we now see.

The one-electron orbitals ξ, η, ζ, θ, ϵ may be classified by their behaviour under the group D_4 about the Z-axis (Tables A 16 and A 17) and they behave as ex, $-ey$, b_2, a_1, b_1, respectively. So the 15 different products ab for the first electron behave as follows:

$$\left.\begin{aligned} \frac{1}{\sqrt{2}}(\xi^2+\eta^2), \zeta^2, \theta^2, \epsilon^2 &\in A_1, \\ \zeta\epsilon &\in A_2, \\ \frac{1}{\sqrt{2}}(\xi^2-\eta^2), \theta\epsilon &\in B_1, \\ \xi\eta, \zeta\theta &\in B_2, \\ \xi\theta, \eta\zeta, \xi\epsilon &\in Ex, \\ \eta\theta, \zeta\xi, \eta\epsilon &\in Ey, \end{aligned}\right\} \tag{9.8}$$

where $\epsilon\, Ex$ means that the function is $\pm$ the x-component of a basis for the representation E. The products cd for the second electron are classified in the same way and we deduce from our general theory (§ 6.5) that $(ab; cd)$ is zero whenever ab and cd belong to a different set in (9.8). Because ξ^2 and η^2 belong to two of the sets, this means that we have five non-interacting sets. $(\zeta\epsilon; cd)$ is always zero unless $cd = \zeta\epsilon$.

Next we ask how many independent parameters are needed to specify the non-vanishing $(ab; cd)$ in the most general case. The ab fall into three sets, according as they belong to the configuration e^2, $t_2 e$ or t_2^2. Clearly no element of O can turn a function of one configuration into a function of another. So there are six independent types of $(ab; cd)$ which may be written $e^2.e^2$; $t_2 e.e^2$; $t_2^2.e^2$; $t_2 e.t_2 e$; $t_2^2.t_2 e$ and $t_2^2.t_2^2$. It is a consequence of our findings of the last paragraph that all integrals of type $t_2 e.e^2$ are zero. The number of independent parameters required, for example, for the integrals of type $t_2^2.e^2$ is equal to the number of times the unit representation A_1 occurs in the direct product of the symmetrized squares $[E^2]$ and $[T_2^2]$. Now $[E^2] = A_1 + E$ and $[T_2^2] = A_1 + E + T_2$ so there are two independent parameters. Similar arguments give the numbers 2, 2, 1, 3 for $e^2.e^2$, $t_2 e.t_2 e$, $t_2^2.t_2 e$, $t_2^2.t_2^2$, respectively.[1]

The last step is simply to apply the generators of O to the $(ab; cd)$ to determine relations between them. Let us take $t_2^2.e^2$ as an example. The classification (9.8) implies that

$$0 = \left(\frac{1}{\sqrt{2}}(\xi^2+\eta^2), \theta\epsilon\right) = \frac{1}{\sqrt{2}}(\xi^2, \theta\epsilon) + \frac{1}{\sqrt{2}}(\eta^2, \theta\epsilon),$$

so

$$(\xi^2, \theta\epsilon) = -(\eta^2, \theta\epsilon) = c, \quad \text{say.}$$

Similarly

$$(\xi^2, \theta^2) = (\eta^2, \theta^2),$$

$$(\xi^2, \epsilon^2) = (\eta^2, \epsilon^2).$$

Then

$$C_4^x(\xi^2, \theta^2) = \tfrac{1}{4}(\xi^2, \theta^2) + \tfrac{3}{4}(\xi^2, \epsilon^2) + \tfrac{1}{2}\sqrt{3}\,(\xi^2, \theta\epsilon).$$

But

$$C_4^x(\xi^2, \theta^2) = (\xi^2, \theta^2)$$

because the latter is just a number, hence

$$(\xi^2, \theta^2) = (\xi^2, \epsilon^2) + \frac{2}{\sqrt{3}}(\xi^2, \theta\epsilon) = d + \frac{2}{\sqrt{3}}c$$

if we put $d = (\xi^2, \epsilon^2)$. c and d are to be the two independent parameters for $t_2^2.e^2$ and we are expressing the other integrals in terms of them. The two remaining non-zero ones are

$$(\zeta^2, \theta^2) = C_4^x(\zeta^2, \theta^2) = \tfrac{1}{4}(\eta^2, \theta^2) + \tfrac{3}{4}(\eta^2, \epsilon^2) + \frac{\sqrt{3}}{2}(\eta^2, \theta\epsilon)$$

$$= d - \frac{1}{\sqrt{3}}c,$$

and

$$(\zeta^2, \epsilon^2) = C_4^x(\zeta^2, \epsilon^2) = \tfrac{3}{4}(\eta^2, \theta^2) + \tfrac{1}{4}(\eta^2, \epsilon^2) - \frac{\sqrt{3}}{2}(\eta^2, \theta\epsilon)$$

$$= d + \sqrt{3}\,c.$$

[1] Precisely the same argument works here for $e^2.e^2$, $t_2 e.t_2 e$, $t_2^2.t_2^2$ but only because there is never a repeated representation in any of e^2, $t_2 e$, t_2^2. Therefore the relationships $(ab; cd) = (cd; ab)$ do not cut down the necessary number of parameters.

The other types of integral are treated in an exactly similar manner and we have finally all the non-vanishing $(ab; cd)$ expressed in terms of 10 independent parameters. The results are in Table A 26, where the 10 parameters are $a, b, \ldots, j$. This is as far as we can go using arguments within the octahedral group O.

When the orbitals are also d orbitals, the integrals may all be expressed in terms of the three Racah parameters A, B and C. As examples we determine c, d and i of Table A 26 in terms of A, B and C:

$$\begin{aligned}(\xi^2; \theta^2) &= \langle \xi\theta \,|\, V \,|\, \xi\theta\rangle \\ &= \tfrac{1}{2}\{\langle 10 \,|\, V \,|\, 10\rangle + \langle -1\,0 \,|\, V \,|\, -1\,0\rangle\} \\ &= A+2B+C = d+\frac{2}{\sqrt{3}}c.\end{aligned}$$

We have used Table 4.5 and the fact that $\langle m_l m_l' \,|\, V \,|\, m_l'' m_l'''\rangle = 0$ unless $m_l + m_l' = m_l'' + m_l'''$. Next,

$$\begin{aligned}(\xi^2; \theta\epsilon) &= \langle \xi\theta \,|\, V \,|\, \xi\epsilon\rangle \\ &= \frac{1}{2\sqrt{2}}\{\langle 10 \,|\, V \,|\, -1\,2\rangle + \langle -1\,0 \,|\, V \,|\, 1\,-2\rangle\} \\ &= 2\sqrt{3}\,B = c.\end{aligned}$$

So $d = A-2B+C$. It is unnecessary to evaluate each of the ten parameters separately. In particular

$$\begin{aligned}2i = (\zeta\theta; \xi\eta) &= C_8^z(\zeta\theta; \xi\eta) = \tfrac{1}{2}(\theta\epsilon; \xi\xi) - \tfrac{1}{2}(\theta\epsilon; \eta\eta) \\ &= -(\theta\epsilon; \eta\eta) = c = 2\sqrt{3}\,B.\end{aligned}$$

All the non-vanishing $(ab; cd)$ are given in Table A 26 in terms of the $a, b, \ldots, j$ and (for d functions) of A, B and C. The coulomb integrals are $J(a, b) = (aa; bb)$ and the exchange integrals $K(a, b) = (ab; ab)$.

We conclude this section by remarking that the matrix of the electrostatic energy of t_2^m is altered only by an additive constant common to all states when the closed shell e^4 is adjoined to the configuration. The same is true for the addition of t_2^6 to e^n. These propositions are special cases of the results established in § 7.2 and enable us to write general formulae for the energies of $t_2^m e^4$ and $t_2^6 e^n$ in terms of t_2^m and e^n, respectively. For d orbitals these formulae become

$$\left.\begin{aligned}E(t_2^m e^4 S\Gamma) &= E(t_2^m S\Gamma) + (4m+6)A + (2m+8)(-2B+C), \\ E(t_2^6 e^n S\Gamma) &= E(e^n S\Gamma) + (6n+15)A + (3n+15)(-2B+C).\end{aligned}\right\} \qquad (9.9)$$

Example

Establish (9.9) directly using Table A 26.

9.3. Examples

We now possess tne basic information necessary to write down the matrix elements of the electrostatic energy between any pair of strong-field terms. For each function of a strong-field term is expressible as a sum of determinantal functions and the matrix element of electrostatic energy between a pair of determinantal functions is a sum of the two-electron integrals discussed in the last section. The calculation of these matrix elements is illustrated in this section by a series of useful examples.

9.3.1. The e^2 configuration. For the configuration e^2 the energies are obtained by taking suitable components of each term from Table A 24. For example

$$|e^2\,{}^3A_2\,1\,a_2\rangle = |\theta^+\epsilon^+\rangle,$$

whence

$$\left.\begin{aligned} E({}^3A_2) &= (\theta^2;\,\epsilon^2)-(\theta\epsilon;\,\theta\epsilon) = e-3f. \\ \text{Similarly}\quad E({}^1A_1) &= \tfrac{1}{2}\{(\theta^2;\,\theta^2)+2(\theta\epsilon;\,\theta\epsilon)+(\epsilon^2;\,\epsilon^2)\} \\ &= e+f, \\ E({}^1E) &= \tfrac{1}{2}\{(\theta^2;\,\theta^2)-2(\theta\epsilon;\,\theta\epsilon)+(\epsilon^2;\,\epsilon^2)\} \\ &= e-f. \end{aligned}\right\} \tag{9.10}$$

In deriving (9.10) we took the θ-component of the ${}^1E$ term but, as we know ((2.73) and (6.30)), the energy is the same for all components of a term. The energy of the ϵ-component is easily seen to be $e-f$ also in accordance with this general theorem. Equations (9.10) show that we predict an interval rule

$$E({}^1A_1)-E({}^1E) = E({}^1E)-E({}^3A_2) \text{ for } e^2.$$

9.3.2. The t_2^m configurations. The energies of the terms of the t_2^m configurations are written down in just the same way as we just saw for e^2, using suitable components from Table A 24. For example, in t_2^3 we have

$$\left.\begin{aligned} E({}^4A_2) &= 3(\xi^2;\,\eta^2)-3(\xi\eta;\,\xi\eta) = 3b-3j, \\ E({}^2E) &= E(\xi^+\eta^-\zeta^+)+(\xi\eta;\,\xi\eta) \\ &= 3(\xi^2;\,\eta^2) = 3b, \\ E({}^2T_1) &= E(\eta^2\xi^+)-(\eta\zeta;\,\eta\zeta) \\ &= (\eta^2;\,\eta^2)+2(\xi^2;\,\eta^2)-2(\eta\zeta;\,\eta\zeta) = a+2b-2j, \\ E({}^2T_2) &= E(\eta^2\xi^+)+(\eta\zeta;\,\eta\zeta) = a+2b. \end{aligned}\right\} \tag{9.11}$$

In the derivations (9.11) I have often replaced an integral such as $(\xi^2;\,\zeta^2)$ by an equal one $(\xi^2;\,\eta^2)$.

All the energies of t_2^m and e^n are given in Table 9.1 both in terms of $a, b, \ldots, j$ and (for d-functions) in terms of A, B and C. It is an interesting fact that for d-functions ${}^1E$ and ${}^1T_2$ of t_2^2 (and of t_2^4) have the same energy and so also do ${}^2E$ and ${}^2T_1$ of t_2^3. This degeneracy is accidental and would be removed in a more accurate treatment which allowed for mixing in of excited configurations involving other atomic orbitals. It is also accidental in the sense that it does not occur for an arbitrary t_2^2 configuration.

We now discuss various further examples and assume that the functions involved are d-functions unless it is stated otherwise.

9.3.3. Ground terms of the ions. If the octahedral field is very large then the ground terms of the ions belong to the configurations t_2^ϵ for $\epsilon \leqslant 6$ and to

Table 9.1. *Electrostatic energies of t_2^m and e^n strong field terms*

$$E(t_2^{6-m}S\Gamma) = E(t_2^m S\Gamma) + (3-m)(a+4b-2j)$$

Configuration	Term	General energy	Energy for d-functions
e^2	$^3A_2$ ^1E $^1A_1$	$e-3f$ $e-f$ $e+f$	$A-8B$ $A+2C$ $A+8B+4C$
e^3	^2E	$3e-5f$	$3A-8B+4C$
e^4	$^1A_1$	$6e-10f$	$6A-16B+8C$
t_2^2	$^3T_1$ $^1A_1$ ^1E $^1T_2$	$b-j$ $a+2j$ $a-j$ $b+j$	$A-5B$ $A+10B+5C$ $A+B+2C$ $A+B+2C$
t_2^3	$^4A_2$ ^2E $^2T_1$ $^2T_2$	$3b-3j$ $3b$ $a+2b-2j$ $a+2b$	$3A-15B$ $3A-6B+3C$ $3A-6B+3C$ $3A+5C$

$t_2^6 e^{\epsilon-6}$ for $\epsilon \geqslant 6$. Therefore Tables A 24 and 9.1 give us the functions and energies of the ground terms for $\epsilon \leqslant 6$. For $\epsilon = 1$ and 3 the ground term is exactly the same as the ground term in the weak-field scheme. For $\epsilon = 2$ both ground terms are $^4T_1$ but the functions and energies are slightly different. This is discussed further in § 9.3.5. For $\epsilon = 4$, 5 and 6, however, the terms are completely different being, respectively, ^5E, ^6S, $^5T_2$ for the weak field and $^3T_1$, $^2T_2$, $^1A_1$ for the strong field. Thus there are at least two possible ground terms, depending on the strength of the field. We shall call a compound in which the metal ion has the weak-field type of ground term a high-spin compound and one with the strong-field type of ground term a low-spin compound. We pursue this matter in § 9.4.

For $\epsilon > 6$ the low-spin ground configurations are $t_2^6 e^{\epsilon-6}$ and so the ground terms are, from Table 9.1 because the t_2^6 closed shell gives only a constant additive energy within a configuration, ^2E, $^3A_2$, ^2E, $^1A_1$ for $\epsilon = 7, 8, 9, 10$, respectively. The functions are derived from Table A 24, for example for d^7

$$|^2E\theta^+\rangle = |\xi^2\eta^2\zeta^2\theta^+\rangle.$$

The energies follow immediately from (9.9) together with Table 9.1. They are all given in Table 9.2 in the column headed low-spin. For $\epsilon = 8, 9, 10$ the functions and electrostatic energies are the same for weak- or strong-field.

9.3.4. The complete matrix for d^2. As well as the terms of t_2^2 and e^2 the four terms $^1T_1$, $^3T_1$, $^1T_2$, $^3T_2$ of $t_2 e$ belong to the atomic configuration d^2. Suitable components of these terms are derived from Table A 20 and are

$$|^3T_1\,1z\rangle = |\zeta^+\epsilon^+\rangle, \quad |^1T_1\,0z\rangle = \frac{1}{\sqrt{2}}(|\zeta^+\epsilon^-\rangle - |\zeta^-\epsilon^+\rangle),$$

$$|^3T_2\,1\zeta\rangle = |\zeta^+\theta^+\rangle, \quad |^1T_2\,0\zeta\rangle = \frac{1}{\sqrt{2}}(|\zeta^+\theta^-\rangle - |\zeta^-\theta^+\rangle).$$

Table 9.2. *Electrostatic energies of ground terms of d^n ions in an octahedral field*

	Low-spin			High-spin		
	Configuration	Term	Energy	Configuration	Term	Energy
d^2	t_2^2	$^3T_1$	$A-5B$	t_2^2	$^3T_1$	$A-5B$
d^3	t_2^3	$^4A_2$	$3A-15B$	t_2^3	$^4A_2$	$3A-15B$
d^4	t_2^4	$^3T_1$	$6A-15B+5C$	t_2^3e	^5E	$6A-21B$
d^5	t_2^5	$^2T_2$	$10A-20B+10C$	$t_2^3e^2$	$^6A_1$	$10A-35B$
d^6	t_2^6	$^1A_1$	$15A-30B+15C$	$t_2^4e^2$	$^5T_2$	$15A-35B+7C$
d^7	t_2^6e	^2E	$21A-36B+18C$	$t_2^5e^2$	$^4T_1$	$21A-40B+14C$
d^8	$t_2^6e^2$	$^3A_2$	$28A-50B+21C$	$t_2^6e^2$	$^3A_2$	$28A-50B+21C$
d^9	$t_2^6e^3$	^2E	$36A-56B+28C$	$t_2^6e^3$	^2E	$36A-56B+28C$
d^{10}	$t_2^6e^4$	$^1A_1$	$45A-70B+35C$	$t_2^6e^4$	$^1A_1$	$45A-70B+35C$

We note that the phase convention implicit in Wigner's formula is used in determining the phases of t_2e. The energies are

$$\left.\begin{aligned} E(^3T_1) &= A+4B, \quad E(^1T_1) = A+4B+2C, \\ E(^3T_2) &= A-8B, \quad E(^1T_2) = A+2C. \end{aligned}\right\} \tag{9.12}$$

The terms $^1A_1$, ^1E, $^1T_2$ and $^3T_1$ each occur twice in the d^2 configuration and we now calculate the off-diagonal matrix elements connecting pairs of them. Between the two $^3T_1$ terms we have

$$\begin{aligned} \langle t_2^2\,{}^3T_1\,1z\,|\,V\,|\,t_2e\,{}^3T_1\,1z\rangle &= \langle \eta^+\xi^+\,|\,V\,|\,\zeta^+\epsilon^+\rangle \\ &= -(\eta\epsilon;\,\xi\zeta)+(\eta\zeta;\,\xi\epsilon) = -6B. \end{aligned}$$

In a similar way we find $\sqrt{6}\,(2B+C)$, $2B\sqrt{3}$ and $-2B\sqrt{3}$ for $^1A_1$, ^1E and $^1T_2$, respectively.

In Table A 27 the matrices of electrostatic energy for d^2 are written out, together with those for d^8. Inspection of these shows that, apart from a constant diagonal energy $27A-42B+21C$ and a few changes of sign in the off-diagonal elements, the matrices for d^8 are identical with those for d^2. This suggests a general relationship between the matrices for d^ϵ and $d^{10-\epsilon}$. The existence of such a relationship is proved in § 9.7, but for the reader who does not wish to work through that proof I state here the exact relationship. With the choice of phases we have adopted, the matrices of electrostatic energy for d^ϵ and $d^{10-\epsilon}$ are the same apart from a constant diagonal energy and some changes of sign in the off-diagonal elements, when we associate $|t_2^m(S_1\Gamma_1)\,e^n(S_2\Gamma_2)\,S\Gamma\rangle$ of d^ϵ with $|t_2^{6-m}(S_1\Gamma_1)\,e^{4-n}(S_2\Gamma_2)\,S\Gamma\rangle$ of $d^{10-\epsilon}$. The signs of the off-diagonal elements would become identical if we were to adopt the following redefinition of the phases of the terms of $d^{10-\epsilon}$

$$|t_2^m(S_1\Gamma_1)\,e^n(S_2\Gamma_2)\,S\Gamma\rangle_R = (-1)^{mn}\,\mu_1\mu_2\,|t_2^m(S_1\Gamma_1)\,e^n(S_2\Gamma_2)\,S\Gamma\rangle,$$

where $\mu_1 = +1$, unless $m = 3$ and $S_1\Gamma_1 = {}^4A_2$, ^2E or $^2T_1$ when it is -1 ($\mu_1 = +1$ when $m = 3$ and $S_1\Gamma_1 = {}^2T_2$). $\mu_2 = +1$, unless $n = 2$ and $S_2\Gamma_2 = {}^3A_2$ or ^1E when it is -1 ($\mu_2 = +1$ for $n = 2$, $S_2\Gamma_2 = {}^1A_1$). The reader will verify for himself that this rule is satisfied for the pair of configurations d^2 and d^8. It also means that it is only necessary to calculate electrostatic matrices for d^ϵ with $\epsilon \leqslant 5$.

The rule is still true for $\epsilon = 5$ and then we obtain relations between different matrix elements of the same configuration. If $m = 3$ and $n = 2$ a term of d^5 goes into the same term of $d^{10-5} = d^5$ and we deduce

$$\langle t_2^3(S_1\Gamma_1)\, e^2(S_2\Gamma_2)\, S\Gamma \,|\, V \,|\, t_2^3(S_1'\Gamma_1')\, e^2(S_2'\Gamma_2')\, S\Gamma\rangle\, \mu_1\mu_2\mu_1'\mu_2'$$
$$= \langle t_2^3(S_1\Gamma_1)\, e^2(S_2\Gamma_2)\, S\Gamma \,|\, V \,|\, t_2^3(S_1'\Gamma_1')\, e^2(S_2'\Gamma_2')\, S\Gamma\rangle.$$

If $\mu_1\mu_2\mu_1'\mu_2' = -1$, then the matrix element is equal to minus itself and therefore is zero. So for these matrix elements we have the rather remarkable selection rule that they are zero unless $\mu_1\mu_2 = \mu_1'\mu_2'$.

For matrix elements between terms, at least one of which has $m \neq 3$, our rule gives relationships between different matrix elements. Typical examples are that

$$\langle t_2^5\, {}^2T_2 \,|\, V \,|\, t_2^4({}^3T_1)\, e\, {}^2T_2\rangle = \langle t_2 e^4\, {}^2T_2 \,|\, V \,|\, t_2^2({}^3T_1)\, e^3\, {}^2T_2\rangle,$$

and $$\langle t_2 e^4\, {}^2T_2 \,|\, V \,|\, t_2 e^4\, {}^2T_2\rangle = \langle t_2^5\, {}^2T_2 \,|\, V \,|\, t_2^5\, {}^2T_2\rangle = 10A - 20B + 10C,$$

where we have also used Table 9.2.

9.3.5. The ${}^4T_1$ terms of d^3. Just as in the weak-field scheme there are two ${}^4T_1$ terms. Their $M_S = \frac{3}{2}$, $M_\Gamma = z$ components are

$$\psi_1 = |t_2^2 e({}^4T_1)\tfrac{3}{2}z\rangle = |\eta^+\xi^+\theta^+\rangle,$$
$$\psi_2 = |t_2 e^2({}^4T_1)\tfrac{3}{2}z\rangle = |\zeta^+\theta^+\epsilon^+\rangle,$$

with a matrix for the electrostatic and crystal-field energies

	ψ_1	ψ_2
ψ_1	$-\frac{1}{5}\Delta + 3A - 3B$	$+6B$
ψ_2	$+6B$	$\frac{4}{5}\Delta + 3A - 12B$

. (9.13)

The corresponding secular equation is identical with the secular equation $f(E, -\Delta) = 0$ for the weak-field scheme (8.35) when allowance is made for the change of origin for energy. The function $|{}^4T_1\frac{3}{2}z\rangle$ which arises, in the weak-field scheme, from ${}^4F$ is equal to $(1/\sqrt{5})(-\psi_1 + 2\psi_2)$. This follows directly on expansion in terms of real orbitals by (9.1).

9.3.6. The quartets of d^5. The quartets arise from t_2^4e, $t_2^3e^2$ and $t_2^2e^3$ which have crystal-field energies $-\Delta$, 0 and Δ, respectively. Both t_2^4e and $t_2^2e^3$ give one ${}^4T_1$ and one ${}^4T_2$. $t_2^3e^2$ gives one ${}^4A_1$, ${}^4A_2$, ${}^4T_1$, ${}^4T_2$ and two ${}^4E$. Hence the crystal field has no matrix elements at all within the ${}^4A_1$, ${}^4A_2$ or ${}^4E$ terms. On the other hand, in the weak-field scheme ${}^4A_1$ and one ${}^4E$ come from ${}^4G$, ${}^4A_2$ comes from ${}^4F$ and the other ${}^4E$ from ${}^4D$. Therefore the energies of these terms in the crystal field are the same as those of the atomic terms from which they arise. Thus

$$\left.\begin{aligned} E({}^4G) &= E({}^4A_1) = E(a\,{}^4E) = 10A - 25B + 5C,\\ E({}^4F) &= E({}^4A_2) = 10A - 13B + 7C,\\ E({}^4D) &= E(b\,{}^4E) = 10A - 18B + 5C \end{aligned}\right\} \qquad (9.14)$$

from Table 4.6.

There are three ${}^4T_2$ terms and their functions with $M_S = \frac{3}{2}$, $M_\Gamma = \zeta$ are, from (9.4):

$$\psi_1 = |\xi^+\eta^+\zeta^2\epsilon^+\rangle,$$

$$\psi_2 = \frac{1}{\sqrt{2}}\{|\xi^2\zeta^+\theta^+\epsilon^+\rangle - |\eta^2\zeta^+\theta^+\epsilon^+\rangle\},$$

$$\psi_3 = |\eta^+\xi^+\theta^2\epsilon^+\rangle,$$

with crystal-field energies $-\Delta$, 0, Δ, respectively. The matrix elements are written straight down, for example

$$V_{12} = \frac{1}{\sqrt{2}}\{-(\xi\zeta;\, \eta\theta) - (\eta\zeta;\, \xi\theta)\}$$

$$= -B\sqrt{6},$$

$$V_{13} = -(\zeta\theta;\, \zeta\theta) = -(4B+C);$$

${}^4T_1$ is treated similarly. The resulting matrices are shown in Table A 30. We again see the consequences of the rule discussed in § 9.3.4.

9.3.7. The first excited configuration of d^6. The strong-field ground configuration of d^6 is t_2^6 and it has just the one term ${}^1A_1$. The first excited configuration is $t_2^5 e$ with the terms ${}^1T_1$, ${}^3T_1$, ${}^1T_2$, ${}^3T_2$. Suitable components are

$$\left.\begin{aligned} \psi_1 &= |{}^1T_1\, z\rangle = \frac{1}{\sqrt{2}}(|\xi^2\eta^2\zeta^+\epsilon^-\rangle - |\xi^2\eta^2\zeta^-\epsilon^+\rangle), \\ \psi_2 &= |{}^3T_1\, 1z\rangle = |\xi^2\eta^2\zeta^+\epsilon^+\rangle, \\ \psi_3 &= |{}^1T_2\, \zeta\rangle = \frac{1}{\sqrt{2}}(|\xi^2\eta^2\zeta^+\theta^-\rangle - |\xi^2\eta^2\zeta^-\theta^+\rangle), \\ \psi_4 &= |{}^3T_2\, 1\zeta\rangle = |\xi^2\eta^2\zeta^+\theta^+\rangle. \end{aligned}\right\} \tag{9.15}$$

The energies relative to ${}^3T_1$ are

$$\left.\begin{aligned} E({}^3T_2) &= \frac{4}{\sqrt{3}}c = 8B, \\ E({}^1T_1) &= 2g - \frac{2}{\sqrt{3}}h = 2C, \\ E({}^1T_2) &= \frac{4}{\sqrt{3}}c + 2g + 2\sqrt{3}\,h = 16B + 2C, \end{aligned}\right\} \tag{9.16}$$

where I have given the energies first in general parameters. They can be worked out directly from the functions given in (9.15) or, in the case of the singlets, from the diagonal-sum rule. For example, $E({}^3T_1) + E({}^1T_1) = 2E(|\xi^2\eta^2\zeta^+\epsilon^-\rangle)$. For future reference we note also that ${}^3T_1$ lies at $-a-4b+2j+5d+\sqrt{3}\,c-3g-\sqrt{3}\,h$ above the ground term.

When we have d orbitals, the four terms of $t_2^5 e$ lie at $\Delta - 3C$ for ${}^3T_1$, $\Delta + 8B - 3C$ for ${}^3T_2$, $\Delta - C$ for ${}^1T_1$ and $\Delta + 16B - C$ for ${}^1T_2$ above the ground term ${}^1A_1$ of t_2^6. Of course, our calculation here is not so complete as those in §§ 9.3.4–9.3.6 because there are other terms of the same type arising from still more excited configurations and we have not yet taken these into account.

9.3.8. Orbital energies. The orbital energy of a state of a strong-field configuration $t_2^m e^n$ is simply $\frac{1}{5}\Delta(3n-2m)$.

9.4. Pairing energies[1]

In Fig. 7.2 we see that the ground term of a d^8 ion has the same spin (in fact $^3A_2$) for all values of the orbital energy separation Δ. As we discussed in § 9.3.3, this is also true for d^1, d^2, d^3, d^9 and d^{10} but not for d^4, d^5, d^6 and d^7. When Δ increases for these latter ions, there ultimately comes a point at which a term with lower spin becomes the ground term. This is illustrated schematically in Fig. 9.1 for a d^5 ion. For small Δ the ground term is $^6A_1$ but, as Δ increases, a $^2T_2$ is progressively stabilized relative to $^6A_1$ and in due course passes below it. We call the point at which the spin of the ground term changes a cross-over point.

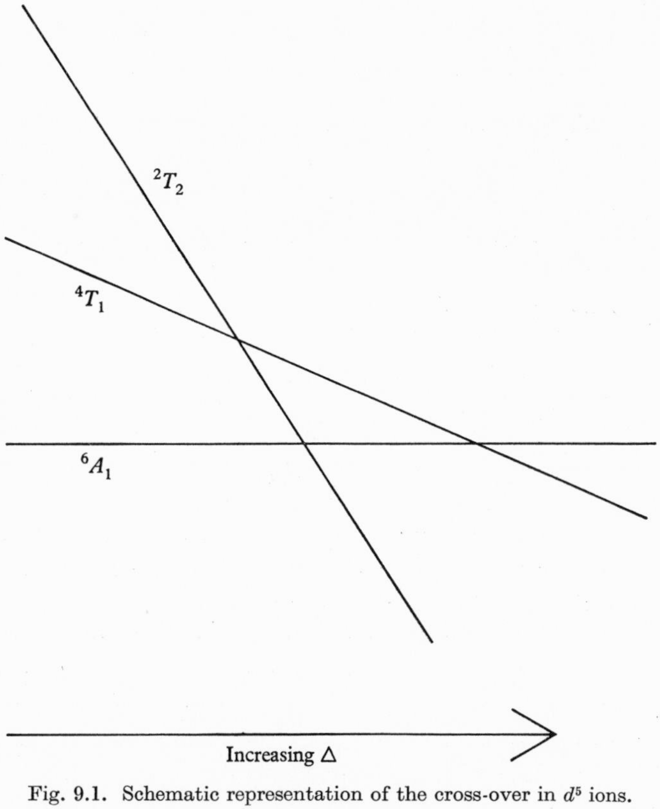

Fig. 9.1. Schematic representation of the cross-over in d^5 ions.

It is easy to obtain an approximate expression for Δ at the cross-over point, using our calculations from the last section. We have

$$\left.\begin{aligned} E_6 &= E(^6A_1) = 10A-35B, \\ E_2 &= E(^2T_2) = 10A-20B+10C-2\Delta, \end{aligned}\right\} \tag{9.17}$$

[1] Discussed independently in Orgel (1955*b*) and Griffith (1956*a*). For *f* electrons see Griffith & Orgel (1957*b*).

and so if $E_6 = E_2$ then $\Delta = 7\frac{1}{2}B+5C$. $^6A_1$ and $^2T_2$ belong respectively to the configurations $t_2^3e^2$ and t_2^5 so the change of ground state involves taking two electrons out of e orbitals and placing them in t_2 orbitals. The states of $^6A_1$ are written $|t_2^3(^4A_2)\,e^2(^3A_2)\,^6A_1M_S\rangle$ and the two t_2 orbitals into which the electrons go have opposite spin to those which are already there. So the change of electrostatic energy, $15B+10C$, may be regarded as a pairing energy and one-half of this as a mean pairing energy per electron transferred from e to t_2. We write Π for the mean pairing energy and define Π to be the calculated value of Δ at the cross-over point (this definition applies to d^4, d^6 and d^7 ions as well). Therefore, in our present approximation, $\Pi = 7\frac{1}{2}B+5C$ for a d^5 ion. When $\Delta < \Pi$ the ground term is $^6A_1$ and when $\Delta > \Pi$ it is $^2T_2$ so Π is a threshold above which Δ must lie in order to lower the spin.

Next we ask if a quartet state could ever be the ground term of d^5. The lowest quartet is $^4T_1$ of t_2^4e with energy

$$E_4 = 10A-25B+6C-\Delta. \tag{9.18}$$

The mean of E_2 and E_6 is

$$\begin{aligned}\tfrac{1}{2}(E_2+E_6) &= 10A-27\tfrac{1}{2}B+5C-\Delta\\ &= E_4-2\tfrac{1}{2}B-C\end{aligned}$$

and so one at least of $^6A_1$ and $^2T_2$ must lie $2\frac{1}{2}B+C$ or more below the lowest quartet.

From Table 9.2 it follows that Π for d^4, d^6 and d^7 is given by

$$\left.\begin{aligned}\Pi(d^4) &= 6B+5C,\\ \Pi(d^6) &= 2\tfrac{1}{2}B+4C,\\ \Pi(d^7) &= 4B+4C.\end{aligned}\right\} \tag{9.19}$$

If B and C were the same for all ions then $\Pi(d^6) < \Pi(d^7) < \Pi(d^4) < \Pi(d^5)$. This is to be correlated with the experimental fact that d^6 ions are often low-spin even though the corresponding d^5 ions are high-spin.

For d^6, just as for d^5, there is a value of the spin intermediate between that of the high-spin and low-spin ground states. This is the $^3T_1$ of t_2^5e and in § 9.3 we saw that it lay $\Delta-3C$ above the $t_2^6\,^1A_1$ term. Hence it lies $2\frac{1}{2}B+C$ above the mean of this $^1A_1$ and the $^5T_2$ term. So for both d^5 and d^6 the lowest intermediate spin term can never be the ground term. This conclusion assumes, of course, a regular octahedral field applied to a pure d^n configuration. In particular it is not necessarily true in a tetragonal field.

9.5. The p^n isomorphism[1]

9.5.1. Definition of correspondence. The three d orbitals which span the t_2 representation can be related to three p orbitals. We write $|p1\rangle$, $|p0\rangle$, $|p-1\rangle$

[1] Based on an idea of Abragam & Pryce (1951*a*).

for the three p orbitals having respectively $m_l = 1, 0$ and -1 and let them correspond to d orbitals according to the rules

$$\left.\begin{aligned} |p1\rangle \;&\sim |2\,-1\rangle &&= |dt_2 1\rangle, \\ |p0\rangle \;&\sim \frac{1}{\sqrt{2}}(|22\rangle - |2\,-2\rangle) &&= |dt_2 0\rangle, \\ |p\,-1\rangle &\sim -|21\rangle &&= |dt_2\,-1\rangle. \end{aligned}\right\} \tag{9.20}$$

Our notation for the t_2 orbitals was chosen with this application in mind.

It is a matter of simple calculation to write down the matrices of the components of $\mathbf{l}$, the orbital angular-momentum vector, within the $|pi\rangle$ and within the $|dt_2 i\rangle$. For example, in units of $\hbar$:

$$\langle pi \,|\, l_z \,|\, pj\rangle = i\delta_{ij}, \quad \langle dt_2 i \,|\, l_z \,|\, dt_2 j\rangle = -i\delta_{ij},$$

($i \neq \sqrt{-1}$ here) and in fact

$$\langle pi \,|\, l_k \,|\, pj\rangle = -\langle dt_2 i \,|\, l_k \,|\, dt_2 j\rangle, \tag{9.21}$$

for all i, j and components l_k of $\mathbf{l}$. We express this by saying that the matrix of $\mathbf{l}$ within the t_2 representation of the d-functions is equal to the matrix of $-\mathbf{l}$ for p-functions when the functions are related according to (9.20).

Passing now to n-electron configurations, any state ψ_p of p^n corresponds to a state ψ_d of $(dt_2)^n$ by simply replacing each one-electron function of ψ_p by its corresponding d-function according to (9.20). We suppose the spin functions to remain unchanged under the correspondence. Thus, for example, the $M_S = M_L = 1$ state of ^3P of p^2 is $\psi_p = |p1^+, p0^+\rangle$ and this corresponds to

$$\psi_d = |dt_2 1^+, dt_2 0^+\rangle = |-1^+ \zeta_1^+\rangle = |t_2^2\, {}^3T_1\, 11\rangle$$

in the notation of Table A 24.

9.5.2. Symmetries of states. Under the octahedral group any term SL of p^n becomes one or more terms $S\Gamma$ of the octahedral group. Therefore the states of p^n get assigned $S\Gamma$ values in a natural way. We now ask how these $S\Gamma$ values are related to those of the corresponding t_2^n states. They are not necessarily the same because (9.20) relates t_1 of p-functions to t_2 of d-functions. Therefore a p-function behaves under the octahedral group as a_2 times the corresponding d-function. Consequently a term $S\Gamma$ of p^n corresponds to the term $S(a_2^n\,\Gamma)$ of t_2^n. For example, $^3T_1$ of p^2 corresponds to $^3T_1$ of $(dt_2)^2$ as we have just seen, but $^4A_1$ of p^3 corresponds to $^4A_2$ of $(dt_2)^3$. In other words, when n is even the correspondence is direct, but when n is odd the correspondence is between a representation and its associated one. The functions listed in Table A 24 are consistent with this correspondence, and with the definitions of Tables A 16 and 19.

9.5.3. Energies of states. Just as $\mathbf{l}$ within t_2 is the same as $-\mathbf{l}$ within p so is the spin-orbit coupling energy $\zeta\mathbf{l}\cdot\mathbf{s}$ within t_2 the same as $-\zeta\mathbf{l}\cdot\mathbf{s}$ within p. Similarly, for an n-electron system, $\mathbf{L}$ and $\Sigma\xi(r_i)\mathbf{l}_i\cdot\mathbf{s}_i$ within $(dt_2)^n$ become $-\mathbf{L}$ and $-\Sigma\xi(r_i)\mathbf{l}_i\cdot\mathbf{s}_i$ within p^n provided we interpret ζ_p as ζ_d.

The electrostatic energies are conveniently discussed by comparing the formulae for p^2 with the formulae for the corresponding terms of t_2^2. Using (4.52–4) and Table 9.1 we have

$$\left.\begin{aligned} E(p^2;\,{}^1S) &= F_0+10F_2, \quad E(dt_2^2;\,{}^1A_1) = A+10B+5C,\\ E(p^2;\,{}^3P) &= F_0-5F_2, \quad E(dt_2^2;\,{}^3T_1) = A-5B,\\ E(p^2;\,{}^1D) &= F_0+F_2, \quad E(dt_2^2;\,{}^1E) = E(dt_2^2;\,{}^1T_2) = A+B+2C, \end{aligned}\right\} \tag{9.22}$$

which are identical if we take $F_0 = A+\frac{5}{3}C$, $F_2 = B+\frac{1}{3}C$. It follows that the electrostatic matrix within t_2^n is then identical with the electrostatic matrix within p^n (or p^{6-n}). The matrix of the spin-orbit coupling within p^{6-n} is minus that for p^n and therefore, under our correspondence, the calculated energies in intermediate coupling are formally identical for t_2^n and p^{6-n}.

When we consider the interaction with an external magnetic field there is a difference between t_2^n and p^{6-n} because $\mathbf{L}+2\mathbf{S}$ for t_2^n corresponds to $-\mathbf{L}+2\mathbf{S}$ for p^{6-n}.[1] When discussing magnetic properties it is more convenient to correlate t_2^n with p^n and replace $\mathbf{L}$, $\mathbf{l}_i$ for p^n by $\gamma\mathbf{L}$, $\gamma\mathbf{l}_i$, where $\gamma = -1$. The calculations of § 5.6 then give us formulae for the magnetic susceptibilities in Russell–Saunders coupling for all dt_2^n ground terms, i.e. for all strong-field d^n ground terms having $n \leqslant 6$.

Example

The g value of the ${}^2P_{\frac{3}{2}}$ level of p^1 is $\frac{4}{3}$ but for the corresponding level of $(dt_2)^5$ it is 0.

9.6. Spin-orbit coupling energies

9.6.1. Introduction. The investigation in the preceding section shows us that the first-order effect of the spin-orbit coupling on the ${}^3T_1$ term of t_2^2 or t_2^4 is the same as its effect on the ${}^3P$ term of p^4 or p^2, respectively. Any ${}^3P$ term splits into three levels ${}^3P_0$, ${}^3P_1$, ${}^3P_2$ with respective degeneracies 1, 3 and 5. It is convenient to write the corresponding states of ${}^3T_1$ as ${}^3T_{10}^M$, ${}^3T_{11}^M$, ${}^3T_{12}^M$, i.e. as ${}^3T_{1J}^M$, where J, M are the quantum numbers of the corresponding states of ${}^3P$. In this particular case we have shown that the ${}^3T_1$ term is split by spin-orbit coupling to first order according to the Landé interval rule. The result is actually true in a much more general form and we will prove this in § 9.6.2.

We use some of the theorems of § 8.7. Let $|\Gamma_1 a\rangle$, $|\Gamma_3 c\rangle$ be functions forming bases for irreducible representations Γ_1, Γ_3 of O and let v_b be the bth component of an operator forming a basis for an irreducible representation Γ_2 of Σ. Then if

$$M_{abc} = \langle \Gamma_1 a \,|\, v_b \,|\, \Gamma_3 c\rangle,$$

it follows from Theorem 8.3 that

$$M_{abc} = \langle \Gamma_1 \,\vdots\, v \,\vdots\, \Gamma_3\rangle \langle \Gamma_2 \Gamma_3 \Gamma_1 a \,|\, \Gamma_2 \Gamma_3 bc\rangle, \tag{9.23}$$

[1] The device of correlating t_2^n with p^{6-n} makes the formal correspondence complete for the spin-orbit coupling $\mathscr{H}_s$ but not for the magnetic energy $\beta\mathbf{H}.(\mathbf{L}+2\mathbf{S})$. Via § 9.7, this is to be related to the fact that $\mathscr{H}_s^* = \mathscr{H}_s$ but $\mathbf{L}^* = -\mathbf{L}$ and $\mathbf{S}^* = -\mathbf{S}$.

where $\langle \Gamma_1 \vdots v \vdots \Gamma_3 \rangle$ is independent of a, b and c. The coupling constants

$$\langle \Gamma_2 \Gamma_3 \Gamma_1 a \mid \Gamma_2 \Gamma_3 bc \rangle$$

are tabulated in Table A 20.

Putting $\Gamma_1 = \Gamma_3 = T_1$ and $v_b = L_b$, the bth component of the orbital angular momentum, we have

$$\langle \alpha T_1 a \mid v_b \mid \alpha' T_1 c \rangle = \gamma \langle Pa \mid L_b \mid Pc \rangle, \tag{9.24a}$$

for some γ independent of a, b, c. Here $|Pa\rangle$ is an eigenket of $\mathbf{L}^2$, L_z for a free atom and having $L = 1, M_L = a$. Equation (9.24a) follows from Theorem 8.4. Finally, our discussion of the correspondence exhibited in (9.20) shows that

$$\langle \alpha T_2 a \mid v_b \mid \alpha' T_2 c \rangle = \gamma \langle Pa \mid L_b \mid Pc \rangle, \tag{9.24b}$$

for some γ independent of a, b, c.

9.6.2. The interval rule. Now let us take a general term $\alpha\,{}^{2S+1}\Gamma$. The matrix elements of the spin-orbit coupling are

$$\langle \alpha S \Gamma M_S a \mid \sum_\mu \xi(\mathbf{r}_\mu) \mathbf{l}_\mu . \mathbf{s}_\mu \mid \alpha S \Gamma M'_S a' \rangle$$
$$= \sum_{\mu, \alpha'} \langle \alpha S \Gamma M_S a \mid \xi(r_\mu) \mathbf{l}_\mu \mid \alpha' S \Gamma M_S a' \rangle . \langle \alpha' S \Gamma M_S a' \mid \mathbf{s}_\mu \mid \alpha S \Gamma M'_S a' \rangle. \tag{9.25}$$

We concentrate attention upon $\langle \alpha S \Gamma M_S a \mid \xi(r_\mu) \mathbf{l}_\mu \mid \alpha' S \Gamma M_S a' \rangle$. The vector $\xi(r_\mu) \mathbf{l}_\mu$ forms a basis for T_1. The square of the irreducible representation Γ contains T_1 if $\Gamma = T_1$ or T_2 but not if $\Gamma = A_1$, A_2 or E. Therefore the first-order effect of the spin-orbit coupling within all ${}^{2S+1}A_1$, ${}^{2S+1}A_2$ and ${}^{2S+1}E$ terms is zero for any S.

When $\Gamma = T_1$ or T_2 we use the two corollaries to the replacement theorem to deduce

$$\langle \alpha S \Gamma M_S a \mid \xi(r_\mu) \mathbf{l}_\mu \mid \alpha' S \Gamma M_S a' \rangle = \gamma \langle S 1 M_S a \mid \mathbf{L} \mid S 1 M_S a' \rangle.$$

Also

$$\langle \alpha' S \Gamma M_S a' \mid \mathbf{s}_\mu \mid \alpha S \Gamma M'_S a' \rangle = \gamma' \langle S 1 M_S a' \mid \mathbf{S} \mid S 1 M'_S a' \rangle,$$

whence

$$\langle \alpha S \Gamma M_S a \mid \sum_\mu \xi(r_\mu) \mathbf{l}_\mu . \mathbf{s}_\mu \mid \alpha S \Gamma M'_S a' \rangle = \lambda \langle S 1 M_S a \mid \mathbf{L} . \mathbf{S} \mid S 1 M'_S a' \rangle.$$

Thus the matrix elements are proportional to those of the spin-orbit coupling within a ${}^{2S+1}P$ atomic term (using the Landé interval rule, (5.4)). Therefore if we define an $ST_i JM_J$ system of quantization by the equation

$$|ST_i JM_J\rangle = \sum_{M_S, a} |ST_i M_S a\rangle \langle S 1 M_S a \mid S 1 J M_J \rangle,$$

where the coefficients $\langle S 1 M_S a \mid S 1 J M_J \rangle$ are Wigner coefficients, then any pair of states belonging to the same 'level' ${}^{2S+1}T_{iJ}$ have the same first-order spin-orbit coupling energy. Furthermore, the three levels ${}^{2S+1}T_{i\,S-1}$, ${}^{2S+1}T_{iS}$, ${}^{2S+1}T_{i\,S+1}$ satisfy a Landé interval rule, to this same approximation.

9.6.3. Matrix elements between different terms. For a pair of terms $\alpha\,{}^{2S+1}\Gamma$ and $\alpha'\,{}^{2S'+1}\Gamma'$ we find

$$\langle \alpha S \Gamma M_S a \mid \sum_\mu \xi(r_\mu) \mathbf{l}_\mu . \mathbf{s}_\mu \mid \alpha' S' \Gamma' M'_S a' \rangle$$
$$= \sum_{\mu, \alpha''} \langle \alpha S \Gamma M_S a \mid \xi(r_\mu) \mathbf{l}_\mu \mid \alpha'' S \Gamma' M_S a' \rangle . \langle \alpha'' S \Gamma' M_S a' \mid \mathbf{s}_\mu \mid \alpha' S' \Gamma' M'_S a' \rangle$$
$$= \sum CC' \langle \alpha S \Gamma \vdots \xi(r_\mu) l_\mu \vdots \alpha'' S \Gamma' \rangle \langle \alpha'' S \Gamma' \vdots s_\mu \vdots \alpha' S' \Gamma' \rangle$$
$$\times \sum_j \langle T_1 \Gamma' \Gamma a \mid T_1 \Gamma' j a' \rangle \langle 1 S' S M_S \mid 1 S' - j M'_S \rangle (-1)^{j+1}, \tag{9.26}$$

where we have used (9.23) and Table A 20. Equation (9.26) is a little complicated but if the reader will look carefully at it he will see that it shows that the entire matrix of spin-orbit coupling between the pair of terms is determined by the coupling coefficients except for a single multiplying constant common to all elements. The same will therefore be true if we use another system of quantization and we shall therefore define one in which the basic states form components of irreducible representations of the spinor group O^* of the regular octahedron.

We adopt the following definitions. For ${}^{2S+1}A_1$, ${}^{2S+1}A_2$ and ${}^{2S+1}E$ we first break up the spin functions under the octahedral group according to Table A 19. Then we couple the resulting irreducible representations with A_1, A_2 or E, putting the spin functions first, according to Table A 20. In general, one would need labels to distinguish different irreducible representations of the spinor group, but for d configurations this is unnecessary as repeated representations do not occur. As an example let us write down the A_1 representation of a ${}^5E$ term. For the spin, $E\theta = |20\rangle$, $E\epsilon = (1/\sqrt{2})(|22\rangle + |2\,-2\rangle)$, so

$$|A_1 a_1\rangle = \frac{1}{\sqrt{2}}(|\theta\theta\rangle + |\epsilon\epsilon\rangle)$$

$$= \frac{1}{\sqrt{2}}|2E0\theta\rangle + \tfrac{1}{2}|2E2\epsilon\rangle + \tfrac{1}{2}|2E\,-2\epsilon\rangle,$$

where the right-hand side uses $S\Gamma M_S M_\Gamma$ quantization.

For ${}^{2S+1}T_1$ and ${}^{2S+1}T_2$ we first combine the states into levels ${}^{2S+1}T_{iJ}$ and then break up the levels according to Table A 19. This completes the definition for ${}^{2S+1}T_1$. For ${}^{2S+1}T_2$ we write

$$|{}^{2S+1}T_{2J}\,\Gamma M\rangle = \sum_{M'} \langle A_2 \Gamma' a_2 M' \mid A_2 \Gamma' \Gamma M\rangle\, |{}^{2S+1}T_{2J}(\Gamma' M')\rangle,$$

where $|{}^{2S+1}T_{2J}(\Gamma' M')\rangle$ is the component and symmetry one obtains by breaking ${}^{2S+1}T_{2J}$ up according to Table A 19. If Γ is not self-associated then we simply change from $\Gamma' M$ to ΓM because $\langle A_2 \Gamma' a_2 M' \mid A_2 \Gamma' \Gamma M\rangle = \delta_{MM'}$.

The definitions are a little complicated and in practice tables of the coefficients in the expansions

$$|S\Gamma\beta\Gamma' M'\rangle = \sum_{M_S M_\Gamma} \langle S\Gamma M_S M_\Gamma \mid S\Gamma\beta\Gamma' M'\rangle\, |S\Gamma M_S M_\Gamma\rangle$$

are all that are necessary. Here β is the value of J when it is necessary to distinguish repeated representations Γ'. For small values of S the coefficients are simply the coupling coefficients of Table A 20. The coefficients for ${}^5E$ and ${}^5T_2$ are also given at the end of that table.

The spin-orbit coupling energy $\mathscr{H}_s$ belongs to the representation A_1 of O^* and therefore

$$\langle \alpha_1 S_1 \Gamma_1 \beta_1 \Gamma_1' M_1' \mid \mathscr{H}_s \mid \alpha_2 S_2 \Gamma_2 \beta_2 \Gamma_2' M_2'\rangle = \lambda c(\beta_1, \Gamma_1', \beta_2, \Gamma_2')\, \delta_{\Gamma_1'\Gamma_2'} \delta_{M_1' M_2'},$$

where λ is independent of β_1, β_2, Γ_1', Γ_2', M_1', M_2'. In other words $\mathscr{H}_s$ is diagonal with respect to Γ', M' and the diagonal elements are

$$\langle \alpha_1 S_1 \Gamma_1 \beta_1 \Gamma' M' \mid \mathscr{H}_s \mid \alpha_2 S_2 \Gamma_2 \beta_2 \Gamma' M'\rangle = \lambda c(\beta_1 \beta_2 \Gamma'), \quad \text{say.} \qquad (9.27)$$

The quantities $c(\beta_1\beta_2\Gamma')$ are completely determined, apart from a multiplying constant which may be absorbed in λ, by the terms concerned and may therefore be calculated once and for all for every pair of terms which can occur in d^n configurations.

As we usually have our states written out in the $S\Gamma M_S M_\Gamma$ scheme it is most convenient to actually calculate spin-orbit coupling matrix elements in this scheme. A typical one is

$$\begin{aligned}
&\langle S_1\Gamma_1 M_{S_1} M_{\Gamma_1} \,|\, \mathscr{H}_s \,|\, S_2\Gamma_2 M_{S_2} M_{\Gamma_2}\rangle \\
&= \sum_{\Gamma'\Gamma''M'M''\beta_1\beta_2} \langle S_1\Gamma_1 M_{S_1} M_{\Gamma_1} \,|\, S_1\Gamma_1\beta_1\Gamma'M'\rangle \langle S_1\Gamma_1\beta_1\Gamma'M' \,|\, \mathscr{H}_s \,|\, S_2\Gamma_2\beta_2\Gamma''M''\rangle \\
&\qquad\qquad \times \langle S_2\Gamma_2\beta_2\Gamma''M'' \,|\, S_2\Gamma_2 M_{S_2} M_{\Gamma_2}\rangle \\
&= \lambda \sum_{\Gamma'M'\beta_1\beta_2} \langle S_1\Gamma_1 M_{S_1} M_{\Gamma_1} \,|\, S_1\Gamma_1\beta_1\Gamma'M'\rangle \langle S_2\Gamma_2\beta_2\Gamma'M' \,|\, S_2\Gamma_2 M_{S_2} M_{\Gamma_2}\rangle\, c(\beta_1\beta_2\Gamma').
\end{aligned} \tag{9.28}$$

Once we know the $c(\beta_1\beta_2\Gamma')$ then, providing the equation is not identically zero on both sides, (9.28) shows how we may calculate λ. We may also use it to calculate the $c(\beta_1\beta_2\Gamma')$ by using the selection rule that

$$\langle S_1\Gamma_1 M_{S_1} M_{\Gamma_1} \,|\, \mathscr{H}_s \,|\, S_2\Gamma_2 M_{S_2} M_{\Gamma_2}\rangle = 0 \quad \text{unless} \quad M_{S_1} - M_{S_2} = 1,\ 0 \text{ or } -1.$$

We illustrate this for the pair of terms ${}^3T_1$, ${}^3T_2$. There are only three c numbers here, namely, $c(E)$, $c(T_1)$ and $c(T_2)$. The parameters β_1, β_2 are not needed as labels. Then we look at the coupling coefficients for $T_1 \times T_1$ and $T_1 \times T_2$ in Table A 20 and note that ${}^3T_1\,1\,1$ differs in its M_S value by 2 units from ${}^3T_2\,-1\,1$ but that both are linear combinations of $A_2 a_2$, $E\epsilon$ and $T_2 0$ in the $S\Gamma JM$ scheme. The left-hand side of (9.28) is therefore zero but the right-hand side is

$$\lambda\left(-\frac{1}{2\sqrt{3}}c(E) - \tfrac{1}{2}c(T_2)\right) = 0.$$

A second equation is obtained from

$$0 = \langle {}^3T_1\,1\,-1 \,|\, \mathscr{H}_s \,|\, {}^3T_2\,-1\,-1\rangle = \lambda\left(\frac{1}{2\sqrt{3}}c(E) - \tfrac{1}{2}c(T_1)\right),$$

whence $c(E) = \sqrt{3}\,c(T_1) = -\sqrt{3}\,c(T_2)$. We choose $c(T_1) = 1$ and then have $c(E) = \sqrt{3}$, $c(T_2) = -1$.

The c numbers between any pair of terms may be calculated and the ones of interest are given in Table A 31. For a given pair of terms, only the relative values of the c numbers are defined and I have chosen them so that no fractions occur. When the two terms are of the same type, for example both ${}^4T_1$, the c numbers can be deduced from the interval rule.

We conclude this subsection by noting the selection rules for spin-orbit coupling $\Gamma\Gamma'$ contains T_1 and $\Delta S = 0, \pm 1$ which follow immediately from (9.26). Because of these rules the matrix elements between ${}^3E$ and ${}^5E$ or between ${}^5A_1$ and ${}^1T_2$ are all zero. c numbers for such pairs of terms are not given in Table A 31 nor for pairs which have only one irreducible representation Γ'' of O^* in common. For pairs of either of these types we always define $c(\Gamma'') = 1$ for the common representations.

9.6.4. Calculation of matrix elements. The use of c numbers is conveniently illustrated by calculating the matrix of spin-orbit coupling within the four triplet terms of d^2. Each term has a T_2 representation in it and we choose a state which includes the $T_2 0$ component. For example

$$\left.\begin{aligned} \psi_1 &= |t_2^2\,{}^3T_1\,11\rangle = |-1^+\zeta_1^+\rangle, \\ \psi_2 &= |t_2 e\,{}^3T_1\,11\rangle = \frac{\sqrt{3}}{2}\,|{}^2T_2\tfrac{1}{2}-1,\,{}^2E\tfrac{1}{2}\theta\rangle - \tfrac{1}{2}\,|{}^2T_2\tfrac{1}{2}1,\,{}^2E\tfrac{1}{2}\epsilon\rangle \\ &= -\frac{\sqrt{3}}{2}\,|1^+\theta^+\rangle - \tfrac{1}{2}\,|-1^+\epsilon^+\rangle, \\ \psi_3 &= |t_2 e\,{}^3T_2\,1-1\rangle = \tfrac{1}{2}\,|1^+\theta^+\rangle - \frac{\sqrt{3}}{2}\,|-1^+\epsilon^+\rangle, \\ \psi_4 &= |e^2\,{}^3A_2\,0a_2\rangle = \frac{1}{\sqrt{2}}\,|\theta^+\epsilon^-\rangle + \frac{1}{\sqrt{2}}\,|\theta^-\epsilon^+\rangle, \end{aligned}\right\} \tag{9.29}$$

where we have used Table A 20 to couple t_2 with e and Table A 24 for the states of t_2, t_2^2 and e^2. Using the suffixes of ψ_i to label the four terms we rewrite (9.27) as

$$\langle i\Gamma'M' \,|\, \mathscr{H}_s \,|\, j\Gamma'M'\rangle = \lambda_{ij}c(\Gamma'), \tag{9.30}$$

and wish to determine the λ_{ij}. This is done by determining the matrix of spin-orbit coupling within the states ψ_i both directly and as a function of λ_{ij}. For λ_{11} we have

$$\langle\bar{\psi}_1 \,|\, \mathscr{H}_s \,|\, \psi_1\rangle = -\tfrac{1}{2}\zeta = \lambda_{11}[\tfrac{1}{2}c(E) + \tfrac{1}{2}c(T_2)] = -\lambda_{11},$$

and for λ_{13}

$$\langle\bar{\psi}_1 \,|\, \mathscr{H}_s \,|\, \psi_3\rangle = -\frac{\sqrt{3}}{2}\zeta = \lambda_{13}\left(-\frac{1}{2\sqrt{3}}c(E) + \tfrac{1}{2}c(T_2)\right) = -\lambda_{13}$$

and similarly for the remaining λ_{ij}. The complete set of λ_{ij} calculated in this way is

$$\begin{aligned} &\lambda_{11} = \tfrac{1}{2}\zeta, \quad && \lambda_{14} = \lambda_{24} = \lambda_{44} = 0, \\ &\lambda_{12} = \tfrac{1}{2}\zeta, && \lambda_{22} = -\tfrac{1}{4}\zeta, \quad \lambda_{33} = -\tfrac{1}{4}\zeta, \\ &\lambda_{13} = \tfrac{1}{2}\sqrt{3}\,\zeta, && \lambda_{23} = \tfrac{1}{4}\sqrt{3}\,\zeta, \quad \lambda_{34} = -\sqrt{2}\,\zeta, \end{aligned}$$

The matrices of spin-orbit coupling in the $S\Gamma JM$ scheme are then written down immediately using (9.30) and the table of c numbers. They are shown in Table A 33 together with the matrix elements connecting the singlet terms with the triplet terms. Of course, Table A 33 could be abbreviated somewhat by just giving the complete matrix of λ_{ij} for the singlet and triplet terms. For applications, however, it is more convenient to have the actual matrix elements of the spin-orbit coupling energy.

Corresponding calculations for the lowest terms of other d^n configurations are easily made and the results are given in Tables A 34–38. The reader would find it good practice to calculate some of these matrix elements for himself.

Example

Think of a straightforward and simple independent method of obtaining the determinants of the matrices given in Table A 33. Confirm that your values agree with those calculated directly from Table A 33.

9.7. Holes and particles[1]

9.7.1. The t_2^n configurations. It is natural to think of the t_2^5 configuration as being a hole in the t_2^6 closed shell and it is the purpose of this section to make such ideas precise. In general, we shall think of t_2^{6-n} as being n holes in t_2^6 and shall relate its states to those of t_2^n.

Unlike a particle a hole has no independent existence and must always be regarded as a hole in something—in our present case in t_2^6. Therefore the properties of t_2^6 itself will play an important role in our discussion. t_2^6 has just one state, ${}^1A_1$, which is the only totally antisymmetric state which can be formed by placing six electrons in the six one-electron kets t_2 (cf. § 2.6; there are six kets because of the two spin functions).

The configuration t_2^n consists of an array of terms but has no repeated terms. t_2^{6-n} consists of the same array. Let us now suppose that we have two different and independent sets of t_2 functions—t_2 and τ_2 say—and try to construct a ${}^1A_1$ state from products of states having the first n electrons in t_2^n and the last $6-n$ in τ_2^{6-n}. We antisymmetrize within t_2^n and within τ_2^{6-n} but not between t_2^n and τ_2^{6-n}. Then we can only obtain a ${}^1A_1$ by coupling a term ${}^{2S+1}\Gamma$ in t_2^n with the same type of term in τ_2^{6-n}. For example, when $n = 2$ we will get four ${}^1A_1$ terms in this way because there are four different terms in t_2^2. These are the only ${}^1A_1$ terms which can be obtained as linear combinations of simple products of the antisymmetrized states of t_2^n and τ_2^{6-n} when the electrons are numbered as indicated previously. We will use a dot to indicate this lack of antisymmetrization across the dot. So we say we have states $\phi.\chi$ of $t_2^n.\tau_2^{6-n}$, where $\phi \in t_2^n$ contains only the first n electrons and $\chi \in \tau_2^{6-n}$ only the last $6-n$.

We now relate the ${}^1A_1$ of t_2^6 to these incompletely antisymmetrized functions. Call it ψ. Then ψ is a sum of simple products, $\psi = \sum_p p$, say. Let P_μ be a permutation of the first n electrons and Q_ν of the last $6-n$. Then

$$n!\,(6-n)!\,\psi = \sum_{\mu,\nu}(-1)^{\mu+\nu}P_\mu Q_\nu \psi = \sum_p \sum_{\mu,\nu}(-1)^{\mu+\nu}P_\mu Q_\nu p. \tag{9.31}$$

Now replace any one-electron function in (9.31) which has one of the last $6-n$ electrons as argument by the corresponding τ_2-function. Let ψ be changed to ψ' by this procedure. Then ψ' is still ${}^1A_1$ and is expressed in the desired form and must, therefore, be a linear combination of the ${}^1A_1$ terms of $t_2^n.\tau_2^{6-n}$. If we now replace τ_2 by t_2 again we have t_2^6 expressed as a linear combination of ${}^1A_1$ terms of $t_2^n.t_2^{6-n}$. Symbolically

$$\psi = |t_2^6\,{}^1A_1\rangle = \Sigma a(S\Gamma)\,|t_2^n S\Gamma.t_2^{6-n}S\Gamma\,{}^1A_1\rangle, \tag{9.32}$$

[1] For the analogous treatment of the free-ion d^n configurations and hence for the weak-field scheme the best classification is in terms of the representations of the full orthogonal group O_5 on the set of five d orbitals (also called the seniority classification; see Racah, 1943, 1949). The electrostatic interaction between the d electrons does not belong to the unit representation of O_5 but would have done had B been zero. This is the reason for the high degeneracies in Figs. 4. 1–4 when $B = 0$. Those figures may be regarded as strictly analogous to the crystal-field diagrams of Fig. 9.2 and in them the 'crystal field' has the three-dimensional rotation group R_3 for its symmetry group.

where the $a(S\Gamma)$ are constants, the sum runs over all terms of t_2^n, and the first n electrons are in t_2^n in each ket in the sum.

Next we determine $a(S\Gamma)$. (9.32) may be slightly rearranged to the form

$$\psi = \Sigma a(S\Gamma)(-1)^{S-M_s}(2S+1)^{-\frac{1}{2}}\langle\Gamma\Gamma M_\Gamma M'_\Gamma \mid \Gamma\Gamma A_1 a_1\rangle \times |t_2^n S\Gamma M_S M_\Gamma\rangle . |t_2^{6-n} S\Gamma - M_S M'_\Gamma\rangle,$$

where the sum is now also over M_S, M_Γ and M'_Γ. For convenience we choose the components of Γ so that they form bases for a real representation (in fact the functions in the left-hand column of Table A 24). Then, using ex. 3, p. 155, we find $\langle\Gamma\Gamma M_\Gamma M'_\Gamma \mid \Gamma\Gamma A_1 a_1\rangle = \lambda(\Gamma)^{-\frac{1}{2}}\delta_{M_\Gamma M_\Gamma}$ and hence

$$\psi = \Sigma a(S\Gamma)(-1)^{S-M_s}(2S+1)^{-\frac{1}{2}}\lambda(\Gamma)^{-\frac{1}{2}}|t_2^n S\Gamma M_S M_\Gamma\rangle . |t_2^{6-n} S\Gamma - M_S M_\Gamma\rangle, \quad (9.33)$$

where $\lambda(\Gamma)$ is the degree of the representation Γ. In (9.33) we have a sum of $q = 6!/[n!(6-n)!]$ products and each state of t_2^n and of t_2^{6-n} occurs just once. If, instead of taking the various components of the terms of t_2^n and t_2^{6-n}, we take single determinant functions we can again write

$$\psi = q^{-\frac{1}{2}}\Sigma\epsilon' |t_2^n d_i\rangle . |t_2^{6-n} d_i\rangle, \quad (9.34)$$

where $\epsilon' = \pm 1$, $|t_2^n d_i\rangle$ is a determinantal function and $|t_2^{6-n} d_i\rangle$ is the (unique apart from sign) determinantal function which has those one-electron functions which do not occur in $|t_2^n d_i\rangle$. Choose the phases so that we always have $\epsilon' = +1$. There are q terms in the sum (9.34). Now the $|t_2^n S\Gamma M_S M_\Gamma\rangle$ are an orthogonal transform, O say, of the $|t_2^n d_i\rangle$. Transform the $|t_2^{6-n} d_i\rangle$ by O also and, because $O'O = OO' = I$, (9.34) becomes transformed into

$$\psi = q^{-\frac{1}{2}}\Sigma |t_2^n S\Gamma M_S M_\Gamma\rangle . |t_2^{6-n} j\rangle, \quad (9.35)$$

and $|t_2^{6-n} j\rangle$ is equal to $|t_2^{6-n} S\Gamma - M_S M_\Gamma\rangle$ apart possibly from phase. This is easily seen by multiplying both (9.33) and (9.35) by $|t_2^n S\Gamma M_S M_\Gamma\rangle$ and integrating. We obtain

$$a(S\Gamma)(-1)^{S-M_s}(2S+1)^{-\frac{1}{2}}\lambda(\Gamma)^{-\frac{1}{2}}|t_2^{6-n} S\Gamma - M_S M_\Gamma\rangle = q^{-\frac{1}{2}}|t_2^{6-n} j\rangle$$

and hence the result because both kets concerned are normalized. It also follows that

$$a(S\Gamma) = \pm(2S+1)^{\frac{1}{2}}\lambda(\Gamma)^{\frac{1}{2}}q^{-\frac{1}{2}}. \quad (9.36)$$

We now define the phases of all t_2^{6-n} terms with $n < 3$ by requiring that the positive sign should hold in (9.36). This definition conforms with the conventions adopted in the theory of atomic structure and is consistent with the phases of the t_2^n functions given in Table A 24 whether we use real or complex t_2 orbitals (see below). Table A 24 is also consistent with the usual phase convention for p^n under the p^n isomorphism discussed in § 9.5.

This now allows us to define a correspondence between individual states of t_2^n and of t_2^{6-n}. We shall say that a state $|t_2^{6-n} S\Gamma - M_S M'_\Gamma\rangle$ is the complementary state to $|t_2^n S\Gamma M_S M_\Gamma\rangle$ if it gets multiplied into it in the relation (9.33) and if, further, $a(S\Gamma)$ is positive. We now call states with $n \leqslant 3$ less-than-half-filled shell states or L-states and those with $n \geqslant 3$ more-than-half-filled shell or R-states.

Equation (9.33) then shows how we derive an L-state from its complementary R-state and conversely. It may now be rewritten

$$\psi = q^{-\frac{1}{2}}\Sigma(-1)^{S-M_S}\,|t_2^n S\Gamma M_S M_\Gamma\rangle_L \, . \, |t_2^{6-n} S\Gamma - M_S M_\Gamma\rangle_R. \qquad (9.37)$$

In case we wish to discuss the case when the components of Γ form bases for a complex representation which is related to the real representations occurring in (9.37) by a unitary matrix A, we have merely to transform the L-states by A and the R-states by $\bar{A}$. This leaves the form of (9.37) unchanged except that the component M_Γ need no longer be the same in the two factors.

States with $n = 3$ can be either L- or R-states, but it may happen that they do not have the same phase in their two manifestations. Take an L-state $|t_2^3 S\Gamma M_S M_\Gamma\rangle_L$. This defines an R-state $|t_2^3 S\Gamma - M_S M_\Gamma\rangle_R$ through (9.37) and then via the standard basis relations between the components of ${}^{2S+1}\Gamma$ an R-state $|t_2^3 S\Gamma M_S M_\Gamma\rangle_R$ which must be the state $|t_2^3 S\Gamma M_S M_\Gamma\rangle_L$ but may differ from it in sign. We call $|t_2^3 S\Gamma M_S M_\Gamma\rangle$ a positive or a negative state according as to whether $|t_2^3 S\Gamma M_S M_\Gamma\rangle_R$ is equal to $+|t_2^3 S\Gamma M_S M_\Gamma\rangle_L$ or $-|t_2^3 S\Gamma M_S M_\Gamma\rangle_L$. All components of a term have the same parity, in this sense, and so terms of t_2^3 get classified as positive or negative. If we change the phase of $|t_2^6\,{}^1A_1\rangle$ we change all positive terms into negative ones and vice versa. The fact that we have two different kinds of terms in the half-filled shell is not, however, merely a matter of definition and is of interest and importance.

The parities of the terms of t_2^3 are easily determined. For example, take

$$|{}^2T_1\,\tfrac{1}{2}x\rangle_L = \frac{1}{\sqrt{2}}(\xi^+\eta^2 - \xi^+\zeta^2)$$

which must be correlated with

$$|{}^2T_1 - \tfrac{1}{2}x\rangle_R = -\frac{1}{\sqrt{2}}(\xi^-\eta^2 - \xi^-\zeta^2) = -|{}^2T_1 - \tfrac{1}{2}x\rangle_L,$$

to get the order $\xi^2\eta^2\zeta^2$ right in t_2^6. Therefore ${}^2T_1$ is negative. ${}^4A_2$ and ${}^2E$ are also negative, but ${}^2T_2$ is positive. We note that the states of t_2^3 in Table A 24 are all regarded as L-states.

e^n configurations are treated in exactly the same way and Table A 24 is consistent with our new-found convention for them also. In e^2, ${}^1A_1$ is positive and ${}^1E$ and ${}^3A_2$ are negative.

It is convenient to extend the definition of L- and R-states so that every state $|t_2^n S\Gamma M_S M_\Gamma\rangle$ and $|e^n S\Gamma M_S M_\Gamma\rangle$ has both an L- and an R-form. We do this simply by taking the states in Table A 24 based on real orbitals as L-states and define the R-states from (9.37) as before. Then we have

$$|t_2^n S\Gamma M_S M_\Gamma\rangle_R = \mu_1\,|t_2^n S\Gamma M_S M_\Gamma\rangle_L,$$

where $\mu_1 = +1$ unless $n = 3$ and $S\Gamma = {}^4A_2$, ${}^2E$ or ${}^2T_1$ when $\mu_1 = -1$, and also

$$|e^n S\Gamma M_S M_\Gamma\rangle_R = \mu_2\,|e^n S\Gamma M_S M_\Gamma\rangle_L,$$

where $\mu_2 = +1$ unless $n = 2$ and $S\Gamma = {}^3A_2$ or ${}^1E$ when $\mu_2 = -1$.

We now derive a result which is useful later. The relation (9.37) shows the dependence of the unique function which forms a basis for the alternating representation of S_6 on the q^2 products $|t_2^n\rangle_L.|t_2^{6-n}\rangle_R$. q^2-1 orthogonal linear combinations of these products can be chosen so that they also form bases for representations of S_6. ψ and these q^2-1 other functions are an orthonormal transform of the $|t_2^n\rangle_L.|t_2^{6-n}\rangle_R$. Inverting this transformation gives

$$(-1)^{S-M_S}|t_2^n S\Gamma M_S M_\Gamma\rangle_L.|t_2^{6-n} S'\Gamma' -M'_S M'_\Gamma\rangle_R$$
$$= \delta(S,S')\,\delta(\Gamma,\Gamma')\,\delta(M_S,M'_S)\,\delta(M_\Gamma,M'_\Gamma)\,q^{-\frac{1}{2}}\psi+\Sigma\psi_i,$$

where the ψ_i all form parts of bases for representations different from the alternating representation of S_6. Therefore they each satisfy $\sum_\nu(-1)^\nu P_\nu\psi_i = 0$ and so we multiply our equation through by $(1/6!)\sum_\nu(-1)^\nu P_\nu$. We want our answer expressed in terms of L-functions so we change to L-functions and drop the suffix L. Then

$$\delta(S,S')\,\delta(\Gamma,\Gamma')\,\delta(M_S,M'_S)\,\delta(M_\Gamma,M'_\Gamma)\,q^{-\frac{1}{2}}\psi$$
$$= \frac{1}{6!}\mu_1(-1)^{S-M_S}\Sigma(-1)^\nu P_\nu|t_2^n S\Gamma M_S M_\Gamma\rangle.|t_2^{6-n} S'\Gamma' -M'_S M'_\Gamma\rangle$$
$$= q^{-\frac{1}{2}}\mu_1(-1)^{S-M_S}|t_2^n S\Gamma M_S M_\Gamma,\, t_2^{6-n} S'\Gamma' -M'_S M'_\Gamma\rangle,$$

whence
$$|t_2^n S\Gamma M_S M_\Gamma,\, t_2^{6-n} S'\Gamma' -M'_S M'_\Gamma\rangle$$
$$= \delta(S,S')\,\delta(\Gamma,\Gamma')\,\delta(M_S,M'_S)\,\delta(M_\Gamma,M'_\Gamma)\,\mu_1(-1)^{S-M_S}|t_2^6\rangle.$$

The functions on both sides of the equation are completely antisymmetrized. Precisely the same form of relation holds for the functions e^n.

9.7.2. One-electron operators. Having defined a relation between L-states of t_2^n and R-states of t_2^{6-n} we next ask for the relationship between the matrices of quantities of interest for t_2^n and t_2^{6-n}. In this section we find this for a one-electron operator $u(\kappa)$.

The proof uses Kramers's star operator. Accordingly, we first investigate the effect of this operator on the kets $|t_2^n\rangle$ and $|e^n\rangle$. We write our equations for t_2^n but they are also directly applicable to e^n with a slight modification which will be indicated as we go along. We write

$$|t_2^n S\Gamma M_S M_\Gamma\rangle_L^* = \epsilon_L(nS\Gamma)(-1)^{M_S}|t_2^n S\Gamma -M_S M_\Gamma\rangle_L,$$
$$|t_2^n S\Gamma M_S M_\Gamma\rangle_R^* = \epsilon_R(nS\Gamma)(-1)^{M_S}|t_2^n S\Gamma -M_S M_\Gamma\rangle_R$$

using real representations Γ. ϵ_L and ϵ_R are independent of M_S and M_Γ. This can be proved by the methods of § 8.4 using the shift operators (see especially ex. 6, p. 208) or verified directly from Table A 24. We have already defined $|t_2^n\rangle$ including phase so the ϵ_L and ϵ_R are therefore implicitly determined. For the purposes of the present proof, however, we will not commit ourselves to quite such a definite choice of phase.

We first establish a relation between ϵ_L and ϵ_R by applying the star operator to (9.37). $\psi^* = -\psi$ and so

$$-\psi = q^{-\frac{1}{2}}\Sigma(-1)^{S-M_S}\epsilon_L(nS\Gamma)\,\epsilon_R(6-n\,S\Gamma)\,|t_2^n\,S\Gamma\,-M_S\,M_\Gamma\rangle_L.|t_2^{6-n}\,S\Gamma M_S M_\Gamma\rangle_R.$$

Comparing coefficients with (9.37) we deduce

$$\epsilon_L(nS\Gamma)\,\epsilon_R(6-n\,S\,\Gamma) = -(-1)^{2M_S} = -(-1)^n.$$

The analogous relation for e^n has $+(-1)^n$. In either case the relation simplifies to $\epsilon_L(nS\Gamma)\,\epsilon_R(nS\Gamma) = 1$ for a half-filled shell. Because of this we may without inconsistency choose $\epsilon_L(nS\Gamma) = \epsilon_R(nS\Gamma) = \pm 1$ always. We do so and drop the suffixes L and R. This is also consistent with the particular choice of phases adopted in Appendix A 24, column 2.

Next we consider a pair of states

$$|\phi\rangle = |t_2^n\,S_1\Gamma_1 M_{S1} M_{\Gamma 1}\rangle_L, \quad |\chi\rangle = |t_2^n\,S_2\Gamma_2 M_{S2} M_{\Gamma 2}\rangle_L$$

and write their complementary states as $|\phi'\rangle$, $|\chi'\rangle$. Suppose $\phi \neq \chi$ so that $|\phi.\chi'\rangle$ does not occur in (9.37). Then when $|\phi.\chi'\rangle$ is expressed as a sum over components of irreducible representations of S_6 the alternating representation does not occur. The sum $\sum_{\kappa=1}^{6} u(\kappa)$ belongs to the unit representation of S_6. Hence by the orthogonality relations for integrals

$$\langle\bar{\phi}.\bar{\chi}'|\sum_{\kappa=1}^{6} u(\kappa)\,|t_2^{6}\,{}^1A_1\rangle = 0,$$

whence

$$\langle\bar{\phi}.\bar{\chi}'|\sum_{\kappa=1}^{n} u(\kappa)\,|t_2^{6}\,{}^1A_1\rangle + \langle\bar{\phi}.\bar{\chi}'|\sum_{n+1}^{6} u(\kappa)\,|t_2^{6}\,{}^1A_1\rangle = 0.$$

So using the expansion (9.37) and writing U for $\Sigma u(\kappa)$ we have

$$(-1)^{S_2-M_{S2}}\langle\bar{\phi}\,|\,U\,|\,\chi\rangle + (-1)^{S_1-M_{S1}}\langle\bar{\chi}'\,|\,U\,|\,\phi'\rangle = 0.$$

We turn the matrix element $\langle\bar{\chi}'\,|\,U\,|\,\phi'\rangle$ over by first barring it and then starring it. The final result of this is to leave its value unaltered. Assume $U = \bar{U} = \eta U^*$ with $\eta = \pm 1$. Also write $\epsilon_1 = \epsilon(nS_1\Gamma_1)$, $\epsilon_2 = \epsilon(nS_2\Gamma_2)$. Then

$$\begin{aligned}\langle\bar{\chi}'\,|\,U\,|\,\phi'\rangle &= \langle\bar{\phi}'\,|\,U\,|\,\chi'\rangle^* \\ &= \langle t_2^{6-n}\,S_1\Gamma_1\,-M_{S1}\,M_{\Gamma 1}\,|\,U\,|\,t_2^{6-n}\,S_2\Gamma_2\,-M_{S2}\,M_{\Gamma 2}\rangle_R^* \\ &= \eta\epsilon_1\epsilon_2(-1)^{M_{S1}-M_{S2}}\langle t_2^{6-n}\,S_1\Gamma_1 M_{S1} M_{\Gamma 1}\,|\,U\,|\,t_2^{6-n}\,S_2\Gamma_2 M_{S2} M_{\Gamma 2}\rangle_R\end{aligned}$$

whence

$$\begin{aligned}&\langle t_2^n\,S_1\Gamma_1 M_{S1} M_{\Gamma 1}\,|\,U\,|\,t_2^n\,S_2\Gamma_2 M_{S2} M_{\Gamma 2}\rangle_L \\ &\quad = -\eta\epsilon_1\epsilon_2(-1)^{S_1-S_2}\langle t_2^{6-n}\,S_1\Gamma_1 M_{S1} M_{\Gamma 1}\,|\,U\,|\,t_2^{6-n}\,S_2\Gamma_2 M_{S2} M_{\Gamma 2}\rangle_R.\end{aligned}$$

This may be abbreviated to

$$U_L = -\eta\epsilon_1\epsilon_2(-1)^{S_1-S_2}\,U_R.$$

We now write $\epsilon(nS\Gamma) = (-i)^{n+2S}$ and have $\epsilon_1\epsilon_2(-1)^{S_1-S_2} = 1$. Therefore, for both t_2^n and e^n, we find

$$U_L = -\eta U_R. \tag{9.38a}$$

Thus when $\eta = +1$, which is true for a ligand field or for the spin-orbit coupling energy, all off-diagonal elements change sign on passing to the complementary scheme. When $\eta = -1$ they remain unchanged. This choice of ϵ values is actually satisfied by the functions based on real orbitals given in Table A 24.

The relation for the diagonal elements is obtained in a similar way. Here

$$(-1)^{S-M_s}|\phi.\phi'\rangle = q^{-\frac{1}{2}}\psi + \Sigma\psi_i$$

and so
$$\langle\bar{\phi}.\bar{\phi}'|U|\phi.\phi'\rangle = \langle\bar{\psi}|U|\psi\rangle = C,$$

where C is a real number independent of ϕ. This yields

$$\langle\bar{\phi}|U|\phi\rangle + \langle\bar{\phi}'|U|\phi'\rangle = C.$$

We find
$$\langle\bar{\phi}'|U|\phi'\rangle = \eta\langle t_2^{6-n}S_1\Gamma_1 M_{S1}M_{\Gamma 1}|U|t_2^{6-n}S_1\Gamma_1 M_{S1}M_{\Gamma 1}\rangle_R$$
$$= \eta U_R,$$

and hence
$$U_L = C - \eta U_R. \tag{9.38b}$$

Thus apart from a constant diagonal energy common to all states we have $U_L = -U_R$ when $\eta = +1$ and $U_L = U_R$ when $\eta = -1$. The constant C is zero unless $u(\kappa)$ contains a component of the unit representation of the symmetry group O^*. For the spin-orbit coupling energy it is zero even though the latter condition is satisfied. The reader should find a group-theoretic reason for this.

Example

The spin-orbit coupling and the ligand field have zero matrix elements between states of t_2^3 having the same 'parity', while the vector **L** has zero matrix elements between states of opposite 'parity'.

9.7.3. The configurations $t_2^m e^n$.

We define complementary states for $t_2^m e^n$ with respect to the (unique) totally antisymmetric $^1A_1$ state $t_2^6 e^4$. The argument used is the same as for t_2^m alone but there is a slight additional complication. A state

$$\phi = |t_2^m(S_1\Gamma_1)\,e^n(S_2\Gamma_2)\,S\Gamma M_S M_\Gamma\rangle$$

is actually complementary to a state

$$\phi' = \pm|t_2^{6-m}(S_1\Gamma_1)\,e^{4-n}(S_2\Gamma_2)\,S\Gamma - M_S M_\Gamma\rangle$$

of $t_2^{6-m}e^{4-n}$. But the term $t_2^{6-m}(S_1\Gamma_1)\,e^{4-n}(S_2\Gamma_2)\,S\Gamma$ is not in general the only term of the type $^{2S+1}\Gamma$ in $t_2^{6-m}e^{4-n}$ and so we require to prove that in the relation corresponding to (9.33) the state ϕ only gets multiplied by the state ϕ' and not by any other states having the same overall symmetry.

The equation corresponding to (9.35) is

$$\psi = |t_2^6 e^4\rangle = q^{-\frac{1}{2}}\Sigma\,|t_2^m(S_1\Gamma_1)\,e^n(S_2\Gamma_2)\,S\Gamma M_S M_\Gamma\rangle.|j\rangle, \tag{9.39}$$

where $q = 10!/\{(m+n)!\,(10-m-n)!\}$, and is still true. In general $|j\rangle$ will be a sum

$$|j\rangle = (-1)^{S-M_s}\sum_\alpha a_\alpha\,|t_2^{m'}(S_1'\Gamma_1')\,e^{n'}(S_2'\Gamma_2')\,S\Gamma - M_S M_\Gamma\rangle$$

with the a_α numbers independent of M_S, M_Γ and satisfying $\Sigma|a_\alpha|^2 = 1$. We let $\alpha = 0$ refer to the term $t_2^{6-m}(S_1\Gamma_1)\,e^{4-n}(S_2\Gamma_2)\,S\Gamma$ and will prove that $a_0 = \pm 1$

which will show that all the other a_α are zero. Before doing this I remark that *all* our kets are built up by the procedure of § 9.1 from $|t_2^m\rangle_L$ and $|e^n\rangle_L$ and are therefore L states.

First, we invert the transformation on q^2 variables of which (9.39) is part to give

$$q^{-\frac{1}{2}}\psi + \Sigma\psi_i = |t_2^m(S_1\Gamma_1)\,e^n(S_2\Gamma_2)\,S\Gamma M_S M_\Gamma\rangle\,.\,|j\rangle$$
$$= (-1)^{S-M_S}\sum_\alpha a_\alpha\,|t_2^m(S_1\Gamma_1)\,e^n(S_2\Gamma_2)\,S\Gamma M_S M_\Gamma\rangle$$
$$.\,|t_2^{m'}(S_1'\Gamma_1')\,e^{n'}(S_2'\Gamma_2')\,S\Gamma\; -M_S M_\Gamma\rangle.$$

Multiply through by $(1/10!)\sum_\nu(-1)^\nu P_\nu$ and we get

$$\psi = (-1)^{S-M_S}\sum_\alpha a_\alpha\,|t_2^m(S_1\Gamma_1)\,e^n(S_2\Gamma_2)\,S\Gamma M_S M_\Gamma,\; t_2^{m'}(S_1'\Gamma_1')\,e^{n'}(S_2'\Gamma_2')\,S\Gamma - M_S M_\Gamma\rangle. \tag{9.40}$$

If we now multiply through (9.40) by the bra corresponding to the αth ket on the right-hand side we obtain an expression for a_α. Because of the orthogonality of the one-electron functions this must be zero unless $m' = 6-m$ and $n' = 4-n$.

In (9.40) both t_2^m and e^n and also t_2^{6-m} and e^{4-n} are coupled together to form the states of the term ${}^{2S+1}\Gamma$. We now uncouple them and alter the order of e^n and t_2^{6-m} to give

$$\psi = (-1)^{S-M_S}\sum_\alpha a_\alpha\,|t_2^m(S_1\Gamma_1)\,e^n(S_2\Gamma_2)\,S\Gamma M_S M_\Gamma,$$
$$t_2^{6-m}(S_1'\Gamma_1')\,e^{4-n}(S_2'\Gamma_2')\,S\Gamma\; -M_S M_\Gamma\rangle$$
$$= (-1)^{mn+S-M_S}\sum_{\substack{\alpha M_{S1}M_{\Gamma 1}M_{S2}M_{\Gamma 2}\\ M'_{S1}M'_{\Gamma 1}M'_{S2}M'_{\Gamma 2}}} a_\alpha\,\langle S_1S_2M_{S1}M_{S2}\,|\,S_1S_2SM_S\rangle \times\langle\Gamma_1\Gamma_2M_{\Gamma 1}M_{\Gamma 2}\,|\,\Gamma_1\Gamma_2\Gamma M_\Gamma\rangle$$
$$\times\langle S_1'S_2'M_{S1}'M_{S2}'\,|\,S_1'S_2'S\; -M_S\rangle\langle\Gamma_1'\Gamma_2'M_{\Gamma 1}'M_{\Gamma 2}'\,|\,\Gamma_1'\Gamma_2'\Gamma M_\Gamma\rangle$$
$$\times\,|\,t_2^m S_1\Gamma_1M_{S1}M_{\Gamma 1},\, t_2^{6-m}S_1'\Gamma_1'M_{S1}'M_{\Gamma 1}',\, e^nS_2\Gamma_2M_{S2}M_{\Gamma 2},\, e^{4-n}S_2'\Gamma_2'M_{S2}'M_{\Gamma 2}'\rangle.$$

We now use the result proved at the end of § 9.7.1. This shows that the only non-zero ket occurring in the sum over α is that for $\alpha = 0$. So ψ becomes expressed as a multiple of itself. Therefore we must have

$$1 = (-1)^{mn+S-M_S}a_0\mu_1\mu_2\sum_{M_{S1}M_{\Gamma 1}M_{S2}M_{\Gamma 2}}(-1)^{S_1+S_2-M_{S1}-M_{S2}}\langle\Gamma_1\Gamma_2M_{\Gamma 1}M_{\Gamma 2}\,|\,\Gamma_1\Gamma_2\Gamma M_\Gamma\rangle^2$$
$$\times\langle S_1S_2M_{S1}M_{S2}\,|\,S_1S_2SM_S\rangle\langle S_1S_2 - M_{S1} - M_{S2}\,|\,S_1S_2S - M_S\rangle$$
$$= (-1)^{mn}\mu_1\mu_2 a_0,$$

where we have used ex. 7, p. 209. This establishes the result and we now define our standard correlation between L and R states from the equation

$$\psi = q^{-\frac{1}{2}}\Sigma(-1)^{S-M_S}\,|t_2^m(S_1\Gamma_1)\,e^n(S_2\Gamma_2)\,S\Gamma M_S M_\Gamma\rangle_L$$
$$.\,|t_2^{6-m}(S_1\Gamma_1)\,e^{4-n}(S_2\Gamma_2)\,S\Gamma\; -M_S M_\Gamma\rangle_R \tag{9.41}$$

and have proved that

$$|t_2^m(S_1\Gamma_1)\,e^n(S_2\Gamma_2)\,S\Gamma M_S M_\Gamma\rangle_R = (-1)^{mn}\,\mu_1\mu_2\,|t_2^m(S_1\Gamma_1)\,e^n(S_2\Gamma_2)\,S\Gamma M_S M_\Gamma\rangle_L, \tag{9.42}$$

where $\mu_1 = +1$ unless $m = 3$ and $S_1\Gamma_1 = {}^4A_2$, ${}^2E$ or ${}^2T_1$, when $\mu_1 = -1$, and $\mu_2 = +1$ unless $n = 2$ and $S\Gamma = {}^3A_2$ or ${}^1E$, when $\mu_2 = -1$.

In order to discuss the matrices of one-electron operators we must know how the star operator operates on the kets of $t_2^m e^n$. In fact the formula

$$\epsilon(aS\Gamma) = (-i)^{a+2S}$$

which holds for t_2^a or e^a separately holds also for any term

$$t_2^m(S_1\Gamma_1)\,e^{a-m}(S_2\Gamma_2)\,S\Gamma.$$

I leave it as an exercise for the reader to prove this using ex. 7, p. 209.

Our previous discussion of one-electron operators can now be taken straight over. There is a slight difference for half-filled shells, however, because not all terms get correlated with themselves. Only the terms of $t_2^3e^2$ get correlated with themselves and their 'parities' are the products of the parities of the constituent terms of t_2^3 and e^2. Thus the matrix elements of the spin-orbit coupling are zero within the terms of $t_2^3e^2$ but not within t_2^5.

9.7.4. An example.

The relationship between the matrix of a two-electron operator in the L-scheme and in the complementary R-scheme is not quite so simple as for a one-electron operator. Therefore it is helpful first to consider an example so that we shall know what to try and prove.

In § 9.3.7 we obtained the energies in general parameters of the terms of t_2^5e relative to ${}^3T_1$ of that configuration. The complementary terms arise from t_2e^3 and their energies are

$$\left.\begin{aligned} E({}^3T_1) &= 3e-5f+3d+\frac{1}{\sqrt{3}}c-2g-\frac{2}{\sqrt{3}}h,\\ E({}^3T_2) &= 3e-5f+3d+\frac{5}{\sqrt{3}}c-2g-\frac{2}{\sqrt{3}}h,\\ E({}^1T_1) &= 3e-5f+3d+\frac{1}{\sqrt{3}}c-\frac{4}{\sqrt{3}}h,\\ E({}^1T_2) &= 3e-5f+3d+\frac{5}{\sqrt{3}}c+\frac{4}{\sqrt{3}}h. \end{aligned}\right\} \tag{9.43}$$

The relative energies here are the same as the relative energies for t_2^5e given in (9.16). The state complementary to the ground term t_2^6 is e^4 with energy

$$E(e^4\,{}^1A_1) = 6e-10f,$$

and

$$E({}^3T_1)-E(e^4\,{}^1A_1) = -3e+5f+3d+\frac{1}{\sqrt{3}}c-2g-\frac{2}{\sqrt{3}}h,$$

which is not the same as the corresponding separation of t_2^5e from t_2^6. In the special case of d-functions, however, both separations become $-3C$.

This example suggests the result we prove in § 9.7.7: for d-functions the matrix of electrostatic energy is the same for an L-scheme as for the complementary R-scheme, except for a constant diagonal energy difference common to all states of the d^e configuration. For functions which are not d-functions the same result is true, except that the diagonal energy difference is only necessarily constant within a configuration $t_2^m e^n$ but may change when we change m and n keeping $m+n$ constant.

9.7.5. Expansion of the electrostatic energy. We saw in § 4.4 that the electrostatic interaction between two electrons can be expressed as a sum of products of one-electron functions. Using the spherical harmonic addition theorem in the form given in (8.11),

$$V(12) = \sum_{k=0}^{\infty} \frac{4\pi e^2}{2k+1} \sum_m \frac{r_<^k}{r_>^{k+1}} Z_{km}(1)\, Z_{km}(2),$$

and hence if we replace the Z_{km} by linear combinations which are components of bases for irreducible representations of O, then

$$V(12) = \Sigma f_{i\Gamma M}(1)\, f_{i\Gamma M}(2). \tag{9.44}$$

In (9.44) $f_{i\Gamma M}$ is a one-electron function which forms the component M of a basis for Γ. The parameter i labels different sets of functions forming bases for the same irreducible representation Γ. We choose all functions to be real and M stands for x, y or z for T_1 and ξ, η or ζ for T_2.

The evaluation of the electrostatic matrix element between a pair of states of $t_2^m e^n$ requires the two-electron integrals $\langle ab \mid V \mid cd\rangle$ shown in (9.5). Because of (9.44),

$$\begin{aligned}\langle ab \mid V \mid cd\rangle &= \Sigma \langle ab \mid f_{i\Gamma M}(1) f_{i\Gamma M}(2) \mid cd\rangle \\ &= \Sigma \langle a \mid f_{i\Gamma M} \mid c\rangle \langle b \mid f_{i\Gamma M} \mid d\rangle\end{aligned}$$

and therefore depends only on the matrix elements of the one-electron operators $f_{i\Gamma M}$ between and amongst the one-electron functions of e and t_2. It is convenient to replace $f_{i\Gamma M}$ by $g_{i\Gamma M}$, where $g_{i\Gamma M}$ has the same matrix elements as $f_{i\Gamma M}$ within the five e and t_2 functions of interest but has zero matrix elements between these functions and any others and also between any other functions. Then the electrostatic energy has become

$$\mathscr{V} = \sum_{\kappa<\lambda} V(\kappa\lambda) = \sum_{i\Gamma M} \sum_{\kappa<\lambda} g_{i\Gamma M}(\kappa)\, g_{i\Gamma M}(\lambda), \tag{9.45}$$

with the $g_{i\Gamma M}$ real and having their components correctly connected in phase.

A one-electron integral $\langle a \mid g_{i\Gamma M} \mid c\rangle$ can only be non-zero if Γ occurs in the product of the representations Γ_1 and Γ_3 to which $|a\rangle$ and $|c\rangle$ belong or, if $\Gamma_1 = \Gamma_3$, to the symmetrized square of Γ_1. So if $\Gamma_1 = \Gamma_3 = e$ then $\Gamma = A_1$ or E; if $\Gamma_1 = e$, $\Gamma_3 = t_2$ then $\Gamma = T_1$ or T_2, and if $\Gamma_1 = \Gamma_3 = t_2$ then $\Gamma = A_1, E$ or T_2. A_2 never occurs. Let us write

$$\mathscr{V}_{i\Gamma} = \sum_M \sum_{\kappa<\lambda} g_{i\Gamma M}(\kappa)\, g_{i\Gamma M}(\lambda), \quad \mathscr{V} = \sum_\Gamma \sum_i \mathscr{V}_{i\Gamma}. \tag{9.46}$$

We do not need $\Gamma = A_2$, so $\mathscr{V}$ is a sum of four contributions, one from each irreducible representation A_1, E, T_1 and T_2.

9.7.6. Two-electron integrals in terms of one-electron operators. For a two-electron system we have $\mathscr{V}_{i\Gamma} = \sum_{M} g_{i\Gamma M}(1)\, g_{i\Gamma M}(2)$. First, let there be just one Γ, of A_1 symmetry, then we may abbreviate this to $\mathscr{V}_{A_1} = g(1)\,g(2)$. The only non-vanishing matrix elements of g are

$$\left.\begin{aligned} &\langle \xi \,|\, g \,|\, \xi\rangle = \langle \eta \,|\, g \,|\, \eta\rangle = \langle \zeta \,|\, g \,|\, \zeta\rangle = k, \\ &\langle \theta \,|\, g \,|\, \theta\rangle = \langle \epsilon \,|\, g \,|\, \epsilon\rangle = l, \quad \text{say.} \end{aligned}\right\} \tag{9.47}$$

The values of the ten parameters $a, \ldots, j$ determining the electrostatic matrix of $\mathscr{V}_{A_1}$ can then be expressed in terms of k and l to give

$$\begin{aligned} &a = (\xi^2;\, \xi^2) = \langle \xi \,|\, g \,|\, \xi\rangle \langle \xi \,|\, g \,|\, \xi\rangle = k^2, \\ &b = (\xi^2;\, \eta^2) = \langle \xi \,|\, g \,|\, \xi\rangle \langle \eta \,|\, g \,|\, \eta\rangle = k^2, \\ &d = (\xi^2;\, \epsilon^2) = kl, \quad e = l^2, \\ &c = f = g = h = i = j = 0. \end{aligned}$$

The treatment of the representations with degree greater than one is only slightly more complicated. For example, if $\mathscr{V}_E = g_\theta(1)\,g_\theta(2) + g_\epsilon(1)\,g_\epsilon(2)$ then g_M has non-zero matrix elements within the e orbitals and within the t_2 orbitals but not between e and t_2. Considering the e orbitals, and using Table A 20, the two functions

$$-\frac{1}{\sqrt{2}} g_\theta \,|\theta\rangle + \frac{1}{\sqrt{2}} g_\epsilon \,|\epsilon\rangle \quad \text{and} \quad \frac{1}{\sqrt{2}} g_\theta \,|\epsilon\rangle + \frac{1}{\sqrt{2}} g_\epsilon \,|\theta\rangle$$

form the θ- and ϵ-components, respectively, of an E-state. Therefore

$$-\frac{1}{\sqrt{2}}\langle \theta \,| g_\theta \,|\, \theta\rangle + \frac{1}{\sqrt{2}}\langle \theta \,|\, g_\epsilon \,|\, \epsilon\rangle = \frac{1}{\sqrt{2}}\langle \epsilon \,|\, g_\theta \,|\, \epsilon\rangle + \frac{1}{\sqrt{2}}\langle \epsilon \,|\, g_\epsilon \,|\, \theta\rangle = p\sqrt{2},$$

say, and $$-\langle \epsilon \,|\, g_\theta \,|\, \theta\rangle + \langle \epsilon \,|\, g_\epsilon \,|\, \epsilon\rangle = \langle \theta \,|\, g_\theta \,|\, \epsilon\rangle + \langle \theta \,|\, g_\epsilon \,|\, \theta\rangle = 0,$$

using the orthogonality properties for integrals. Similar relations follow on coupling g_θ, g_ϵ with $|\theta\rangle$, $|\epsilon\rangle$ to form A_1 and A_2 states. Solution of the resulting linear equations (or use (9.23)) yields

$$\left.\begin{aligned} &\langle \epsilon \,|\, g_\theta \,|\, \epsilon\rangle = -\langle \theta \,|\, g_\theta \,|\, \theta\rangle = \langle \theta \,|\, g_\epsilon \,|\, \epsilon\rangle = \langle \epsilon \,|\, g_\epsilon \,|\, \theta\rangle = p, \\ &\langle \epsilon \,|\, g_\epsilon \,|\, \epsilon\rangle = \quad \langle \theta \,|\, g_\epsilon \,|\, \theta\rangle = \langle \theta \,|\, g_\theta \,|\, \epsilon\rangle = \langle \epsilon \,|\, g_\theta \,|\, \theta\rangle = 0. \end{aligned}\right\} \tag{9.48}$$

Similarly $$\begin{aligned} &\langle \xi \,|\, g_\theta \,|\, \xi\rangle = \quad \langle \eta \,|\, g_\theta \,|\, \eta\rangle = -\tfrac{1}{2}\langle \zeta \,|\, g_\theta \,|\, \zeta\rangle = q, \\ &\langle \xi \,|\, g_\epsilon \,|\, \xi\rangle = -\langle \eta \,|\, g_\epsilon \,|\, \eta\rangle = -q\sqrt{3}, \end{aligned}$$

and all other elements are zero. Hence

$$a = 4q^2, \quad b = -2q^2, \quad c = -pq\sqrt{3}, \quad d = pq, \quad e = f = p^2, \quad g = h = i = j = 0.$$

For both T_1 and T_2, $a = b = c = d = e = f = 0$. $g = 3r^2$, $h = -r^2\sqrt{3}$, $i = j = 0$ for T_1 and $g = s^2$, $h = s^2\sqrt{3}$, $i = -st$, $j = t^2$ for T_2 for suitable parameters r, s, t.

Armed with these results it is easy to determine equivalent one-electron operators for the electrostatic interaction between d electrons. We consider the parts arising from F_0, F_2 and F_4 separately. Take F_0 first, then, from Table A 26, we see that $a = b = d = e = F_0$, $c = f = g = h = i = j = 0$ so the equivalent one-electron operator (of A_1 symmetry) has $k = l = F_0^{\frac{1}{2}}$. The coefficient of F_2 has one-electron functions of E and of T_2 symmetry so we solve the equation system

$$a = 4F_2 = 4q^2, \quad b = -2F_2 = -2q^2, \quad c = 2\sqrt{3}\,F_2 = -pq\sqrt{3},$$
$$d = -2F_2 = pq, \quad e = f = 4F_2 = p^2, \quad g = F_2 = s^2,$$
$$h = \sqrt{3}\,F_2 = s^2\sqrt{3}, \quad i = \sqrt{3}\,F_2 = -st, \quad j = 3F_2 = t^2,$$

which are satisfied by

$$q = s = F_2^{\frac{1}{2}}, \quad p = -2F_2^{\frac{1}{2}}, \quad t = -\sqrt{3}\,F_2^{\frac{1}{2}}.$$

In the same way the part arising from F_4 is replaced by one-electron operators of A_1, E, T_1 and T_2 symmetries with parameters satisfying

$$k' = \frac{2\sqrt{7}}{\sqrt{3}}F_4^{\frac{1}{2}}, \quad l' = -\sqrt{21}\,F_4^{\frac{1}{2}}, \quad p' = \sqrt{15}\,F_4^{\frac{1}{2}}, \quad q' = \frac{2\sqrt{5}}{\sqrt{3}}F_4^{\frac{1}{2}},$$
$$r' = \tfrac{1}{2}\sqrt{35}\,F_4^{\frac{1}{2}}, \quad s' = \tfrac{1}{2}\sqrt{15}\,F_4^{\frac{1}{2}}, \quad t' = 2\sqrt{5}\,F_4^{\frac{1}{2}},$$

where the dashes distinguish them from the parameters for F_0 and F_2.

9.7.7. Electrostatic matrices for complementary schemes. We are now in a position to derive the relationship between the electrostatic matrices for the L- and R-schemes. First, we treat a fully symmetric part $\mathscr{V}_{iA_1}$ of $\mathscr{V}$. From (9.47), $\mathscr{V}_{iA_1}$ is diagonal within a scheme in which the basic states are determinantal functions built up from e and t_2. The diagonal element for a state of $t_2^m e^n$ is

$$E(t_2^m e^n) = \tfrac{1}{2}m(m-1)\,k_i^2 + mnk_i l_i + \tfrac{1}{2}n(n-1)\,l_i^2, \tag{9.49}$$

and depends only on m and n. Therefore the same formula holds for the energy of a state $|t_2^m(S_1\Gamma_1)\,e^n(S_2\Gamma_2)\,S\Gamma M_S M_\Gamma\rangle$ and we may write down an explicit formula for the energy difference between complementary states as a function of m and n. For d-functions there are just two $\mathscr{V}_{iA_1}$ and if we sum over i the formula for the energy becomes

$$E(t_2^m e^n) = \tfrac{1}{2}m(m-1)\,(F_0 + 9\tfrac{1}{3}F_4) + mn(F_0 - 14F_4) + \tfrac{1}{2}n(n-1)\,(F_0 + 21F_4) \tag{9.50}$$

and the energy difference between complementary states is

$$E(t_2^{6-m}e^{4-n}) - E(t_2^m e^n) = (45 - 9m - 9n)\,F_0 + (-70 + 9\tfrac{1}{3}m + 21n)\,F_4. \tag{9.51}$$

Next let $\Gamma \neq A_1$, and write $\epsilon = m+n$. Then

$$\mathscr{V}_{i\Gamma} = \tfrac{1}{2}\sum_M \left\{\sum_{\kappa=1}^{\epsilon} g_{i\Gamma M}(\kappa)\right\}^2 - \tfrac{1}{2}\sum_M \sum_{\kappa=1}^{\epsilon} [g_{i\Gamma M}(\kappa)]^2. \tag{9.52}$$

The sum $\sum_{\kappa=1}^{\epsilon} g_{i\Gamma M}(\kappa)$ is a one-electron quantity and, because $\Gamma \neq A_1$

$$\langle t_2^6 e^4 |\sum_{\kappa=1}^{10} g_{i\Gamma M}(\kappa)\,| t_2^6 e^4\rangle = 0.$$

Hence by the discussion of § 9.7.3 the matrix of $\Sigma g_{i\Gamma M}$ changes its sign throughout on passing from an L-scheme to an R-scheme. Now

$$\langle\bar{\phi}|\left\{\sum_{\kappa=1}^{\epsilon} g_{i\Gamma M}(\kappa)\right\}^2|\phi\rangle = \sum_{\psi}\langle\bar{\phi}|\sum_{\kappa=1}^{\epsilon} g_{i\Gamma M}(\kappa)|\psi\rangle\langle\bar{\psi}|\sum_{\kappa=1}^{\epsilon} g_{i\Gamma M}(\kappa)|\phi\rangle,$$

where ψ runs over the other (fully antisymmetric) states of $(t_2 e)^\epsilon$. Therefore the matrix of the first term of $\mathscr{V}_{i\Gamma}$ in (9.52) is identical for the L-scheme and the complementary R-scheme.

The second term of (9.52) is a sum of one-electron operators and may be rewritten

$$\mathscr{V}_{i\Gamma}^2 = -\tfrac{1}{2}\sum_{\kappa=1}^{\epsilon}\sum_{M}[g_{i\Gamma M}(\kappa)]^2,$$

where $\sum_M [g_{i\Gamma M}(\kappa)]^2$ is a function which forms a basis for A_1. So $\sum_M g_{i\Gamma M}^2$ is diagonal within the five one-electron functions e and t_2, though it may have a different value for e from its value for t_2. Let these values be Y_1, Y_2, respectively. Then the matrix of $\mathscr{V}_{i\Gamma}^2$ is diagonal within $t_2^m e^n$ with the constant diagonal energy

$$E(t_2^m e^n) = -\tfrac{1}{2}mY_2 - \tfrac{1}{2}nY_1.$$

This completes the proof that complementary schemes have identical matrices of electrostatic energy, apart, possibly, from a change in the diagonal elements. The change in the diagonal elements is constant for all L-states having the same m and n.

For d-functions we may use the results of § 9.7.6 and express Y_1 and Y_2 in terms of F_2 and F_4. We find $Y_1 = 14F_2 + 105F_4$ and $Y_2 = 14F_2 + 116\tfrac{2}{3}F_4$. Combining this with (9.50) and (9.51) for the A_1 part of $\mathscr{V}$, the total difference of diagonal energy between $t_2^{6-m}e^{4-n}$ and $t_2^m e^n$ is

$$E(t_2^{6-m}e^{4-n}) - E(t_2^m e^n) = (45-9\epsilon)A + (-70+14\epsilon)B + (35-7\epsilon)C, \quad (9.53)$$

where $\epsilon = m+n$ and we have expressed the result in terms of Racah parameters. For d-functions, therefore, the energy difference is a function only of ϵ and is independent of the individual values of m and n.

For clarity of presentation, I have proved the results in a rather special form. It is unnecessary to choose real orbitals, and, also the results can be extended to any group of the first kind under which the set of five d orbitals contains no repeated representation. If the set breaks up into three or more irreducible representations the order of coupling is important and must be in an opposite sense for L and R configurations (§ 9.8.3).

9.8. Fractional parentage

9.8.1. Introduction. The concept of fractional parentage is important generally in theoretical physics and was used by Tanabe & Sugano (1954) in their original calculation of the complete matrix of electrostatic energy for d^n in the strong-field coupling scheme. The idea, in a fairly general form, is to regard the states of a configuration $a^{m+n}b^{p+q}$ as combinations of states of $a^m b^p$ and $a^n b^q$

and to relate the properties of the former states, which are the 'children', to those of the latter, which are their parents. I do not use this method much in this book and, accordingly, give a short, self-contained account in the present section, which is unnecessary for the understanding of later chapters in the book.

9.8.2. Fractional parentage in t_2^n configurations. If we take two t_2 orbitals and build up composite states, including spin, by coupling the two t_2 representations, we obtain some states which are antisymmetric and some which are symmetric. The former only are allowed as states of t_2^2. For example

$$|{}^3T_1\,1\,x\rangle = \frac{1}{\sqrt{2}}|\zeta^+.\eta^+\rangle - \frac{1}{\sqrt{2}}|\eta^+.\zeta^+\rangle$$

is the allowed state $|\zeta^+\eta^+\rangle$ of Table A 24, but

$$|{}^3A_1\,1\rangle = \frac{1}{\sqrt{3}}|\xi^+.\xi^+\rangle + \frac{1}{\sqrt{3}}|\eta^+.\eta^+\rangle + \frac{1}{\sqrt{3}}|\zeta^+.\zeta^+\rangle$$

is not allowed. As in § 9.7, there is no antisymmetrization across a dot. In this particular case the classification in terms inevitably implies a classification as symmetric or antisymmetric.

The next step is to build states of t_2^3 by coupling those of t_2^2 with t_2. We can only obtain one ${}^4A_2$ in this way

$$\begin{aligned}|{}^4A_2\,\tfrac{3}{2}\rangle &= \frac{1}{\sqrt{3}}|{}^3T_1\,1x.\xi^+\rangle + \frac{1}{\sqrt{3}}|{}^3T_1\,1y.\eta^+\rangle + \frac{1}{\sqrt{3}}|{}^3T_1\,1z.\zeta^+\rangle \\ &= |\zeta^+\eta^+\xi^+\rangle,\end{aligned}$$

where the factor $\sqrt{3}$ has disappeared because of the normalization factor of $|\zeta^+\eta^+\xi^+\rangle$. However, a ${}^2T_1$ term can be obtained by adding a t_2 electron to ${}^3T_1$, ${}^1E$ or ${}^1T_2$. The $\frac{1}{2}x$ components of these ${}^2T_1$ terms are

$$\begin{aligned}\psi_1 = |t_2^2({}^3T_1).t_2\,{}^2T_1\,\tfrac{1}{2}x\rangle &= -\frac{1}{\sqrt{3}}|{}^3T_1\,1y.\zeta^-\rangle - \frac{1}{\sqrt{3}}|{}^3T_1\,1z.\eta^-\rangle \\ &\qquad + \frac{1}{\sqrt{6}}|{}^3T_1\,0y.\zeta^+\rangle + \frac{1}{\sqrt{6}}|{}^3T_1\,0z.\eta^+\rangle \\ &= -\frac{1}{\sqrt{3}}|\xi^+\zeta^+.\zeta^-\rangle - \frac{1}{\sqrt{3}}|\eta^+\xi^+.\eta^-\rangle + \frac{1}{2\sqrt{3}}|\xi^+\zeta^-.\zeta^+\rangle + \frac{1}{2\sqrt{3}}|\xi^-\zeta^+.\zeta^+\rangle \\ &\qquad + \frac{1}{2\sqrt{3}}|\eta^+\xi^-.\eta^+\rangle + \frac{1}{2\sqrt{3}}|\eta^-\xi^+.\eta^+\rangle,\end{aligned}$$

$$\psi_2 = |t_2^2({}^1E).t_2\,{}^2T_1\,\tfrac{1}{2}x\rangle = -\frac{1}{\sqrt{2}}|\eta^2.\xi^+\rangle + \frac{1}{\sqrt{2}}|\zeta^2.\xi^+\rangle,$$

$$\begin{aligned}\psi_3 = |t_2^2({}^1T_2).t_2\,{}^2T_1\,\tfrac{1}{2}x\rangle &= \tfrac{1}{2}|\xi^+\zeta^-.\zeta^+\rangle + \tfrac{1}{2}|\zeta^+\xi^-.\zeta^+\rangle \\ &\qquad - \tfrac{1}{2}|\xi^+\eta^-.\eta^+\rangle - \tfrac{1}{2}|\eta^+\xi^-.\eta^+\rangle,\end{aligned}$$

and the ${}^2T_1$ term which actually occurs in Table A 24 is equal to

$$\frac{1}{\sqrt{2}}\psi_1 - \frac{1}{\sqrt{3}}\psi_2 + \frac{1}{\sqrt{6}}\psi_3.$$

We say that it has each of ψ_1, ψ_2, ψ_3 as a partial, or fractional, parent and write

$$|t_2^3\,{}^2T_1 M_S M_\Gamma\rangle = \sum_{S_1\Gamma_1} \langle t_2^2(S_1\Gamma_1)\,t_2\,{}^2T_1|\}\,t_2^3\,{}^2T_1\rangle\,|t_2^2(S_1\Gamma_1)\,.\,t_2\,{}^2T_1 M_S M_\Gamma\rangle. \quad (9.54)$$

The numbers $\langle t_2^2(S_1\Gamma_1)\,t_2\,{}^2T_1|\}\,t_2^3\,{}^2T_1\rangle$, which are the three coefficients of the ψ_i, are called coefficients of fractional parentage. They are, of course, independent of M_S and M_Γ. The coefficients of fractional parentage for t_2^n and e^n configurations are given in Table A 39.[1]

A knowledge of the coefficients of fractional parentage enables us to calculate matrix elements for a configuration containing n electrons in terms of those with $n-1$ electrons. Consider the electrostatic energy of a state $|X\rangle = |t_2^n S\Gamma M_S M_\Gamma\rangle$. Because the state is totally antisymmetric we have

$$E(t_2^n S\Gamma) = \langle X|\sum_{\kappa<\lambda\leqslant n} V(\kappa\lambda)\,|X\rangle = \tfrac{1}{2}n(n-1)\,\langle X|\,V(12)\,|X\rangle$$

$$= \frac{n}{n-2}\langle X|\sum_{\kappa<\lambda\leqslant n-1} V(\kappa\lambda)\,|X\rangle. \quad (9.55)$$

Writing $\mathscr{V}_{n-1} = \sum_{\kappa<\lambda\leqslant n-1} V(\kappa\lambda)$ and using the fractional parentage equation

$$|X\rangle = \sum_{S_1\Gamma_1} \langle t_2^{n-1}(S_1\Gamma_1)\,t_2 S\Gamma|\}\,t_2^n S\Gamma\rangle\,|t_2^{n-1}(S_1\Gamma_1)\,.\,t_2 S\Gamma M_S M_\Gamma\rangle,$$

remembering that the electron after the dot is always the nth and therefore does not occur in the sum in the last part of (9.55), we find

$$E(t_2^n S\Gamma) = \frac{n}{n-2}\sum_{S_1\Gamma_1 S_1'\Gamma_1'} \langle t_2^n S\Gamma\{|t_2^{n-1}(S_1'\Gamma_1')\,t_2 S\Gamma\rangle\langle t_2^{n-1}(S_1\Gamma_1)\,t_2 S\Gamma|\}\,t_2^n S\Gamma\rangle$$

$$\times\langle t_2^{n-1}(S_1'\Gamma_1')\,.\,t_2 S\Gamma M_S M_\Gamma\,|\,\mathscr{V}_{n-1}\,|\,t_2^{n-1}(S_1\Gamma_1)\,.\,t_2 S\Gamma M_S M_\Gamma\rangle$$

$$= \frac{n}{n-2}\sum_{\substack{S_1 S_1'\Gamma_1\Gamma_1' M_{S1} M_{S2} M_{S1}' M_{S2}'\\ M_{\Gamma1} M_{\Gamma2} M_{\Gamma1}' M_{\Gamma2}'}} \langle t_2^n S\Gamma\{|t_2^{n-1}(S_1'\Gamma_1')\,t_2 S\Gamma\rangle\langle t_2^{n-1}(S_1\Gamma_1)\,t_2 S\Gamma|\}\,t_2^n S\Gamma\rangle$$

$$\times\langle S_1'\tfrac{1}{2}SM_S\,|\,S_1'\tfrac{1}{2}M_{S1}'M_{S2}'\rangle\langle\Gamma_1' T_2\Gamma M_\Gamma\,|\,\Gamma_1' T_2 M_{\Gamma1}' M_{\Gamma2}'\rangle$$

$$\times\langle t_2^{n-1}S_1'\Gamma_1' M_{S1}' M_{\Gamma1}'|\,.\,\langle t_2 M_{S2}' M_{\Gamma2}'\,|\,\mathscr{V}_{n-1}\,|\,t_2^{n-1}S_1\Gamma_1 M_{S1} M_{\Gamma1}\rangle\,.\,|\,t_2 M_{S2} M_{\Gamma2}\rangle$$

$$\times\langle S_1\tfrac{1}{2}M_{S1}M_{S2}\,|\,S_1\tfrac{1}{2}SM_S\rangle\langle\Gamma_1 T_2 M_{\Gamma1} M_{\Gamma2}\,|\,\Gamma_1 T_2\Gamma M_\Gamma\rangle$$

$$= \frac{n}{n-2}\sum_{S_1\Gamma_1 M_{S1} M_{S2} M_{\Gamma1} M_{\Gamma2}} |\langle t_2^{n-1}(S_1\Gamma_1)\,t_2 S\Gamma|\}\,t_2^n S\Gamma\rangle|^2\,|\langle S_1\tfrac{1}{2}SM_S\,|\,S_1\tfrac{1}{2}M_{S1}M_{S2}\rangle|^2$$

$$\times|\langle\Gamma_1 T_2\Gamma M_\Gamma\,|\,\Gamma_1 T_2 M_{\Gamma1} M_{\Gamma2}\rangle|^2\langle t_2^{n-1}S_1\Gamma_1 M_{S1} M_{\Gamma1}\,|\,\mathscr{V}_{n-1}\,|\,t_2^{n-1}S_1\Gamma_1 M_{S1} M_{\Gamma1}\rangle$$

$$= \frac{n}{n-2}\sum_{S_1\Gamma_1} |\langle t_2^{n-1}(S_1\Gamma_1)\,t_2 S\Gamma|\}\,t_2^n S\Gamma\rangle|^2\,E(t_2^{n-1}S_1\Gamma_1). \quad (9.56)$$

Contrary to our usual convention we have used capital letters for one-electron quantum numbers. In spite of the complexity of the intermediate steps, we have arrived at a very simple formula relating the energies of the terms of t_2^n to those of t_2^{n-1}. In the particular case of the ${}^2T_1$ term of t_2^3 we take the energies of the three relevant terms of t_2^2 from Table 9.1, in general parameters,

$$E(t_2^2\,{}^3T_1) = b-j, \quad E(t_2^2\,{}^1E) = a-j, \quad E(t_2^2\,{}^1T_2) = b+j;$$

[1] As calculated by Tanabe & Sugano (1954), but with their phases altered. I have checked $\langle\gamma^{n-1}(S'\Gamma')\,\gamma S\Gamma|\}\,\gamma^n S\Gamma\rangle$ but not $\langle\gamma^{n-2}(S_1\Gamma_1)\,\gamma^2(S_2\Gamma_2)\,S\Gamma|\}\,\gamma^n S\Gamma\rangle$.

and (9.56) becomes

$$E(t_2^3\,{}^2T_1) = \tfrac{3}{2}E(t_2^2\,{}^3T_1) + E(t_2^2\,{}^1E) + \tfrac{1}{2}E(t_2^2\,{}^1T_2)$$
$$= a + 2b - 2j,$$

in agreement with the energy we found earlier for this term.

We call this method of calculation the inductive method. Of course, for t_2^n it is not nearly such an easy way of calculating energies as our original straightforward set of calculations. However, it illustrates the method and shows clearly what is going on. We shall see in the next section how Tanabe & Sugano used this method to obtain the complete electrostatic matrix for d^n in the strong-field scheme although a considerable part of that matrix, too, can be obtained more easily by the elementary methods we used in § 9.3.

The idea of fractional parentage can be generalized somewhat. Suppose now that we build up the terms of t_2^{m+n} by coupling terms of t_2^m with those of t_2^n. In analogy with (9.54) we write

$$|t_2^{m+n}S\Gamma\rangle = \sum_{S_1\Gamma_1S_2\Gamma_2} \langle t_2^m(S_1\Gamma_1)\,t_2^n(S_2\Gamma_2)|\}t_2^{m+n}S\Gamma\rangle |t_2^m(S_1\Gamma_1)\,.\,t_2^n(S_2\Gamma_2)\,S\Gamma\rangle, \quad (9.57)$$

where the labels $M_S M_\Gamma$ are omitted for brevity. Reversing the order of coupling t_2^m to t_2^n gives the relation

$$\langle t_2^n(S_2\Gamma_2)\,t_2^m(S_1\Gamma_1)|\}t_2^{m+n}S\Gamma\rangle$$
$$= (-1)^{mn+S_1+S_2-S+\eta}\langle t_2^m(S_1\Gamma_1)\,t_2^n(S_2\Gamma_2)|\}t_2^{m+n}S\Gamma\rangle, \quad (9.58)$$

where $\eta = 0$ unless $(\Gamma_1, \Gamma_2, \Gamma) = (E, E, A_2)$ or (T_1, T_1, T_1) or (T_2, T_2, T_1) when $\eta = 1$ (use ex. 1, p. 24, and Table A 20). This relation depends only on the coupling coefficients and therefore still holds when t_2 is replaced by e or t_1.

For a filled shell the coefficients in (9.57) follow immediately from (9.37) and are

$$\langle t_2^m(S_1\Gamma_1)\,t_2^n(S_2\Gamma_2)|\}t_2^6\,{}^1A_1\rangle = \frac{\mu(2S_1+1)^{\frac{1}{2}}\lambda(\Gamma_1)^{\frac{1}{2}}}{\sqrt{q}}\,\delta_{S_1S_2}\delta_{\Gamma_1\Gamma_2},$$

$$\langle e^2(S_1\Gamma_1)\,e^2(S_2\Gamma_2)|\}e^4\,{}^1A_1\rangle = \frac{\mu(2S_1+1)^{\frac{1}{2}}\lambda(\Gamma_1)^{\frac{1}{2}}}{\sqrt{6}}\,\delta_{S_1S_2}\delta_{\Gamma_1\Gamma_2}, \quad (9.59)$$

where μ is the 'parity' of ${}^{2S_1+1}\Gamma_1$ and $\lambda(\Gamma_1)$ the degree of Γ_1. The only remaining coefficients are for t_2 and $m = 2, 3$; $n = 2$. They were calculated by Tanabe & Sugano and are also in Table A 39 with their phases in accord with our basis conventions.

We can now give a slightly different fractional parentage method of calculating energies. Take

$$E(t_2^n S\Gamma) = \tfrac{1}{2}n(n-1)\langle X|\,V(n-1,n)\,|X\rangle$$

in the derivation (9.56) and also expand $|X\rangle$ in terms of

$$|t_2^{n-2}(S_1\Gamma_1)\,.\,t_2^2(S_2\Gamma_2)\,S\Gamma\rangle.$$

Then

$$E(t_2^n S\Gamma) = \tfrac{1}{2}n(n-1)\sum_{S_1S_2\Gamma_1\Gamma_2}|\langle t_2^{n-2}(S_1\Gamma_1)\,t_2^2(S_2\Gamma_2)|\}t_2^n S\Gamma\rangle|^2\,E(t_2^2 S_2\Gamma_2). \quad (9.60)$$

This may be called a direct method of calculation because it calculates $E(t_2^n S\Gamma)$ directly in terms of the energies of two-electron terms. For example, for $t_2^4\,{}^1A_1$ we have

$$E(t_2^4\,{}^1A_1) = \tfrac{4}{3}E(t_2^2\,{}^1A_1) + 3E(t_2^2\,{}^3T_1) + \tfrac{2}{3}E(t_2^2\,{}^1E) + E(t_2^2\,{}^1T_2)$$
$$= 2a + 4b$$

in agreement with Table 9.1. This method can be used to give an explicit formula for certain off-diagonal matrix elements d^n in the strong-field scheme (see the examples at the end of the section).

9.8.3. Coupling of three representations. If we have functions forming bases for three irreducible representations Γ_1, Γ_2, Γ_3 and couple them to form a basis for Γ we may either couple Γ_1 to Γ_2 and then couple the resultant functions to Γ_3, or we may couple Γ_2 to Γ_3 first and then couple Γ_1 to these resultants. In symbols

$$|\Gamma_1\Gamma_2\Gamma'M'\rangle = \sum_{M_1M_2} \langle\Gamma_1\Gamma_2 M_1 M_2 \,|\, \Gamma_1\Gamma_2\Gamma'M'\rangle\,|\Gamma_1 M_1\rangle\,|\Gamma_2 M_2\rangle,$$

$$|\Gamma_1\Gamma_2(\Gamma')\,\Gamma_3\Gamma M\rangle = \sum_{M'M_3} \langle\Gamma'\Gamma_3 M'M_3 \,|\, \Gamma'\Gamma_3\Gamma M\rangle\,|\Gamma_1\Gamma_2\Gamma'M'\rangle\,|\Gamma_3 M_3\rangle$$
$$= \sum_{M_1M_2M'M_3} \langle\Gamma'\Gamma_3 M'M_3 \,|\, \Gamma'\Gamma_3\Gamma M\rangle\langle\Gamma_1\Gamma_2 M_1 M_2 \,|\, \Gamma_1\Gamma_2\Gamma'M'\rangle \times |\Gamma_1 M_1\rangle\,|\Gamma_2 M_2\rangle\,|\Gamma_3 M_3\rangle$$

and also

$$\Gamma_1, \Gamma_2\Gamma_3(\Gamma'')\,\Gamma M\rangle = \sum_{M_1M_2M_3M''} \langle\Gamma_1\Gamma'' M_1 M'' \,|\, \Gamma_1\Gamma''\Gamma M\rangle\langle\Gamma_2\Gamma_3 M_2 M_3 \,|\, \Gamma_2\Gamma_3\Gamma''M''\rangle \times |\Gamma_1 M_1\rangle\,|\Gamma_2 M_2\rangle\,|\Gamma_3 M_3\rangle.$$

The transformation matrix between these two systems of coupling is

$$\langle\Gamma_1\Gamma_2(\Gamma')\,\Gamma_3\Gamma \,|\, \Gamma_1, \Gamma_2\Gamma_3(\Gamma'')\,\Gamma\rangle$$
$$= \sum_{M_1M_2M_3M'M''} \langle\Gamma'\Gamma_3\Gamma M \,|\, \Gamma'\Gamma_3 M'M_3\rangle\langle\Gamma_1\Gamma_2\Gamma'M' \,|\, \Gamma_1\Gamma_2 M_1 M_2\rangle \times \langle\Gamma_1\Gamma'' M_1 M'' \,|\, \Gamma_1\Gamma''\Gamma M\rangle\langle\Gamma_2\Gamma_3 M_2 M_3 \,|\, \Gamma_2\Gamma_3\Gamma''M''\rangle. \quad (9.61)$$

It is convenient to write $\delta = S\Gamma$ in § 9.8.4 and then we define

$$\langle\delta_1\delta_2(\delta')\,\delta_3\delta \,|\, \delta_1, \delta_2\delta_3(\delta'')\,\delta\rangle$$
$$= \langle\Gamma_1\Gamma_2(\Gamma')\,\Gamma_3\Gamma \,|\, \Gamma_1, \Gamma_2\Gamma_3(\Gamma'')\,\Gamma\rangle\,\langle S_1S_2(S')\,S_3S \,|\, S_1, S_2S_3(S'')\,S\rangle. \quad (9.62)$$

9.8.4. Tanabe & Sugano's induction formula for $t_2^m e^n$. The complete matrices of electrostatic energy for all d^n configurations in the strong-field coupling scheme were first calculated by Tanabe & Sugano using an induction formula which is a generalization of (9.56). Their results, altered to accord with our phase conventions, are given in Tables A 28–30 and their energy level diagrams for $2 \leqslant n \leqslant 7$ are shown in Fig. 9.2. Tanabe & Sugano did not publish a proof of their formula and so I give a proof here.

The object is to express a matrix element

$$M = \langle t_2^n(\delta_1)\,e^p(\delta_2)\,\delta m \,|\, \mathscr{V}_N \,|\, t_2^{n'}(\delta_3)\,e^{p'}(\delta_4)\,\delta m\rangle,$$

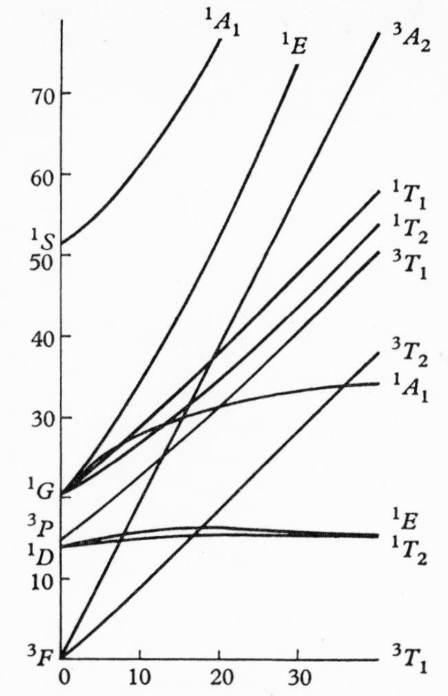

(*a*) d^2, V^{3+}, $\gamma = 4{\cdot}428$, $B = 860$ cm^{-1}.

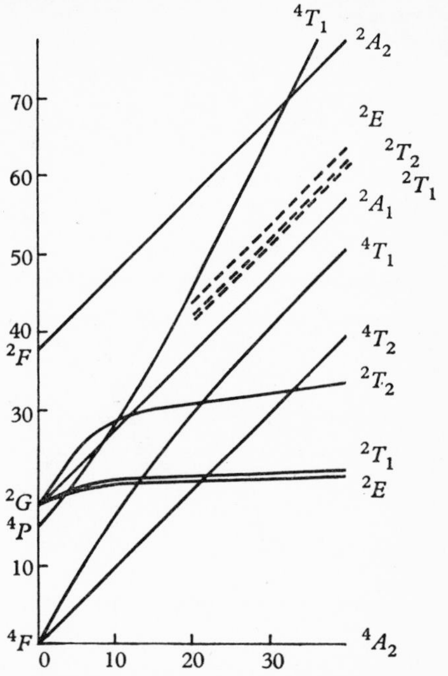

(*b*) d^3, Cr^{3+}, $\gamma = 4{\cdot}502$, $B = 918$ cm^{-1}.

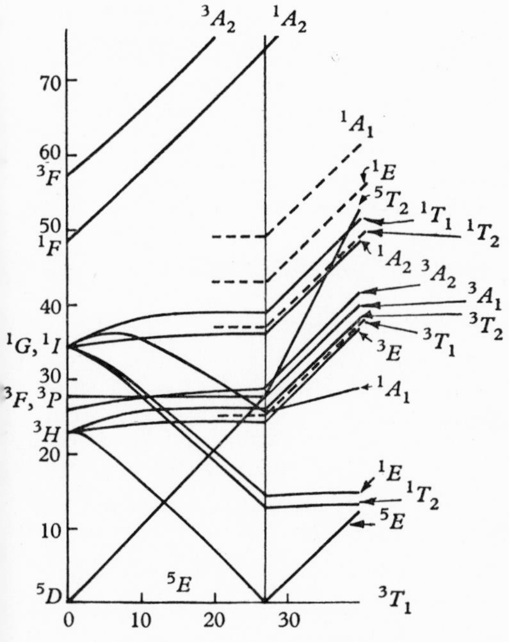

(*c*) d^4, Mn^{3+}, $\gamma = 4{\cdot}611$, $B = 965$ cm^{-1}.

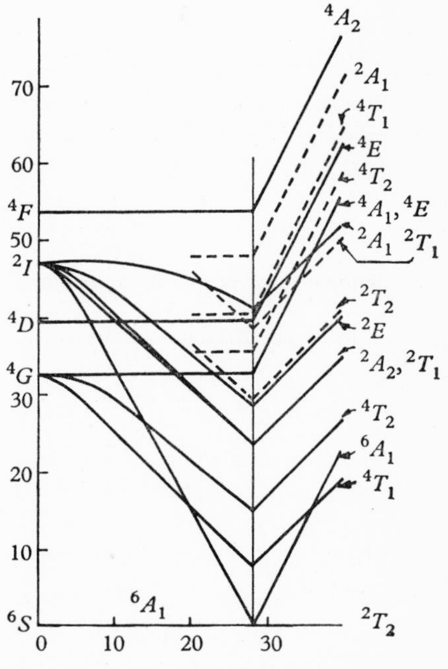

(*d*) d^5, Mn^{++}, $\gamma = 4{\cdot}477$, $B = 860$ cm^{-1}.

Fig. 9.2 For legend see over.

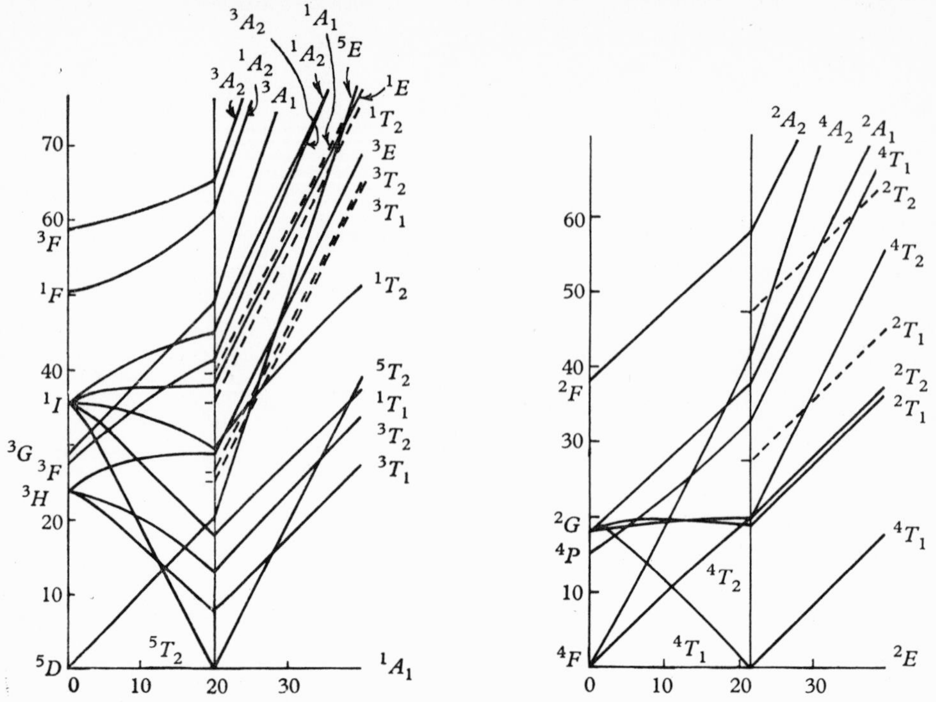

(*e*) d^6, Co^{3+}, $\gamma = 4{\cdot}808$, $B = 1{,}065\ \text{cm}^{-1}$.

(*f*) d^7, Co^{++}, $\gamma = 4{\cdot}633$, $B = 971\ \text{cm}^{-1}$.

Figs. 9.2*a–f*. Energy plotted against Δ (both in units of B) for d^n ions, as calculated by Tanabe & Sugano (1954). The ion for which the calculation was actually made is shown in each case, together with their values for $\gamma = CB^{-1}$ and B. Except for d^2, not all the highly excited states are plotted.

where $N = n+p = n'+p'$, in terms of matrix elements for the $N-1$ electron system. Then just as in the derivation of (9.56) we put

$$M = \frac{N}{N-2}\langle t_2^n(\delta_1)\, e^p(\delta_2)\, \delta m \mid \mathscr{V}_{N-1} \mid t_2^{n'}(\delta_3)\, e^{p'}(\delta_4)\, \delta m\rangle, \tag{9.63}$$

where $\mathscr{V}_{N-1}$ is a sum over the first $N-1$ electrons only. We also write

$$|t_2^n(\delta_1)\, e^p(\delta_2)\, \delta m\rangle = (N!\,n!\,p!)^{-\frac{1}{2}} \sum_{\nu m_1 m_2} (-1)^\nu P_\nu(T_{NN} - T_{1N} - T_{2N} - \ldots - T_{N-1N})$$
$$\times \langle \delta_1 \delta_2 m_1 m_2 \mid \delta_1 \delta_2 \delta m\rangle\, |t_2^n \delta_1 m_1\rangle \,.\, |e^p \delta_2 m_2\rangle, \tag{9.64}$$

where ν runs over all permutations of the first $N-1$ electrons and T_{iN} is that transposition which exchanges the ith electron with the last electron. The functions t_2^n and e^p on the right-hand side of (9.64) are antisymmetrized but t_2^n involves only the first n electrons and e^p only the last p. Every permutation of N electrons occurs in (9.64) and with its right parity (see § 2.6). The normalizing factor is because the permutations may be written as products, P^2P^1 say, where the $n!\,p!$ permutations P^1 are that subgroup of S_N which is the direct product of the symmetric group on the first n electrons with the symmetric group on the last p. There are $N!/n!\,p!$ permutations P^2 which are chosen so that P^2P^1 exhaust

S_N (ex. 3, p. 141). Then the $(-1)^{\nu_1}P^1$ give the same state $t_2^n.e^p$, while $(-1)^{\nu_2}P^2$ give different states and the normalization factor is

$$\left(\frac{n!\,p!}{N!}\right)^{\frac{1}{2}}\frac{1}{n!\,p!} = (N!\,n!\,p!)^{-\frac{1}{2}}.$$

In evaluating (9.63) we are interested in the position of the Nth electron. Equation (9.64) shows that n transpositions assign it to t_2^n and p to e^p and we break (9.64) up into two parts, $|X_t\rangle+|X_e\rangle$ say. Take $|X_e\rangle$ first and consider the term which arises from the transposition T_{iN}. We now perform this transposition again to get the Nth electron back to its natural position, giving

$$|X_e\rangle = \frac{p}{(N!\,n!\,p!)^{\frac{1}{2}}}\sum_{\nu m_1 m_2}(-1)^\nu P_\nu\langle\delta_1\delta_2 m_1 m_2\,|\,\delta_1\delta_2\delta m\rangle\,|t_2^n\delta_1 m_1\rangle\,.\,|e^p\delta_2 m_2\rangle$$

$$= \frac{p}{(N!\,n!\,p!)^{\frac{1}{2}}}\sum_{\nu m_1 m_2 m' m_e \delta'}(-1)^\nu P_\nu\langle\delta_1\delta_2 m_1 m_2\,|\,\delta_1\delta_2\delta m\rangle\langle\delta' e m' m_e\,|\,\delta' e\delta_2 m_2\rangle$$
$$\times\langle e^{p-1}(\delta')\,e|\}e^p\delta_2\rangle\,|t_2^n\delta_1 m_1\rangle\,.\,|e^{p-1}\delta' m'\rangle\,.\,|e m_e\rangle$$

$$= \left(\frac{p}{N}\right)^{\frac{1}{2}}\sum_{\substack{m_1 m_2 m' m_e\\ \delta'\bar{\delta}\bar{m}}}\langle\delta_1\delta_2 m_1 m_2\,|\,\delta_1\delta_2\delta m\rangle\langle\delta' e m' m_e\,|\,\delta' e\delta_2 m_2\rangle\langle\delta_1\delta'\bar{\delta}\bar{m}\,|\,\delta_1\delta' m_1 m'\rangle$$
$$\times\langle e^{p-1}(\delta')\,e|\}e^p\delta_2\rangle\,|t_2^n(\delta_1)\,e^{p-1}(\delta')\,\bar{\delta}\bar{m}\rangle\,.\,|e m_e\rangle. \quad (9.65)$$

We now write

$$|t_2^n(\delta_1)\,e^{p-1}(\delta')\,\bar{\delta}\bar{m}\rangle\,.\,|e m_e\rangle = \sum_{\delta_a m_a}\langle\bar{\delta}e\delta_a m_a\,|\,\bar{\delta}e\bar{m}m_e\rangle\,|t_2^n(\delta_1)\,e^{p-1}(\delta')\,\bar{\delta}\,.\,e\delta_a m_a\rangle \quad (9.66)$$

and observe that the only terms in the sum over $\delta_a m_a$ which can give a non-zero contribution to (9.63) are those for which $\delta_a = \delta$, $m_a = m$. Substituting (9.66) in (9.65) and using (9.61) and (9.62) we have

$$|X_e\rangle = \left(\frac{p}{N}\right)^{\frac{1}{2}}\sum_{\delta'\bar{\delta}}\langle\delta_1\delta'(\bar{\delta})\,e\delta\,|\,\delta_1,\delta' e(\delta_2)\,\delta\rangle\langle e^{p-1}(\delta')\,e|\}e^p\delta_2\rangle\,|t_2^n(\delta_1)\,e^{p-1}(\delta')\,\bar{\delta}\,.\,e\delta m\rangle. \quad (9.67)$$

Because the e orbitals are orthogonal to the t_2 orbitals, M breaks up into two parts

$$M = \langle X_t\,|\,\mathscr{V}\,|\,X_t'\rangle+\langle X_e\,|\,\mathscr{V}\,|\,X_e'\rangle = M_t+M_e, \quad \text{say.}$$

Using (9.67) and the corresponding expression for $|X_e'\rangle$ we find M_e. We integrate the Nth electron out by applying the transformation inverse to (9.66) and summing over $\bar{m}m_e$. Then, finally

$$M_e = \frac{\sqrt{pp'}}{N-2}\sum_{\delta'\delta''\bar{\delta}}\langle e^p\delta_2\{|e^{p-1}(\delta')\,e\rangle$$
$$\times\langle\delta_1,\delta' e(\delta_2)\,\delta\,|\,\delta_1\delta'(\bar{\delta})\,e\delta\rangle\langle t_2^n(\delta_1)\,e^{p-1}(\delta')\,\bar{\delta}\,|\,\mathscr{V}_{N-1}\,|\,t_2^{n'}(\delta_3)\,e^{p'-1}(\delta'')\,\bar{\delta}\rangle$$
$$\times\langle\delta_3\delta''(\bar{\delta})\,e\delta\,|\,\delta_3,\delta'' e(\delta_4)\,\delta\rangle\langle e^{p'-1}(\delta'')\,e|\}e^{p'}\delta_4\rangle. \quad (9.68)$$

The treatment of M_t is exactly the same except for a trivial difference in the discussion of the effect of the transpositions T_{iN}. One obtains

$$M_t = \frac{\sqrt{nn'}}{N-2}\sum_{\delta'\delta''\bar{\delta}}\langle t_2^n\delta_1\{|t_2,t_2^{n-1}(\delta')\,\delta_1\rangle\langle t_2\delta'(\delta_1)\,\delta_2\delta\,|\,t_2,\delta'\delta_2(\bar{\delta})\,\delta\rangle$$
$$\times\langle t_2^{n-1}(\delta')\,e^p(\delta_2)\,\bar{\delta}\,|\,\mathscr{V}_{N-1}\,|\,t_2^{n'-1}(\delta'')\,e^p(\delta_4)\,\bar{\delta}\rangle$$
$$\times\langle t_2,\delta''\delta_4(\bar{\delta})\,\delta\,|\,t_2\delta''(\delta_3)\,\delta_4\delta\rangle\langle t_2,t_2^{n'-1}(\delta'')\,\delta_3|\}t_2^n\delta_3\rangle. \quad (9.69)$$

Examples

1. Prove that

$$\langle t_2^{n+2}(\delta_1)\, e^p(\delta_2)\, \delta \,|\, \mathscr{V}_N \,|\, t_2^n(\delta_3)\, e^{p+2}(\delta_4)\, \delta\rangle = M(^1A_1) + M(^1E),$$

where $N = n+p+2$ and

$$M(^1A_1) = \chi\delta(\delta_1,\delta_3)\,\delta(\delta_2,\delta_4)\,\sqrt{6}\,(2B+C)\,\langle t_2^{n+2}\delta_1\{|t_2^n(\delta_1)\,t_2^2(^1A_1)\rangle\,\langle e^p(\delta_2)\,e^2(^1A_1)\,\delta_2|\}\,e^{p+2}\delta_2\rangle,$$

$$M(^1E) = \chi(-1)^\eta\,\delta(S_1,S_3)\,\delta(S_2,S_4)\,(2\sqrt{3}\,B)$$

$$\times\langle\Gamma_3 E(\Gamma_1)\,\Gamma_2\Gamma\,|\,\Gamma_3, E\Gamma_2(\Gamma_4)\,\Gamma\rangle\,\langle t_2^{n+2}\delta_1\{|t_2^n(\delta_3)\,t_2^2(^1E)\rangle\,\langle e^p(\delta_2)\,e^2(^1E)\,\delta_4|\}\,e^{p+2}\delta_4\rangle$$

in which $\chi = \frac{1}{2}\{(n+1)(n+2)(p+1)(p+2)\}^{\frac{1}{2}}$ and $\eta = \eta(\Gamma_2 E\Gamma_4)$ is as in (9.58).

2. Prepare a table of the $\langle\Gamma_3 E(\Gamma_1)\,\Gamma_2\Gamma\,|\,\Gamma_3, E\Gamma_2(\Gamma_4)\,\Gamma\rangle$ and hence calculate the matrix elements $\langle t_2^{n+2}e^p\,|\,\mathscr{V}_N\,|\,t_2^n e^{p+2}\rangle$ for all d^n configurations.

CHAPTER 10

PARAMAGNETIC SUSCEPTIBILITIES[1]

10.1. Van Vleck's theorem

We saw in § 9.4 that for some ions there is more than one possible value for the spin of the ground term. Therefore we wish to know how to distinguish different values of the spin experimentally. In this chapter we consider paramagnetic susceptibilities and show that their measurement usually gives us fairly definite information on this point. We shall find that, to quite a good approximation, room temperature paramagnetic susceptibilities of d^n ions having spins of S are functions of n and S alone. In other words a particular susceptibility for the ground state of a d^n ion is characteristic of a particular spin state for the ion, and usually enables us to infer that spin state with some confidence. First, we shall show why this is so and then later see what limitations the assertion has.

In § 5.6 we calculated the susceptibility χ of an aggregate of metal ions, each of which was supposed to be in a spherically symmetric electrostatic field due to the rest of the crystal. In § 10.2 we will calculate χ assuming a regular octahedral field around each ion. In fact, fields around ions are rarely exactly octahedral and there are usually lower symmetry components with associated matrix elements large compared with the energy of interaction with the external magnetic field ($\sim$ 1 cm^{-1}). Therefore our correct method of calculation should be to diagonalize these low-symmetry elements and only after this to consider the effect of the magnetic field. However, it was shown by Van Vleck that provided the low-symmetry matrix elements are small compared with kT ($\sim$ 200 cm^{-1} at room temperature) the susceptibility, calculated by taking a statistical average over the occupied states, is practically the same as the susceptibility calculated with neglect of the low-symmetry matrix elements. This means that we can make fairly definite *a priori* calculations of χ and is the reason why χ is approximately the same for all ions with the same n, S and a similar environment.

We take the general expression (5.62) for the susceptibility

$$\chi = -\frac{N}{H}\frac{\Sigma(\partial E/\partial H)\, e^{-E/kT}}{\Sigma\, e^{-E/kT}}. \tag{10.1}$$

In § 9.6 we saw that in a regular octahedral field the ground term is split by spin-orbit coupling into a number of degenerate, or near-degenerate, sets of states. For example, the $^3T_1$ ground term of t_2^2 splits in first order into $^3T_{10}$, $^3T_{11}$ and $^3T_{12}$ with respective degeneracies 1, 3 and 5. In higher order the $^3T_{12}$ separates slightly into states belonging to the two irreducible representations E and T_2.

[1] For early and important work in the field of paramagnetism one naturally refers to Van Vleck (1932*a*,*b*) and his co-workers (Schlapp & Penney, 1932; Jordahl, 1934; Howard, 1935) on the one hand and to Kramers (see § 8.4) on the other.

Let us ask the following question. Suppose we have a set of n states of an ion which are degenerate in the absence of a magnetic field and that there are no other states near (using the thermal energy kT as a yard-stick). Then this set of states has a calculated susceptibility

$$\chi = -\frac{N}{nH}\sum_{i=1}^{n}\frac{\partial E_i}{\partial H}, \tag{10.2}$$

from (10.1). Now we add a low-symmetry perturbation V having matrix elements V_{ij} within these states, where the V_{ij} are all small compared with kT, and show that the calculated value of χ is not greatly altered.

The energy of interaction with the external magnetic field will, in general, have matrix elements between our set of n states and states outside, but we neglect these at first for simplicity of exposition. For our n states we choose a basic set $|\phi_i\rangle$ within which the energy of interaction with the external magnetic field is diagonal. Without loss of generality we take the field to be $(0, 0, H)$. Then if a state $|\phi_i\rangle$ has magnetic energy $n_i\beta H$ the conjugate state $|\phi_i\rangle^*$ has energy $-n_i\beta H$. On the other hand, as V is real $\langle\overline{\phi}_i^* \,|\, V \,|\, \phi_i^*\rangle = \langle\overline{\phi}_i \,|\, V \,|\, \phi_i\rangle$ (see § 8.4). So we may suppose the $|\phi_i\rangle$ chosen so that they can be grouped in pairs having diagonal energies $V_{ii} \pm n_i\beta H$ and, perhaps, some self-conjugate states having diagonal energies V_{ii}. Then if E_i are the energies when the matrix M_{ij} of the sum of the magnetic energy and V is diagonalized, the E_j are the n solutions of the secular equation

$$f(E) \equiv |M_{ij} - E\delta_{ij}| = 0. \tag{10.3}$$

Hence $\Sigma E_i = \Sigma M_{ii} = \Sigma V_{ii}$ is independent of H and we take it to be zero.

Now

$$\chi = \frac{NkT}{HZ}\frac{\partial Z}{\partial H},$$

where the function Z is defined by

$$Z = \Sigma\, e^{-E_i/kT}$$

and in our case we expand Z as a power series to give

$$Z = n - \frac{1}{kT}\Sigma E_i + \frac{1}{2k^2T^2}\Sigma E_i^2 + \ldots,$$

where $\Sigma E_i = 0$ and

$$\Sigma E_i^2 = (\Sigma E_i)^2 - 2\sum_{i<j} E_i E_j$$

$$= -2\sum_{i<j}(M_{ii}M_{jj} - M_{ij}^2) = 2\sum_{i<j} M_{ij}^2 - (\sum_i M_{ii})^2 + \Sigma M_{ii}^2$$

from (10.3). M_{ij} is independent of H, and $\Sigma M_{ii} = 0$, so

$$\frac{\partial Z}{\partial H} = \frac{1}{2k^2T^2}\frac{\partial}{\partial H}(\Sigma M_{ii}^2) = \frac{1}{2k^2T^2}\frac{\partial}{\partial H}(\Sigma n_i^2\beta^2H^2)$$

$$= \frac{\beta^2 H}{k^2T^2}\Sigma n_i^2$$

and as $Z = n + O[(E/kT)^2]$ we have finally

$$\chi = \frac{N\beta^2 \Sigma n_i^2}{nkT}. \tag{10.4}$$

In (10.4) the sum over n_i^2 counts each of the n states separately. We notice that all dependence on the matrix elements of the low-symmetry field has disappeared.

If the matrix elements of the magnetic field energy to states outside the set of n are included then (10.3) is modified slightly. Suppose we choose a scheme in which the energy is diagonal except for the interaction with these other states and that there are m excited states which interact directly. Then we must solve

$$\begin{vmatrix} E_1 - E & 0 & \dots & 0 & a_{11}\beta H & \dots & a_{1m}\beta H \\ 0 & E_2 - E & \dots & 0 & a_{21}\beta H & \dots & a_{2m}\beta H \\ \\ 0 & 0 & \dots & E_n - E & a_{n1}\beta H & \dots & a_{nm}\beta H \\ \bar{a}_{11}\beta H & \bar{a}_{21}\beta H & \dots & \bar{a}_{n1}\beta H & \Delta_1 - E & \dots & 0 \\ \\ \bar{a}_{1m}\beta H & \bar{a}_{2m}\beta H & \dots & \bar{a}_{nm}\beta H & 0 & \dots & \Delta_m - E \end{vmatrix} = 0. \tag{10.5}$$

In calculating χ we sum over those values of E which are close to zero and so $\Delta_i - E \simeq \Delta_i$. The $a_{ij}\beta H$ are small quantities of order of magnitude $1\,\mathrm{cm}^{-1}$. Then (10.5) is approximately

$$\prod_{i=1}^{n} (E_i - E) \prod_{j=1}^{m} (\Delta_j - E) \left(1 - \sum_{p,l} \frac{|a_{pl}|^2 \beta^2 H^2}{(E_p - E)(\Delta_l - E)}\right) = 0.$$

We may divide through by $\Pi(\Delta_j - E)$ and then have

$$f(E) + (-1)^n E^{n-1} \beta^2 H^2 \sum_{p,l} \frac{|a_{pl}|^2}{\Delta_l} + \dots = 0. \tag{10.6}$$

This is the desired modification to (10.3). Higher terms are of the same order as those neglected in deriving (10.4). From (10.6) we derive

$$\Sigma E_i = \beta^2 H^2 \sum_{p,l} \frac{|a_{pl}|^2}{\Delta_l},$$

but ΣE_i^2 is unchanged from its previous expression to this order. Hence

$$\frac{\partial Z}{\partial H} = -\frac{2\beta^2 H}{kT} \sum_{p,l} \frac{|a_{pl}|^2}{\Delta_l} + \frac{\beta^2 H}{k^2 T^2} \sum_i n_i^2,$$

$$\chi = \frac{N\beta^2}{n} \Sigma \left(\frac{n_i^2}{kT} - 2\frac{|a_{il}|^2}{\Delta_l}\right). \tag{10.7}$$

The quantity $\quad \sum_i |a_{il}|^2 = \sum_i |\langle \bar{\phi}_i | (L_z + 2S_z) | \psi_l \rangle|^2,$

where **L** and **S** are measured in units of $\hbar$, is invariant to an orthonormal transformation of the ground functions $|\phi_i\rangle$ so (10.7) also is completely independent of the matrix elements V_{ij} of the low-symmetry field. This establishes the theorem even when there is more than one set of occupied states for we then simply sum over the sets, weighting each set with its appropriate Boltzmann factor.

In the form in which it is proved here the theorem is not true to the next order in V/kT and χ then involves the values of V_{ij} explicitly. We will not investigate this point here but refer the reader to Van Vleck (1932*a*) for discussion of this and of yet higher approximations. Next I remark that it is usual to prove the theorem by expanding E as a power series in H. This is unsatisfactory, however, because it is not necessarily possible to do this. Consider a pair of states having energies $\pm 2\beta H$ and a matrix element V between them. Then (10.3) becomes $E = \pm(V^2+4\beta^2H^2)^{\frac{1}{2}}$ which is a power series in H if $4\beta^2H^2 < V^2$ and in H^{-1} if $V^2 < 4\beta^2H^2$. In practice, the elements V_{ij} may be larger than, smaller than, or comparable with the magnetic energies, and it is for this reason that I have given a proof which does not require this power series expansion. The possibility of expanding $e^{-E_i/kT}$ as a fairly rapidly convergent series is, however, fundamental to the theorem.

The theorem is also relevant to the influence of the nuclear moments. Here the coupling between the nuclear moment and the electronic moment has no appreciable effect on the susceptibility until very low temperatures and therefore the magnetic moments of nuclei give simply an extra term in the susceptibility. But the nuclear magneton β_N is approximately $10^{-3}\beta$ and therefore this extra term is only of the order of 10^{-6} of the electronic term.

It is probably best to think of Van Vleck's theorem as asserting the approximate irrelevance of low-symmetry elements rather than to worry too much about its quantitative aspects. In accurate work one would include low-symmetry elements explicitly and find their effect. However, these elements can really only be obtained by fitting theoretical formulae to experimental data (also electron resonance and thermodynamic data, chs. 11 and 12), whilst if we neglect them we can make fairly definite *a priori* calculations of susceptibilities using only the parameter ζ obtained from free-ion data. We do this in the next section and then compare the resulting formulae with experimental susceptibilities.

Now I remark that though we cannot, in general, expand E as a power series in H we have shown that we can replace a system by one having the same calculated susceptibility (here to a very good approximation) and for which such an expansion is possible. The expansion is possible for all the examples discussed in the next section. Then, as in § 5.6, we write

$$E = W_0 + HW_1 + H^2W_2 + \dots \tag{10.8}$$

and have
$$\frac{\partial E}{\partial H} = W_1 + 2HW_2 + \dots,$$

$$\chi = N\frac{\Sigma\left(\frac{W_1^2}{kT} - 2W_2\right)e^{-W_0/kT}}{\Sigma\, e^{-W_0/kT}}. \tag{10.9}$$

which are (5.61) and (5.63). Equation (10.7) is just the same, apart from notation, as the last equation of (10.9) for the case in which there is just one value of W_0 for the occupied states. In (10.8) and (10.9)

$$W_1 = \langle \bar{\psi}_0 | \beta(L_{\parallel} + 2S_{\parallel}) | \psi_0 \rangle,$$

$$W_2 = \sum_n \frac{|\langle \bar{\psi}_n | \beta(L_{\parallel} + 2S_{\parallel}) | \psi_0 \rangle|^2}{E_0 - E_n},$$

where $L_{\parallel} + 2S_{\parallel}$ is the component of $\mathbf{L} + 2\mathbf{S}$ parallel to the external field $\mathbf{H}$ ($|\mathbf{H}| = H$). For an octahedral field, therefore, χ referred to one axis is equal to

$$\chi = \frac{N\beta^2}{3} \frac{\sum \left\{ \frac{\langle \bar{\psi}_0 | (\mathbf{L} + 2\mathbf{S}) | \psi_0 \rangle^2}{kT} + 2 \sum_n \frac{|\langle \bar{\psi}_n | (\mathbf{L} + 2\mathbf{S}) | \psi_0 \rangle|^2}{E_n - E_0} \right\} e^{-W_0/kT}}{\sum e^{-W_0/kT}}. \tag{10.10}$$

In the particular case that the ground term is orbitally non-degenerate (strictly we should say spatially non-degenerate), is an eigenstate of the total spin and we neglect matrix elements of $\mathbf{L}$ between the ground states and excited states, (10.9) becomes

$$\chi = \frac{N\beta^2}{kT} \frac{\sum \langle \bar{\psi}_0 | 2S_z | \psi_0 \rangle^2}{2S+1} = \frac{4N\beta^2}{(2S+1)kT} \sum_{n=-S}^{+S} n^2$$

$$= \frac{4N\beta^2 S(S+1)}{3kT}. \tag{10.11}$$

Equation (10.11) is called the spin-only formula for the susceptibility. We see the close analogy with the classical Langevin–Debye formula $\chi = N\mu^2/3kT$, where μ is the magnetic moment.

Lastly, we ask what is the group-theoretic behaviour of χ? In general, the induced moment $\boldsymbol{\mu}$ is not necessarily parallel to the field $\mathbf{H}$ and we have $\mu_i = \sum_j \chi_{ij} H_j$. $\boldsymbol{\mu} \cdot \mathbf{H} = \Sigma H_i \chi_{ij} H_j$ must belong to the unit representation of the site-symmetry group G. This is because the energy of the system is invariant to group operations provided we apply these operations to $\mathbf{H}$ also. If G is the octahedral group then $(H_i) \in T_1$. Without loss of generality we suppose χ_{ij} symmetric and then $(\chi_{ij}) \in A_1 + E + T_2$. χ_{ij} itself must be invariant to operations of G applied to the electronic states alone and therefore χ_{ij} only consists of the A_1 state and hence $\chi_{ij} = \chi \delta_{ij}$, using the coupling coefficients for $T_1 \times T_1 = A_1$. This means that the susceptibility is a scalar for the regular octahedral sites, although the coefficients of higher powers of H in the expansion of μ as a function of H are not necessarily scalars (the coefficient of $H_j H_k H_l$ is not, in general). In particular, formula (10.10) for χ is true for the magnetic field along any axis.

10.2. Susceptibilities of ground terms of d^n configurations

10.2.1. Susceptibility of a term. A convenient first approximation in the calculation of susceptibilities is obtained by considering only the matrix elements of spin-orbit coupling and magnetic-field energy within the ground term. When the site symmetry is octahedral we have five possible types of term: ${}^{2S+1}A_1$,

${}^{2S+1}A_2$, ${}^{2S+1}E$, ${}^{2S+1}T_1$ and ${}^{2S+1}T_2$. Because $\mathbf{L}$ transforms as T_1 its matrix elements within the first three types are all zero and we already know that the spin-orbit coupling has zero matrix elements within them. Therefore, the susceptibility is given by the spin-only formula.

The spin-orbit coupling splits ${}^{2S+1}T_i$ according to a Landé interval rule and we associate a term ${}^{2S+1}T_i$ with a ${}^{2S+1}P$ term. Under this correspondence $\mathbf{L}+2\mathbf{S}$ passes over into $\gamma\mathbf{L}+2\mathbf{S}$ calculated within this ${}^{2S+1}P$ term. So we have to calculate the susceptibility for a ${}^{2S+1}P$ term having spin-orbit coupling energy $\nu\mathbf{L}\,.\,\mathbf{S}$ and an interaction $\gamma\mathbf{L}+2\mathbf{S}$ with the external magnetic field. But we already did this in § 5.6 and found for the level ${}^{2S+1}P_J$ (and hence for ${}^{2S+1}T_{iJ}$)

$$\chi_J = \frac{N\beta^2 g^2 J(J+1)}{3kT} + \frac{2N\beta^2(g-\gamma)(g-2)}{3\nu}, \tag{10.12}$$

where

$$g = \gamma + (2-\gamma)\frac{J(J+1)+S(S+1)-2}{2J(J+1)} \tag{10.13}$$

unless $J = 0$ when $g = 4-\gamma$. The susceptibility of the whole term is

$$\chi = \frac{\sum\limits_{J=|S-1|}^{S+1} (2J+1)\chi_J \exp[-\nu J(J+1)/2kT]}{\sum\limits_{J=|S-1|}^{S+1} (2J+1)\exp[-\nu J(J+1)/2kT]}. \tag{10.14}$$

We now rewrite (10.12) in terms of a dimensionless quantity. This is the Bohr magneton number μ_{eff}. It is positive and defined in terms of the susceptibility by the relation

$$\chi = \frac{N\beta^2\mu_{\text{eff}}^2}{3kT}. \tag{10.15}$$

Corresponding to theoretical and experimental susceptibilities we have theoretical and experimental μ_{eff} values. Then (10.12) becomes

$$\mu_{\text{eff}}^2 = g^2 J(J+1) + \frac{2kT}{\nu}(g-\gamma)(g-2). \tag{10.16}$$

μ_{eff}^2 for the term is the statistical average (10.14) of μ_{eff}^2 for the individual levels. For a d^n term ν is a multiple of ζ, the spin-orbit constant for a single d electron, so we may put

$$\mu_{\text{eff}}^2 = A + Bx^{-1}, \tag{10.17}$$

where $x = \zeta/kT$ and A and B are functions of S, J and γ according to (10.13) and (10.16).

This is as far as we can go in general because μ and γ depend on the particular term concerned. They may both be obtained by evaluating the spin-orbit and magnetic field energies of just one state of the term and so apart from this necessary extra calculation we have completely calculated the susceptibilities of all terms.

10.2.2. Ground terms of t_2^n. From the spin-only formula for χ we deduce $\mu_{\text{eff}}^2 = 4S(S+1)$ and therefore $\mu_{\text{eff}} = \sqrt{15}$ for the ${}^4A_2$ ground term of t_2^3. $\mu_{\text{eff}} = 0$ for the ${}^1A_1$ ground term of t_2^6 (but see § 10.3). The other t_2^n ground terms are ${}^2T_2$ or ${}^3T_1$ and so we must use formulae (10.12)–(10.16).

Take $t_2^2\,{}^3T_1$ as an example and use the ${}^3T_{12}^2$ state to determine γ and μ. Then

$$|{}^3T_{12}^2\rangle = |{}^3T_1\,11\rangle = |-1^+\zeta_1^+\rangle,$$

and hence $\gamma = -1$, $\nu = -\frac{1}{2}\zeta$. Similarly for $t_2\,{}^2T_2$ we find $\gamma = -1$, $\nu = -\zeta$. A, B and μ_{eff}^2 now follow on substitution in our previous formulae. The values of A and B are given in Table 10.1 and the resulting Bohr magneton numbers satisfy

$$\left.\begin{aligned} \mu_{\text{eff}}^2(dt_2^1) &= \frac{3x + (16-3x)\tanh\frac{3}{4}x}{x(3+\tanh\frac{3}{4}x)}, \\ \mu_{\text{eff}}^2(dt_2^2) &= \frac{5(3x+18)\,e^{\frac{3}{2}x} + 3(x+18)\,e^{\frac{1}{2}x} - 144}{2x(1+3\,e^{\frac{1}{2}x}+5\,e^{\frac{3}{2}x})}, \\ \mu_{\text{eff}}^2(dt_2^3) &= 15. \end{aligned}\right\} \qquad (10.18)$$

Table 10.1. *The constants for the paramagnetic properties of d-electron systems*

Term	J	g	A	B
$d^1\,t_2^1\,{}^2T_2$	$\frac{1}{2}$	-2	3	-8
	$1\frac{1}{2}$	0	0	4
$d^2\,t_2^2\,{}^3T_1$	0	5	0	-72
	1	$\frac{1}{2}$	$\frac{1}{2}$	9
	2	$\frac{1}{2}$	$1\frac{1}{2}$	9
$d^2\,{}^3F\,{}^3T_1$	0	$5\frac{1}{2}$	0	$-65\frac{1}{3}$
	1	$\frac{1}{4}$	$\frac{1}{8}$	$8\frac{1}{6}$
	2	$\frac{1}{4}$	$\frac{3}{8}$	$8\frac{1}{6}$
$d^6\,t_2^4e^2\,{}^5T_2$	1	$3\frac{1}{2}$	$24\frac{1}{2}$	54
	2	$1\frac{1}{2}$	$13\frac{1}{2}$	-10
	3	1	12	-16
$d^7\,t_2^5e^2\,{}^4T_1$	$\frac{1}{2}$	4	12	60
	$1\frac{1}{2}$	$1\frac{1}{5}$	$5\frac{2}{5}$	$-10\frac{14}{25}$
	$2\frac{1}{2}$	$\frac{4}{5}$	$5\frac{3}{5}$	$-12\frac{24}{25}$
$d^7\,{}^4F\,{}^4T_1$	$\frac{1}{2}$	$4\frac{1}{3}$	$14\frac{1}{12}$	$54\frac{4}{9}$
	$1\frac{1}{2}$	$1\frac{1}{15}$	$4\frac{4}{15}$	$-9\frac{131}{225}$
	$2\frac{1}{2}$	$\frac{3}{5}$	$3\frac{3}{20}$	$-11\frac{19}{25}$

Both the spin-orbit coupling energy and the magnetic field energy are one-electron operators. It follows from the results of § 9.7 that the Bohr magneton numbers for t_2^4 and t_2^5 are obtained from those of t_2^2 and t_2^1, respectively, by changing the sign of x throughout (10.18). These formulae were first obtained by Kotani (1949) by direct calculation and are plotted showing μ_{eff}^2 as a function of $x^{-1} = kT/\zeta$ in Fig. 10.1.

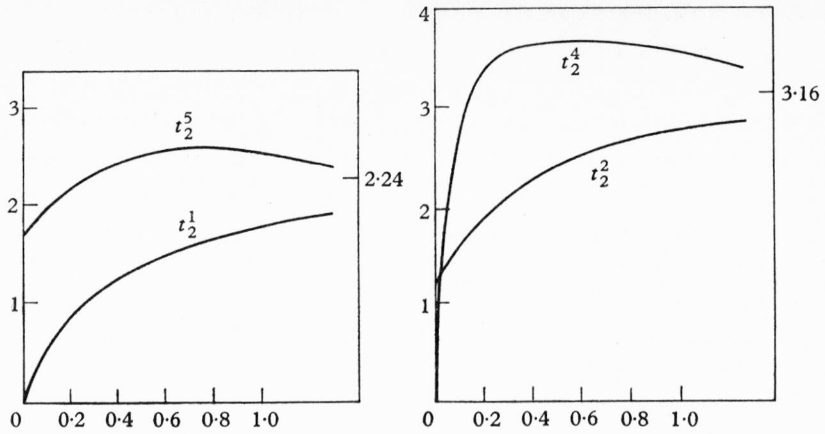

Fig. 10.1. μ_{eff} plotted against kT/ζ for t_2^n ions ($n = 1, 2, 4, 5$). The limiting values of μ_{eff} as $T \to \infty$ are marked at the right of the graphs (after Kotani, 1949).

10.2.3. Other ground terms. The high-spin ground term of d^6 is $^5T_2$ and the $^5T_{23}^3$ state is

$$|^5T_{23}^3\rangle = |t_2^4(^3T_1)\, e^2(^3A_2)\, ^5T_2\, 21\rangle$$
$$= -|1^+ -1^2\, \zeta_1^+\, \theta^+\, \epsilon^+\rangle$$

and $\nu = \frac{1}{4}\zeta$, $\gamma = -1$. g, A and B are given in Table 10.1 and

$$\mu_{\text{eff}}^2(d^6) = \frac{3(49x+108)+5(27x-20)\,e^{-\frac{1}{2}x}+56(3x-4)\,e^{-\frac{5}{4}x}}{2x(3+5\,e^{-\frac{1}{2}x}+7\,e^{-\frac{5}{4}x})}. \tag{10.19}$$

The ground term of d^7 is $^4T_1$ but differs slightly according to whether we use a strong field, weak field or an intermediate field. For the strong-field ground term $|t_2^5 e^2\, ^4T_1\rangle$, $\nu = \frac{1}{3}\zeta$ and $\gamma = -1$ whence

$$\mu_{\text{eff}}^2(d^7) = \frac{300(x+5)+6(45x-88)\,e^{-\frac{1}{2}x}+12(35x-81)\,e^{-\frac{4}{3}x}}{25x(1+2\,e^{-\frac{1}{2}x}+3\,e^{-\frac{4}{3}x})}. \tag{10.20}$$

Formulae (10.19) and (10.20) are a little complicated and to facilitate writing similar formulae we introduce the notation

$$\mu_{\text{eff}}^2 = \frac{\Sigma W_J\, \mu_{\text{eff}}^2(J)}{\Sigma W_J}, \tag{10.21}$$

where W_J is a weighting factor for the level J. The sum in (10.21) is over at most three values of J and the W_J are proportional to $(2J+1)\exp[-\nu J(J+1)/2kT]$. For the weak-field ground term $|d^7\, ^4F\, ^4T_1\rangle$, $\nu = \frac{1}{2}\zeta$ and $\gamma = -\frac{3}{2}$ whence we derive the parameters g, A and B given in Table 10.1. For an intermediate field we will work in the strong-field scheme. The $^4T_1\frac{3}{2}1$ state is a linear combination ψ of

$$\psi_1 = |t_2^5(^2T_2)\, e^2(^3A_2)\, ^4T_1\tfrac{3}{2}1\rangle = -|1^+ -1^2\, \zeta_1^2\, \theta^+\, \epsilon^+\rangle,$$

$$\psi_2 = |t_2^4(^3T_1)\, e^3(^2E)\, ^4T_1\tfrac{3}{2}1\rangle = \tfrac{1}{2}|1^+ -1^2\, \zeta_1^+\, \theta^+\, \epsilon^2\rangle + \tfrac{1}{2}\sqrt{3}\,|1^2 -1^+\, \zeta_1^+\, \theta^2\, \epsilon^+\rangle,$$

and so

$$\psi = \psi_1 \cos\theta + \psi_2 \sin\theta.$$

The matrices of spin-orbit coupling and of L_z within (ψ_1, ψ_2) are

$$\langle \overline{\psi}_1 | L_z | \psi_1 \rangle = \langle \overline{\psi}_1 | L_z | \psi_2 \rangle = -2 \langle \overline{\psi}_2 | L_z | \psi_2 \rangle = -1$$

and

$$\langle \overline{\psi}_i | \mathscr{H}_s | \psi_j \rangle = -\tfrac{1}{2}\zeta \langle \overline{\psi}_i | L_z | \psi_j \rangle,$$

whence $\gamma = -\cos^2\theta - 2\sin\theta\cos\theta + \frac{1}{2}\sin^2\theta$ and $\nu = -\frac{1}{3}\gamma\zeta$. One easily proves that $-\frac{3}{2} \leqslant \gamma \leqslant -1$, using the secular equation for the electrostatic energies of the $^4T_1$ terms to determine the possible range of θ.

Then

$$\left.\begin{aligned} g(\tfrac{1}{2}) &= \tfrac{2}{3}(5-\gamma), \quad \mu^2_{\text{eff}}(\tfrac{1}{2}) = \tfrac{1}{3}(5-\gamma)^2 - \frac{20}{3x\gamma}(2-\gamma)^2, \quad W_{\frac{1}{2}} = 1, \\ g(\tfrac{3}{2}) &= \tfrac{2}{15}(11+2\gamma), \quad \mu^2_{\text{eff}}(\tfrac{3}{2}) = \tfrac{1}{15}(11+2\gamma)^2 + \frac{88}{75x\gamma}(2-\gamma)^2, \quad W_{\frac{3}{2}} = 2\,e^{\frac{1}{2}\gamma x}, \\ g(\tfrac{5}{2}) &= \tfrac{2}{3}(3+\gamma), \quad \mu^2_{\text{eff}}(\tfrac{5}{2}) = \tfrac{7}{5}(3+\gamma)^2 + \frac{36}{25x\gamma}(2-\gamma)^2, \quad W_{\frac{5}{2}} = 3\,e^{\frac{4}{3}\gamma x}. \end{aligned}\right\} \quad (10.22)$$

For d^2, also, there is an excited term of the same symmetry as the ground term. The constants, g, A and B are given in Table 10.1 for the weak-field d^2 ground term.[1] I leave it as an exercise for the reader to show that for an intermediate field $\nu = \frac{1}{2}\gamma\zeta$, $-\frac{3}{2} \leqslant \gamma \leqslant -1$ and

$$\left.\begin{aligned} g(0) &= 4-\gamma, \quad \mu^2_{\text{eff}}(0) = \frac{8}{x\gamma}(2-\gamma)^2, \quad W_0 = 1, \\ g(1) &= 1+\tfrac{1}{2}\gamma, \quad \mu^2_{\text{eff}}(1) = \tfrac{1}{2}(2+\gamma)^2 - \frac{1}{x\gamma}(2-\gamma)^2, \quad W_1 = 3\,e^{-\frac{1}{2}\gamma x}, \\ g(2) &= 1+\tfrac{1}{2}\gamma, \quad \mu^2_{\text{eff}}(2) = \tfrac{3}{2}(2+\gamma)^2 - \frac{1}{x\gamma}(2-\gamma)^2, \quad W_2 = 5\,e^{-\frac{3}{2}\gamma x}. \end{aligned}\right\} \quad (10.23)$$

The remaining ground terms of d^n configurations in octahedral fields have spin-only susceptibilities to the present approximation, so we have found formulae for μ^2_{eff} for all high- and low-spin ground terms.

10.2.4. Ground terms in tetrahedral fields. Page 222 and § 9.7 show the matrix for a d^n ion in a tetrahedral field is simply related to the matrix of a d^{10-n} ion in an octahedral field. Therefore we may use the formulae we have just found to deduce formulae for tetrahedral fields. Specifically the formula for μ^2_{eff} for a tetrahedral ground term of d^n is the same as that for the corresponding octahedral ground term of d^{10-n} of the same type (i.e. also high spin or also low spin) except for a change of sign of $x = \zeta/kT$. γ does not change sign.

A well-known example here is the cobaltous ion, which appears in octahedral and tetrahedral coordination. The octahedral ground term is $^4T_1$ and the tetrahedral one $^4A_2$. To our present approximation the room-temperature calculated values of μ_{eff} are 4·68[2] for $^4T_1$ and 3·87 for $^4A_2$ and it has been suggested that the

[1] Formulae (10.19) and (10.20) were given in Griffith (1958*a*) and confirmed by Figgis (1958) who also gave formulae in the weak-field limit for d^7 and d^2. The latter agree with (10.22) and (10.23) in the text on putting $\gamma = -\frac{3}{2}$ and with Table 10.1.

[2] For the strong-field case.

experimental susceptibilities might be used as a diagnostic test for the coordination number, and ligand arrangement, of the ion. The large calculated difference is, however, deceptive and when the mixing of the $^4T_{2\frac{3}{2}}$ states into the ground term in tetrahedral Co^{++} is taken into account the difference is largely removed. In Cs_3CoCl_5 we expect $\mu_{\text{eff}} = 4{\cdot}51$ (§ 10.4.1) which is so close to the strong field value for an octahedral compound that it is doubtful whether any reliable test could be based upon the difference.[1] Perhaps I should emphasize that it is not only the experimental difficulty which is relevant here but also the uncertainty of the expected values. Apart from their dependence on ζ and Δ, small changes from compound to compound are to be expected both because of the varying effect of lower symmetry perturbations and (§ 10.2.8) because of varying exchange interactions between ions. The octahedral and tetrahedral cobaltic compounds differ much more in the anisotropy of their susceptibility, however, as discussed in § 10.2.6. They also differ in their temperature dependence, the tetrahedral but not the octahedral ones satisfying a Curie law (p. 277) at normal temperatures.

10.2.5. Comparison with experiment. Generally speaking, the experimental susceptibilities agree rather well with the theoretical formulae. Values of μ_{eff} are shown in Table 10.2 for first transition-series metal ions at room temperature.

Table 10.2. *Comparison of experimental room temperature susceptibilities with those calculated within the ground terms for octahedral symmetry*

Experimental data (*a*) for aqueous ions (Pauling, 1940), (*b*) typical values for octahedral coordination (for data see Selwood (1943), Nyholm (1956)). $\mu = \mu_{\text{eff}}$.

Ions	μ spin only	μ calc.	μ exp. (*a*)	μ exp. (*b*)
d^1: V^{4+}, Ti^{3+}	1·73	1·95	—	1·7–1·9
d^2: V^{3+}	2·83	2·72	2·4	2·7–2·9
d^3: V^{++}, Cr^{3+}	3·87	3·87	3·8–3·9	3·8–3·9
d^4: Cr^{++}, Mn^{3+}	4·90	4·90	4·8–4·9	4·8–4·9
d^5: Mn^{++}, Fe^{3+}	5·92	5·92	5·9	5·8–5·9
d^6: Fe^{++}	4·90	5·64	5·3	5·2–5·5
d^7: Co^{++}	3·87	4·68	5·0–5·2	4·8–5·1
d^8: Ni^{++}	2·83	2·83	3·2	2·8–3·3
d^9: Cu^{++}	1·73	1·73	1·9–2·0	1·8–2·0

The calculated values use (10.18) for d^1, d^2 and d^3, (10.19) for d^6, (10.20) for d^7 and the spin-only formula for the remainder and take $kT = 200\ \text{cm}^{-1}$. The fourth column has the susceptibilities of the aqueous ions and the last column has values selected as typical from a wide range of measured susceptibilities. The agreement is satisfactory, the deviations for nickel and copper being explicable when we include the matrix elements of the spin-orbit coupling energies with excited terms—$^3T_2$ for nickel and $^2T_2$ for copper (see §§ 10.4.1 and 12.4.2, 3).

[1] However, Nyholm (1956) reports that for some classes of cobaltous compounds, at least, the octahedral compounds have consistently higher susceptibilities than the tetrahedral ones. It is far from certain that this would always be true.

Kotani's formulae for d^4 and d^5 low-spin compounds also give fairly good agreement with experiment. χ was measured in the temperature range 20–300° K for the compound $K_3Mn(CN)_6$ by Cooke & Duffus (1955). Here the manganese ion is d^4. Comparison with experiment on the assumption that the ground term is $dt_2^4\,{}^3T_1$ is shown in Fig. 10.2 and is seen to be good but not perfect. The deviation

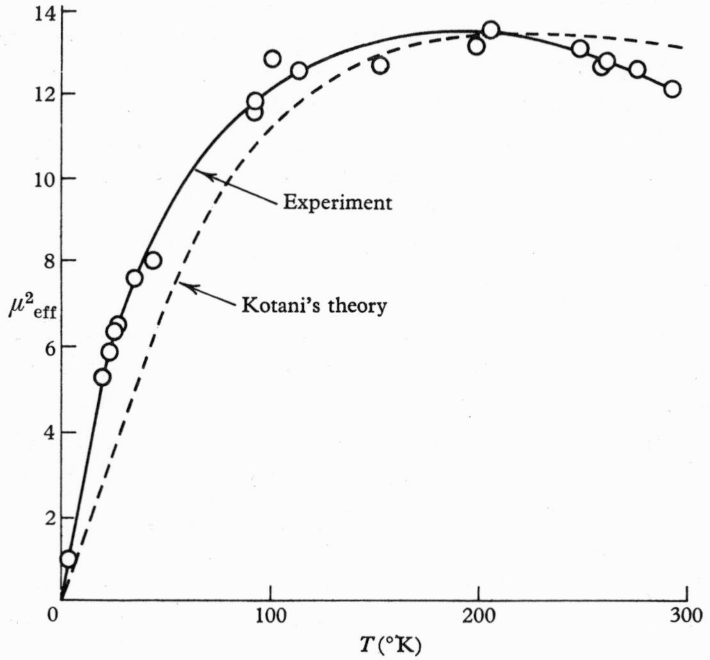

Fig. 10.2. Theoretical and experimental μ_{eff} plotted against temperature for $K_3Mn(CN)_6$ (from Cooke & Duffus, 1955).

is at least partly explicable in terms of lower symmetry perturbations (Kamimura, 1956). χ for $K_3Fe(CN)_6$, which is d^5, has been measured carefully over a large temperature range, chiefly by Jackson (1938). It is very anisotropic at low temperatures with approximately tetragonal symmetry. The mean susceptibility is in fair agreement with Kotani's formula and the individual values of χ have been largely, but not completely, explained in terms of a tetragonal perturbation with a small rhombic component (Howard, 1935; Bleaney & O'Brien, 1956).

10.2.6. Magnetic anisotropy of high-spin compounds. Let us now consider the effect of a field of lower symmetry on the susceptibilities of high-spin compounds. This can only produce a first-order splitting of the ground term if the ground term is spatially degenerate. In the latter case the Jahn–Teller theorem leads us to expect that there will actually be a splitting. This will produce some effect on the mean susceptibility, especially at low temperatures, but will mainly produce an anisotropy of susceptibility. A number of interesting detailed calculations have been made of this but here I merely make the general observation, due originally to Van Vleck, that we should expect the anisotropy to be very much smaller for ions with ${}^{2S+1}A_1$ or ${}^{2S+1}A_2$ ground terms than for those having ${}^{2S+1}E$,

${}^{2S+1}T_1$ or ${}^{2S+1}T_2$ ground terms. In an octahedral field the high-spin ions of the former type are d^3 (Cr^{3+}), d^5 (Mn^{++}, Fe^{3+}) and d^8 (Ni^{++}) and in a tetrahedral field they are d^2 (Ti^{++}, V^{3+}), d^5 (Mn^{++}, Fe^{3+}) and d^7 (Co^{++}). The latter type are d^1 (Ti^{3+}), d^4 (Cr^{++}), d^6(Fe^{++}, Co^{3+} in K_3CoF_6), d^9 (Co^{++}) and also d^2, d^7 for an octahedral and d^3, d^8 for a tetrahedral field. Van Vleck (1958) has given a table of relative anisotropies for typical octahedral high-spin ions with the values 0·25, 0·10, 0·20, 1·50 for Cr^{3+}, Mn^{++}, Fe^{3+}, Ni^{++}, respectively, and 16, 30, 20 for Fe^{++}, Co^{++}, Cu^{++}, respectively, which is a nice illustration of the general proposition. Finally, we note that not merely the magnitude of the susceptibility but also of the anisotropy can serve to distinguish an octahedral from a tetrahedral field in the cases of d^2, d^3, d^7 and d^8. As an example of this, tetrahedral Co^{++} compounds have low anisotropy unlike the highly anisotropic octahedral Co^{++} compounds and this is likely to furnish a more reliable magnetic criterion to distinguish these two symmetries.

10.2.7. Correction for diamagnetism. The reader may have noticed that we have completely neglected the diamagnetic term in the susceptibility. This is practically negligible for strongly paramagnetic ions in concentrated salts, but allowance must be made for it when the ions are very weakly paramagnetic or are diluted in isomorphous diamagnetic compounds. The latter is done in order to avoid the exchange effects discussed in the next section.

In the kind of compounds we consider, the diamagnetic contributions χ_d are additive functions of the components to quite a good approximation. For example

$$\chi_d(ZnSO_4 . 5H_2O) \approx \chi_d(Zn^{++}) + \chi_d(SO_4^{--}) + 5\chi_d(H_2O),$$

and even

$$\chi_d(H_2O) \approx 2\chi_d(H) + \chi_d(O).$$

These relations are known as Pascal's rules. As examples, the χ_d for Mg^{++}, Ca^{++}, Zn^{++}, Ag^+, Cd^{++} are respectively $-10{\cdot}1$, $-16{\cdot}0$, $-13{\cdot}5$, $-31{\cdot}0$ and $-20{\cdot}0$ in units of 10^{-6}. NH_3 and H_2O have $\chi_d = -16{\cdot}5 \times 10^{-6}$ and $\chi_d = -13{\cdot}0 \times 10^{-6}$, respectively. The paramagnetic contribution to χ for an ion with $S = \frac{1}{2}$ is 1260×10^{-6} at room temperature and proportionally greater at lower temperatures. These numbers illustrate the general remarks at the beginning of this subsection.

10.2.8. Exchange and saturation effects. One of the most serious barriers to the accurate experimental knowledge of that part of the susceptibility which belongs to the metal ions considered as isolated members of an aggregate lies in the existence of exchange effects. The assumption that there is no coupling between the spins of different paramagnetic ions is never strictly correct and is often very untrue indeed. The theory of aggregates of ions with interionic couplings is outside the scope of this book. In other words we do not treat ferromagnetic or antiferromagnetic materials. However, as small couplings always occur it is desirable to consider briefly their effect and we do this by taking the simplest possible example.

We suppose we have an aggregate of $\frac{1}{2}N$ pairs of metal ions, each metal ion having a spin of $\frac{1}{2}$ being orbitally non-degenerate and with a spin-only interaction with any external magnetic field. Let the interaction between the ions of a pair be given by $-J\mathbf{s}_1 . \mathbf{s}_2$, where $\mathbf{s}_1$ and $\mathbf{s}_2$ are the spin vectors for the two ions and J is a constant. $\mathbf{s}_1$ and $\mathbf{s}_2$ are measured in units of $\hbar$. The Hamiltonian for a pair is then

$$\left.\begin{aligned}\mathscr{H} &= -J\mathbf{s}_1 . \mathbf{s}_2 + 2\beta \mathbf{H} . \mathbf{S}, \\ &= -\tfrac{1}{2}JS(S+1) + 2\beta\mathbf{H} . \mathbf{S} + \tfrac{3}{4}J.\end{aligned}\right\} \tag{10.24}$$

Equation (10.24) shows that in zero magnetic field the pair has a singlet and a triplet state separated in energy by J and with the triplet lying lower if J is positive. When the latter is true we say we have a ferromagnetic interaction and when J is negative an antiferromagnetic one. $J\mathbf{s}_1 . \mathbf{s}_2$ has the form of a magnetic interaction but, just as in atoms singlets and triplets get separated in energy as a consequence of the combined action of the Pauli antisymmetry requirement and the electrostatic interaction, so here the origin of J is electric and not magnetic.

The Hamiltonian $\mathscr{H}$ is isotropic, so without loss of generality we take $\mathbf{H}$ along OZ. Then $\mathscr{H}$ commutes with S_z and we have four states with energies

$$-\tfrac{1}{4}J - 2\beta H, \quad -\tfrac{1}{4}J, \quad -\tfrac{1}{4}J + 2\beta H \quad \text{and} \quad \tfrac{3}{4}J.$$

Hence the susceptibility is

$$\chi = \frac{\tfrac{1}{2}N\beta^2}{kT}\frac{8}{3+e^{-J/kT}}. \tag{10.25}$$

Now suppose $J \ll kT$ and replace $e^{-J/kT}$ by $1-(J/kT)$. Then we find

$$\chi = \frac{N\beta^2}{k[T-(J/4k)]}. \tag{10.26}$$

If the susceptibility varies with temperature according to the law $\chi = N\beta^2\mu^2/kT$ with constant μ we say it satisfies a Curie law after Pierre Curie who discovered the law experimentally. If it satisfies

$$\chi = \frac{N\beta^2\mu^2}{k(T+\Delta)}, \tag{10.27}$$

with μ constant and a non-zero constant Δ we say it satisfies a Curie–Weiss law, commemorating also Weiss who discussed ferromagnetic exchange couplings of the above type in terms of a classical model. Δ is called the Weiss constant and in our model $\Delta = -J/4k$. Δ is positive for antiferromagnetic and negative for ferromagnetic couplings. The behaviour of actual materials as T approaches $|\Delta|$ is hardly ever usefully approximated by our model and is often very complicated.

The actual values of the part of Δ which is due to exchange coupling vary greatly with dilution. Δ is usually in the range 20–200° K for concentrated salts without water of hydration and is generally smaller for hydrated and more dilute salts. J itself has been measured directly in a few salts where coupled pairs of ions occur. In copper acetate, where the pair of ions is separated by 2·6 Å, $J = -300\ \text{cm}^{-1}$, while in copper sulphate where they are 5·2 Å apart $|J| = 0{\cdot}15\ \text{cm}^{-1}$. These were obtained by an analysis of electron resonance measurements.[1]

[1] See Bagguley & Owen (1957).

It is evident that in our particular model if we plot kT against $N\beta^2\chi^{-1}$ we get a straight line (for $kT \gg J$) with slope μ^2, with $\mu^2 = 3$, which is correct for non-interacting metal ions with spin $\frac{1}{2}$. Therefore in certain cases we can eliminate the effects of exchange interactions by making measurements over a range of temperature. Unfortunately this is not always easy because if an individual ion has thermally occupied states which are not completely degenerate its susceptibility satisfies a Curie–Weiss law at temperatures high compared with the separations between the states. Therefore in such a case one would get an incorrect estimate for μ^2 if one ascribes the Curie–Weiss temperature dependence to exchange effects and attempts to correct for them.

Let us illustrate this latter point. Take Kotani's formula (10.18) for the susceptibility of d^1 and let $\zeta \ll kT$. Then to first order in ζ it becomes

$$\chi = \frac{5N\beta^2}{3k[T+(2\zeta/5k)]}, \tag{10.28}$$

having the form of an antiferromagnetic interaction with the Bohr magneton number $\sqrt{5}$. The latter is actually only correct for $T = \infty$.

Until now we have supposed χ to be independent of H. This is very accurately true at normal temperatures and normally accessible magnetic fields. The relevant relation here is $\beta H \ll kT$ and so at very low temperatures χ will depend on H. However, for most compounds Δ is much larger than the temperature corresponding to βH (about 1° K for $H = 10{,}000$ gauss). Δ is particularly small for the rare-earth compound gadolinium sulphate which contains trivalent gadolinium Gd^{3+} and saturation effects have been observed here. Gd^{3+} is a particularly favourable ion for this purpose because it has the half-filled shell ground term ^8S of f^7 and ligand fields are in any case small in rare-earth compounds.

10.3. Residual paramagnetism

So far we have considered only ions whose ground state is degenerate in the absence of the external magnetic field and the low-symmetry components of the crystal field. All these ions can have, and do have, non-zero matrix elements of the magnetic field energy amongst the occupied states. However, there is one octahedral d^n ground term which is non-degenerate, so has no first-order interaction with the magnetic field. This is the $^1A_1$ term which is the low-spin ground state of d^6. We assume the strong-field case and then have the ground term $|\psi_0\rangle = |1^2-1^2\zeta_1^2\rangle$.

For such an ion $W_1 = 0$ and so the susceptibility χ arises solely from the second-order term W_2 and is, from (10.10),

$$\chi = 2N\beta^2 \sum_n \frac{|\langle\psi_n|(L_z+2S_z)|\psi_0\rangle|^2}{E_n-E_0}. \tag{10.29}$$

$|\psi_0\rangle$ is a singlet and so the matrix elements of S_z to any other term are zero. The selection rules for L_z show that its matrix elements are zero, except to z-components of $^1T_{1g}$ states and, in the strong-field scheme, except to states of t_2^5e. Therefore the sum in (10.29) involves just the one state

$$|\psi_n\rangle = |t_2^5 e\, ^1T_1\, 0\rangle = -\frac{1}{\sqrt{2}}|1^2 - 1^2 \zeta_1^+ \epsilon^-\rangle + \frac{1}{\sqrt{2}}|1^2 - 1^2 \zeta_1^- \epsilon^+\rangle$$

from Table A 20 and 24. Hence $\langle\overline{\psi}_n|(L_z+2S_z)|\psi_0\rangle = -2\sqrt{2}$ and

$$\chi = \frac{16N\beta^2}{E_n - E_0} = \frac{4{\cdot}085}{E_n - E_0}, \tag{10.30}$$

where, in the second formula for χ, $E_n - E_0$ is in wave-numbers.

$E_n - E_0$ is known experimentally from the spectra for many cobaltic salts. In the hexammine, for example, it is 21,000 cm^{-1}. Substituting this value in (10.30) gives $\chi = 195 \times 10^{-6}$. The experimental values are rather discordant but it appears that χ is probably somewhere in the range 60–150 $\times 10^{-6}$. Theoretically, one expects χ to be reduced through orbital reduction and we discuss the theory of this in § 10.4.4.

10.4. More accurate treatments

10.4.1. High-frequency elements from outside the ground configuration. It is customary to refer to that part of χ which arises from matrix elements of $\beta\mathbf{H}\,.\,(\mathbf{L}+2\mathbf{S})$ between states which are thermally occupied and those which are not appreciably so as arising from high-frequency elements. These elements contribute to χ through W_2 in (10.9) and hence give a temperature-independent contribution apart from the Boltzmann factor dependence. The residual paramagnetism considered in the last section is a typical example. We now derive a general approximate expression for the part of χ which comes from high-frequency elements to excited configurations in the case of octahedral symmetry.

Let us suppose first that we have a determinantal function $|\psi_0\rangle$ belonging to e^m. Then, from (10.10), the relevant high-frequency contribution to χ is

$$\chi_h = \frac{2N\beta^2}{3}\sum_n \frac{|\langle\overline{\psi}_n|(\mathbf{L}+2\mathbf{S})|\psi_0\rangle|^2}{E_n - E_0}. \tag{10.31}$$

In (10.31) the sum runs over all excited configurations. Now $\mathbf{L}+2\mathbf{S}$ is a one-electron operator and therefore the only one-vanishing elements arise from $|\psi_n\rangle$ in te^{m-1} (writing $t = t_2$ for short). Further, $\mathbf{S}$ has no matrix elements between configurations because of the orthogonality of the spatial functions. Our approximation is to be the assumption that we may replace $E_n - E_0$ by the same mean energy $\bar{E}$ for all relevant states. Doing this, we have

$$\chi_h = \frac{2N\beta^2}{3\bar{E}}\sum_n |\langle\overline{\psi}_n|\mathbf{L}|\psi_0\rangle|^2. \tag{10.32}$$

The sum in (10.32) is invariant to an orthonormal transform of the states $|\psi_n\rangle$ and we choose these latter states to be determinantal functions also. Then the ground state may be written $|e_1 \ldots e_m\rangle$, where the symbol e_i includes a specification of spin, and an excited state as $|t_j e_2 \ldots e_m\rangle$, $|e_1 t_j \ldots e_m\rangle$, ... or $|e_1 e_2 \ldots t_j\rangle$ for varying t_j. Other states of te^{m-1} do not contribute. We now evaluate the part of the sum in (10.32) depending on L_z. It is (with real spatial functions)

$$\sum_n |\langle \overline{\psi}_n | L_z | \psi_0 \rangle|^2 = \sum_{i,j} |\langle t_j | l_z | e_i \rangle|^2$$
$$= \sum_{i,j} (|\langle t_j | l_z | e_i \rangle|^2 + |\langle e_j | l_z | e_i \rangle|^2) = \sum_i \langle e_i | l_z^2 | e_i \rangle \quad (10.33)$$

and similarly for the parts of (10.32) depending on L_x and L_y. We now have

$$\chi_h(e^m) = \frac{2N\beta^2}{3\bar{E}} \sum_{i,\epsilon} \langle e_i | l_\epsilon^2 | e_i \rangle = \frac{2N\beta^2}{3\bar{E}} \sum_i \langle e_i | \mathbf{l}^2 | e_i \rangle$$
$$= \frac{4mN\beta^2}{\bar{E}}. \quad (10.34)$$

A similar analysis holds for a state of t_2^n except that one step in (10.33) is different. Here
$$\langle t_i | l_z^2 | t_i \rangle = \sum_j |\langle e_j | l_z | t_i \rangle|^2 + \sum_j |\langle t_j | l_z | t_i \rangle|^2$$
as before, but $\langle t_j | l_z | t_i \rangle$, unlike $\langle e_j | l_z | e_i \rangle$, is not necessarily zero. We recall, however, the isomorphism between t_2-orbitals and p-orbitals and remark that the sum $\sum_j |\langle t_j | l_z | t_i \rangle|^2$ is the same as it would be for p-orbitals. On adding it to the corresponding sum for l_x and l_y we obtain $\langle p | \mathbf{l}^2 | p \rangle = 2$. Hence

$$\sum_{j,\epsilon} |\langle e_j | l_\epsilon | t_i \rangle|^2 = 6 - 2 = 4,$$

and
$$\chi_h(t^n) = \frac{8nN\beta^2}{3\bar{E}}. \quad (10.35)$$

Formulae (10.34) and (10.35) are independent of our choice of ground state and therefore are true for any states of e^m or t^n and not only for determinantal functions. $\bar{E}$ is now the mean energy of the excited configuration above the ground term. Also, because of our relationship between states and matrices of d^n and d^{10-n}, they are true for the complementary configurations too. So $\chi_h(t^6e^{4-m}) = \chi_h(e^m)$ and $\chi_h(t^{6-n}e^4) = \chi_h(t^n)$. Putting $m = 4$ and $n = 6$ we find

$$\chi_h(e^4) = \frac{16N\beta^2}{\bar{E}} = \chi_h(t^6)$$

showing the consistency of the two formulae. For e^4 and t^6 there is only one excited term which contributes, so (10.34) and (10.35) are exact and become identical with formula (10.30) for t^6 which we found by a direct calculation.

For e^m or t^6e^{4-m} configurations we may now write the general expression

$$\chi = \frac{N\beta^2\overline{g^2}S(S+1)}{3kT} + \frac{4mN\beta^2}{\bar{E}}, \quad (10.36)$$

for the susceptibility, where $\overline{g^2} = \frac{1}{3}(g_x^2 + g_y^2 + g_z^2)$. We illustrate this first for the aqueous nickel ion. Here the only term of t^5e^3 which contributes to the high-frequency part is $^3T_2$. Therefore

$$\mu_{\text{eff}}^2(\text{Ni}^{++}) = 2\overline{g^2} + \frac{24kT}{E(^3T_2) - E(^3A_2)}. \quad (10.37)$$

We anticipate to say that g is available from electron resonance experiments (§ 12.4) and $E(^3T_2)-E(^3A_2)$ from optical measurements (§ 11.4). g is isotropic at 2·25 and $E(^3T_2)-E(^3A_2) = 8500\,\text{cm}^{-1}$. Taking $kT = 200\,\text{cm}^{-1}$, which is approximately room temperature, we obtain $\mu_{\text{eff}} = 3{\cdot}27$ compared with the experimental value of 3.2.

For the cupric ion we have

$$\mu_{\text{eff}}^2(\text{Cu}^{++}) = \tfrac{3}{4}\overline{g^2}+\frac{12kT}{E(^2T_2)-E(^2E)}. \tag{10.38}$$

An interesting example here is solid copper sulphate, $CuSO_4.5H_2O$. The copper ion has four near neighbours which form a planar complex $(Cu(H_2O)_4)^{++}$ with oxygen atoms from the sulphate groups completing a very distorted octahedron. Our analysis still applies on replacing $\overline{g^2}$ by g^2 and we have different values for μ_{eff}^2 parallel and perpendicular to the fourfold axis of the planar complex. Comparison with experiment is shown in Table 10.3.

Table 10.3. *Comparison of experimental susceptibilities with those calculated from formulae* (10.36)–(10.39)

Experimental values at room temperature for aqueous Cr^{3+}, Ni^{++} and for solid $CuSO_4.5H_2O$. $\|$ and $\perp$ refer to the plane of the water molecules. $\mu = \mu_{\text{eff}}$.

Ion	g	$\bar{E}$ (cm^{-1})	μ calc. ($kT = 200$ cm^{-1})	μ exp.
Cr^{3+}	1·98	17,400	3·87	3·8–3·9
Ni^{++}	2·25	8,500	3·27	3·2
$Cu^{++}\|$	2·46	12,600	2·17	2·12
$Cu^{++}\perp$	2·08	12,600	1·85	1·80

A third example is the chromic ion for which

$$\mu_{\text{eff}}^2(\text{Cr}^{3+}) = \tfrac{15}{4}\overline{g^2}+\frac{24kT}{E(^4T_2)-E(^4A_2)}. \tag{10.39}$$

This is a t_2^3 ion but (10.36) still applies, with m replaced by $\frac{2}{3}n$, so long as the ground term $^4A_2$ is not appreciably mixed with other terms of t_2^3 having different multiplicity. For the whole high-frequency part of the susceptibility arises from t_2^2e and in fact just from $^4T_2$. Here g is isotropic at 1·98 and again the experimental susceptibility is in good agreement with the theoretical formula. In the nickel, cupric and chromic ions, however, the high-frequency part of the susceptibility is a small proportion of the whole and so the experimental agreement is not really quite so impressive at it looks at first sight.

Because of the relationship between octahedral d^3 and tetrahedral d^7 we may use (10.39) for the tetrahedral cobaltous ion also. Electron resonance gives a nearly isotropic g with a root-mean-square value of 2·29 for Cs_3CoCl_5. $E(^4T_2)-E(^4A_2) = 6300\,\text{cm}^{-1}$ (§ 11.5) and we deduce $\mu_{\text{eff}} = 4{\cdot}51$ which as we remarked earlier is not far different from the value for an octahedral compound. The qualitative reason why the deviation from the value of 3·87 calculated for the term $^4A_2$ is so large is partly that we are at the end of the transition series, so ζ is large, but also because the separation $^4T_2 - {}^4A_2$ is much smaller for tetrahedral than octahedral compounds (see § 8.3).

In our experimental comparisons in this subsection we used experimentally determined g values. It is possible to give theoretical formulae for these values and, often, to calculate them. We will see how this is done in ch. 12.

10.4.2. The low-spin d^4 ions. In the second and especially in the third transition series ζ becomes so large that it is not always satisfactory to treat its effects by first-order perturbation theory. We now consider t_2^4 in an octahedral field and place no restriction on the magnitude of ζ. We take the four terms $^3T_1$, ^1E, $^1T_2$, $^1A_1$ as our basic states. Including spin-orbit coupling we have $(^3T_{10}, {}^1A_1) \in A_1$, $(^3T_{12}E, {}^1E) \in E$, $^3T_{11} \in T_1$ and $(^3T_{12}T_2, {}^1T_2) \in T_2$ and we already possess the matrix elements of the electrostatic energy and the spin-orbit coupling energy between these states (Tables 9.1 and A 35). Relative to the centre of gravity of $^3T_1$ they are given in Table 10.4. We notice that the matrices for E and for T_2 are identical.

Table 10.4. *Matrix elements of electrostatic and spin-orbit coupling energy within* t_2^4

A_1	$^3T_1$	$^1A_1$
$^3T_1$	$-\zeta$	$-\zeta\sqrt{2}$
$^1A_1$	$-\zeta\sqrt{2}$	$15B+5C$

E	$^3T_1$	^1E
$^3T_1$	$\frac{1}{2}\zeta$	$\frac{1}{2}\zeta\sqrt{2}$
^1E	$\frac{1}{2}\zeta\sqrt{2}$	$6B+2C$

T_2	$^3T_1$	$^1T_2$
$^3T_1$	$\frac{1}{2}\zeta$	$\frac{1}{2}\zeta\sqrt{2}$
$^1T_2$	$\frac{1}{2}\zeta\sqrt{2}$	$6B+2C$

T_1	$^3T_1$
$^3T_1$	$-\frac{1}{2}\zeta$

This is because of the isomorphism between t_2^4 and p^2. When the electrostatic parameters are zero we obtain three equally spaced sets of states, $(t_{2\frac{3}{2}})^4$ with energy -2ζ and symmetry A_1; $(t_{2\frac{3}{2}})^3(t_{2\frac{1}{2}})^1$ with energy $-\frac{1}{2}\zeta$ and symmetries E, T_1, T_2; $(t_{2\frac{3}{2}})^2(t_{2\frac{1}{2}})^2$ with energy $+\zeta$ and symmetries A_1, E, T_2. This corresponds to the jj-coupling limit in p^2 and the transition from

$$\zeta/(3B+C) = 0 \quad \text{to} \quad \zeta/(3B+C) = \infty$$

is shown schematically in Fig. 10.3. The J values appropriate for p^2 are shown.

We now calculate the susceptibility for temperatures low enough for the ground A_1 state to be the only one which is appreciably occupied. Let the ground state be

$$|\psi_0\rangle = c_1|^3T_{10}\rangle + c_2|^1A_1\rangle, \tag{10.40}$$

where c_1, c_2 are determined by solving the quadratic secular equation associated with the A_1 matrix of Table 10.4. The susceptibility is given by (10.31) with $|\psi_n\rangle$ running over states of t_2^4. The state $|\psi_0\rangle \in A_1$ and $\mathbf{L}$ and $2\mathbf{S} \in T_1$ so only states $|\psi_n\rangle$ belonging to T_1 can have non-zero matrix elements of $\mathbf{L}+2\mathbf{S}$

with $|\psi_0\rangle$. ${}^3T_{11}$ is the only possibility and it has zero matrix elements with ${}^1A_1$ because it differs in spin. Let us put $m = E({}^3T_{11}) - E(\psi_0)$ and then we have

$$\begin{aligned}\chi &= \frac{2N\beta^2}{3m}\sum_{\epsilon}|\langle {}^3T_{11}T_1\epsilon|(L_\epsilon+2S_\epsilon)|\psi_0\rangle|^2\\ &= \frac{2N\beta^2c_1^2}{m}|\langle {}^3T_{11}T_1z|(L_z+2S_z)|{}^3T_{10}\rangle|^2\\ &= \frac{2N\beta^2c_1^2}{m}|\langle {}^3P_1 0|(-L_z+2S_z)|{}^3P_0\rangle|^2 = \frac{12N\beta^2c_1^2}{m}. \end{aligned} \qquad (10.41)$$

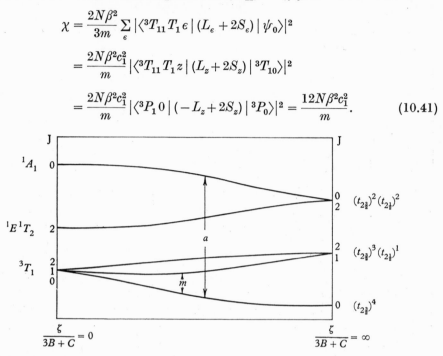

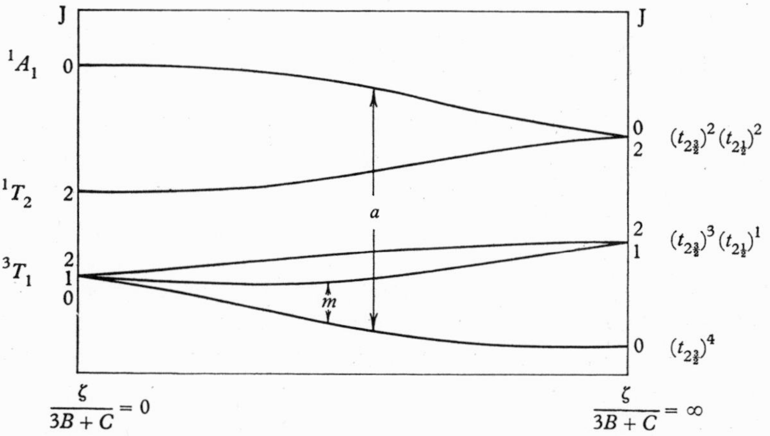

Fig. 10.3. Schematic representation of the behaviour of the energies of the states of t_2^4 as a function of $\zeta/(3B+C)$, exhibiting also the isomorphism with p^2 in intermediate coupling (after Condon & Shortley, 1953, and Griffith, 1958*b*).

We have used the p^2 isomorphism and (5.68) but the matrix element

$$\langle {}^3T_{11}T_1z|(L_z+2S_z)|{}^3T_{10}\rangle$$

can of course be worked out directly from the definitions of ${}^3T_{10}$, ${}^3T_{11}$ (Tables A 20 and 24) with little extra trouble. We now solve the secular equation for A_1 and find

$$\left.\begin{aligned} m &= \tfrac{1}{2}\zeta + \delta(\sec\theta - 1),\\ \chi &= \frac{6N\beta^2(1+\cos\theta)}{m\delta} + \frac{32N\beta^2}{3\bar{E}}, \end{aligned}\right\} \qquad (10.42)$$

where $2\delta = 15B + 5C + \zeta$ and $\delta\tan\theta = \zeta\sqrt{2}$ and we have included also the contribution (10.35) from the high-frequency elements to t_2^3e.

A comparison of the theoretical formula (10.42) with experimental measurements by Figgis, Lewis, Nyholm & Peacock (1958) is given in Table 10.5. For ruthenium and iridium the individual values are close to the average values but for osmium the experimental values vary in an apparently haphazard manner for a series of osmihalides M_2OsHal_6 (M = K, Cs; Hal = F, Cl, Br, I). If the field contains a lower symmetry component this would split the ${}^3T_{11}$ level and have some effect on the susceptibility. m would then become anisotropic. However, it seems unlikely that any large effect could arise in this way.

Table 10.5. *Comparison of theoretical with experimental susceptibilities for d^4 ions*

χ is in units of 10^{-6}. The experimental χ is to be compared with the sum of the theoretical quantities $\chi(t_2^4)$ and χ_h.

Ion	ζ	δ	$\chi(t_2^4)$ Kotani	$\chi(t_2^4)$ New theory	Exp. average χ	χ_h $\bar{E} = 4 \times 10^4$	χ_h $\bar{E} = 5 \times 10^4$
Ru^{4+}	1,400	13,200	4,375	3,593	3,695	68	54
Os^{4+}	4,500	11,000	1,362	722	936	68	54
Ir^{5+}	5,000	11,625	1,226	634	725	68	54

There is a more serious way in which our calculation is incomplete although it does not appear to affect the calculated values of χ very much. This the so-called orbital reduction of the magnetic moment and will be discussed in 10.4.4 (see ex. 2, p. 285).

10.4.3. K_2ReCl_6 and related compounds. For the d^3 configuration in the later transition series K_2ReCl_6 is especially interesting. The g value here is 1·8 for the salt diluted in K_2PtCl_6 and if we introduce this into (10.39) using a value for $E(^4T_2)-E(^4A_2)$, which is unknown experimentally, in the range 10,000–60,000 cm^{-1} we obtain $3{\cdot}50 \leqslant \mu_{\text{eff}} \leqslant 3{\cdot}55$. The susceptibility of pure K_2ReCl_6 has been measured by two groups of workers over a large temperature range. There is a considerable antiferromagnetic interaction (Curie–Weiss constant of 90°) and when this is allowed for $\mu \simeq 3{\cdot}6$–$3{\cdot}7$, which is satisfactory (see Klemm & Steinberg, 1936; Griffiths, Owen & Ward, 1953; Figgis *et al.* 1958).

10.4.4. Orbital reduction. We have carried since § 7.3, although mainly at the back of our minds, the realization that our t_2 and e-orbitals do not necessarily approximate very closely to d-orbitals and it is time to see how the appropriate modification to our theory is introduced. It is simplest to use the replacement theorem and hypothesize orbitals t_2' and e' which are actually d-orbitals. Then $\mathbf{l}$ transforms as T_1 and therefore

$$\left.\begin{aligned} \langle t_2\kappa \,|\, l_\lambda \,|\, t_2\mu\rangle &= k\,\langle t_2'\kappa \,|\, l_\lambda \,|\, t_2'\mu\rangle, \\ \langle t_2\kappa \,|\, l_\lambda \,|\, e\mu\rangle &= k'\,\langle t_2'\kappa \,|\, l_\lambda \,|\, e'\mu\rangle, \\ \langle e\kappa \,|\, l_\lambda \,|\, e\mu\rangle &= 0. \end{aligned}\right\} \tag{10.43}$$

k and k' are numbers, not necessarily equal. They are called orbital reduction factors. In practice they are less than unity and closer to unity for the first transition series than for the later ones. $k = k' = 1$ for d-orbitals, so we may regard $1-k$ and $1-k'$ as some measure of the deviation of the actual orbitals from being pure d-orbitals.

In a similar way there will be two spin-orbit coupling constants, ζ for use among t_2-orbitals and ζ' for use between t_2- and e-orbitals.

The orbital reduction affects the matrix elements of **l**, **L** but not of **s**, **S** and therefore will have most effect on the magnetic susceptibilities when the matrix elements of the spin moment are small compared with that of the orbital moment. The most obvious example here is the low-spin t_2^6 ion where the matrix elements of **S** are zero. All of χ in (10.30) arises from matrix elements of **l** between t_2- and e-orbitals. So we now have

$$\chi = \frac{16N\beta^2k'^2}{E_n - E_0}, \tag{10.44}$$

giving a new and, in general, lower value for χ. In comparing (10.44) with experiment we should remember also that it is calculated assuming the ground state is t_2^6. Actually there will be four excited $d^6\,{}^1A_1$ terms mixed in in varying amounts (Table A 29). These are unlikely, however, to have a large effect nor are the spin-orbit matrix elements of $t_2^6\,{}^1A_1$ with excited terms of different multiplicity (Table A 37).

Examples

Assume orbital reduction according to (10.43).

1. Equations (10.34) and (10.35) each acquire a factor k'^2.
2. Show that Kotani's formulae (10.18) become replaced by

$$\mu^2_{\text{eff}}(t_2^1) = \frac{3x(2k+1)^2 - 8(k+2)^2 + 2(15x(k-1)^2 + 4(k+2)^2)\,e^{\frac{3}{2}x}}{9x(1+2e^{\frac{3}{2}x})},$$

$$\mu^2_{\text{eff}}(t_2^2) = \frac{2(2+k)^2(-8+3e^{\frac{1}{2}x}+5e^{\frac{3}{2}x}) + 12x(1-\frac{1}{2}k)^2(e^{\frac{1}{2}x}+5e^{\frac{3}{2}x})}{2(1+3e^{\frac{1}{2}x}+5e^{\frac{3}{2}x})},$$

$$\mu^2_{\text{eff}}(t_2^3) = 15.$$

3. In the notation of § 10.4.8 show that (10.42) becomes

$$\chi = \frac{2N\beta^2(1+\cos\theta)(k+2)^2}{3m\delta} + \frac{32N\beta^2k'^2}{3\overline{E}}.$$

Dr Jørgensen finds experimentally that the energy separation a in Fig. 10.3 is 17,240 cm^{-1} for $OsCl_6^{--}$ and 16,100 cm^{-1} for $OsBr_6^{--}$. Reducing ζ, B, C and k all in the same proportion I calculate the susceptibilities to be 848×10^{-6} and 878×10^{-6}, respectively, using the first part of the formula. Including the high-frequency part this then gives very close agreement with the experimental average in Table 10.5.

CHAPTER 11

OPTICAL SPECTRA AND THERMODYNAMIC PROPERTIES[1]

11.1. Introduction to d^n spectra

In this chapter we are going to interpret a number of spectra of transition-metal ions as due to transitions between states of d^n configurations. It will clarify matters if we first consider a specific example and we take the tris ethylenediamine nickel ion in aqueous solution. This ion has the formula $Ni^{++}(en)_3$ where *en* is the ethylenediamine molecule $NH_2CH_2CH_2NH_2$. Each *en* molecule is joined to the nickel ion by its nitrogen atoms, thus:

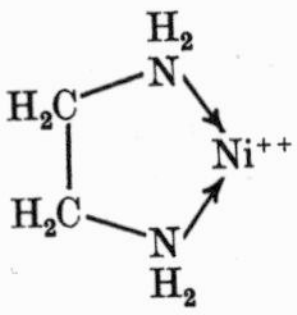

The arrow between a nitrogen atom and the nickel ion is the chemical symbol for a coordinate bond or, alternatively, we may regard it as merely signifying an electrostatic attraction. There are six nitrogen atoms around each nickel ion approximately at the vertices of a regular octahedron. Therefore we may expect them to give rise to a predominantly octahedral field and the energies of the various states of the d^8 nickel ion to lie on a vertical line of Fig. 7.2. We know that electric-dipole, magnetic-dipole and electric-quadrupole matrix elements all satisfy the selection rule $\Delta S = 0$, so at most the spin-allowed transitions can occur with appreciable intensity. The ground term is $^3A_2$ and there are just three excited triplet terms, $^3T_2$, $a\,^3T_1$ and $b\,^3T_1$ say. Therefore we expect just three absorption bands.

The absorption spectrum of the $Ni^{++}(en)_3$ ion has been measured by Jørgensen and is shown in Fig. 11.1 (the open circles). There are just three bands and we tentatively assign them to transitions to the three excited triplet terms. The oscillator strengths can be estimated and we obtain

$$\left.\begin{aligned} E(^3T_2) &= 11{,}200\,\text{cm}^{-1}, & f(^3T_2) &= 1{\cdot}0\times10^{-4},\\ E(a\,^3T_1) &= 18{,}350\,\text{cm}^{-1}, & f(a\,^3T_1) &= 1{\cdot}0\times10^{-4},\\ E(b\,^3T_1) &= 29{,}000\,\text{cm}^{-1}, & f(b\,^3T_1) &= 1{\cdot}4\times10^{-4},\end{aligned}\right\} \quad (11.1)$$

where the energies are measured relative to the ground state. According to our theory, $E(^3T_2) = \Delta$, the octahedral field parameter. So $\Delta = 11{,}200\,\text{cm}^{-1}$ and B

[1] General references on spectra: Tanabe & Sugano (1954); Hartmann & Schläfer (1954); Orgel (1955*a*); Owen (1955); Jørgensen (1956, 1957); Holmes & McClure (1957); Runciman (1958).

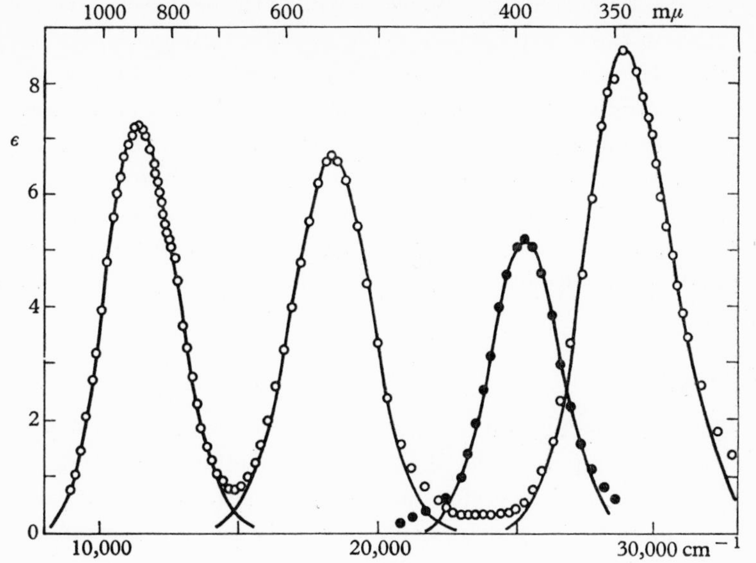

Fig. 11.1. The absorption spectrum of $Ni^{++}(en)_3$, open circles, and the violet band of $Ni(H_2O)_6^{++}$, filled circles. The full curves are the components separated with the assumption of Gaussian form (from Jørgensen, 1954*a*).

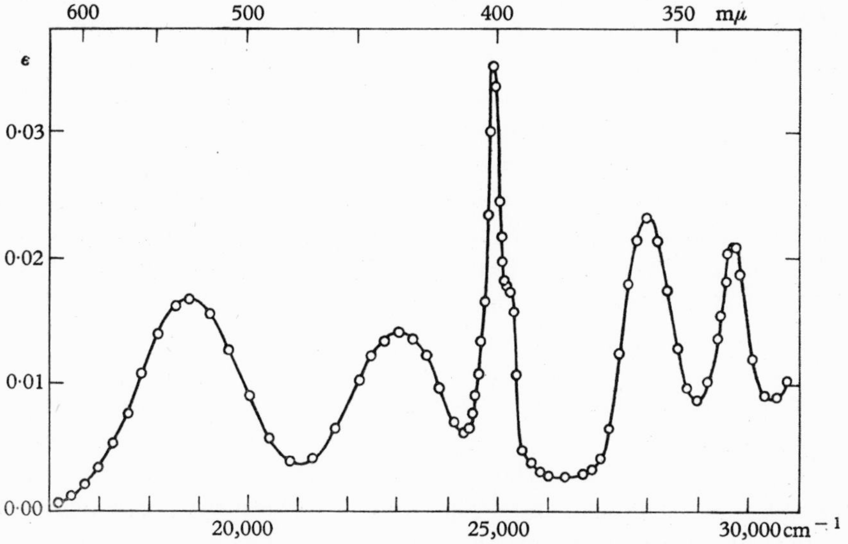

Fig. 11.2. The absorption spectrum of the aqueous manganous ion (from Jørgensen, 1954*b*).

was obtained from atomic spectra (Appendix 6) to be 1084 cm^{-1}. The energies of the two $^3T_1$ terms are the solutions of the secular equation

$$\begin{vmatrix} 12B+\Delta-E & -6B \\ -6B & 3B+2\Delta-E \end{vmatrix} = 0 \qquad (11.2)$$

and are now calculated to be $E = 18{,}400\,\text{cm}^{-1}$ and $31{,}400\,\text{cm}^{-1}$ in pleasing agreement with (11.1) There is no reasonable doubt that this interpretation of the $Ni^{++}(en)_3$ spectrum is correct and we pass on in the next section to a discussion of the question of the interpretation of the oscillator strengths.

We conclude this section by seeing what a typical spin-forbidden d^n spectrum looks like. The ground term of the d^5 configuration is ^6S and there is no other sextet. In the gaseous Mn^{++} ion the first excited term is ^4G of d^5 lying 26,800 cm^{-1} above the ground term. This is split by an octahedral field into $^4A_1$, ^4E, $^4T_1$ and $^4T_2$ (for details, see §§ 9.3 and 11.4). The absorption spectrum of the Mn^{++} ion in aqueous solution, as measured by Jørgensen, is shown in Fig. 11.2. The intensity here is much lower, each band having an oscillator strength of $1-2\times10^{-7}$.

11.2. Intensities

11.2.1. General discussion of spectra. We have supposed a complex ion to be made up of two parts—the filled shells S_f together with the nuclei and the partly-filled shells S_p say. Therefore as well as transitions within S_p there will be transitions from S_f to S_p and from both of these to empty orbitals which lie higher still. Generally these extra transitions all lie at higher energies than many at least of the transitions within S_p, but this is not always true.

I think it will be helpful to introduce a more familiar classification of these transitions based on a separation of the complex ion into parts, S_p together with the metal ion nucleus and inner shells, pertaining to the metal ion, and the remainder of S_f, which pertains to the ligands. Then there are four kinds of transition possible—excitation within the metal ion, excitation within the ligands, transfer of an electron from ligands to metal, transfer of an electron from metal to ligands. We can call the first kind a metal-ion transition, the second a ligand transition and both the last two charge-transfer transitions. Then the likelihood of low-energy transitions may be estimated by referring to the free metal ion and the free ligands. Broadly speaking, unless the ligands are coloured, we do not get any visible absorption due to the second kind of transition. Charge-transfer transitions may occur with colourless ligands; the blood-red $Fe(CNS)_3$ complex of ferric iron with the colourless thiocyanate ion CNS^- is a good example here. We remember, however, that it is only an approximation to divide up our complex ion in this way and may be a very bad one for the excited states.

The lowest state of one of these excited configurations will usually have a higher spin than the ground state. The higher states can have the same spin and opposite parity. Therefore we may get intense bands preceded on the long wavelength side by some weak spin-forbidden bands. It may sometimes be difficult to distinguish these latter bands from transitions within S_p.

We now return to the transitions within S_p, calling them *d-d* transitions. This is a convenient but not entirely accurate terminology.

11.2.2. Intensities of *d-d* transitions. We now consider the ways in which transitions can take place between states of a d^n configuration. Evidently magnetic-dipole and electric-quadrupole transitions provide two mechanisms. At a site whose symmetry group does not contain the inversion, electric-dipole transitions are also possible. This is because atomic states of odd parity are mixed into the d^n states, which means that the orbitals of the partly-filled shells are no longer eigenstates of the parity operator. Another, less accurate, way of describing this is to say that $4p$ atomic orbitals are mixed into the metal $3d$-orbitals. If we write each orbital as $f_g + f_u$, where f_g and f_u are even and odd, respectively, one expects f_u to be small and we may expand the functions of the partly-filled shell configuration and retain only the first power in f_u. Typical states are now

$$\left.\begin{aligned}\psi_0 &= \phi_0 + \epsilon_0 \chi_0,\\ \psi_1 &= \phi_1 + \epsilon_1 \chi_1,\end{aligned}\right\} \tag{11.3}$$

where ϕ_0, ϕ_1 are d^n states, χ_0, χ_1 have odd parity and ϵ_0, ϵ_1 are small coefficients. Then the electric-dipole moment has non-zero matrix elements between ϕ_0 and χ_1 and between χ_0 and ϕ_1 even though it has no non-zero ones between ϕ_0 and ϕ_1. We shall say that these transitions are electric-dipole transitions due to an unsymmetrical environment.

Electric-dipole transitions can arise in another way, however. So far we have assumed the nuclei in the environment of our metal ions to be fixed. This is obviously not strictly true—the nuclei will have states of motion. These correspond to the classical concepts of vibration and rotation. The states of vibration are the relevant ones here and the system will have excited vibrational states. The existence of these is responsible for electric-dipole transitions even when the site symmetry group contains the inversion. The formal possibility of this is easily demonstrated. Suppose the ground and excited states are $\phi_0 f_0$ and $\phi_1 f_1$, respectively, where ϕ_0, ϕ_1 are electronic wave-functions and f_0, f_1 represent vibrational states of the environment. Then if f_0 has opposite parity to f_1 we have a transition with change of parity, which is therefore possible for electric-dipole radiation (there may, of course, be other selection rules which forbid a particular transition of this type). The theory of these vibrationally-induced transitions rests essentially on the fact that, even if the environment has inversion symmetry on the average, momentarily it will usually be in an unsymmetrical configuration. In this unsymmetrical configuration states of odd parity are mixed into the d^n electronic functions and the transitions become weakly allowed. The overall parity of states remains unchanged because, for example, $\phi_0 f_0$ becomes $\phi_0 f_0 + \Sigma \epsilon \phi_0' f_0'$ where ϕ_0', f_0' both have opposite parity respectively to ϕ_0, f_0.

In the rest of this section, I describe the theory of three of the four mechanisms in greater detail reserving the vibration-induced mechanism till the next section. It will become apparent that the two electric-dipole types of transition are the only important sources of intensity in d-electron spectra. As illustrations I shall give explicit calculations of the triplet-triplet intensities of magnetic dipole, and electric quadrupole transitions in the Ni^{++} ion.

11.2.3. Magnetic-dipole transitions. We recall from § 3.2.3 that the oscillator strength is given approximately by $f = (4\pi m\nu/h)\,|M|^2$, where

$$|M|^2 = \tfrac{1}{3}\,|\langle a'|\,\frac{1}{2mc}(\mathbf{L}+2\mathbf{S})\,|a\rangle|^2 \tag{11.4}$$

for magnetic-dipole transitions. If we assume that the states are eigenstates of the total spin, then the spin vector **S** has no matrix elements from the ground term (term here refers to the site symmetry group, e.g. $^3A_2$ for Ni^{++}) to any excited term and we may put $\langle a' \mid \mathbf{S} \mid a\rangle = 0$ in (11.4) to obtain

$$f = \frac{\pi\nu}{3hmc^2}|\langle a' \mid \mathbf{L} \mid a\rangle|^2. \tag{11.5}$$

In order to obtain the total oscillator strength from a state of the ground term to all states of a given excited term we must sum $|a'\rangle$ over the excited term. However, if we sum $|\langle a' \mid \mathbf{L} \mid a\rangle|^2$ over all states $|a'\rangle$ we obtain

$$\Sigma |\langle a' \mid \mathbf{L} \mid a\rangle|^2 = \langle a' \mid \mathbf{L}^2 \mid a\rangle = L(L+1)\hbar^2$$

and hence
$$f \leqslant \frac{\pi\nu}{3hmc^2}L(L+1)\hbar^2 = \frac{h\nu L(L+1)}{12\pi mc^2} \leqslant \frac{h\nu}{\pi mc^2} \tag{11.6}$$

because $L \leqslant 3$ for the terms of d^n configurations having highest spin. Restricting attention to light which is at the long wave-length side of 30,000 cm^{-1} we deduce that $f \leqslant 1{\cdot}4\times 10^{-5}$ (for $L = 3$; $f \leqslant 0{\cdot}7\times 10^{-5}$ for $L = 2$). The observed oscillator strengths correspond to an average over the states of the ground term, weighted according to their occupancy. Therefore, if they are due to magnetic-dipole transitions, we deduce that for them also $f \leqslant 1{\cdot}4\times 10^{-5}$. It is clear, therefore, that when oscillator strengths as large as 10^{-4} are observed, as in $Ni^{++}(en)_3$, their origin must be mainly due to some other cause.

The oscillator strengths for magnetic-dipole transitions between the ground term and the three excited triplet terms of d^8 ions in an octahedral field can be deduced immediately from the preceding discussion. The components of **L** transform as T_1 under the octahedral group and have zero matrix elements within the $^3A_2$ ground term and between that term and either of the excited $^3T_1$ terms. Therefore the whole of the sum $\Sigma|\langle a' \mid \mathbf{L} \mid a\rangle|^2$ comes from matrix elements of **L** to $^3T_2$. So

$$f(^3A_2 \rightarrow a\,^3T_1) = f(^3A_2 \rightarrow b\,^3T_1) = 0$$

and
$$f(^3A_2 \rightarrow {}^3T_2) = \frac{h\nu}{\pi mc^2} = 5{\cdot}4\times 10^{-6} \quad \text{for} \quad Ni^{++}(en)_3.$$

The observed oscillator strengths for transitions from the ground state of Mn^{++} to the excited states are $1\text{–}2\times 10^{-7}$. However, here the magnetic-dipole strengths are much lower also. We have seen (5.11) that the ground level is approximately
$$\psi_M = \psi(^6S, M0) - \epsilon\psi(^4P^M_{\frac{5}{2}}),$$

where $\epsilon = \zeta\sqrt{5}/(E(^4P_{\frac{5}{2}}) - E(^6S)) = 2{\cdot}66\times 10^{-2}$. The only excited level to which there could be appreciable magnetic-dipole intensity is

$$\phi_M = \psi(^4P^M_{\frac{5}{2}}) + \epsilon\psi(^6S,\ M0).$$

It is immediate that
$$\langle ^6S, \tfrac{5}{2}0|\,(L_z + 2S_z)\,|^6S, \tfrac{5}{2}0\rangle = 5\hbar,$$
$$\langle ^4P^{\frac{5}{2}}_{\frac{5}{2}}|\,(L_z + 2S_z)\,|^4P^{\frac{5}{2}}_{\frac{5}{2}}\rangle = 4\hbar,$$

and hence that
$$\langle \psi_{\frac{5}{2}}|\,(L_z + 2S_z)\,|\phi_{\frac{5}{2}}\rangle = \epsilon\hbar.$$

It then follows from (2.80) that

$$\langle \psi_{\frac{3}{2}} | (L_z + 2S_z) | \phi_{\frac{3}{2}} \rangle = \tfrac{3}{5}\epsilon\hbar,$$

and

$$\langle \psi_{\frac{1}{2}} | (L_z + 2S_z) | \phi_{\frac{1}{2}} \rangle = \tfrac{1}{5}\epsilon\hbar.$$

Therefore, taking the mean over the ground level,

$$f = \frac{\pi\nu}{3hmc^2}\tfrac{1}{6}\Sigma\, |\langle a' | (\mathbf{L} + 2\mathbf{S}) | a \rangle|^2 = \frac{\pi\nu}{6hmc^2}\Sigma\, |\langle a' | (L_z + 2S_z) | a \rangle|^2$$

$$= \frac{\pi\nu}{6hmc^2}\tfrac{14}{5}\epsilon^2\hbar^2 = 1{\cdot}2 \times 10^{-9},$$

which is much too small.

The preceding calculations are quite definite, involve no arbitrary parameters which were not determined independently, and should give reliable upper limits to the magnetic-dipole intensities. Further refinements of the theory slightly lower the calculated intensities through reductions of the matrix elements of **L** and the value of ζ. We may therefore completely ignore magnetic-dipole intensities except possibly in cases where the oscillator strength for a transition is, anomalously, extremely low.

11.2.4. Electric-quadrupole transitions. Here the intensities are even smaller and to demonstrate this it will suffice to calculate them for the triplet-triplet spectrum of a d^8 ion in an octahedral field. It is reasonable to suppose this example to be fairly typical.

The quadrupole tensor $\mathcal{N}_{ij}$ for a single electron has for its components xy, yz, zx and $x^2 - \frac{1}{3}r^2$, $y^2 - \frac{1}{3}r^2$, $z^2 - \frac{1}{3}r^2$. The former three form a basis for t_{2g} and the latter are linear combinations of the pair $x^2 - y^2$, $z^2 - \frac{1}{3}r^2$ which form a basis for e_g. The quadrupole tensor for n electrons, therefore, has off-diagonal elements which form a basis for T_{2g} and diagonal elements which are linear combinations of a basis for E_g. It follows that the diagonal elements have zero matrix elements between the ground term, $^3A_2$, of a d^8 ion and all the excited terms. The whole tensor has zero matrix elements to the $^3T_2$ term. If we write $\boldsymbol{\eta}^i$ for the pseudo-vector $(y_i z_i, z_i x_i, x_i y_i)$ referring to the ith electron then we deduce from § 3.2.3 that

$$f = \frac{\pi m \nu^3}{5hc^2} \sum_i |\langle a' | \boldsymbol{\eta}^i | a \rangle|^2. \tag{11.7}$$

We actually calculate f for the complementary configuration d^2 and use the strong-field coupling scheme. Then, because $\boldsymbol{\eta}^i$ is a one-electron quantity, there are no matrix elements to the upper $^3T_1$ term. Using Table A 20 and the discussion following (6.30)

$$f = \frac{3\pi m \nu^3}{5hc^2} |\langle a\, {}^3T_1\, 1\, \zeta | \, \Sigma x_i y_i \, | {}^3A_2\, 1\, a_2 \rangle|^2$$

whence, as $|{}^3A_2\, 1\, a_2\rangle = |\theta^+ \epsilon^+\rangle$ and $|{}^3T_1\, 1\, \zeta\rangle = |\zeta^+ \epsilon^+\rangle$,

$$f = \frac{3\pi m \nu^3}{5hc^2} |\langle \zeta \,|\, xy \,|\, \theta \rangle|^2.$$

$\langle \zeta \,|\, xy \,|\, \theta \rangle = -(2/7\sqrt{3})\, \overline{r^2}$ (use Tables 4.1 and A 21) and so

$$f = \frac{4\pi m \nu^3}{245hc^2} (\overline{r^2})^2.$$

For light at 18,000 cm^{-1} and $\overline{r^2} = 10^{-16}\,cm^2$ this gives $f = 3{\cdot}1\times10^{-9}$. So the calculated electric-quadrupole intensity is quite negligible and would remain so even if we decided that we have underestimated $\overline{r^2}$ considerably. One should bear in mind that with very extensive delocalization it would be possible to get an appreciable contribution, but we would need $\overline{r^2} = 10^{-14}\,cm^2$, i.e. $\bar{r} \approx 10$ Å to get $f \approx 0{\cdot}3\times10^{-4}$.

11.2.5. Unsymmetrical environments. Here we content ourselves with making a rough order of magnitude estimate. We suppose that in (11.3) the transition from ϕ_0 to χ_1 or from χ_0 to ϕ_1 is fully allowed. Then the transition $\psi_0 \rightarrow \psi_1$ has an oscillator strength of the order of magnitude of the quantities ϵ_0^2 and ϵ_1^2, unless there is an accidental cancellation. These quantities are M^2/E^2 where M is the matrix element of the non-centrosymmetric part of the field between ϕ and χ while E is the energy difference $E(\chi)-E(\phi)$. In the free ion the lowest states χ would belong to the $3d^{n-1}4p$ configuration and $E \approx 5\times10^4$–10^5 but in compounds there may be lower-lying charge-transfer states. With $M = 10^3$–10^4 we find f between 4×10^{-2} and 10^{-4}. Thus providing M is fairly large we can get a substantial intensity in this way.

11.3. Coupling of nuclear and electronic motions

11.3.1. Normal co-ordinates for nuclear motion. It is convenient to use coordinates, X_i say, which measure the displacement of the nuclei from their equilibrium positions. Let us examine this first from the point of view of classical mechanics and suppose that the amplitude of the nuclear motion is small. Then the potential energy of the nuclei may be expanded in a power series in the X_i, thus

$$V = V_0 + \sum_{i=1}^{n} a_i X_i + \sum_{i,j=1}^{n} b_{ij} X_i X_j + \dots, \tag{11.8}$$

where n is the total number of coordinates. We choose our energy zero so that $V_0 = 0$. At equilibrium, $\partial V/\partial X_i = 0$, so $a_i = 0$ for each i. At first we neglect third-power and higher orders of X_i and have the quadratic form

$$V = \sum_{i,j=1}^{n} b_{ij} X_i X_j,$$

for the potential energy. We define b_{ij} so that $b_{ij} = b_{ji}$.

We now choose linear combinations Q_j of the X_i so that V is diagonal. Then

$$V = \sum_{j=1}^{n} \nu_j^2 Q_j^2. \tag{11.9}$$

Now the equilibrium configuration of the nuclei can only be stable if $V > 0$ for all small displacements X_i or, equivalently, for all small displacements Q_j. Therefore $\nu_j^2 > 0$ for each j.

In discussing the Jahn–Teller effect in § 8.5 we found that the presence of a degenerate electronic state often distorted a system away from an expected symmetrical configuration into a less symmetrical one. We wish to translate this into our present scheme and do so by dividing our complex ion into two parts—the filled shells together with the nuclei and the partly-filled shells. Then the Hamiltonian breaks up into three parts. First, there is the motion of the nuclei under the field due to the other nuclei and to the filled shells. This is the motion which we are first treating classically and the equilibrium positions for this motion have the symmetry of the expected symmetrical configuration. Then there is the interaction between the nuclear motion and the motion of the electrons in the partly-filled shells. It arises via the ligand field and is invariant under rotations applied to all particles of the system but not necessarily to

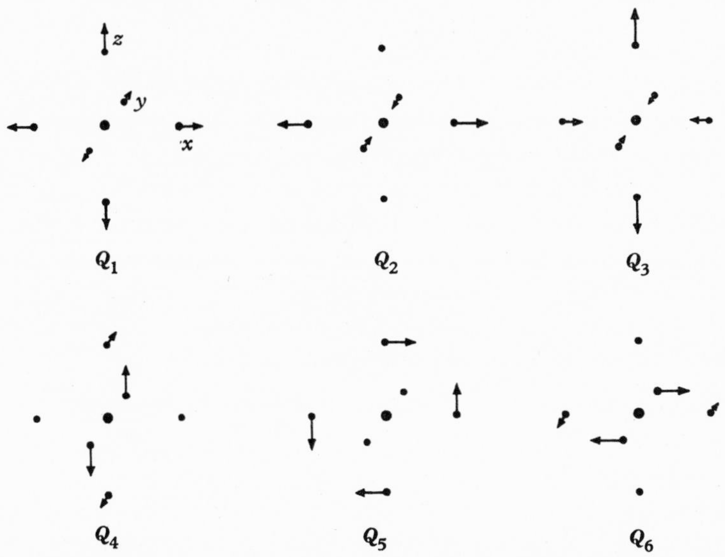

Fig. 11.3. Normal coordinates for centrosymmetric distortions of a regular octahedral molecule (from Öpik & Pryce, 1956).

rotations of nuclear coordinates or partly-filled shell coordinates alone. The parts which are separately invariant to elements of the symmetry group of the filled shells in their equilibrium positions may be conveniently included in the potential energy for the nuclear motion. Finally, there is the interaction between the electrons in the partly-filled shells. We are already familiar with this.

Now let us fix our ideas and consider an isolated ML_6 complex ion. There are 21 coordinates necessary to specify the positions of the seven nuclei. However, not all of them describe vibrations. Three describe a translational motion of the centre of mass and three an overall rotation. This leaves 15. We may classify these according to their behaviour under the symmetry group O_h relevant to the nuclei fixed in an equilibrium configuration. The usual way to do this is to attach little arrows to each nucleus showing the way in which they move under the displacement. The simplest displacements are the X_i which each refer to only one nucleus. Examples of Q_j are shown in Fig. 11.3. Q_1 is totally symmetric and the arrangement of arrows, i.e. displacements, is invariant under all operations of O_h. So it belongs to the unit representation A_{1g} and we say the coordinate Q_1 has A_{1g} symmetry. (Q_2, Q_3) have E_g symmetry and (Q_4, Q_5, Q_6) have T_{2g} symmetry.

V itself must have A_{1g} symmetry and so if we choose our Q_j so that they are components of irreducible representations of O_h there can be no cross-terms in V between Q_j which are different components (we use § 8.6 and ex. 3, p. 155 here). We therefore choose the Q_j for (11.9) in this way. These Q_j are called normal co-ordinates. As an example, the actual positions of the nuclei in Fig. 11.3 after a nuclear displacement (Q_2, Q_3) are

$$\left.\begin{aligned}&[\pm(R-\tfrac{1}{2}Q_3+\tfrac{1}{2}\sqrt{3}\,Q_2),0,0],\\&[0,\pm(R-\tfrac{1}{2}Q_3-\tfrac{1}{2}\sqrt{3}\,Q_2),0],\\&[0,0,\pm(R+Q_3)],\end{aligned}\right\} \tag{11.10}$$

from (9.1), where R is the equilibrium ML internuclear distance.

We find which irreducible representations the set of Q_j spans by determining the character of the representation whose basis is the set of 21 little arrows attached in orthogonal trios to the 7 nuclei. This character shows that the representation is $A_{1g}+E_g+T_{1g}+3T_{1u}+T_{2g}+T_{2u}$. The translations are T_{1u} and the rotations T_{1g}. Hence the Q_j span $A_{1g}+E_g+2T_{1u}+T_{2g}+T_{2u}$. The normal coordinates are all defined, apart from some constant factors, by group-theoretic requirements except for the repeated representation T_{1u}.

The kinetic energy of the nuclei is a sum $\frac{1}{2}\Sigma m_i \dot{X}_i^2$ where m_i is the mass of the nucleus to which X_i refers. If we now write $Y_i = m_i^{\frac{1}{2}} X_i$, then the kinetic energy assumes the simple form $\frac{1}{2}\Sigma \dot{Y}_i^2$. Suppose that we had already done this at the beginning of the section and had diagonalized $\Sigma b_{ij} Y_i Y_j$ to obtain (11.9). Then as the Q_j are an orthonormal transform of the Y_i both the kinetic and the potential energy is diagonal in Q_j. The total energy of the system is

$$\mathscr{H} = \sum_{j=1}^{n} (\tfrac{1}{2}\dot{Q}_j^2 + \nu_j^2 Q_j^2). \tag{11.11}$$

This system goes over in quantum theory to n independent harmonic oscillators. Inclusion of higher terms in the series for V introduces a coupling between them.

We discussed oscillators having the Hamiltonian (11.11) in § 3.1.3, but with a different zero for their energies. The energies are now $(n+\frac{1}{2})\hbar\nu_j$ for the jth one, where n is a non-negative integer. Take a particular one and drop the suffix j. Then we write its eigenstates $|n\rangle$ and have

$$\langle n-1\,|\,Q\,|\,n\rangle = \langle n\,|\,Q\,|\,n-1\rangle = \left(\frac{n\hbar}{2\nu}\right)^{\frac{1}{2}}. \tag{11.12}$$

All other matrix elements of Q are zero.

For compounds ML_6 there are six independent frequencies, one for each irreducible representation. They are usually numbered in the order A_{1g}, E_g, T_{1u}, T_{1u}, T_{2g}, T_{2u} with the convention that $\nu_3 \leqslant \nu_4$. In case the ligands L contain more than one nucleus or ML_6 forms part of a more extended system there will be other frequencies, but $\nu_1, \ldots, \nu_6$ can still often be distinguished. In $Mn(H_2O)_6^{++}$, Koide & Pryce (1958) estimate $\nu_1 = 250$–$300\,\text{cm}^{-1}$, $\nu_3 = 170\,\text{cm}^{-1}$, $\nu_4 = 320$–$400\,\text{cm}^{-1}$, $\nu_6 = 85\,\text{cm}^{-1}$ which gives an idea of orders of magnitude. ν_i would be larger in more tightly bound compounds.

Having described the Hamiltonian for the nuclear motion alone we pass on to describe the nature of the coupling to the electronic motion. Before doing so I remark that we are deliberately adopting an unsophisticated and semi-classical approach to the problem. This is logically unsatisfactory but makes for clarity of presentation. It is possible to approach the problem from the quantum Hamiltonian for the whole complex ion and one arrives at the same model if one makes suitable approximations.

11.3.2. The interaction. The coupling between the nuclear and electronic motions arises because the ligand field changes as the nuclear positions change. This may be represented by the crystal-field expressions (8.12) and (8.13). We write it more generally as $f(Q_i)$, where the Q_i are normal coordinates and f is also a function of electronic coordinates. To a first approximation it is, however, a one-electron operator. Let us take a one-electron system and expand f as a power series

$$f(Q_i) = p + \sum_i q_i Q_i + \ldots, \tag{11.13}$$

p gives no coupling so we put $p = 0$. Neglecting terms quadratic in the Q_i we have the interaction

$$f(Q_i) = \sum_i q_i Q_i. \tag{11.14}$$

For an n-electron system (11.14) becomes

$$f(Q_i) = \sum_i \sum_j q_i(j)\, Q_i. \tag{11.15}$$

We suppose we have a symmetry group G with representations all of category one. Now $f(Q_i)$ is left invariant by any operation g of G. g operates on the Q_i as well as on the electrons. Therefore each q_i forms the same component of an irreducible representation as Q_i does (use §§ 8.6, 8.7). This completes our general description of the interaction. With particular models for the field of the ligands, for example a point charge or point dipole model, the q_i can be calculated explicitly. Even the assumption that the partly-filled orbitals are d-orbitals introduces considerable restrictions on the q_i which must then be sums $\Sigma \eta_{k\alpha} r^k Z_{k\alpha}$ over spherical harmonics with k even and not greater than four.

Arguing from a qualitative point of view we can say something about the expected magnitudes of the matrix elements of the interaction. Naturally the size of these depends upon the extent to which a change in nuclear configuration can affect and distinguish between the different components of the initially degenerate state. The primary influence is electrostatic and can therefore separate space components, M_Γ, but not spin components, M_S. It follows immediately that the Jahn–Teller effect will be smallest for spatially non-degenerate states with $S \neq 0$ and largest for spatially degenerate states. Among the latter it will be larger, for octahedral coordination, when the degeneracy arises from electrons in e-orbitals, e.g. $t_2^3 e\, {}^5E$ of Cr^{++}, than when it arises from those in t_2-orbitals, e.g. $t_2^4 e^2\, {}^5T_2$ of Fe^{++}. This is because the e-orbitals have a distribution which is closer to the adjacent nuclei than that of the t_2-orbitals (from an L.C.A.O. point of view because they involve σ-bonding to the ligands, which is expected to be strong, rather than π-bonding which should be weak, see Appendix 8).[1] Analogous remarks apply to other site symmetry groups.

We now have a nuclear system with vibrational states classified by the occupancy numbers n_i of the normal modes of vibration Q_i, the linear interaction Hamiltonian (11.15), and the states of the electronic system, whose energies and eigenfunctions are to be evaluated for the equilibrium nuclear arrangement. The basic states may be written $|X\rangle = |\psi_j, n_1, n_2, \ldots, n_r\rangle$ and the Hamiltonian as

$$\mathscr{H} = \mathscr{H}_e + \mathscr{H}_v + \mathscr{H}_n. \tag{11.16}$$

$\mathscr{H}_e$ refers to the electronic system. $\mathscr{H}_n$ refers to the nuclear vibrations and is a sum of harmonic oscillator Hamiltonians. Ideally the functions $|X\rangle$ are eigenstates of $\mathscr{H}_e$ and $\mathscr{H}_n$ but not of $\mathscr{H}_v$, though in practice we have to be content with approximate eigenstates of $\mathscr{H}_e$. $\mathscr{H}_v$ is not only a one-electron operator, but also has zero matrix elements between states differing in the occupancy of more than one of the n_i. It also satisfies the selection rule $\Delta n_i = \pm 1$ and hence has no non-zero diagonal elements within our basic states.

The reader should note the very close analogy between (11.16) and (3.8) giving the Hamiltonian for an atomic system in interaction with a radiation field.

11.3.3. Intensities in centrosymmetric complex ions. Let the site symmetry group contain the inversion. Then q_i and Q_i have definite parities and, in fact, the same parity. The first-order increments to the basic states under the influence of the perturbation (11.15) have the same parities as the states they modify, but the electronic parts of the increments may have opposite parities. It is because of this possibility that we can get electric-dipole radiation. From a formal point of view, therefore, our next step is clear. We calculate the modifications and then determine the matrix elements of components of the electric-dipole moment $\mathbf{P} = \Sigma e\mathbf{r}_j$ between them.[2] We use the notation $P_\epsilon = \Sigma p(j)$ for a typical component and proceed.

The modified ground and excited states are

$$\left.\begin{aligned} |X_g\rangle &= |X,0\rangle - \Sigma E^{-1}\langle Y,n\,|\,\mathscr{H}_v\,|\,X,0\rangle\,|\,Y,n\rangle, \\ |X_e\rangle &= |X',m\rangle - \Sigma E_1^{-1}\langle Y,n\,|\,\mathscr{H}_v\,|\,X',m\rangle\,|\,Y,n\rangle. \end{aligned}\right\} \tag{11.17}$$

We have assumed no quanta of vibration to be present in the ground state. This restriction is unnecessary and is for simplicity only. The sum is over all excited electronic states Y and over all vibrational states. n really stands for $n_1, n_2, \ldots, n_r$. Because of the selection rules for $\mathscr{H}_v$, we must have $n = 1$ in the first equation (i.e. an occupancy of 1 for one of the n_i). Also n and m differ by one in the second.

[1] The relevance of this to interatomic distances ML in the grouping ML_6 is that these distances do not vary smoothly with atomic number as M passes along a transition series, but are anomalously long according to the extent to which the ground state of the d^n ion contains e-electrons. See Van Santen & Van Wieringen (1952); Orgel (1956); Hush & Pryce (1958).

[2] We omit the part depending on the nuclear coordinates. Its effect is probably small but does not appear to have been discussed quantitatively.

The matrix element of P_ϵ is

$$\begin{aligned}\langle X_e \,|\, P_\epsilon \,|\, X_g \rangle &= -\Sigma E^{-1} \langle X', m \,|\, P_\epsilon \,|\, Y, 1 \rangle \langle Y, 1 \,|\, \mathscr{H}_v \,|\, X, 0 \rangle \\ &\quad - \Sigma E_1^{-1} \langle X', m \,|\, \mathscr{H}_v \,|\, Y, n \rangle \langle Y, n \,|\, P_\epsilon \,|\, X, 0 \rangle \\ &= -\Sigma E^{-1} \langle X', 1 \,|\, P_\epsilon \,|\, Y, 1 \rangle \langle Y, 1 \,|\, \mathscr{H}_v \,|\, X, 0 \rangle \\ &\quad - \Sigma E_1^{-1} \langle X', 1 \,|\, \mathscr{H}_v \,|\, Y, 0 \rangle \langle Y, 0 \,|\, P_\epsilon \,|\, X, 0 \rangle. \end{aligned} \tag{11.18}$$

We have again used the selection rules for $\mathscr{H}_v$. $\langle X_e \,|\, P_\epsilon \,|\, X_g \rangle$ is given by (11.18) when $m = 1$. Otherwise it is zero. Thus only one vibrational quantum can be excited at a time.

Next we factorize out the nuclear part of (11.18). Write $\mathscr{H}_v = \Sigma q(j)\, Q$. The suffix i has been dropped from Q and q for clarity and the sum is actually over j and over the normal coordinates. Therefore

$$\left.\begin{aligned}\langle X', 1 \,|\, P_\epsilon \,|\, Y, 1 \rangle &= \langle X' \,|\, P_\epsilon \,|\, Y \rangle, \\ \langle Y, 1 \,|\, \mathscr{H}_v \,|\, X, 0 \rangle &= \langle Y |\, \Sigma q(j) \,| X \rangle \langle 1 \,|\, Q \,|\, 0 \rangle \\ &= \left(\frac{\hbar}{2\nu}\right)^{\frac{1}{2}} \langle Y |\, \Sigma q(j) \,| X \rangle, \end{aligned}\right\} \tag{11.19}$$

whence

$$\begin{aligned}\langle X_e \,|\, P_\epsilon \,|\, X_g \rangle &= -\Sigma E^{-1} \left(\frac{\hbar}{2\nu}\right)^{\frac{1}{2}} \langle X' |\, \Sigma p(j) \,| Y \rangle \langle Y |\, \Sigma q(j) \,| X \rangle \\ &\quad - \Sigma E_1^{-1} \left(\frac{\hbar}{2\nu}\right)^{\frac{1}{2}} \langle X' |\, \Sigma q(j) \,| Y \rangle \langle Y |\, \Sigma p(j) \,| X \rangle. \end{aligned} \tag{11.20}$$

It will be apparent that the theory is simple and straightforward from a formal point of view. The difficulty is to make reasonable estimates of the quantities occurring in (11.19). Some further general progress may be made by assuming that all the important excited states Y lie at the same energies E and E_1 above X_e and X_g. We may then take the factors $E^{-1}(\hbar/2\nu)^{\frac{1}{2}}$ and $E_1^{-1}(\hbar/2\nu)^{\frac{1}{2}}$ outside the summation over $|\, Y \rangle$. Following the earlier work of Van Vleck (1937) and Tanabe & Sugano (1954), two methods have been used to perform this summation. Both suppose the $|\, Y \rangle$ to be $3d^{n-1}\, 4p$ states, although they are essentially unaltered by taking molecular orbitals instead of $4p$. The first calculates the matrix elements of $\Sigma p(j)$ and $\Sigma q(j)$ from the $| X \rangle$ to the $|\, Y \rangle$ in terms of one-electron matrix elements between $3d$ and $4p$ (Liehr & Ballhausen, 1957, 1959). The second uses a closure procedure assuming that in the relation

$$\Sigma \,|\, Y \rangle \langle Y \,| = 1, \tag{11.21}$$

only our important excited states occur (Koide & Pryce, 1958; Koide, 1959). This is, however, not consistent with assuming that the $|\, Y \rangle$ are $3d^{n-1}\, 4p$ states. The two methods can be interrelated and the inconsistencies removed. They both also neglect to consider excited states in which an electron has been transferred from a filled odd parity orbital to a d-orbital rather than the other way round—they consider odd parity orbitals which are particles but not those which are holes, relative to the filled shells of the complex ion. When these matters are put right the theory becomes considerably more complicated and although in principle it can be parametrized it may prove difficult to do so in a way which does not merely slavishly reproduce the experimental results.[1]

The calculations made by the aforementioned authors generally agree rather well with experiment.[2] In view of the theoretical criticism which can be levelled at their schemes, together with the fact that the parameters involved can only be estimated by making drastic and unrealistic assumptions, we should be gratified only at the order-of-magnitude agreement. Closer agreement cannot reasonably be expected and when it occurs should not be taken to be in itself evidence that the physical mechanism assumed to be responsible for the intensities is the only important one.

11.3.4. Further discussion of the coupling. If we have a basic state with its electronic part belonging to an irreducible representation Γ_1 and having just one vibrational mode, belonging to Γ_2, excited then the representation $\Gamma_1\Gamma_2$ may be reducible. Its components will be split apart by the interaction $\mathscr{H}_v$ but, because of the selection rules for the vibrational quantum numbers, not to first order. We recall from § 8.5 that $\mathscr{H}_v$ is the perturbation which is responsible for the Jahn–Teller effect. Assuming that the matrix elements of $\mathscr{H}_v$ are small compared with the relevant vibrational separations, the splitting of $\Gamma_1\Gamma_2$ may be calculated by second-order perturbation theory. The Jahn–Teller effect, then, is not so much to distort the complex as to introduce a direct coupling between the electronic and vibrational motions. It is usual to call this the dynamical Jahn–Teller effect.

There is not only a splitting between the levels of $\Gamma_1\Gamma_2$ but also a modification of the eigenfunctions. This results in the nuclei having most probable configurations slightly away from the original equilibrium positions. However, there are generally several of these most probable configurations.

The assumption that $\mathscr{H}_v$ is small compared with the vibrational separations is only sometimes true and often $\mathscr{H}_v$ is large compared with them. Let us examine what happens as $\mathscr{H}_v$ increases. More and more excited vibrational states get mixed in until ultimately the eigenstates are all sums over large numbers of them. The lowest state having electronic part Γ_1 now has most probable configurations far removed from regular symmetry. There may be several such configurations with equal energy, but with the energy minima separated by such large maxima that they are virtually inaccessible from one another. If the ions form part of a solid lattice, as in a crystal, because of packing considerations they will usually all distort in a phased manner. The whole crystal then stays near one particular distorted configuration and each individual ion always occupies just one of the configurations of minimum energy for the free ion. This illustrates the passage from the dynamic to the static Jahn–Teller effect.

Even apart from the cooperative interaction there may be a preferred mini-

[1] See the series referred to on p. 446 for criticism of Liehr, Ballhausen, Koide & Pryce's work.

[2] Englmann (1960, *Molecular Physics*, **3**, 48) has given an alternative analysis for d^n ions assuming the $|Y\rangle$ to be charge transfer states.

mum. Suppose an octahedral ion has six ligands which are nearly but not quite identical. Then the symmetry group will be smaller than O_h although a model assuming O_h symmetry may be quite a good approximation. Then there can be several relatively inaccessible minima with nearly but not quite the same energies. At low enough temperatures the system only occupies one of them. Both this and the cooperative effect may combine.

11.3.5. The Jahn–Teller effect in the cupric ion. The ground term of the cupric ion in an octahedral field is ^2E. The Jahn–Teller effect here is due to an electron in an e-orbital and would be expected to be large. Empirically it appears that $\mathscr{H}_v$ is large compared with the vibrational spacings.

We return to the classical model for the nuclei and contemplate finding the electronic energies for particular values of the Q_i. Of the various vibrations enumerated in § 11.3.1 for a complex ML_6 only those with A_{1g} and E_g symmetry get multiplied by q_i which have non-zero matrix elements within ^2E. The A_{1g} is not a distortion from octahedral symmetry and so we discuss only the E_g displacements. They are represented by Q_2 and Q_3. Q_2 is the ϵ-component and Q_3 the θ-component and the displacements corresponding to them are given in (11.10).

Apart from a constant energy, the Hamiltonian is

$$\mathscr{H} = \nu^2(Q_2^2+Q_3^2)+q_2Q_2+q_3Q_3, \tag{11.22}$$

where q_3, q_2 are respectively the θ- and ϵ-components of a basis for E. Put $Q_3 = \rho\cos\theta$, $Q_2 = \rho\sin\theta$ and then

$$\mathscr{H} = \nu^2\rho^2+\rho q_2\sin\theta+\rho q_3\cos\theta. \tag{11.23}$$

The matrix elements of q_2 and q_3 are determined completely by group-theoretic arguments except for one parameter and, in effect, we already determined them in § 9.7.6. Using (9.48), the matrix of $\mathscr{H}$ is

$\mathscr{H}$	θ	ϵ
θ	$\nu^2\rho^2-\rho A\cos\theta$	$\rho A\sin\theta$
ϵ	$\rho A\sin\theta$	$\nu^2\rho^2+\rho A\cos\theta$

with energies $\nu^2\rho^2\pm\rho\,|A|$. The minimum energy is obtained by differentiating this, occurs at $\rho = \frac{1}{2}|A|\,\nu^{-2}$ and is $-\frac{1}{4}A^2\nu^{-2}$. The minimum is independent of θ but this is a consequence of retaining only second-order terms in the expression for V.

Passing to the next approximation, we add a term of the third degree in Q_2, Q_3 to V and having symmetry A_1. There is only one such because A_1 occurs in E^3 only once. Using the coupling coefficients A 20 we readily find

$$V = \nu^2\rho^2+B\rho^3\cos 3\theta. \tag{11.24}$$

The next order in the interaction has the three products Q_2^2, Q_3^2, Q_2Q_3 which form a basis for the symmetrized square of E, i.e. A_1+E. As a result, the matrix of $\mathscr{H}$ now becomes

$\mathscr{H}$	θ	ϵ
θ	$(\nu^2+a)\rho^2+B\rho^3\cos 3\theta-A\rho\cos\theta+C\rho^2\cos 2\theta$	$A\rho\sin\theta+C\rho^2\sin 2\theta$
ϵ	$A\rho\sin\theta+C\rho^2\sin 2\theta$	$(\nu^2+a)\rho^2+B\rho^3\cos 3\theta+A\rho\cos\theta-C\rho^2\cos 2\theta$

with energies

$$E = (\nu^2+a)\rho^2+B\rho^3\cos 3\theta \pm (A^2\rho^2+C^2\rho^4+2AC\rho^3\cos 3\theta)^{\frac{1}{2}}$$

$$= (\nu^2+a)\rho^2+B\rho^3\cos 3\theta \pm (A\rho+C\rho^2\cos 3\theta). \tag{11.25}$$

E now has minima at $\theta = 0$, $\frac{2}{3}\pi$, $\frac{4}{3}\pi$ if $B\rho^3 < C\rho^2$ and at $\frac{1}{3}\pi$, π, $\frac{5}{3}\pi$ if $B\rho^3 > C\rho^2$. When $\theta = 0$ the two ligands along the $\pm OZ$-axis have moved out and the other four have moved in. The same happens for $\theta = \frac{2}{3}\pi, \frac{4}{3}\pi$ except that the two ligands are now along the x-, y-axes, respectively. In the other three cases two ligands move in and the other four out.

The preceding analysis is due to Öpik & Pryce (1956), except that they omitted the second-order terms in the interaction.[1] It is not obvious that this is allowable. Assuming a reasonable model, but with $a = C = 0$, they found $B < 0$ which suggests that two ligands move out.

Orgel & Dunitz (1957) independently discussed the Jahn–Teller effect in cupric compounds and gave a survey of the then available crystal structures. It appears that the two ligands do move out in every genuinely octahedral case.[2]

Although in most cupric compounds, only one minimum is populated, there are a few in which all three are. In the former case the electronic part of the wave-function for the ground state is $^2E\epsilon^{\pm}$ (or $^2E\theta^{\pm}$) to a good approximation, where the Z-axis is the axis of the distortion. In the latter case it is a mixture of $^2E\epsilon^{\pm}$ and $^2E\theta^{\pm}$, an equal mixture if the population of each of the minima is the same. $CuSiF_6$ is an example of this latter type and we discuss it further in § 12.4.1.

Belford, Calvin & Belford (1957), and also Holmes & McClure (1957) analysed optical spectra of cupric compounds and came to the conclusion that not only transitions to $^2T_2$, but also those to the other component of ^2E occur at visible or near visible wave-lengths. Putting all the results of this section together we arrive at the schematic energy-level diagram for the cupric ion shown in Fig. 11.4. The relative positions of the upper states are uncertain. The main absorption proceeds vertically upwards in the diagram, not to the minimum of the $^2T_2$ term. This is because the transition involves a change of the electronic state but very little change of the nuclear configuration. This kind of consideration is called the Franck–Condon principle. Our previous analysis now shows that the trough lies a distance below the undistorted configuration which is approximately a quarter of the energy of transition to the upper state of ^2E. The aqueous cupric ion has a broad band centred at 12,600 cm^{-1} so the trough is about 3000 cm^{-1} deep. The ratios implied in Fig. 11.4 should only be taken as a rough guide.

[1] First pointed out by Liehr & Ballhausen (1958), *Ann. Phys.* (*N.Y.*), **3**, 304.
[2] Although see Knox (1959), *J. Chem. Phys.* **30**, 991.

I have chosen the cupric ion as an example,[1] but a similar analysis applies to the chromous ion whose ground term is ^5E and which also contains an odd e-electron. For the other high-spin ions, the Jahn–Teller effect is due to t_2-electrons and would be much smaller. This point is incorporated into Fig. 11.4

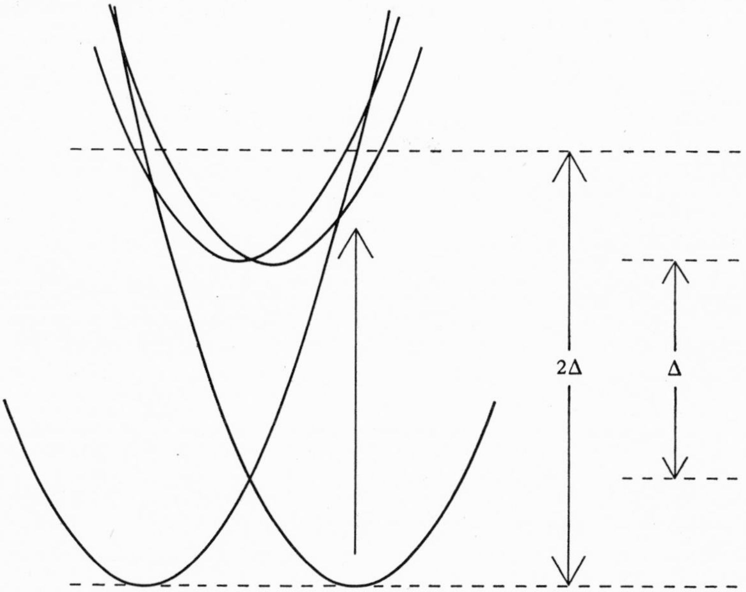

Fig. 11.4. Schematic representation of the Jahn–Teller distortion in the hydrated cupric ion. (Arising from a discussion over coffee with H. C. Longuet-Higgins, L. E. Orgel and M. H. L. Pryce.) $\Delta \approx 6300$ cm^{-1}.

because the trough for the upper term is very small. The spin-orbit coupling is comparable with the Jahn–Teller energy in ions distorting because of t_2-electrons, but is quite negligible for Cr^{++} and Cu^{++} partly because the Jahn–Teller energies are larger but mainly because it has no first-order matrix elements within a ^{2S+1}E term.

11.4. Spectra of the aqueous aquo-ions

A convenient introduction to the spectra is through those ions having six water molecules as their ligands. We take first the configurations d^1, d^4, d^6 and d^9 having a high-spin ground term and consider only the terms of highest multiplicity. Then each ion has one excited term of the same multiplicity as the ground term and with energy Δ above the ground term. Therefore we expect one allowed transition from each. These are:

$$d^1: {}^2T_2 \to {}^2E, \qquad d^4: {}^5E \to {}^5T_2,$$
$$d^6: {}^5T_2 \to {}^5E, \qquad d^9: {}^2E \to {}^2T_2.$$

For each of the ions $Ti^{3+}(d^1)$, $Cr^{++}(d^4)$, $Mn^{3+}(d^4)$, $Fe^{++}(d^6)$ and $Cu^{++}(d^9)$, a single fairly broad band is observed in the visible or near infra-red region. Taking the

[1] For further treatment of the Jahn–Teller effect see Öpik & Pryce (1956); Orgel (1957); Dunitz & Orgel (1957); Moffitt & Liehr (1957) and Longuet-Higgins, Öpik, Pryce & Sack (1958).

centres of these bands as reasonable estimates of the energy of the electronic transitions we obtain the Δ values shown in column 2 of Table 11.1. It is probably quite a good approximation to take the centre of the band in calculating from spectra measured at room temperature because although each transition is accompanied, for a centrosymmetric ligand grouping, by a change in the vibrational quantum number this change is almost as likely to represent absorption

Table 11.1. *Parameters calculated from spectra for transition metal aquo-ions (see* § 11.4)

Ion	Δ	Ion	Δ	B (mol.)	B (atomic)	Calc. δT_1	Obs. δT_1
Ti^{3+}	20,300	V^{3+}	16,700	787	862	—	—
Cr^{++}	13,900	V^{++}	12,600	460	755	10,100	8,300
Mn^{3+}	21,000	Cr^{3+}	17,400	633	918	13,950	12,300
Fe^{++}	10,400	Co^{++}	7,900	883	971	19,060	19,400
Cu^{++}	12,600	Ni^{++}	8,500	887	1,056	10,660	11,800

as emission. In any case the vibrational quanta are small compared with Δ. Our discussion of the Jahn–Teller effect shows that the Δ in Table 11.1 for d^4 and d^9 is about twice the true Δ, i.e. that appropriate to the undistorted configuration. Corresponding to this, the uncorrected Δ measured from the spectra for Cr^{++}, Mn^{3+} and especially Cu^{++} is noticeably higher than for adjacent ions with the same charge and ligands.

The next set of high-spin configurations which we take is d^2, d^3, d^7 and d^8 and again we restrict attention to the terms of highest multiplicity. For each configuration there is one A_2, one T_2 and two T_1 terms with calculated energies:

$$\left.\begin{aligned} E(T_1) &= \tfrac{1}{2}(15B+\Delta) \pm \tfrac{1}{2}(225B^2+18B\Delta+\Delta^2)^{\frac{1}{2}},\\ E(T_2) &= \Delta, \quad E(A_2) = 2\Delta \end{aligned}\right\} \tag{11.26}$$

for d^2, d^7 and

$$\left.\begin{aligned} E(T_1) &= \tfrac{1}{2}(15B-\Delta) \pm \tfrac{1}{2}(225B^2-18B\Delta+\Delta^2)^{\frac{1}{2}},\\ E(T_2) &= -\Delta, \quad E(A_2) = -2\Delta \end{aligned}\right\} \tag{11.27}$$

for d^3, d^8. We have dropped the spin affix. The positions of the terms as a function of Δ are shown schematically in Fig. 11.5. For $V^{3+}(d^2)$ only the excited T_1 and T_2 terms are known (for collection of spectra see Table A 40) and we then calculate $B = 787\ \text{cm}^{-1}$, $\Delta = 16{,}700\ \text{cm}^{-1}$ from (11.26). For the other aquo-ions for which data are available, all three excited terms are known and then we can use T_2, A_2 and the mean of the T_1 terms to determine B and Δ and calculate

$$\delta T_1 = (225B^2 \pm 18B\Delta + \Delta^2)^{\frac{1}{2}}$$

which is the difference between the T_1 terms. The results are in Table 11.1 and show fair agreement with experiment. B is in the column headed B (mol.). This B determined from molecular spectra is to be compared with B determined from atomic spectra by taking $\frac{1}{15}$ of the separation between the centre of gravity of the ground term and that of the excited P term of the same multiplicity. This quantity appears in the column headed B (atomic) and is consistently larger than B (mol.). This reduction of the parameters of electrostatic interaction in passing from the gaseous ion to the compound appears to be of universal occurrence and we already learned to expect it in § 7.3.

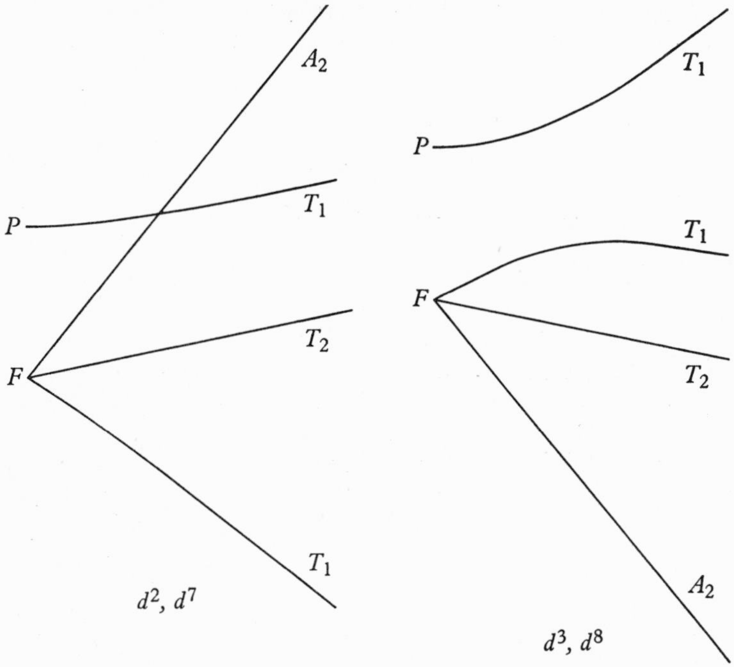

Fig. 11.5. The behaviour in an octahedral crystal field of the states of highest multiplicity of d^2, d^3, d^7 and d^8 (after Orgel).

The remaining type of high-spin aqueous ion is d^5 with a ${}^6A_1$ (${}^6S$) ground term, as typified by Mn^{++} or Fe^{3+}. We already mentioned this in § 11.1 and saw that the intensities of the d-d transitions are anomalously low ($f \sim 10^{-7}$). This is because there are no excited terms of the same multiplicity. The spectrum of the aqueous Mn^{++} ion is in Fig. 11.2 and the bands are ascribed to sextet-quartet transitions within d^5. The quartets comprise ${}^4A_1$, ${}^4A_2$, two ${}^4E$ and three each of ${}^4T_1$ and ${}^4T_2$. The energies of the first four terms were the same (9.14) as those of certain terms of the free ion and if we take the energy of the ground sextet as zero we have

$$\left.\begin{aligned} E({}^4A_1) &= E(a\,{}^4E) = E({}^4G), \\ E({}^4A_2) &= E({}^4F), \quad E(b\,{}^4E) = E({}^4D). \end{aligned}\right\} \tag{11.28}$$

The energies of the ${}^4T_1$ and ${}^4T_2$ terms must be obtained by solving the cubic secular equations. It is most convenient to work in the weak-field scheme and take the observed energies for the free ion rather than calculated ones. This was done by Orgel who solved the equations as a function of Δ and drew the energy-level diagrams for the quartets shown in Fig. 11.6. Comparing this with the absorption spectrum we see the two lowest terms appearing as fairly broad bands (${}^6A_1 \rightarrow {}^4T_1$ at 18,800 cm^{-1} and ${}^6A_1 \rightarrow {}^4T_2$ at 23,000 cm^{-1}). Next come two nearly coincident narrow bands which are probably the two terms ${}^4E$ and ${}^4A_1$ which should, theoretically, be coincident. It is natural to guess that the more intense band centred at about 24,900 cm^{-1} is the doubly spatially degenerate term ^4E and that the less intense one centred at 25,150 cm^{-1} is $^4A_1$.[1] The next two bands are presumably $^4T_2$ at 28,000 cm^{-1} and ^4E at 29,700 cm^{-1}. The spectrum has been investigated further into the ultra-violet and more terms identified (see Table A 40, giving more recent data, and also for the aqueous Fe^{3+} ion).

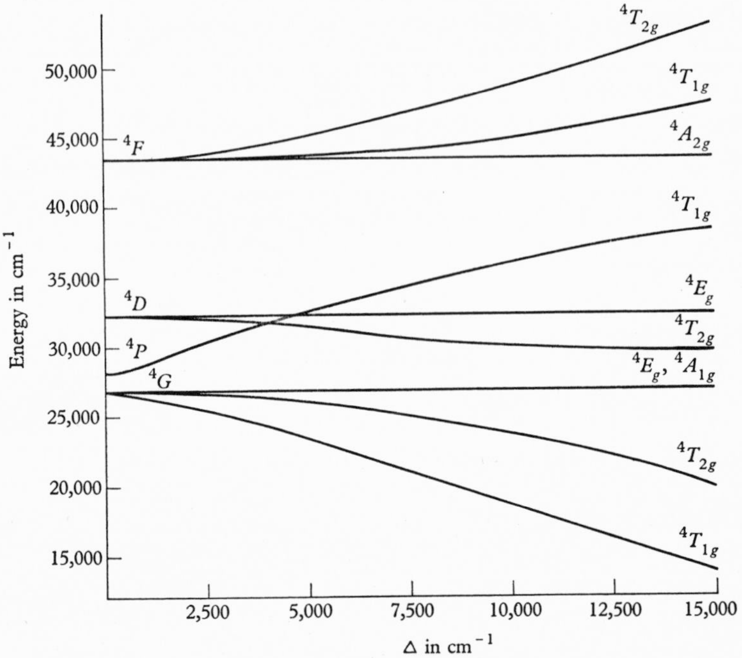

Fig. 11.6. Energy-level diagram for the Mn^{++} ion (from Orgel, 1955*a*).

There are two points of particular interest about this spectrum. The first is the sharpness of the bands at 25,000 cm^{-1}, which is interpreted in the following way.[2] The environment of every manganous ion is not identical but there are variations due to thermal fluctuations in manganese-ligand distances and in the precise arrangement of the water molecules around the ion. The energies of $^6A_1$ to $^4T_1$ or $^4T_2$ transitions depend on Δ and therefore have a range because of these varying ionic environments. However, the transitions to $^4A_1$, $^4A_2$ or ^4E are completely independent of Δ, in our approximation, and are therefore not susceptible to this line-broadening mechanism. Consequently, they should be sharper and Fig. 11.2 shows just this, especially for the bands at 25,000 cm^{-1} but also to some extent for the ^4E at 29,700 cm^{-1}.

The second point is that in the d^5 ions we have the possibility of a particularly direct comparison between term separations in free ions and in compounds. If the term separations are unchanged, the equalities of (11.28) should hold. Actually $E(^4A_1, a\,^4E) = 25{,}000\ \text{cm}^{-1}$, $E(^4G) = 26{,}846\ \text{cm}^{-1}$, and $E(b\,^4E) = 29{,}700\ \text{cm}^{-1}$,

[1] It is difficult to tell what this small separation is due to—second-order spin-orbit coupling, vibrations or changes in the relations between the parameters $a, \ldots, j$ of § 9.2 from those in the free ion are obvious suggestions. Some discussion is in Koide & Pryce (1958).

[2] Originally by Orgel.

$E(^4D) = 32{,}375\,\text{cm}^{-1}$, where we have lumped $^4A_1$ together with $a\,^4E$ in case the assignment of the two components of the sharp band is incorrect, and so there is a definite reduction. See Table 11.2 for data for other manganous compounds.

Table 11.2. *The energy in* cm^{-1} *corresponding to the free-ion energy difference* ^6S–^4G *of* $3d^5$ *for some gaseous ions and for various* Mn^{++} *and* Fe^{3+} *compounds (from Jørgensen,* 1958)

(py = pyridine; ox = oxalate; mal = malonate; en = ethylenediamine; enta = ethylenediaminetetra-acetate.)

Manganese	Energy	Iron	Energy
Mn $3d^5\,4s^2$	25,279	Fe^{3+} $3d^5$	(32,800)
Mn^{++} $3d^5$	26,846	FeF_6^{3-}	25,350
$Mn(H_2O)_6^{++}$	25,000	$Fe(H_2O)_6^{3+}$	24,450
$MnCl_2(H_2O)_4$	24,650	Fe $urea_6^{3+}$	23,250
Mn $enta^{--}$	24,000	Fe mal_6^{3-}	22,800
Mn en_3^{++}	23,800	Fe $formate_6^{3-}$	22,750
Mn py_2Cl_2	23,800	Fe ox_3^{3-}	22,200
$MnBr_4^{--}$	22,300	$FeCl_4^-$	18,800
MnS—ZnS	~ 20,000		

The only stable aquo-ion in the first transition series to have a low-spin ground term is the cobaltic ion $d^6\,{}^1A_1$ and we know that this fact is to be correlated with the lower pairing energy for d^6 than for d^4, d^5 or d^7. We discuss this ion in the strong-field limit. Then the ground term is t_2^6 and the first excited configuration is t_2^5e containing $^3T_1$, $^3T_2$, $^1T_1$, $^1T_2$. The two singlet bands are observed at 16,600 and $24{,}900\,\text{cm}^{-1}$. The calculated energies are (§ 9.3.7) $\Delta - C$ for $^1T_1$ and $\Delta + 16B - C$ for $^1T_2$ so we must assign the band at $16{,}600\,\text{cm}^{-1}$ to $^1T_1$, whence $\Delta - C = 16{,}600\,\text{cm}^{-1}$ and $B = 520\,\text{cm}^{-1}$. So $\Delta \approx 21{,}000\,\text{cm}^{-1}$ which is comparable with the other values for trivalent aquo-ions, but B is very low compared with the probable value of B for the gaseous Co^{3+} ion (calculated value $1150\,\text{cm}^{-1}$ from (4.79)).

11.5. Further examples of spectra

11.5.1. Spin-forbidden absorption bands. Spin-forbidden bands occur not only for the d^5 high-spin ions where they are the only d-d transitions but also in all other d^n configurations for $2 \leqslant n \leqslant 8$. In order to find their expected positions we look at the more complete energy-level diagrams of Figs. 7.2 and 9.2. A transition between a term ^{2S+1}X and $^{2S'+1}Y$ is made allowed by the admixture into ^{2S+1}X of a term $^{2S'+1}Z$ with the same spin as $^{2S'+1}Y$ under the influence of the spin-orbit coupling energy, or alternatively by the admixture of ^{2S+1}W into $^{2S'+1}Y$. The transition is now allowed but with an intensity proportional to the square of the coefficient of admixture. As the spin-orbit coupling operator has matrix elements only between terms differing at most by one unit in S it follows that transitions with $\Delta S \geqslant 2$ will normally be very weak indeed. The coefficient of admixture depends not only on the matrix element between the terms but also on the proximity of those terms and therefore formally forbidden transitions can achieve quite large intensities in cases of near degeneracy of one term (usually the upper) with a term having the same multiplicity as the other.

An extremely interesting example of a spin-forbidden band is the lowest $^3A_2 \rightarrow {}^1E$ transition in Ni^{++} which was observed by Jørgensen in a series of nickel compounds. As we see from Fig. 7.2, as Δ increases the ^1E term passes

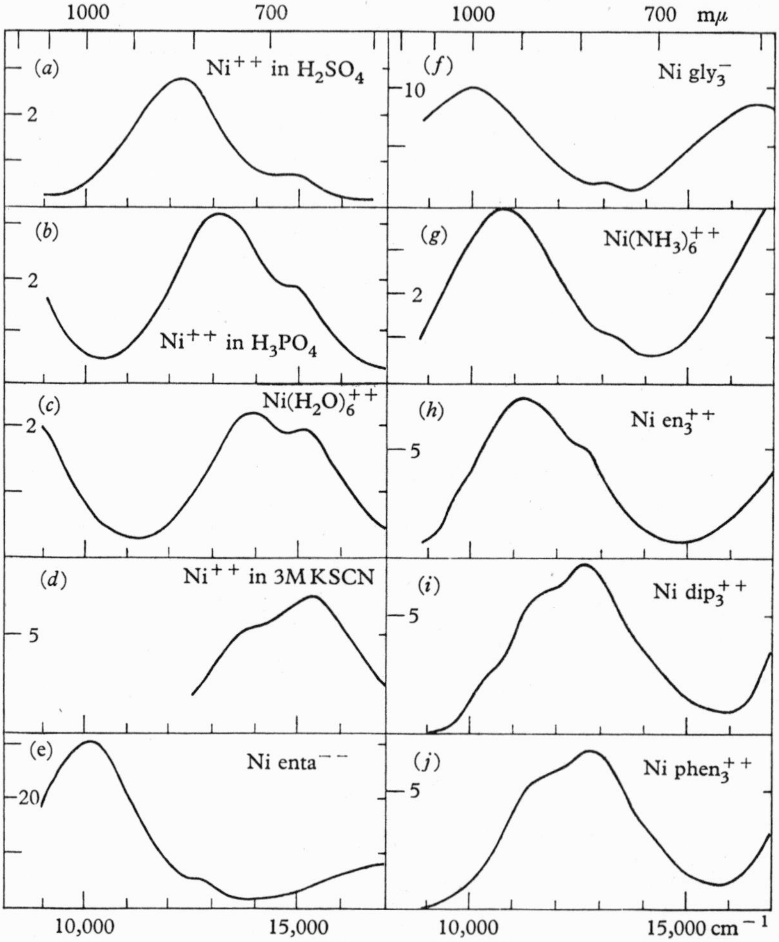

Fig. 11.7. Absorption spectra of Ni^{++} complex ions in the range 8000–17,000 cm^{-1} (from Jørgensen, 1955). The Δ values are: (*a*) 7000, (*b*) 7900, (*c*) 8500, (*d*) 8600, (*e*) 10,000, (*f*) 10,100, (*g*) 10,800, (*h*) 11,600, (*i*) 12,200, (*j*) 12,200, all in cm^{-1}.

through first a $^3T_1$ and then the $^3T_2$ term. In the strong-field limit these terms are, respectively, $t_2^6e^2\,{}^1E$, $t_2^5e^3\,{}^3T_1$ and $t_2^5e^3\,{}^3T_2$. They are correlated in d^2 with the terms $e^2\,{}^1E$, $t_2e\,{}^3T_1$ and $t_2e\,{}^3T_2$ and then Table A 33 shows that ^1E has non-zero matrix elements with the E-components of $^3T_1$ and $^3T_2$. In the strong-field limit these matrix elements are, apart from sign, $\frac{1}{2}\sqrt{6}\,\zeta$ for both.

Let us now consider what happens when the ^1E would have nearly the same energy as the E-component of one of these two triplets if the matrix element between ^1E and the triplet were neglected. Then to a first approximation we need only consider the matrix elements within the pair of levels ${}^1E$ and the E-component of the triplet in question. These are just

$$\begin{bmatrix} \epsilon & \frac{1}{2}\zeta\sqrt{6} \\ \frac{1}{2}\zeta\sqrt{6} & \epsilon' \end{bmatrix},$$

where ϵ is the energy of ${}^1E$ and ϵ' of the other level. The correct energies are now given by

$$E^2-(\epsilon+\epsilon')E+\epsilon\epsilon'-\tfrac{3}{2}\zeta^2 = 0$$

and are

$$E = \tfrac{1}{2}(\epsilon+\epsilon')\pm\tfrac{1}{2}\{(\epsilon-\epsilon')^2+6\zeta^2\}^{\frac{1}{2}},$$

so the difference in energy between the pair of levels is $\{(\epsilon-\epsilon')^2+6\zeta^2\}^{\frac{1}{2}}$ which is never smaller than $\zeta\sqrt{6}$ and is $\zeta\sqrt{6}$ when $\epsilon = \epsilon'$. Using $\zeta = 649\,\text{cm}^{-1}$ from Appendix 6 we have $\zeta\sqrt{6} = 1590\,\text{cm}^{-1}$. The crossing of the ${}^3T_2$ term of Ni^{++} by the ${}^1E$ is shown in Fig. 11.8.

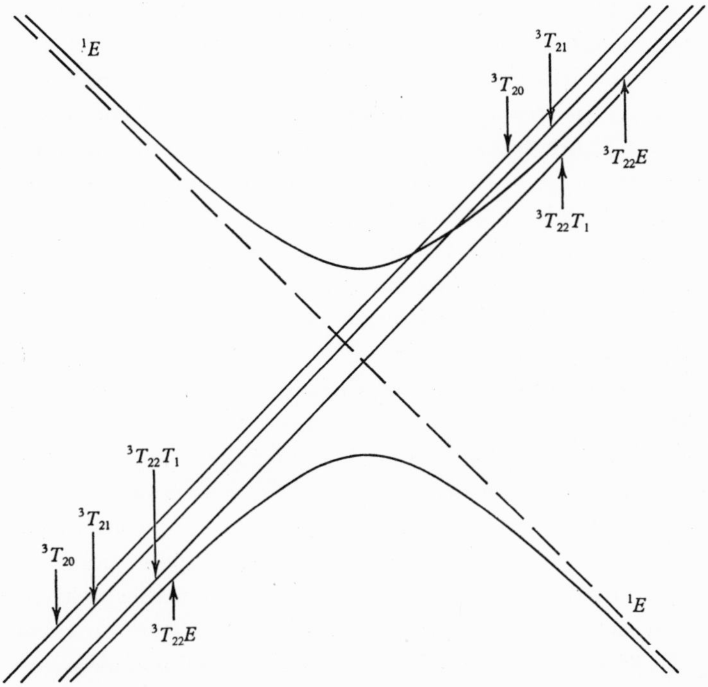

Fig. 11.8. The crossing of the ${}^3T_2$ term by ${}^1E$ in the Ni^{++} ion.

Fig. 11.7 shows the actual spectra for a series of nickel compounds with various values of Δ in the range 7000 to 12,200 cm^{-1} and we see that ${}^1E$ passes from the short wave-length side of ${}^3T_1$ in the top left-hand corner through the ${}^3T_1$ and practically through the ${}^3T_2$ also. Fig. 11.8 suggests that there should be five distinct transitions, but this finer structure cannot be resolved in Jørgensen's spectra. The minimum separation between the peaks has the right order of magnitude, but the precise value cannot be compared with the theory because we cannot see enough structure to be sure which parts of the bands correspond to E symmetry when they are crossing. (See also top of Table A 40.)

11.5.2. Tetrahedral ions. There is not much data available for tetrahedrally coordinated compounds. As we know, in the absence of spin-orbit coupling, the energy-level diagram for d^n ions in a tetrahedral field is the same function of Δ as is the energy-level diagram for d^{10-n} ions in an octahedral field. In each case Δ is the (positive) separation between the t_2- and e-orbitals. Hence d^5 has the same energy-level diagram in tetrahedral and octahedral environments.

Interesting compounds which have had their spectra measured include VCl_4 with one d-electron and several cobaltous compounds with seven. VCl_4 is tetrahedral in the gas phase and also, presumably, in solution in carbon tetrachloride. Its spectrum was measured in solution in the latter and has a band with a maximum at about 9000 cm^{-1} and $\epsilon_{max.} = 110$. The ground term is expected to be ^2E and VCl_4 has a nearly spin-only susceptibility which supports this view. Then the transition is to $^2T_2$ and so $\Delta = 9000\ cm^{-1}$ which is very low for an ion of such a high valency (see Table 11.3). This accords with the general expectation of low values of Δ for tetrahedral environments (§ 8.3). Then the ϵ_{max} is rather high reflecting the natural asymmetry of the environment.

The tetrahedral cobaltous compounds turn up as their anions with the general formula CoX_4^{--}. Spectra are available for X = Cl, Br and I (and also for X = CNS). Each halide has an absorption band in the infra-red, at 6300, 5300 and 5000 cm^{-1} with $\epsilon_{max} = 15$, 100 and 160, respectively, for chloride, bromide and iodide. There is also a strong band in the visible, at 15,000, 13,700 and 12,500 cm^{-1}, respectively. The intensity of this is high, for example $\epsilon_{max} = 600$ in the chloride. At first sight this is all very satisfactory because we take the infra-red band to be $^4A_2 \rightarrow {}^4T_2$ with energy Δ and again have low Δ and high intensity. However, then we predict another band ($^4A_2 \rightarrow {}^4T_1$) much too low (11,000 cm^{-1} in the chloride). This was discussed by Orgel (1955*a*) who suggested that the alternative of assigning the infra-red bands to $^4A_2 \rightarrow {}^4T_1$ was more probable. This assignment leads to predicted transitions $^4A_2 \rightarrow$ upper $^4T_1$ which are too high. It does have the advantage, though, of assigning both the relatively high intensity observed bands to $A_2 \rightarrow T_1$ transitions. These are formally allowed because under the tetrahedral group the electric-dipole moment transforms as T_2. It leads to the prediction that there should be a rather weak band far into the infra-red in each of these ions (between 3000 and 3500 cm^{-1}). If this is correct, then $\Delta = 3750$, 3100 and 2900 cm^{-1} for chloride, bromide and iodide, respectively. We shall see in § 12.4.9 that electron resonance measurements make it almost certain that this latter set of assignments are actually correct.

11.5.3. Other compounds. The majority of the data is for octahedrally coordinated ions and there is a great deal of it so we cannot go through the spectrum of each compound separately. Broadly speaking the general theoretical interpretation we gave for the aquo-ions is universally applicable. Therefore I content myself now with giving a collection of data (collected, and often measured, by Jørgensen). This is in Table A 40 together with the suggested assignments. From these spectra we can derive values for Δ and B. This, as we have seen, is straightforward if we only take into account the terms of maximum multiplicity for high-spin ions. Otherwise we should solve the secular equations derived from the electrostatic matrices of Tables A 27–30. This is a little tedious as, for example, we must solve equations of degree 4 and 7 to get the relative positions of $^1T_1$ and $^1T_2$ in d^6. So we often use second-order perturbation theory or even, as in § 11.4, first-order perturbation theory to give estimates of parameter values. A table of parameters determined by Jørgensen is given in Table 11.3. They do not always agree with those given in the text, but I thought it better to leave them different in order to underline the fact that different acceptable methods of estimation lead to slightly different values and therefore the exact numbers should never be taken too seriously. However, Table 11.3 illustrates a number of general points which we now enumerate.

For a given metal ion Δ always increases as we pass from the left to the right of the table. Thus the ligands can be placed in order of their Δ-forming power. This order is virtually independent of the central metal ion—that is, so long as we keep the latter fixed, the order of Δ is the same, no matter which metal ion we are considering. For this purpose we must regard two ions of the same metal as different if they differ in charge. The set of ligands in order is called the spectrochemical series. It was discovered empirically in terms of spectral shifts by Tsuchida (1938) and independently by Orgel (1955*a*) who related it to ligand-field theory. A version of the series is I^-, Br^-, Cl^-, F^-, oxalate, H_2O, enta^{4-}, NH_3, en, NO_2^-, CN^-, where en = ethylenediamine, enta^{4-} = ethylenediamine-tetra-acetate, and the Δ values increase towards the right. Table 11.3 shows that the order is nearly, but not quite, independent of central metal ion.

Δ does not vary nearly so much along a transition series for fixed ionic charge and ligand. However, it tends to be high when obtained from d^4 or d^6 spectra because of the tetragonal splitting of the ground term. The 'true' value of Δ would be lower but is difficult to estimate very precisely (see § 11.3.5). It is low for high-spin d^5 ions, possibly because the ligands are held less rigidly octahedrally around the spherically symmetric ^6S ions.

Next we remark that Δ is virtually independent of groups which are not directly attached to the metal ion.

The variation of Δ among iso-electronic ions with the same ligands is interesting also. Here we have an increase of about 50 % when we pass from divalent to trivalent ions in the same transition series (e.g. $V^{++} \to Cr^{3+}$) and also in passing from the first to the second transition series keeping the charge fixed. Δ for the third series is a little higher than in the second. Remembering that the electrostatic expression (8.23) for Δ contains no explicit dependence on the charge of the central metal ion, the increase of 50 % between divalent and trivalent ions shows a deficiency in the electrostatic theory.

11.5.4. The nephelauxetic series. Jørgensen has found that the empirical data for Δ may be well represented by the formula $\Delta = fg$, where f depends on the ligand only and g on the central ion only. This is a further refinement of the idea of spectrochemical series. His values of f and g for various common ligands are shown in Table 11.4.

Table 11.3. *The values of* Δ *are given in units of* 1000 *cm*$^{-1}$. *Those for strongly distorted Jahn–Teller systems such as* Cr (II) *and* Cu (II) *have been omitted*

(The values of B are given in cm^{-1} and estimated from the energy difference between states with the maximum value of the total spin S. They are given in parentheses for gaseous ions, when they are obtained by extrapolation (assuming the ratios for corresponding $3d$, $4d$, and $5d$ ions to be 1 : 0·66 : 0·60); also for complexes when obtained from linear combinations of B and C, i.e. from spin-forbidden bands (e.g. $3B+C$ in d^3 and $10B+5C$ in d^5 systems), assuming $C = 4B$; also for the tetrahedral ions $FeCl_4^-$, $CoBr_4^{--}$ and $CoCl_4^{--}$. (I am indebted to C. K. Jørgensen for preparing this table for me.))

		Gaseous	6Br⁻		6Cl⁻		6F⁻		$6H_2O$		3ox⁻⁻		$6NH_3$		3en		6CN⁻	
		B	Δ	B	Δ	B	Δ	B	Δ	B	Δ	B	Δ	B	Δ	B	Δ	B
$3d^2$	V (III)	860	—	—	—	—	—	—	19·0	620	17·8	550	—	—	—	—	—	—
$3d^3$	V (II)	750	—	—	—	—	—	—	12·4	690	—	—	—	—	—	—	—	—
	Cr (III)	920	—	—	13·8	510	15·2	820	17·4	725	17·5	620	21·6	650	21·9	620	26·7	530
	Mn (IV)	1060	—	—	—	—	21·8	600	—	—	—	—	—	—	—	—	—	—
$3d^5$	Mn (II)	(895)	—	—	—	(785)	—	(850)	8·3	(835)	—	—	—	—	9·9	(785)	—	—
	Fe (III)	(1090)	—	—	—	(625)	13·9	(845)	14·2	(815)	13·6	(740)	—	—	—	—	—	—
$3d^6$	Fe (II)	(940)	—	—	—	—	—	—	10·4	—	—	—	—	—	—	—	31·4	400
	Co (III)	(1100)	—	—	—	—	—	—	18·5	800	18·1	630	23·0	680	23·2	650	33·5	440
$3d^7$	Co (II)	980	—	(660)	—	(690)	—	—	9·3	—	—	—	10·1	—	—	—	—	—
$3d^8$	Ni (II)	1040	7·0	740	7·2	760	7·3	960	8·5	940	—	—	10·8	890	11·5	850	—	—
$4d^3$	Mo (III)	(610)	—	—	19·2	440	—	—	—	—	—	—	—	—	—	—	—	—
	Tc (IV)	(700)	—	(390)	—	(420)	—	—	—	—	—	—	—	—	—	—	—	—
$4d^6$	Rh (III)	(720)	19·0	290	20·4	350	—	—	27·0	530	26·4	—	34·1	430	34·6	420	—	—
$5d^3$	Re (IV)	(640)	—	(370)	27·5	(400)	32	(500)	—	—	—	—	—	—	—	—	—	—
	Ir (VI)	(810)	—	—	—	—	—	(320)	—	—	—	—	—	—	—	—	—	—
$5d^4$	Os (IV)	(670)	—	(350)	—	(320)	—	—	—	—	—	—	—	—	—	—	—	—
	Pt (VI)	(840)	—	—	—	—	—	(260)	—	—	—	—	—	—	—	—	—	—
$5d^6$	Ir (III)	(660)	23·1	250	25·0	300	—	—	—	—	—	—	—	—	41·4	—	—	—
	Pt (IV)	(750)	—	—	29	—	33·0	380	—	—	—	—	—	—	—	—	—	—

Table 11.4. *The parameters for the approximate factorizations (from Jørgensen*, 1958)

($\Delta = fg$ and $(1-\beta) = hk$)

Ligands	f	h	Ion	g (cm^{-1})	k
$6H_2O$	1·00	1·0	Mn^{++}	7,600	0·07
$6NH_3$	1·25	1·4	Ni^{++}	8,900	0·12
3en	1·28	1·5	Fe^{3+}	14,400	0·24
$3ox^{--}$	0·98	1·4	Cr^{3+}	17,400	0·21
$6Cl^-$	0·80	2·0	Co^{3+}	19,000	0·30
$6CN^-$	1·7	2·0	Rh^{3+}	27,000	0·30
$6Br^-$	0·76	2·3	Ir^{3+}	32,000	0·3

The parameters of electrostatic interaction in a compound may be written βB, βC, where B and C are the free ion values, providing that they are all reduced in the same proportion. In the absence of sufficient data one is forced to make some such assumption in order to make any comparison between experiment and theory. One will often go further and assume that ζ and **L** are reduced in the same proportion. For example, we did this in ex. 3, p. 285. Naturally, however, such assumptions are far from satisfactory.

Taking a single parameter β to represent the reduction, Schaffer & Jørgensen proposed a new series to give the variation of β as a function of ligand for a fixed metal ion. This is, at least approximately, independent of metal ion. They call it a nephelauxetic series, or cloud-increasing series, from νεφέλη = cloud and αὐξάνω = I increase. Here again one can factorize the effect. Jørgensen writes $1-\beta = hk$ where h refers to ligand and k to metal ion. His values for h and k are also given in Table 11.4. f and h, however, do not follow the same sequence.

The series is conveniently illustrated by the ^6S–^4G energies for the Mn^{++} ion which, theoretically, depend on B and C but not on Δ. They are shown in Table 11.2 for a number of compounds.

11.5.5. The assignment of bands. The reader naturally wonders how we decide which band corresponds to which transition. I will list the sort of evidence one can use for octahedral complexes.

(1) The observed spectrum is fitted to the energy-level diagram. This is the main method of assignment and can be justified on the grounds that it gives a plausible and consistent interpretation of a large number of spectra and leads to no serious difficulties. It is supplemented by other, some potentially more reliable, methods.

(2) The behaviour of bands under low-symmetry perturbations is observed and interpreted in terms of the wave-functions.

(3) The polarizations of absorption bands split under low symmetry are measured and interpreted.

(4) The rotation of the plane of polarized light is analysed. This depends on $E^{-1}\langle 0 | \mathbf{L} | n\rangle \langle n | \Sigma \mathbf{r}_i | 0\rangle$ where $|n\rangle$ is an excited state and E its energy above $|0\rangle$.

The selection rules for **L** are then the main source of information (see Moffitt, 1956).

(5) Transitions are the sharper the less their sensitivity to changes in Δ, i.e. the smaller $\partial E/\partial\Delta$ where E is the energy of the transition.

(6) In the strong-field case two-electron jumps (e.g. $t_2^2\,{}^3T_1 \rightarrow e^2\,{}^3A_2$) should be weaker than one-electron ones.

(7) The selection rule for spin is: $\Delta S = 1$ usually weak; $\Delta S \rangle 1$ usually extremely weak (see § 11.5.1).

(8) The Zeeman effect on the transitions—that is we observe the splitting of narrow bands in strong magnetic fields. This is only possible at very low temperatures.

(9) The splitting due to spin-orbit coupling could, in principle, be used.

Methods (2) and (3) are described in the next section. A very detailed analysis for V^{3+} in corundum has been given by Pryce & Runciman (1958) and for Cr^{3+} in corundum (ruby) by Sugano, Tanabe & Tsujikawa (1958). The latter gives the theory of the Zeeman effect in d^3 subject to a large octahedral field with a trigonal component.

Finally, we emphasize that in some cases there is doubt about an assignment. This is especially true for weak bands or for bands near to charge-transfer or ligand bands for they may always not be *d-d* transitions at all but spin-forbidden charge-transfer or ligand bands.

11.6. The cobaltic ion

11.6.1. Singlet-singlet transitions in the aquo-ion. In § 11.4 we assigned the visible absorption bands of the Co^{3+} ion in aqueous solution to the transitions $t_2^6\,{}^1A_1 \rightarrow t_2^5 e\,{}^1T_1$ and $t_2^6\,{}^1A_1 \rightarrow t_2^5 e\,{}^1T_2$, with respective energies 16,600 and 24,900 cm^{-1}. This then led to the relation $16B = 8300$ cm^{-1}, $B = 520$ cm^{-1} which is really extremely low compared either with $B = 883$ cm^{-1} which we obtained for the aqueous Co^{++} ion or with $B = 1150$ cm^{-1} for the gaseous Co^{3+} ion from (4.79). We now investigate this ion in more detail and show that the anomalously low value is due entirely to our having used an unrealistically approximate theory.

Suppose $C = 4B = 4000$ cm^{-1} which is a 13 % reduction for B below the theoretical expression (4.79) for a gaseous cobaltic ion and therefore represents a realistic estimate of B for the aqueous cobaltic ion. Next we contemplate correcting the energies of the lowest ${}^1A_1$, ${}^1T_1$, ${}^1T_2$ in the strong-field scheme by second-order perturbation theory using the matrices A 29. This means correcting the energies by an amount $-\Sigma M^2/E$, where M is the matrix element to an excited term and E its energy above the lower term in question and is a reasonable thing to do providing $M/E \ll 1$ for all excited terms. Taking a trial $\Delta = 21{,}000$ cm^{-1} we find $M/E \leqslant \frac{1}{5}$ for all excited terms except $t_2^2({}^3T_1)\,e^2({}^3A_2)\,{}^1T_2$, for which $M/E \approx 1$. This, then, is a reason for the small distance between the lowest ${}^1T_1$ and ${}^1T_2$—there is a nearby ${}^1T_2$ which pushes the lowest ${}^1T_2$ down substantially.

It is interesting to consider also the effect of the polarization correction discussed in §4.5. This added a term $\alpha L(L+1)$ to the calculated energies of terms, $\alpha = 76\,\text{cm}^{-1}$ is a reasonable estimate for Co^{3+}. We evaluate

$$\alpha \mathbf{L}^2 = 36\alpha + 2\alpha \sum_{\kappa<\lambda} \mathbf{l}_\kappa . \mathbf{l}_\lambda$$

for the strong-field terms of t_2^6 and $t_2^5 e$ obtaining 24, 16, 16, 32, 16 for $^1A_1$, $^3T_1$, $^3T_2$, $^1T_1$, $^1T_2$, respectively. In the simplest approximation, then, $^1T_1$ is raised relative to $^1T_2$ by $1220\,\text{cm}^{-1}$, which is not entirely negligible. This particular correction is likely to be lowered when the correct $^1T_2$ function is taken.

With the above values for α, B, C we now choose Δ so that the relative energy of $^1A_1$ and $^1T_1$ calculated by second-order perturbation theory agrees with the experimental value of $16{,}600\,\text{cm}^{-1}$. This $\Delta = 19{,}210\,\text{cm}^{-1}$. Using it we find the term $t_2^4(^3T_1)\,e^2(^3A_2)\,^1T_2$ lies $6210\,\text{cm}^{-1}$ above $t_2^5 e\,^1T_2$. $M/E \leqslant \frac{1}{5}$ for the higher terms. We then solve the 2×2 secular equation for the lowest two $^1T_2$ terms to obtain

$$|a\,^1T_2\rangle = 0{\cdot}833\,|t_2^5 e\,^1T_2\rangle - 0{\cdot}553\,|t_2^4(^3T_1)\,e^2(^3A_2)\,^1T_2\rangle, \tag{11.29}$$

with energy $-4873\,\text{cm}^{-1}$ relative to $t_2^5 e\,^1T_2$. We now use second-order perturbation theory to correct the energy of the function $|a\,^1T_2\rangle$. It is depressed by a further $923\,\text{cm}^{-1}$. The lowest $^1T_1$ is depressed by $1240\,\text{cm}^{-1}$ hence, apart from polarization, we predict the energy difference to be $11{,}440\,\text{cm}^{-1}$. In view of the large coefficient of the excited term in (11.29), the polarization correction is probably quite a way from $16\alpha = 1220\,\text{cm}^{-1}$, but if we take this value we get finally $E(^1T_2) - E(^1T_1) = 10{,}220\,\text{cm}^{-1}$ to be compared with the observed value of $8330\,\text{cm}^{-1}$. This is already within the range of deviation shown in Table 11.1 for the calculated T_1-T_1 separations in high-spin ions and the agreement could obviously be improved by taking B rather smaller (850–$900\,\text{cm}^{-1}$, which is acceptable).

In §9.4 we gave a simple account of pairing energies and cross-over points. Clearly, however, the real cross-over point occurs not where the lowest terms in the strong-field coupling scheme cross each other but rather where the energies of the actual terms cross. Using our values for Δ, B, C and α we now calculate the energy separation $\delta = E(^5D) - E(^1A_1)$ where $E(^1A_1)$ is the energy of the lowest $^1A_1$ calculated as described above. Then

$$\begin{aligned}\delta &= 2\Delta - 5B - 8C - 18\alpha + 3236\\ &= 3290\,\text{cm}^{-1},\end{aligned}$$

and would be increased to $6990\,\text{cm}^{-1}$ if $B = \frac{1}{4}C = 900\,\text{cm}^{-1}$. As the ground term is a singlet, this result is naturally gratifying.

Before leaving this topic I remind the reader that as $\Delta \to \infty$, $|a\,^1T_2\rangle \to |t_2^5 e\,^1T_2\rangle$ and for the cyanide where $\Delta \approx 35{,}000\,\text{cm}^{-1}$ the lowest $^1T_2$ does belong to the configuration $t_2^5 e$ to quite a good approximation and there B really is substantially reduced ($B \approx 400\,\text{cm}^{-1}$).

11.6.2. Tetragonal perturbations and optical dichroism. Consider now the compound trans $[Co(en)_2Cl_2]ClO_4$ of the cobaltic ion. The structure of the complex ion in one plane is illustrated and has a chloride ion above and below the cobalt. Thus the cobaltic ion is octahedrally coordinated and the symmetry is approximately D_{4h}. The orbital pattern is shown in Fig. 12.3. $\mu > 0$ because, according to Table 11.3, ethylenediamine splits the d-orbitals more than the chloride ion. δ is also expected to be positive because of π-bonding from the chlorine ions (Appendix 8). The lowest excited singlet is $^1T_1$ and we have just seen that the strong-field function $t_2^5 e\, ^1T_1$ approximates to it fairly well. In the strong-field scheme the result of the perturbation in Fig. 12.3 is to split the $^1T_1$, as shown in Fig. 11.9, by $\delta + \frac{3}{4}\mu$ with the doubly degenerate level lower but without altering the position of the centre of gravity (the calculation is not quite so simple for the $^1T_2$, cf. (11.29)).

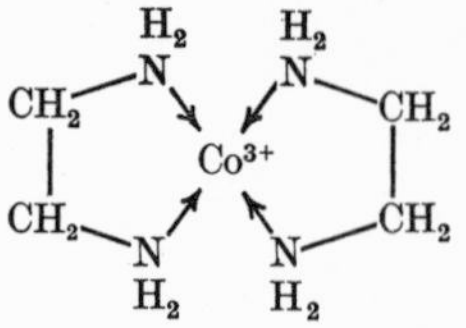

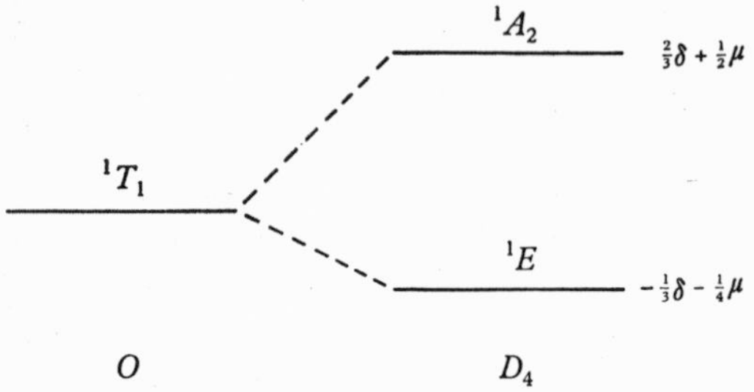

Fig. 11.9. Effect of tetragonal perturbation on the $t_2^5 e\, ^1T_1$ term of d^6.

In our compound there are bands at 16,130 and at 23,260 cm^{-1} and they show interesting dichroic properties. Before considering these, however, let us look at the energies. If they both arise from the $^1T_1$ of a strictly octahedral complex then their weighted mean energy might be expected to be related to the weighted mean of the energies of the $^1A_1 \to {}^1T_1$ transitions in $Co(en)_3^{3+}$ and $CoCl_6^{3-}$. The former is $E_1 = 21{,}400$ cm^{-1} but the latter is unknown. They are both known for the chromic ion and so by reducing Δ for Co^{3+} in the same proportion as for Cr^{3+} we obtain an estimated $E_2 = 12{,}600$ cm^{-1} and $\frac{2}{3}E_1 + \frac{1}{3}E_2 = 18{,}500$ cm^{-1}. The weighted mean of the observed bands is 18,500 cm^{-1} also so we appear to be on the right lines.

The dichroism is as follows: when the electric vector is parallel to the fourfold axis absorption only occurs at 16,130 cm^{-1}; when it is perpendicular to the fourfold axis absorption occurs at both 16,130 and at 23,260 cm^{-1} (Yamada, Nakahara, Shimura & Tsuchida, 1955).

We use the notation appropriate to D_{4h} and find selection rules for absorption from the vibrationally unexcited ground state. The ground state is $|^1A_1\rangle$ and an excited state is $|\Gamma a \Gamma' b\rangle$ where this represents the component a of an electronic term of symmetry Γ, together with one unit of the component b of a vibration of symmetry Γ'. Then electric-dipole absorption can only take place for the electric vector along the fourfold axis (OZ) if $A_1\Gamma\Gamma'$ contains A_{2u}, i.e. if $A_{2u}\Gamma'$ contains Γ (use 11.3.3 and App. 3.2 a), because $z \in A_{2u}$ under D_{4h}. For an octahedrally coordinated compound the six nearest neighbour atoms can execute odd vibrations of A_{2u}, B_{2u} or E_u symmetry but not of A_{1u} or B_{1u}. We see this by decomposing the odd vibrations appropriate to O_h according to Table A 11. Therefore the even part of $A_{2u}\Gamma'$ contains A_{1g}, B_{1g} and E_g but not A_{2g} or B_{2g}. So the transition to ${}^1A_2$ in Fig. 11.9 is forbidden but not the one to ${}^1E$ (the transition to ${}^1B_{2g}$ from ${}^1T_{2g}$ is also forbidden). For perpendicular polarization the electric vector transforms as E_u and then all even symmetries occur in $E_u\Gamma'$. Hence the selection rules match the observed dichroism.

The above interpretation of optical dichroism was given by Ballhausen & Moffitt (1956) and by Yamatera (1958) and is presumably correct. However, it depends upon the assumption that all the absorption (or at least the great majority) occurs from the ground vibrational state. The measurements were at room temperature (Yamada, private communication). In their calculations on intensities of the hydrated manganous ion Koide & Pryce (1958) concluded that the majority of the intensity arose from vibrations with energies $\nu_3 = 170$ and $\nu_6 = 85\,\text{cm}^{-1}$. The energies should be higher in the more strongly bound cobaltic complexes so it may be reasonable to assume there that the majority of the absorption occurs from a vibrationally unexcited state even at room temperature.

11.7. Thermodynamic properties[1]

11.7.1. Ligand-field stabilization of octahedral complexes. We discuss now the heat of formation ΔH of a complex ion in solution from the free gaseous ion. ΔH is defined by

$$M^{n+}_{\text{gas}} + (\text{H}_2\text{O})_{\text{liq}} = M^{n+}_{\text{aq}} + (\text{H}_2\text{O})_{\text{liq}} + \Delta H_1, \tag{11.30}$$

for the aquo-ions and $$M^{n+}_{\text{gas}} + 6L_{\text{aq}} = (ML_6)_{\text{aq}} + \Delta H_2, \tag{11.31}$$

for octahedral coordination of six ligands L. The observed value of ΔH_1 for divalent metal ions of the first transition series is plotted in curve (a) of Fig. 11.10. This curve has a two-humped form with minima at $\text{Ca}^{++}(d^0)$, $\text{Mn}^{++}(d^5)$ and $\text{Zn}^{++}(d^{10})$. Much more data is available for the second half of the series than for the first and an extensive survey of data by George & McClure (1959) shows that the characteristic humped form occurs with minor modifications for ΔH_2 for any ligand L, for trivalent ions and also both for solid compounds containing discrete complex ions and those containing infinite lattices (the metal halides, for example).

There is a simple interpretation of these facts in terms of an electrostatic crystal-field theory. In the free ions the ground configuration is d^n and the d-orbitals are degenerate. In the compound the dt_2-orbitals have energy $-\frac{2}{5}\Delta$ and the de-orbitals $+\frac{3}{5}\Delta$. As a consequence the ground terms of the ions are, in general, split and the ground state of the complex has an energy below the mean energy of the free-ion ground term. We may say that in the free ion the t_2- and e-orbitals are equally likely to be occupied, but in the crystal field the t_2-orbitals get more occupied than the e-orbitals. This change in mean occupancy confers an extra

[1] For the ligand-field interpretation see Orgel (1952, 1956); George (1956); Bjerrum & Jørgensen (1956); Griffith (1956); George & McClure (1959).

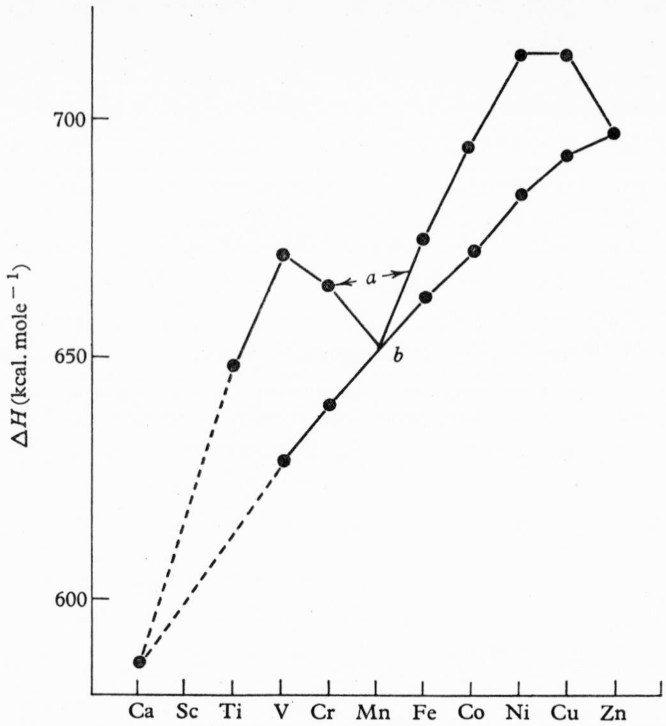

Fig. 11.10. Heats of formation of divalent aqueous ions of the first transition series from the gaseous ions. (*a*) uncorrected; (*b*) corrected (from Orgel, 1956; Griffith & Orgel, 1957).

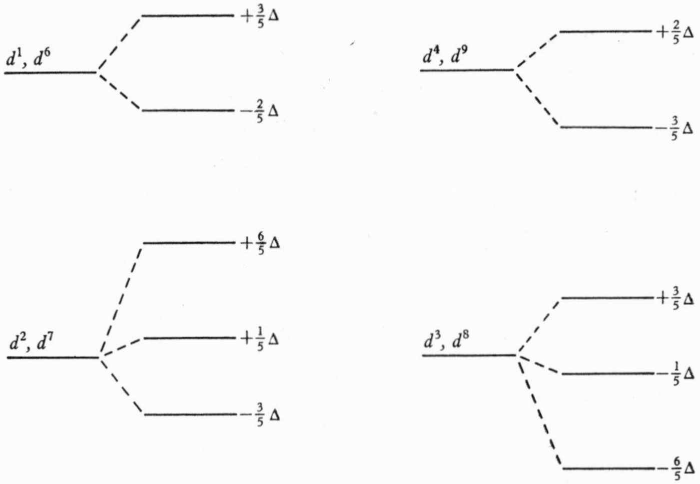

Fig. 11.11. Ligand-field stabilizations for d^n ions with high-spin ground terms and $L \neq 0$, in an octahedral field.

stabilization on the complex which is larger the greater the lowering of the ground state below the mean energy of the free-ion ground term. These crystal-field stabilizations are shown in Fig. 11.11 for the weak-field case and high-spin ground terms. They are zero for d^0, d^5 and d^{10}. Apart from the possibility of change of spin, it is only for d^2 and d^7 that they are not accurate for a large Δ.

For those two the mixing of the ground FT_1 with the excited PT_1 leads to a stabilization lying between the weak-field limit $-\frac{3}{5}\Delta$ and the strong-field limit $-\frac{4}{5}\Delta$.

One might expect the mean heat of formation taken over all the components of the free-ion ground term to vary smoothly as a function of atomic number for fixed metal valency and ligand. Therefore, if we subtract what we have called the crystal-field stabilizations from the observed heats of formation we should obtain a smooth curve. Using values of Δ obtained from spectra, we then get curve (*b*) in Fig. 11.10 and see that this expectation is borne out.

The electrostatic crystal-field theory is a good way to get a general idea of the nature of the stabilizations—which we will now call ligand-field stabilizations. We can obtain them equally easily from a simple molecular orbital model. Here we expect everything that determines the energy to vary smoothly including the orbital energies $E(t_2)$ and $E(e)$ of the two kinds of d-electrons. However, the occupancy of these orbitals does not do so and if we write $E(t_2) = a - \frac{2}{5}\Delta$ and $E(e) = a + \frac{3}{5}\Delta$ we have the orbital energy

$$E(t_2^m e^n) = (m+n)\,a + \tfrac{3}{5}n\Delta - \tfrac{2}{5}m\Delta. \tag{11.32}$$

According to our hypothesis, $(m+n)\,a$ varies smoothly as a function of $(m+n)$. So the 'unsmooth' part is $\frac{1}{5}\Delta(3n-2m)$ which is just the extra stabilization predicted by the electrostatic theory except for d^2 and d^7 where it gives the strong-field limiting value $-\frac{4}{5}\Delta$. Our results are summed up in Table 11.5.

Table 11.5. *The extra stabilizations in* (*a*) *the weak-field and* (*b*) *the strong-field limits for high-spin ions. The units are* Δ; *the latter varies from ion to ion*

Ion	d^0	d^1	d^2	d^3	d^4	d^5	d^6	d^7	d^8	d^9	d^{10}
(*a*)	0	$-\frac{2}{5}$	$-\frac{3}{5}$	$-\frac{6}{5}$	$-\frac{3}{5}$	0	$-\frac{2}{5}$	$-\frac{3}{5}$	$-\frac{6}{5}$	$-\frac{3}{5}$	0
(*b*)	0	$-\frac{2}{5}$	$-\frac{4}{5}$	$-\frac{6}{5}$	$-\frac{3}{5}$	0	$-\frac{2}{5}$	$-\frac{4}{5}$	$-\frac{6}{5}$	$-\frac{3}{5}$	0

The essential feature of our derivation is that all ions except S-state ions can achieve extra stability by orienting their spatial distributions to minimize their repulsive interaction with the surroundings. This is possible when the ion has spatial degeneracy but not otherwise (to first order). This more general formulation now applies to any site symmetry. The S-state ions (d^0, d^5, d^{10}) should always be anomalously unstable and the F ions (d^2, d^3, d^7, d^8) probably more stable than the D ions (d^1, d^4, d^6, d^9). These remarks should all be interpreted in terms of stability relative to a smoothly varying background. The relation of F to D also assumes a smooth variation of the parameters determining the orbital pattern. For octahedral coordination we saw earlier that this is not entirely correct and that d^4 and d^9 ions tend to distort strongly. This gives a large extra stabilization due to distortion. Referring to Fig. 11.4 for the aqueous cupric ion we find an extra stabilization of $\frac{1}{2}\Delta$, giving a total of $1{\cdot}1\Delta = 6930\,\text{cm}^{-1} = 20\,\text{kcal}$. This is very little different from the stabilization of $7560\,\text{cm}^{-1} = 22\,\text{kcal}$ which we would have obtained if we had assumed the cupric ion to be in a regular octahedral field and that the observed visible band is due to an $e \rightarrow t_2$ transition.

In tetrahedral symmetry all the diagrams in Fig. 11.11 are turned upside down. The t_2-orbitals now contain $3d$- and $4p$-orbitals from the metal atom, as well as other admixtures, and we assume the coefficients of the $4p$-orbitals in t_2 to vary smoothly as a function of atomic number.

11.7.2. Possible refinements to the theory. There is little doubt that the general form of variation shown in Fig. 11.10 for octahedral complexes can be ascribed mainly to the changes in relative occupancy of the t_2- and e-orbitals combined in the case of d^4 and d^9 with a substantial influence of Jahn–Teller distortion. It is instructive, however, to examine in a qualitative way how the theory might be further refined. We return to the discussion of § 7.2 and remark that we must compare the energy

$$E_0 = E_0^{(f)} + (m+n)\,E_0^{(i)} + E_0^{(p)}, \tag{11.33}$$

for the free d^{m+n} ion with

$$E_1 = E_1^{(f)} + mE_1^{(i)}(t_2) + nE_1^{(i)}(e) + E_1^{(p)}, \tag{11.34}$$

for the $t_2^m e^n$ complex ion. Here $E_0^{(f)}$ and $E_1^{(f)}$ are the energies of the filled shells and $E_0^{(p)}$ and $E_1^{(p)}$ of the partly-filled shells, without any interaction between them. $E_0^{(i)}$, $E_1^{(i)}(t_2)$ and $E_1^{(i)}(e)$ arose from the interaction between them and we remember that our assumption about electron configurations led to each of the $E^{(i)}$ being just numbers independent of m and n. They are the orbital energies in the field of the filled shells.

The first point which emerges from (11.33) and (11.34) is that we have assumed $E_0^{(p)} = E_1^{(p)}$, in other words that the parameters A, B and C of electrostatic interaction are the same in the compound as in the free ion. This is not strictly true. As $E_0^{(p)}$ is positive, this effect should be an extra stabilization which is largest relatively for those ions which, when free, have the largest $E_0^{(p)}$.

Next we note that Fig. 4.6 is good evidence that $E_0^{(f)} + (m+n)\,E_0^{(i)}$ really does vary smoothly as a function of $(m+n)$.

Equation (11.34) has been used assuming that $E_1^{(f)}$, $E_1^{(i)}(t_2)$ and $E_1^{(i)}(e)$ are independent of the occupancy of t_2 and e, i.e. that they are functions of $(m+n)$ but are independent of $(m-n)$. This will not be true in general and the distribution of nuclei, and of the electrons in the filled shells, will adjust itself in a way which will somewhat decrease the energy separations between pairs of configurations $t_2^m e^n$ and $t_2^{m+1} e^{n-1}$. However, this also affects our empirical estimate of Δ and so our semi-empirical corrections to ΔH already contain some allowance for this polarizability of the filled shells.

I conclude this section by remarking that the experimental chemist is often interested in differences between heats of formation because these occur in replacement reactions such as

$$(M^{++}L_6)_{\text{aq}} + 6N_{\text{aq}} = (M^{++}N_6)_{\text{aq}} + 6L_{\text{aq}}, \tag{11.35}$$

and even more in the free energy changes in these reactions. The latter include an entropy contribution $-T\Delta S$ which is not necessarily either small or exactly the same for all metal ions with the same ligand. ΔS for formation includes contributions from possible change in degeneracy of the ground state and a variable contribution which is a function of the rigidity of attachment of the ligands. The rigidity is likely to be least for the S-state ions d^0, d^5 and d^{10}.

The variations in ΔH as a function of cation in (11.35) are often rather small and then when they deviate from the predictions of the simplest version of ligand-field theory it is difficult to assign the origin of the deviation with any confidence.

11.7.3. The significance of the free-ion ionization potentials. Before the ligand-field interpretation of Fig. 11.10 by Orgel (1952) had appeared, it had been noticed empirically by Mellor & Maley (1947, 1948) that the free-energy changes ΔF for (11.35) for a given pair of ligands seemed always to be numerically in the same order when considered as a function of cation. ΔF was particularly large for Cu^{++} in the first transition series. This order was confirmed by Irving & Williams (1948, 1953) and also found by Calvin & Melchior (1948). Both these groups independently suggested that it was to be correlated with the order of the sum Q of the first two ionization potentials of the free metal ion. The argument here is that the larger are these ionization potentials, the greater is the affinity of the metal ion for ligand electrons and hence the more readily will the metal ion bond with the ligands. Hence Q, ΔH and probably ΔF are all correlated.

However, it is difficult to make such an idea quantitative, which restricts its utility. Further, although when it was originally proposed it was perfectly reasonable to suppose the whole of the extra stabilization shown in Fig. 11.10 to be due to this irregular electron affinity, with the understanding we now have of the ligand-field theory this is no longer true. However, while the electron affinity argument makes no allowance for ligand-field effects the converse is not true. We saw in § 11.7.2 that the spectroscopically determined Δ includes some allowance for the polarization of the filled shells or, amongst other things and in other words, for electron affinity. It will also be apparent to the reader that there will be several kinds of electron affinity which must be considered if we adopt the affinity point of view—these are the affinity of t_2, e and higher metal orbitals for electrons from the various ligand orbitals. Such ideas can be made precise most easily in terms of some kind of ligand-field theory.

CHAPTER 12

PARAMAGNETIC RESONANCE

12.1. Electron resonance

We have discussed the optical spectra of transition-metal compounds. We now consider radio spectra—that is the absorption and emission of electromagnetic radiation in the radio-frequency region. There are two kinds of spectra of interest to us here: electron spin resonance spectra and nuclear magnetic resonance spectra. For completeness one should also mention nuclear quadrupole resonance but this is not important for transition-metal compounds. Of the other two, electron spin resonance has by far the wider application and interest in the study of transition-metal compounds. Accordingly, we treat electron spin resonance first and leave nuclear magnetic resonance until § 12.5.

Let us illustrate electron resonance by considering an ion having a ground term which is orbitally non-degenerate and has spin S and the spin-only interaction

$$\mathscr{H} = 2\beta \mathbf{H} . \mathbf{S} \tag{12.1}$$

with an external magnetic field $\mathbf{H}$. $\mathbf{S}$ is in units of $\hbar$. If $\mathbf{H}$ is along OZ the term splits into $2S+1$ equally spaced states with energies $2\beta H M_S$ and separation $2\beta H$ between adjacent ones. Electron resonance studies the absorption (and occasionally emission) of radiation accompanying transitions between these states. The transitions are magnetic dipole allowed and have intensities proportional to

$$P_\epsilon = |\langle \bar{\phi} | (L_\epsilon + 2S_\epsilon) | \psi \rangle|^2, \tag{12.2}$$

for transitions between states ϕ and ψ where the suffix ϵ denotes the component along the direction of the magnetic vector of the incident radiation. With the Hamiltonian (12.1) and $\mathbf{H}$ along OZ we have

$$P_x = P_y = S(S+1) - M_S(M_S+1), \tag{12.3}$$

for transitions between the states $|SM_S\rangle$ and $|SM_{S+1}\rangle$ and $P_x = P_y = P_z = 0$ for all other transitions (see Fig. 12.1a). In particular we have the selection rule $\Delta M_S = \pm 1$ so the energy change can only be $2\beta H$. For $H = 10^4$ gauss, $2\beta H \approx 1$ cm^{-1} so we are concerned here with centimetre wave-length radio waves.

After these preliminary remarks I now describe briefly the electron resonance experimental arrangement. A suitable specimen containing unpaired electrons is placed in a static external magnetic field. Plane polarized radio waves are sent through the specimen with their direction of propagation at right-angles to the static field and with their magnetic vector usually also at right-angles to the static field, but occasionally parallel to it. The net absorption of radiation by the specimen from the radiation field is measured. The static magnetic field is then varied until a resonance (i.e. a maximum absorption) is obtained. The radio frequency is not usually changed during an experiment.

In our example no radiation would be absorbed if the radio-frequency magnetic field were parallel to the static field, but this is not always true because there are extra terms in the Hamiltonian which can mix states with different M_S values and then P_z is not necessarily zero between the mixed states. When the two fields are at right-angles we obtain absorption only when the so-called resonance condition $h\nu = 2\beta H$ is satisfied. This is shown, for $S = 1$, in Fig. 12.1 *a*,

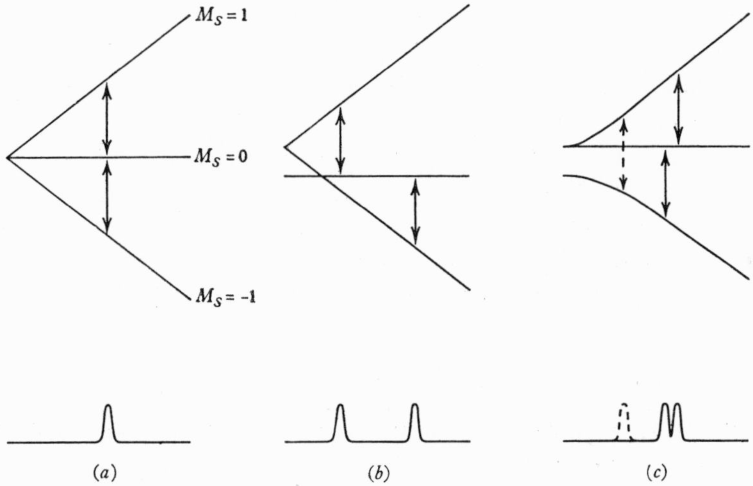

Fig. 12.1. Fine structure for a spatially non-degenerate ion with $S = 1$, as a function of the magnetic field **H**. (*a*) No zero-field splitting. (*b*) Zero-field splitting with axis parallel to static magnetic field **H**. (*c*) Axis at right-angles to **H**. The broken curve is not to scale with the full ones.

where the energy levels and the absorption line are plotted as a function of the magnetic-field strength. In general, the first-order interaction with the magnetic field may be more complicated than (12.1) and we write the resonance condition $h\nu = g\beta H$ and regard a measurement of ν and H together with an observation of a maximum absorption as giving us a value for g. For our simple model we predict $g = 2$ independently of the direction of the static magnetic field. We will see in §§ 12.3 and 12.4 how g values different from 2 and anisotropic g values can arise.

Next we notice that the situation here is not quite identical with the model experiment we discussed in § 3.2.3. There the incident light induced transitions to excited states which were not thermally occupied. Here our excited states are practically as much occupied thermally as the ground state. Therefore transitions are induced both ways. It is, however, just the small difference in occupancy which still persists (the occupancy is, of course, proportional to $e^{-E/kT}$ where $E = 2\beta H M_S$ in our example) which gives rise to a net absorption of radiation. Therefore the intensity is enhanced at low temperatures.

This is one advantage in working at lower temperatures but a more important one is that the line width is reduced at low temperatures. The main causes of line width are spin-lattice interaction, magnetic spin-spin interaction and exchange interaction. The first of these arises from a coupling of lattice vibrations to the spin moment of the ion through the intermediary of the spin-orbit coupling. It is reduced by lowering the temperature and consequently in many salts electron resonance is not observed at room temperature but only at low temperatures (liquid air temperature is often necessary and sometimes even liquid helium temperature). The magnetic spin-spin interaction is the classical magnetic coupling between the electrons of different ions and between one ion and the nuclear spin of another. As there is usually no strong correlation between the orientations of these spins this leads to a broadening. The exchange interaction was mentioned in § 10.2.8 in connexion with the Curie–Weiss law. Its effect on line width is rather complicated and in some cases can actually narrow the line. Both the spin-spin and the exchange interactions are reduced by diluting the paramagnetic salt in an isomorphous diamagnetic one.

The existence of line width places a lower limit on the useful frequency of the radiation and the difficulty of making the appropriate microwave equipment an upper limit. Most measurements are made with the radiation having energy in the range 0·1–2·0 cm^{-1}. For more detail about the experimental methods of electron resonance I refer the reader to a book by Ingram (1955).

12.2. Refinements to the Hamiltonian

Under the usual experimental conditions in electron resonance experiments the energy of magnetic interaction with an external static magnetic field is of the order of 1 cm^{-1}, and hitherto neglected terms in the Hamiltonian now become important. They are of two kinds. The first consists of small extra interactions between the electrons, mainly magnetic in character. The second consists of interactions between the electrons and the nucleus other than the spherically symmetric Coulomb interaction Ze^2/r.

Just as we discussed the spin-orbit coupling energy from the point of view of Dirac's relativistic wave-equation, so also should we treat the small extra interactions between the electrons from the point of view of a suitable relativistic equation. The equation here is the Breit equation or preferably the Bethe–Salpeter equation. However, to do this properly is considerably more difficult than the corresponding derivation for the one-electron case and as we make comparatively little use of these extra terms it is more suitable to treat them from a semi-classical and phenomenological view-point which is what we will do.

The new terms in the Hamiltonian include a spin-spin magnetic coupling having the same form as a classical interaction between two small bar magnets, also a magnetic coupling between the two orbital moments (orbit-orbit coupling) of a pair of electrons and between the spin moment of one electron and the orbital moment of the other (spin-other-orbit coupling). As one might expect there also energes from the treatment via the Breit equation a Fermi-type delta function interaction. I discuss the theory of the spin-spin interaction in § 12.2.4. The other terms, however, we shall entirely neglect in this book because they are not only small but also lead to no new effects to a good approximation. This is because they mimic on a smaller scale the first-order effects of terms already present in the Hamiltonian. With the methods that we used to derive the Landé interval rule in § 5.1 we may show that the first-order effect of the magnetic spin-other-orbit coupling within a Russell–Saunders term is formally the same as the effect of the spin-orbit coupling, i.e. it is proportional to **L . S**. The orbit-orbit coupling commutes with both **L** and **S** and, therefore, only gives rise to a uniform shift in all levels of a term. Lastly, the Fermi-type interaction between a pair of electrons is proportional to $\mathbf{s}_1 . \mathbf{s}_2\, \delta(r_{12})$ and again gives rise only to an overall shift of any term.

The extra interactions with the nucleus include a magnetic interaction between the electrons and the magnetic moment of the nucleus and also a Coulomb interaction between the electrons and the nuclear quadrupole moment, provided that the nucleus has a moment of the appropriate kind. We have already discussed the theory of the magnetic interaction in § 5.5.3 and we will use those results to derive the theory of the magnetic hyperfine structure of electron resonance. The quadrupole interaction arises because the nucleus is not necessarily spherically symmetric, and therefore its Coulomb field will contain a non-spherically symmetric component. This produces very small energies and, among the transition-metal ions, the effects have only been analysed in detail and published for the cupric ion. We discuss the general theory of the quadrupole coupling in § 12.2.3 and apply it to Cu^{++} and Ti^{3+} in §§ 12.4.1 and 12.4.2.

12.2.1. Two useful identities. I now describe a technique which greatly simplifies the theory of the nuclear hyperfine structure and electron spin-spin interaction. If we have a pair of vectors $\mathbf{T}_1$ and $\mathbf{T}_2$ each of type T with respect to an angular momentum **J** we know (§§ 2.8, 8.10) that we may regard them as operator eigenstates of **J**, each having $J = 1$. From the nine products T_{1x}, T_{2x}, etc., of the components of $\mathbf{T}_1$ and $\mathbf{T}_2$ taken with $\mathbf{T}_1$ preceding $\mathbf{T}_2$ we may construct three sets of operator eigenstates of **J** having $J = 0, 1, 2$, respectively. In case $\mathbf{T}_1 = \mathbf{T}_2$ and the components of $\mathbf{T}_1$ commute the states having $J = 1$ become identically zero. Linear combinations giving $J = 2$ are

$$\left.\begin{aligned} V_1(\mathbf{T}_1, \mathbf{T}_2) &= (1/\sqrt{3})\,(2T_{1z}T_{2z} - T_{1x}T_{2x} - T_{1y}T_{2y}), \\ V_2(\mathbf{T}_1, \mathbf{T}_2) &= T_{1x}T_{2x} - T_{1y}T_{2y}, \quad V_3(\mathbf{T}_1, \mathbf{T}_2) = T_{1x}T_{2y} + T_{1y}T_{2x}, \\ V_4(\mathbf{T}_1, \mathbf{T}_2) &= T_{1y}T_{2z} + T_{1z}T_{2y}, \quad V_5(\mathbf{T}_1, \mathbf{T}_2) = T_{1z}T_{2x} + T_{1x}T_{2z}. \end{aligned}\right\} \tag{12.4}$$

Let us now suppose that we have another vector **T** which is of type T with respect to an angular momentum $\mathbf{J}'$ and that $\mathbf{T}_1$, $\mathbf{T}_2$ and **J** each commute with **T** and $\mathbf{J}'$. Then the quantity

$$W = \sum_{k=1}^{5} V_k(\mathbf{T}, \mathbf{T})\, V_k(\mathbf{T}_1, \mathbf{T}_2), \tag{12.5}$$

commutes with $\mathbf{j} = \mathbf{J} + \mathbf{J}'$ as one verifies by direct computation (see the example at the end of the subsection). W is the operator eigenstate (unique apart from a constant factor) of **j** having $j = 0$ and formed from products of the $V_k(\mathbf{T}, \mathbf{T})$ with $V_{k'}(\mathbf{T}_1, \mathbf{T}_2)$. Equation (12.5) may also be regarded as a spherical harmonic addition theorem for operators (compare (8.11)). On multiplying out the terms of W and simplifying we obtain, after quite a bit of algebra,

$$W = -\tfrac{2}{3}\mathbf{T}^2(\mathbf{T}_1 . \mathbf{T}_2) + 2(\mathbf{T} . \mathbf{T}_1)(\mathbf{T} . \mathbf{T}_2) - \Sigma[T_x, T_y](\mathbf{T}_1 \wedge \mathbf{T}_2)_z, \tag{12.6}$$

where the sum is over the cyclic permutations of x, y, z. If the components of $\mathbf{T}$ commute among themselves we have

$$W = -\tfrac{2}{3}[\mathbf{T}^2(\mathbf{T}_1 . \mathbf{T}_2) - 3(\mathbf{T} . \mathbf{T}_1)(\mathbf{T} . \mathbf{T}_2)]. \tag{12.7}$$

If $\mathbf{T}$ satisfies the commutation rules for an angular momentum, then $[T_x, T_y] = iT_z$ (we write angular momenta in units of $\hbar$ throughout this chapter) and the third term in (12.6) becomes the triple scalar product $-i\mathbf{T} . \mathbf{T}_1 \wedge \mathbf{T}_2$. If, now, $\mathbf{T}_1 = \mathbf{T}_2 = \mathbf{U}$ is also an angular momentum then

$$W = 2[(\mathbf{T} . \mathbf{U})^2 + \tfrac{1}{2}(\mathbf{T} . \mathbf{U}) - \tfrac{1}{3}\mathbf{T}^2\mathbf{U}^2]. \tag{12.8}$$

In (12.8), $\mathbf{T}$ commutes with $\mathbf{U}$.

The value of the identities is that they connect (12.5), which can easily be replaced by operator equivalents, with the forms (12.7) and (12.8) which occur directly in the theory. In the next three sections we use these results to discuss nuclear magnetic hyperfine and quadrupole interactions and electronic spin-spin couplings. This method is due to Araki (1948) and was used by him to discuss spin-spin coupling in atoms.[1]

It is often convenient to take instead of the V_k an orthonormal transform of them, V'_k say. W is unchanged by this if we write

$$W = \sum_{k=1}^{5} V'_k(\mathbf{T}, \mathbf{T})\, \bar{V}'_k(\mathbf{T}_1, \mathbf{T}_2). \tag{12.9}$$

A choice of V'_k which we use later is

$$\left.\begin{aligned} V'_1 &= V_1, \quad V'_2 = \frac{1}{\sqrt{2}}(V_2 + iV_3) = \frac{1}{\sqrt{2}} T_1^+ T_2^+, \\ V'_3 &= \frac{1}{\sqrt{2}}(V_2 - iV_3) = \frac{1}{\sqrt{2}} T_1^- T_2^-, \\ V'_4 &= \frac{1}{\sqrt{2}}(V_5 + iV_4) = \frac{1}{\sqrt{2}}(T_1^+ T_{2z} + T_{1z} T_2^+), \\ V'_5 &= \frac{1}{\sqrt{2}}(V_5 - iV_4) = \frac{1}{\sqrt{2}}(T_1^- T_{2z} + T_{1z} T_2^-), \end{aligned}\right\} \tag{12.10}$$

where $T_1^{\pm} = T_{1x} \pm iT_{1y}, \quad T_2^{\pm} = T_{2x} \pm iT_{2y}.$

Example

Show that if $\mathbf{T}_1$, $\mathbf{T}_2$ are both of type T with respect to $\mathbf{J}$, then

$$[V_1, J_z] = 0, \quad [V_2, J_z] = -2iV_3, \quad [V_3, J_z] = 2iV_2, \quad [V_4, J_z] = iV_5, \quad [V_5, J_z] = -iV_4.$$

$\mathbf{T}_3$, $\mathbf{T}_4$ are of type T with respect to $\mathbf{J}'$. Deduce that

$$W = \sum_{k=1}^{5} V_k(\mathbf{T}_1, \mathbf{T}_2)\, V_k(\mathbf{T}_3, \mathbf{T}_4) \quad \text{commutes with } \mathbf{J} + \mathbf{J}'.$$

[1] Although the majority of the so-called spin-spin coupling is a second-order effect of spin-orbit coupling (see Trees, 1951*a*).

12.2.2. The magnetic hyperfine structure. We already discussed the hyperfine interaction between an electron and the nuclear magnetic moment in § 5.5.3. The interaction added a term

$$\Delta = 2\gamma\beta\beta_N \mathbf{I}\,.\left\{f(r^{-3}\mathbf{l}-r^{-3}\mathbf{s}+3r^{-5}(\mathbf{s}\,.\,\mathbf{r})\,\mathbf{r})+r^{-2}\frac{df}{dr}(\mathbf{s}-r^{-2}(\mathbf{s}\,.\,\mathbf{r})\,\mathbf{r})\right\}, \quad (12.11)$$

to the Hamiltonian for a one-electron system. We now replace the parts of Δ by equivalent operators in order to facilitate calculations. It is convenient to do this in stages and we take the first bracket first and let $f = 1$. Then the z-component of the vector $-r^{-3}\mathbf{s}+3r^{-5}(\mathbf{s}\,.\,\mathbf{r})\,\mathbf{r}$ is

$$-r^{-3}s_z+3r^{-5}(xzs_x+yzs_y+z^2 s_z) = 3r^{-5}(xzs_x+yzs_y+\tfrac{1}{3}(3z^2-r^2)\,s_z).$$

By (8.36) an operator equivalent to this within a term is

$$-3\xi\overline{r^{-3}}(\tfrac{1}{2}(l_x l_z+l_z l_x)\,s_x+\tfrac{1}{2}(l_y l_z+l_z l_y)\,s_y+\tfrac{1}{3}(2l_z^2-l_x^2-l_y^2)\,s_z), \quad (12.12)$$

where ξ is a dimensionless constant. Equation (12.12) is then rearranged to read

$$\xi\overline{r^{-3}}(l(l+1)\,s_z-\tfrac{3}{2}(\mathbf{l}\,.\,\mathbf{s})\,l_z-\tfrac{3}{2}l_z(\mathbf{l}\,.\,\mathbf{s})).$$

Thus the first three terms of (12.11) may be replaced by

$$2\gamma\beta\beta_N\overline{r^{-3}}\{\mathbf{l}\,.\,\mathbf{I}+\xi[l(l+1)\,\mathbf{s}\,.\,\mathbf{I}-\tfrac{3}{2}(\mathbf{l}\,.\,\mathbf{s})\,(\mathbf{l}\,.\,\mathbf{I})-\tfrac{3}{2}(\mathbf{l}\,.\,\mathbf{I})\,(\mathbf{l}\,.\,\mathbf{s})]\}. \quad (12.13)$$

The value of ξ is determined by evaluating

$$a = -r^{-3}\mathbf{s}\,.\,\mathbf{I}+3r^{-5}(\mathbf{s}\,.\,\mathbf{r})\,(\mathbf{I}\,.\,\mathbf{r})$$

and
$$b = \overline{r^{-3}}\xi[l(l+1)\,\mathbf{s}\,.\,\mathbf{I}-\tfrac{3}{2}(\mathbf{l}\,.\,\mathbf{s})\,(\mathbf{l}\,.\,\mathbf{I})-\tfrac{3}{2}(\mathbf{l}\,.\,\mathbf{I})\,(\mathbf{l}\,.\,\mathbf{s})], \quad (12.14)$$

for a state $|nl\tfrac{1}{2}l, I\rangle$ having the maximum possible values for m_s, m_l and m_I. For such a state the mean value of a is

$$\bar{a} = \overline{r^{-3}}(-\tfrac{1}{2}I+\tfrac{3}{2}I\,\overline{\cos^2\theta}) = -\frac{lI\overline{r^{-3}}}{2l+3},$$

where we have used $\overline{\cos^2\theta} = (2l+3)^{-1}$ from ex. 5, p. 62. On the other hand

$$\bar{b} = \overline{r^{-3}}\xi(\tfrac{1}{2}l(l+1)\,I-\tfrac{3}{2}l^2 I)$$
$$= -\tfrac{1}{2}\overline{r^{-3}}\xi lI(2l-1),$$

and so $\xi = 2/[(2l-1)\,(2l+3)]$. For d electrons $l = 2$, so $\xi = \frac{2}{21}$.

For an n-electron system the operator equivalent to $A = \Sigma a_i$ within a term ^{2S+1}L is

$$B = \xi\overline{r^{-3}}\mathbf{N}_1\,.\,\mathbf{I} = \xi\overline{r^{-3}}\{L(L+1)\,\mathbf{S}\,.\,\mathbf{I}-\tfrac{3}{2}(\mathbf{L}\,.\,\mathbf{S})\,(\mathbf{L}\,.\,\mathbf{I})-\tfrac{3}{2}(\mathbf{L}\,.\,\mathbf{I})\,(\mathbf{L}\,.\,\mathbf{S})\}, \quad (12.15)$$

where ξ is yet to be determined. (12.15) is the natural generalization of (12.14) to an n-electron system. We note first that it is obvious that $A' = \Sigma b_i$ is an operator equivalent to A and we shall always use this alternative form for the nuclear hyperfine interaction, both to establish (12.15), and also to calculate the interaction for electronic states which are not parts of free-ion terms. To prove that B is correct we remark that b of (12.14) satisfies

$$b = \overline{r^{-3}}\xi(l(l+1)\,\mathbf{s}\,.\,\mathbf{I}-3(\mathbf{l}\,.\,\mathbf{s})\,(\mathbf{l}\,.\,\mathbf{I})+\tfrac{3}{2}i\mathbf{l}\,.\,\mathbf{s}\wedge\mathbf{I}),$$

which follows from the commutation rules for $\mathbf{l}$. But (see the discussion after (12.6) and (12.7)) this means that

$$b = -\tfrac{3}{2}\xi\overline{r^{-3}} \sum_{k=1}^{5} V_k(\mathbf{l},\mathbf{l})\, V_k(\mathbf{s},\mathbf{I}). \tag{12.16}$$

We now evaluate Σb_λ for typical states of a term of an n-electron l^n system:

$$\begin{aligned}
&\langle SLM_SM_L \,|\, \Sigma b_\lambda \,|\, SLM'_SM'_L\rangle \\
&= -\tfrac{3}{2}\xi\overline{r^{-3}} \sum_{\lambda,k,\beta} \langle SLM_SM_L \,|\, V_k(\mathbf{l}_\lambda,\mathbf{l}_\lambda) \,|\, \beta SLM_SM'_L\rangle\langle \beta SLM_SM'_L \,|\, V_k(\mathbf{s}_\lambda,\mathbf{I}) \,|SLM'_SM'_L\rangle \\
&= -\tfrac{3}{2}\xi\overline{r^{-3}} \sum_{\lambda,k,\beta} \alpha_1\alpha_2 \langle SLM_SM_L|\, V_k(\mathbf{L},\mathbf{L}) \,|SLM_SM'_L\rangle\langle SLM_SM'_L|\, V_k(\mathbf{S},\mathbf{I}) \,|SLM'_SM'_L\rangle \\
&= -\tfrac{3}{2}\xi\overline{r^{-3}} \sum_{\lambda,k,\beta} \alpha_1\alpha_2 \langle SLM_SM_L|\, V_k(\mathbf{L},\mathbf{L})\, V_k(\mathbf{S},\mathbf{I}) \,|SLM'_SM'_L\rangle.
\end{aligned} \tag{12.17}$$

ξ in (12.17) is the ξ appropriate to a single electron. We have replaced $V_k(\mathbf{l}_\lambda,\mathbf{l}_\lambda)$ by $\alpha_1 V_k(\mathbf{L},\mathbf{L})$ with a coefficient α_1 which depends on β but not on k. Similarly we have replaced $\mathbf{s}_\lambda$ by $\alpha_2\mathbf{S}$ because the $V_k(\mathbf{s}_\lambda,\mathbf{I})$ are homogeneous linear functions of the components of $\mathbf{s}_\lambda$. The last form of (12.17) is a multiple of B of (12.15) which proves the result.

Having derived (12.15) we now determine ξ for the ground term of an l^n configuration having $n < 2l+1$ by evaluating A' and B for the state

$$|X\rangle = |l^+, l-1^+, \quad \ldots, \quad l-n+1^+, I\rangle$$

and equating the results. This is a matter of straightforward algebra and we find

$$\langle X \,|\, A' \,|\, X\rangle = -\frac{2IL\overline{r^{-3}}(l+\tfrac{1}{2}-2S)}{(2l-1)(2l+3)},$$

$$\langle X \,|\, B \,|\, X\rangle = -\xi LSI(2L-1),$$

whence
$$\xi = \frac{2l+1-4S}{S(2l-1)(2l+3)(2L-1)}. \tag{12.18}$$

We have used $S = \tfrac{1}{2}n$ and $L = \tfrac{1}{2}n(2l+1-n)$, which are true for the ground terms. Both A' and B are zero for the ground term when $n = 2l+1$.

The formulae for the hyperfine energies in l^n and l^{4l+2-n} are the same and (12.18) holds for a more-than-half-filled shell also. This is because the hyperfine interaction is a one-electron operator n which satisfies $n = \overline{n} = -n^*$ and follows immediately from the discussion given in § 9.7.

The last part of (12.11) may be replaced by an expression due to Fermi involving a delta function. This expression is $\tfrac{1}{3}(16\pi)\gamma\beta\beta_N\delta(r_k)\,\mathbf{s}\,.\,\mathbf{I}$, where

$$\int g(x,y,z)\,\delta(r)\,d\tau = g(0,0,0),$$

for any continuous function $g(x,y,z)$. It has the same matrix elements as

$$\Delta_2 = 2\gamma\beta\beta_N r^{-2}\frac{df}{dr}(\mathbf{s}\,.\,\mathbf{I} - r^{-2}(\mathbf{s}\,.\,\mathbf{r})(\mathbf{r}\,.\,\mathbf{I})) \tag{12.19}$$

to the first approximation in the small quantities occurring in (12.19). To prove this let us consider the part of Δ_2 involving I_z. This is

$$\Delta_{2z} = 2\gamma\beta\beta_N I_z r^{-2}\frac{df}{dr}(s_z\sin^2\theta - s_x\sin\theta\cos\theta\cos\phi - s_y\sin\theta\cos\theta\sin\phi), \tag{12.20}$$

where are we using spherical polar coordinates for the electron. Then use an $lm_s m_l m_I$ scheme and calculate a matrix element of Δ_{2z} between two states having radial functions R_1, R_2, respectively. This matrix element contains as a factor the integral over radial functions

$$\int_0^\infty R_1 r^{-2} \frac{df}{dr} R_2 r^2 dr = \int_0^\infty R_1 R_2 \frac{df}{dr} dr$$

$$= -\int_0^\infty f \frac{d(R_1 R_2)}{dr} dr \approx R_1(0) R_2(0),$$

where we have integrated by parts and then set $f = 1$. The reader may care to satisfy himself that the vanishing of f at the origin does not invalidate our approximation of replacing f by unity in the derivation. Therefore the matrix element can only be non-zero between two s-states, in which case the coefficients

$$\sin\theta \cos\theta \cos\phi \quad \text{and} \quad \sin\theta \cos\theta \sin\phi$$

of s_x and s_y, respectively, are zero when we integrate over θ and ϕ. For an s-state the mean value of $\sin^2\theta$ is

$$\overline{\sin^2\theta} = \overline{\left(\frac{x^2+y^2}{r^2}\right)} = \frac{2}{3}\overline{\left(\frac{x^2+y^2+z^2}{r^2}\right)} = \frac{2}{3},$$

whence the matrix element of Δ_{2z} is

$$2\gamma\beta\beta_N m_I^1 R_1(0) R_2(0) \,.\, \tfrac{2}{3} \,.\, m_s^1 \delta(m_I^1, m_I^2)\, \delta(m_s^1, m_s^2),$$

where m_I^1, m_s^1 refer to one state and m_I^2, m_s^2 to the other. But the part of $\frac{1}{3}(16\pi)\gamma\beta\beta_N \delta(r_k)\, \mathbf{s} \,.\, \mathbf{I}$ involving I_z also has this matrix element. This establishes our result. The reader should realize the strength of this result. We have obtained a simple operator equivalent to the Fermi part of Δ which has the same matrix elements as that part for all pairs of one-electron states, not merely for those which are both part of the same one-electron term.

There is no Fermi energy within a d^n configuration but in spite of this the analysis of the hyperfine structure of supposedly d-electron terms requires the assumption of an extra isotropic interaction proportional to $\mathbf{S} \,.\, \mathbf{I}$. This extra interaction is, of course, almost certainly due to the inadequacy of the assumption that our states are entirely d^n states. Let us take an example. The ground term of the manganous ion is

$$1s^2\, 2s^2\, 2p^6\, 3s^2\, 3p^6\, 3d^5\, {}^6S.$$

Configuration interaction mixes in a little of various excited configurations including

$$1s^2\, 2s^2\, 2p^6\, 3s\, 4s\, ({}^3S)\, 3p^6\, 3d^5\, ({}^6S)\, {}^6S$$

in which the $3s$ and $4s$ electrons are coupled to form a ${}^3S$ term and this is then coupled with $3d^5\,{}^6S$ to form a ${}^6S$ term for the ion. If the ground term contains some of this particular excited term, then it contains unpaired s electrons and we obtain a Fermi energy proportional to $R_{3s}(0)\, R_{4s}(0)$ times the coefficient of the excited term in the ground term. Many other excited terms also contribute in this way. As a consequence, it is customary to take the hyperfine Hamiltonian for a single d electron as

$$\Delta = 2\gamma\beta\beta_N \overline{r^{-3}}\{\mathbf{l}.\mathbf{I}+\tfrac{1}{7}(4\mathbf{s}.\mathbf{I}-(\mathbf{l}.\mathbf{s})(\mathbf{l}.\mathbf{I})-(\mathbf{l}.\mathbf{I})(\mathbf{l}.\mathbf{s}))-\kappa\mathbf{s}.\mathbf{I}\}, \quad (12.21)$$

where κ is a dimensionless constant. It is usual to put $P = 2\gamma\beta\beta_N\overline{r^{-3}}$.

Another way of looking at the occurrence of κ is to stick to the configuration assignment $1s^2\,2s^2\,2p^6\,3s^2\,3p^6\,3d^5\,{}^6S$, but to allow the $1s$, $2s$ and $3s$ electrons to have different radial functions depending on their spin-quantum number. Then if the $3d^5$ shell has $M_S = \frac{5}{2}$, an s electron with $m_s = +\frac{1}{2}$ has a different exchange interaction with the $3d^5$ shell from that of an s electron with $m_s = -\frac{1}{2}$. So the minimum energy will in general be achieved for different radial functions. Therefore the Fermi energy for an s^2 shell is not zero because $\psi^2(0)$ is different for the two s electrons. It is not easy to make either idea quantitative or even to calculate the sign of κ.

These two ways of looking at κ are essentially equivalent. A third, and different, possibility is that the ligand field mixes $3d^{n-1}\,4s$ into $3d^n$. This cannot happen for regular octahedral or tetrahedral compounds but can happen in tetragonal symmetry for then $3d_{z^2}$ and $4s$ both belong to the unit representation. This time κ is proportional to the square of the coefficient of the $4s$ orbital and to $\psi_{4s}(0)^2$. This effect is probably not very important unless there is a large deviation from octahedral symmetry, for example, in planar compounds.[1]

We conclude this subsection with a few remarks designed to assist the reader to calculate hyperfine energies. The n-electron hyperfine Hamiltonian for a d^n system may be taken as

$$\mathscr{H}_m = P\left(\mathbf{L}.\mathbf{I}-\kappa\mathbf{S}.\mathbf{I}+\frac{1}{7}\sum_{k=1}^{n}\mathbf{a}_k.\mathbf{I}\right),$$

where $$\mathbf{a}_k = 4\mathbf{s}_k-(\mathbf{l}_k.\mathbf{s}_k)\mathbf{l}_k-\mathbf{l}_k(\mathbf{l}_k.\mathbf{s}_k) \quad\text{and}\quad P = 2\gamma\beta\beta_N\overline{r^{-3}}. \quad (12.22)$$

The matrix elements of the components a_z and $a^- = a_x - ia_y$ of the vector $\mathbf{a}$ are given in Table A 41. This means that we can work out the matrix elements of $\mathscr{H}_m$ between any pair of determinantal functions. $\mathscr{H}_m$ is a one-electron operator, so they vanish when the determinantal functions differ by more than one constituent one-electron function. In the particular case that we wish to evaluate $\mathscr{H}_m$ only between functions which form part of the ground term of the ion we can take

$$\mathscr{H}_m = P\{\mathbf{L}.\mathbf{I}-\kappa\mathbf{S}.\mathbf{I}+\xi[L(L+1)\,\mathbf{S}.\mathbf{I}-\tfrac{3}{2}(\mathbf{L}.\mathbf{S})(\mathbf{L}.\mathbf{I})-\tfrac{3}{2}(\mathbf{L}.\mathbf{I})(\mathbf{L}.\mathbf{S})]\}, \quad (12.23)$$

where ξ is given by (12.18).

Examples

1. (Fermi.) For a one-electron atom in free space, in an s-state and with nuclear spin I, show that the two hyperfine components have energies

$$E(I+\tfrac{1}{2}) = \tfrac{8}{3}\pi\gamma\beta\beta_N I\psi^2(0) \quad\text{and}\quad E(I-\tfrac{1}{2}) = -\tfrac{8}{3}\pi\gamma\beta\beta_N(I+1)\,\psi^2(0).$$

2. (Breit.) For the atom of ex. 1 in a state with $l \neq 0$ and $j = l\pm\frac{1}{2}$ show that the energy as a function of the total angular momentum f is

$$E(f) = \gamma\beta\beta_N\overline{r^{-3}}\frac{l(l+1)}{j(j+1)}\{f(f+1)-I(I+1)-j(j+1)\}.$$

[1] References to: first way (Abragam, Horowitz & Pryce, 1955); second way (Heine, 1957); third way (Griffith, 1958*b*).

12.2.3. The nuclear quadrupole interaction.[1] The energy of Coulomb interaction between electrons with position vectors $\mathbf{r}_\lambda$ and the protons of the nucleus with position vectors $\mathbf{R}_\mu$ is

$$E = -\sum_{\lambda,\mu} \frac{e^2}{|\mathbf{r}_\lambda - \mathbf{R}_\mu|}. \tag{12.24}$$

If we put $R_\mu = 0$ we get the central-field energy $-\sum_\lambda Ze^2/r_\lambda$. Now expand V as a power series in R_μ/r_λ which is allowable so long as the electron is outside the nucleus. Then

$$E = -\sum_{\lambda,\mu} e^2 \sum_k \frac{R_\mu^k}{r_\lambda^{k+1}} P_k(\cos\omega). \tag{12.25}$$

The first term of (12.25) is the central-field energy, the second term is zero because the nucleus does not have a permanent dipole moment and the third term is the interaction with the nuclear quadrupole moment. This third term is

$$\begin{aligned} E_q &= -\sum_{\lambda,\mu} \frac{e^2}{r_\lambda^5} r_\lambda^2 R_\mu^2 P_2(\cos\omega) \\ &= -\sum_{\lambda,\mu} \frac{e^2}{4r_\lambda^5} \sum_{m=1}^{5} V_m(\mathbf{r}_\lambda, \mathbf{r}_\lambda) V_m(\mathbf{R}_\mu, \mathbf{R}_\mu), \end{aligned} \tag{12.26}$$

where we have used the spherical harmonic addition theorem to express E_q in terms of the V_m of (12.4). Now the nucleus is in an eigenstate of its angular momentum and the interaction E_q does not appreciably mix in any excited nuclear states. Therefore we may replace the $V_m(\mathbf{R}_\mu, \mathbf{R}_\mu)$, which form operator eigenstates of the nuclear angular momentum $\mathbf{I}$ having $I = 2$, by multiples of $V_m(\mathbf{I}, \mathbf{I})$. It is customary to choose the numerical factors in such a way that

$$E_q = -\frac{3e^2Q}{4I(2I-1)} \sum_\lambda r_\lambda^{-5} \sum_{m=1}^{5} V_m(\mathbf{r}_\lambda, \mathbf{r}_\lambda) V_m(\mathbf{I}, \mathbf{I}) \tag{12.27}$$

$$= \frac{e^2Q}{2I(2I-1)} \sum_\lambda \left(\frac{I(I+1)}{r_\lambda^3} - \frac{3(\mathbf{r}_\lambda \cdot \mathbf{I})^2}{r_\lambda^5}\right), \tag{12.28}$$

where we have used (12.7). Q is the nuclear quadrupole moment as it is usually defined. Because the $V_m(\mathbf{R}, \mathbf{R})$ have $I = 2$ the nucleus cannot have a non-zero quadrupole interaction unless it has $I \geqslant 1$.

Let us now look for an operator equivalent to E_q within an l^n term ${}^{2S+1}L$. Taking (12.27) for E_q we have a form linear in the electronic operators $r_\lambda^{-5} V_m(\mathbf{r}_\lambda, \mathbf{r}_\lambda)$ and (§ 8.10, especially (8.36)) can replace these operators by multiples of the equivalent ones $V_m(\mathbf{L}, \mathbf{L})$. We set $q' = e^2Q\overline{r^{-3}}/[2I(2I-1)]$ and then the multiples are chosen so that

$$\begin{aligned} E_q &= \tfrac{3}{2}\eta q' \sum_{m=1}^{5} V_m(\mathbf{L}, \mathbf{L}) V_m(\mathbf{I}, \mathbf{I}) \\ &= \eta q'[3(\mathbf{L} \cdot \mathbf{I})^2 + \tfrac{3}{2}(\mathbf{L} \cdot \mathbf{I}) - L(L+1)I(I+1)]. \end{aligned} \tag{12.29}$$

For one electron we determine η by evaluating the alternative expressions (12.28) and (12.29) for E_q over the state $|X\rangle$ having $m_s = \frac{1}{2}$, $m_l = l$, $m_I = I$. We use ex. 5, p. 62 and obtain $\eta = 2/\{(2l-1)(2l+3)\}$. For d electrons, $\eta = \frac{2}{21}$.

[1] After Bleaney & Stevens (1953).

The value of η for an n-electron system is obtained by evaluating

$$C = \sum_\lambda \frac{2}{(2l-1)(2l+3)}[3(\mathbf{l}_\lambda . \mathbf{I})^2 + \tfrac{3}{2}(\mathbf{l}_\lambda . \mathbf{I}) - l(l+1)I(I+1)]$$

and $$D = \eta[3(\mathbf{L}.\mathbf{I})^2 + \tfrac{3}{2}\mathbf{L}.\mathbf{I} - L(L+1)I(I+1)]$$

for the state $|l^+, l-1^+, ..., l-n+1^+, I\rangle$, as in § 12.2.2.

$$\bar{C} = \sum_{m=l-n+1}^{l} \frac{2}{(2l-1)(2l+3)}(I^2 - \tfrac{1}{2}I)(3m^2 - l(l+1)) = -\frac{(2I-1)\bar{A}'}{\overline{r^{-3}}},$$

$$\bar{D} = \tfrac{1}{2}\eta LI(2I-1)(2L-1) = -\frac{(2I-1)\bar{B}\eta}{2\xi S\overline{r^{-3}}},$$

whence $\eta = 2S\xi$ where ξ is given by (12.18).

For a more-than-half-filled shell the matrix of the quadrupole energy changes sign because E_q is a one-electron operator satisfying $E_q = \bar{E}_q = E_q^*$ (§ 9.7). Therefore $\eta = -2S\xi$ for $n > 2l+1$. Both $\bar{C}$ and $\bar{D}$ are zero for the ground term when $n = 2l+1$.

12.2.4. Spin-spin coupling energies.[1] The energy of interaction between the spin-magnetic moments of a pair of electrons is given by

$$E_{12} = \beta^2\left(\frac{\boldsymbol{\sigma}_1 . \boldsymbol{\sigma}_2}{r_{12}^3} - \frac{3(\boldsymbol{\sigma}_1 . \mathbf{r}_{12})(\boldsymbol{\sigma}_2 . \mathbf{r}_{12})}{r_{12}^5}\right)$$

$$= -\tfrac{3}{2}\beta^2 \sum_{k=1}^{5} V_k(\hat{\mathbf{r}}_{12}, \hat{\mathbf{r}}_{12}) V_k(\boldsymbol{\sigma}_1, \boldsymbol{\sigma}_2), \qquad (12.30)$$

where we have used (12.7) and $\hat{\mathbf{r}}_{12} = \mathbf{r}_{12}/r_{12}^{\frac{5}{2}}$. We may therefore replace $\Sigma E_{\kappa\lambda}$ within a term ^{2S+1}L by

$$-\tfrac{1}{2}\rho \sum_{k=1}^{5} V_k(\mathbf{L},\mathbf{L}) V_k(\mathbf{S},\mathbf{S})$$

$$= -\rho[(\mathbf{L}.\mathbf{S})^2 + \tfrac{1}{2}\mathbf{L}.\mathbf{S} - \tfrac{1}{3}L(L+1)S(S+1)], \qquad (12.31)$$

using (12.8).

The value of ρ in transition-metal ions is uncertain, but at least it is probably not greater than 0·1 cm^{-1} in the ground terms of the divalent or trivalent ions of the first transition series. Larger values have been derived in the past from deviations from the Landé interval rule but these deviations appear to be mainly a consequence of second-order effects of the spin-orbit coupling energy (Trees, 1951*a*).

12.3. The spin-Hamiltonian

The spin-Hamiltonian is a convenient resting place during the long trek from fundamental theory to the squiggles on an oscilloscope which are the primary result of electron resonance experiments. For us it represents nearly the last outpost in our land of theoretical physics and it is there that we meet most of

[1] Araki (1948); Pryce (1950*a*).

our experimental results already prepared and packaged for us by experimental workers. We now approach this concept, but cautiously by way of an illuminating example.

12.3.1. An ion with $S = 1$ and the concept of a spin-Hamiltonian. Consider now an ion with spin $S = 1$ and orbitally non-degenerate. If we neglect spin-orbit coupling it will be threefold degenerate in the absence of a magnetic field and strictly so. The combined action of the spin-orbit coupling and the ligand field may lift this degeneracy and we will then have three states which are nearly but not necessarily quite degenerate. Let us suppose first that each of these three states can be written, to first order in perturbation theory, as $|M_S\rangle - \epsilon|X\rangle$ where ϵ is small and $|X\rangle$ is a linear combination of excited states and is different for each of the $|M_S\rangle$. It is convenient to classify these three states by the M_S values of their parent unperturbed states. Thus we may write

$$|a(M_S)\rangle = |M_S\rangle - \epsilon|X\rangle, \tag{12.32}$$

remembering that $|a(M_S)\rangle$ is not necessarily an eigenstate of S_z. If $|a(M_S)\rangle$ has energy $E(M_S)$ then $E(1) = E(-1)$ (use the Kramers operator) and so there are just the two independent energies $E(0)$ and $E(1)$.

The idea of the spin-Hamiltonian is to produce an operator which is a polynomial in the components of the spin-vector $\mathbf{S}$ and having an effect equivalent to these energies. We take $\mathscr{H}(S) = [E(1)-E(0)]S_z^2 + E(0)$ and calculate its matrix elements within the $|a(M_S)\rangle$ as if these states were eigenstates of S_z (and correctly connected in phase, though this is not relevant yet). The matrix elements of $\mathscr{H}(S)$ are

$$\langle a(1)|\mathscr{H}(S)|a(1)\rangle = \langle a(-1)|\mathscr{H}(S)|a(-1)\rangle = E(1)$$

and
$$\langle a(0)|\mathscr{H}(S)|a(0)\rangle = E(0)$$

showing that we have achieved our object.

A much more general form for the spin-Hamiltonian would be $L = \sum_{\kappa,\lambda} D_{\kappa\lambda} S_\kappa S_\lambda$. We show in §12.3.3 that for an ion with $S = 1$ without a magnetic field any perturbation may be represented by the operator equivalent L with a symmetric matrix ($D_{\kappa\lambda} = D_{\lambda\kappa}$). It is natural to expect that we can also find spin operators equivalent to the interaction with the external magnetic field and to the magnetic hyperfine and quadrupole interactions with the nucleus, and we can in fact do so. The sum of all these spin-operator equivalents is called the spin-Hamiltonian.

It is probably helpful to anticipate and state that the expression

$$\mathscr{H}(S) = \sum_{\kappa,\lambda}(\beta g_{\kappa\lambda} H_\kappa S_\lambda + D_{\kappa\lambda} S_\kappa S_\lambda + A_{\kappa\lambda} S_\kappa I_\lambda + Q_{\kappa\lambda} I_\kappa I_\lambda) - \gamma\beta_N \mathbf{H}.\mathbf{I}, \tag{12.33}$$

is an adequate spin-operator equivalent in many, but not all, circumstances. All the constants in (12.33) are real, $D_{\kappa\lambda} = D_{\lambda\kappa}$ and $Q_{\kappa\lambda} = Q_{\lambda\kappa}$. In the case that the site has tetragonal or trigonal symmetry about the Z-axis (12.33) reduces to

$$\mathscr{H}(S) = \beta g_\parallel H_z S_z + \beta g_\perp (H_x S_x + H_y S_y) + D[S_z^2 - \tfrac{1}{3}S(S+1)]$$
$$+ A S_z I_z + B(S_x I_x + S_y I_y) + Q[I_z^2 - \tfrac{1}{3}I(I+1)] - \gamma\beta_N \mathbf{H}.\mathbf{I}. \tag{12.34}$$

We may add a constant diagonal energy to $\mathscr{H}(S)$ without making any significant difference and this constant energy has been chosen in (12.34) so that the mean energy of the set of states is zero.

The spin in an equation such as (12.33) is not the same as the true spin as applied to the perturbed kets and it is customary to refer to it as a fictitious spin. The discussion in 12.3.2 and 12.3.3 will convince the reader that it may not even be very closely related to it. As a consequence there is considerable freedom of choice for the kets $|a(M_S)\rangle$ and hence for the actual values of the constants in the spin-Hamiltonian. This problem has not been completely cleared up, but some progress has been made (Pryce, 1959). A result of importance for our calculations of g values in this chapter is that, when $g_{ij} = 0$ unless $i = j$, the values of g_{xx}^2, g_{yy}^2, g_{zz}^2 and $\det g_{ij}$ are always well defined.

12.3.2. Kramers doublets. In an odd-electron system the combined effect of spin-orbit coupling and low-symmetry fields often leaves a single Kramers doublet lying an amount large compared with βH below any other. I now show that the spin-Hamiltonian (12.33) is completely adequate to represent the matrix elements of the magnetic and quadrupole interactions within this doublet. As the Kramers doublet must be degenerate in the absence of an external magnetic field and non-zero nuclear moment, we may take $D_{\kappa\lambda} = 0$.

The proof is simple. Let ψ and ψ^* be the states of the Kramers doublet and ρ be a real linear operator (i.e. $\bar{\rho} = \rho$). First suppose $\rho = \rho^*$. Then

$$\langle\bar{\psi}\,|\,\rho\,|\,\psi\rangle = \langle\bar{\psi}\,|\,\rho\,|\,\psi\rangle^* = \langle\bar{\psi}^*\,|\,\rho\,|\,\psi^*\rangle$$

and $$\langle\bar{\psi}\,|\,\rho\,|\,\psi^*\rangle = \overline{\langle\bar{\psi}^*\,|\,\rho\,|\,\psi\rangle} = \langle\bar{\psi}^*\,|\,\rho\,|\,\psi\rangle^* = -\langle\bar{\psi}\,|\,\rho\,|\,\psi^*\rangle = 0,$$

where we have used the properties of ρ and the fact that $\bar{c} = c^*$ when c is a complex number. Hence if $\rho = \bar{\rho} = \rho^*$, ρ has a real diagonal matrix within the Kramers doublet and the two diagonal elements are equal. Therefore it can be replaced by a number as its operator equivalent, the number being equal to the diagonal elements.

We now apply this result to the quadrupole interaction. It has the form $\Sigma\rho_k V_k(\mathbf{I},\mathbf{I})$ with $\rho_k = \bar{\rho}_k = \rho_k^*$ (see (12.27)) and therefore has an operator equivalent $\Sigma a_k V_k(\mathbf{I},\mathbf{I})$, where the a_k are real numbers. On expanding this we obtain an expression of the form $\sum\limits_{\kappa,\lambda} Q_{\kappa\lambda} I_\kappa I_\lambda$ with $Q_{\kappa\lambda} = Q_{\lambda\kappa}$ and $\sum\limits_{\kappa=1}^{3} Q_{\kappa\kappa} = 0$.

Next suppose $\rho = \bar{\rho} = -\rho^*$. Then

$$\langle\bar{\psi}\,|\,\rho\,|\,\psi\rangle = \langle\bar{\psi}\,|\,\rho\,|\,\psi\rangle^* = -\langle\bar{\psi}^*\,|\,\rho\,|\,\psi^*\rangle$$

and so ρ has a matrix of the form

ρ	ψ	ψ^*
ψ	c	$a-ib$
ψ^*	$a+ib$	$-c$

with a, b and c real. There is, in general, no further restriction on the matrix of ρ. We now seek a spin-Hamiltonian with $S = \frac{1}{2}$ and basic states $|\frac{1}{2}\rangle$ correlated with ψ and $|-\frac{1}{2}\rangle$ correlated with ψ^*. The operator $2aS_x + 2bS_y + 2cS_z$ has exactly the same matrix within $|\frac{1}{2}\rangle$, $|-\frac{1}{2}\rangle$ as ρ has within ψ, ψ^*. Therefore any operator ρ satisfying $\rho = \bar{\rho} = -\rho^*$ may be replaced by a linear form in the components of **S** having real coefficients. It follows immediately from this that the forms $\Sigma\beta g_{\kappa\lambda}H_\kappa S_\lambda$ and $\Sigma A_{\kappa\lambda}S_\kappa I_\lambda$ may always be used as operators equivalent within the ground doublet to the interaction with the external magnetic field and the magnetic moment of the nucleus, respectively.

An important example of an isolated Kramers doublet occurs in the high-spin cobaltous ion in an octahedral field. The ground term is $^4T_1$ and the $^4T_{1\frac{1}{2}}$ doublet lies $\frac{1}{2}\zeta$ below the next level of the term. Therefore we may use the spin-Hamiltonian (12.33) for it. We notice that the actual spin of the ground doublet is $S = \frac{3}{2}$ but the spin to be used with the spin-Hamiltonian is $S' = \frac{1}{2}$. The latter is sometimes called a fictitious spin. This example also underlines the fact that a spin-Hamiltonian is not completely defined until the fictitious spin is given. The degeneracy of the set of states for which $\mathscr{H}(S)$ is being used is, of course, $2S'+1$.

12.3.3. Kramers triplets. If we have an even electron system with, in the absence of magnetic fields, three near degenerate states lying lowest, ψ_1, ψ_2, ψ_3 say, and satisfying $\psi_i^* = \sum_{j=1}^{3} a_{ji}\psi_j$ for each i then we may reasonably call (ψ_1, ψ_2, ψ_3) a Kramers triplet. Ex. 5, p. 208, shows that we may, without loss of generality, suppose $\psi_i^* = \psi_i$ for each i.

Just as in the last section we determine the implicit restrictions on the operator ρ. There is no extra complication here and we deduce that if $\rho_1 = \bar{\rho}_1 = \rho_1^*$ and $\rho_2 = \bar{\rho}_2 = -\rho_2^*$ their matrices have the form

ρ_1	ψ_1	ψ_2	ψ_3
ψ_1	a	d	e
ψ_2	d	b	f
ψ_3	e	f	c

ρ_2	ψ_1	ψ_2	ψ_3
ψ_1	0	$i\alpha$	$i\beta$
ψ_2	$-i\alpha$	0	$i\gamma$
ψ_3	$-i\beta$	$-i\gamma$	0

where $a, b, c, d, e, f, \alpha, \beta$ and γ are real.

The spin-Hamiltonian is chosen to have $S = 1$ and states $|M_S\rangle$ which are eigenstates of S_z. ψ_1, ψ_2, ψ_3 get correlated with $-\frac{1}{2}\sqrt{2}\,|1\rangle+\frac{1}{2}\sqrt{2}\,|-1\rangle$, $|0\rangle$, $\frac{1}{2}i\sqrt{2}\,|1\rangle+\frac{1}{2}i\sqrt{2}\,|-1\rangle$, respectively. It is a matter of simple computation to show ρ_1 can be replaced by $\Sigma a_{\kappa\lambda}S_\kappa S_\lambda$ and ρ_2 by $\Sigma b_\kappa S_\kappa$, where $a_{\kappa\lambda} = a_{\lambda\kappa}$ and both the $a_{\kappa\lambda}$ and the b_κ are real. It then follows at once that $\mathscr{H}(S)$ of (12.33), with $S = 1$, is an adequate spin-Hamiltonian for a Kramers triplet, apart from the nuclear quadrupole interaction. For comparison with experiment we only calculate the latter for Kramers doublets and so do not consider it further.

This method of obtaining spin-Hamiltonians may be extended to higher degeneracies but then, in general, we shall need higher powers of the components of **S**. If the degeneracy is fourfold for an odd-electron system, i.e. a pair of Kramers doublets, the quadratic form $\Sigma D_{\kappa\lambda}S_\kappa S_\lambda$ is still adequate but, in general, we need higher powers of the S_κ for the two magnetic interactions. The reader will find it good practice to verify these assertions.

Examples

1. Show that any eigenstate ψ of the Hamiltonian $\mathscr{H}$ for an even-electron system with no magnetic fields present either satisfies $\psi^* = e^{i\alpha}\psi$ with α a real number or the pair (ψ, ψ^*) are linear combinations of an orthonormal pair of real functions f_1, f_2 which are each eigenstates of $\mathscr{H}$ with the same eigenvalues.

(This shows that our assumption in the present section that ψ_1, ψ_2, ψ_3 form a Kramers triplet is unnecessary. We can prove it.)

2. V_k of (12.4) are symmetric to exchange of $\mathbf{T}_1$, $\mathbf{T}_2$ and are related in an obvious way to d-functions. Define the seven quantities $W_k(\mathbf{T}_1, \mathbf{T}_2, \mathbf{T}_3)$ which correspond in the same way to f-functions. Show that the first-order interaction of an external magnetic field with a Kramers quartet may always be represented by

$$\beta \sum_{\kappa\lambda} g_{\kappa\lambda} H_\kappa S_\lambda + \beta \sum h_{\kappa\lambda} H_\kappa W_\lambda(\mathbf{S}, \mathbf{S}, \mathbf{S}).$$

12.3.4. The use of perturbation theory. We now treat the spin-orbit coupling as a small perturbation and derive the spin-Hamiltonian for a spatially non-degenerate ground state, but for general S, correct to first order in the modification of the ground states by the spin-orbit coupling. The main terms in the true Hamiltonian may be written

$$\mathscr{H} = \mathscr{H}_0 + \mathscr{H}_1 + \mathscr{H}_2,$$

where $\mathscr{H}_0$ is the electrostatic interaction between and among the electrons and nuclei, $\mathscr{H}_1$ is the spin-orbit coupling and $\mathscr{H}_2$ is the sum of the other magnetic interactions and the nuclear quadrupole interactions.

The zero-order states are eigenstates of $\mathscr{H}_0$. We take them also as eigenstates of $\mathbf{S}^2$ and S_z (which commute with $\mathscr{H}_0$) and write then $|Mj\rangle$. Here M is the eigenvalue of S_z and j is a parameter which includes the specification of the eigenvalue of $\mathbf{S}^2$. $|M0\rangle$ give the unperturbed ground states and we let their spin be S. It is convenient also to define the effect of the Kramers operator and we do this so that $|Mj\rangle^* = (-1)^M|-Mj\rangle$. This is always possible by ex. 6, p. 208. The first-order modification to the $|M0\rangle$ is

$$|\alpha M0\rangle = |M0\rangle - \sum_{j, M'} \frac{\langle M'j|\,\mathscr{H}_1\,|M0\rangle\,|M'j\rangle}{E_j}, \tag{12.35}$$

and our energies are obtained by evaluating $\mathscr{H}$ within the states $|\alpha M0\rangle$ neglecting terms of order $\mathscr{H}_1^3$ or $\mathscr{H}_1^2\mathscr{H}_2$. It is convenient to define the arbitrary additive constant energy in such a way that $\langle M0|\,\mathscr{H}_0\,|M0\rangle = 0$. (For justification of this perturbation procedure, see Appendix 3.)

We first consider the interaction with the external magnetic field. This is $\beta\mathbf{H}.(\mathbf{L}+2\mathbf{S})$. If L_a is a component of $\mathbf{L}$, then

$$\langle M0|\,L_a\,|M0\rangle = \langle M0|\,L_a\,|M0\rangle^* = -\langle -M0|\,L_a\,|-M0\rangle = -\langle M0|\,L_a\,|M0\rangle = 0.$$

So the zero-order contribution is simply the matrix of $2\beta\mathbf{H}.\mathbf{S}$. The first-order contribution arises from the matrix elements

$$\langle M0|\,(L_a+2S_a)\,|M'j\rangle = \langle M0|\,L_a\,|M'j\rangle$$

because our zero-order states are eigenstates of $\mathbf{S}^2$ and S_z. This latter matrix element is also zero unless the spin S' of the excited states is the same as that of the ground state and also $M = M'$. Write $\langle M0 \,|\, L_a \,|\, Mj \rangle = i\Lambda_a^j$. Λ_a^j is independent of M. Further $(i\Lambda_a^j)^* = \langle M0 \,|\, L_a \,|\, Mj \rangle^* = -\langle -M0 \,|\, L_a \,|\, -Mj \rangle = -i\Lambda_a^j$, and so Λ_a^j is real.

We now make the assumption that the $|M0\rangle$ form part of the ground term of the free ion. In this case the only excited states of the same multiplicity, having non-zero matrix elements of $\mathbf{L}$ with the ground states, also belong to the ground term. Therefore we may use the special form $\lambda\mathbf{L}\,.\,\mathbf{S}$ for the spin-orbit coupling energy. Then for these excited states

$$\sum_{M'} \langle M'j \,|\, \mathscr{H}_1 \,|\, M0 \rangle \,|M'j\rangle = \lambda \sum_{M'} \langle M'j|\,(L_z S_z + \tfrac{1}{2}L^+S^- + \tfrac{1}{2}L^-S^+)\,|M0\rangle\,|M'j\rangle$$

$$= -i\lambda\Lambda_z^j M\,|Mj\rangle - \tfrac{1}{2}i\lambda(\Lambda_x^j + i\Lambda_y^j)\,[S(S+1) - M(M-1)]^{\frac{1}{2}}\,|M-1j\rangle$$

$$-\tfrac{1}{2}i\lambda(\Lambda_x^j - i\Lambda_y^j)\,[S(S+1) - M(M+1)]^{\frac{1}{2}}\,|M+1j\rangle$$

$$= -i\lambda\Lambda_z^j S_z\,|Mj\rangle - \tfrac{1}{2}i\lambda(\Lambda_x^j + i\Lambda_y^j)\,S^-\,|Mj\rangle$$

$$-\tfrac{1}{2}i\lambda(\Lambda_x^j - i\Lambda_y^j)\,S^+\,|Mj\rangle$$

$$= -i\lambda\mathbf{\Lambda}^j\,.\,\mathbf{S}\,|Mj\rangle, \tag{12.36}$$

and

$$|\alpha M0\rangle = |M0\rangle + i\lambda \sum_j E_j^{-1}\mathbf{\Lambda}^j\,.\,\mathbf{S}\,|Mj\rangle. \tag{12.37}$$

In (12.37) the components of the vector $\mathbf{\Lambda}^j$ are just numbers. Evaluating $\beta \sum_a H_a L_a$ to first order among the $|\alpha M0\rangle$ of (12.37) we obtain two equal expressions, one being

$$i\lambda\beta \sum_j E_j^{-1} \sum_{a,b} \langle M'0 \,|\, H_a L_a \Lambda_b^j S_b \,|\, Mj \rangle$$

$$= i\lambda\beta \sum_j E_j^{-1} \sum_{a,b} H_a \Lambda_b^j \langle M'0 \,|\, S_b \,|\, M0 \rangle \langle M0 \,|\, L_a \,|\, Mj \rangle$$

$$= -\lambda\beta \sum_j E_j^{-1} \sum_{a,b} H_a \Lambda_b^j \Lambda_a^j \langle M'0 \,|\, S_b \,|\, M0 \rangle.$$

This shows that the first-order interaction with the external magnetic field may be represented in the spin-Hamiltonian by $\beta \sum_{a,b} g_{ab} H_a S_b$, where

$$g_{ab} = 2\delta_{ab} - 2\lambda \sum_j \frac{\Lambda_a^j \Lambda_b^j}{E_j}. \tag{12.38}$$

This result and the concept of the spin-Hamiltonian is due to Pryce (1950*b*). Pryce also derived expressions for the other terms in the spin-Hamiltonian with the assumption that only the excited states from the same free-ion term as the ground states need be considered. The quadratic form $\Sigma D_{\kappa\lambda} S_\kappa S_\lambda$ and the form $\Sigma A_{\kappa\lambda} S_\kappa I_\lambda$ follow at once from (12.37) using the expressions $\mathscr{H}_0 + \lambda\mathbf{L}\,.\,\mathbf{S}$ and the magnetic hyperfine operator equivalent given in (12.23).

We can establish the form (12.33), apart from the quadrupole term, to our present order in perturbation theory without Pryce's restricting assumption.

For excited states having $S' = S$, this can be done by a slight adaptation of the preceding proof. We do this first. In (12.36), $\mathscr{H}_1$ can no longer be replaced by $\lambda\mathbf{L}\,.\,\mathbf{S}$ but we must use $\zeta\Sigma\mathbf{l}_\kappa\,.\,\mathbf{s}_\kappa$. So we require matrix elements like

$$\langle M'j\,|\,l_{\kappa z}s_{\kappa z}\,|\,M0\rangle = \sum_k \langle M'j\,|\,l_{\kappa z}\,|\,M'k\rangle\langle M'k\,|\,s_{\kappa z}\,|\,M0\rangle$$
$$= \sum_k \langle M'j\,|\,l_{\kappa z}\,|\,M'k\rangle\langle k\,\vdots\,s_\kappa\,\vdots\,0\rangle M\delta_{MM'},$$

where we have used the fact that $\mathbf{s}_\kappa$ is of type T with respect to $\mathbf{S}$. Similarly for $l_\kappa^+ s_\kappa^-$ and $l_\kappa^- s_\kappa^+$. But now we can derive (12.37) again and the only difference is in the formulae for the real numbers Λ_a^j. For this reason Λ_b^j but not Λ_a^j is modified in (12.38). The derivation of $\Sigma D_{\kappa\lambda}S_\kappa S_\lambda$ and $\Sigma A_{\kappa\lambda}S_\kappa I_\lambda$ carries over also and is straightforward but is a little more complicated. Let us illustrate the method briefly for the part of the magnetic hyperfine structure arising from matrix elements between $|M0\rangle$ and $|M'j\rangle$. Apart from $P\mathbf{L}\,.\,\mathbf{I}$ the operator for the hyperfine structure is a sum of terms of the form $\Sigma\rho_{kl}s_k I_l$, where ρ_{kl} is real and a function of space alone and s_k is the kth component of a spin vector for a single electron. (12.37) may be written

$$|\alpha M0\rangle = |M0\rangle + i\Sigma a_m^j S_m\,|Mj\rangle$$

and then a typical contribution to the hyperfine interaction is

$$i\sum_{klm} a_m^j I_l\langle M'0\,|\,\rho_{kl}s_k S_m\,|\,Mj\rangle - i\sum_{klm} a_m^j I_l\langle M'j\,|\,S_m\rho_{kl}s_k\,|\,M0\rangle$$
$$= i\sum_{klm} a_m^j I_l\langle M'0\,|\,\rho_{kl}[s_k, S_m]\,|\,Mj\rangle,$$

where we have exchanged 0 and j in the second term by a double expansion and the properties of the matrix elements of $\rho_{\kappa\lambda}$ and s_κ which follow from $\rho_{\kappa\lambda}^* = \rho_{\kappa\lambda}$ and $s_\kappa^* = -s_\kappa$ (i.e. $\langle j\,\vdots\,s_\kappa\,\vdots\,p\rangle = \langle p\,\vdots\,s_\kappa\,\vdots\,j\rangle$). The commutator $[s_k, S_m]$ is zero or a pure imaginary multiple of a component of $\mathbf{s}$, which can then be replaced by the corresponding component of $\mathbf{S}$ as its operator equivalent.

As $\mathbf{L}+2\mathbf{S}$ has no matrix elements between states with differing multiplicity we have established the expression $\beta\Sigma g_{\kappa\lambda}H_\kappa S_\lambda$ completely to our order of approximation. The second and third terms of (12.33), however, need also contributions from excited terms with spins $S' = S \pm 1$.

It is possible to show that the excited terms with spins $S' \neq S$ give contributions which can also be represented by quadratic forms in the S_κ and I_λ. I will merely show how this can be done for the second-order contribution to the spin-orbit coupling energy.[1] This is given by

$$\langle \alpha M0\,|\,\mathscr{H}_1\,|\,\alpha M'0\rangle = -\sum_{j,\,M''} E_j^{-1}\langle M0\,|\,\mathscr{H}_1\,|\,M''j\rangle\langle M''j\,|\,\mathscr{H}_1\,|\,M'0\rangle.$$

The contribution for a particular j is $-E_j^{-1}X_{MM'}$ where

$$X_{MM'} = \sum_{M''}\langle M0\,|\,\mathscr{H}_1\,|\,M''j\rangle\langle M''j\,|\,\mathscr{H}_1\,|\,M'0\rangle$$

[1] For more detail, see Griffith (1960), *Molecular Physics*, **3**, 79.

and we evaluate $X_{MM'}$. If we can then find an operator equivalent whose matrix elements within the $|M0\rangle$ are the $X_{MM'}$ we have merely to multiply by $-E_j^{-1}$ and sum over j to obtain an operator equivalent to $\mathscr{H}_1$.

Write $l^{\pm} = \frac{1}{2}(l_x \pm il_y)$, $s^{\pm} = s_x \pm is_y$ and $l^0 = l_z$, $s^0 = s_z$ and let the affixes μ, ν stand for $+$, $-$ or 0. Then we have

$$\begin{aligned}\langle M'0 \,|\, \mathscr{H}_1 \,|\, M''j\rangle &= \sum_{\kappa,\nu} \langle M'0 \,|\, \xi_\kappa l_\kappa^{-\nu} s_\kappa^{\nu} \,|\, M''j\rangle \\ &= \sum_{\kappa,\nu,n} \langle M'0 \,|\, \xi_\kappa l_\kappa^{-\nu} \,|\, M'n\rangle \langle M'n \,|\, s_\kappa^{\nu} \,|\, M''j\rangle \\ &= \sum_{\kappa,\nu,n} \langle M'0 \,|\, \xi_\kappa l_\kappa^{-\nu} \,|\, M'n\rangle \langle n \,\vdots\, s_\kappa \,\vdots\, j\rangle f(SM', \nu, S'M'') \\ &= \sum_{\nu} Y_{-\nu} f(SM', \nu, S'M''), \end{aligned} \tag{12.39}$$

where we have used (2.80), $-\nu = +,\ 0,\ -$ when $\nu = -,\ 0,\ +$, respectively. Using the star operator we deduce $\overline{Y}_{-\nu} = Y_\nu$ for $S' = S \pm 1$. $f(SM', \nu, S'M'')$ is the coefficient of $\langle S \,\vdots\, T \,\vdots\, S'\rangle$ in the equation for $\langle SM' \,|\, T_\nu \,|\, S'M''\rangle$ among (2.80). Then

$$X_{MM'} = \sum_{\mu,\nu,M''} Y_{-\mu} Y_\nu f(SM, \mu, S'M'') f(SM', \nu, S'M''),$$

and it is a mere matter of putting in the values of the f to show that $X_{MM'}$ can be obtained from an operator equivalent $\sum_{\kappa,\lambda} D_{\kappa\lambda} S_\kappa S_\lambda$.

I illustrate this for $S' = S+1$. Both $X_{MM'}$ and our suggested operator equivalent vanish when $|M-M'| > 2$. Let $M' = M+2$. Then the only non-vanishing term in $X_{MM'}$ arises from $M'' = M+1$ and is

$$\begin{aligned} X_{M,\,M+2} &= Y_+^2 f(SM, -, S+1, M+1) f(S, M+2, +, S+1, M+1) \\ &= -Y_+^2 \{(S+M+1)(S+M+2)(S-M-1)(S-M)\}^{\frac{1}{2}}. \end{aligned}$$

Evidently $-Y_+^2 (S^-)^2$ is an operator equivalent with the matrix elements $X_{MM'}$ when $M' = M+2$ and with zero elements otherwise. It is convenient to define the real vector $\mathbf{Y}$ from the equations $Y_x \pm iY_y = 2Y_\pm$, $Y_0 = Y_z$ and then, in the notation of 12.2.1, $-Y_+^2(S^-)^2 = -\frac{1}{2} V_2'(\mathbf{Y}, \mathbf{Y})\, \overline{V_2'}(\mathbf{S}, \mathbf{S})$.

Next let $M' = M+1$. Here we have two non-zero terms in $X_{MM'}$, corresponding to $M'' = M$ or $M+1$. Then

$$\begin{aligned} X_{M,\,M+1} &= Y_0 Y_+ f(SM, 0, S+1, M) f(S, M+1, +, S+1, M) \\ &\quad + Y_+ Y_0 f(SM, -, S+1, M+1) f(S, M+1, 0, S+1, M+1) \\ &= -Y_0 Y_+ (2M+1) \{(S+M+1)(S-M)\}^{\frac{1}{2}}. \end{aligned}$$

Here our operator equivalent is

$$-Y_0 Y_+ (S_z S^- + S^- S_z) = -\tfrac{1}{2} V_4'(\mathbf{Y}, \mathbf{Y})\, \overline{V_4'}(\mathbf{S}, \mathbf{S}).$$

When $M' = M$ there are three terms in the sum and after a certain amount of manipulation we find as our equivalent

$$-\tfrac{1}{2} V_1'(\mathbf{Y}, \mathbf{Y})\, \overline{V_1'}(\mathbf{S}, \mathbf{S}) + \tfrac{1}{3}(S+1)(2S+3)\, \mathbf{Y}^2.$$

Finally, for $M' < M$ we may deduce the expressions by barring the matrix elements we already possess. We now replace the V'_k by the V_k and have

$$-\tfrac{1}{2}\sum_{k=1}^{5} V_k(\mathbf{Y},\mathbf{Y})\,V_k(\mathbf{S},\mathbf{S})+\tfrac{1}{3}(S+1)\,(2S+3)\,\mathbf{Y}^2$$

as a symmetric quadratic operator equivalent with real coefficients which has the $X_{MM'}$ as matrix elements. This easily simplifies, using (12.7), to

$$\frac{S+1}{S}\mathbf{Y}^2\mathbf{S}^2-(\mathbf{Y}\,.\,\mathbf{S})^2.$$

12.3.5. Discussion of the spin-Hamiltonian. We have now established (12.33) in a number of cases. (For a more detailed discussion of these matters, see the paper referred to in the footnote on p. 336.) It is well to emphasize again that it is not true, in general, except for Kramers doublets of odd-electron systems and, apart from the quadrupole interaction, for triply degenerate states. In practice the deviations from (12.33) for the two magnetic interactions have not been definitely observed although for $S' \geqslant \frac{3}{2}$ they will of course be there. The term $\Sigma D_{\kappa\lambda}S_\kappa S_\lambda$, which is called the fine structure term, is adequate for $S' \leqslant \frac{3}{2}$ but not for larger S'. The inadequacy of most interest occurs when $S' = \frac{5}{2}$, in the sextet ground term of the high-spin Mn^{++} or Fe^{3+} ion. In an octahedral field the ground term belongs to $E''+U'$ and E'' will, in general, have a different energy from U'. But in an octahedral field $\Sigma D_{\kappa\lambda}S_\kappa S_\lambda$ becomes $DS(S+1)$ and cannot separate E'' and U'. The simplest expression which can is $S_x^4+S_y^4+S_z^4$ and this appears in the spin-Hamiltonian as

$$\tfrac{1}{6}a(S_x^4+S_y^4+S_z^4-\tfrac{707}{16}),$$

the notation being conventional. I do not discuss these higher power spin-Hamiltonian components further in this book but mentioned them for completeness (see Bleaney & Stevens, 1953).

In the next two subsections we briefly discuss the experimental significance of the terms in $\mathscr{H}(S)$ and then return to the theoretical calculation of $\mathscr{H}(S)$.

Example

Put $W_x = 5S_x^3-3S(S+1)\,S_x$, $W_y = 5S_y^3-3S(S+1)\,S_y$, $W_z = 5S_z^3-3S(S+1)\,S_z$. Show that a sufficiently general spin-Hamiltonian for a Kramers quartet in an octahedral field, corresponding to the first three terms of (12.33), is

$$\mathscr{H}(S) = \beta g\mathbf{H}\,.\,\mathbf{S}+\beta h\mathbf{W}\,.\,\mathbf{H}+A\mathbf{S}\,.\,\mathbf{I}+A'\mathbf{W}\,.\,\mathbf{I}.$$

12.3.6. The fine structure. We return to our ion with $S = 1$ and suppose it is placed in a field tetragonal about the OZ-axis and that its nucleus has $I = 0$. Then the spin-Hamiltonian is

$$\mathscr{H}(S) = D(S_z^2-\tfrac{1}{3}S(S+1))+g\beta\mathbf{H}\,.\,\mathbf{S}. \qquad (12.40)$$

Writing $H^{\pm} = H_x \pm iH_y$, the matrix of $\mathscr{H}(S)$ is

$\mathscr{H}(S)$	1	0	-1
1	$\frac{1}{3}D+g\beta H_z$	$\frac{1}{2}\sqrt{2}\,g\beta H^-$	0
0	$\frac{1}{2}\sqrt{2}\,g\beta H^+$	$-\frac{2}{3}D$	$\frac{1}{2}\sqrt{2}\,g\beta H^-$
-1	0	$\frac{1}{2}\sqrt{2}\,g\beta H^+$	$\frac{1}{3}D-g\beta H_z$

If **H** is along the direction (θ, ϕ) we have

$$\mathbf{H} = (H \sin\theta \cos\phi,\ H\sin\theta \sin\phi,\ H\cos\theta)$$

and then the secular equation for the energies is derived from the matrix for $\mathscr{H}(S)$ as

$$(E-\tfrac{1}{3}D)^2(E+\tfrac{2}{3}D)-\beta^2g^2H^2(E-\tfrac{1}{3}D)\sin^2\theta$$
$$-\beta^2g^2H^2(E+\tfrac{2}{3}D)\cos^2\theta = 0. \qquad (12.41)$$

If the incident radiation has frequency $h\nu$ resonance is possible for those values of H for which there exist a pair of roots of (12.41) differing by $h\nu$. The relative intensities are then derived from (12.2) and a consideration of Boltzmann factors. The latter are only relevant at low temperatures because $h\nu$ is the same for each resonance and hence the relative Boltzmann factors of ground and excited states are the same for each resonance.

The computation becomes very simple for the two special angles $\theta = 0$ and $\theta = \frac{1}{2}\pi$ for the external field. For these angles we can solve (12.41) explicitly to obtain

$$\left.\begin{aligned} &E_1(0) = \tfrac{1}{3}D-g\beta H, && E_1(\tfrac{1}{2}\pi) = -\tfrac{1}{6}D-\tfrac{1}{2}(D^2+4g^2\beta^2H^2)^{\frac{1}{2}},\\ &E_2(0) = -\tfrac{2}{3}D, && E_2(\tfrac{1}{2}\pi) = \tfrac{1}{3}D,\\ &E_3(0) = \tfrac{1}{3}D+g\beta H, && E_3(\tfrac{1}{2}\pi) = -\tfrac{1}{6}D+\tfrac{1}{2}(D^2+4g^2\beta^2H^2)^{\frac{1}{2}}. \end{aligned}\right\} \qquad (12.42)$$

These solutions are plotted schematically as functions of H in Figs. 12.1*b* and *c*, with the resonance spectrum plotted below as a full line when the radio-frequency magnetic field is perpendicular to the static magnetic field and as a dotted line when it is parallel to it. The reader may care to verify the existence of a non-zero intensity for the dotted resonance in Fig. 12.1*c* and show that it tends to zero with D. The main features of fine structure are now clear from Fig. 12.1: a zero-field splitting; a complicated behaviour when $g\beta H$ is comparable with the zero-field splitting; a spectrum for $h\nu$ large or D small practically the same as without any zero-field splitting, and finally the breakdown of the selection rules appropriate for no zero-field splitting. The last would have been more complete had we chosen a lower site symmetry or a general angle for θ.

12.3.7. The hyperfine structure. We illustrate this for the isotropic Hamiltonian

$$\mathscr{H}(S) = g\beta\mathbf{H}\,.\,\mathbf{S}+A\mathbf{I}\,.\,\mathbf{S},$$

and at first arbitrary electronic spin S and nuclear spin I. Then there are $(2S+1)(2I+1)$ basic states $|M_S M_I\rangle$ which for **H** parallel to OZ are eigenstates of $\mathbf{H}\,.\,\mathbf{S}$ but not of $\mathbf{I}\,.\,\mathbf{S}$. However, A is usually so small compared with $g\beta H$ that we may neglect the matrix elements of $A\mathbf{I}\,.\,\mathbf{S}$ between states differing in their M_S values. This is equivalent to replacing $A\mathbf{I}\,.\,\mathbf{S}$ by AI_zS_z and the energy of $|M_S M_I\rangle$ becomes $g\beta HM_S+AM_SM_I$. The allowed transitions are now those for which $\Delta M_S = \pm 1$, $\Delta M_I = 0$. (The neglected interaction $-\gamma\beta_N\mathbf{H}\,.\,\mathbf{I}$ has a completely negligible effect on intensities.) The energy of the transition

$$|M_S M_I\rangle \to |M_{S+1} M_I\rangle \quad \text{is} \quad g\beta H+AM_I$$

and therefore each resonance line splits up into $(2I+1)$ hyperfine components. This is illustrated for $S = I = \frac{1}{2}$ in Fig. 12.2.

The effect of the quadrupole interaction has been analysed in only a very few cases. Consider the Hamiltonian (12.34) appropriate to axial symmetry. Then if **H** is parallel to OZ the quadrupole interaction changes the energy of $|M_S M_I\rangle$ to $g\beta HM_S + AM_S M_I + Q[M_I^2 - \frac{1}{3}I(I+1)]$ but does not affect the energies of the allowed transitions because for these $\Delta M_I = 0$. When **H** is in any other direction, there is competition between the demand of the magnetic hyperfine interaction

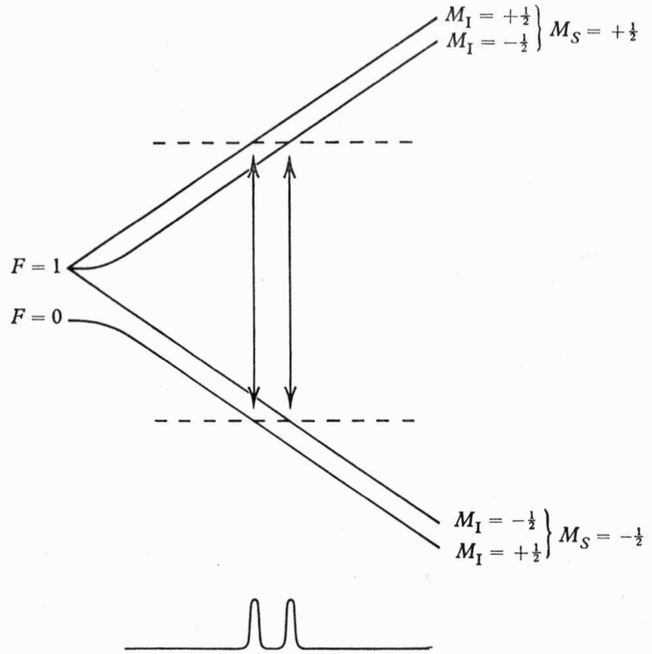

Fig. 12.2. Hyperfine structure for a spatially non-degenerate ion with $S = I = \frac{1}{2}$ and $A > 0$, plotted as a function of H.

that the nuclear states should be quantized with reference to the direction of the magnetic moment of the electrons and hence to that of **H** and that of the quadrupole interaction that they should be quantized along OZ. The stationary states are then intermediate between those two limits, the behaviour is in general complicated (as a matter of solving secular equations, not in principle) and the selection rule $\Delta M_I = 0$ is no longer valid. The effects here are small but enable one, in theory and sometimes in practice, to obtain Q.

12.4. Survey of electron resonance measurements

In this section we make a general survey of electron resonance using as simple a version of the theory as it is at all plausible and trying to look at examples of all the main points of interest. The majority of compounds have a symmetry which is predominantly octahedral with a smaller admixture of lower symmetry components. The spin-Hamiltonian parameters are usually very sensitive to this admixture and its presence is often necessary in order to secure even approximate agreement with experiment. We cannot make a reliable *a priori* calculation of the parameters characterizing these lower symmetry fields and therefore, to some extent, our interpretations are a mere translation of one set of incalculable numbers into another. However, we do really do more than this and so I enumerate here some of these satisfactory features of our theoretical interpretation: the interpretation is seldom, if ever, forced and provides a unified and internally consistent picture of the electronic structure of all transition-metal ions (and in fact also of rare earth ions)—a picture which is broadly consistent with the optical spectra, the thermodynamic properties and the bulk paramagnetic susceptibilities; the parameters characterizing the low-symmetry fields turn out to have values plausible on physical grounds; using optical data we can make *a priori* calculations for those ions for which, on theoretical grounds, we should expect low-symmetry fields to have little effect (octahedral d^3, d^5, d^8, for example) and for these theory usually agrees well with experiment, while the exceptions can be explained convincingly by the assumption of an unusually strong low-symmetry field; for other ions we can often make *a priori* calculations of some constants on the basis of a general and plausible hypothesis about the nature of the ground state (e.g. Cr^{++}) and obtain fair agreement with experiment; finally the theory often predicts relationships between constants and ranges outside which constants cannot lie, in general agreement with experiment (see, for example, Co^{++} in § 12.4.8 and Fig. 12.10).

It is convenient to give here a general definition of the names of the parameters we shall use in describing a tetragonal field. We put

$$\left.\begin{aligned} E(\zeta) &= E(d_{xy}) = -\tfrac{2}{5}\Delta - \tfrac{2}{3}\delta, \\ E(\xi,\eta) &= E(d_{yz}, d_{zx}) = -\tfrac{2}{5}\Delta + \tfrac{1}{3}\delta, \\ E(\theta) &= E(d_{z^2}) = \tfrac{3}{5}\Delta - \tfrac{1}{2}\mu, \\ E(\epsilon) &= E(d_{x^2-y^2}) = \tfrac{3}{5}\Delta + \tfrac{1}{2}\mu, \end{aligned}\right\} \tag{12.43}$$

and shall use these symbols for the relative energies although we will feel free to choose the centre of gravity appropriately to the particular problem concerned. The orbital pattern (12.43) is illustrated in Fig. 12.3.

In learning about electron resonance I have found the article by Abragam & Pryce (1951*a*) and the reviews of Bleaney & Stevens (1953) and Bowers & Owen (1955) especially helpful. Data quoted in this chapter without reference come from the latter review.

12.4.1. d^9: copper,[1] Cu^{++}. The ground term of the cupric ion is $d^9\,{}^2E$ in an octahedral field and in a tetragonal field this breaks up into two terms ${}^2A_1 + {}^2B_1$ (Table A 11). We use the notation for the groups D_4 and D_4^* in this section and first note that the spin-orbit coupling energy has zero matrix elements within and between ${}^2A_1$ and ${}^2B_1$ because they are derived from ${}^2E$ of the octahedral group. ${}^2A_1$ and ${}^2B_1$ form respectively the representations E' and E'' of the spinor group D_4^*.

It is convenient to work in the complementary scheme and treat the ion as a single hole in the d^{10} shell. Then we evaluate the g values and hyperfine constants

[1] Abragam & Pryce (1951*b*); Bleaney, Bowers & Ingram (1955); Bleaney, Bowers & Trenam (1955); Bleaney, Bowers & Pryce (1955).

for d^1 and change the signs of ζ and q, but not of A or B. Because the Hamiltonian commutes with all elements of D_4^*, the two possible ground doublets form bases for E' and E'' when calculated correct to any order of perturbation

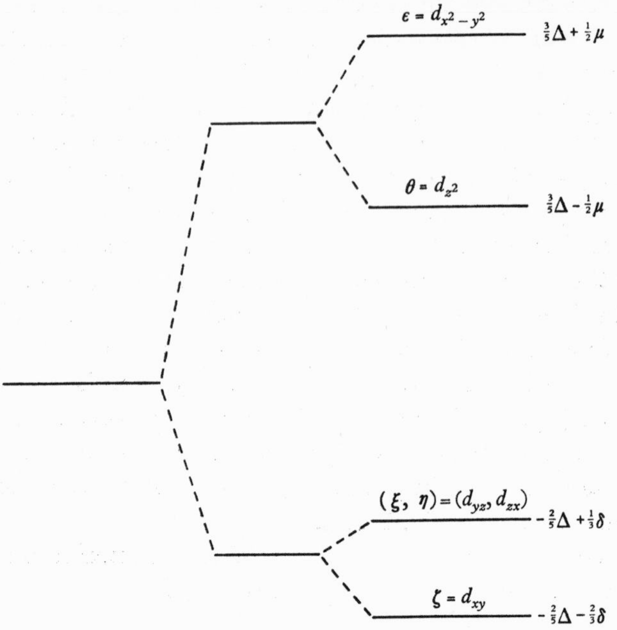

Fig. 12.3. The orbital pattern for a single d-electron in a field of tetragonal symmetry.

theory. The perturbation we consider here is the mixing in of the excited doublets of d^1 by the spin-orbit coupling energy. Having made our calculation of the ground doublet to a suitable order in perturbation theory we then evaluate

$$\mathscr{H} = \beta\mathbf{H}\,.\,(\mathbf{L}+2\mathbf{S})+P(\mathbf{l}\,.\,\mathbf{I}-\kappa\mathbf{s}\,.\,\mathbf{I}+\tfrac{1}{7}(4\mathbf{s}\,.\,\mathbf{I}-(\mathbf{l}\,.\,\mathbf{s})(\mathbf{l}\,.\,\mathbf{I})-(\mathbf{l}\,.\,\mathbf{I})(\mathbf{l}\,.\,\mathbf{s})))$$
$$+\tfrac{1}{7}q'\sum_{m=1}^{5} V_m(\mathbf{l},\mathbf{l})\,V_m(\mathbf{I},\mathbf{I}), \quad (12.44)$$

for that doublet (see (12.21) and (12.29)) and match this with a spin-Hamiltonian

$$\mathscr{H}(S) = \beta g_{\parallel}H_zS_z+\beta g_{\perp}(H_xS_x+H_yS_y)+AI_zS_z+B(I_xS_x+I_yS_y)$$
$$+Q[I_z^2-\tfrac{1}{3}I(I+1)]. \quad (12.45)$$

We proved in §12.3.2 that this is always possible, and we naturally associate $|E'\alpha'\rangle$ with $|\frac{1}{2}\rangle$, $|E'\beta'\rangle$ with $|-\frac{1}{2}\rangle$ for the E' doublet and $|E''\alpha''\rangle$ with $|\frac{1}{2}\rangle$, $E''\beta''\rangle$ with $|-\frac{1}{2}\rangle$ for the E'' doublet. Let us examine this in more detail. Take the interaction with the external magnetic field first. L_z+2S_z and S_z each transform as A_2, therefore as the symmetrized square $[E'^2]$ contains A_2 just once, the replacement theorem shows that the matrix of L_z+2S_z within E' is proportional under our correspondence to that of S_z within $|\frac{1}{2}\rangle, |-\frac{1}{2}\rangle$. (L_x+2S_x, L_y+2S_y) transform as E which again occurs just once in $[E'^2]$. Therefore their matrix elements are proportional to those of (S_x, S_y) within $|\frac{1}{2}\rangle, |-\frac{1}{2}\rangle$. The proportionality also holds for E'' because if $f \in B_1$ then the pair $|fE''\alpha''\rangle$, $|fE''\beta''\rangle$ form respectively the α', β' components of an E' doublet. But $f^2 \in A_1$ and therefore as f appears twice in any matrix element the result is established. Exactly the same argument applies to the nuclear magnetic hyperfine interaction.

For the quadrupole interaction we remark that as the $V_m(\mathbf{1},\mathbf{1})$ are real they have non-zero matrix elements within the ground doublet only when they are part of a basis for the antisymmetrized square $(E'^2) = (E''^2) = A_1$. But the V_m have $L = 2$ and therefore (Table A 15) contain A_1 only once. Actually $V_1 = (1/\sqrt{3})(3l_z^2-6)$ belongs to A_1 and therefore the only part of the quadrupole term of $\mathscr{H}$ which can have non-zero matrix elements is

$$\mathscr{H}' = \tfrac{1}{7}q'.\frac{1}{\sqrt{3}}(3l_z^2-6).\frac{1}{\sqrt{3}}[3I_z^2-I(I+1)]$$

and therefore
$$Q = -\tfrac{3}{7}q'\langle\overline{\psi}|\,(l_z^2-2)\,|\psi\rangle, \tag{12.46}$$

where $|\psi\rangle$ is the electronic part of any state of the doublet concerned and we have changed the sign of Q because we are really concerned with d^9. It also follows from our remarks about the other operators in $\mathscr{H}$ that if we define $g_{L\parallel}$, $g_{L\perp}$ and $g_{S\parallel}$, $g_{S\perp}$ to be the parts of g which arise from **L** and **S**, respectively, in (12.44) we have

$$\left.\begin{aligned} g_\parallel &= g_{L\parallel}+g_{S\parallel}, \qquad g_\perp = g_{L\perp}+g_{S\perp},\\ g_{L\parallel} &= 2\langle E'\alpha'\,|\,L_z\,|\,E'\alpha'\rangle, \quad g_{S\parallel} = 4\langle E'\alpha'\,|\,S_z\,|\,E'\alpha'\rangle,\\ g_{L\perp} &= \langle E'\beta'\,|\,L^-\,|\,E'\alpha'\rangle, \quad g_{S\perp} = 2\langle E'\beta'\,|\,S^-\,|\,E'\alpha'\rangle,\\ A &= P\{g_{L\parallel}-\tfrac{1}{2}\kappa g_{S\parallel}+\tfrac{2}{7}\langle E'\alpha'|\,[4s_z-(\mathbf{l}\,.\,\mathbf{s})\,l_z-l_z(\mathbf{l}\,.\,\mathbf{s})]\,|E'\alpha'\rangle\},\\ B &= P\{g_{L\perp}-\tfrac{1}{2}\kappa g_{S\perp}+\tfrac{1}{7}\langle E'\beta'|\,[4s^- -(\mathbf{l}\,.\,\mathbf{s})\,l^- -l^-(\mathbf{l}\,.\,\mathbf{s})]\,|E'\alpha'\rangle\}, \end{aligned}\right\} \tag{12.47}$$

with a precisely analogous set of expressions for E''.

Let us now take the zero-order approximation in which $E'\alpha' = \theta^+$, $E'\beta' = \theta^-$ and $E''\alpha'' = \epsilon^+$, $E''\beta'' = \epsilon^-$. Then in either case $g_\parallel = g_\perp = 2$, the spin-only value. From (12.46), $Q = \pm\frac{6}{7}q'$ and from (12.47), $A = P(\pm\frac{4}{7}-\kappa)$ and $B = P(\mp\frac{2}{7}-\kappa)$ where the upper sign is for E'. The actual g values for copper salts in tetragonal fields deviate significantly from 2 and are anisotropic so it is clear that we must go to the next order of approximation.

The states which now become mixed in are those which form $^2T_2$ of the octahedral group and under D_4 they become classified as $^2B_2$ and ^2E. Suppose they have energies E_1 and E_2, respectively, above the ground doublet. We expect that usually $E_1 \approx E_2$, i.e. $\delta \ll \Delta$ in Fig. 12.3. Now put $x_1 = \zeta/E_1$, $x_2 = \zeta/E_2$ and the two possible ground doublets correct to first order in the spin-orbit coupling energy are

$$\left.\begin{aligned} |E'\alpha'\rangle &= |\theta^+\rangle+\tfrac{1}{2}\sqrt{6}\,x_2\,|1^-\rangle,\\ |E'\beta'\rangle &= |\theta^-\rangle+\tfrac{1}{2}\sqrt{6}\,x_2\,|-1^+\rangle, \end{aligned}\right\} \tag{12.48}$$

and

$$\left.\begin{aligned} |E''\alpha''\rangle &= |\epsilon^+\rangle+x_1\,|\zeta_1^+\rangle+\tfrac{1}{2}\sqrt{2}\,x_2\,|-1^-\rangle,\\ |E''\beta''\rangle &= |\epsilon^-\rangle-x_1\,|\zeta_{\bar{1}}^-\rangle+\tfrac{1}{2}\sqrt{2}\,x_2\,|1^+\rangle, \end{aligned}\right\} \tag{12.49}$$

where we have already changed the sign of the spin-orbit coupling constant to make these functions give the right g values for d^9, not d^1. Then

$$\left.\begin{aligned} g_\parallel &= 2, \qquad\quad g_\perp = 2+6x_2 \quad \text{for} \quad E',\\ g_\parallel &= 2+8x_1, \quad g_\perp = 2+2x_2 \quad \text{for} \quad E'', \end{aligned}\right\} \tag{12.50}$$

where $g_{S\parallel} = g_{S\perp} = 2$ in each case, the g_L being the parts involving the x_i.

For evaluating nuclear magnetic hyperfine parameters it is useful to have the complete matrices for one electron of the quantities $[4s_z-(\mathbf{l}\,.\,\mathbf{s})\,l_z-l_z(\mathbf{l}\,.\,\mathbf{s})]$ and $4s^- - (\mathbf{l}\,.\,\mathbf{s})\,l^- - l^-(\mathbf{l}\,.\,\mathbf{s})$ occurring in (12.47) and these are easily worked out. They are in Table A 41. Using them we obtain

$$\left.\begin{aligned} A &= P(-\kappa-\tfrac{4}{7}+g_{L\parallel}\mp\tfrac{6}{7}x_2),\\ B &= P(-\kappa+\tfrac{2}{7}+g_{L\perp}\pm\tfrac{3}{7}x_2),\end{aligned}\right\} \tag{12.51}$$

where the upper sign applies to E' and the lower to E''. Q is still $\pm\frac{6}{7}q'$ in this approximation.

A typical example of a strong tetragonal field occurs in copper phthalocyanine. Here

$$\left.\begin{aligned} g_{\parallel} &= 2{\cdot}165\pm0{\cdot}003, & A &= 0{\cdot}022\pm0{\cdot}001\ \text{cm}^{-1},\\ g_{\perp} &= 2{\cdot}045\pm0{\cdot}003, & B &= 0{\cdot}003\pm0{\cdot}001\ \text{cm}^{-1},\end{aligned}\right\} \tag{12.52}$$

the effect of the quadrupole moment of the nucleus is not resolved and the relative sign of A and B have not been determined (Gibson, Ingram & Schonland, 1958). There are actually two naturally occurring isotopes of copper and the spin I is $\frac{3}{2}$ for each. It is clear from the g values that we must choose E'' for the ground doublet, a choice that is reasonable on other grounds (see § 12.4.13). Then $x_1 = 0{\cdot}0206$, $x_2 = 0{\cdot}0225$ and so $x_1 \approx x_2$ as expected. Substituting these values in (12.51) for A and B we obtain

$$\left.\begin{aligned} A &= P(-\kappa-0{\cdot}387),\\ B &= P(-\kappa+0{\cdot}321),\end{aligned}\right\} \tag{12.53}$$

and these are satisfied by taking $P = 0{\cdot}0353$, $\kappa = 0{\cdot}237$ or $P = 0{\cdot}0268$, $\kappa = 0{\cdot}433$ according to the choice of signs for A and B.

Resonance has been observed in a large number of compounds of copper. There are three main points of interest. First, many compounds have a symmetry lower than tetragonal and then we have a ground doublet which consists of linear combinations of E' and E'' in (12.48) and (12.49). The interaction with the external magnetic field, referred to suitable axes, now becomes

$$\beta g_x H_x S_x + \beta g_y H_y S_y + \beta g_z H_z S_z$$

with g_x, g_y, g_z in general all different.

Secondly, there are often strong exchange interactions between pairs of adjacent copper ions. This was mentioned in § 10.2.8. In copper acetate each pair of ions gives rise to a singlet and a triplet by coupling the individual doublets. The singlet lies 300 cm^{-1} below the triplet and resonance has been observed in the triplet and interpreted in terms of a spin-Hamiltonian with $S = 1$.

Thirdly, we recall from the discussion in § 11.3 that whilst the Jahn–Teller effect in cupric compounds usually results in a practically static distortion with a well-defined electronic wave-function E' or E'', it sometimes gives rise to a dynamical distortion travelling round the trough in Fig. 11.4. The electronic part of the wave-function is then equally E' and E'' and the g values are the mean of those appropriate to E' and E'' and, hence, isotropic. From (12.50) we find $g = 2+4x$.

At sufficiently low temperatures the ion becomes frozen into one of the minima of the trough and then we get g values satisfying (12.50). Take Cu^{++} partly replacing the zinc in $ZnSiF_6.6H_2O$ as a typical example. At 90° K, $g_\parallel = 2{\cdot}221$, $g_\perp = 2{\cdot}230$, while at 20° and 12° K, $g_\parallel = 2{\cdot}46$, $g_\perp = 2{\cdot}10$. However, the mean $\bar{g} = \frac{1}{3}g_\parallel + \frac{2}{3}g_\perp$ is 2·227 at 90° K and 2·22 at the lower temperatures. The slight anisotropy at 90° K is probably to be interpreted as a slight difference in occupancy of the three Jahn–Teller valleys corresponding to distortions along the x-, y- and z-axes.

Example

Using group-theoretic arguments show that, when $\kappa = 0$, the only non-vanishing part of the nuclear magnetic hyperfine interaction within ${}^2A_1$, ${}^2A_2$, ${}^2B_1$ or ${}^2B_2$ of D_4 in a one-electron system can be $-\frac{1}{7}PV_1(\mathbf{l},\mathbf{l})\,V_1(\mathbf{s},\mathbf{I})$. Deduce that the spin-Hamiltonian constants satisfy

$$A:B:Q = 2P:-P:3q'.$$

12.4.2. d^1: titanium, Ti^{3+}. We calculate the constants in the spin-Hamiltonian, working only within the ground doublet ${}^2T_2$. Titanium has five naturally occurring isotopes and two, with total abundance 13 %, have non-zero nuclear spin—^{47}Ti with $I = \frac{5}{2}$ and ^{49}Ti with $I = \frac{7}{2}$—but hyperfine structure has not yet been resolved.

In the absence of a tetragonal field ${}^2T_2$ splits into the two levels ${}^2T_{2\frac{1}{2}}$ with energy ζ and ${}^2T_{2\frac{3}{2}}$ with energy $-\frac{1}{2}\zeta$. ${}^2T_{2\frac{3}{2}}$ is therefore lower. It has no interaction with the external magnetic field ($g = 0$ in Table 10.1). Under the group D_4 the upper level has symmetry E'' and the lower one splits into $E'+E''$. The tetragonal field then separates the lower pair and we have a Kramers doublet E' or E'' lowest and the situation is formally similar to that of the cupric ion. Therefore we may use (12.47) to calculate the constants.

The functions appropriate to zero tetragonal field are

$$\left.\begin{aligned}
\psi_1^+ &= |{}^2T_{2\frac{1}{2}}E''\alpha''\rangle = \frac{1}{\sqrt{3}}|\zeta_1^+\rangle - \frac{\sqrt{2}}{\sqrt{3}}|-1^-\rangle,\\
\psi_1^- &= |{}^2T_{2\frac{1}{2}}E''\beta''\rangle = -\frac{1}{\sqrt{3}}|\zeta_1^-\rangle - \frac{\sqrt{2}}{\sqrt{3}}|1^+\rangle,\\
\psi_2^+ &= |{}^2T_{2\frac{3}{2}}E''\alpha''\rangle = \frac{\sqrt{2}}{\sqrt{3}}|\zeta_1^+\rangle + \frac{1}{\sqrt{3}}|-1^-\rangle,\\
\psi_2^- &= |{}^2T_{2\frac{3}{2}}E''\beta''\rangle = -\frac{\sqrt{2}}{\sqrt{3}}|\zeta_1^-\rangle + \frac{1}{\sqrt{3}}|1^+\rangle,\\
\phi^+ &= |{}^2T_{2\frac{3}{2}}E'\alpha'\rangle = -|1^-\rangle,\\
\phi^- &= |{}^2T_{2\frac{3}{2}}E'\beta'\rangle = |-1^+\rangle,
\end{aligned}\right\}\qquad (12.54)$$

and are written straight down using Tables A 20 and A 24. The matrix of the tetragonal field and the spin-orbit coupling within the ϕ and ψ breaks up into the two parts:

E''	ψ_1	ψ_2
ψ_1	ζ	$-\frac{\sqrt{2}}{3}\delta$
ψ_2	$-\frac{\sqrt{2}}{3}\delta$	$-\frac{1}{2}\zeta-\frac{1}{3}\delta$

E': $E(\phi) = \frac{1}{3}\delta-\frac{1}{2}\zeta$.

The energies of the E'' doublets are

$$E_{\pm} = -\tfrac{1}{6}\delta+\tfrac{1}{4}\zeta\pm\tfrac{1}{2}(\delta^2+\zeta\delta+\tfrac{9}{4}\zeta^2)^{\frac{1}{2}}. \tag{12.55}$$

The E' doublet lies below the lower E'' doublet when $\delta < 0$ and above when $\delta > 0$. It is instructive to give a formal proof because the method used is of wide application in establishing inequalities between roots of secular equations. The secular equation for E'' is

$$f(E) \equiv (\zeta-E)(-\tfrac{1}{2}\zeta-\tfrac{1}{3}\delta-E)-\tfrac{2}{9}\delta^2 = 0, \tag{12.56}$$

and $f(-\infty) = f(+\infty) = \infty$. If, therefore, we can find a value of E, E_0 say, for which $f(E_0) < 0$ we can deduce that the two roots $E_{\pm}$ of (12.56) satisfy $E_1 < E_0 < E_2$. Take $E_0 = \frac{1}{3}\delta-\frac{1}{2}\zeta$ which is the energy of the E' doublet. Then $f(E_0) = -\zeta\delta$ and so if $\delta > 0$ an E'' doublet is the ground state. If δ is negative either both or neither of the E'' doublets lie above $\frac{1}{3}\delta-\frac{1}{2}\zeta$. That it is both we see by taking $E_0 = -\frac{1}{3}\delta-\frac{1}{2}\zeta$ which is now greater than $\frac{1}{3}\delta-\frac{1}{2}\zeta$ and, from (12.56), $f(-\frac{1}{3}\delta-\frac{1}{2}\zeta) = -\frac{2}{9}\delta^2$ which establishes the result.

The two possible ground doublets are $\phi^{\pm}$ and

$$\left.\begin{aligned}\psi^+ &= c_1|\zeta_1^+\rangle+c_2|-1^-\rangle,\\ \psi^- &= -c_1|\zeta_1^-\rangle+c_2|1^+\rangle.\end{aligned}\right\} \tag{12.57}$$

The constants c_1 and c_2 are determined from the secular equations and satisfy the relations

$$c_1^2 = \tfrac{1}{2}+S^{-1}(\tfrac{1}{4}\zeta+\tfrac{1}{2}\delta),\quad c_2^2 = \tfrac{1}{2}-S^{-1}(\tfrac{1}{4}\zeta+\tfrac{1}{2}\delta),\quad c_1c_2 = \tfrac{1}{2}\sqrt{2}\,S^{-1}\zeta, \tag{12.58}$$

where

$$S = (\delta^2+\zeta\delta+\tfrac{9}{4}\zeta^2)^{\frac{1}{2}}. \tag{12.59}$$

The g values for $\phi^{\pm}$ are both zero in our approximation because $\phi^{\pm}$ are part of $^2T_{2\frac{3}{2}}$. Actually the measured g values are never very close to zero and so do not correspond to E' as the ground doublet. For E'' we obtain

$$\left.\begin{aligned}g_{\parallel} &= -1+3S^{-1}(\delta+\tfrac{1}{2}\zeta),\\ g_{\perp} &= 1+S^{-1}(\delta-\tfrac{3}{2}\zeta).\end{aligned}\right\} \tag{12.60}$$

and for the parameters of the hyperfine interaction (12.46) and (12.47)

$$\left.\begin{aligned}A &= P\{-\kappa S^{-1}(\tfrac{1}{2}\zeta+\delta)-\tfrac{10}{7}+\tfrac{6}{7}S^{-1}(\delta-\tfrac{1}{2}\zeta)\},\\ B &= P\{\kappa[\tfrac{1}{2}+S^{-1}(\tfrac{1}{4}\zeta+\tfrac{1}{2}\delta)]-\tfrac{1}{7}+\tfrac{3}{2}\zeta S^{-1}-\tfrac{1}{7}\delta S^{-1}\},\\ Q &= \tfrac{3}{14}q'[1+3S^{-1}(\tfrac{1}{2}\zeta+\delta)].\end{aligned}\right\} \tag{12.61}$$

Resonance has been observed in caesium titanium alum with $g_\parallel = 1{\cdot}25 \pm 0{\cdot}02$, $g_\perp = 1{\cdot}14 \pm 0{\cdot}02$. The formulae (12.60) are functions of the single parameter ζ/δ and these g values can be roughly but not accurately fitted to the theoretical formulae.

Under a trigonal perturbation the t_2-orbitals split into a_1 and e of the group D_3 (Table A 11). We write δ for the orbital energy difference $E(e) - E(a_1)$. Because of the structural isomorphism between t_2- and p-orbitals both the theory for a tetragonal field and the theory for a trigonal field are correlated with the theory for a ^2P term in which one p-orbital has a different energy from the other two. Therefore, as functions of δ and ζ, they are formally identical and we may use (12.60) and (12.61) for them also. Resonance has been observed in titanium trisacetylacetonate with $g_\parallel = 2{\cdot}00$, $g_\perp = 1{\cdot}93$ (Jarrett, 1957). One easily proves that the only way in which $g_\parallel$ of (12.60) can be near 2 is for ζ/δ to be very small. When that is so we have $g_\parallel = 2 + 0(\zeta^2/\delta^2)$, $g_\perp = 2 - (2\zeta/\delta) + 0(\zeta^2/\delta_2)$, so here $\zeta/\delta \approx 0{\cdot}035$, and if $\zeta = 150\,\text{cm}^{-1}$, then $\delta \approx 4000\,\text{cm}^{-1}$.

However, at the limit of small ζ/δ we must clearly also include the effects of the E'' doublet arising from ^2E. We then have $g_\parallel = 2$ and $g_\perp = 2 - (2\zeta/\delta) - (4\zeta/\delta_1)$, where δ_1 is the separation between the ground doublet and the E'' of ^2E. δ now lies somewhere nearer to $8000\,\text{cm}^{-1}$ depending on the position of this excited doublet.[1] This estimate would, of course, be reduced with a reduction of ζ.

12.4.3. d^8: nickel, Ni^{++}. The ground term of d^8 in an octahedral field is $^3A_2$. We work in the complementary scheme and change the sign of ζ. We possess the complete matrix of the spin-orbit coupling energy within d^2 (Table A 33) and a reference to that shows that the only two levels to be mixed into the ground level in first approximation come from the terms $t_2e\,^1T_2$ and $t_2e\,^3T_2$. Change the sign of ζ and the ground level now becomes

$$|i\rangle = |e^2\,{}^3A_2T_2i\rangle - \frac{\zeta\sqrt{2}}{\Delta}|t_2e\,{}^3T_2T_2i\rangle - \frac{\zeta}{\Delta+8B+2C}|t_2e\,{}^1T_2i\rangle, \qquad (12.62)$$

in this approximation. It has symmetry T_2 and the antisymmetrized square of T_2 contains T_1 just once. Therefore the matrix of $\mathbf{L}+2\mathbf{S}$ is completely determined except for one numerical factor and is therefore proportional to the matrix of $\mathbf{S}$ within a spatially non-degenerate triplet if we correlate the state $|i\rangle$ of (12.62) with a state having $M_S = i$. The spin-Hamiltonian, then, has $S = 1$ and $g = \langle 1|\,(L_z + 2S_z)\,|1\rangle$ is to be determined from (12.62). We do so to first order in ζ/Δ.

Before doing this we repeat two general points. First, as the original ground term $^3A_2$ is an eigenstate of $\mathbf{S}^2$ there is no non-vanishing matrix element of $\mathbf{S}$ between it and any other zero-order state whatsoever. Therefore $\langle 1\,|\,2S_z\,|\,1\rangle = 2$. Secondly, as L_z does not involve spin it has no non-vanishing matrix elements with any excited term of different multiplicity from the ground term. So in a

[1] With $\delta = 7500\,\text{cm}^{-1}$ if $\zeta = 150\,\text{cm}^{-1}$ and $\delta_1 = 20{,}000\,\text{cm}^{-1}$. I disagree with Jarrett (1957), p. 1301, because he incorrectly assigns 'cubic energy' to only one of his ^2E terms.

first-order calculation of g values one may ignore such excited terms. It is not true, however, that they can be ignored in an approximate calculation of the zero-field splitting. Here they are often the most important terms because in many cases, although not here, they lie closer to the ground term than the first excited term of the same multiplicity. Erroneous conclusions have sometimes been drawn in the literature through a failure to appreciate this point.

Our digression leaves us with the expression

$$g = 2 - \frac{2\zeta\sqrt{2}}{\Delta} \langle e^2\,{}^3A_2\,T_2\,1 \,|\, L_z \,|\, t_2 e\,{}^3T_2\,T_2\,1\rangle, \tag{12.63}$$

and using Table A 20 we have

$$|{}^3T_2\,T_2\,1\rangle = \frac{1}{\sqrt{2}}|{}^3T_2\,10\rangle - \frac{1}{\sqrt{2}}|{}^3T_2\,01\rangle,$$

$$|{}^3A_2\,T_2\,1\rangle = |{}^3A_2\,1a_2\rangle.$$

Now $|{}^3A_2\,1a_2\rangle = |\theta^+\epsilon^+\rangle$ and $|{}^3T_2\,10\rangle = |\zeta_1^+\theta^+\rangle$ and so

$$g = 2 - \frac{2\zeta}{\Delta}\langle e^2\,{}^3A_2\,1a_2 \,|\, L_z \,|\, t_2 e\,{}^3T_2\,10\rangle = 2 + \frac{4\zeta}{\Delta}. \tag{12.64}$$

Data is available for numerous nickel compounds, mainly containing $Ni(H_2O)_6^{++}$ groups, and g is nearly isotropic and usually in the range 2·2–2·3. For those compounds containing $Ni(H_2O)_6^{++}$, $g = 2{\cdot}25 \pm 0{\cdot}05$. Using $\zeta = 649$ cm^{-1} from atomic spectra and $\Delta = 8500\,cm^{-1}$ in (12.64) we get $g = 2.31$ which is a little higher and indicates a slight reduction of ζ in the compound.

A fairly satisfactory spin-Hamiltonian is

$$\mathscr{H}(S) = g\beta\mathbf{H}\,.\,\mathbf{S} + D(S_z^2 - \tfrac{2}{3}) + E(S_x^2 - S_y^2), \tag{12.65}$$

and D is usually of the order of $-2\,cm^{-1}$ and E about $-0{\cdot}5\,cm^{-1}$. Both are very variable[1] and probably mainly arise through the combined effect of spin-orbit coupling and fields of low symmetry within the states from ${}^3A_2$ and ${}^3T_2$. There is a nickel isotope, ^{61}Ni, which should have a magnetic moment and probable spin $I = \frac{3}{2}$ but its natural abundance is only 1·25 % and no hyperfine structure has been observed.

12.4.4. d^3**: chromium,** Cr^{3+}**; vanadium,** V^{2+}**; molybdenum,** Mo^{3+}**; rhenium,** Re^{4+}**.** The ground term of d^3 in an octahedral field is ${}^4A_2$. The only excited term of the same multiplicity which can be mixed in is ${}^4T_2$. Referring to Table A 34 we see that the lower-lying doublet $t_2^3\,{}^2T_2$ also interacts directly with the ground level and must be taken into account in an accurate treatment. Neglecting this and proceeding to first order only, the calculation is very similar to the one for nickel. There are a few differences however. The term ${}^4A_2$ can have

[1] An interesting pressure dependence of the zero-field splitting in $NiSiF_6.6H_2O$ has been observed by Walsh & Bloembergen (1957). Here D actually changes sign at 6200 Kg/cm².

its states classified as $|^4A_2\, ia_2\rangle$ or as $|^4A_2\, U'\, i\rangle$, the two being connected via the coupling coefficients:

$$|^4A_2\, U'i\rangle = \sum_j \langle U'A_2 ja_2 \,|\, U'A_2 U'i\rangle\, |^4A_2\, ja_2\rangle$$

according to Table A 20. In forming a spin-Hamiltonian it is, of course, the $|^4A_2\, ia_2\rangle$ which we associate with the spin-Hamiltonian states with $M_S = i$ and $S = \frac{3}{2}$. Next we remark that as the symmetrized square of U' contains T_1 twice, the matrix of $\mathbf{L}+2\mathbf{S}$ within U' depends, in general, on two parameters. We could also have deduced this result from the example in § 12.3.5. However, as we proved in § 12.3.4, at least as long as we calculate the g values to first order only we get the first-order interaction with the magnetic field in the form $\beta g\mathbf{H}\,.\,\mathbf{S}$. So we need only evaluate the matrix element of $L_z + 2S_z$ for one state. We do this for $|U'\mu\rangle$ which gets correlated with the $M_S = \frac{3}{2}$ state for the spin-Hamiltonian. Then as $|^4A_2\, \frac{3}{2}a_2\rangle = -\,|1^+ - 1^+ \zeta_1^+\rangle$, $|^4T_2\, \frac{3}{2}0\rangle = |1^+ - 1^+ \epsilon^+\rangle$ we deduce

$$\left.\begin{aligned} \tfrac{3}{2}g &= \langle U'\mu|\,(L_z+2S_z)\,|U'\mu\rangle = 3-\frac{4\zeta}{\Delta}, \\ g &= 2-\frac{8\zeta}{3\Delta}. \end{aligned}\right\} \qquad (12.66)$$

In the first transition series data is available for V^{++} and Cr^{3+}. g is practically isotropic and slightly less than 2. The chromic ion has been more investigated and when surrounded by six water molecules has $g = 1{\cdot}98$. Using $\zeta = 273\,\mathrm{cm}^{-1}$ from atomic spectra and $\Delta = 17{,}400\,\mathrm{cm}^{-1}$ from molecular spectra we calculate $g = 1{\cdot}96$. As with nickel there is probably some reduction of ζ in the compound. In the cyanides $K_4V(CN)_6$ and $K_3Cr(CN)_6$ (diluted in isomorphous diamagnetic cyanides) g is isotropic at $1{\cdot}992$ and the smaller deviation from 2 is to be correlated with the fact that Δ is much larger for Cr^{3+} here and is expected to be for V^{++} also (see Table 11.3). ζ also may be further reduced (see Table 11.4). Vanadium has isotopes ^{50}V and ^{51}V with nuclear spins $I = 6$ and $I = \frac{7}{2}$ and abundances $0{\cdot}25$ and $99{\cdot}75\,\%$, respectively. Chromium has an isotope ^{53}Cr with $I = \frac{3}{2}$ and abundance $9{\cdot}6\,\%$. The spin-Hamiltonian contains the isotropic hyperfine interaction $A\mathbf{I}\,.\,\mathbf{S}$ which arises mainly from the s-electron Fermi interaction. There is usually also a zero-field splitting with D of the order of $0{\cdot}05$–$0{\cdot}1\,\mathrm{cm}^{-1}$. In a few compounds it is considerably larger, being $-0{\cdot}36\,\mathrm{cm}^{-1}$ in ruby and $0{\cdot}59\,\mathrm{cm}^{-1}$ in chromium trisacetylacetonate. In both these latter compounds there is a trigonal field along a threefold axis of the octahedron and the zero-field splitting is along this axis. It appears probable that these large splittings arise mainly from an anisotropy of the spin-orbit coupling constant rather than from the combined effect of an isotropic spin-orbit coupling and the trigonal field (Sugano & Tanabe, 1958).

In molybdenum compounds with the d^3 configuration the zero-field splitting is still larger and for normal magnetic fields is much larger than the energy of interaction with the external magnetic field. This means that the states with $M_S = \pm\frac{1}{2}$ are separated from those with $M_S = \pm\frac{3}{2}$ by $2D$, say. $M_S = \pm\frac{1}{2}$ lies lower. It is now possible to use a new spin-Hamiltonian for this lower doublet with an $S' = \frac{1}{2}$. Consider just the parts depending on the external magnetic field, then the two spin-Hamiltonians may be written

$$\mathscr{H}(\tfrac{3}{2}) = g\beta\mathbf{H}\,.\,\mathbf{S}, \quad \text{with four basic states } |\tfrac{3}{2}M_S\rangle,$$

$$\mathscr{H}(\tfrac{1}{2}) = g'_{\parallel}\beta H_z S_z + g'_{\perp}\beta(H_x S_x + H_y S_y), \quad \text{with two basic states } |\tfrac{1}{2}M_S\rangle.$$

Then
$$\langle\tfrac{1}{2}\tfrac{1}{2}|\,g'_{\parallel}\beta H_z S_z\,|\tfrac{1}{2}\tfrac{1}{2}\rangle = \langle\tfrac{3}{2}\tfrac{1}{2}|\,g\beta H_z S_z\,|\tfrac{3}{2}\tfrac{1}{2}\rangle,$$

$$\langle\tfrac{1}{2}\,-\tfrac{1}{2}|\,g'_{\perp}\beta H^+S^-\,|\tfrac{1}{2}\tfrac{1}{2}\rangle = \langle\tfrac{3}{2}\,-\tfrac{1}{2}|\,g\beta H^+S^-\,|\tfrac{3}{2}\tfrac{1}{2}\rangle,$$

whence
$$g'_{\parallel} = g \approx 2 - \frac{8\zeta}{3\Delta}, \quad g'_{\perp} = 2g \approx 4 - \frac{16\zeta}{3\Delta}. \tag{12.67}$$

For K_3MoCl_6, diluted in an isomorphous diamagnetic compound, $g = 1{\cdot}93$. Hyperfine structure is also observed. ζ from atomic spectra is 817 cm^{-1} and Δ from molecular spectra is 19,200 cm^{-1} whence we predict $g = 1{\cdot}89$ from (12.66). Resonance is also observed in K_2ReCl_6 with $g = 1{\cdot}8$ corresponding, presumably, to a big increase in ζ compensated to some extent by an increase in Δ.

12.4.5. d^5: manganese, Mn^{++}; **iron,** Fe^{3+}. The high-spin ground term is $^6A_1$ in an octahedral field. There are no other terms of the same or higher multiplicity in d^5 and the selection rules for the spin-orbit coupling energy then show that only $^4T_1$ terms can be mixed into the ground term to first order. The ground term consists of the two levels E'' and U'. However, although they can be separated by going to high order in perturbation theory they cannot be by merely diagonalizing the spin-orbit coupling within the terms $^6A_1$ and the three $^4T_1$ terms of d^5. This is because the c-numbers (Table A 31) force the matrices for E'' and U' to be identical. Alternatively, one may see this by letting $^6A_1$, $^4T_1$ correspond respectively to atomic ^6S, ^4P terms and then E'', U' are the two parts of $J = \frac{5}{2}$ levels and hence strictly degenerate. Actually our ground $^6A_1$ is a ^6S term to a good approximation and this makes it even more difficult for E'' to get separated from U'. The spin-spin coupling energy may be partly responsible for the small separations which are observed but it can only act, to first order, on the deviations of the actual ground term from spherical symmetry.

As we remarked in § 12.3.5, a spin-Hamiltonian with only quadratic forms in the spin components cannot give any zero-field splitting in an octahedral field and we must take a more general form. Including also an axial field but assuming g and the interaction with the nucleus to be isotropic we have

$$\mathscr{H}(S) = g\beta\mathbf{H}\,.\,\mathbf{S} + \tfrac{1}{6}a(S_x^4 + S_y^4 + S_z^4 - \tfrac{707}{16}) + D(S_z^2 - \tfrac{35}{12})$$
$$+ \tfrac{7}{36}F(S_z^4 - \tfrac{95}{14}S_z^2 + \tfrac{81}{16}) + A\mathbf{S}\,.\,\mathbf{I}, \tag{12.68}$$

where the names of the constants are conventional. Experimentally g is isotropic and very close to 2, A is isotropic and a, D are usually very small and F practically negligible. Actually A is only known for Mn^{++}, whose only naturally occurring isotope has $I = \frac{5}{2}$, because hyperfine structure has not been observed for Fe^{3+}.

For ferric compounds a and D are often comparable but for manganous compounds a is usually much smaller than D. Then the main features of the resonance spectrum are obtained with the simple Hamiltonian

$$\mathscr{H}(S) = g\beta \mathbf{H}\,.\,\mathbf{S} + D(S_z^2 - \tfrac{35}{12}) + A\mathbf{S}\,.\,\mathbf{I}. \tag{12.69}$$

The zero-field splitting means that the transitions $M_S \rightarrow M_S + 1$ do not all have the same energy and we get five lines. Each of these is split into six components by the nuclear hyperfine interaction and the resulting spectrum, especially for **H** not parallel to OZ, is very complicated. This complexity is added to in some ions by the presence of two inequivalent manganese ions each giving separately a spin-Hamiltonian of the form (12.68) or (12.69) but with different axes for the

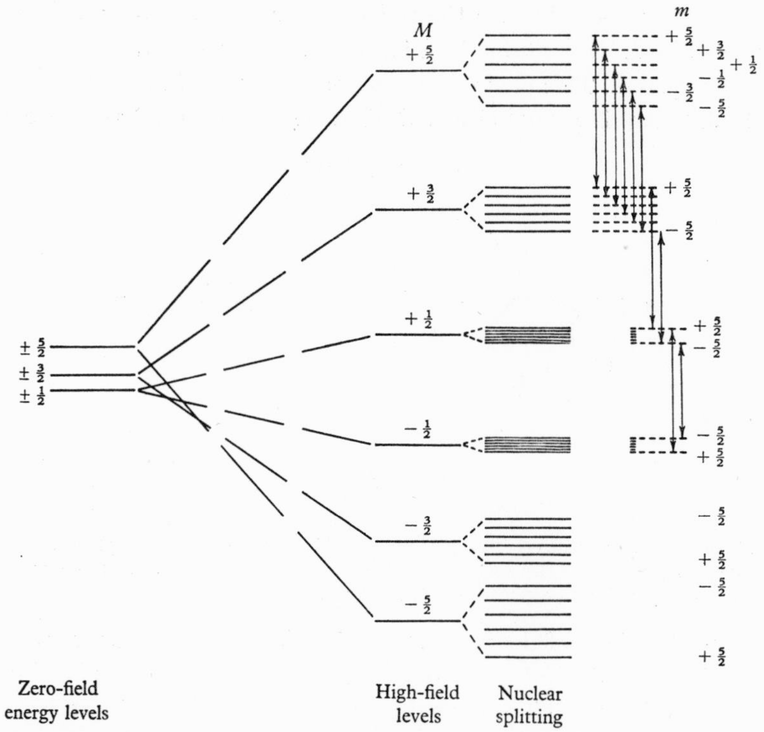

Fig. 12.4. A schematic energy-level diagram for the high-spin manganous ion, showing the electronic levels in zero and in strong magnetic field, together with the splitting due to the nuclear spin. Some of the typical transitions are indicated by arrows (from Bleaney & Ingram, 1951).

zero-field splitting. As an example let us take manganese ammonium sulphate for which $D = 2{\cdot}77 \times 10^{-2}\,\text{cm}^{-1}$ and a nearly isotropic $A = 9{\cdot}3 \times 10^{-3}\,\text{cm}^{-1}$ at 20° K. A schematic energy-level diagram for this salt with the magnetic field parallel to the zero-field splitting axis was given by Bleaney & Ingram (1951) and is shown in Fig. 12.4. For **H** at other angles the behaviour is more complicated but the energy levels are still at $g\beta HM_S$, where M_S refers to quantization along the magnetic field direction, to a good approximation. The effects of D and of A are second-order ones.

Now suppose D to be much larger and actually considerably larger than the magnetic field energy. In this case the ground doublet is not appreciably mixed with the excited doublets under the influence of the external magnetic field. This is just the situation we met in the d^3 molybdenum compounds and we do the same thing as we did there and take a new spin-Hamiltonian with $S' = \frac{1}{2}$. We leave aside the nuclear interaction and have

$$\mathscr{H}(\tfrac{1}{2}) = g'_{\parallel}\beta H_z S_z + g'_{\perp}\beta(H_x S_x + H_y S_y), \quad \text{with basic states } |\tfrac{1}{2}M_S\rangle, \tag{12.70}$$

where

$$\left.\begin{aligned} g'_{\parallel} &= 2\langle\tfrac{5}{2}\tfrac{1}{2}|gS_z|\tfrac{5}{2}\tfrac{1}{2}\rangle = g \approx 2, \\ g'_{\perp} &= \langle\tfrac{5}{2}-\tfrac{1}{2}|gS^-|\tfrac{5}{2}\tfrac{1}{2}\rangle = 3g \approx 6. \end{aligned}\right\} \tag{12.71}$$

This behaviour has not been observed in manganous compounds but has been observed in a few ferric compounds. These are all derivatives of the protein haemoglobin which is responsible for the red colour and oxygen-carrying ability of mammalian blood. It contains iron ions in the middle of a larger planar conjugated ring system. This part of haemoglobin is called haem and is illustrated in Fig. 12.5. There is probably a fifth ligand attached (perhaps by a nitrogen

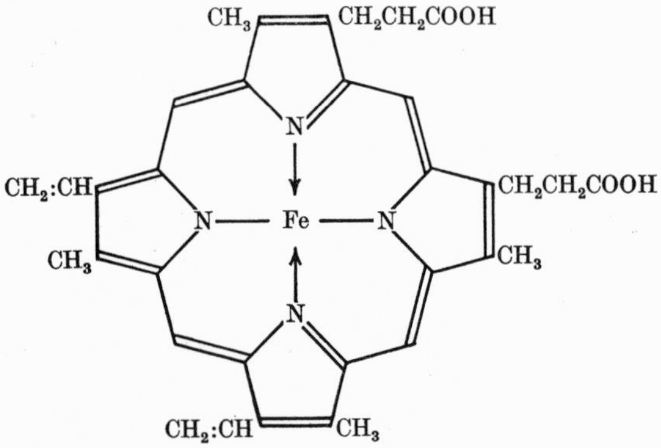

Fig. 12.5. Ferrous protoporphyrin (haem), the prosthetic group of haemoglobin. One classical chemical structure is shown: the ring system is actually expected to have the symmetry of its nuclei to a good approximation (from Griffith, 1956*b*).

atom) below the plane of the paper. This fifth ligand forms part of the protein. Then the sixth octahedral coordination position can be occupied by various ligands—in the living cell it is there that the oxygen molecule is carried. Natural haemoglobin has the iron in the ferrous state, but in the laboratory it can be oxidized to the ferric state and then electron resonance is observed. This oxidized form, ferrihaemoglobin, occurs in both the high-spin and the low-spin form depending on the nature of the occupant of the sixth coordination position. In the former case resonance is observed from a single doublet and when fitted to the spin-Hamiltonian (12.70) gives $g_{\parallel} = 2$, $g_{\perp} = 6$, where $\parallel$ refers to the fourfold axis, i.e. the axis at right-angles to the plane of the paper in Fig. 12.5. The origin of this large zero-field splitting is not completely certain but is probably mainly due to a combination of the isotropic spin-orbit coupling with a rather strong tetragonal field (see § 12.4.12).

12.4.6. d^4: chromium, Cr^{++}. Electron resonance has been observed in $CrSO_4 . 5H_2O$ and was interpreted in terms of the spin-Hamiltonian

$$\mathscr{H}(S) = \beta g_{\parallel} H_z S_z + \beta g_{\perp}(H_x S_x + H_y S_y) + D(S_z^2 - 2) + E(S_x^2 - S_y^2) \tag{12.72}$$

with $S = 2$ and $g_\parallel = 1{\cdot}95$, $g_\perp = 1{\cdot}99$, $|D| = 2{\cdot}24\,\text{cm}^{-1}$, $|E| = 0{\cdot}10\,\text{cm}^{-1}$. No hyperfine structure was observed. $\parallel$ and $\perp$ refer to a fourfold axis of the octahedron formed by the water molecules around the chromous ion. E is very small and we neglect it. It can be interpreted by including a small term of less than tetragonal symmetry in the ligand field.

The ground term of the high-spin chromous ion is $t_2^3 e\,{}^5E$ and this breaks up into ${}^5A_1 + {}^5B_1$ under a tetragonal field. As the splitting arises from the effect of the field on the e-orbitals we expect it to be fairly large. There is no non-zero spin-orbit coupling energy within ${}^5E$ so if we calculate only within ${}^5E$ we get a spin-only isotropic g value of 2 for ${}^5A_1$ and ${}^5B_1$. In the next approximation various excited terms are mixed into ${}^5E$ under the influence of the spin-orbit coupling energy, but to calculate the g values we need only consider ${}^5T_2$ to this order. We do this first and can then use $\frac{1}{4}\zeta\mathbf{L}\,.\,\mathbf{S}$ for the spin-orbit coupling energy. The ground term is either ${}^5A_1$ with states $|M_S\theta\rangle$ or ${}^5B_1$ with states $|M_S\epsilon\rangle$ and it is convenient to take the excited term states as $|M_S\zeta_1\rangle$, $|M_S 1\rangle$, $|M_S -1\rangle$, where the second number in each bracket is the M_L value or indicates it

$$(|M_S\theta\rangle = |M_S 0\rangle, \quad |M_S\epsilon\rangle = \frac{1}{\sqrt{2}}|M_S 2\rangle + \frac{1}{\sqrt{2}}|M_S -2\rangle,$$

$$|M_S\zeta_1\rangle = \frac{1}{\sqrt{2}}|M_S 2\rangle - \frac{1}{\sqrt{2}}|M_S -2\rangle,$$

i.e. $|M_S 1\rangle$ is not $|{}^5T_2 M_S 1\rangle$).

Writing $x = \zeta/\Delta$, the ground states become

$$\left.\begin{aligned}
|2\theta\rangle' &= |2\theta\rangle - \frac{\sqrt{3}}{2\sqrt{2}}x\,|11\rangle,\\
|2\epsilon\rangle' &= |2\epsilon\rangle - x\,|2\zeta_1\rangle - \frac{1}{2\sqrt{2}}x\,|1\,-1\rangle,\\
|1\theta\rangle' &= |1\theta\rangle - \frac{\sqrt{3}}{2\sqrt{2}}x\,|2\,-1\rangle - \tfrac{3}{4}x\,|01\rangle,\\
|1\epsilon\rangle' &= |1\epsilon\rangle - \tfrac{1}{2}x\,|1\zeta_1\rangle - \frac{1}{2\sqrt{2}}x\,|21\rangle - \frac{\sqrt{3}}{4}x\,|0\,-1\rangle,
\end{aligned}\right\} \tag{12.73}$$

for $M_S = 2$ or 1 and hence the g values are

$$g_\parallel = 2, \qquad g_\perp = 2 - \tfrac{3}{2}x \quad \text{for} \quad {}^5A_1,$$

$$g_\parallel = 2 - 2x, \quad g_\perp = 2 - \tfrac{1}{2}x \quad \text{for} \quad {}^5B_1.$$

Using $\zeta = 230\,\text{cm}^{-1}$, $\Delta = 13{,}900\,\text{cm}^{-1}$ we calculate $g_\parallel = 2$, $g_\perp = 1{\cdot}98$ for ${}^5A_1$ and $g_\parallel = 1{\cdot}97$, $g_\perp = 1{\cdot}99$ for ${}^5B_1$. Therefore the ground term is ${}^5B_1$.

Next we calculate the zero-field splitting due to second-order effects of the spin-orbit coupling energy. There are two important excited terms here—the ${}^5T_2$ and also $t_2^4\,{}^3T_1$. The effect of the former is to lower the energy of a state by Δ times the sum of the squares of the coefficients of those states of ${}^5T_2$ which are mixed in (12.73). For example, $|2\theta\rangle'$ is lowered by $\frac{3}{8}x^2\Delta$ and $|1\theta\rangle'$ by $\frac{15}{16}x^2\Delta$. Therefore

$$3D = -\tfrac{3}{8}x^2\Delta + \tfrac{15}{16}x^2\Delta = \tfrac{9}{16}x^2\Delta \quad \text{and} \quad D = \tfrac{3}{16}(\zeta^2/\Delta) \quad \text{for} \quad {}^5A_1$$

and $$D = -(3\zeta^2/16\Delta) \quad \text{for} \quad {}^5B_1.$$

With the above values of ζ and Δ this gives $D = \pm 0{\cdot}71\,\text{cm}^{-1}$.

The position of $t_2^4\,{}^3T_1$, which is the only term of t_2^4 having a non-zero matrix element with ${}^5E$, is unknown. In the strong-field scheme its energy is calculated to be $E = 6B+5C-\Delta = 8230\,\text{cm}^{-1}$ using B and C from Appendix 6. This will be an overestimate both because B and C should be reduced somewhat in the compound and because we have neglected interactions with more excited strong-field ${}^3T_1$ terms.

We possess the matrix of spin-orbit interaction between $t_2^4\,{}^3T_1$ and $t_2^3e\,{}^5E$ in Table A 35. I now show how we deduce the matrix elements of $|M_S\theta\rangle$ and $|M_S\epsilon\rangle$ with $t_2^4\,{}^3T_1$. Take

$$|{}^5E\,2\theta\rangle = -\tfrac{1}{2}\,|{}^5E\,A_2a_2\rangle + \tfrac{1}{2}\,|{}^5E\,E\epsilon\rangle + \frac{1}{\sqrt{2}}\,|{}^5E\,T_20\rangle,$$

from the coupling coefficients in Table A 20. Then A_2 does not occur in ${}^3T_1$ but E and T_2 do and from Table A 35

$$\langle{}^5E\,E\epsilon\,|\,\mathscr{H}_s\,|\,t_2^4\,{}^3T_1\,E\epsilon\rangle = -\zeta,$$

$$\langle{}^5E\,T_20\,|\,\mathscr{H}_s\,|\,t_2^4\,{}^3T_1\,T_20\rangle = -\frac{1}{\sqrt{2}}\zeta,$$

whence $$-\sum_{X\in t_2^4\,{}^3T_1}\frac{|\langle{}^5E\,2\theta\,|\,\mathscr{H}_s\,|\,X\rangle|^2}{E} = -\frac{\zeta^2}{2E}$$

which is the second-order correction to the energy of $|{}^5E\,2\theta\rangle$ arising from the term $t_2^4\,{}^3T_1$. The second-order corrections to the other states are written down just as easily and the net result is, apart from shifts of the centres of gravity, to add a zero-field splitting $D(S_z^2-2)$ with $D = \pm\zeta^2/4E$ the upper sign holding for ${}^5A_1$ and the lower for ${}^5B_1$. With $E = 8230\,\text{cm}^{-1}$ we find $D = \pm 1{\cdot}61\,\text{cm}^{-1}$. Combining this with our earlier contribution to D we have a total $D = \pm 2{\cdot}32\,\text{cm}^{-1}$ in very good agreement with the observed $|D| = 2{\cdot}24\,\text{cm}^{-1}$.

As we remarked earlier, $8230\,\text{cm}^{-1}$ is probably an overestimate for E and so D should be increased. However, ζ will be reduced in the ion and this will reduce D again. Therefore the close agreement between calculated and experimental D is partly a consequence of cancellation of two not very large errors. As an example, if we decrease ζ, B and C all in the same proportion $\frac{20}{23}$, i.e. about 13 %, the calculated value of D only changes to $\pm 2{\cdot}41\,\text{cm}^{-1}$ with contributions 0·54 and 1·87 from ${}^5T_2$ and $t_2^4\,{}^3T_1$, respectively. The separation to ${}^5T_2$, of course, is an experimental quantity and cannot be changed.

Next, the influence of the triplets of t_2^3e is not necessarily negligible, but it happens that there is a cancellation of their effect and they are unlikely to affect D by more than about $0{\cdot}03\,\text{cm}^{-1}$ at most and probably less (see the example below).

Lastly, there is the spin-spin coupling energy. This is given by (12.31) within ${}^5E$ and is easily worked out. The term in $\mathbf{L}\,.\,\mathbf{S}$ is zero within ${}^5E$ so we need only consider $-\rho[(\mathbf{L}\,.\,\mathbf{S})^2-12]$. For $|{}^5E\,2\theta\rangle$ this is

$$12\rho-\rho\,\langle{}^5E\,2\theta\,|\,L_z^2S_z^2\,|\,{}^5E\,2\theta\rangle-\tfrac{1}{4}\rho\,\langle{}^5E\,2\theta\,|\,L^-L^+S^+S^-\,|\,{}^5E\,2\theta\rangle = 12\rho-6\rho = 6\rho,$$

the other terms vanishing because $S^+|^5E\,2\theta\rangle = 0$. $|^5E\,2\epsilon\rangle$ is treated similarly and we conclude that the contribution to D is $\pm 3\rho$ with the upper sign for $^5A_1$ and the lower for $^5B_1$. The value of ρ is not known but is probably not larger than $0{\cdot}1\ \mathrm{cm}^{-1}$. As we obtain a satisfactory interpretation of the observed D with $\rho = 0$ it would be merely an embarrassment to have a large ρ. Sadly we must admit that this is no strong argument that ρ is actually small, but merely say that it is unnecessary to include spin-spin energy in order to explain the electron resonance data for $CrSO_4.5H_2O$.

I should perhaps say a word about the theoretical interpretation given for D by Ono, Koide, Sekiyama & Abe (1954) who measured the electron resonance spectrum. They ascribe a large proportion of D to spin-spin energy and still obtain agreement between theory and experiment. This is because they neglected $t_2^4\,{}^3T_1$. The large spin-spin constant $\rho = 0{\cdot}4\ \mathrm{cm}^{-1}$ which they used was obtained by Pryce (1950) from the optical spectrum of gaseous Cr^{++} by incorrectly ascribing all the deviation from the Landé interval rule for the ground ^5D term to spin-spin energies.

Examples

1. Writing the matrix of spin-orbit interaction between $t_2^3e\,{}^5E$ and a $^3T_1$ term as $\lambda\zeta c(\Gamma)$ according to Table A 31 show that the $^3T_1$ term produces a zero-field splitting $D(S_z^2-2)$ and $D = \pm(\lambda^2\zeta^2/2\delta)$, where δ is the difference in energy between $^3T_1$ and ^5E and the upper sign is for the $^5A_1$ part of ^5E and the lower for $^5B_1$. With the same notation show that the signs change place if we replace $^3T_1$ by $^3T_2$.

2. In the strong-field scheme show that there are only two triplet terms, one $^3T_1$ and one $^3T_2$, of the configuration t_2^3e which have non-zero matrix elements of spin-orbit coupling energy with $t_2^3e\,{}^5E$. Deduce that the entire contribution to D from the triplets of t_2^3e is approximately $\mp\zeta^2B/[16(8B+3C)(9B+3C)] \approx \mp 0{\cdot}01\ \mathrm{cm}^{-1}$ with the upper sign for $^5A_1$ and the lower for $^5B_1$.

12.4.7. d^6: iron, Fe^{++}. The ground term of the high-spin ferrous ion in an octahedral field is $^5T_2$. This is split by the spin-orbit coupling energy into the three levels $^5T_{21}$, $^5T_{22}$ and $^5T_{23}$ with $^5T_{21}$ lying lowest. So long as we stay within this ground term we may use the structural isomorphism of $^5T_2$ with ^5P to calculate our energies within ^5P. Then we must use $\frac{1}{4}\zeta\mathbf{L}\,.\,\mathbf{S}$ for the spin-orbit coupling energy and $\beta\mathbf{H}\,.\,(-\mathbf{L}+2\mathbf{S})$ for the magnetic energy (see ch. 10).

Now suppose we have a lower-symmetry component in the ligand field which separates the triple orbital degeneracy of $^5T_2$ into a single and a double degeneracy. By choosing axes appropriately[1] the single degeneracy can be made to correspond to the $M_L = 0$ states of ^5P. Doing this we can calculate the effect of the low-symmetry component within ^5P by taking a perturbation $\delta(L_z^2-\frac{2}{3})$. From a purely formal point of view our calculation is the same no matter what are the relative orientations of the axes of the octahedron and of the low-symmetry perturbation. In particular it is the same for a distortion along the fourfold or the threefold axes of the octahedron. The Z-axis we use in our calculation

[1] The angular dependence of the spatial part of ^5P is as (x, y, z), therefore that of the singly degenerate spatial part is $lx+my+nz$, say, where l, m, n can be taken real (apply the Kramers operator). Now take the new Z-axis along the vector (l, m, n).

is the axis of the low-symmetry perturbation, not necessarily the fourfold axis of the octahedron.

Our Hamiltonian is now

$$\mathscr{H} = \tfrac{1}{4}\zeta \mathbf{L}\,.\,\mathbf{S} + \delta(L_z^2 - \tfrac{2}{3}) + \beta \mathbf{H}\,.\,(-\mathbf{L} + 2\mathbf{S}), \tag{12.74}$$

to be evaluated within the 15 states $|M_S M_L\rangle$ of a ^5P term. We do this in the absence of a magnetic field first. Then M_J is a good quantum number and the matrix of $\mathscr{H}$ breaks up into seven submatrices whose elements can be written straight down. They are shown in Table 12.1. There is essentially only one parameter, the ratio of δ to ζ. We write $\omega = \delta/\zeta$. The secular equations for the energies may now be written down and are a quadratic and a linear equation for

Table 12.1. *Matrices of spin-orbit coupling energy and pseudo-axial perturbation within the ground $^5T_2$ term of the ferrous ion, classified by the parameters M_S, M_L, M_J of the correlated ^5P term*

(The matrices for $|-M_S, -M_L\rangle$ are the same as for $|M_S, M_L\rangle$.)

$M_J = 0$	$\|1, -1\rangle$	$\|0, 0\rangle$	$\|-1, 1\rangle$
$\|1, -1\rangle$	$\frac{1}{3}\delta - \frac{1}{4}\zeta$	$\frac{\sqrt{3}}{4}\zeta$	0
$\|0, 0\rangle$	$\frac{\sqrt{3}}{4}\zeta$	$-\frac{2}{3}\delta$	$\frac{\sqrt{3}}{4}\zeta$
$\|-1, 1\rangle$	0	$\frac{\sqrt{3}}{4}\zeta$	$\frac{1}{3}\delta - \frac{1}{4}\zeta$

$M_J = 1$	$\|2, -1\rangle$	$\|1, 0\rangle$	$\|0, 1\rangle$
$\|2, -1\rangle$	$\frac{1}{3}\delta - \frac{1}{2}\zeta$	$\frac{\sqrt{2}}{4}\zeta$	0
$\|1, 0\rangle$	$\frac{\sqrt{2}}{4}\zeta$	$-\frac{2}{3}\delta$	$\frac{\sqrt{3}}{4}\zeta$
$\|0, 1\rangle$	0	$\frac{\sqrt{3}}{4}\zeta$	$\frac{1}{3}\delta$

$M_J = 2$	$\|2, 0\rangle$	$\|1, 1\rangle$
$\|2, 0\rangle$	$-\frac{2}{3}\delta$	$\frac{\sqrt{2}}{4}\zeta$
$\|1, 1\rangle$	$\frac{\sqrt{2}}{4}\zeta$	$\frac{1}{3}\delta + \frac{1}{4}\zeta$

$M_J = 3$	$\|2, 1\rangle$
$\|2, 1\rangle$	$\frac{1}{3}\delta + \frac{1}{2}\zeta$

$M_J = 0$, a cubic for $M_J = 1$, a quadratic for $M_J = 2$ and a linear equation for $M_J = 3$. I have solved them numerically[1] and plotted the energies, in units of ζ, as functions of ω in Fig. 12.6. We see the interval rule satisfied for $\omega = 0$. For $\omega > 0$ the ground state is singly degenerate with $M_J = 0$ and for $\omega < 0$ it is doubly degenerate with $M_J = \pm 1$.

It may perhaps be helpful to point out that Fig. 12.6 can only be used in a straightforward way to give relative energies for a chosen fixed value of ω. This is because we defined our low-symmetry perturbation so that it left the centre of gravity of the term unaltered. In other words we have omitted the potential energy, quadratic in the normal coordinates (see § 11.3).

[1] Let E be the energy. In such a calculation it is actually easier to calculate ω as a function of E rather than vice versa as one readily sees by inspecting the secular equations. This also applies to the calculation of the relation between g and ω.

Let us now determine the g value of the lowest doublet using a spin-Hamiltonian with $S' = \frac{1}{2}$. The doublet may be written

$$\left.\begin{aligned} \psi &= c_1\,|2,-1\rangle + c_2\,|1,0\rangle + c_3\,|0,1\rangle, \\ \psi^* &= -c_1\,|-2,1\rangle - c_2\,|-1,0\rangle - c_3\,|0,-1\rangle, \end{aligned}\right\} \qquad (12.75)$$

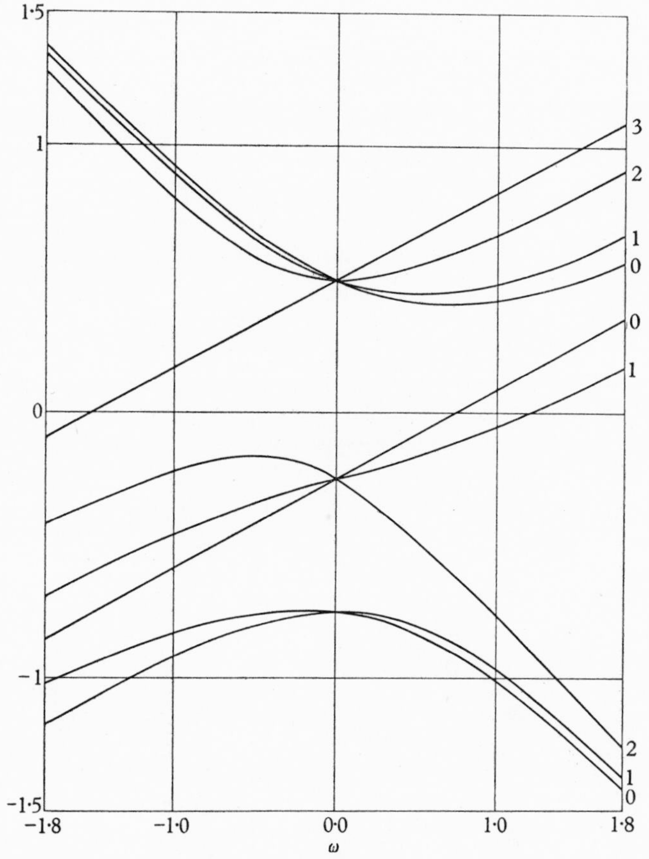

Fig. 12.6. The splitting of the ground term of the high-spin ferrous ion in a pseudo-axial field. Energy in units of ζ marked at left and M_J values shown at right.

where the c_i are real and ψ^* is the Kramers conjugate of ψ. Then we deduce

$$g_{\parallel} = 10c_1^2 + 4c_2^2 - 2c_3^2 \quad (g_{\perp} = 0), \qquad (12.76)$$

c_1, c_2, c_3 are determined, via the secular equation for $|M_J| = 1$, from ω and hence we can find $g_{\parallel}$ as a function of ω. This is plotted in Fig. 12.7. $g_{\parallel}$ satisfies $4 \leqslant g_{\parallel} \leqslant 10$. The intensity for magnetic-dipole transitions between ψ and ψ^* in (12.75) is necessarily identically zero. A non-zero intensity could arise from mixing of ψ with ψ^* and with the lowest state with $M_J = 0$ under the influence of yet lower symmetry components of the field.

Electron resonance has been observed (Tinkham, 1955, 1956) and analysed in detail for the ferrous ion present substitutionally in zinc fluoride. Zinc fluoride

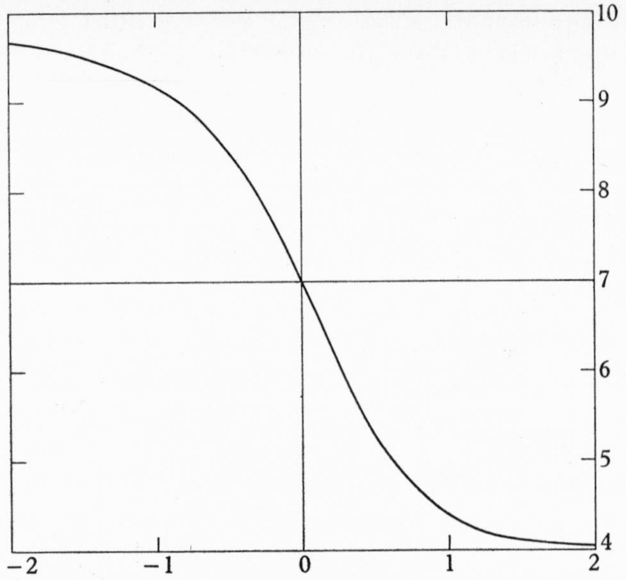

Fig. 12.7. Calculated ferrous ion $g_{\parallel}$ as a function of the low-symmetry field parameter $\omega = \delta/\zeta$.

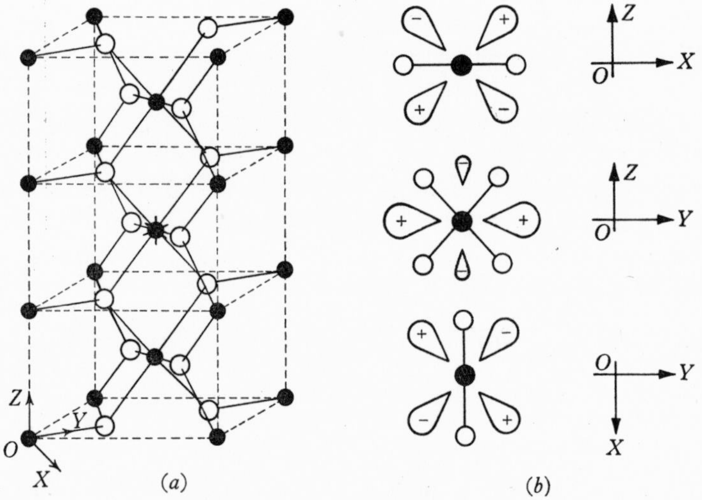

Fig. 12.8. (*a*) A ferrous ion (centre) present substitutionally in a zinc fluoride lattice. Open circles are fluorine atoms and filled ones are metal atoms (after Wells, 1956). (*b*) The three lowest *d*-orbitals projected into planes of bilateral symmetry.

has the rutile structure. This is shown in Fig. 12.8 where three unit cells are drawn. The experimental results are interpreted in terms of a spin-Hamiltonian

$$\mathscr{H}(S') = g_{\parallel}\beta H_z S_z + aS_x,$$

with $S' = \frac{1}{2}$, $g_{\parallel} = 8{\cdot}97 \pm 0{\cdot}02$, $a = 0{\cdot}203\ \text{cm}^{-1}$ at 20° K and $a = 0{\cdot}224\ \text{cm}^{-1}$ at 90° K. The axes are drawn in Fig. 12.8. The resonance occurs from the ground doublet.

It would be quite natural to try to interpret these results in terms of a near degeneracy for a pair of the dt_2 orbitals and $\omega = -0{\cdot}82$. This, however, appears to be incorrect[1] and I now outline some interesting parts of Tinkham's interpretation. For more detail I refer the reader to his paper (Tinkham, 1956).

The four fluorine ions surrounding a ferrous ion in the YZ-plane are all at 2·03 Å from it but at the vertices of a rectangle with sides 2·59 and 3·13 Å and elongated parallel to the Z-axis. The two fluorine neighbours in the directions of the X-axis are at 2·04 Å. Classifying the d-functions under the group D_2 we find that $d_{z^2-y^2}$ and d_{x^2} both belong to A_1 and therefore can be mixed by the ligand field. We write $\xi, \eta, \zeta, \theta, \epsilon$ referred as usual to the x-, y-, z-axes and have an orbital pattern of the form

$$\left.\begin{aligned} \psi_1 &= \zeta, \\ \psi_2 &= -(\tfrac{1}{2}\sqrt{3}\,\theta+\tfrac{1}{2}\epsilon)\cos\alpha-(-\tfrac{1}{2}\theta+\tfrac{1}{2}\sqrt{3}\,\epsilon)\sin\alpha, \\ \psi_3 &= \eta, \\ \psi_4 &= -(\tfrac{1}{2}\sqrt{3}\,\theta+\tfrac{1}{2}\epsilon)\sin\alpha+(-\tfrac{1}{2}\theta+\tfrac{1}{2}\sqrt{3}\,\epsilon)\cos\alpha, \\ \psi_5 &= \xi, \end{aligned}\right\} \qquad (12.77)$$

with energies E_i. Using a crystal field model, Tinkham calculates $\tan\alpha \approx 0{\cdot}2$ and $E_2-E_1 \approx E_3-E_2 \approx 1000\,\text{cm}^{-1}$. The orbitals ψ_1, ψ_2, ψ_3 are shown in Fig. 12.8 in order of increasing energy and projected into their planes of bilateral symmetry. Because of the large separation between the states of the orbital triplet we have a non-degenerate orbital level lowest and may conveniently use a spin-Hamiltonian with $S' = 2$. The fine-structure Hamiltonian then becomes

$$D_x S_x^2 + D_y S_y^2 + D_z S_z^2,$$

where we choose $D_x+D_y+D_z = 0$. There are essentially two parameters in the fine-structure Hamiltonian. One of these is found experimentally from a in the spin-Hamiltonian and the other by comparing the calculated with the experimental anisotropy of susceptibility. The latter involves the D_ϵ through the Boltzmann factor in the formula for the susceptibility. I merely state the results.

$$D_z = -4{\cdot}9 \pm 0{\cdot}5\,\text{cm}^{-1}, \quad \eta = \frac{D_x-D_y}{D_z\sqrt{3}} = 0{\cdot}167 \pm 0{\cdot}020.$$

The interesting thing here is that the results appear to conflict with Fig. 12.6 which suggests that when an orbitally non-degenerate state is lowest we must have a fine-structure Hamiltonian $D[S_z^2-\frac{1}{3}S(S+1)]$ with $D > 0$. It is difficult to see how a lower symmetry perturbation could change the sign of D. The resolution of this apparent paradox is that the lowest orbital in Tinkham's interpretation is not the one corresponding to p_z in the p^n isomorphism but the one corresponding to p_y. So if $E_2 = E_3$ and there is no mixing in of ψ_4 or ψ_5 we have the fine-structure Hamiltonian $D[S_y^2-\frac{1}{3}S(S+1)] = \frac{2}{3}D[S_y^2-\frac{1}{2}(S_x^2+S_z^2)]$. Therefore the sign of D_z is already correct. But D_z mainly arises from

$$\frac{|\langle 1\,|\,L_z\,|\,2\rangle|^2}{E_2-E_1} = \frac{(\cos\alpha+\sqrt{3}\sin\alpha)^2}{E_2-E_1},$$

[1] It appears that the case of a pseudo-axial field does occur in $FeSiF_6.6H_2O$. See Palumbo (1958); Jackson (1959).

while D_x arises from $$\frac{|\langle 1|L_x|3\rangle|^2}{E_3-E_1} = \frac{1}{E_3-E_1},$$

and as $E_3-E_1 \approx 2(E_2-E_1)$ and $(\cos\alpha+\sqrt{3}\sin\alpha)^2 \approx 1{\cdot}8$, D_z is increased fourfold relatively. The change in the axis of apparent tetragonality has changed the sign of D, so the $M_S = \pm 2$ doublet is lowest and the g value is largely due to spin.

We have here also an example of the principle in crystallography elaborated by Dunitz & Orgel (1957). The direction of the distortion of the rutile structure and the position of the odd d electron collaborate so as to achieve a minimum energy. The direction of causal implication is uncertain; the presence of the collaboration is clear enough.

12.4.8. d^7: cobalt, Co^{++}, in an octahedral field. The theory of the high-spin cobaltous ion in an octahedral field is formally very similar to that of the ferrous ion. The ground term is now $^4T_1$ giving the three levels $^4T_{1\frac{1}{2}}$, $^4T_{1\frac{3}{2}}$ and $^4T_{1\frac{5}{2}}$ with the $^4T_{1\frac{1}{2}}$ level lowest. This lowest level is a Kramers doublet and so cannot be split except by magnetic fields. The exact nature of the $^4T_1$ ground term is determined by the magnitude of the cubic component of the ligand field. In the weak-field limit the ground doublet $^4T_{1\frac{1}{2}}$ has an isotropic $g = 4\frac{1}{3}$ and in the strong-field limit $g = 4$ (Table 10.1).

Recalling our findings in § 10.2.3, the calculation of the energies and g values when we apply a perturbation which separates the triple spatial degeneracy into a single and a double degeneracy is accomplished by using the Hamiltonian

$$\mathscr{H} = -\tfrac{1}{3}\gamma\zeta\mathbf{L}\,.\,\mathbf{S}-\delta(L_z^2-\tfrac{2}{3})+\beta\mathbf{H}(\gamma\mathbf{L}+2\mathbf{S}), \tag{12.78}$$

where the Z-axis is the axis of the low-symmetry perturbation (see § 12.4.7) and γ satisfies $-\frac{3}{2} \leqslant \gamma \leqslant -1$. $\gamma = -\frac{3}{2}$ in the weak-field and -1 in the strong-field limit. The calculation is to be performed within a ^4P term. If our orbitals are not d-orbitals we do not use the free-ion value of ζ and elsewhere in $\mathscr{H}$ we replace $\mathbf{L}$ by $k\mathbf{L}$ (and $\frac{2}{3}$ by $\frac{2}{3}k^2$).

The calculation of the energies in zero magnetic field is a straightforward matter and the appropriate matrices are given in Table 12.2. The energy levels are shown schematically in Fig. 12.9. The ground doublet is

$$\left.\begin{aligned}\psi &= c_1|\tfrac{3}{2},-1\rangle+c_2|\tfrac{1}{2},0\rangle+c_3|-\tfrac{1}{2},1\rangle,\\ \psi' &= c_1|-\tfrac{3}{2},1\rangle+c_2|-\tfrac{1}{2},0\rangle+c_3|\tfrac{1}{2},-1\rangle,\end{aligned}\right\} \tag{12.79}$$

where c_1, c_2, c_3 are real and $\psi' = -i\psi^*$. c_1, c_2, c_3 are determined from the secular equation. The g values are

$$\left.\begin{aligned}g_\parallel &= 2\langle\bar{\psi}|(\gamma L_z+2S_z)|\psi\rangle = (6-2\gamma)c_1^2+2c_2^2+(2\gamma-2)c_3^2,\\ g_\perp &= \langle\bar{\psi}'|(\gamma L^-+2S^-)|\psi\rangle = 4c_2^2+4\sqrt{3}\,c_1c_3+2\sqrt{2}\,\gamma c_2c_3.\end{aligned}\right\} \tag{12.80}$$

When $\delta = 0$ we have $c_1 = (1/\sqrt{2})$, $c_2 = -(1/\sqrt{3})$, $c_3 = (1/\sqrt{6})$ and so g is isotropic at $\frac{2}{3}(5-\gamma)$ in agreement with (10.22). In the limit $\delta = +\infty$ we see by inspecting

Table 12.2. *Matrices of spin-orbit coupling energy and pseudo-axial perturbation within the ground* $^4T_1$ *term of the cobaltous ion, classified by the parameters* M_S, M_L, M_J *of the correlated* ^4P *term*

($\gamma = -1$ for the strong-field ground term and $\gamma = -\frac{3}{2}$ for the weak-field ground term. The matrices for $|-M_S, -M_L\rangle$ are the same as for $|M_S, M_L\rangle$.)

$M_J = \frac{1}{2}$	$\|\frac{3}{2}, -1\rangle$	$\|\frac{1}{2}, 0\rangle$	$\|-\frac{1}{2}, 1\rangle$
$\|\frac{3}{2}, -1\rangle$	$-\frac{1}{3}\delta + \frac{1}{2}\gamma\zeta$	$-\frac{1}{\sqrt{6}}\gamma\zeta$	0
$\|\frac{1}{2}, 0\rangle$	$-\frac{1}{\sqrt{6}}\gamma\zeta$	$\frac{2}{3}\delta$	$-\frac{\sqrt{2}}{3}\gamma\zeta$
$\|-\frac{1}{2}, 1\rangle$	0	$-\frac{\sqrt{2}}{3}\gamma\zeta$	$-\frac{1}{3}\delta + \frac{1}{6}\gamma\zeta$

$M_J = \frac{3}{2}$	$\|\frac{3}{2}, 0\rangle$	$\|\frac{1}{2}, 1\rangle$
$\|\frac{3}{2}, 0\rangle$	$\frac{2}{3}\delta$	$-\frac{1}{\sqrt{6}}\gamma\zeta$
$\|\frac{1}{2}, 1\rangle$	$-\frac{1}{\sqrt{6}}\gamma\zeta$	$-\frac{1}{3}\delta - \frac{1}{6}\gamma\zeta$

$M_J = \frac{5}{2}$	$\|\frac{3}{2}, 1\rangle$
$\|\frac{3}{2}, 1\rangle$	$-\frac{1}{3}\delta - \frac{1}{2}\gamma\zeta$

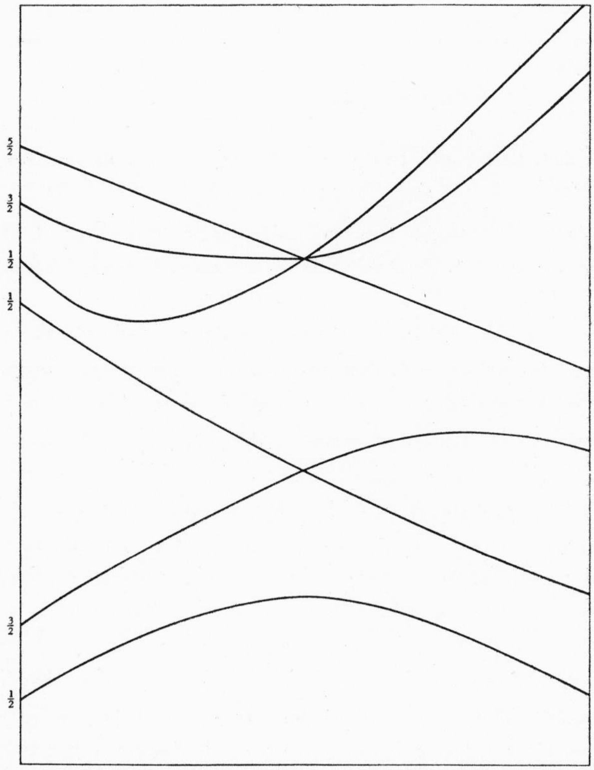

Fig. 12.9. Schematic representation of the splitting of the $^4T_1$ ground term of the high-spin cobaltous ion in a pseudo-axial field. M_J values shown at left.

Table 12.2 that $c_1 = 1, c_2 = c_3 = 0$ and similarly when $\delta = -\infty, c_2 = 1, c_1 = c_3 = 0$. These limiting cases lead to the g values

$$\left.\begin{aligned} \delta = +\infty: \quad g_\parallel = 2(3-\gamma), \quad g_\perp = 0, \\ \delta = -\infty: \quad g_\parallel = 2, \quad g_\perp = 4. \end{aligned}\right\} \tag{12.81}$$

For a given γ, the two g values are functions of the single parameter δ/ζ and so they bear a functional relationship to each other. Therefore we can plot a curve showing this relation between the calculated values of $g_\parallel$ and $g_\perp$. The observed g values are for octahedral environments with either tetragonal or trigonal fields and should therefore all lie on such a curve. The curve is drawn in Fig. 12.10 for $\gamma = -1$ and $\gamma = -\frac{3}{2}$ (strong-field and weak-field cases). For intermediate cases

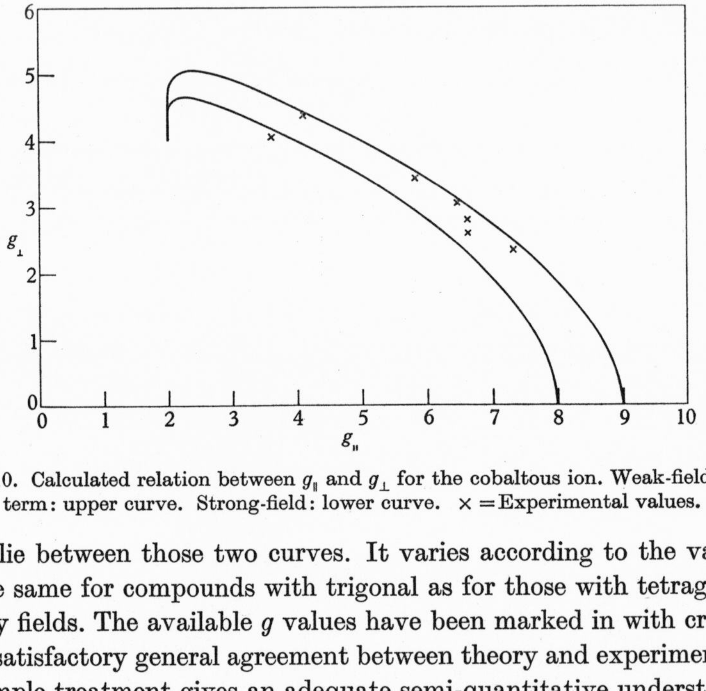

Fig. 12.10. Calculated relation between $g_\parallel$ and $g_\perp$ for the cobaltous ion. Weak-field ground term: upper curve. Strong-field: lower curve. × =Experimental values.

it would lie between those two curves. It varies according to the value of γ, but is the same for compounds with trigonal as for those with tetragonal low-symmetry fields. The available g values have been marked in with crosses and we see a satisfactory general agreement between theory and experiment.

Our simple treatment gives an adequate semi-quantitative understanding of the g values. A more detailed treatment was given by Abragam & Pryce (1951*c*) taking into account the mixing in of the upper $^4T_2$, $^4A_2$ and $^4T_1$ terms by the low-symmetry field, but not including orbital reduction or any terms of lower multiplicity. The extra refinement modifies the calculated values slightly but the qualitative features of the theory are unaltered. There is, however, then a very small difference between the calculated curve of $g_\parallel$ against $g_\perp$ for tetragonal and trigonal symmetry. My treatment in the present section owes, of course, all its essentials to a careful perusal of Abragam & Pryce's paper.

Finally, the cobalt nucleus has a spin $I = \frac{7}{2}$ and therefore there is a hyperfine structure. In their detailed calculations Abragam & Pryce found fair but not perfect agreement between the calculated and observed hyperfine parameters for various compounds. As they remark, there are two parameters, P and κ, to be determined empirically so the satisfactory nature of the agreement may be deceptive. The hyperfine structure is calculated using (12.47) with the hyperfine-structure Hamiltonian (12.22). In the weak-field limit the ground $^4T_1$ is part of the free-ion ^4F term and then we can use the operator equivalent (12.23).

A point of general interest about the hyperfine interaction concerns the possibility of replacing it within a $^4T_1$ term by an operator whose matrix elements are to be calculated within ^4P. The hyperfine operator is

$$\mathscr{H}_m = P\left\{\mathbf{L}\,.\,\mathbf{I}-\kappa\mathbf{S}\,.\,\mathbf{I}-\tfrac{1}{7}\sum_{\lambda=1}^{n}\sum_{k=1}^{5} V_k(\mathbf{l}_\lambda,\mathbf{l}_\lambda)\,V_k(\mathbf{s}_\lambda,\mathbf{I})\right\}. \quad (12.82)$$

$\mathbf{L}$ is replaced by $\gamma\mathbf{L}$ and $\mathbf{S}$ is left unchanged, as before. $V_k(\mathbf{l}_\lambda,\mathbf{l}_\lambda)$, however, forms a basis for the reducible representation $E+T_2$ of the octahedral group. Therefore the matrix elements of $V_k(\mathbf{l}_\lambda,\mathbf{l}_\lambda)$ are determined group-theoretically apart from two parameters. An operator equivalent within ^4P to $\mathscr{H}_m$ is

$$\mathscr{H}'_m = P\left\{\gamma\mathbf{L}\,.\,\mathbf{I}-\kappa\mathbf{S}\,.\,\mathbf{I}+\alpha\sum_{k=1}^{2} V_k(\mathbf{L},\mathbf{L})\,V_k(\mathbf{S},\mathbf{I})+\beta\sum_{k=3}^{5} V_k(\mathbf{L},\mathbf{L})\,V_k(\mathbf{S},\mathbf{I})\right\}. \quad (12.83)$$

Unless $\alpha = \beta$, then, the nuclear hyperfine interaction behaves differently for the tetragonal and trigonal axes. Actually for cobalt Abragam & Pryce found the contribution from the last term of (12.82) to be even different in sign for the trigonal and tetragonal axes.

12.4.9. Cs_3CoCl_5. In Cs_3CoCl_5 the cobaltous ion is in the tetrahedral complex ion $CoCl_4^{--}$. Therefore the theory is the same as for an octahedral d^3 ion provided that we change the sign of the spin-orbit coupling constant. We have already treated this d^3 ion in § 12.4.4. and use those results now. The ground term is $^4A_2$, is orbitally non-degenerate, and has

$$g = 2+\frac{8\zeta}{3\Delta}. \quad (12.84)$$

The observed g values are $g_{\parallel} = 2{\cdot}32 \pm 0{\cdot}04$ and $g_{\perp} = 2{\cdot}27 \pm 0{\cdot}04$ with a large zero-field splitting represented in the spin-Hamiltonian with $S' = \frac{3}{2}$ by $D(S_z^2-\frac{5}{4})$ and $D \approx -4{\cdot}5\ \text{cm}^{-1}$.

We recall from § 11.5.2 that there is some doubt about the assignments of the bands of tetrahedral cobaltous compounds. The infra-red band of $CoCl_4^{--}$ can be assigned either to $^4A_2 \rightarrow {}^4T_2$ or to $^4A_2 \rightarrow {}^4T_1$ leading to alternative values for Δ of 6300 or 3750 cm^{-1}, respectively. ζ for the free ion is 533 cm^{-1} and then (12.84) gives $g = 2.23$ or $2{\cdot}38$, respectively. If ζ is reduced to 90 % of its free-ion value then $g = 2{\cdot}20$ or $2{\cdot}34$ and if 80 % then $g = 2{\cdot}18$ or $2{\cdot}30$. These figures definitely favour the lower value for Δ and the correctness of Orgel's assignment of the near infra-red bands of tetrahedral cobaltous halides to the transition $^4A_2 \rightarrow {}^4T_1$.

12.4.10. Low-spin d^5: iron, Fe^{3+}; manganese, Mn^{++}; ruthenium, Ru^{3+}; iridium, Ir^{4+}. The ground term is $^2T_2$ and in the strong-field limit is $t_2^5\,{}^2T_2$ which is the complementary term of $t_2^1\,{}^2T_2$ with respect to t_2^6. Therefore for tetragonal symmetry we may use (12.60) and (12.61) with change of sign of δ, ζ and q'. In most low-spin d^5 compounds, however, the symmetry is lower than tetragonal and there are three quite distinct g values.

Let us suppose, then, that we have a low-symmetry perturbation which separates the energies of the real functions ξ, η, ζ but has no matrix elements between them. We work in the scheme complementary to t_2^5 so that, including spin, we have six one-electron functions altogether. Let the energies of ξ, η, ζ be $\frac{1}{2}V$, $-\frac{1}{2}V$, Δ, respectively. Then the matrix of the low-symmetry perturbation and the spin-orbit coupling breaks up into two submatrices which are shown in Table 12.3. The secular equations associated with them give the energies of the

Table 12.3. *Matrix of spin-orbit coupling and low-symmetry field within* $t_2^1\,{}^2T_2$

	$\vert 1^+\rangle$ (or $\vert -1^-\rangle$)	$\vert \zeta_1^-\rangle$ (or $-\vert \zeta_1^+\rangle$)	$\vert -1^+\rangle$ (or $\vert 1^-\rangle$)
$\vert 1^+\rangle$ (or $\vert -1^-\rangle$)	$\frac{1}{2}\zeta$	$\frac{1}{2}\zeta\sqrt{2}$	$\frac{1}{2}V$
$\vert \zeta_1^-\rangle$ (or $-\vert \zeta_1^+\rangle$)	$\frac{1}{2}\zeta\sqrt{2}$	Δ	0
$\vert -1^+\rangle$ (or $\vert 1^-\rangle$)	$\frac{1}{2}V$	0	$-\frac{1}{2}\zeta$

three Kramers doublets. If we know $V\zeta^{-1}$ and $\Delta\zeta^{-1}$ we could solve them, obtain the ground doublet and predict the parameters of the resonance spectrum. In practice we do not, so we work in reverse and assume a general form for the ground doublet

$$\left.\begin{aligned} \psi &= A\,|1^+\rangle + B\,|\zeta_1^-\rangle + C\,|-1^+\rangle, \\ \psi' &= A\,|-1^+\rangle - B\,|\zeta_1^+\rangle + C\,|1^-\rangle, \end{aligned}\right\} \tag{12.85}$$

with A, B and C real. Then $\psi' = i\psi^*$. Take the interaction with the external magnetic field to be $\beta\mathbf{H}\,.\,(k\mathbf{L}+2\mathbf{S})$.

Consider first kL_z+2S_z. This has zero matrix elements between ψ and ψ' and

$$\langle\bar{\psi}|\,(kL_z+2S_z)\,|\psi\rangle = k(A^2-C^2)+A^2-B^2+C^2. \tag{12.86}$$

$$\langle\bar{\psi}'|\,(kL_z+2S_z)\,|\psi'\rangle = -\langle\bar{\psi}|\,(kL_z+2S_z)\,|\psi\rangle$$

because (ψ, ψ') form a Kramers doublet. Hence, if we associate $|\psi\rangle$ with $|\frac{1}{2}\rangle$ and $|\psi'\rangle$ with $|-\frac{1}{2}\rangle$ in a spin-Hamiltonian with $S = \frac{1}{2}$, the part involving H_z is

$$2\beta H_z(k(A^2-C^2)+A^2-B^2+C^2).$$

In a similar way we examine the matrix elements of kL_x+2S_x and kL_y+2S_y and find that the entire first-order interaction with the external magnetic field is represented by $\beta(g_xH_x+g_yH_y+g_zH_z)$ where (Stevens, 1953)

$$\left.\begin{aligned} g_x &= 2[2AC-B^2+kB(C-A)\sqrt{2}], \\ g_y &= 2[2AC+B^2+kB(C+A)\sqrt{2}], \\ g_z &= 2[A^2-B^2+C^2+k(A^2-C^2)]. \end{aligned}\right\} \tag{12.87}$$

If we put $k = 1$ we get the simpler expressions

$$\left.\begin{aligned} g_x &= 2(\sqrt{2}\,A+B)\,(\sqrt{2}\,C-B), \\ g_y &= 2(\sqrt{2}\,A+B)\,(\sqrt{2}\,C+B), \\ g_z &= 2(2A^2-B^2). \end{aligned}\right\} \tag{12.88}$$

Writing $p = \sqrt{2}\,A+B$ we find $g_x^2+g_y^2+g_z^2 = 4p^2(1-p^2)$.

As $A^2+B^2+C^2 = 1$, (12.87) depends on the three parameters k and $A:B:C$. Thus if we know the experimental g values (they are the moduli of g_x, g_y, g_z of (12.88)) we can deduce A, B, C and k from (12.87) and then as ψ and ψ' are solutions of the secular equations associated with Table 12.3, we can calculate $V\zeta^{-1}$ and $\Delta\zeta^{-1}$. The secular equations are

$$\left.\begin{aligned} (\tfrac{1}{2}\zeta-E)\,A+\tfrac{1}{2}\zeta\sqrt{2}\,B+\tfrac{1}{2}VC &= 0,\\ \tfrac{1}{2}\zeta\sqrt{2}\,A+(\Delta-E)\,B &= 0,\\ \tfrac{1}{2}VA+(-\tfrac{1}{2}\zeta-E)\,C &= 0,\end{aligned}\right\} \qquad (12.89)$$

and are soluble for E, $V\zeta^{-1}$ and $\Delta\zeta^{-1}$ in terms of A, B and C. Next, knowing $V\zeta^{-1}$ and $\Delta\zeta^{-1}$ we can calculate the positions and wave-functions of the excited pair of Kramers doublets and, knowing all these things, calculate the susceptibility as a function of $T\zeta^{-1}$. By comparing this with the observed susceptibility one calculates ζ.

For d^5 we must change the signs of ζ, V and Δ. In potassium ferricyanide, diluted in diamagnetic cobalt ferricyanide, $g_x = 2{\cdot}35$, $g_y = 2{\cdot}10$, $g_z = 0{\cdot}915$. The susceptibility has been measured by Jackson (1938) over a wide temperature range and down to 14° K, but for the concentrated salt. Electron resonance has been observed in the concentrated salt, but exchange effects are important and the results are not understood in detail. They are in general agreement with the values given above. Combining Jackson's measurements with the g values from the dilute salt, Bleaney & O'Brien (1956) find $k = 0{\cdot}87$ and $\zeta = 278\,\text{cm}^{-1}$ (free-ion value $410\,\text{cm}^{-1}$). The g values could also be fitted with $k = 0{\cdot}56$ but the agreement with susceptibility measurements is then worse.

Another interesting point about the susceptibilities is that as $T \to 0$ only the ground doublet is thermally occupied and the temperature independent contribution to the susceptibility becomes negligible compared with the part arising from matrix elements within the ground doublet. Therefore if χ is the susceptibility along an axis, $\lim_{T\to 0} kT\chi = \frac{1}{4}N\beta^2g^2$ giving a relation between χ and g. The mean susceptibility $\bar{\chi}$ gives $\lim_{T\to 0} kT\bar{\chi} = \frac{1}{4}N\beta^2\overline{g^2}$. These relations are quite well satisfied experimentally for potassium ferricyanide.

Resonance has also been observed in $K_4Mn(CN)_6$ diluted in potassium ferrocyanide. It is low-spin with $g_x = 2{\cdot}62$, $g_y = 2{\cdot}18$, $g_z = 0{\cdot}72$ and an anisotropic hyperfine structure. No hyperfine structure has been observed for the ferric compounds. Resonance has been found in low-spin Ru^{3+} and Ir^{4+} compounds. In the $IrCl_6^{--}$ and $IrBr_6^{--}$ anions hyperfine structure from the chlorine and bromine nuclei has been observed.[1] This hyperfine structure from an ion not at the centre of the complex was in fact first observed in the $IrCl_6^{--}$ anion. For this particular anion the g values are usually isotropic with g slightly less than $1{\cdot}78$. If we assume octahedral symmetry and use (10.13) we obtain $g = \frac{2}{3}(2k+1)$, whence $k = 0{\cdot}83$.

[1] In accordance with the general discussion in ch. 7. For more detail see Stevens (1953).

12.4.11. Monovalent ions. Sodium fluoride crystallizes in a simple cubic lattice in which each sodium ion is surrounded octahedrally by six fluorine ions and conversely. By irradiating sodium fluoride containing transition-metal impurities it is possible to replace the sodium ions with monovalent transition-metal ions.[1] Resonance has been observed (Bleaney & Hayes, 1957; Hayes, 1958) for Cr^+, Fe^+, Co^+ and Ni^+. The free gaseous ions have ground terms arising from the respective configurations d^5, $d^6 s$, d^8 and d^9 and it appears that the same is true for the ions in the crystal except for Fe^+ which is probably d^7. Our previous theoretical interpretation applies to the g values and we now discuss these briefly.

Cr^+ has an isotropic $g = 2{\cdot}000$, and this arises from $d^5\,{}^6S$. Fe^+ has a single g value of 4·344 to be compared with the isotropic g lying between 4 and $4\frac{1}{3}$ expected for the ground ${}^4T_{1\frac{1}{2}}$ level of d^7 in an octahedral field.

Co^+ has an isotropic g value of 2·31 which, using (12.64) for d^8 and $\zeta = 456\ \text{cm}^{-1}$, gives $\Delta \simeq 5900\ \text{cm}^{-1}$. Ni^+ has nearly tetragonal symmetry with $g_\parallel = 2{\cdot}766$, $g_\perp = 2{\cdot}114$. This means the ground doublet is $d^9\,E''$ and, using (12.50) and $\zeta = 603\ \text{cm}^{-1}$, we deduce $E(\epsilon) - E(\zeta) = 6300\ \text{cm}^{-1}$ and $E(\epsilon) - E(1, -1) = 10{,}600$ cm^{-1} for the orbital energies. The orbital separations are fairly low here both because the ligands are fluoride ions (see Table 11.3) and because we have monovalent ions. We have no independent evidence for the values of ζ; if ζ is much lower than in the free ion, the orbital separations will be much lower than our calculated values.

Another feature of special interest about these experiments is the presence, for Cr^+, Fe^+ and Ni^+, of a resolved hyperfine structure from the fluorine nuclear magnetic moments. This is due almost entirely to the e_g magnetic electrons being in orbitals extending over the fluorine ions. Fluorine hyperfine structure has also been observed for transition-metal ions in a zinc fluoride lattice (Tinkham, 1956).

Example

Show that the ground level of $d^6 s$ in an octahedral field should belong to the irreducible representation U' and has $g = \frac{2}{5}(7+k)$, where k is an orbital reduction factor.

12.4.12. Zero-field splitting for high-spin d^5 in tetragonal symmetry. We know that in an octahedral field for d^5 the only two possible ground terms are ${}^6A_1$ and ${}^2T_2$; the ${}^4T_1$ can never be lowest. Let us now consider the effect of a perturbation which does not separate the t_2-orbitals but splits the e-orbitals by a large amount μ leaving their centre of gravity unaltered. This will be a reasonable approximation to the situation in haemoglobin and $e\epsilon$ which is directed towards the nitrogen atoms of the porphyrin ring is expected to lie above $e\theta$, at least for ligands near or to the weak-field side of the crossover. Fig. 12.3 is still relevant, but it is no longer obvious that θ lies above (ξ, η) or ζ.

The perturbation has no first-order effect on ${}^6A_1$ and, at least in the strong-field limit, none on ${}^2T_2$ either. This is because these terms have either a half-filled or an empty shell of e-electrons. Not so ${}^4T_1$ which is $t_2^4 e$. With $\mu = E(\epsilon) - E(\theta)$, and positive, ${}^4T_1$ splits into two terms separated by $\frac{3}{4}\mu$. Using the notation

[1] For resonance from impurity ions in MgO see Low (1958).

adequate to D_4, ${}^4A_2$ lies lower and ${}^4E$ higher. If μ is large enough we obtain the behaviour shown in Fig. 12.11*a*, and for a range of values of Δ a quartet will be the ground term. There will also be a range of Δ for which the sextet is the ground term, but the quartet is very close above. This latter may occur even if there is no Δ for which the quartet is actually the ground term (Fig. 12.11*b*).

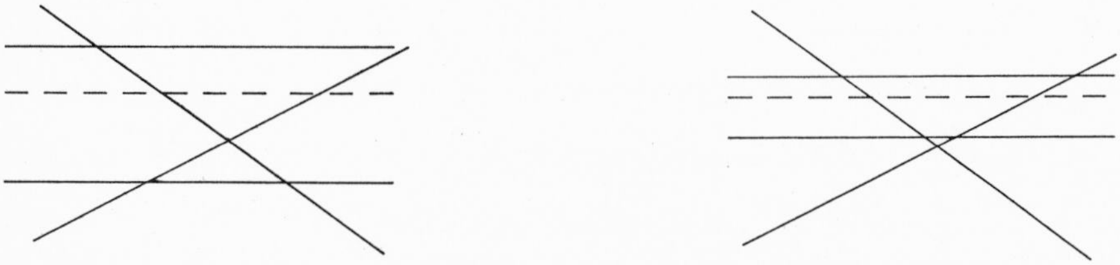

(*a*) (*b*)

Fig. 12.11. The electrostatic part of the crossover in planar compounds. Dotted lines for ${}^4T_1$ in an octahedral field. It is supposed that ferric porphyrin chloride has its energy levels on a vertical line of (*a*) and ferrihaemoglobin derivatives on vertical lines of (*b*).

Examining Fig. 12.11 suggests we should obtain for our lowest approximation the matrix of spin-orbit coupling energy just within the three terms ${}^6A_1$, ${}^2T_2$ (both classified by O) and ${}^4A_2$ (classified by D_4). We can write this matrix down immediately using the tables given in Appendix 2. I shall do this in a leisurely manner so that the reader will understand this kind of use of the tables.

The group classifying the states including spin-orbit coupling is D_4^* which has the two possible irreducible representations E' and E'' for odd-electron systems. Under O^*, ${}^6A_1$ is $E''+U'$ (Table A 14) and ${}^2T_2$ is $E'\times T_2 = E''+U'$ (Table A 9). Now, from Table A 11, E'' and U' under O^* become respectively E'' and $E'+E''$ under D_4^*. So both the doublet and the sextet span $E'+2E''$ under D_4^*. The ${}^4A_2$ is $A_2\times(E'+E'') = E'+E''$. Hence we have found

$$\left.\begin{aligned} {}^6A_1 &\to E'+2E'', \\ {}^4A_2({}^4T_1) &\to E'+E'', \\ {}^2T_2 &\to E'+2E''. \end{aligned}\right\} \qquad (12.90)$$

So there will be a 3×3 matrix for E' and a 5×5 one for E''.

We possess spin-orbit matrices for O^* (Table A 36) but not for D_4^*. Therefore we must express all our functions as sums over components of irreducible representations of O^*. Take the states of ${}^4A_2$ as an example, writing ϕ_2 for the $E'\alpha'$ component in (12.90) and ψ_3 for the $E''\alpha''$ component. Then (use Table A 16 to see the effect of the group generators on ${}^4A_2$) we have

$$\phi_2 = |{}^4T_1\tfrac{1}{2}0\rangle, \quad \psi_3 = |{}^4T_1 -\tfrac{3}{2}0\rangle,$$

and using the coupling coefficients in Table A 20

$$\left.\begin{aligned} \phi_2 &= -\frac{1}{\sqrt{3}}|{}^4T_1E'\alpha'\rangle + \frac{1}{\sqrt{15}}|{}^4T_{1\frac{3}{2}}\,U'\lambda\rangle + \frac{\sqrt{3}}{\sqrt{5}}|{}^4T_{1\frac{5}{2}}\,U'\lambda\rangle, \\ \psi_3 &= -\frac{1}{\sqrt{3}}|{}^4T_1E''\alpha''\rangle - \frac{\sqrt{3}}{\sqrt{5}}|{}^4T_{1\frac{3}{2}}\,U'\nu\rangle + \frac{1}{\sqrt{15}}|{}^4T_{1\frac{5}{2}}\,U'\nu\rangle, \end{aligned}\right\} \qquad (12.91)$$

where the symbols in the kets are all octahedral group classifiers.

To determine the other $E'\alpha'$ and $E''\alpha''$ components (under D_4^*) we merely note that E' and E'' under O^* become E' and E'' under D_4^*, while $U'\lambda$ and $-U'\nu$ become $E'\alpha'$ and $E''\alpha''$, respectively, under D_4^* (Table A 16). Therefore the other two $E'\alpha'$ components under D_4^* are

$$\phi_1 = |{}^6A_1\, U'\lambda\rangle \quad \text{and} \quad \phi_3 = |{}^2T_2\, U'\lambda\rangle.$$

The $E''\alpha''$ components are

$$\psi_1 = -|{}^6A_1\, U'\nu\rangle, \quad \psi_2 = |{}^6A_1\, E''\alpha''\rangle,$$
$$\psi_4 = -|{}^2T_2\, U'\nu\rangle, \quad \psi_5 = |{}^2T_2\, E''\alpha''\rangle.$$

The matrices of spin-orbit coupling are now written straight down using Table A 36. We write E_6, E_4, E_2 for the electrostatic energies of ${}^6A_1$, ${}^4T_1$, ${}^2T_2$, respectively, and define the linear combinations of the ψ_i

$$\psi_1' = -\frac{\sqrt{5}}{\sqrt{6}}\psi_1 + \frac{1}{\sqrt{6}}\psi_2, \quad \psi_4' = \frac{1}{\sqrt{3}}\psi_4 - \frac{\sqrt{2}}{\sqrt{3}}\psi_5,$$
$$\psi_2' = -\frac{1}{\sqrt{6}}\psi_1 - \frac{\sqrt{5}}{\sqrt{6}}\psi_2, \quad \psi_5' = \frac{\sqrt{2}}{\sqrt{3}}\psi_4 + \frac{1}{\sqrt{3}}\psi_5.$$

They have the advantage that ψ_1' has no non-zero matrix element with any other function in our set and ψ_5' has no non-zero matrix element with ψ_3. So $E(\psi_1') = E_6$ and the matrices for the remaining states are given in Table 12.4. This completes our formal derivation and we can only proceed by making assumptions about the values of E_2, E_4, E_6 and ζ.

Table 12.4. *Matrices of electrostatic energy and spin-orbit coupling energy between the lowest terms of a planar d^5 compound (symmetry group D_4^*). For notation for the functions see* § 12.4.12

E'	ϕ_1	ϕ_2	ϕ_3
ϕ_1	E_6	$-\frac{\sqrt{6}}{\sqrt{5}}\zeta$	0
ϕ_2	$-\frac{\sqrt{6}}{\sqrt{5}}\zeta$	E_4	$-\frac{1}{\sqrt{2}}\zeta$
ϕ_3	0	$-\frac{1}{\sqrt{2}}\zeta$	$E_2+\frac{1}{2}\zeta$

E''	ψ_1'
ψ_1'	E_6

E''	ψ_2'	ψ_3	ψ_4'	ψ_5'
ψ_2'	E_6	$-\frac{2}{\sqrt{5}}\zeta$	0	0
ψ_3	$-\frac{2}{\sqrt{5}}\zeta$	E_4	$\frac{\sqrt{3}}{\sqrt{2}}\zeta$	0
ψ_4'	0	$\frac{\sqrt{3}}{\sqrt{2}}\zeta$	$E_2-\frac{1}{2}\zeta$	$\frac{1}{\sqrt{2}}\zeta$
ψ_5'	0	0	$\frac{1}{\sqrt{2}}\zeta$	E_2

First, suppose the sextet lies lowest and that the other states lie sufficiently far above for us to be able to use perturbation theory. Using Table A 19 we discover that ϕ_1, ψ'_2, ψ'_1 are respectively the $M_S = \frac{1}{2}, -\frac{3}{2}, \frac{5}{2}$ states of $^6A_1$. This is why ψ'_1 does not interact with $^4T_1$: in the related ^6S and ^4P terms, ψ'_1 has a larger M_J value than any state in ^4P correlated with the $^4A_2$ part of $^4T_1$. The second-order corrections to the energies of the $M_S = \frac{1}{2}, -\frac{3}{2}, \frac{5}{2}$ states are, respectively, $6\zeta^2/[5(E_6-E_4)]$, $4\zeta^2/[5(E_6-E_4)]$, 0 which can be represented by a quadratic spin-Hamiltonian, as indeed we saw that it must in § 12.3.4. Apart from a shift in the centre of gravity the fine structure part of the spin-Hamiltonian is $D(S_z^2-\frac{35}{12})$ with $D = \zeta^2/5(E_4-E_6)$. D is positive, which is necessary in order to understand the experimental resonance spectrum.

The theory I have just given to account for the g values of haemoglobin fluoride is, of course, oversimplified. The neglected upper part of $^4T_1$ partly cancels the influence of the lower part (exactly if $\mu = 0$). But if Fig. 12.11 is correct this effect will be small except far from the cross-over point, and then E_4-E_6 is too large anyway to produce an appreciable D. I think the majority of the zero-field splitting arises in the way I have described. It is necessary for D to be at least about 5 cm^{-1} to observe a true $g_\perp$ of 6, because otherwise the matrix elements of $\beta\mathbf{H}.(\mathbf{L}+2\mathbf{S})$ to the $M_S = \pm\frac{3}{2}$ states cause $g_\perp$ to be field dependent, and it is difficult to see how else such a large D could arise.

From susceptibility measurements it was suggested by Pauling (1940) that ferrihaemoglobin hydroxide has $S = \frac{3}{2}$. Actually $\mu_{\text{eff}} = 4{\cdot}47$ at room temperature. It appears now to be much more probable that it is very near the cross-over point between $S = \frac{1}{2}$ and $S = \frac{5}{2}$ and is a thermal mixture of these latter two spin states (Griffith, 1956*b*; Gibson, Ingram & Schonland, 1958).

However, $S = \frac{3}{2}$ probably does occur in ferric porphyrin chloride (Gibson *et al.* 1958; Griffith, 1958*b*). Here $g_\parallel = 2$ and $g_\perp = 3{\cdot}8$. We recall our discussion of zero-field splittings in d^3 (§ 12.4.4) and naturally expect these g values to arise from resonance within the $M_S = \pm\frac{1}{2}$ ground doublet of a spatially non-degenerate state with $S = \frac{3}{2}$. For our $^4A_2$ from $^4T_1$ of d^5, this ground doublet is E', the $M_S = \pm\frac{3}{2}$ doublet being E''. We now ask what happens if we suppose $E_4 < E_2$, E_6 and use perturbation theory. The second-order correction to the energies is

$$M_S = \pm\tfrac{1}{2}: \quad E(\phi_2) = -\frac{6\zeta^2}{5(E_6-E_4)}-\frac{\zeta^2}{2(E_2-E_4)},$$

$$M_S = \pm\tfrac{3}{2}: \quad E(\psi_3) = -\frac{4\zeta^2}{5(E_6-E_4)}-\frac{3\zeta^2}{2(E_2-E_4)},$$

and so the $\pm\frac{1}{2}$ doublet lies lower if $2(E_2-E_4) > 5(E_6-E_4)$. This equality is clearly satisfied near the $^6A_1$, $^4T_1$ cross-over point and not satisfied near the $^4T_1$, $^2T_2$ cross-over point. We therefore have a reasonable interpretation of the g values and can deduce that $E_2-E_4 > \frac{5}{2}(E_6-E_4)$.

I have shown in Fig. 12.12 a schematic representation of the cross-over. The proportions are not correct, the ratio of ζ to E_2-E_4 and E_6-E_4 is likely to be less than that implied by the figure. The ground pair of doublets passes from being pure $^6A_1$ through being nearly pure $^4T_1$, at which part E'' passes below E', and finishes up pure $^2T_{2\frac{1}{2}}$. Fig. 12.12 is most inadequate at the strong-field limit because the neglected splittings between the t_2-orbitals then become important in determining the magnetic properties of the ground term.

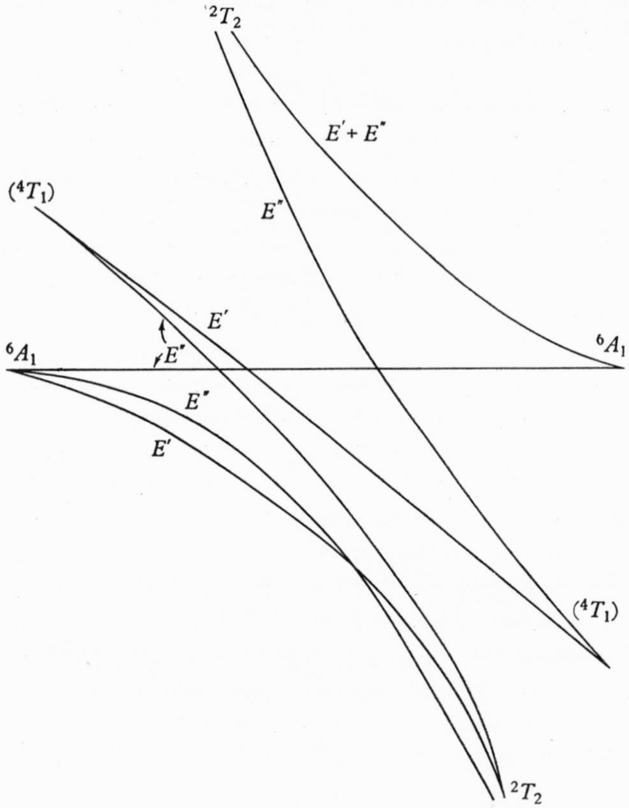

Fig. 12.12. Schematic representation of the details of the cross-over in planar compounds classified according to the representations of O and of D_4^*. The $M_S = \pm\frac{1}{2}$ ground doublet of $^6A_1$ in haemoglobin fluoride and the $M_S = \pm\frac{1}{2}$ ground doublet of ferric porphyrin chloride are both E'.

12.4.13. Planar compounds. As well as haemoglobin and its derivatives there are numerous other compounds in which the symmetry is octahedral but with a large tetragonal component superposed. These are generally compounds which have four ligands coplanar with the metal ion and strongly linked, while the fifth and sixth coordination positions are either empty or contain only more weakly linked groups. In these circumstances it is natural to expect the strongly linked ligands in the plane (XY-plane say) to raise the energy of the $\epsilon = d_{x^2-y^2}$ orbital far above any of the others. The position of the θ-orbital should depend markedly on the presence and nature of the ligands in the fifth and sixth positions. It is difficult to predict its position *a priori*—simple-minded approaches lead to its calculated position being below or above ξ, η, ζ according as to whether one supposes the electrostatic crystal field or molecular orbital resonance integrals to be more important. Under the group D_4, (ξ, η) belong to E and ζ to B_2. Again it is difficult to decide the relative order from *a priori* considerations.

The experimental data from which we might infer the order are electron resonance, paramagnetic susceptibilities and, sometimes, optical spectra. I shall discuss also metal phthalocyanines which are structurally rather similar to metal porphyrins. The experimental difficulties here may be stated briefly: the organic ring systems have intense (electric dipole allowed) transitions which swamp the *d-d* transitions; the compounds tend to aggregate whereupon there are strong metal-metal exchange couplings. Because of this is it very difficult to measure the susceptibilities.

Susceptibilities and resonance data are, however, available for several of them and we discuss these shortly. Before doing so, let us appreciate that it is almost impossible to produce a logically compelling interpretation because there are so many unknown parameters in the theory. The orbital pattern has three, whilst unless we insist that our orbitals are actually *d*-orbitals the spin-orbit coupling has five and there are five separate orbital reduction factors. These last remarks follow from Theorem 8.5 and $(E^2) = A_2$. Nevertheless, there is an orbital pattern which is strongly suggested by general intuition and the available facts. This is as illustrated in Fig. 12.3 but with θ down amongst (ξ, η) and ζ.

My suggestion is that this orbital pattern occurs in all metal porphyrins and phthalocyanines and in all haemoglobin derivatives. The relative positions of ϵ, (ξ, η) and ζ are always the same with energy spacings in the ranges $2\text{–}5 \times 10^4$ and $1\text{–}5 \times 10^3\,\text{cm}^{-1}$, respectively. The position of θ depends upon the nature of the out-of-plane ligands but is near ξ, η and ζ if there are no out-of-plane ligands and above them otherwise. My suggested pattern has an appealing simplicity about it but, like all simple ideas, may well need to be modified considerably when more data are available. We now consider individual compounds.

The g values of copper phthalocyanine were interpreted to show that ϵ lay highest and not θ (§ 12.4.1). The reader may care to verify that, arguing from pure atomic orbitals, they are not reasonably consistent with the top orbital being the pair (ξ, η) or $4p_z$. The latter has been considered in the past to be a possibility. We may be fairly confident that the top orbital is ϵ.

Nickel phthalocyanine and nickel porphyrin are diamagnetic and show no electron resonance. This is interpreted as meaning that the eight electrons are crammed into ξ, η, ζ and θ, while ϵ is empty. The lowest triplets are obtained, in a strong-field scheme, by taking one electron from ξ, η, ζ or θ and putting it in ϵ. They are ^3E, $^3A_2$ and $^3B_1$ and, working in the scheme complementary to d^8 with respect to d^{10}, we immediately calculate their electrostatic energies as $-9B-3C$, $-3C$ and $-12B-3C$ relative to the lowest singlet. Using free-ion values for B and C these are, $-24{,}250$, $-14{,}500$ and $-27{,}500\,\text{cm}^{-1}$, respectively. Hence the diamagnetism is consistent with the orbital pattern.

Cobalt phthalocyanine has $\mu = 2{\cdot}16$ at room temperature and electron resonance gives

$$g_{\parallel} = 1{\cdot}92 \pm 0{\cdot}01, \qquad g_{\perp} = 2{\cdot}90 \pm 0{\cdot}01,$$
$$A = 0{\cdot}017 \pm 0{\cdot}001, \qquad B = 0{\cdot}027 \pm 0{\cdot}001.$$

g is slightly anisotropic in the XY-plane but we ignore this here. Clearly cobalt phthalocyanine is low-spin, both because of the susceptibility and the g values, which would be difficult to understand if it were high-spin (see § 12.4.8). We work in the complementary, d^3, scheme. The three possibilities for the low-spin ground term are $|\epsilon^2\theta\,{}^2A_1\rangle$, $|\epsilon^2(\xi,\eta)\,{}^2E\rangle$ and $|\epsilon^2\zeta\,{}^2B_2\rangle$ with electrostatic energies

$$\left.\begin{aligned} E(^2A_1) &= 3A-8B+4C,\\ E(^2E) &= 3A-3B+4C,\\ E(^2B_2) &= 3A+12B+4C.\end{aligned}\right\} \qquad (12.92)$$

The ${}^2E$ term splits under the first-order influence of spin-orbit coupling to give a pair of Kramers doublets, forming bases for the representations E' and E''

$$\left.\begin{aligned} E'\alpha' &= |\epsilon^2\,1^-\rangle, & E'\beta' &= -|\epsilon^2\,-1^+\rangle,\\ E''\alpha'' &= |\epsilon^2\,-1^-\rangle, & E''\beta'' &= -|\epsilon^2\,1^+\rangle,\end{aligned}\right\} \qquad (12.93)$$

with spin-orbit coupling energies $-\frac{1}{2}\zeta$ for E' and $\frac{1}{2}\zeta$ for E''. This means that E'' lies lower in the complementary scheme and has g values $g_\parallel = 4$, $g_\perp = 0$. Lower symmetry perturbations would alter these somewhat but they are clearly incompatible with the observed g values.

In order for ${}^2B_2$ to be the ground term it would be necessary for the orbital energy of ζ to be at least $15B$ above (ξ,η) and $20B$ above θ. This seems improbable and would imply a very different order from copper phthalocyanine where the g values suggest that ζ has about the same energy as (ξ,η) and is probably below them.

We now assume ${}^2A_1$ to be the ground term and calculate the g values. It is a Kramers doublet forming a basis for the representation E'

$$E'\alpha' = |\epsilon^2\theta^+\rangle, \quad E'\beta' = |\epsilon^2\theta^-\rangle. \qquad (12.94)$$

To first-order in the spin-orbit coupling, the only important excited term is the ${}^2E$ of (12.93). Let it lie at an energy $y^{-1}\zeta$ above ${}^2A_1$, then the first-order modification to the ground-state doublet gives

$$\left.\begin{aligned} E'\alpha' &= |\epsilon^2\theta^+\rangle-\tfrac{1}{2}\sqrt{6}\,y\,|\epsilon^2 1^-\rangle,\\ E'\beta' &= |\epsilon^2\theta^-\rangle-\tfrac{1}{2}\sqrt{6}\,y\,|\epsilon^2-1^+\rangle,\end{aligned}\right\} \qquad (12.95)$$

and with the interaction $\mathbf{L}+2\mathbf{S}$ with the external magnetic field, from which we calculate

$$\left.\begin{aligned} g_\parallel &= 2,\\ g_\perp &= 2-6y.\end{aligned}\right\} \qquad (12.96)$$

In the complementary configuration we must change the sign of y. The experimental $g_\perp$ is fitted by $y = 0{\cdot}15$. The deviation for $g_\parallel$ appears to be due to higher-order terms.

The hyperfine constants A and B follow from (12.96) and the one-electron matrix elements in Table A 41 and are

$$\begin{aligned} A &= P(-\kappa+\tfrac{4}{7}-\tfrac{6}{7}y) &&= P(0{\cdot}44-\kappa),\\ B &= P(-\kappa-\tfrac{2}{7}+\tfrac{45}{7}y) &&= P(0{\cdot}68-\kappa).\end{aligned}$$

The experimental A and B are fitted with $P = 0{\cdot}04$ and $\kappa \approx 0$.

A comment on the low value of κ is appropriate here. κ is normally about $+0{\cdot}3$ for first transition series ions. So κ is anomalously low here by 0·3. This means that in cobalt phthalocyanine there is an extra isotropic interaction with the nuclear moment which can be represented in the Hamiltonian by $a\mathbf{S}\,.\,\mathbf{I}$ with $a = +0{\cdot}012$. This probably arises from the fact that the s-orbitals belong to the representation A_1, as well as $3d_{z^2}$. Therefore the orbital θ will be a mixture of, at least, $3d_{z^2}$ and $4s$. We may write it

$$\theta = 3d_{z^2}\cos\alpha + 4s\sin\alpha.$$

Therefore there will be a Fermi-type interaction of $\theta^{\pm}$ with the nuclear magnetic moment. It is known from optical measurements on the $3d^{10}\,4s\,{}^2S$ ground term of the neutral copper atom that the interaction is $A\mathbf{I}\,.\,\mathbf{S}$ with $A = 0{\cdot}197\,\text{cm}^{-1}$. If we take this value as an order of magnitude estimate for the interaction of a $4s$ orbital in our cobaltous compound, we get $a = A\sin^2\alpha \approx 0{\cdot}2\sin^2\alpha$ and $\sin^2\alpha \approx 0{\cdot}06$. This type of contribution to κ only occurs, of course, when there is a d-function belonging to the unit representation of the site symmetry group and not, therefore, for regular octahedral or tetrahedral symmetry. It is to be expected also for D_3 symmetry.

We have already discussed resonance in ferric porphyrin chloride. Apparently $S = \frac{3}{2}$ and calculations of the electrostatic energies show this to be consistent with Fig. 12.11 *a*. Susceptibility measurements suggest $S = \frac{5}{2}$ so there is a conflict. I think the resonance result is more likely to be reliable here. Finally, ferrous porphyrin (and phthalocyanine) has $\mu = 4{\cdot}0$ at room temperature. This lowers to $\mu = 3{\cdot}3$ at 90° K. There may be a thermal equilibrium between $S = 0$ and $S = 2$ here, or the compound may be in an $S = 1$ ground term. Electrostatic calculations based on my assumed orbital pattern favour the latter.

For general survey, more detail and earlier references see: Gibson, Ingram & Schonland (1958); Griffith (1958*b*).

12.4.14. The specific heat tail. The thermodynamic properties of an aggregate of identical independent metal ions are calculated from the partition function

$$Z = \Sigma\, e^{-E_i/kT}, \tag{12.97}$$

where the sum is over all states of one of the ions. Naturally only those states which are appreciably occupied contribute significantly to Z. Taking the Ni^{++} ion at low temperatures, in zero magnetic field and with the spin-Hamiltonian (12.65) with $E = 0$, the sum (12.97) is over the threefold near-degenerate ground state. Then $E_i = -\frac{2}{3}D,\ +\frac{1}{3}D,\ +\frac{1}{3}D.$

The specific heat is given by

$$C_v = 2kT\frac{\partial \log Z}{\partial T} + kT^2\frac{\partial^2 \log Z}{\partial T^2} \tag{12.98}$$

for each ion and for a gram-ion is NC_v where N is Avogadro's number. Evidently if we know the fine-structure Hamiltonian we can calculate C_v. C_v tends to zero both as $T \to 0$ and as $T \to \infty$ and is only appreciable when kT is comparable with the splitting between the lowest state and one of the higher ones. The specific heat tail refers to the high-temperature side of the region where C_v is appreciable (in our case to the high-temperature side of $T = k^{-1}D$; in general, of course, this will be the low-temperature side of some other contribution to C_v although for the nickel ion this will not normally appear until very high temperatures). The leading term in an expansion of C_v in powers of T^{-1} is the T^{-2} term. The coefficient of this term characterizes the high-temperature behaviour of the specific heat tail and can be determined experimentally. We now see how to calculate it from the fine-structure Hamiltonian.

Expand $e^{-E_i/kT}$ in powers of T^{-1}, retaining the terms up to T^{-2}. Then if n is the number of occupied states

$$Z = n(1-\bar{E}k^{-1}T^{-1}+\tfrac{1}{2}k^{-2}T^{-2}\overline{E^2}),$$

and
$$C_v = \frac{2kT}{Z}\frac{\partial Z}{\partial T}+kT^2\frac{\partial}{\partial T}\left(Z^{-1}\frac{\partial Z}{\partial T}\right),$$

whence after a little algebra

$$C_v = k^{-1}(\overline{E^2}-\bar{E}^2)\,T^{-2},$$

where $\bar{E}$ and $\overline{E^2}$ are, respectively, the mean and mean squares of the E_i. It is customary to list the quantity $t = (T^2C_v/k)_{T=\infty}$ and to measure the parameters occurring in E_i in units of temperature, not energy. In these units $t = \overline{E^2}-\bar{E}^2$. Using the fact that the E_i are the solutions of the secular equation associated with a fine-structure matrix, V_{ij} say, we deduce $t = (1/n)\sum_{i,j} V_{ij}^2$, when $\bar{E} = 0$.

To return to the nickel ion, $\bar{E} = 0$ and $\overline{E^2} = \frac{2}{9}D^2$. Therefore $t = \frac{2}{9}D^2$. If we had included the term $E(S_x^2-S_y^2) = \frac{1}{2}E\,[(S^+)^2-(S^-)^2]$ it would then have been easier to use the alternative expression for t. The V_{ij}^2 are worked out at once and are

	1	0	−1
1	$\frac{1}{3}D$	0	E
0	0	$-\frac{2}{3}D$	0
−1	E	0	$\frac{1}{3}D$

Hence $t = \frac{2}{9}(D^2+3E^2)$.

In our discussion we have neglected the influence of hyperfine interactions. They are usually important at lower temperatures.

Example

Using the spin-Hamiltonian (12.72), show that $t = \frac{14}{5}(D^2+3E^2)$ for the chromous ion.

12.5. Nuclear magnetic resonance in cobaltic compounds

Nuclear magnetic resonance is paramagnetic resonance using the nuclear spin **I** and magnetic moment $\boldsymbol{\mu} = \gamma\beta_N\mathbf{I}$. **I** refers to the lowest nuclear level and is the total angular momentum vector for the nucleus. It is customary to do nuclear resonance on compounds which have singlet electronic ground states and we discuss this case here for octahedrally coordinated cobaltic compounds.

The direct interaction of the nuclear magnetic moment with an external magnetic field **H** is $-\boldsymbol{\mu}.\mathbf{H}$. However, the actual interaction is different because the field at the nucleus is modified by the magnetic moment induced in the electronic system by the external field. It is conventional to write the actual first-order interaction with **H** as $-(1-\sigma)\boldsymbol{\mu}.\mathbf{H}$. So $\sigma\boldsymbol{\mu}.\mathbf{H}$ represents the modification and σ is called the chemical screening constant.

σ has two parts. One comes from the nuclear hyperfine interaction described in § 12.2.2 and the other from the diamagnetic part of the Hamiltonian for the interaction of the electronic system with magnetic fields. We now evaluate these in order.

We know from § 10.3 that the strong-field ground term, correct to first order in the magnetic field H along OZ, is

$$\psi = |t_2^6\,{}^1A_1\rangle + \frac{2\sqrt{2}\,\beta H}{E}|t_2^5 e\,{}^1T_1\,0\rangle, \tag{12.99}$$

where E is the energy separation between t_2^6 and $t_2^5e\,{}^1T_1$. As both the ground and excited terms are singlets, the only part of the nuclear hyperfine interaction (12.22) which contributes is $P\mathbf{L}.\mathbf{I}$. Therefore, to first order in H, we find from (12.99) that the induced coupling to the nuclear moment is

$$\langle\overline{\psi}|\,P\mathbf{L}.\mathbf{I}\,|\psi\rangle = -\frac{16PI_z\,\beta H}{E}. \tag{12.100}$$

By writing down ψ for an arbitrary direction of **H**, (12.100) is immediately generalized to

$$\langle\overline{\psi}|\,P\mathbf{L}.\mathbf{I}\,|\psi\rangle = -\frac{16P\beta}{E}\mathbf{I}.\mathbf{H}$$

$$= -\frac{32\beta^2}{E}\overline{r^{-3}}\boldsymbol{\mu}.\mathbf{H}, \tag{12.101}$$

using $P = 2\gamma\beta\beta_N\overline{r^{-3}}$. The contribution to the chemical screening constant is the coefficient of $\boldsymbol{\mu}.\mathbf{H}$ and is

$$\sigma_1 = -\frac{32\beta^2}{E}\overline{r^{-3}}. \tag{12.102}$$

To obtain the diamagnetic contribution we go back to (5.43), with $f = 1$. The diamagnetic term in the Hamiltonian for the electronic system is $(e^2/2mc^2)\mathbf{A}^2$, where **A** is the sum of the vector potential $\mathbf{A}_1 = \frac{1}{2}\mathbf{H}\wedge\mathbf{r}$ for the static external field and $\mathbf{A}_2 = (\boldsymbol{\mu}\wedge\mathbf{r}/r^3)$ for the magnetic field due to the nuclear moment. It is only the cross-terms in $(\mathbf{A}_1+\mathbf{A}_2)^2$ which give a coupling between the nuclear moment and the external field. Hence the coupling is

$$\frac{e^2}{mc^2}\mathbf{A}_1.\mathbf{A}_2 = \frac{e^2}{2mc^2r^3}(\mathbf{H}\wedge\mathbf{r}).(\boldsymbol{\mu}\wedge\mathbf{r})$$

$$= \frac{e^2}{2mc^2r^3}[r^2\mathbf{H}.\boldsymbol{\mu}-(\mathbf{H}.\mathbf{r})(\boldsymbol{\mu}.\mathbf{r})]. \tag{12.103}$$

Now if we evaluate (12.103) over the $t_2^6\,{}^1A_1$ ground term only those parts which belong to the spatial representation A_1 contribute. In $(\mathbf{H}.\mathbf{r})(\boldsymbol{\mu}.\mathbf{r})$ we have a sum of products like $H_x \mu_x x^2$ and $H_x \mu_y xy$. The only combination of x^2, xy, etc., which transforms as A_1 is $x^2+y^2+z^2 = r^2$ and hence the mean values over angle satisfy

$$\overline{xy} = \overline{yz} = \overline{zx} = 0, \quad \overline{x^2} = \overline{y^2} = \overline{z^2} = \tfrac{1}{3}r^2.$$

Introducing this finding, (12.103) simplifies to

$$\frac{e^2}{2mc^2r^3}(r^2\mathbf{H}\,.\,\boldsymbol{\mu} - \tfrac{1}{3}r^2\mathbf{H}\,.\,\boldsymbol{\mu}) = \frac{e^2}{3mc^2r}\mathbf{H}\,.\,\boldsymbol{\mu},$$

giving a contribution
$$\sigma_2 = \frac{e^2}{3mc^2}\overline{r^{-1}}, \tag{12.104}$$

to the chemical screening constant.

If we now take the Z-axis along the direction of the external field, the energy levels of the nucleus are at $-(1-\sigma)\gamma\beta_N HM_I$ and for absorption and emission of electromagnetic radiation we have the usual selection rule $\Delta M_I = \pm 1$. Therefore we may expect one line at a frequency ν satisfying

$$h\nu = (1-\sigma)\gamma\beta_N H, \tag{12.105}$$

where
$$\sigma = -\frac{32\beta^2}{E}\overline{r^{-3}} + \frac{e^2}{3mc^2}\overline{r^{-1}}. \tag{12.106}$$

Assuming that $\overline{r^{-3}}$ and $\overline{r^{-1}}$ remain the same for all cobaltic compounds we deduce that for a constant external field H there is a linear relation between ν and the wave-length of the lowest ${}^1A_1$-${}^1T_1$ transition (Griffith & Orgel, 1957*a*). Experiments have been made by Proctor & Yu (1951) and Freeman, Murray & Richards (1957) and agree with the theoretical conclusion. In Fig. 12.13 are plotted results obtained by Freeman *et al.* (1957) for compounds for each of which there are six identical adjacent atoms arranged octahedrally.[1] It is seen that a good straight line is obtained. The slope gives $32\beta^2\overline{r^{-3}} = 450 \pm 30\,\text{cm}^{-1}$.

From (12.105) and (12.106) we see that if we know $\overline{r^{-1}}$ as well as $\overline{r^{-3}}$ we can calculate γ from ν and H. An approximate theoretical formula for $\overline{r^{-1}}$ for atoms was given by Lamb (1941) and leads to $\sigma_2 = 0{\cdot}0021$. This is fairly small but it is rather difficult to assess its reliability (perhaps 10–20 %).

Using this value Freeman *et al.* (1957) calculated γ. Because of the difficulty of measuring the absolute value of H accurately they actually compared γ for cobalt with γ for ^{23}Na, which is accurately known,[2] using a ratio of ν for ^{60}Co and ^{23}Na in $K_3Co(CN)_6$ solution measured by Proctor & Yu (1951). This latter ratio can be measured accurately by using the same solution and hence the same magnetic field for both ions. Their result was $\mu = \gamma I = \frac{7}{2}\gamma = 4{\cdot}583$ and the accuracy is probably to within 0·01.

The reader may ask what effect will the replacement of dt_2 orbitals by more general t_2-orbitals have here. First, it will introduce orbital reduction of σ_1 by a factor k'^2 in the notation of (10.43). This will not affect the preceding analysis

[1] Data are also given for non-octahedral compounds. Here ν is different for different directions but molecular rotation in the liquid averages it to a single observed line which fits Fig. 12.13 quite well for the measured compounds.

[2] Using also an estimate of σ_2 for sodium (and $\sigma_1 = 0$). Note also that there is a dependence on temperature for σ, via (12.106) because E depends slightly on temperature.

and derivation of μ at all except that $\overline{r^{-3}}$ becomes replaced by $k'^2\overline{r^{-3}}$. Secondly, there will now be matrix elements between the t_2-orbitals of the ground term and more than one excited set of orbitals. So (12.99) now involves a sum over excited

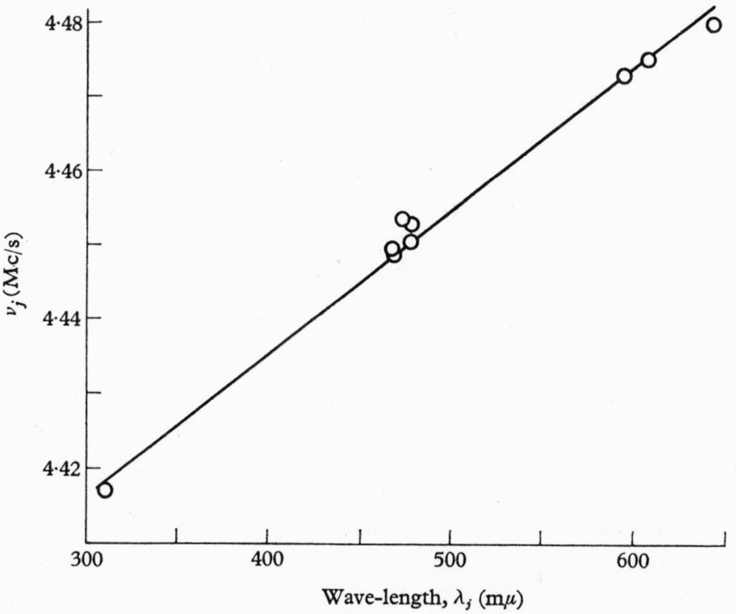

Fig. 12.13. Nuclear magnetic resonance frequency plotted against wave-length for a number of octahedral low-spin cobaltic compounds (after Freeman, Murray & Richards, 1957).

$t_2^5 b\, {}^1T_1$ terms where b is now an orbital belonging to the A_2, E, T_1 or T_2 representation of the octahedral group. These extra terms, however, will have high energies and small coefficients and be unlikely to contribute very much to σ_1. Apart from this type of refinement, however, a more accurate theory would include the departure of the ground electronic wave-function from being $t_2^6\, {}^1A_1$. The most important departure is through electrostatic interaction (Table A 29); there would also be a departure through spin-orbit interaction (Table A 37).

APPENDIX 1

THE ELEMENTS

A1.1. Names, symbols and atomic numbers

1	H	Hydrogen	35	Br	Bromine	69	Tm	Thulium
2	He	Helium	36	Kr	Krypton	70	Yb	Ytterbium
3	Li	Lithium	37	Rb	Rubidium	71	Lu	Lutetium
4	Be	Beryllium	38	Sr	Strontium	72	Hf	Hafnium
5	B	Boron	39	Y	Yttrium	73	Ta	Tantalum
6	C	Carbon	40	Zr	Zirconium	74	W	Tungsten
7	N	Nitrogen	41	Nb	Niobium	75	Re	Rhenium
8	O	Oxygen	42	Mo	Molybdenum	76	Os	Osmium
9	F	Fluorine	43	Tc	Technetium	77	Ir	Iridium
10	Ne	Neon	44	Ru	Ruthenium	78	Pt	Platinum
11	Na	Sodium	45	Rh	Rhodium	79	Au	Gold
12	Mg	Magnesium	46	Pd	Palladium	80	Hg	Mercury
13	Al	Aluminium	47	Ag	Silver	81	Tl	Thallium
14	Si	Silicon	48	Cd	Cadmium	82	Pb	Lead
15	P	Phosphorus	49	In	Indium	83	Bi	Bismuth
16	S	Sulphur	50	Sn	Tin	84	Po	Polonium
17	Cl	Chlorine	51	Sb	Antimony	85	At	Astatine
18	A	Argon	52	Te	Tellurium	86	Rn	Radon
19	K	Potassium	53	I	Iodine	87	Fr	Francium
20	Ca	Calcium	54	Xe	Xenon	88	Ra	Radium
21	Sc	Scandium	55	Cs	Caesium	89	Ac	Actinium
22	Ti	Titanium	56	Ba	Barium	90	Th	Thorium
23	V	Vanadium	57	La	Lanthanum	91	Pa	Protoactinium
24	Cr	Chromium	58	Ce	Cerium	92	U	Uranium
25	Mn	Manganese	59	Pr	Praseodymium	93	Np	Neptunium
26	Fe	Iron	60	Nd	Neodymium	94	Pu	Plutonium
27	Co	Cobalt	61	Pm	Promethium	95	Am	Americium
28	Ni	Nickel	62	Sm	Samarium	96	Cm	Curium
29	Cu	Copper	63	Eu	Europium	97	Bk	Berkelium
30	Zn	Zinc	64	Gd	Gadolinium	98	Cf	Californium
31	Ga	Gallium	65	Tb	Terbium	99	Es	Einsteinium
32	Ge	Germanium	66	Dy	Dysprosium	100	Fm	Fermium
33	As	Arsenic	67	Ho	Holmium	101	Md	Mendelevium
34	Se	Selenium	68	Er	Erbium	102	No	Nobelium

A1.2. Properties of transition-metal and neighbouring atoms and ions

(The closed shells comprise $1s^2\,2s^2\,2p^6\,3s^2\,3p^6$ for the first series, $1s^2\,2s^2\,2p^6\,3s^2\,3p^6\,3d^{10}\,4s^2\,4p^6$ for the second and $1s^2\,2s^2\,2p^6\,3s^2\,3p^6\,3d^{10}\,4s^2\,4p^6\,4d^{10}\,4f^{14}\,5s^2\,5p^6$ for the third. C = configuration outside the closed shells. T = ground term. I = ionization potential.)

	M I			M II			M III		
	C	*T*	*I*	*C*	*T*	*I*	*C*	*T*	*I*
First series									
Calcium	$4s^2$	^1S	6·11	$4s$	^2S	11·87	—	^1S	51·21
Scandium	$3d\,4s^2$	^2D	6·54	$3d\,4s$	^3D	12·80	$3d$	^2D	24·75
Titanium	$3d^2\,4s^2$	^3F	6·82	$3d^2\,4s$	^4F	13·57	$3d^2$	^3F	27·47
Vanadium	$3d^3\,4s^2$	^4F	6·74	$3d^4$	^5D	14·65	$3d^3$	^4F	29·31
Chromium	$3d^5\,4s$	^7S	6·76	$3d^5$	^6S	16·49	$3d^4$	^5D	30·95
Manganese	$3d^5\,4s^2$	^6S	7·43	$3d^5\,4s$	^7S	15·64	$3d^5$	^6S	33·69
Iron	$3d^6\,4s^2$	^5D	7·87	$3d^6\,4s$	^6D	16·18	$3d^6$	^5D	30·64
Cobalt	$3d^7\,4s^2$	^4F	7·86	$3d^8$	^3F	17·05	$3d^7$	^4F	33·49
Nickel	$3d^9\,4s$	^3D	7·63	$3d^9$	^2D	18·15	$3d^8$	^3F	35·16
Copper	$3d^{10}\,4s$	^2S	7·72	$3d^{10}$	^1S	20·29	$3d^9$	^2D	36·83
Zinc	$3d^{10}\,4s^2$	^1S	9·39	$3d^{10}\,4s$	^2S	17·96	$3d^{10}$	^1S	39·70
Second series									
Strontium	$5s^2$	^1S	5·69	$5s$	^2S	11·03	—	—	—
Yttrium	$4d\,5s^2$	^2D	6·38	$5s^2$	^1S	12·23	$4d$	^2D	20·5
Zirconium	$4d^2\,5s^2$	^3F	6·84	$4d^2\,5s$	^4F	13·13	$4d^2$	^3F	22·98
Niobium	$4d^4\,5s$	^6D	6·88	$4d^4$	^5D	14·32	$4d^3$	^4F	25·04
Molybdenum	$4d^5\,5s$	^7S	7·10	$4d^5$	^6S	16·15	$4d^4$	^5D	27·13
Technetium	$4d^5\,5s^2$	^6S	7·28	$4d^5\,5s$	^7S	15·26	—	—	—
Ruthenium	$4d^7\,5s$	^5F	7·36	$4d^7$	^4F	16·76	$4d^6$	^5D	28·46
Rhodium	$4d^8\,5s$	^4F	7·46	$4d^8$	^3F	18·07	$4d^7$	^4F	31·05
Palladium	$4d^{10}$	^1S	8·33	$4d^9$	^2D	19·42	$4d^8$	^3F	32·92
Silver	$4d^{10}\,5s$	^2S	7·57	$4d^{10}$	^1S	21·48	$4d^9$	^2D	34·82
Cadmium	$4d^{10}\,5s^2$	^1S	8·99	$4d^{10}\,5s$	^2S	16·90	$4d^{10}$	^1S	37·47
Third series									
Hafnium	$5d^2\,6s^2$	^3F	(7)	$5d\,6s^2$	^2D	14·9	—	—	—
Tantalum	$5d^3\,6s^2$	^4F	7·88	$5d^3\,6s$	^5F	(16·2)	—	—	—
Tungsten	$5d^4\,6s^2$	^5D	7·98	$5d^4\,6s$	^6D	(17·7)	—	—	—
Rhenium	$5d^5\,6s^2$	^6S	7·87	$5d^5\,6s$	^7S	(16·6)	—	—	—
Osmium	$5d^6\,6s^2$	^5D	8·7	$5d^6\,6s$	^6D	(17)	—	—	—
Iridium	$5d^7\,6s^2$	^4F	(9)	—	—	—	—	—	—
Platinum	$5d^9\,6s$	^3D	9·0	$5d^9$	^2D	18·56	—	—	—
Gold	$5d^{10}\,6s$	^2S	9·22	$5d^{10}$	^1S	20·5	—	—	—
Mercury	$5d^{10}\,6s^2$	^1S	10·43	$5d^{10}\,6s$	^2S	18·75	$5d^{10}$	^1S	34·2

The data are from Moore, Atomic energy levels, *National Bureau of Standards*, no. 467 (1949, 1952, 1958). Values in brackets are rather doubtful. The term assignments assume Russell–Saunders coupling. The ions M III which are missing from the above tabulation should all belong to the configuration d^n. The same is true for all higher states of ionization having electrons outside the closed shells.

A1.3. Some properties of nuclei

(Natural abundances, spins and magnetic moments are given for nuclei which occur in important ligands and for the first three transition series. It is probable that all the nuclei with an even number of protons and neutrons (even-even nuclei) have spin zero. The spin is given in brackets for such nuclei when direct experimental proof is lacking. The magnetic moments are as directly measured, without allowance for electronic shielding (see § 12.5). Naturally occurring elements which are very weakly radio-active are marked with one, and other radio-active elements with two, asterisks. Reference for all data: Strominger, Hollander & Seaborg, *Rev. Mod. Phys.* (1958), **30**, 585.)

Element	%	I	μ	Element	%	I	μ
^{1}H	99·985	½	+2·79270	^{54}Fe	5·84	(0)	—
^{2}H	0·015	1	+0·857393	^{56}Fe	91·68	(0)	—
^{3}H**	—	½	+2·9788	^{57}Fe	2·17	½	<0·05
^{12}C	98·892	0	—	^{58}Fe*	0·31	(0)	—
^{13}C	1·108	½	+0·702205	^{59}Co	100	7/2	+4·6399
^{14}C**	—	0	—	^{58}Ni	67·76	(0)	—
^{14}N	99·635	1	+0·40357	^{60}Ni	26·16	(0)	—
^{15}N	0·365	½	−0·2830	^{61}Ni	1·25	—	~0
^{16}O	99·759	0	—	^{62}Ni	3·66	(0)	—
^{17}O	0·037	5/2	−1·8930	^{64}Ni	1·16	(0)	—
^{18}O	0·204	0	—	^{63}Cu	69·1	3/2	+2·221
^{19}F	100	½	+2·6275	^{65}Cu	30·9	3/2	+2·380
^{28}Si	92·2	(0)	—	^{64}Zn*	48·89	(0)	~0
^{29}Si	4·7	½	±0·5548	^{66}Zn	27·81	(0)	~0
^{30}Si	3·1	(0)	—	^{67}Zn	4·11	5/2	+0·8735
^{31}P	100	½	+1·1305	^{68}Zn	18·56	(0)	~0
^{32}S	95·0	0	—	^{70}Zn*	0·62	(0)	—
^{33}S	0·76	3/2	+0·6427	^{89}Y	100	½	−0·13683
^{34}S	4·22	(0)	—	^{90}Zr	51·46	(0)	—
^{36}S	0·02	(0)	—	^{91}Zr	11·23	5/2	−1·29803
^{35}Cl	75·5	3/2	+0·82091	^{92}Zr	17·11	(0)	—
^{37}Cl	24·5	3/2	+0·6833	^{94}Zr	17·40	(0)	—
^{79}Br	50·5	3/2	+2·0992	^{96}Zr*	2·80	(0)	—
^{81}Br	49·5	3/2	+2·2625	^{93}Nb	100	9/2	+6·144
^{127}I	100	5/2	+2·7935	^{92}Mo*	15·86	(0)	~0
^{45}Sc	100	7/2	+4·749	^{94}Mo	9·12	(0)	~0
^{46}Ti	8·0	(0)	—	^{95}Mo	15·70	(5/2)	−0·9290
^{47}Ti	7·5	5/2	−0·7871	^{96}Mo	16·50	(0)	~0
^{48}Ti	73·7	(0)	—	^{97}Mo	9·45	(5/2)	−0·9485
^{49}Ti	5·5	7/2	−1·1023	^{98}Mo	23·75	(0)	~0
^{50}Ti	5·3	(0)	—	^{100}Mo*	9·62	(0)	~0
^{50}V	0·25	6	+3·3413	^{99}Tc**	—	9/2	+5·657
^{51}V	99·75	7/2	+5·139	^{96}Ru	5·6	(0)	—
^{50}Cr	4·31	(0)	—	^{98}Ru	2·0	(0)	—
^{52}Cr	83·76	(0)	—	^{99}Ru	12·7	5/2	−0·6
^{53}Cr	9·55	3/2	−0·4735	^{100}Ru	12·7	(0)	—
^{54}Cr	2·38	(0)	—	^{101}Ru	17·0	5/2	−0·7
^{55}Mn	100	5/2	+3·4614	^{102}Ru	31·5	(0)	—

Element	%	I	μ	Element	%	I	μ
^{104}Ru	18·5	(0)	—	^{183}W	14·4	$\frac{1}{2}$	+0·115
^{103}Rh	100	$\frac{1}{2}$	−0·0879	^{184}W	30·6	(0)	—
^{102}Pd	0·9	(0)	—	^{186}W*	28·4	(0)	—
^{104}Pd	10·1	(0)	—	^{185}Re	37·07	$\frac{5}{2}$	+3·144
^{105}Pd	22·4	$\frac{5}{2}$	−0·57	^{187}Re*	62·93	$\frac{5}{2}$	+3·176
^{106}Pd	27·3	(0)	—	^{184}Os	0·018	(0)	—
^{108}Pd	26·7	(0)	—	^{186}Os	1·59	(0)	—
^{110}Pd	12·6	(0)	—	^{187}Os*	1·64	$\frac{1}{2}$	+0·12
^{107}Ag	51·35	$\frac{1}{2}$	−0·1130	^{188}Os	13·3	(0)	—
^{109}Ag	48·65	$\frac{1}{2}$	−0·1299	^{189}Os	16·1	$\frac{3}{2}$	+0·6507
^{106}Cd*	1·22	(0)	—	^{190}Os	26·4	(0)	—
^{108}Cd	0·88	(0)	—	^{192}Os*	41·0	(0)	—
^{110}Cd	12·39	(0)	~0	^{191}Ir	38·5	$\frac{3}{2}$	+0·2
^{111}Cd	12·75	$\frac{1}{2}$	−0·5922	^{193}Ir	61·5	$\frac{3}{2}$	+0·2
^{112}Cd	24·07	(0)	~0	^{190}Pt*	0·012	(0)	—
^{113}Cd	12·26	$\frac{1}{2}$	−0·6195	^{192}Pt*	0·78	(0)	—
^{114}Cd	28·86	(0)	~0	^{194}Pt	32·8	(0)	~0
^{116}Cd*	7·58	(0)	~0	^{195}Pt	33·8	$\frac{1}{2}$	+0·6004
^{174}Hf	0·2	(0)	—	^{196}Pt	25·3	(0)	~0
^{176}Hf	5·2	(0)	—	^{198}Pt*	7·2	(0)	—
^{177}Hf	18·5	$\frac{7}{2}$	+0·61	^{197}Au	100	$\frac{3}{2}$	+0·14
^{178}Hf	27·1	(0)	~0	^{196}Hg	0·146	0	—
^{179}Hf	13·7	$\frac{9}{2}$	−0·47	^{198}Hg	10·02	(0)	~0
^{180}Hf	35·2	(0)	~0	^{199}Hg	16·84	$\frac{1}{2}$	+0·4993
^{180}Ta*	0·0123	—	—	^{200}Hg	23·13	(0)	~0
^{181}Ta	99·9877	$\frac{7}{2}$	+2·1	^{201}Hg	13·22	$\frac{3}{2}$	−0·607
^{180}W	0·135	(0)	—	^{202}Hg	29·80	(0)	~0
^{182}W	26·4	(0)	—	^{204}Hg	6·85	(0)	~0

APPENDIX 2

TABLES

Table A 1. *The character table of the group D_2*

D_2	1	C_2^x	C_2^y	C_2^z
A_1	1	1	1	1
B_1	1	-1	-1	1
B_2	1	-1	1	-1
B_3	1	1	-1	-1

Table A 2. *The character table for D_{3v} and D_6*

(The classes of D_6 are written above those of D_{3v} and its representations to the left. P represents inversion. The top left-hand corner gives the character table for D_3 when the suffix g is omitted. $2C_3$ means that there are two elements in the class of rotations of type C_3.)

	D_6	1	$2C_3$	$3C_2'$	C_2	$2C_6$	$3C_2''$
D_6	D_{3v}	1	$2C_3$	$3C_2'$	P	$2S_6$	$3\sigma_v$
A_1	A_{1g}	1	1	1	1	1	1
A_2	A_{2g}	1	1	-1	1	1	-1
E_2	E_g	2	-1	0	2	-1	0
B_1	A_{1u}	1	1	1	-1	-1	-1
B_2	A_{2u}	1	1	-1	-1	-1	1
E_1	E_u	2	-1	0	-2	1	0

Table A 3. *The character table for D_4 and C_{4v}*

C_{4v}	1	C_2^z	$2C_4$	$\sigma_v^{yz}, \sigma_v^{zx}$	$2\sigma_v'$
D_4	1	C_2^z	$2C_4$	C_2^x, C_2^y	$2C_2'$
A_1	1	1	1	1	1
A_2	1	1	1	-1	-1
B_1	1	1	-1	1	-1
B_2	1	1	-1	-1	1
E	2	-2	0	0	0

Table A 4. *The character table for D_5*

D_5	1	$2C_5$	$2C_5^2$	$5C_2$
A_1	1	1	1	1
A_2	1	1	1	-1
E_1	2	$\frac{1}{2}(\sqrt{5}-1)$	$-\frac{1}{2}(\sqrt{5}+1)$	0
E_2	2	$-\frac{1}{2}(\sqrt{5}+1)$	$\frac{1}{2}(\sqrt{5}-1)$	0

Table A 5. *The character table for T_d and O. Mulliken's notation for the representations is to the left of Bethe's*

T_d	1	$8C_3$	$3C_2$	$6\sigma_d$	$6S_4$
O	1	$8C_3$	$3C_2$	$6C_2'$	$6C_4$
A_1, Γ_1	1	1	1	1	1
A_2, Γ_2	1	1	1	-1	-1
E, Γ_3	2	-1	2	0	0
T_1, Γ_4	3	0	-1	-1	1
T_2, Γ_5	3	0	-1	1	-1

Table A 6. *The character table for T. The representation E is irreducible in terms of real matrices.* $\omega = \frac{1}{2}(-1+i\sqrt{3})$

T		1	$3C_2$	$4C_3$	$4C_3'$
A_1	Γ_1	1	1	1	1
E	Γ_2	1	1	ω	ω^2
	Γ_3	1	1	ω^2	ω
T	Γ_4	3	-1	0	0

Table A 7. *The character table for the icosahedral group K*

K	1	$20C_3$	$15C_2$	$12C_5$	$12C_5^2$
A	1	1	1	1	1
T_1	3	0	-1	$\frac{1}{2}(1+\sqrt{5})$	$\frac{1}{2}(1-\sqrt{5})$
T_2	3	0	-1	$\frac{1}{2}(1-\sqrt{5})$	$\frac{1}{2}(1+\sqrt{5})$
U	4	1	0	1	-1
V	5	-1	1	0	0

Table A 8. *The 'two-valued' representations of the most important rotation groups (and of T_d^*)*

(The representation spanned by the two spin functions α, β is always written E'. These supplement Tables A 1–7 and the notation corresponds. We write R for the unique rotation through 360°. Bethe's notation for representations is put in brackets. Note that $C_n^n = R$, not 1.)

D_2^*	1	R	$2C_2^x$	$2C_2^y$	$2C_2^z$
$E'(\Gamma_5)$	2	-2	0	0	0

D_3^*		1	R	$2C_3$	$2C_3R$	$3C_2'$	$3C_2'R$
E'		2	-2	1	-1	0	0
E''	ρ_1	1	-1	-1	1	i	$-i$
	ρ_2	1	-1	-1	1	$-i$	i

D_4^*	1	R	$2C_2^z$	$2C_4$	$2C_4R$	$4C_2^{x,y}$	$4C_2'$
$E'(\Gamma_6)$	2	-2	0	$\sqrt{2}$	$-\sqrt{2}$	0	0
$E''(\Gamma_7)$	2	-2	0	$-\sqrt{2}$	$\sqrt{2}$	0	0

D_5^*		1	R	$2C_5$	$2C_5R$	$2C_5^2$	$2C_5^2R$	$5C_2$	$5C_2R$
E'		2	-2	$\frac{1}{2}(\sqrt{5}+1)$	$-\frac{1}{2}(\sqrt{5}+1)$	$\frac{1}{2}(\sqrt{5}-1)$	$-\frac{1}{2}(\sqrt{5}-1)$	0	0
E''		2	-2	$-\frac{1}{2}(\sqrt{5}-1)$	$\frac{1}{2}(\sqrt{5}-1)$	$-\frac{1}{2}(\sqrt{5}+1)$	$\frac{1}{2}(\sqrt{5}+1)$	0	0
E'''	ρ_1	1	-1	-1	1	1	-1	i	$-i$
	ρ_2	1	-1	-1	1	1	-1	$-i$	i

D_6^*	1	R	$2C_3$	$2C_3R$	$6C_2'$	$2C_2$	$2C_6$	$2C_6R$	$6C_2''$
$E'(\Gamma_7)$	2	-2	1	-1	0	0	$\sqrt{3}$	$-\sqrt{3}$	0
$E''(\Gamma_8)$	2	-2	1	-1	0	0	$-\sqrt{3}$	$\sqrt{3}$	0
$E'''(\Gamma_9)$	2	-2	-2	2	0	0	0	0	0

Table A 8 (*cont.*)

T_d^*	1	R	$8C_3$	$8C_3R$	$6C_2$	$12\sigma_d$	$6S_4$	$6S_4R$
O^*	1	R	$8C_3$	$8C_3R$	$6C_2$	$12C_2'$	$6C_4$	$6C_4R$
$E'(\Gamma_6)$	2	-2	1	-1	0	0	$\sqrt{2}$	$-\sqrt{2}$
$E''(\Gamma_7)$	2	-2	1	-1	0	0	$-\sqrt{2}$	$\sqrt{2}$
$U'(\Gamma_8)$	4	-4	-1	1	0	0	0	0

T^*	1	R	$6C_2$	$4C_3$	$4C_3R$	$4C_3'$	$4C_3'R$
E'	2	-2	0	1	-1	-1	1
U' { E''	2	-2	0	ω	$-\omega$	$-\omega^2$	ω^2
U' { E'''	2	-2	0	ω^2	$-\omega^2$	$-\omega$	ω

K^*	1	R	$20C_3$	$20C_3R$	$30C_2$	$12C_5$	$12C_5R$	$12C_5^2$	$12C_5^2R$
E'	2	-2	1	-1	0	$\frac{1}{2}(\sqrt{5}+1)$	$-\frac{1}{2}(\sqrt{5}+1)$	$\frac{1}{2}(\sqrt{5}-1)$	$-\frac{1}{2}(\sqrt{5}-1)$
E''	2	-2	1	-1	0	$-\frac{1}{2}(\sqrt{5}-1)$	$\frac{1}{2}(\sqrt{5}-1)$	$-\frac{1}{2}(\sqrt{5}+1)$	$\frac{1}{2}(\sqrt{5}+1)$
U'	4	-4	-1	1	0	1	-1	-1	1
W'	6	-6	0	0	0	-1	1	1	-1

Table A 9. *Reduction of the direct products of one- and two-valued representations of D_4, O and K*

D_4	A_1	A_2	B_1	B_2	E	E'	E''
A_1	A_1	A_2	B_1	B_2	E	E'	E''
A_2	A_2	A_1	B_2	B_1	E	E'	E''
B_1	B_1	B_2	A_1	A_2	E	E''	E'
B_2	B_2	B_1	A_2	A_1	E	E''	E'
E	E	E	E	E	$A_1+A_2+B_1+B_2$	$E'+E''$	$E'+E''$
E'	E'	E'	E''	E''	$E'+E''$	A_1+A_2+E	B_1+B_2+E
E''	E''	E''	E'	E'	$E'+E''$	B_1+B_2+E	A_1+A_2+E

O	A_1	A_2	E	T_1	T_2	E'	E''	U'
A_1	A_1	A_2	E	T_1	T_2	E'	E''	U'
A_2	A_2	A_1	E	T_2	T_1	E''	E'	U'
E	E	E	A_1+A_2+E	T_1+T_2	T_1+T_2	U'	U'	$E'+E''+U'$
T_1	T_1	T_2	T_1+T_2	$A_1+E+T_1+T_2$	$A_2+E+T_1+T_2$	$E'+U'$	$E''+U'$	$E'+E''+2U'$
T_2	T_2	T_1	T_1+T_2	$A_2+E+T_1+T_2$	$A_1+E+T_1+T_2$	$E''+U'$	$E'+U'$	$E'+E''+2U'$
E'	E'	E''	U'	$E'+U'$	$E''+U'$	A_1+T_1	A_2+T_2	$E+T_1+T_2$
E''	E''	E'	U'	$E''+U'$	$E'+U'$	A_2+T_2	A_1+T_1	$E+T_1+T_2$
U'	U'	U'	$E'+E''+U'$	$E'+E''+2U'$	$E'+E''+2U'$	$E+T_1+T_2$	$E+T_1+T_2$	$A_1+A_2+E+2T_1+2T_2$

K	A	T_1	T_2	U	V
A	A	T_1	T_2	U	V
T_1	T_1	$A+T_1+V$	$U+V$	T_2+U+V	T_1+T_2+U+V
T_2	T_2	$U+V$	$A+T_2+V$	T_1+U+V	T_1+T_2+U+V
U	U	T_2+U+V	T_1+U+V	$A+T_1+T_2+U+V$	T_1+T_2+U+2V
V	V	T_1+T_2+U+V	T_1+T_2+U+V	T_1+T_2+U+2V	$A+T_1+T_2+2U+2V$

K	E'	E''	U'	W'
A	E'	E''	U'	W'
T_1	$E'+U'$	W'	$E'+U'+W'$	$E''+U'+2W'$
T_2	W'	$E''+U'$	$E''+U'+W'$	$E'+U'+2W'$
U	$E''+W'$	$E'+W'$	$U'+2W'$	$E'+E''+2U'+2W'$
V	$U'+W'$	$U'+W'$	$E'+E''+U'+2W'$	$E'+E''+2U'+3W'$
E'	$A+T_1$	U	T_1+V	T_2+U+V
E''	U	$A+T_2$	T_2+V	T_1+U+V
U'	T_1+V	T_2+V	$A+T_1+T_2+U+V$	$T_1+T_2+2U+2V$
W'	[illegible]	[illegible]	[illegible]	[illegible]

Table A 10. *An enumeration of all the distinct proper subgroups of certain important symmetry groups*

Group	Subgroups
O_h	0, T_d, T_h, D_{4h}, T, D_{3v}, D_4, D_{2v}, C_{4h}, C_{4v}, $2D_{2h}$, D_3, C_{3v}, S_6, C_4, S_4, $3C_{2v}$, $2D_2$, $2C_{2h}$, C_3, $2C_2$, $2C_{1h}$, S_2, C_1
O	T, D_4, D_3, $2D_2$, C_4, C_3, $2C_2$, C_1
T_d	T, D_{2v}, C_{3v}, D_2, C_{2v}, S_4, C_3, C_{1h}, C_2, C_1
D_{4h}	D_4, D_{2v}, C_{4h}, C_{4v}, $2D_{2h}$, C_4, S_4, $4C_{2v}$, $2D_2$, $3C_{2h}$, $3C_2$, $3C_{1h}$, S_2, C_1
D_{3v}	D_3, S_6, C_{3v}, C_{2h}, C_3, C_2, C_{1h}, S_2, C_1
T_h	T, D_{2h}, S_6, D_2, C_{2v}, C_{2h}, C_3, C_2, C_{1h}, S_2, C_1
K	T, D_5, D_3, C_5, D_2, C_3, C_2, C_1
K_h	K, T_h, D_{5v}, T, D_{3v}, D_5, C_{5v}, S_{10}, D_{2h}, D_3, S_6, C_{3v}, C_5, D_2, C_{2v}, C_{2h}, C_3, C_2, C_{1h}, S_2, C_1

Table A 11. *Decomposition of the representations of O_h relative to its subgroups T_d, D_{4h}, D_{3v} and T_h. The dashed symbols refer to the two-valued representations*

O_h	T_d	D_{4h}	D_{3v}	T_h
A_{1g}	A_1	A_{1g}	A_{1g}	A_g
A_{1u}	A_2	A_{1u}	A_{1u}	A_u
A_{2g}	A_2	B_{1g}	A_{2g}	A_g
A_{2u}	A_1	B_{1u}	A_{2u}	A_u
E_g	E	$A_{1g}+B_{1g}$	E_g	E_g
E_u	E	$A_{1u}+B_{1u}$	E_u	E_u
T_{1g}	T_1	$A_{2g}+E_g$	$A_{2g}+E_g$	T_g
T_{1u}	T_2	$A_{2u}+E_u$	$A_{2u}+E_u$	T_u
T_{2g}	T_2	$B_{2g}+E_g$	$A_{1g}+E_g$	T_g
T_{2u}	T_1	$B_{2u}+E_u$	$A_{1u}+E_u$	T_u
E'_g	E'	E'_g	E'_g	E'_g
E'_u	E''	E'_u	E'_u	E'_u
E''_g	E''	E''_g	E'_g	E'_g
E''_u	E'	E''_u	E'_u	E'_u
U'_g	U'	$E'_g+E''_g$	$E'_g+E''_g$	$E''_g+E'''_g$
U'_u	U'	$E'_u+E''_u$	$E'_u+E''_u$	$E''_u+E'''_u$

Table A 12. *Decomposition of the representations of K relative to its subgroups D_5, D_3 and D_2. The dashed symbols refer to the two-valued representations*

K	D_5	D_3	D_2
A	A_1	A_1	A_1
T_1	A_2+E_1	A_2+E	$B_1+B_2+B_3$
T_2	A_2+E_2	A_2+E	$B_1+B_2+B_3$
U	E_1+E_2	A_1+A_2+E	$A_1+B_1+B_2+B_3$
V	$A_1+E_1+E_2$	A_1+2E	$2A_1+B_1+B_2+B_3$
E'	E'	E'	E'
E''	E''	E'	E'
U'	$E'+E''$	$E'+E''$	$2E'$
W'	$E'+E''+E'''$	$2E'+E''$	$3E'$

Table A 13. *Table of values of the character χ^J of various rotations C_n and their powers C_n^m for atomic levels with $J \leqslant 6$. For the meaning of C_n for J non-integral, see* §§ 6.9–11

J	C_1	C_2	C_3	C_3^2	C_4	C_5	C_5^2	C_6
0	1	1	1	1	1	1	1	1
$\frac{1}{2}$	-2	0	1	-1	$\sqrt{2}$	$\frac{1}{2}(\sqrt{5}+1)$	$\frac{1}{2}(\sqrt{5}-1)$	$\sqrt{3}$
1	3	-1	0	0	1	$\frac{1}{2}(\sqrt{5}+1)$	$\frac{1}{2}(1-\sqrt{5})$	2
$1\frac{1}{2}$	-4	0	-1	1	0	1	-1	$\sqrt{3}$
2	5	1	-1	-1	-1	0	0	1
$2\frac{1}{2}$	-6	0	0	0	$-\sqrt{2}$	-1	1	0
3	7	-1	1	1	-1	$-\frac{1}{2}(\sqrt{5}+1)$	$\frac{1}{2}(\sqrt{5}-1)$	-1
$3\frac{1}{2}$	-8	0	1	-1	0	$-\frac{1}{2}(\sqrt{5}+1)$	$\frac{1}{2}(1-\sqrt{5})$	$-\sqrt{3}$
4	9	1	0	0	1	-1	-1	-2
$4\frac{1}{2}$	-10	0	-1	1	$\sqrt{2}$	0	0	$-\sqrt{3}$
5	11	-1	-1	-1	1	1	1	-1
$5\frac{1}{2}$	-12	0	0	0	0	$\frac{1}{2}(\sqrt{5}+1)$	$\frac{1}{2}(\sqrt{5}-1)$	0
6	13	1	1	1	-1	$\frac{1}{2}(\sqrt{5}+1)$	$\frac{1}{2}(1-\sqrt{5})$	1

Table A 14. *Reduction to irreducible representations of O, O^* and of K, K^* of the $(2J+1)$ states with given J*

(If $b(J)$ is the breakdown for J under the group O^* (or O for J integral), then

$$b(J+12m) = mA_1+mA_2+2mE+3mT_1+3mT_2+b(J)$$

when J is integral and

$$b(J+12m) = 2mE'+2mE''+4mU'+b(J)$$

otherwise.)

J	O, O^*	K, K^*
0	A_1	A
1	T_1	T_1
2	$E+T_2$	V
3	$A_2+T_1+T_2$	T_2+U
4	$A_1+E+T_1+T_2$	$U+V$
5	$E+2T_1+T_2$	T_1+T_2+V
6	$A_1+A_2+E+T_1+2T_2$	$A+T_1+U+V$
7	$A_2+E+2T_1+2T_2$	T_1+T_2+U+V
8	$A_1+2E+2T_1+2T_2$	T_2+U+2V
9	$A_1+A_2+E+3T_1+2T_2$	T_1+T_2+2U+V
10	$A_1+A_2+2E+2T_1+3T_2$	$A+T_1+T_2+U+2V$
11	$A_2+2E+3T_1+3T_2$	$2T_1+T_2+U+2V$
12	$2A_1+A_2+2E+3T_1+3T_2$	$A+T_1+T_2+2U+2V$

J	O^*	K^*
$\frac{1}{2}$	E'	E'
$1\frac{1}{2}$	U'	U'
$2\frac{1}{2}$	$E''+U'$	W'
$3\frac{1}{2}$	$E'+E''+U'$	$E''+W'$
$4\frac{1}{2}$	$E'+2U'$	$U'+W'$
$5\frac{1}{2}$	$E'+E''+2U'$	$E'+U'+W'$
$6\frac{1}{2}$	$E'+2E''+2U'$	$E'+E''+U'+W'$
$7\frac{1}{2}$	$E'+E''+3U'$	$U'+2W'$
$8\frac{1}{2}$	$2E'+E''+3U'$	$E''+U'+2W'$
$9\frac{1}{2}$	$2E'+2E''+3U'$	$E'+E''+U'+2W'$
$10\frac{1}{2}$	$E'+2E''+4U'$	$E'+2U'+2W'$
$11\frac{1}{2}$	$2E'+2E''+4U'$	$E'+E''+2U'+2W'$
$12\frac{1}{2}$	$3E'+2E''+4U'$	$E'+E''+U'+3W'$

Table A 15. *Reduction to irreducible representations of D_4, D_4^* and of D_3, D_3^* of the $(2J+1)$ states with given J*

In the notation of Table A 14:

$b(J+4m) = mA_1+mA_2+mB_1+mB_2+2mE+b(J)$ for D_4, D_4^*, for J integral;

$b(J+6m) = 2mA_1+2mA_2+4mE+b(J)$ for D_3, D_3^*, for J integral;

$b(J+4m) = 2mE'+2mE''+b(J)$ for D_4^*, for $J-\frac{1}{2}$ integral;

$b(J+3m) = 2mE'+mE''+b(J)$ for D_3^*, for $J-\frac{1}{2}$ integral.

J	D_4, D_4^*	D_3, D_3^*
0	A_1	A_1
1	A_2+E	A_2+E
2	$A_1+B_1+B_2+E$	A_1+2E
3	$A_2+B_1+B_2+2E$	A_1+2A_2+2E
4	$2A_1+A_2+B_1+B_2+2E$	$2A_1+A_2+3E$
5	$A_1+2A_2+B_1+B_2+3E$	A_1+2A_2+4E
6	$2A_1+A_2+2B_1+2B_2+3E$	$3A_1+2A_2+4E$

J	D_4^*	D_3^*
$\frac{1}{2}$	E'	E'
$1\frac{1}{2}$	$E'+E''$	$E'+E''$
$2\frac{1}{2}$	$E'+2E''$	$2E'+E''$
$3\frac{1}{2}$	$2E'+2E''$	$3E'+E''$
$4\frac{1}{2}$	$3E'+2E''$	$3E'+2E''$

Table A 16. *Behaviour of functions belonging to irreducible representations of O, O^*, D_4, D_4^*, D_3, D_3^* under useful group generators and definition of the standard basis relations*

($C_3^{(111)}$ is the rotation D for the octahedron in Fig. 6.4. Remember that the group operations rotate the functions, not the coordinate system.)

O, O^*		C_4^z	C_4^x	$C_3^{(111)}$
A_1	a_1	a_1	a_1	a_1
A_2	a_2	$-a_2$	$-a_2$	a_2
E	θ	θ	$-\frac{1}{2}\theta-\frac{\sqrt{3}}{2}\epsilon$	$-\frac{1}{2}\theta+\frac{\sqrt{3}}{2}\epsilon$
	ϵ	$-\epsilon$	$-\frac{\sqrt{3}}{2}\theta+\frac{1}{2}\epsilon$	$-\frac{\sqrt{3}}{2}\theta-\frac{1}{2}\epsilon$
T_1	x	y	x	y
	y	$-x$	z	z
	z	z	$-y$	x
T_2	ξ	$-\eta$	$-\xi$	η
	η	ξ	$-\zeta$	ζ
	ζ	$-\zeta$	η	ξ

Table A 16 (*cont.*)

D_4, D_4^*		C_4^z	C_2^x
A_1	a_1	a_1	a_1
A_2	a_2	a_2	$-a_2$
B_1	b_1	$-b_1$	b_1
B_2	b_2	$-b_2$	$-b_2$
E	x	y	x
	y	$-x$	$-y$

D_3, D_3^*		C_3^z	C_2^x
A_1	a_1	a_1	a_1
A_2	a_2	a_2	$-a_2$
E	x	$-\frac{1}{2}x+\frac{\sqrt{3}}{2}y$	x
	y	$-\frac{\sqrt{3}}{2}x-\frac{1}{2}y$	$-y$

$$T_1 1 = -\frac{i}{\sqrt{2}}(T_1x+iT_1y) \qquad T_2 1 = -\frac{i}{\sqrt{2}}(T_2\xi+iT_2\eta) \qquad E\,1 = -\frac{i}{\sqrt{2}}(Ex+iEy)$$

$$T_1 0 = iT_1z \qquad T_2 0 = iT_2\zeta \qquad E\,-1 = -\frac{i}{\sqrt{2}}(Ex-iEy)$$

$$T_1 -1 = \frac{i}{\sqrt{2}}(T_1x-iT_1y) \qquad T_2 -1 = \frac{i}{\sqrt{2}}(T_2\xi-iT_2\eta)$$

O^*		C_4^z	C_4^x
E'	α'	$\frac{1}{\sqrt{2}}(1-i)\,\alpha'$	$\frac{1}{\sqrt{2}}\alpha'-\frac{i}{\sqrt{2}}\beta'$
	β'	$\frac{1}{\sqrt{2}}(1+i)\,\beta'$	$-\frac{i}{\sqrt{2}}\alpha'+\frac{1}{\sqrt{2}}\beta'$
E''	α''	$\frac{1}{\sqrt{2}}(-1+i)\,\alpha''$	$-\frac{1}{\sqrt{2}}\alpha''+\frac{i}{\sqrt{2}}\beta''$
	β''	$\frac{1}{\sqrt{2}}(-1-i)\,\beta''$	$\frac{i}{\sqrt{2}}\alpha''-\frac{1}{\sqrt{2}}\beta''$
U'	κ	$\frac{1}{\sqrt{2}}(-1-i)\,\kappa$	$\frac{1}{2\sqrt{2}}\kappa-\frac{i\sqrt{3}}{2\sqrt{2}}\lambda-\frac{\sqrt{3}}{2\sqrt{2}}\mu+\frac{i}{2\sqrt{2}}\nu$
	λ	$\frac{1}{\sqrt{2}}(1-i)\,\lambda$	$-\frac{i\sqrt{3}}{2\sqrt{2}}\kappa-\frac{1}{2\sqrt{2}}\lambda-\frac{i}{2\sqrt{2}}\mu-\frac{\sqrt{3}}{2\sqrt{2}}\nu$
	μ	$\frac{1}{\sqrt{2}}(1+i)\,\mu$	$-\frac{\sqrt{3}}{2\sqrt{2}}\kappa-\frac{i}{2\sqrt{2}}\lambda-\frac{1}{2\sqrt{2}}\mu-\frac{i\sqrt{3}}{2\sqrt{2}}\nu$
	ν	$\frac{1}{\sqrt{2}}(-1+i)\,\nu$	$\frac{i}{2\sqrt{2}}\kappa-\frac{\sqrt{3}}{2\sqrt{2}}\lambda-\frac{i\sqrt{3}}{2\sqrt{2}}\mu+\frac{1}{2\sqrt{2}}\nu$

D_4^*		C_4^z	C_2^x
E'	α'	$\frac{1}{\sqrt{2}}(1-i)\,\alpha'$	$-i\beta'$
	β'	$\frac{1}{\sqrt{2}}(1+i)\,\beta'$	$-i\alpha'$
E''	α''	$\frac{1}{\sqrt{2}}(-1+i)\,\alpha''$	$-i\beta''$
	β''	$\frac{1}{\sqrt{2}}(-1-i)\,\beta''$	$-i\alpha''$

D_3^*		C_3^z	C_2^x
E'	α'	$\frac{1}{2}(1-i\sqrt{3})\,\alpha'$	$-i\beta'$
	β'	$\frac{1}{2}(1+i\sqrt{3})\,\beta'$	$-i\alpha'$
E''	ρ_1	$-\rho_1$	$i\rho_1$
	ρ_2	$-\rho_2$	$-i\rho_2$

Table A 17. *The way in which functions belonging to irreducible representations of O span irreducible representations of the subgroups D_4 (about OZ) and D_3 (about the $x = y = z > 0$ axis and C_2 along $x+y = z = 0$)*

O	D_4
A_1a_1, $E\theta$	$\rightarrow A_1a_1$
T_1z	$\rightarrow A_2a_2$
A_2a_2, $E\epsilon$	$\rightarrow B_1b_1$
$T_2\zeta$	$\rightarrow B_2b_2$
(T_1x, T_1y) and $(T_2\xi, -T_2\eta)$	$\rightarrow (Ex, Ey)$

O	D_3
A_1a_1, $\frac{1}{\sqrt{3}}(T_2\xi+T_2\eta+T_2\zeta)$	$\rightarrow A_1a_1$
A_2a_2, $\frac{1}{\sqrt{3}}(T_1x+T_1y+T_1z)$	$\rightarrow A_2a_2$
$(E\theta, E\epsilon)$, $\left[\frac{1}{\sqrt{2}}(T_1x-T_1y), \frac{1}{\sqrt{6}}(T_1x+T_1y-2T_1z)\right]$ and $\left[-\frac{1}{\sqrt{6}}(T_2\xi+T_2\eta-2T_2\zeta), \frac{1}{\sqrt{2}}(T_2\xi-T_2\eta)\right]$	$\rightarrow (Ex, Ey)$

Table A 18. *The behaviour of functions quantized with respect to* $\mathbf{L}^2$ *under the groups D_3 and D_4 about OZ. In each case the OX-axis is a twofold axis*

Representation of D_3	Function		M (a integral)
	$M+L$ even	$M+L$ odd	
A_1	Z^c_{LM}	Z^s_{LM}	$3a$
A_2	Z^s_{LM}	Z^c_{LM}	$3a$
E	(Z^c_{LM}, Z^s_{LM})		$3a\pm 1$

Representation of D_4	Component	Function		M (a integral)
		L even	L odd	
A_1	a_1	Z^c_{LM}	Z^s_{LM}	$4a$
A_2	a_2	Z^s_{LM}	Z^c_{LM}	$4a$
B_1	b_1	Z^c_{LM}	Z^s_{LM}	$4a+2$
B_2	b_2	Z^s_{LM}	Z^c_{LM}	$4a+2$
E	x	Z^s_{LM}	$(-1)^{\frac{1}{2}(M-1)}Z^c_{LM}$	$2a+1$
	y	$(-1)^{\frac{1}{2}(M+1)}Z^c_{LM}$	Z^s_{LM}	

Table A 19. *Behaviour of the kets $|JM\rangle$ under the octahedral group O^* (or O for the J integral)*

(The bases for O^*, O are chosen in accord with Table A 16. When $|JM\rangle$ are the Y_{JM} the linear combinations given are real for A_1, A_2 and E and give real T_1x, T_1y, T_1z, $T_2\xi, T_2\eta, T_2\zeta$ according to A 16, when J is even. When J is odd they are all pure imaginary.)

$J = 0$ $\quad |A_1 a_1\rangle = |0\,0\rangle$

$J = 1$

$$|T_1\,1\rangle = |1\,1\rangle$$
$$|T_1\,0\rangle = |1\,0\rangle$$
$$|T_1\,-1\rangle = |1\,-1\rangle$$

$J = 2$

$$|E\theta\rangle = |2\,0\rangle$$
$$|E\epsilon\rangle = \frac{1}{\sqrt{2}}|2\,2\rangle + \frac{1}{\sqrt{2}}|2\,-2\rangle$$
$$|T_2\,1\rangle = |2\,-1\rangle$$
$$|T_2\,0\rangle = \frac{1}{\sqrt{2}}|2\,2\rangle - \frac{1}{\sqrt{2}}|2\,-2\rangle$$
$$|T_2\,-1\rangle = -|2\,1\rangle$$

$J = 3$

$$|A_2 a_2\rangle = \frac{1}{\sqrt{2}}|3\,2\rangle - \frac{1}{\sqrt{2}}|3\,-2\rangle$$
$$|T_1\,1\rangle = -\frac{\sqrt{5}}{2\sqrt{2}}|3\,-3\rangle - \frac{\sqrt{3}}{2\sqrt{2}}|3\,1\rangle$$
$$|T_1\,0\rangle = |3\,0\rangle$$
$$|T_1\,-1\rangle = -\frac{\sqrt{5}}{2\sqrt{2}}|3\,3\rangle - \frac{\sqrt{3}}{2\sqrt{2}}|3\,-1\rangle$$
$$|T_2\,1\rangle = -\frac{\sqrt{3}}{2\sqrt{2}}|3\,3\rangle + \frac{\sqrt{5}}{2\sqrt{2}}|3\,-1\rangle$$
$$|T_2\,0\rangle = \frac{1}{\sqrt{2}}|3\,2\rangle + \frac{1}{\sqrt{2}}|3\,-2\rangle$$
$$|T_2\,-1\rangle = -\frac{\sqrt{3}}{2\sqrt{2}}|3\,-3\rangle + \frac{\sqrt{5}}{2\sqrt{2}}|3\,1\rangle$$

$J = 4$

$$|A_1 a_1\rangle = \frac{\sqrt{7}}{2\sqrt{3}}|4\,0\rangle + \frac{\sqrt{5}}{2\sqrt{6}}|4\,4\rangle + \frac{\sqrt{5}}{2\sqrt{6}}|4\,-4\rangle$$
$$|E\theta\rangle = -\frac{\sqrt{5}}{2\sqrt{3}}|4\,0\rangle + \frac{\sqrt{7}}{2\sqrt{6}}|4\,4\rangle + \frac{\sqrt{7}}{2\sqrt{6}}|4\,-4\rangle$$
$$|E\epsilon\rangle = \frac{1}{\sqrt{2}}|4\,2\rangle + \frac{1}{\sqrt{2}}|4\,-2\rangle$$
$$|T_1\,1\rangle = -\frac{1}{2\sqrt{2}}|4\,-3\rangle - \frac{\sqrt{7}}{2\sqrt{2}}|4\,1\rangle$$
$$|T_1\,0\rangle = \frac{1}{\sqrt{2}}|4\,4\rangle - \frac{1}{\sqrt{2}}|4\,-4\rangle$$
$$|T_1\,-1\rangle = \frac{1}{2\sqrt{2}}|4\,3\rangle + \frac{\sqrt{7}}{2\sqrt{2}}|4\,-1\rangle$$
$$|T_2\,1\rangle = \frac{\sqrt{7}}{2\sqrt{2}}|4\,3\rangle - \frac{1}{2\sqrt{2}}|4\,-1\rangle$$
$$|T_2\,0\rangle = \frac{1}{\sqrt{2}}|4\,2\rangle - \frac{1}{\sqrt{2}}|4\,-2\rangle$$
$$|T_2\,-1\rangle = \frac{1}{2\sqrt{2}}|4\,1\rangle - \frac{\sqrt{7}}{2\sqrt{2}}|4\,-3\rangle$$

$J = 5$

$$|E\theta\rangle = \frac{1}{\sqrt{2}}|5\,4\rangle - \frac{1}{\sqrt{2}}|5\,-4\rangle$$

$$|E\epsilon\rangle = -\frac{1}{\sqrt{2}}|5\,2\rangle + \frac{1}{\sqrt{2}}|5\,-2\rangle$$

$$|aT_1 1\rangle = \frac{3\sqrt{7}}{8\sqrt{2}}|5\,5\rangle + \frac{\sqrt{35}}{8\sqrt{2}}|5\,-3\rangle + \frac{\sqrt{15}}{8}|5\,1\rangle$$

$$|aT_1 0\rangle = |5\,0\rangle$$

$$|aT_1 -1\rangle = \frac{3\sqrt{7}}{8\sqrt{2}}|5\,-5\rangle + \frac{\sqrt{35}}{8\sqrt{2}}|5\,3\rangle + \frac{\sqrt{15}}{8}|5\,-1\rangle$$

$$|bT_1 1\rangle = \frac{\sqrt{5}}{8\sqrt{2}}|5\,5\rangle - \frac{9\sqrt{2}}{16}|5\,-3\rangle + \frac{\sqrt{21}}{8}|5\,1\rangle$$

$$|bT_1 0\rangle = \frac{1}{\sqrt{2}}|5\,4\rangle + \frac{1}{\sqrt{2}}|5\,-4\rangle$$

$$|bT_1 -1\rangle = \frac{\sqrt{5}}{8\sqrt{2}}|5\,-5\rangle - \frac{9\sqrt{2}}{16}|5\,3\rangle + \frac{\sqrt{21}}{8}|5\,-1\rangle$$

$$|T_2 1\rangle = \frac{\sqrt{15}}{4\sqrt{2}}|5\,-5\rangle - \frac{\sqrt{3}}{4\sqrt{2}}|5\,3\rangle - \frac{\sqrt{7}}{4}|5\,-1\rangle$$

$$|T_2 0\rangle = \frac{1}{\sqrt{2}}|5\,2\rangle + \frac{1}{\sqrt{2}}|5\,-2\rangle$$

$$|T_2 -1\rangle = \frac{\sqrt{15}}{4\sqrt{2}}|5\,5\rangle - \frac{\sqrt{3}}{4\sqrt{2}}|5\,-3\rangle - \frac{\sqrt{7}}{4}|5\,1\rangle$$

$J = 6$

$$|A_1 a_1\rangle = \frac{1}{2\sqrt{2}}|6\,0\rangle - \frac{\sqrt{7}}{4}|6\,4\rangle - \frac{\sqrt{7}}{4}|6\,-4\rangle$$

$$|A_2 a_2\rangle = \frac{\sqrt{11}}{4\sqrt{2}}|6\,2\rangle + \frac{\sqrt{11}}{4\sqrt{2}}|6\,-2\rangle - \frac{\sqrt{5}}{4\sqrt{2}}|6\,6\rangle - \frac{\sqrt{5}}{4\sqrt{2}}|6\,-6\rangle$$

$$|E\theta\rangle = \frac{\sqrt{7}}{2\sqrt{2}}|6\,0\rangle + \tfrac{1}{4}|6\,4\rangle + \tfrac{1}{4}|6\,-4\rangle$$

$$|E\epsilon\rangle = \frac{\sqrt{5}}{4\sqrt{2}}|6\,2\rangle + \frac{\sqrt{5}}{4\sqrt{2}}|6\,-2\rangle + \frac{\sqrt{11}}{4\sqrt{2}}|6\,6\rangle + \frac{\sqrt{11}}{4\sqrt{2}}|6\,-6\rangle$$

$$|T_1 1\rangle = \frac{\sqrt{3}}{4}|6\,1\rangle - \frac{\sqrt{30}}{8}|6\,-3\rangle - \frac{\sqrt{22}}{8}|6\,5\rangle$$

$$|T_1 0\rangle = \frac{1}{\sqrt{2}}|6\,4\rangle - \frac{1}{\sqrt{2}}|6\,-4\rangle$$

$$|T_1 -1\rangle = -\frac{\sqrt{3}}{4}|6\,-1\rangle + \frac{\sqrt{30}}{8}|6\,3\rangle + \frac{\sqrt{22}}{8}|6\,-5\rangle$$

$$|aT_2 1\rangle = \frac{\sqrt{10}}{16}|6\,-1\rangle - \tfrac{9}{16}|6\,3\rangle + \frac{\sqrt{165}}{16}|6\,-5\rangle$$

$$|aT_2 0\rangle = \frac{1}{\sqrt{2}}|6\,2\rangle - \frac{1}{\sqrt{2}}|6\,-2\rangle$$

$$|aT_2 -1\rangle = -\frac{\sqrt{10}}{16}|6\,1\rangle + \tfrac{9}{16}|6\,-3\rangle - \frac{\sqrt{165}}{16}|6\,5\rangle$$

$$|bT_2\,1\rangle = \frac{3\sqrt{22}}{16}|6\,-1\rangle + \frac{\sqrt{55}}{16}|6\,3\rangle + \frac{\sqrt{3}}{16}|6\,-5\rangle$$

$$|bT_2\,0\rangle = \frac{1}{\sqrt{2}}|6\,6\rangle - \frac{1}{\sqrt{2}}|6\,-6\rangle$$

$$|bT_2\,-1\rangle = -\frac{3\sqrt{22}}{16}|6\,1\rangle - \frac{\sqrt{55}}{16}|6\,-3\rangle - \frac{\sqrt{3}}{16}|6\,5\rangle$$

$J = \frac{1}{2}$

$$|E'\alpha'\rangle = |\tfrac{1}{2}\,\tfrac{1}{2}\rangle$$

$$|E'\beta'\rangle = |\tfrac{1}{2}\,-\tfrac{1}{2}\rangle$$

$J = \frac{3}{2}$

$$|U'\kappa\rangle = |\tfrac{3}{2}\,\tfrac{3}{2}\rangle$$

$$|U'\lambda\rangle = |\tfrac{3}{2}\,\tfrac{1}{2}\rangle$$

$$|U'\mu\rangle = |\tfrac{3}{2}\,-\tfrac{1}{2}\rangle$$

$$|U'\nu\rangle = |\tfrac{3}{2}\,-\tfrac{3}{2}\rangle$$

$J = \frac{5}{2}$

$$|E''\alpha''\rangle = \frac{1}{\sqrt{6}}|\tfrac{5}{2}\,\tfrac{5}{2}\rangle - \frac{\sqrt{5}}{\sqrt{6}}|\tfrac{5}{2}\,-\tfrac{3}{2}\rangle$$

$$|E''\beta''\rangle = \frac{1}{\sqrt{6}}|\tfrac{5}{2}\,-\tfrac{5}{2}\rangle - \frac{\sqrt{5}}{\sqrt{6}}|\tfrac{5}{2}\,\tfrac{3}{2}\rangle$$

$$|U'\kappa\rangle = -\frac{1}{\sqrt{6}}|\tfrac{5}{2}\,\tfrac{3}{2}\rangle - \frac{\sqrt{5}}{\sqrt{6}}|\tfrac{5}{2}\,-\tfrac{5}{2}\rangle$$

$$|U'\lambda\rangle = |\tfrac{5}{2}\,\tfrac{1}{2}\rangle$$

$$|U'\mu\rangle = -|\tfrac{5}{2}\,-\tfrac{1}{2}\rangle$$

$$|U'\nu\rangle = \frac{\sqrt{5}}{\sqrt{6}}|\tfrac{5}{2}\,\tfrac{5}{2}\rangle + \frac{1}{\sqrt{6}}|\tfrac{5}{2}\,-\tfrac{3}{2}\rangle$$

$J = \frac{7}{2}$

$$|E'\alpha'\rangle = \frac{\sqrt{5}}{\sqrt{12}}|\tfrac{7}{2}\,-\tfrac{7}{2}\rangle + \frac{\sqrt{7}}{\sqrt{12}}|\tfrac{7}{2}\,\tfrac{1}{2}\rangle$$

$$|E'\beta'\rangle = -\frac{\sqrt{5}}{\sqrt{12}}|\tfrac{7}{2}\,\tfrac{7}{2}\rangle - \frac{\sqrt{7}}{\sqrt{12}}|\tfrac{7}{2}\,-\tfrac{1}{2}\rangle$$

$$|E''\alpha''\rangle = \frac{\sqrt{3}}{2}|\tfrac{7}{2}\,\tfrac{5}{2}\rangle - \tfrac{1}{2}|\tfrac{7}{2}\,-\tfrac{3}{2}\rangle$$

$$|E''\beta''\rangle = -\frac{\sqrt{3}}{2}|\tfrac{7}{2}\,-\tfrac{5}{2}\rangle + \tfrac{1}{2}|\tfrac{7}{2}\,\tfrac{3}{2}\rangle$$

$$|U'\kappa\rangle = \tfrac{1}{2}|\tfrac{7}{2}\,-\tfrac{5}{2}\rangle + \frac{\sqrt{3}}{2}|\tfrac{7}{2}\,\tfrac{3}{2}\rangle$$

$$|U'\lambda\rangle = \frac{\sqrt{7}}{\sqrt{12}}|\tfrac{7}{2}\,-\tfrac{7}{2}\rangle - \frac{\sqrt{5}}{\sqrt{12}}|\tfrac{7}{2}\,\tfrac{1}{2}\rangle$$

$$|U'\mu\rangle = \frac{\sqrt{7}}{\sqrt{12}}|\tfrac{7}{2}\,\tfrac{7}{2}\rangle - \frac{\sqrt{5}}{\sqrt{12}}|\tfrac{7}{2}\,-\tfrac{1}{2}\rangle$$

$$|U'\nu\rangle = \tfrac{1}{2}|\tfrac{7}{2}\,\tfrac{5}{2}\rangle + \frac{\sqrt{3}}{2}|\tfrac{7}{2}\,-\tfrac{3}{2}\rangle$$

$J = \frac{9}{2}$

$$|E'\alpha'\rangle = \frac{\sqrt{3}}{2\sqrt{2}}|\tfrac{9}{2}\,\tfrac{9}{2}\rangle + \frac{\sqrt{7}}{2\sqrt{3}}|\tfrac{9}{2}\,\tfrac{1}{2}\rangle + \frac{1}{2\sqrt{6}}|\tfrac{9}{2}\,-\tfrac{7}{2}\rangle$$

$$|E'\beta'\rangle = \frac{\sqrt{3}}{2\sqrt{2}}|\tfrac{9}{2}\,-\tfrac{9}{2}\rangle + \frac{\sqrt{7}}{2\sqrt{3}}|\tfrac{9}{2}\,-\tfrac{1}{2}\rangle + \frac{1}{2\sqrt{6}}|\tfrac{9}{2}\,\tfrac{7}{2}\rangle$$

$$|aU'\kappa\rangle = |\tfrac{9}{2}\,-\tfrac{5}{2}\rangle$$

$$|aU'\lambda\rangle = -\frac{\sqrt{3}}{4}|\tfrac{9}{2}\,\tfrac{9}{2}\rangle + \frac{\sqrt{14}}{4\sqrt{3}}|\tfrac{9}{2}\,\tfrac{1}{2}\rangle - \frac{5}{4\sqrt{3}}|\tfrac{9}{2}\,-\tfrac{7}{2}\rangle$$

$$|aU'\mu\rangle = \frac{\sqrt{3}}{4}|\tfrac{9}{2}\,-\tfrac{9}{2}\rangle - \frac{\sqrt{14}}{4\sqrt{3}}|\tfrac{9}{2}\,-\tfrac{1}{2}\rangle + \frac{5}{4\sqrt{3}}|\tfrac{9}{2}\,\tfrac{7}{2}\rangle$$

$$|aU'\nu\rangle = -|\tfrac{9}{2}\,\tfrac{5}{2}\rangle$$

$$|bU'\kappa\rangle = |\tfrac{9}{2}\,\tfrac{3}{2}\rangle$$

$$|bU'\lambda\rangle = -\frac{\sqrt{7}}{4}|\tfrac{9}{2}\,\tfrac{9}{2}\rangle + \frac{1}{2\sqrt{2}}|\tfrac{9}{2}\,\tfrac{1}{2}\rangle + \frac{\sqrt{7}}{4}|\tfrac{9}{2}\,-\tfrac{7}{2}\rangle$$

$$|bU'\mu\rangle = \frac{\sqrt{7}}{4}|\tfrac{9}{2}\,-\tfrac{9}{2}\rangle - \frac{1}{2\sqrt{2}}|\tfrac{9}{2}\,-\tfrac{1}{2}\rangle - \frac{\sqrt{7}}{4}|\tfrac{9}{2}\,\tfrac{7}{2}\rangle$$

$$|bU'\nu\rangle = -|\tfrac{9}{2}\,-\tfrac{3}{2}\rangle$$

Table A 20. *Coupling coefficients for the octahedral group*

(We define $\langle \Gamma_2\Gamma_1\, ba \mid \Gamma_2\Gamma_1\Gamma c\rangle = \langle \Gamma_1\Gamma_2\, ab \mid \Gamma_1\Gamma_2\Gamma c\rangle$ when $\Gamma_1 \neq \Gamma_2$.)

$A_2 \times A_2$		A_1
		a_1
a_2	a_2	1

$A_2 \times E$		E	
		θ	ϵ
a_2	θ	0	-1
a_2	ϵ	1	0

$A_2 \times T_1$		T_2		
		ξ	η	ζ
a_2	x	1	0	0
a_2	y	0	1	0
a_2	z	0	0	1

$A_2 \times T_2$		T_1		
		x	y	z
a_2	ξ	1	.	.
a_2	η	.	1	.
a_2	ζ	.	.	1

$E \times E$		A_1	A_2	E	
		a_1	a_2	θ	ϵ
θ	θ	$\frac{1}{\sqrt{2}}$	.	$-\frac{1}{\sqrt{2}}$	.
ϵ	ϵ	$\frac{1}{\sqrt{2}}$	.	$\frac{1}{\sqrt{2}}$	.
θ	ϵ	.	$\frac{1}{\sqrt{2}}$	.	$\frac{1}{\sqrt{2}}$
ϵ	θ	.	$-\frac{1}{\sqrt{2}}$	.	$\frac{1}{\sqrt{2}}$

$T_1 \times T_1$		or $T_2 \times T_2$		A_1	E	
				a_1	θ	ϵ
x	x	ξ	ξ	$\frac{1}{\sqrt{3}}$	$\frac{1}{\sqrt{6}}$	$-\frac{1}{\sqrt{2}}$
y	y	η	η	$\frac{1}{\sqrt{3}}$	$\frac{1}{\sqrt{6}}$	$\frac{1}{\sqrt{2}}$
z	z	ζ	ζ	$\frac{1}{\sqrt{3}}$	$-\frac{2}{\sqrt{6}}$	.

$T_1 \times T_2$		A_2	E	
		a_2	θ	ϵ
x	ξ	$\frac{1}{\sqrt{3}}$	$-\frac{1}{\sqrt{2}}$	$-\frac{1}{\sqrt{6}}$
y	η	$\frac{1}{\sqrt{3}}$	$\frac{1}{\sqrt{2}}$	$-\frac{1}{\sqrt{6}}$
z	ζ	$\frac{1}{\sqrt{3}}$	.	$\frac{2}{\sqrt{6}}$

Table A 20 (*cont.*)

$E \times T_1$		T_1			T_2		
		x	y	z	ξ	η	ζ
θ	x	$-\frac{1}{2}$	.	.	$-\frac{1}{2}\sqrt{3}$	.	.
θ	y	.	$-\frac{1}{2}$	.	.	$\frac{1}{2}\sqrt{3}$	.
θ	z	.	.	1	.	.	.
ϵ	x	$\frac{1}{2}\sqrt{3}$	.	.	$-\frac{1}{2}$	.	.
ϵ	y	.	$-\frac{1}{2}\sqrt{3}$	.	.	$-\frac{1}{2}$	.
ϵ	z	.	.	.	.	.	1

$E \times T_2$		T_1			T_2		
		x	y	z	ξ	η	ζ
θ	ξ	$-\frac{1}{2}\sqrt{3}$	.	.	$-\frac{1}{2}$	.	.
θ	η	.	$\frac{1}{2}\sqrt{3}$	.	.	$-\frac{1}{2}$	.
θ	ζ	.	.	.	.	.	1
ϵ	ξ	$-\frac{1}{2}$	.	.	$\frac{1}{2}\sqrt{3}$	.	.
ϵ	η	.	$-\frac{1}{2}$	.	.	$-\frac{1}{2}\sqrt{3}$	.
ϵ	ζ	.	.	1	.	.	.

$T_1 \times T_1$ or $T_2 \times T_2$				T_1			T_2		
				x	y	z	ξ	η	ζ
x	y	ξ	η	.	.	$-\frac{1}{\sqrt{2}}$	.	.	$-\frac{1}{\sqrt{2}}$
x	z	ξ	ζ	.	$\frac{1}{\sqrt{2}}$	.	.	$-\frac{1}{\sqrt{2}}$	.
y	x	η	ξ	.	.	$\frac{1}{\sqrt{2}}$	.	.	$-\frac{1}{\sqrt{2}}$
y	z	η	ζ	$-\frac{1}{\sqrt{2}}$	.	.	$-\frac{1}{\sqrt{2}}$	.	.
z	x	ζ	ξ	.	$-\frac{1}{\sqrt{2}}$	.	.	$-\frac{1}{\sqrt{2}}$	.
z	y	ζ	η	$\frac{1}{\sqrt{2}}$	.	.	$-\frac{1}{\sqrt{2}}$	.	.

Table A 20 (*cont.*)

$T_1 \times T_2$	T_1			T_2		
	x	y	z	ξ	η	ζ
$x \quad \eta$	.	.	$-\frac{1}{\sqrt{2}}$	.	.	$-\frac{1}{\sqrt{2}}$
$x \quad \zeta$	.	$-\frac{1}{\sqrt{2}}$	.	.	$\frac{1}{\sqrt{2}}$	.
$y \quad \xi$	.	.	$-\frac{1}{\sqrt{2}}$	.	.	$\frac{1}{\sqrt{2}}$
$y \quad \zeta$	$-\frac{1}{\sqrt{2}}$	.	.	$-\frac{1}{\sqrt{2}}$	.	.
$z \quad \xi$	.	$-\frac{1}{\sqrt{2}}$	.	.	$-\frac{1}{\sqrt{2}}$	.
$z \quad \eta$	$-\frac{1}{\sqrt{2}}$	.	.	$\frac{1}{\sqrt{2}}$	.	.

$\langle A_2 T_1 a_2 i \,|\, A_2 T_1 T_2 j\rangle = \langle A_2 T_2 a_2 i \,|\, A_2 T_2 T_1 j\rangle = \delta_{ij}$, for both systems of quantization for the pair T_1, T_2. All coupling coefficients are consistent with the relation between the systems given in Table A 16.

$E \times T_1$	T_1			T_2		
	1	0	-1	1	0	-1
$\theta \quad 1$	$-\frac{1}{2}$	.	.	.	.	$\frac{1}{2}\sqrt{3}$
$\theta \quad 0$	.	1	.	.	.	.
$\theta \quad -1$	.	.	$-\frac{1}{2}$	$\frac{1}{2}\sqrt{3}$	.	.
$\epsilon \quad 1$	.	.	$-\frac{1}{2}\sqrt{3}$	$-\frac{1}{2}$	.	.
$\epsilon \quad 0$	.	.	.	.	1	.
$\epsilon \quad -1$	$-\frac{1}{2}\sqrt{3}$	.	.	.	.	$-\frac{1}{2}$

$E \times T_2$	T_1			T_2		
	1	0	-1	1	0	-1
$\theta \quad 1$	.	.	$\frac{1}{2}\sqrt{3}$	$-\frac{1}{2}$	.	.
$\theta \quad 0$	.	.	.	.	1	.
$\theta \quad -1$	$\frac{1}{2}\sqrt{3}$	.	.	.	.	$-\frac{1}{2}$
$\epsilon \quad 1$	$-\frac{1}{2}$	.	.	.	.	$-\frac{1}{2}\sqrt{3}$
$\epsilon \quad 0$	.	1	.	.	.	.
$\epsilon \quad -1$	.	.	$-\frac{1}{2}$	$-\frac{1}{2}\sqrt{3}$	.	.

Table A 20 (*cont.*)

$T_1 \times T_1$ and $T_2 \times T_2$		A_1	E		T_1			T_2		
T_1 / T_2	T_1 / T_2	a_1	θ	ϵ	1	0	-1	1	0	-1
1	1	.	.	$\frac{1}{\sqrt{2}}$	.	.	.	.	$\frac{1}{\sqrt{2}}$	.
1	0	.	.	.	$\frac{1}{\sqrt{2}}$	.	.	.	.	$-\frac{1}{\sqrt{2}}$
1	-1	$\frac{1}{\sqrt{3}}$	$\frac{1}{\sqrt{6}}$	.	.	$\frac{1}{\sqrt{2}}$	.	.	.	.
0	1	.	.	.	$-\frac{1}{\sqrt{2}}$	.	.	.	.	$-\frac{1}{\sqrt{2}}$
0	0	$-\frac{1}{\sqrt{3}}$	$\frac{2}{\sqrt{6}}$	.	.	.	.	.	.	.
0	-1	.	.	.	.	.	$\frac{1}{\sqrt{2}}$	$\frac{1}{\sqrt{2}}$	.	.
-1	1	$\frac{1}{\sqrt{3}}$	$\frac{1}{\sqrt{6}}$	.	.	$-\frac{1}{\sqrt{2}}$	.	.	.	.
-1	0	.	.	.	.	.	$-\frac{1}{\sqrt{2}}$	$\frac{1}{\sqrt{2}}$	.	.
-1	-1	.	.	$\frac{1}{\sqrt{2}}$	.	.	.	.	$-\frac{1}{\sqrt{2}}$	.

$T_1 \times T_2$		A_2	E		T_1			T_2		
		a_2	θ	ϵ	1	0	-1	1	0	-1
1	1	.	$\frac{1}{\sqrt{2}}$	.	.	$\frac{1}{\sqrt{2}}$	.	.	.	.
1	0	.	.	.	.	.	$-\frac{1}{\sqrt{2}}$	$\frac{1}{\sqrt{2}}$	.	.
1	-1	$\frac{1}{\sqrt{3}}$	.	$-\frac{1}{\sqrt{6}}$	.	.	.	.	$\frac{1}{\sqrt{2}}$	.
0	1	.	.	.	.	.	$-\frac{1}{\sqrt{2}}$	$-\frac{1}{\sqrt{2}}$	.	.
0	0	$-\frac{1}{\sqrt{3}}$	.	$-\frac{2}{\sqrt{6}}$	.	.	.	.	.	.
0	-1	.	.	.	$\frac{1}{\sqrt{2}}$	.	.	.	.	$\frac{1}{\sqrt{2}}$
-1	1	$\frac{1}{\sqrt{3}}$	.	$-\frac{1}{\sqrt{6}}$	.	.	.	.	$-\frac{1}{\sqrt{2}}$	.
-1	0	.	.	.	$\frac{1}{\sqrt{2}}$	.	.	.	.	$-\frac{1}{\sqrt{2}}$
-1	-1	.	$\frac{1}{\sqrt{2}}$	.	.	$-\frac{1}{\sqrt{2}}$	.	.	.	.

Table A 20 (*cont.*)

$A_2 \times U'$		U'			
		κ	λ	μ	ν
a_2	κ	.	.	1	.
a_2	λ	.	.	.	-1
a_2	μ	-1	.	.	.
a_2	ν	.	1	.	.

$E \times E'$		U'			
		κ	λ	μ	ν
θ	α'	.	-1	.	.
θ	β'	.	.	1	.
ϵ	α'	.	.	.	-1
ϵ	β'	1	.	.	.

$E \times E''$		U'			
		κ	λ	μ	ν
θ	α''	.	.	.	-1
θ	β''	1	.	.	.
ϵ	α''	.	1	.	.
ϵ	β''	.	.	-1	.

$E \times U'$		E'		E''		U'			
		α'	β'	α''	β''	κ	λ	μ	ν
θ	κ	.	.	.	$\frac{1}{\sqrt{2}}$	$\frac{1}{\sqrt{2}}$	.	.	.
θ	λ	$\frac{1}{\sqrt{2}}$	.	.	.	.	$-\frac{1}{\sqrt{2}}$	.	.
θ	μ	.	$-\frac{1}{\sqrt{2}}$	.	.	.	.	$-\frac{1}{\sqrt{2}}$	.
θ	ν	.	.	$-\frac{1}{\sqrt{2}}$	.	.	.	.	$\frac{1}{\sqrt{2}}$
ϵ	κ	.	$-\frac{1}{\sqrt{2}}$	.	.	.	.	$\frac{1}{\sqrt{2}}$	.
ϵ	λ	.	.	$\frac{1}{\sqrt{2}}$	.	.	.	.	$\frac{1}{\sqrt{2}}$
ϵ	μ	.	.	.	$-\frac{1}{\sqrt{2}}$	$\frac{1}{\sqrt{2}}$	.	.	.
ϵ	ν	$\frac{1}{\sqrt{2}}$	.	.	.	.	$\frac{1}{\sqrt{2}}$	.	.

$T_1 \times E'$		E'		U'			
		α'	β'	κ	λ	μ	ν
1	α'	.	.	1	.	.	.
0	α'	$\frac{1}{\sqrt{3}}$	.	.	$\frac{\sqrt{2}}{\sqrt{3}}$	.	.
-1	α'	.	$\frac{\sqrt{2}}{\sqrt{3}}$	.	.	$\frac{1}{\sqrt{3}}$	.
1	β'	$-\frac{\sqrt{2}}{\sqrt{3}}$	.	.	$\frac{1}{\sqrt{3}}$	.	.
0	β'	.	$-\frac{1}{\sqrt{3}}$	.	.	$\frac{\sqrt{2}}{\sqrt{3}}$	.
-1	β'	.	.	.	.	.	1

$T_2 \times E'$		E''		U'			
		α''	β''	κ	λ	μ	ν
1	α'	.	.	.	.	1	.
0	α'	$\frac{1}{\sqrt{3}}$	.	.	.	.	$-\frac{\sqrt{2}}{\sqrt{3}}$
-1	α'	.	$\frac{\sqrt{2}}{\sqrt{3}}$	$-\frac{1}{\sqrt{3}}$	.	.	.
1	β'	$-\frac{\sqrt{2}}{\sqrt{3}}$	.	.	.	.	$-\frac{1}{\sqrt{3}}$
0	β'	.	$-\frac{1}{\sqrt{3}}$	$-\frac{\sqrt{2}}{\sqrt{3}}$	.	.	.
-1	β'	.	.	.	1	.	.

Table A 20 (*cont.*)

$T_1 \times U'$	E'		E''		$\frac{3}{2}U'$				$\frac{5}{2}U'$			
	α'	β'	α''	β''	κ	λ	μ	ν	κ	λ	μ	ν
$1\ \kappa$	.	.	$\frac{1}{\sqrt{6}}$	.	.	.	.	.	.	.	.	$\frac{\sqrt{5}}{\sqrt{6}}$
$0\ \kappa$	.	.	.	$-\frac{1}{\sqrt{3}}$	$\frac{\sqrt{3}}{\sqrt{5}}$	.	.	.	$-\frac{1}{\sqrt{15}}$	.	.	.
$-1\ \kappa$	$\frac{1}{\sqrt{2}}$	.	.	.	.	$\frac{\sqrt{2}}{\sqrt{5}}$	.	.	.	$\frac{1}{\sqrt{10}}$	.	.
$1\ \lambda$	.	.	.	$-\frac{1}{\sqrt{2}}$	$-\frac{\sqrt{2}}{\sqrt{5}}$	.	.	.	$-\frac{1}{\sqrt{10}}$	.	.	.
$0\ \lambda$	$-\frac{1}{\sqrt{3}}$	.	.	.	.	$\frac{1}{\sqrt{15}}$	.	.	.	$\frac{\sqrt{3}}{\sqrt{5}}$	.	.
$-1\ \lambda$	.	$\frac{1}{\sqrt{6}}$	.	.	.	.	$\frac{2\sqrt{2}}{\sqrt{15}}$	.	.	.	$-\frac{\sqrt{3}}{\sqrt{10}}$	.
$1\ \mu$	$\frac{1}{\sqrt{6}}$	.	.	.	.	$-\frac{2\sqrt{2}}{\sqrt{15}}$	.	.	.	$\frac{\sqrt{3}}{\sqrt{10}}$	.	.
$0\ \mu$	.	$-\frac{1}{\sqrt{3}}$	.	.	.	.	$-\frac{1}{\sqrt{15}}$	.	.	.	$-\frac{\sqrt{3}}{\sqrt{5}}$	.
$-1\ \mu$	.	.	$-\frac{1}{\sqrt{2}}$	.	.	.	.	$\frac{\sqrt{2}}{\sqrt{5}}$	.	.	.	$\frac{1}{\sqrt{10}}$
$1\ \nu$	.	$\frac{1}{\sqrt{2}}$	.	.	.	.	$-\frac{\sqrt{2}}{\sqrt{5}}$	.	.	.	$-\frac{1}{\sqrt{10}}$	.
$0\ \nu$	.	.	$-\frac{1}{\sqrt{3}}$	.	.	.	.	$-\frac{\sqrt{3}}{\sqrt{5}}$	.	.	.	$\frac{1}{\sqrt{15}}$
$-1\ \nu$	.	.	.	$\frac{1}{\sqrt{6}}$	.	.	.	.	$-\frac{\sqrt{5}}{\sqrt{6}}$	.	.	.

Table A 20 (*cont.*)

$T_2 \times U'$	E'		E''		$\frac{3}{2}U'$				$\frac{5}{2}U'$			
	α'	β'	α''	β''	κ	λ	μ	ν	κ	λ	μ	ν
$1\ \kappa$	$\frac{1}{\sqrt{6}}$	.	.	.	.	.	.	.	.	$\frac{\sqrt{5}}{\sqrt{6}}$	.	.
$0\ \kappa$	.	$-\frac{1}{\sqrt{3}}$	.	.	.	.	$\frac{\sqrt{3}}{\sqrt{5}}$	.	.	.	$-\frac{1}{\sqrt{15}}$	.
$-1\ \kappa$	.	.	$\frac{1}{\sqrt{2}}$	.	.	.	.	$-\frac{\sqrt{2}}{\sqrt{5}}$	.	.	.	$-\frac{1}{\sqrt{1\ldots}}$
$1\ \lambda$	.	$-\frac{1}{\sqrt{2}}$	.	.	.	.	$-\frac{\sqrt{2}}{\sqrt{5}}$	.	.	.	$-\frac{1}{\sqrt{10}}$	.
$0\ \lambda$	.	.	$-\frac{1}{\sqrt{3}}$	.	.	.	.	$-\frac{1}{\sqrt{15}}$	.	.	.	$-\frac{\sqrt{\ldots}}{\sqrt{\ldots}}$
$-1\ \lambda$	.	.	.	$\frac{1}{\sqrt{6}}$	$-\frac{2\sqrt{2}}{\sqrt{15}}$	.	.	.	$\frac{\sqrt{3}}{\sqrt{10}}$	.	.	.
$1\ \mu$	.	.	$\frac{1}{\sqrt{6}}$	.	.	.	.	$\frac{2\sqrt{2}}{\sqrt{15}}$	.	.	.	$-\frac{\sqrt{\ldots}}{\sqrt{1\ldots}}$
$0\ \mu$	.	.	.	$-\frac{1}{\sqrt{3}}$	$\frac{1}{\sqrt{15}}$	.	.	.	$\frac{\sqrt{3}}{\sqrt{5}}$	.	.	.
$-1\ \mu$	$-\frac{1}{\sqrt{2}}$	.	.	.	.	$\frac{\sqrt{2}}{\sqrt{5}}$	.	.	.	$\frac{1}{\sqrt{10}}$	.	.
$1\ \nu$	.	.	.	$\frac{1}{\sqrt{2}}$	$\frac{\sqrt{2}}{\sqrt{5}}$	.	.	.	$\frac{1}{\sqrt{10}}$	.	.	.
$0\ \nu$	$-\frac{1}{\sqrt{3}}$	.	.	.	.	$-\frac{\sqrt{3}}{\sqrt{5}}$	.	.	.	$\frac{1}{\sqrt{15}}$	.	.
$-1\ \nu$	.	$\frac{1}{\sqrt{6}}$	.	.	.	.	.	.	.	.	$-\frac{\sqrt{5}}{\sqrt{6}}$	.

^5E	A_1	A_2	E		T_1			T_2		
	a_1	a_2	θ	ϵ	1	0	-1	1	0	-1
2θ	.	$-\frac{1}{2}$	.	$\frac{1}{2}$	.	.	.	.	$\frac{1}{\sqrt{2}}$	.
2ϵ	$\frac{1}{2}$	.	$\frac{1}{2}$	.	.	$\frac{1}{\sqrt{2}}$	.	.	.	.
1θ	.	.	.	.	$-\frac{1}{2}\sqrt{3}$	.	.	.	.	$\frac{1}{2}$
1ϵ	.	.	.	.	.	.	$\frac{1}{2}$	$\frac{1}{2}\sqrt{3}$	.	.
0θ	$\frac{1}{\sqrt{2}}$	.	$-\frac{1}{\sqrt{2}}$	.	.	.	.	.	.	.
0ϵ	.	$\frac{1}{\sqrt{2}}$	.	$\frac{1}{\sqrt{2}}$	.	.	.	.	.	.
-1θ	.	.	.	.	.	.	$\frac{1}{2}\sqrt{3}$	$-\frac{1}{2}$	.	.
-1ϵ	.	.	.	.	$-\frac{1}{2}$	.	.	.	.	$-\frac{1}{2}\sqrt{3}$
-2θ	.	$-\frac{1}{2}$	.	$\frac{1}{2}$	.	.	.	.	$-\frac{1}{\sqrt{2}}$	.
-2ϵ	$\frac{1}{2}$	.	$\frac{1}{2}$	.	.	$-\frac{1}{\sqrt{2}}$	.	.	.	.

Table A 20 (*cont.*)

$^5T_2$	$(J=1)$			$(J=2)$					$(J=3)$						
	$1T_2$			E		$2T_1$			A_1	$3T_1$			$3T_2$		
	1	0	-1	θ	ϵ	1	0	-1	a_1	1	0	-1	1	0	-1
2 1	.	.	.	.	.	.	.	.	.	$-\frac{\sqrt{3}}{2\sqrt{2}}$	.	.	.	.	$-\frac{\sqrt{5}}{2\sqrt{2}}$
2 0	.	.	.	$\frac{1}{\sqrt{3}}$	.	.	$\frac{1}{\sqrt{3}}$	.	$\frac{1}{\sqrt{6}}$	.	$\frac{1}{\sqrt{6}}$	.	.	.	.
2 -1	$\frac{\sqrt{3}}{\sqrt{5}}$	.	.	.	.	.	.	$-\frac{1}{\sqrt{3}}$	.	.	.	$\frac{1}{2\sqrt{6}}$	$-\frac{1}{2\sqrt{10}}$	.	.
1 1	.	.	.	$-\frac{1}{\sqrt{6}}$	.	.	$-\frac{1}{\sqrt{6}}$	.	$\frac{1}{\sqrt{3}}$	.	$\frac{1}{\sqrt{3}}$	.	.	.	.
1 0	$-\frac{\sqrt{3}}{\sqrt{10}}$	.	.	.	.	.	.	$-\frac{1}{\sqrt{6}}$	.	.	.	$\frac{1}{\sqrt{3}}$	$-\frac{1}{\sqrt{5}}$	.	.
1 -1	.	$\frac{\sqrt{3}}{\sqrt{10}}$	.	.	$-\frac{1}{\sqrt{2}}$	.	.	.	.	.	.	.	.	$\frac{1}{\sqrt{5}}$	.
0 1	$\frac{1}{\sqrt{10}}$	.	.	.	.	.	.	$-\frac{1}{\sqrt{2}}$	.	.	.	$\frac{1}{2}$	$-\frac{\sqrt{3}}{2\sqrt{5}}$	.	.
0 0	.	$-\frac{\sqrt{2}}{\sqrt{5}}$	.	.	.	.	.	.	.	.	.	.	.	$\frac{\sqrt{3}}{\sqrt{5}}$	.
0 -1	.	.	$\frac{1}{\sqrt{10}}$	.	.	$\frac{1}{\sqrt{2}}$	.	.	.	$\frac{1}{2}$	.	.	.	.	$-\frac{\sqrt{3}}{2\sqrt{5}}$
-1 1	.	$\frac{\sqrt{3}}{\sqrt{10}}$	.	.	$\frac{1}{\sqrt{2}}$	.	.	.	.	.	.	.	.	$\frac{1}{\sqrt{5}}$	.
-1 0	.	.	$-\frac{\sqrt{3}}{\sqrt{10}}$	.	.	$-\frac{1}{\sqrt{6}}$	.	.	.	$\frac{1}{\sqrt{3}}$	.	.	.	.	$-\frac{1}{\sqrt{5}}$
-1 -1	.	.	.	$\frac{1}{\sqrt{6}}$	.	.	$-\frac{1}{\sqrt{6}}$	.	$-\frac{1}{\sqrt{3}}$	.	$\frac{1}{\sqrt{3}}$	.	.	.	.
-2 1	.	.	$\frac{\sqrt{3}}{\sqrt{5}}$	.	.	$-\frac{1}{\sqrt{3}}$	.	.	.	$\frac{1}{2\sqrt{6}}$	.	.	.	.	$-\frac{1}{2\sqrt{10}}$
-2 0	.	.	.	$-\frac{1}{\sqrt{3}}$	.	.	$\frac{1}{\sqrt{3}}$	.	$-\frac{1}{\sqrt{6}}$	.	$\frac{1}{\sqrt{6}}$	.	.	.	.
-2 -1	.	.	.	.	.	.	.	.	.	.	.	$-\frac{\sqrt{3}}{2\sqrt{2}}$	$-\frac{\sqrt{5}}{2\sqrt{2}}$	.	.

Table A 21. *The products of $Z_{2\alpha'}Z_{2\alpha''}$ expressed in terms of $Z_{k\alpha}$*

$$(Z_{22}^c)^2\sqrt{4\pi} = \frac{\sqrt{5}}{\sqrt{7}}Z_{44}^c + \tfrac{1}{7}Z_{40} - \frac{2\sqrt{5}}{7}Z_{20} + Z_{00}$$

$$(Z_{21}^c)^2\sqrt{4\pi} = \frac{2\sqrt{5}}{7}Z_{42}^c - \tfrac{4}{7}Z_{40} + \frac{\sqrt{15}}{7}Z_{22}^c + \frac{\sqrt{5}}{7}Z_{20} + Z_{00}$$

$$Z_{20}^2\sqrt{4\pi} = \tfrac{6}{7}Z_{40} + \frac{2\sqrt{5}}{7}Z_{20} + Z_{00}$$

$$(Z_{21}^s)^2\sqrt{4\pi} = -\frac{2\sqrt{5}}{7}Z_{42}^c - \tfrac{4}{7}Z_{40} - \frac{\sqrt{15}}{7}Z_{22}^c + \frac{\sqrt{5}}{7}Z_{20} + Z_{00}$$

$$(Z_{22}^s)^2\sqrt{4\pi} = -\frac{\sqrt{5}}{\sqrt{7}}Z_{44}^c + \tfrac{1}{7}Z_{40} - \frac{2\sqrt{5}}{7}Z_{20} + Z_{00}$$

$$Z_{22}^c Z_{21}^c\sqrt{4\pi} = \frac{\sqrt{5}}{\sqrt{14}}Z_{43}^c - \frac{\sqrt{5}}{7\sqrt{2}}Z_{41}^c + \frac{\sqrt{15}}{7}Z_{21}^c$$

$$Z_{22}^c Z_{20}\sqrt{4\pi} = \frac{\sqrt{15}}{7}Z_{42}^c - \frac{2\sqrt{5}}{7}Z_{22}^c$$

$$Z_{22}^c Z_{21}^s\sqrt{4\pi} = \frac{\sqrt{5}}{7\sqrt{2}}Z_{41}^s + \frac{\sqrt{5}}{\sqrt{14}}Z_{43}^s - \frac{\sqrt{15}}{7}Z_{21}^s$$

$$Z_{22}^c Z_{22}^s\sqrt{4\pi} = \frac{\sqrt{5}}{\sqrt{7}}Z_{44}^s$$

$$Z_{21}^c Z_{20}\sqrt{4\pi} = \frac{\sqrt{30}}{7}Z_{41}^c + \frac{\sqrt{5}}{7}Z_{21}^c$$

$$Z_{21}^c Z_{21}^s\sqrt{4\pi} = \frac{2\sqrt{5}}{7}Z_{42}^s + \frac{\sqrt{15}}{7}Z_{22}^s$$

$$Z_{21}^c Z_{22}^s\sqrt{4\pi} = -\frac{\sqrt{5}}{7\sqrt{2}}Z_{41}^s + \frac{\sqrt{5}}{\sqrt{14}}Z_{43}^s + \frac{\sqrt{15}}{7}Z_{21}^s$$

$$Z_{20} Z_{21}^s\sqrt{4\pi} = \frac{\sqrt{30}}{7}Z_{41}^s + \frac{\sqrt{5}}{7}Z_{21}^s$$

$$Z_{20} Z_{22}^s\sqrt{4\pi} = \frac{\sqrt{15}}{7}Z_{42}^s - \frac{2\sqrt{5}}{7}Z_{22}^s$$

$$Z_{21}^s Z_{22}^s\sqrt{4\pi} = -\frac{\sqrt{5}}{\sqrt{14}}Z_{43}^c - \frac{\sqrt{5}}{7\sqrt{2}}Z_{41}^c + \frac{\sqrt{15}}{7}Z_{21}^c$$

Table A 22. *Representations of finite rotation groups according to Frobenius and Schur's categories*

Group	Representation categories	'Two-valued' representation categories
C_1	1	1
C_2	1	3
C_{2n}	Two of 1. The rest 3	3
C_{2n+1}	One of 1. The rest 3	One of 1. The rest 3
D_{2n}	1	2
D_{2n+1}	1	One pair 3. The rest 2
T	Two of 1. One pair 3	One of 2. One pair 3
O, K	1	2

Table A 23. *Symmetrized,* $[\Gamma^2]$, *and antisymmetrized,* (Γ^2), *squares of irreducible representations of* O^*, D_4^* *and* D_3^*. E'' *of* D_3^* *is reducible over the complex field*

The group O^*		
Γ	$[\Gamma^2]$	(Γ^2)
A_1	A_1	—
A_2	A_1	—
E	A_1+E	A_2
T_1	A_1+E+T_2	T_1
T_2	A_1+E+T_2	T_1
E'	T_1	A_1
E''	T_1	A_1
U'	$A_2+2T_1+T_2$	A_1+E+T_2

The group D_4^*		
Γ	$[\Gamma^2]$	(Γ^2)
A_1	A_1	—
A_2	A_1	—
B_1	A_1	—
B_2	A_1	—
E	$A_1+B_1+B_2$	A_2
E'	A_2+E	A_1
E''	A_2+E	A_1

The group D_3^*		
Γ	$[\Gamma^2]$	(Γ^2)
A_1	A_1	—
A_2	A_1	—
E	A_1+E	A_2
E'	A_2+E	A_1
(E'')	A_1+2A_2	A_1

Table A 24. *The functions of t_2^m and e^n*

(The bases given are consistent with Table A 20 and Wigner's formula and also with the hole particle convention of § 9.7. They correspond under the p^n isomorphism to p^n functions defined w the usual phase convention (see Racah, 1943), although there is an ambiguity of sign for $t_2^3\,{}^2E \to p^3$ The functions in the column headed 'complex orbitals' below are related to the corresponding functions via the definitions in Table A 20. $|\zeta_1\rangle = i\,|\zeta\rangle = \frac{1}{\sqrt{2}}|2\,2\rangle - \frac{1}{\sqrt{2}}|2\,-2\rangle$. We write $|m_l^\pm\rangle$ inst of $|2m_l^\pm\rangle$, and usually omit the ket signs for the functions based on real orbitals. $\xi^2 = \xi^+\xi^-$, etc.)

Configuration	Real orbitals	Complex orbitals
t_2^1	$\lvert {}^2T_2\,\tfrac{1}{2}\xi\rangle = \lvert\xi^+\rangle$	$\lvert {}^2T_2\,\tfrac{1}{2}\,1\rangle = \lvert -1^+\rangle$
	$\lvert {}^2T_2\,\tfrac{1}{2}\eta\rangle = \lvert\eta^+\rangle$	$\lvert {}^2T_2\,\tfrac{1}{2}\,0\rangle = \lvert\zeta_1^+\rangle$
	$\lvert {}^2T_2\,\tfrac{1}{2}\zeta\rangle = \lvert\zeta^+\rangle$	$\lvert {}^2T_2\,\tfrac{1}{2}\,-1\rangle = -\lvert 1^+\rangle$
t_2^2	${}^3T_1\,1x = \zeta^+\eta^+$	$\lvert {}^3T_1\,1\,1\rangle = \lvert -1^+\zeta_1^+\rangle$
	${}^3T_1\,1y = \xi^+\zeta^+$	$\lvert {}^3T_1\,1\,0\rangle = \lvert 1^+\,-1^+\rangle$
	${}^3T_1\,1z = \eta^+\xi^+$	$\lvert {}^3T_1\,1\,-1\rangle = \lvert 1^+\zeta_1^+\rangle$
	${}^1A_1 = \frac{1}{\sqrt{3}}(\xi^2+\eta^2+\zeta^2)$	$= -\frac{1}{\sqrt{3}}(\lvert -1^+1^-\rangle + \lvert 1^+\,-1^-\rangle + \lvert\zeta_1^2\rangle)$
	${}^1E\theta = \frac{1}{\sqrt{6}}(\xi^2+\eta^2-2\zeta^2)$	$= \frac{1}{\sqrt{6}}(-\lvert -1^+1^-\rangle - \lvert 1^+\,-1^-\rangle + 2\lvert\zeta_1^2\rangle)$
	${}^1E\epsilon = \frac{1}{\sqrt{2}}(\eta^2-\xi^2)$	$= \frac{1}{\sqrt{2}}(\lvert 1^2\rangle + \lvert -1^2\rangle)$
	${}^1T_2\,\xi = -\frac{1}{\sqrt{2}}(\eta^+\zeta^- + \zeta^+\eta^-)$	$\lvert {}^1T_2\,1\rangle = -\frac{1}{\sqrt{2}}(\lvert 1^+\zeta_1^-\rangle + \lvert\zeta_1^+\,1^-\rangle)$
	${}^1T_2\,\eta = -\frac{1}{\sqrt{2}}(\xi^+\zeta^- + \zeta^+\xi^-)$	$\lvert {}^1T_2\,0\rangle = -\frac{1}{\sqrt{2}}(\lvert 1^2\rangle - \lvert -1^2\rangle)$
	${}^1T_2\,\zeta = -\frac{1}{\sqrt{2}}(\xi^+\eta^- + \eta^+\xi^-)$	$\lvert {}^1T_2\,-1\rangle = -\frac{1}{\sqrt{2}}(\lvert -1^+\zeta_1^-\rangle + \lvert\zeta_1^+\,-1^-\rangle)$
t_2^3	${}^4A_2\,\tfrac{3}{2}a_2 = -\xi^+\eta^+\zeta^+$	$= -\lvert 1^+\,-1^+\zeta_1^+\rangle$
	${}^2E\,\tfrac{1}{2}\theta = \frac{1}{\sqrt{2}}(\xi^+\eta^-\zeta^+ - \xi^-\eta^+\zeta^+)$	$= -\frac{1}{\sqrt{2}}(\lvert 1^2\zeta_1^+\rangle - \lvert -1^2\zeta_1^+\rangle)$
	${}^2E\,\tfrac{1}{2}\epsilon = \frac{1}{\sqrt{6}}(2\xi^+\eta^+\zeta^- - \xi^+\eta^-\zeta^+ - \xi^-\eta^+\zeta^+)$	$= \frac{1}{\sqrt{6}}(2\lvert 1^+\,-1^+\zeta_1^-\rangle - \lvert 1^+\,-1^-\zeta_1^+\rangle - \lvert 1^-\,-1^+\zeta_1$
	${}^2T_1\,\tfrac{1}{2}x = \frac{1}{\sqrt{2}}(\xi^+\eta^2 - \xi^+\zeta^2)$	$\lvert {}^2T_1\,\tfrac{1}{2}\,1\rangle = \frac{1}{\sqrt{2}}(\lvert 1^+\zeta_1^2\rangle + \lvert 1^2\,-1^+\rangle)$
	${}^2T_1\,\tfrac{1}{2}y = \frac{1}{\sqrt{2}}(\eta^+\zeta^2 - \xi^2\eta^+)$	$\lvert {}^2T_1\,\tfrac{1}{2}\,0\rangle = -\frac{1}{\sqrt{2}}(\lvert 1^2\zeta_1^+\rangle + \lvert -1^2\zeta_1^+\rangle)$
	${}^2T_1\,\tfrac{1}{2}z = \frac{1}{\sqrt{2}}(\xi^2\zeta^+ - \eta^2\zeta^+)$	$\lvert {}^2T_1\,\tfrac{1}{2}\,-1\rangle = -\frac{1}{\sqrt{2}}(\lvert -1^+\zeta_1^2\rangle + \lvert 1^+\,-1^2\rangle)$
	${}^2T_2\,\tfrac{1}{2}\xi = \frac{1}{\sqrt{2}}(\xi^+\zeta^2 + \xi^+\eta^2)$	$\lvert {}^2T_2\,\tfrac{1}{2}\,1\rangle = \frac{1}{\sqrt{2}}(\lvert 1^+\,-1^2\rangle - \lvert -1^+\zeta_1^2\rangle)$
	${}^2T_2\,\tfrac{1}{2}\eta = \frac{1}{\sqrt{2}}(\eta^+\zeta^2 + \xi^2\eta^+)$	$\lvert {}^2T_2\,\tfrac{1}{2}\,0\rangle = \frac{1}{\sqrt{2}}(\lvert 1^-\,-1^+\zeta_1^+\rangle - \lvert 1^+\,-1^+\zeta_1^+\rangle)$
	${}^2T_2\,\tfrac{1}{2}\zeta = \frac{1}{\sqrt{2}}(\xi^2\zeta^+ + \eta^2\zeta^+)$	$\lvert {}^2T_2\,\tfrac{1}{2}\,-1\rangle = \frac{1}{\sqrt{2}}(\lvert 1^+\zeta_1^2\rangle - \lvert 1^2\,-1^+\rangle)$

Table A 24 (*cont.*)

Configuration	Real orbitals	Complex orbitals
t_2^4	${}^3T_1\,1x = \xi^2\eta^+\zeta^+$	$\lvert {}^3T_1\,1\,1\rangle = -\lvert 1^+\,-1^2\,\zeta_1^+\rangle$
	${}^3T_1\,1y = -\xi^+\eta^2\zeta^+$	$\lvert {}^3T_1\,1\,0\rangle = \lvert 1^+\,-1^+\,\zeta_1^2\rangle$
	${}^3T_1\,1z = \xi^+\eta^+\zeta^2$	$\lvert {}^3T_1\,1\,-1\rangle = -\lvert 1^2\,-1^+\,\zeta_1^+\rangle$
	${}^1A_1 = \dfrac{1}{\sqrt{3}}(\xi^2\eta^2+\eta^2\zeta^2+\zeta^2\xi^2)$	$= \dfrac{1}{\sqrt{3}}(-\lvert 1^2\,-1^2\rangle - \lvert 1^-\,-1^+\,\zeta_1^2\rangle + \lvert 1^+\,-1^-\,\zeta_1^2\rangle)$
	${}^1E\theta = \dfrac{1}{\sqrt{6}}(\eta^2\zeta^2+\zeta^2\xi^2-2\xi^2\eta^2)$	$= \dfrac{1}{\sqrt{6}}(2\lvert 1^2\,-1^2\rangle - \lvert 1^-\,-1^+\,\zeta_1^2\rangle + \lvert 1^+\,-1^-\,\zeta_1^2\rangle)$
	${}^1E\epsilon = \dfrac{1}{\sqrt{2}}(\zeta^2\xi^2-\eta^2\zeta^2)$	$= \dfrac{1}{\sqrt{2}}(\lvert 1^2\,\zeta_1^2\rangle + \lvert -1^2\,\zeta_1^2\rangle)$
	${}^1T_2\,\xi = \dfrac{1}{\sqrt{2}}(\xi^2\eta^+\zeta^- - \xi^2\eta^-\zeta^+)$	$\lvert {}^1T_2\,1\rangle = \dfrac{1}{\sqrt{2}}(\lvert 1^2\,-1^+\,\zeta_1^-\rangle - \lvert 1^2\,-1^-\,\zeta_1^+\rangle)$
	${}^1T_2\,\eta = \dfrac{1}{\sqrt{2}}(\xi^+\eta^2\zeta^- - \xi^-\eta^2\zeta^+)$	$\lvert {}^1T_2\,0\rangle = \dfrac{1}{\sqrt{2}}(\lvert -1^2\,\zeta_1^2\rangle - \lvert 1^2\,\zeta_1^2\rangle)$
	${}^1T_2\,\zeta = \dfrac{1}{\sqrt{2}}(\xi^+\eta^-\zeta^2 - \xi^-\eta^+\zeta^2)$	$\lvert {}^1T_2\,-1\rangle = \dfrac{1}{\sqrt{2}}(-\lvert 1^-\,-1^2\,\zeta_1^+\rangle + \lvert 1^+\,-1^2\,\zeta_1^-\rangle)$
t_2^5	${}^2T_2\,\tfrac{1}{2}\xi = \xi^+\eta^2\zeta^2$	$\lvert {}^2T_2\,\tfrac{1}{2}\,1\rangle = -\lvert 1^+\,-1^2\,\zeta_1^2\rangle$
	${}^2T_2\,\tfrac{1}{2}\eta = \xi^2\eta^+\zeta^2$	$\lvert {}^2T_2\,\tfrac{1}{2}\,0\rangle = -\lvert 1^2\,-1^2\,\zeta_1^+\rangle$
	${}^2T_2\,\tfrac{1}{2}\zeta = \xi^2\eta^2\zeta^+$	$\lvert {}^2T_2\,\tfrac{1}{2}\,-1\rangle = \lvert 1^2\,-1^+\,\zeta_1^2\rangle$
t_2^6	${}^1A_1 = \xi^2\eta^2\zeta^2$	$= \lvert 1^2\,-1^2\,\zeta_1^2\rangle$
e^2	${}^3A_2\,1a_2 = \theta^+\epsilon^+$	
	${}^1A_1 = \dfrac{1}{\sqrt{2}}(\theta^2+\epsilon^2)$	
	${}^1E\theta = \dfrac{1}{\sqrt{2}}(\epsilon^2-\theta^2)$	
	${}^1E\epsilon = \dfrac{1}{\sqrt{2}}(\theta^+\epsilon^- - \theta^-\epsilon^+)$	
e^3	${}^2E\,\tfrac{1}{2}\theta = \theta^+\epsilon^2$	
	${}^2E\,\tfrac{1}{2}\epsilon = \theta^2\epsilon^+$	
e^4	${}^1A_1 = \theta^2\epsilon^2$	

Table A 25. *Numbers of allowed terms for octahedral strong-field configurations*

Configuration	Representation					Spin
	A_1	A_2	E	T_1	T_2	
e^1, e^3	0	0	1	0	0	Doublet
e^2	0	1	0	0	0	Triplet
	1	0	1	0	0	Singlet
t_2^1, t_2^5	0	0	0	0	1	Doublet
t_2^2, t_2^4	0	0	0	1	0	Triplet
	1	0	1	0	1	Singlet
t_2^3	0	1	0	0	0	Quartet
	0	0	1	1	1	Doublet
$t_2^1e^1$, $t_2^5e^1$,	0	0	0	1	1	Triplet
$t_2^1e^3$, $t_2^5e^3$	0	0	0	1	1	Singlet
$t_2^2e^1$, $t_2^4e^1$,	0	0	0	1	1	Quartet
$t_2^2e^3$, $t_2^4e^3$	1	1	2	2	2	Doublet
$t_2^3e^1$, $t_2^3e^3$	0	0	1	0	0	Quintet
	1	1	2	2	2	Triplet
	1	1	1	2	2	Singlet
$t_2^1e^2$, $t_2^5e^2$	0	0	0	1	0	Quartet
	0	0	0	2	2	Doublet
$t_2^2e^2$, $t_2^4e^2$	0	0	0	0	1	Quintet
	0	1	1	3	2	Triplet
	2	1	3	1	3	Singlet
$t_2^3e^2$	1	0	0	0	0	Sextet
	1	1	2	1	1	Quartet
	2	1	3	4	4	Doublet

Table A 26. *The expressions for the non-zero* $(ab; cd) \equiv \langle ac \mid V \mid bd \rangle$, *classified according to the representation of* D_4. *They are given first in general parameters and then, for d electrons, in Racah parameters*

A_1, B_1	ξ^2	η^2	ζ^2
ξ^2	$a = A+4B+3C$	$b = A-2B+C$	$b = A-2B+C$
η^2	$b = A-2B+C$	$a = A+4B+3C$	$b = A-2B+C$
ζ^2	$b = A-2B+C$	$b = A-2B+C$	$a = A+4B+3C$
θ^2	$d+\frac{2}{\sqrt{3}}c = A+2B+C$	$d+\frac{2}{\sqrt{3}}c = A+2B+C$	$d-\frac{1}{\sqrt{3}}c = A-4B+C$
ϵ^2	$d = A-2B+C$	$d = A-2B+C$	$d+c\sqrt{3} = A+4B+C$
$\theta\epsilon$	$c = 2B\sqrt{3}$	$-c = -2B\sqrt{3}$	0

A_1, B_1	θ^2	ϵ^2	$\theta\epsilon$
ξ^2	$d+\frac{2}{\sqrt{3}}c = A+2B+C$	$d = A-2B+C$	$c = 2B\sqrt{3}$
η^2	$d+\frac{2}{\sqrt{3}}c = A+2B+C$	$d = A-2B+C$	$-c = -2B\sqrt{3}$
ζ^2	$d-\frac{1}{\sqrt{3}}c = A-4B+C$	$d+c\sqrt{3} = A+4B+C$	0
θ^2	$e = A+4B+3C$	$e-2f = A-4B+C$	0
ϵ^2	$e-2f = A-4B+C$	$e = A+4B+3C$	0
$\theta\epsilon$	0	0	$f = 4B+C$

A_2	$\epsilon\zeta$
$\epsilon\zeta$	$g-\frac{1}{\sqrt{3}}h = C$

B_2	$\theta\zeta$	$\xi\eta$
$\theta\zeta$	$g+h\sqrt{3} = 4B+C$	$-2i = -2B\sqrt{3}$
$\xi\eta$	$-2i = -2B\sqrt{3}$	$j = 3B+C$

Ex	$\theta\xi$	$\epsilon\xi$	$\eta\zeta$
$\theta\xi$	$g = B+C$	$-h = -B\sqrt{3}$	$i = B\sqrt{3}$
$\epsilon\xi$	$-h = -B\sqrt{3}$	$g+\frac{2}{\sqrt{3}}h = 3B+C$	$-i\sqrt{3} = -3B$
$\eta\zeta$	$i = B\sqrt{3}$	$-i\sqrt{3} = -3B$	$j = 3B+C$

Ey	$\theta\eta$	$\epsilon\eta$	$\zeta\xi$
$\theta\eta$	$g = B+C$	$h = B\sqrt{3}$	$i = B\sqrt{3}$
$\epsilon\eta$	$h = B\sqrt{3}$	$g+\frac{2}{\sqrt{3}}h = 3B+C$	$i\sqrt{3} = 3B$
$\zeta\xi$	$i = B\sqrt{3}$	$i\sqrt{3} = 3B$	$j = 3B+C$

Table A 27. *Electrostatic matrices for d^2 and d^8*

$^1A_1(d^2)$	e^2	t_2^2
e^2	$A+8B+4C$	$(2B+C)\sqrt{6}$
t_2^2	$(2B+C)\sqrt{6}$	$A+10B+5C$

$^1A_1(d^8)$	$t_2^6e^2$	$t_2^4e^4$
$t_2^6e^2$	$28A-34B+25C$	$(2B+C)\sqrt{6}$
$t_2^4e^4$	$(2B+C)\sqrt{6}$	$28A-32B+26C$

$^1E(d^2)$	e^2	t_2^2
e^2	$A+2C$	$2B\sqrt{3}$
t_2^2	$2B\sqrt{3}$	$A+B+2C$

$^1E(d^8)$	$t_2^6e^2$	$t_2^4e^4$
$t_2^6e^2$	$28A-42B+23C$	$-2B\sqrt{3}$
$t_2^4e^4$	$-2B\sqrt{3}$	$28A-41B+23C$

$^1T_2(d^2)$	t_2^2	t_2e
t_2^2	$A+B+2C$	$-2B\sqrt{3}$
t_2e	$-2B\sqrt{3}$	$A+2C$

$^1T_2(d^8)$	$t_2^4e^4$	$t_2^5e^3$
$t_2^4e^4$	$28A-41B+23C$	$2B\sqrt{3}$
$t_2^5e^3$	$2B\sqrt{3}$	$28A-42B+23C$

$^3T_1(d^2)$	t_2^2	t_2e
t_2^2	$A-5B$	$-6B$
t_2e	$-6B$	$A+4B$

$^3T_1(d^8)$	$t_2^4e^4$	$t_2^5e^3$
$t_2^4e^4$	$28A-47B+21C$	$6B$
$t_2^5e^3$	$6B$	$28A-38B+21C$

$d^2\quad {}^1T_1 = A+4B+2C$

$d^2\quad {}^3T_2 = A-8B$

$d^2\quad {}^3A_2 = A-8B$

$d^8\quad {}^1T_1 = 28A-38B+23C$

$d^8\quad {}^3T_2 = 28A-50B+21C$

$d^8\quad {}^3A_2 = 28A-50B+21C$

Table A 28. *Electrostatic matrices for d^3. To obtain the matrices for d^7 add $18A-28B+14C$ to all diagonal elements (for the phases of d^7 see* §§ 9.3.4 and 9.7)

^2E	t_2^3	$t_2^2(^1A_1)\,e$	$t_2^2(^1E)\,e$	e^3
t_2^3	$3A-6B+3C$	$-6B\sqrt{2}$	$3B\sqrt{2}$	0
$t_2^2(^1A_1)\,e$	$-6B\sqrt{2}$	$3A+8B+6C$	$-10B$	$(2B+C)\sqrt{3}$
$t_2^2(^1E)\,e$	$3B\sqrt{2}$	$-10B$	$3A-B+3C$	$-2B\sqrt{3}$
e^3	0	$(2B+C)\sqrt{3}$	$-2B\sqrt{3}$	$3A-8B+4C$

$^2T_1$	t_2^3	$t_2^2(^3T_1)\,e$	$t_2^2(^1T_2)\,e$	$t_2\,e^2(^3A_2)$	$t_2\,e^2(^1E)$
t_2^3	$3A-6B+3C$	$3B$	$-3B$	0	$-2B\sqrt{3}$
$t_2^2(^3T_1)\,e$	$3B$	$3A+3C$	$-3B$	$3B$	$-3B\sqrt{3}$
$t_2^2(^1T_2)\,e$	$-3B$	$-3B$	$3A-6B+3C$	$-3B$	$B\sqrt{3}$
$t_2\,e^2(^3A_2)$	0	$3B$	$-3B$	$3A-6B+3C$	$-2B\sqrt{3}$
$t_2\,e^2(^1E)$	$-2B\sqrt{3}$	$-3B\sqrt{3}$	$B\sqrt{3}$	$-2B\sqrt{3}$	$3A-2B+3C$

$^2T_2$	t_2^3	$t_2^2(^3T_1)\,e$	$t_2^2(^1T_2)\,e$	$t_2\,e^2(^1A_1)$	$t_2\,e^2(^1E)$
t_2^3	$3A+5C$	$-3B\sqrt{3}$	$5B\sqrt{3}$	$4B+2C$	$2B$
$t_2^2(^3T_1)\,e$	$-3B\sqrt{3}$	$3A-6B+3C$	$-3B$	$-3B\sqrt{3}$	$-3B\sqrt{3}$
$t_2^2(^1T_2)\,e$	$5B\sqrt{3}$	$-3B$	$3A+4B+3C$	$B\sqrt{3}$	$-B\sqrt{3}$
$t_2e^2(^1A_1)$	$4B+2C$	$-3B\sqrt{3}$	$B\sqrt{3}$	$3A+6B+5C$	$10B$
$t_2e^2(^1E)$	$2B$	$-3B\sqrt{3}$	$-B\sqrt{3}$	$10B$	$3A-2B+3C$

$^4T_1$	$t_2^2(^3T_1)\,e$	$t_2\,e^2(^3A_2)$
$t_2^2(^3T_1)\,e$	$3A-3B$	$6B$
$t_2\,e^2(^3A_2)$	$6B$	$3A-12B$

$t_2^2(^1E)\,e$ $\quad ^2A_1 = 3A-11B+3C$

$t_2^2(^1E)\,e$ $\quad ^2A_2 = 3A+9B+3C$

t_2^3 $\quad ^4A_2 = 3A-15B$

$t_2^2(^3T_1)\,e$ $\quad ^4T_2 = 3A-15B$

Table A 29. *Electrostatic matrices for d^4. To obtain the matrices for d^6 add $9A-14B+7C$ to all diagonal elements*

$^3T_1$	t_2^4	$t_2^3(^2T_1)\,e$	$t_2^3(^2T_2)\,e$	$t_2^2(^3T_1)\,e^2(^1A_1)$	$t_2^2(^3T_1)\,e^2(^1E)$	$t_2^2(^1T_2)\,e^2(^3A_2)$	t_2e^3
t_2^4	$6A-15B+5C$	$-B\sqrt{6}$	$-3B\sqrt{2}$	$-(2B+C)\sqrt{2}$	$2B\sqrt{2}$	0	0
$t_2^3(^2T_1)\,e$	$-B\sqrt{6}$	$6A-11B+4C$	$5B\sqrt{3}$	$-B\sqrt{3}$	$B\sqrt{3}$	$3B$	$B\sqrt{6}$
$t_2^3(^2T_2)\,e$	$-3B\sqrt{2}$	$5B\sqrt{3}$	$6A-3B+6C$	$3B$	$3B$	$5B\sqrt{3}$	$(B+C)\sqrt{2}$
$t_2^2(^3T_1)\,e^2(^1A_1)$	$-(2B+C)\sqrt{2}$	$-B\sqrt{3}$	$3B$	$6A-B+6C$	$-10B$	0	$-3B\sqrt{2}$
$t_2^2(^3T_1)\,e^2(^1E)$	$2B\sqrt{2}$	$B\sqrt{3}$	$3B$	$-10B$	$6A-9B+4C$	$2B\sqrt{3}$	$3B\sqrt{2}$
$t_2^2(^1T_2)\,e^2(^3A_2)$	0	$3B$	$5B\sqrt{3}$	0	$2B\sqrt{3}$	$6A-11B+4C$	$B\sqrt{6}$
t_2e^3	0	$B\sqrt{6}$	$(B+C)\sqrt{2}$	$-3B\sqrt{2}$	$3B\sqrt{2}$	$B\sqrt{6}$	$6A-16B+5C$

$^1T_2$	t_2^4	$t_2^3(^2T_1)\,e$	$t_2^3(^2T_2)\,e$	$t_2^2(^3T_1)\,e^2(^3A_2)$	$t_2^2(^1T_2)\,e^2(^1E)$	$t_2^2(^1T_2)\,e^2(^1A_1)$	t_2e^3
t_2^4	$6A-9B+7C$	$-3B\sqrt{2}$	$-5B\sqrt{6}$	0	$2B\sqrt{2}$	$-(2B+C)\sqrt{2}$	0
$t_2^3(^2T_1)\,e$	$-3B\sqrt{2}$	$6A-9B+6C$	$5B\sqrt{3}$	$3B$	$-3B$	$-3B$	$B\sqrt{6}$
$t_2^3(^2T_2)\,e$	$-5B\sqrt{6}$	$5B\sqrt{3}$	$6A+3B+8C$	$3B\sqrt{3}$	$-5B\sqrt{3}$	$5B\sqrt{3}$	$(3B+C)\sqrt{2}$
$t_2^2(^3T_1)\,e^2(^3A_2)$	0	$3B$	$3B\sqrt{3}$	$6A-9B+6C$	$-6B$	0	$3B\sqrt{6}$
$t_2^2(^1T_2)\,e^2(^1E)$	$2B\sqrt{2}$	$-3B$	$-5B\sqrt{3}$	$-6B$	$6A-3B+6C$	$-10B$	$-B\sqrt{6}$
$t_2^2(^1T_2)\,e^2(^1A_1)$	$-(2B+C)\sqrt{2}$	$-3B$	$5B\sqrt{3}$	0	$-10B$	$6A+5B+8C$	$-B\sqrt{6}$
t_2e^3	0	$B\sqrt{6}$	$(3B+C)\sqrt{2}$	$3B\sqrt{6}$	$-B\sqrt{6}$	$-B\sqrt{6}$	$6A+7C$

$^1A_1$	t_2^4	$t_2^3(^2E)\,e$	$t_2^2(^1A_1)\,e^2(^1A_1)$	$t_2^2(^1E)\,e^2(^1E)$	e^4
t_2^4	$6A+10C$	$-12B\sqrt{2}$	$(4B+2C)\sqrt{2}$	$-2B\sqrt{2}$	0
$t_2^3(^2E)\,e$	$-12B\sqrt{2}$	$6A+6C$	$-12B$	$6B$	0
$t_2^2(^1A_1)\,e^2(^1A_1)$	$(4B+2C)\sqrt{2}$	$-12B$	$6A+14B+11C$	$-20B$	$(2B+C)\sqrt{6}$
$t_2^2(^1E)\,e^2(^1E)$	$-2B\sqrt{2}$	$6B$	$-20B$	$6A-3B+6C$	$-2B\sqrt{6}$
e^4	0	0	$(2B+C)\sqrt{6}$	$-2B\sqrt{6}$	$6A-16B+8C$

^1E	t_2^4	$t_2^3(^2E)e$	$t_2^2(^1E)\,e^2(^1A_1)$	$t_2^2(^1A_1)\,e^2(^1E)$	$t_2^2(^1E)\,e^2(^1E)$
t_2^4	$6A-9B+7C$	$6B$	$-(2B+C)\sqrt{2}$	$-2B$	$4B$
$t_2^3(^2E)\,e$	$6B$	$6A-6B+6C$	$3B\sqrt{2}$	$-12B$	0
$t_2^2(^1E)\,e^2(^1A_1)$	$-(2B+C)\sqrt{2}$	$3B\sqrt{2}$	$6A+5B+8C$	$-10B\sqrt{2}$	$-10B\sqrt{2}$
$t_2^2(^1A_1)\,e^2(^1E)$	$-2B$	$-12B$	$-10B\sqrt{2}$	$6A+6B+9C$	0
$t_2^2(^1E)\,e^2(^1E)$	$4B$	0	$-10B\sqrt{2}$	0	$6A-3B+6C$

Table A 29 (*cont.*)

$^1T_1$	$t_2^3(^2T_1)\,e$	$t_2^3(^2T_2)\,e$	$t_2^2(^1T_2)\,e^2(^1E)$	t_2e^3
$t_2^3(^2T_1)\,e$	$6A-3B+6C$	$5B\sqrt{3}$	$-3B$	$B\sqrt{6}$
$t_2^3(^2T_2)\,e$	$5B\sqrt{3}$	$6A-3B+8C$	$5B\sqrt{3}$	$(B+C)\sqrt{2}$
$t_2^2(^1T_2)\,e^2(^1E)$	$-3B$	$5B\sqrt{3}$	$6A-3B+6C$	$B\sqrt{6}$
t_2e^3	$B\sqrt{6}$	$(B+C)\sqrt{2}$	$B\sqrt{6}$	$6A-16B+7C$

$^3T_2$	$t_2^3(^2T_1)\,e$	$t_2^3(^2T_2)\,e$	$t_2^2(^3T_1)\,e^2(^3A_2)$	$t_2^2(^3T_1)\,e^2(^1E)$	t_2e^3
$t_2^3(^2T_1)\,e$	$6A-9B+4C$	$5B\sqrt{3}$	$B\sqrt{6}$	$-B\sqrt{3}$	$B\sqrt{6}$
$t_2^3(^2T_2)\,e$	$5B\sqrt{3}$	$6A-5B+6C$	$3B\sqrt{2}$	$3B$	$(3B+C)\sqrt{2}$
$t_2^2(^3T_1)\,e^2(^3A_2)$	$B\sqrt{6}$	$3B\sqrt{2}$	$6A-13B+4C$	$2B\sqrt{2}$	$6B$
$t_2^2(^3T_1)\,e^2(^1E)$	$-B\sqrt{3}$	$3B$	$2B\sqrt{2}$	$6A-9B+4C$	$3B\sqrt{2}$
t_2e^3	$B\sqrt{6}$	$(3B+C)\sqrt{2}$	$6B$	$3B\sqrt{2}$	$6A-8B+5C$

^3E	$t_2^3(^4A_2)\,e$	$t_2^3(^2E)\,e$	$t_2^2(^1E)\,e^2(^3A_2)$
$t_2^3(^4A_2)\,e$	$6A-13B+4C$	$-4B$	0
$t_2^3(^2E)\,e$	$-4B$	$6A-10B+4C$	$3B\sqrt{2}$
$t_2^2(^1E)\,e^2(^3A_2)$	0	$3B\sqrt{2}$	$6A-11B+4C$

$^3A_2$	$t_2^3(^2E)\,e$	$t_2^2(^1A_1)\,e^2(^3A_2)$
$t_2^3(^2E)\,e$	$6A-8B+4C$	$-12B$
$t_2^2(^1A_1)\,e^2(^3A_2)$	$-12B$	$6A-2B+7C$

$^1A_2$	$t_2^3(^2E)\,e$	$t_2^2(^1E)\,e^2(^1E)$
$t_2^3(^2E)\,e$	$6A-12B+6C$	$-6B$
$t_2^2(^1E)\,e^2(^1E)$	$-6B$	$6A-3B+6C$

$t_2^3(^2E)\,e \qquad {}^3A_1 = 6A-12B+4C$

$t_2^3(^4A_2)\,e \qquad {}^5E = 6A-21B$

$t_2^2(^3T_1)\,e^2(^3A_2) \qquad {}^5T_2 = 6A-21B$

Table A 30. *Electrostatic matrices for d^5*

Terms of $t_2^3e^2$ which have negative 'parity' in the sense of § 9.7 have been marked with an asterisk.

$^2A_1$	$t_2^4(^1E)\,e$	$t_2^3(^2E)\,e^2(^1E)$	$t_2^3(^4A_2)\,e^2(^3A_2)$	$t_2^2(^1E)\,e^3$
$t_2^4(^1E)\,e$	$10A-3B+9C$	$-3B\sqrt{2}$	0	$-(6B+C)$
$t_2^3(^2E)\,e^2(^1E)$	$-3B\sqrt{2}$	$10A-12B+8C$	$4B\sqrt{3}$	$-3B\sqrt{2}$
$t_2^3(^4A_2)\,e^2(^3A_2)$	0	$4B\sqrt{3}$	$10A-19B+8C$	0
$t_2^2(^1E)\,e^3$	$-(6B+C)$	$-3B\sqrt{2}$	0	$10A-3B+9C$

$^2A_2$	$t_2^4(^1E)\,e$	$t_2^3(^2E)\,e^2(^1E)$	$t_2^2(^1E)\,e^3$
$t_2^4(^1E)\,e$	$10A-23B+9C$	$3B\sqrt{2}$	$2B-C$
$t_2^3(^2E)\,e^2(^1E)$	$3B\sqrt{2}$	$10A-12B+8C$	$3B\sqrt{2}$
$t_2^2(^1E)\,e^3$	$2B-C$	$3B\sqrt{2}$	$10A-23B+9C$

^2E	$t_2^4(^1A_1)\,e$	$t_2^4(^1E)\,e$	$t_2^3(^2E)\,e^2(^1A_1)$*	$t_2^3(^2E)\,e^2(^3A_2)$	$t_2^3(^2E)\,e^2(^1E)$	$t_2^2(^1E)\,e^3$	$t_2^2(^1A_1)\,e^3$
$t_2^4(^1A_1)\,e$	$10A-4B+12C$	$10B$	$6B$	$6B\sqrt{3}$	$6B\sqrt{2}$	$2B$	$4B+2C$
$t_2^4(^1E)\,e$	$10B$	$10A-13B+9C$	$-3B$	$3B\sqrt{3}$	0	$-(2B+C)$	$2B$
$t_2^3(^2E)\,e^2(^1A_1)$*	$6B$	$-3B$	$10A-4B+10C$	0	0	$3B$	$-6B$
$t_2^3(^2E)\,e^2(^3A_2)$	$6B\sqrt{3}$	$3B\sqrt{3}$	0	$10A-16B+8C$	$2B\sqrt{6}$	$3B\sqrt{3}$	$6B\sqrt{3}$
$t_2^3(^2E)\,e^2(^1E)$	$6B\sqrt{2}$	0	0	$2B\sqrt{6}$	$10A-12B+8C$	0	$6B\sqrt{2}$
$t_2^2(^1E)\,e^3$	$2B$	$-(2B+C)$	$3B$	$3B\sqrt{3}$	0	$10A-13B+9C$	$10B$
$t_2^2(^1A_1)\,e^3$	$4B+2C$	$2B$	$-6B$	$6B\sqrt{3}$	$6B\sqrt{2}$	$10B$	$10A-4B+12C$

^4E	$t_2^3(^2E)\,e^2(^3A_2)$	$t_2^3(^4A_2)\,e^2(^1E)$
$t_2^3(^2E)\,e^2(^3A_2)$	$10A-22B+5C$	$-2B\sqrt{3}$
$t_2^3(^4A_2)\,e^2(^1E)$	$-2B\sqrt{3}$	$10A-21B+5C$

$^4T_1$	$t_2^4(^3T_1)\,e$	$t_2^3(^2T_2)\,e^2(^3A_2)$*	$t_2^2(^3T_1)\,e^3$
$t_2^4(^3T_1)\,e$	$10A-25B+6C$	$3B\sqrt{2}$	$-C$
$t_2^3(^2T_2)\,e^2(^3A_2)$*	$3B\sqrt{2}$	$10A-16B+7C$	$-3B\sqrt{2}$
$t_2^2(^3T_1)\,e^3$	$-C$	$-3B\sqrt{2}$	$10A-25B+6C$

Table A 30 (*cont.*)

$^4T_2$	$t_2^4(^3T_1)\,e$	$t_2^3(^2T_1)\,e^2(^3A_2)$	$t_2^2(^3T_1)\,e^3$
$t_2^4(^3T_1)\,e$	$10A-17B+6C$	$-B\sqrt{6}$	$-(4B+C)$
$t_2^3(^2T_1)\,e^2(^3A_2)$	$-B\sqrt{6}$	$10A-22B+5C$	$-B\sqrt{6}$
$t_2^2(^3T_1)\,e^3$	$-(4B+C)$	$-B\sqrt{6}$	$10A-17B+6C$

$t_2^3(^4A_2)\,e^2(^3A_2)$ $\quad {}^4A_1 = 10A-25B+5C$

$t_2^3(^4A_2)\,e^2(^1A_1)^*$ $\quad {}^4A_2 = 10A-13B+7C$

$t_2^3(^4A_2)\,e^2(^3A_2)$ $\quad {}^6A_1 = 10A-35B$

$^2T_1$	$t_2^4(^3T_1)\,e$	$t_2^4(^1T_2)\,e$	$t_2^3(^2T_1)\,e^2(^1A_1)^*$	$t_2^3(^2T_1)\,e^2(^1E)$	$t_2^3(^2T_2)\,e^2(^3A_2)^*$	$t_2^3(^2T_2)\,e^2(^1E)^*$	$t_2^2(^1T_2)\,e^3$	$t_2^2(^3T_1)\,e^3$
$t_2^4(^3T_1)\,e$	$10A-22B+9C$	$-3B$	$-\tfrac{3}{2}B\sqrt{2}$	$\tfrac{3}{2}B\sqrt{2}$	$\tfrac{3}{2}B\sqrt{2}$	$-\tfrac{3}{2}B\sqrt{6}$	0	$-C$
$t_2^4(^1T_2)\,e$	$-3B$	$10A-8B+9C$	$\tfrac{3}{2}B\sqrt{2}$	$\tfrac{3}{2}B\sqrt{2}$	$-\tfrac{15}{2}B\sqrt{2}$	$\tfrac{5}{2}B\sqrt{6}$	$-(4B+C)$	0
$t_2^3(^2T_1)\,e^2(^1A_1)^*$	$-\tfrac{3}{2}B\sqrt{2}$	$\tfrac{3}{2}B\sqrt{2}$	$10A-4B+10C$	0	0	$10B\sqrt{3}$	$-\tfrac{3}{2}B\sqrt{2}$	$\tfrac{3}{2}B\sqrt{2}$
$t_2^3(^2T_1)\,e^2(^1E)$	$\tfrac{3}{2}B\sqrt{2}$	$\tfrac{3}{2}B\sqrt{2}$	0	$10A-12B+8C$	0	0	$\tfrac{3}{2}B\sqrt{2}$	$\tfrac{3}{2}B\sqrt{2}$
$t_2^3(^2T_2)\,e^2(^3T_2)^*$	$\tfrac{3}{2}B\sqrt{2}$	$-\tfrac{15}{2}B\sqrt{2}$	0	0	$10A-10B+10C$	$-2B\sqrt{3}$	$\tfrac{15}{2}B\sqrt{2}$	$-\tfrac{3}{2}B\sqrt{2}$
$t_2^3(^2T_2)\,e^2(^1E)^*$	$-\tfrac{3}{2}B\sqrt{6}$	$\tfrac{5}{2}B\sqrt{6}$	$10B\sqrt{3}$	0	$-2B\sqrt{3}$	$10A-6B+10C$	$-\tfrac{5}{2}B\sqrt{6}$	$\tfrac{3}{2}B\sqrt{6}$
$t_2^2(^1T_2)\,e^3$	0	$-(4B+C)$	$-\tfrac{3}{2}B\sqrt{2}$	$\tfrac{3}{2}B\sqrt{2}$	$\tfrac{15}{2}B\sqrt{2}$	$-\tfrac{5}{2}B\sqrt{6}$	$10A-8B+9C$	$-3B$
$t_2^2(^3T_1)\,e^3$	$-C$	0	$\tfrac{3}{2}B\sqrt{2}$	$\tfrac{3}{2}B\sqrt{2}$	$-\tfrac{3}{2}B\sqrt{2}$	$\tfrac{3}{2}B\sqrt{6}$	$-3B$	$10A-22B+9C$

$^2T_2$	t_2^5	$t_2^4(^3T_1)\,e$	$t_2^4(^1T_2)\,e$	$t_2^3(^2T_1)\,e^2(^3A_2)$	$t_2^3(^2T_1)\,e^2(^1E)$	$t_2^3(^2T_2)\,e^2(^1A_1)$	$t_2^3(^2T_2)\,e^2(^1E)^*$	$t_2^2(^1T_2)\,e^3$	$t_2^2(^3T_1)\,e^3$	t_2e^4
t_2^5	$10A-20B+10C$	$-3B\sqrt{6}$	$B\sqrt{6}$	0	$2B\sqrt{3}$	$4B+2C$	$2B$	0	0	0
$t_2^4(^3T_1)\,e$	$-3B\sqrt{6}$	$10A-8B+9C$	$-3B$	$-\tfrac{1}{2}B\sqrt{6}$	$-\tfrac{3}{2}B\sqrt{2}$	$-\tfrac{3}{2}B\sqrt{6}$	$-\tfrac{3}{2}B\sqrt{6}$	0	$-(4B+C)$	0
$t_2^4(^1T_2)\,e$	$B\sqrt{6}$	$-3B$	$10A-18B+9C$	$\tfrac{3}{2}B\sqrt{6}$	$\tfrac{3}{2}B\sqrt{2}$	$\tfrac{5}{2}B\sqrt{6}$	$-\tfrac{5}{2}B\sqrt{6}$	$-C$	0	0
$t_2^3(^2T_1)\,e^2(^3A_2)$	0	$-\tfrac{1}{2}B\sqrt{6}$	$\tfrac{3}{2}B\sqrt{6}$	$10A-16B+8C$	$-2B\sqrt{3}$	0	0	$\tfrac{3}{2}B\sqrt{6}$	$-\tfrac{1}{2}B\sqrt{6}$	0
$t_2^3(^2T_1)\,e^2(^1E)$	$2B\sqrt{3}$	$-\tfrac{3}{2}B\sqrt{2}$	$\tfrac{3}{2}B\sqrt{2}$	$-2B\sqrt{3}$	$10A-12B+8C$	$10B\sqrt{3}$	0	$\tfrac{3}{2}B\sqrt{2}$	$-\tfrac{3}{2}B\sqrt{2}$	$2B\sqrt{3}$
$t_2^3(^2T_2)\,e^2(^1A_1)$	$4B+2C$	$-\tfrac{3}{2}B\sqrt{6}$	$\tfrac{5}{2}B\sqrt{6}$	0	$10B\sqrt{3}$	$10A+2B+12C$	0	$\tfrac{5}{2}B\sqrt{6}$	$-\tfrac{3}{2}B\sqrt{6}$	$4B+2C$
$t_2^3(^2T_2)\,e^2(^1E)^*$	$2B$	$-\tfrac{3}{2}B\sqrt{6}$	$-\tfrac{5}{2}B\sqrt{6}$	0	0	0	$10A-6B+10C$	$\tfrac{5}{2}B\sqrt{6}$	$\tfrac{3}{2}B\sqrt{6}$	$-2B$
$t_2^2(^1T_2)\,e^3$	0	0	$-C$	$\tfrac{3}{2}B\sqrt{6}$	$\tfrac{3}{2}B\sqrt{2}$	$\tfrac{5}{2}B\sqrt{6}$	$\tfrac{5}{2}B\sqrt{6}$	$10A-18B+9C$	$-3B$	$B\sqrt{6}$
$t_2^2(^3T_1)\,e^3$	0	$-(4B+C)$	0	$-\tfrac{1}{2}B\sqrt{6}$	$-\tfrac{3}{2}B\sqrt{2}$	$-\tfrac{3}{2}B\sqrt{6}$	$\tfrac{3}{2}B\sqrt{6}$	$-3B$	$10A-8B+9C$	$-3B\sqrt{6}$
t_2e^4	0	0	0	0	$2B\sqrt{3}$	$4B+2C$	$-2B$	$B\sqrt{6}$	$-3B\sqrt{6}$	$10A-20B+10C$

Table A 31. *c numbers for use in the calculation of spin-orbit coupling energies. For definition and notation see* § 9.6.3

Terms	c numbers
^5E $^5T_2$	$c(A_1) = 2\sqrt{5}$, $c(E) = \sqrt{10}$, $c(2T_1) = \sqrt{5}$, $c(3T_1) = \sqrt{10}$, $c(1T_2) = -\sqrt{3}$, $c(3T_2) = -\sqrt{2}$
^3E $^5T_2$	$c(2T_1) = -\sqrt{15}$, $c(3T_1) = \sqrt{30}$, $c(1T_2) = 1$, $c(3T_2) = -3\sqrt{6}$
$^3T_1$ $^5T_2$	$c(A_1) = -2\sqrt{15}$, $c(E) = -\sqrt{15}$, $c(2T_1) = -\sqrt{5}$, $c(3T_1) = -2\sqrt{10}$, $c(1T_2) = 1$, $c(3T_2) = 2\sqrt{6}$
$^3T_2$ $^5T_2$	$c(E) = c(2T_1) = 3$, $c(1T_2) = \sqrt{5}$, $c(3T_1) = c(3T_2) = 0$
$^3T_1$ ^5E	$c(A_1) = 2$, $c(E) = \sqrt{2}$, $c(T_1) = \sqrt{3}$, $c(T_2) = 1$
$^3T_2$ ^5E	$c(A_2) = 2$, $c(E) = \sqrt{2}$, $c(T_1) = 1$, $c(T_2) = -\sqrt{3}$
$^3T_1$ $^3T_2$	$c(E) = \sqrt{3}$, $c(T_1) = 1$, $c(T_2) = -1$
^3E $^3T_2$	$c(T_1) = \sqrt{3}$, $c(T_2) = -1$
^3E $^3T_1$	$c(T_1) = 1$, $c(T_2) = \sqrt{3}$
$^3T_1$ $^3T_1$	$c(A_1) = 2$, $c(T_1) = 1$, $c(E) = c(T_2) = -1$
$^3T_2$ $^3T_2$	$c(A_2) = 2$, $c(T_2) = 1$, $c(E) = c(T_1) = -1$
$^4T_1$ $^6A_1$	$c(\frac{3}{2}U') = 0$, $c(\frac{5}{2}U') = 1$
$^4A_2$ $^4T_2$	$c(\frac{3}{2}U') = 1$, $c(\frac{5}{2}U') = 0$
$^4T_1$ ^4E	$c(E') = \sqrt{5}$, $c(E'') = 3\sqrt{5}$, $c(\frac{3}{2}U') = -4$, $c(\frac{5}{2}U') = 3$
$^4T_2$ ^4E	$c(E') = -3\sqrt{5}$, $c(E'') = \sqrt{5}$, $c(\frac{3}{2}U') = -4$, $c(\frac{5}{2}U') = 3$
$^4T_1$ $^4T_1$	$c(E') = 5$, $c(\frac{3}{2}U') = 2$, $c(E'') = c(\frac{5}{2}U') = -3$
$^4T_2$ $^4T_2$	$c(E'') = 5$, $c(\frac{3}{2}U') = 2$, $c(E') = c(\frac{5}{2}U') = -3$
$^4T_1$ $^4T_2$	$c(E') = -c(E'') = 5$, $c(\frac{3}{2}\frac{3}{2}U') = -4$, $c(\frac{3}{2}\frac{5}{2}U') = c(\frac{5}{2}\frac{3}{2}U') = 8$, $c(\frac{5}{2}\frac{5}{2}U') = 9$
$^2T_1$ ^4E	$c(E') = \sqrt{2}$, $c(U') = -1$
$^2T_2$ ^4E	$c(E'') = \sqrt{2}$, $c(U') = -1$
$^2T_1$ $^4T_2$	$c(E') = \sqrt{10}$, $c(\frac{3}{2}U') = -1$, $c(\frac{5}{2}U') = 2$
$^2T_2$ $^4T_1$	$c(E'') = -\sqrt{10}$, $c(\frac{3}{2}U') = -1$, $c(\frac{5}{2}U') = 2$
$^2T_2$ $^4T_2$	$c(E'') = \sqrt{2}$, $c(\frac{3}{2}U') = \sqrt{5}$, $c(\frac{5}{2}U') = 0$
$^2T_1$ $^4T_1$	$c(E') = \sqrt{2}$, $c(\frac{3}{2}U') = \sqrt{5}$, $c(\frac{5}{2}U') = 0$
^2E $^4T_1$	$c(\frac{3}{2}U') = 1$, $c(\frac{5}{2}U') = 3$
^2E $^4T_2$	$c(\frac{3}{2}U') = 1$, $c(\frac{5}{2}U') = 3$
$^2T_1$ $^2T_1$	$c(E') = 2$, $c(U') = -1$
$^2T_2$ $^2T_2$	$c(E'') = 2$, $c(U') = -1$

Table A 32. *Matrices of spin-orbit coupling within $d^2\,{}^3F$ and $d^3\,{}^4F$ in the weak-field coupling scheme*

(Units of ζ. E, T_1 and T_2 refer to d^2 and E', E'' and U' to d^3. ${}^3T_1A_1 = \frac{3}{2}$ and ${}^3T_2A_2 = -\frac{1}{2}$.)

E	${}^3T_1$	${}^3T_2$
${}^3T_1$	$-\frac{3}{4}$	$\frac{3\sqrt{5}}{4}$
${}^3T_2$	$\frac{3\sqrt{5}}{4}$	$\frac{1}{4}$

T_1	${}^3T_1$	${}^3T_2$
${}^3T_1$	$\frac{3}{4}$	$\frac{\sqrt{15}}{4}$
${}^3T_2$	$\frac{\sqrt{15}}{4}$	$\frac{1}{4}$

T_2	${}^3A_2$	${}^3T_1$	${}^3T_2$
${}^3A_2$	0	0	$\sqrt{2}$
${}^3T_1$	0	$-\frac{3}{4}$	$-\frac{\sqrt{15}}{4}$
${}^3T_2$	$\sqrt{2}$	$-\frac{\sqrt{15}}{4}$	$-\frac{1}{4}$

E'	${}^4T_1$	${}^4T_2$
${}^4T_1$	$\frac{5}{4}$	$\frac{\sqrt{5}}{4}$
${}^4T_2$	$\frac{\sqrt{5}}{4}$	$\frac{1}{4}$

E''	${}^4T_1$	${}^4T_2$
${}^4T_1$	$-\frac{3}{4}$	$-\frac{\sqrt{5}}{4}$
${}^4T_2$	$-\frac{\sqrt{5}}{4}$	$-\frac{5}{12}$

U'	${}^4A_2$	${}^4T_{1\frac{3}{2}}$	${}^4T_{1\frac{5}{2}}$	${}^4T_{2\frac{3}{2}}$	${}^4T_{2\frac{5}{2}}$
${}^4A_2$	0	0	0	$\frac{\sqrt{5}}{\sqrt{3}}$	0
${}^4T_{1\frac{3}{2}}$	0	$\frac{1}{2}$	0	$-\frac{1}{\sqrt{5}}$	$\frac{2}{\sqrt{5}}$
${}^4T_{1\frac{5}{2}}$	0	0	$-\frac{3}{4}$	$\frac{2}{\sqrt{5}}$	$\frac{9}{4\sqrt{5}}$
${}^4T_{2\frac{3}{2}}$	$\frac{\sqrt{5}}{\sqrt{3}}$	$-\frac{1}{\sqrt{5}}$	$\frac{2}{\sqrt{5}}$	$-\frac{1}{6}$	0
${}^4T_{2\frac{5}{2}}$	0	$\frac{2}{\sqrt{5}}$	$\frac{9}{4\sqrt{5}}$	0	$\frac{1}{4}$

Table A 33. *Spin-orbit coupling matrices, in units of* ζ, *for* d^2 *in the strong-field coupling scheme*

(Change the signs of the matrices throughout for d^8 (see §§ 9.3.4 and 9.7). d^2 terms are given in brackets.)

A_1	$t_2^2(^1A_1)$	$e^2(^1A_1)$	$t_2^2(^3T_1)$	$t_2e(^3T_1)$
$t_2^2(^1A_1)$	0	0	$\sqrt{2}$	$-\sqrt{2}$
$e^2(^1A_1)$	0	0	0	$-\sqrt{3}$
$t_2^2(^3T_1)$	$\sqrt{2}$	0	1	1
$t_2e(^3T_1)$	$-\sqrt{2}$	$-\sqrt{3}$	1	$-\frac{1}{2}$

A_2	$t_2e(^3T_2)$
$t_2e(^3T_2)$	$-\frac{1}{2}$

E	$t_2^2(^1E)$	$e^2(^1E)$	$t_2^2(^3T_1)$	$t_2e(^3T_1)$	$t_2e(^3T_2)$
$t_2^2(^1E)$	0	0	$-\frac{1}{\sqrt{2}}$	$-\sqrt{2}$	0
$e^2(^1E)$	0	0	0	$\frac{1}{2}\sqrt{6}$	$-\frac{1}{2}\sqrt{6}$
$t_2^2(^3T_1)$	$-\frac{1}{\sqrt{2}}$	0	$-\frac{1}{2}$	$-\frac{1}{2}$	$\frac{3}{2}$
$t_2e(^3T_1)$	$-\sqrt{2}$	$\frac{1}{2}\sqrt{6}$	$-\frac{1}{2}$	$\frac{1}{4}$	$\frac{3}{4}$
$t_2e(^3T_2)$	0	$-\frac{1}{2}\sqrt{6}$	$\frac{3}{2}$	$\frac{3}{4}$	$\frac{1}{4}$

T_1	$t_2e(^1T_1)$	$t_2^2(^3T_1)$	$t_2e(^3T_1)$	$t_2e(^3T_2)$
$t_2e(^1T_1)$	0	$-\frac{1}{\sqrt{2}}$	$-\frac{1}{2\sqrt{2}}$	$-\frac{\sqrt{3}}{2\sqrt{2}}$
$t_2^2(^3T_1)$	$-\frac{1}{\sqrt{2}}$	$\frac{1}{2}$	$\frac{1}{2}$	$\frac{\sqrt{3}}{2}$
$t_2e(^3T_1)$	$-\frac{1}{2\sqrt{2}}$	$\frac{1}{2}$	$-\frac{1}{4}$	$\frac{\sqrt{3}}{4}$
$t_2e(^3T_2)$	$-\frac{\sqrt{3}}{2\sqrt{2}}$	$\frac{\sqrt{3}}{2}$	$\frac{\sqrt{3}}{4}$	$\frac{1}{4}$

T_2	$t_2^2(^1T_2)$	$t_2e(^1T_2)$	$t_2^2(^3T_1)$	$t_2e(^3T_1)$	$t_2e(^3T_2)$	$e^2(^3A_2)$
$t_2^2(^1T_2)$	0	0	$-\frac{1}{\sqrt{2}}$	$\frac{1}{\sqrt{2}}$	$-\frac{1}{2}\sqrt{6}$	0
$t_2e(^1T_2)$	0	0	$-\frac{1}{2}\sqrt{6}$	$\frac{\sqrt{3}}{2\sqrt{2}}$	$-\frac{1}{2\sqrt{2}}$	-1
$t_2^2(^3T_1)$	$-\frac{1}{\sqrt{2}}$	$-\frac{1}{2}\sqrt{6}$	$-\frac{1}{2}$	$-\frac{1}{2}$	$-\frac{\sqrt{3}}{2}$	0
$t_2e(^3T_1)$	$\frac{1}{\sqrt{2}}$	$\frac{\sqrt{3}}{2\sqrt{2}}$	$-\frac{1}{2}$	$\frac{1}{4}$	$-\frac{\sqrt{3}}{4}$	0
$t_2e(^3T_2)$	$-\frac{1}{2}\sqrt{6}$	$-\frac{1}{2\sqrt{2}}$	$-\frac{\sqrt{3}}{2}$	$-\frac{\sqrt{3}}{4}$	$-\frac{1}{4}$	$-\sqrt{2}$
$e^2(^3A_2)$	0	-1	0	0	$-\sqrt{2}$	0

Table A 34. *Spin-orbit coupling matrices for the lowest terms of d^3 in units of ζ*

E'	$t_2^3(^2T_1)$	$t_2^2e(^4T_2)$	$t_2^2e(^4T_1)$
$t_2^3(^2T_1)$	0	$\frac{1}{\sqrt{2}}$	$-\frac{1}{\sqrt{2}}$
$t_2^2e(^4T_2)$	$\frac{1}{\sqrt{2}}$	$\frac{1}{4}$	$-\frac{1}{4}$
$t_2^2e(^4T_1)$	$-\frac{1}{\sqrt{2}}$	$-\frac{1}{4}$	$-\frac{5}{12}$

E''	$t_2^3(^2T_2)$	$t_2^2e(^4T_2)$	$t_2^2e(^4T_1)$
$t_2^3(^2T_2)$	0	$\frac{1}{\sqrt{6}}$	$-\frac{1}{2}\sqrt{6}$
$t_2^2e(^4T_2)$	$\frac{1}{\sqrt{6}}$	$-\frac{5}{12}$	$\frac{1}{4}$
$t_2^2e(^4T_1)$	$-\frac{1}{2}\sqrt{6}$	$\frac{1}{4}$	$\frac{1}{4}$

U'	$t_2^3(^4A_2)$	$t_2^3(^2E)$	$t_2^3(^2T_1)$	$t_2^3(^2T_2)$	$t_2^2e(^4T_{2\frac{3}{2}})$	$t_2^2e(^4T_{2\frac{5}{2}})$	$t_2^2e(^4T_{1\frac{3}{2}})$	$t_2^2e(^4T_{1\frac{5}{2}})$
$t_2^3(^4A_2)$	0	0	0	1	$-\frac{\sqrt{5}}{\sqrt{3}}$	0	0	0
$t_2^3(^2E)$	0	0	0	$\frac{1}{\sqrt{2}}$	$-\frac{\sqrt{2}}{\sqrt{15}}$	$-\frac{\sqrt{6}}{\sqrt{5}}$	0	0
$t_2^3(^2T_1)$	0	0	0	$-\frac{1}{2}\sqrt{3}$	$-\frac{1}{2\sqrt{5}}$	$\frac{1}{\sqrt{5}}$	$-\frac{1}{2}\sqrt{5}$	0
$t_2^3(^2T_2)$	1	$\frac{1}{\sqrt{2}}$	$-\frac{1}{2}\sqrt{3}$	0	$\frac{\sqrt{5}}{2\sqrt{3}}$	0	$-\frac{\sqrt{3}}{2\sqrt{5}}$	$\frac{\sqrt{3}}{\sqrt{5}}$
$t_2^2e(^4T_{2\frac{3}{2}})$	$-\frac{\sqrt{5}}{\sqrt{3}}$	$-\frac{\sqrt{2}}{\sqrt{15}}$	$-\frac{1}{2\sqrt{5}}$	$\frac{\sqrt{5}}{2\sqrt{3}}$	$-\frac{1}{6}$	0	$\frac{1}{5}$	$-\frac{2}{5}$
$t_2^2e(^4T_{2\frac{5}{2}})$	0	$-\frac{\sqrt{6}}{\sqrt{5}}$	$\frac{1}{\sqrt{5}}$	0	0	$\frac{1}{4}$	$-\frac{2}{5}$	$-\frac{9}{20}$
$t_2^2e(^4T_{1\frac{3}{2}})$	0	0	$-\frac{1}{2}\sqrt{5}$	$-\frac{\sqrt{3}}{2\sqrt{5}}$	$\frac{1}{5}$	$-\frac{2}{5}$	$-\frac{1}{6}$	0
$t_2^2e(^4T_{1\frac{5}{2}})$	0	0	0	$\frac{\sqrt{3}}{\sqrt{5}}$	$-\frac{2}{5}$	$-\frac{9}{20}$	0	$\frac{1}{4}$

Table A 35. *Spin-orbit coupling matrices for the lowest terms of d^4 in units of ζ*

(The terms given are ^5E of t_2^3e, $^5T_2$ of $t_2^2e^2$ and $^3T_1$, ^1E, $^1T_2$ and $^1A_1$ of t_2^4.)

A_1	$^3T_1$	^5E	$^5T_2$	$^1A_1$
$^3T_1$	-1	$-\sqrt{2}$	0	$-\sqrt{2}$
^5E	$-\sqrt{2}$	0	$-\frac{1}{2}\sqrt{6}$	0
$^5T_2$	0	$-\frac{1}{2}\sqrt{6}$	$-\frac{1}{2}$	0
$^1A_1$	$-\sqrt{2}$	0	0	0

T_1	$^3T_1$	^5E	$^5T_{22}$	$^5T_{23}$
$^3T_1$	$-\frac{1}{2}$	$\frac{1}{2}\sqrt{6}$	0	0
^5E	$\frac{1}{2}\sqrt{6}$	0	$-\frac{1}{4}\sqrt{6}$	$-\frac{1}{2}\sqrt{3}$
$^5T_{22}$	0	$-\frac{1}{4}\sqrt{6}$	$\frac{1}{4}$	0
$^5T_{23}$	0	$-\frac{1}{2}\sqrt{3}$	0	$-\frac{1}{2}$

A_2	^5E
^5E	0

E	$^3T_1$	^1E	^5E	$^5T_2$
$^3T_1$	$\frac{1}{2}$	$\frac{1}{\sqrt{2}}$	-1	0
^1E	$\frac{1}{\sqrt{2}}$	0	0	0
^5E	-1	0	0	$-\frac{1}{2}\sqrt{3}$
$^5T_2$	0	0	$-\frac{1}{2}\sqrt{3}$	$\frac{1}{4}$

T_2	$^3T_1$	$^1T_2$	^5E	$^5T_{21}$	$^5T_{23}$
$^3T_1$	$\frac{1}{2}$	$\frac{1}{\sqrt{2}}$	$-\frac{1}{\sqrt{2}}$	0	0
$^1T_2$	$\frac{1}{\sqrt{2}}$	0	0	0	0
^5E	$-\frac{1}{\sqrt{2}}$	0	0	$\frac{3}{2\sqrt{10}}$	$\frac{\sqrt{3}}{2\sqrt{5}}$
$^5T_{21}$	0	0	$\frac{3}{2\sqrt{10}}$	$\frac{3}{4}$	0
$^5T_{23}$	0	0	$\frac{\sqrt{3}}{2\sqrt{5}}$	0	$-\frac{1}{2}$

Table A 36. *Spin-orbit coupling matrices for d^5 in units of ζ*

(The terms given are $^2T_2$ of t_2^5, $^4T_1$ and $^4T_2$ of t_2^4e and $^6A_1$ of $t_2^3e^2$.)

E'	$^4T_1$	$^4T_2$
$^4T_1$	$\frac{5}{12}$	$\frac{1}{4}$
$^4T_2$	$\frac{1}{4}$	$-\frac{1}{4}$

E''	$^2T_2$	$^4T_1$	$^4T_2$	$^6A_1$
$^2T_2$	-1	$\sqrt{3}$	$-\frac{1}{\sqrt{3}}$	0
$^4T_1$	$\sqrt{3}$	$-\frac{1}{4}$	$-\frac{1}{4}$	$-\sqrt{2}$
$^4T_2$	$-\frac{1}{\sqrt{3}}$	$-\frac{1}{4}$	$\frac{5}{12}$	0
$^6A_1$	0	$-\sqrt{2}$	0	0

U'	$^2T_2$	$^4T_{1\frac{3}{2}}$	$^4T_{1\frac{5}{2}}$	$^4T_{2\frac{3}{2}}$	$^4T_{2\frac{5}{2}}$	$^6A_1$
$^2T_2$	$\frac{1}{2}$	$\frac{\sqrt{6}}{2\sqrt{5}}$	$-\frac{\sqrt{6}}{\sqrt{5}}$	$-\frac{\sqrt{5}}{\sqrt{6}}$	0	0
$^4T_{1\frac{3}{2}}$	$\frac{\sqrt{6}}{2\sqrt{5}}$	$\frac{1}{6}$	0	$-\frac{1}{5}$	$\frac{2}{5}$	0
$^4T_{1\frac{5}{2}}$	$-\frac{\sqrt{6}}{\sqrt{5}}$	0	$-\frac{1}{4}$	$\frac{2}{5}$	$\frac{9}{20}$	$-\sqrt{2}$
$^4T_{2\frac{3}{2}}$	$-\frac{\sqrt{5}}{\sqrt{6}}$	$-\frac{1}{5}$	$\frac{2}{5}$	$\frac{1}{6}$	0	0
$^4T_{2\frac{5}{2}}$	0	$\frac{2}{5}$	$\frac{9}{20}$	0	$-\frac{1}{4}$	0
$^6A_1$	0	0	$-\sqrt{2}$	0	0	0

Table A 37. *Spin-orbit coupling matrices for d^6 in units of ζ*

(The terms given are $^1A_1$ of t_2^6, $^3T_1$, $^1T_1$, $^3T_2$, and $^1T_2$ of t_2^5e, $^5T_2$ of $t_2^4e^2$ and ^5E of $t_2^3e^3$.)

A_1	$^1A_1$	$^3T_1$	$^5T_2$	^5E
$^1A_1$	0	$-\sqrt{6}$	0	0
$^3T_1$	$-\sqrt{6}$	$\frac{1}{2}$	$\sqrt{3}$	0
$^5T_2$	0	$\sqrt{3}$	$\frac{1}{2}$	$-\frac{1}{2}\sqrt{6}$
^5E	0	0	$-\frac{1}{2}\sqrt{6}$	0

A_2	$^3T_2$	^5E
$^3T_2$	$\frac{1}{2}$	0
^5E	0	0

E	$^3T_1$	$^3T_2$	$^5T_2$	^5E
$^3T_1$	$-\frac{1}{4}$	$-\frac{3}{4}$	$\frac{\sqrt{3}}{2}$	0
$^3T_2$	$-\frac{3}{4}$	$-\frac{1}{4}$	$\frac{\sqrt{3}}{2}$	0
$^5T_2$	$\frac{\sqrt{3}}{2}$	$\frac{\sqrt{3}}{2}$	$-\frac{1}{4}$	$-\frac{\sqrt{3}}{2}$
^5E	0	0	$-\frac{\sqrt{3}}{2}$	0

T_1	$^1T_1$	$^3T_1$	$^3T_2$	$^5T_{22}$	$^5T_{23}$	^5E
$^1T_1$	0	$\frac{1}{2\sqrt{2}}$	$\frac{1}{4}\sqrt{6}$	0	0	0
$^3T_1$	$\frac{1}{2\sqrt{2}}$	$\frac{1}{4}$	$-\frac{\sqrt{3}}{4}$	$\frac{1}{2}$	$\sqrt{2}$	0
$^3T_2$	$\frac{1}{4}\sqrt{6}$	$-\frac{\sqrt{3}}{4}$	$-\frac{1}{4}$	$\frac{\sqrt{3}}{2}$	0	0
$^5T_{22}$	0	$\frac{1}{2}$	$\frac{\sqrt{3}}{2}$	$-\frac{1}{4}$	0	$-\frac{1}{4}\sqrt{6}$
$^5T_{23}$	0	$\sqrt{2}$	0	0	$\frac{1}{2}$	$-\frac{\sqrt{3}}{2}$
^5E	0	0	0	$-\frac{1}{4}\sqrt{6}$	$-\frac{\sqrt{3}}{2}$	0

T_2	$^1T_2$	$^3T_1$	$^3T_2$	$^5T_{21}$	$^5T_{23}$	^5E
$^1T_2$	0	$-\frac{1}{4}\sqrt{6}$	$\frac{1}{2\sqrt{2}}$	0	0	0
$^3T_1$	$-\frac{1}{4}\sqrt{6}$	$-\frac{1}{4}$	$\frac{\sqrt{3}}{4}$	$-\frac{1}{2\sqrt{5}}$	$-\frac{\sqrt{6}}{\sqrt{5}}$	0
$^3T_2$	$\frac{1}{2\sqrt{2}}$	$\frac{\sqrt{3}}{4}$	$\frac{1}{4}$	$\frac{\sqrt{5}}{2\sqrt{3}}$	0	0
$^5T_{21}$	0	$-\frac{1}{2\sqrt{5}}$	$\frac{\sqrt{5}}{2\sqrt{3}}$	$-\frac{3}{4}$	0	$\frac{3}{2\sqrt{10}}$
$^5T_{23}$	0	$-\frac{\sqrt{6}}{\sqrt{5}}$	0	0	$\frac{1}{2}$	$\frac{\sqrt{3}}{2\sqrt{5}}$
^5E	0	0	0	$\frac{3}{2\sqrt{10}}$	$\frac{\sqrt{3}}{2\sqrt{5}}$	0

Table A 38. *Spin-orbit coupling matrices for d^7 in units of ζ*

(The terms given are ^2E of t_2^6e and $^4T_1$ of $t_2^5e^2$.)

U'	^2E	$^4T_{1\frac{3}{2}}$	$^4T_{1\frac{5}{2}}$
^2E	0	$\frac{1}{\sqrt{5}}$	$\frac{3}{\sqrt{5}}$
$^4T_{1\frac{3}{2}}$	$\frac{1}{\sqrt{5}}$	$-\frac{1}{3}$	0
$^4T_{1\frac{5}{2}}$	$\frac{3}{\sqrt{5}}$	0	$\frac{1}{2}$

$$^4T_{1\frac{1}{2}}E' = -\tfrac{5}{6}$$
$$^4T_{1\frac{5}{2}}E'' = \tfrac{1}{2}$$

Table A 39. *Coefficients of fractional parentage for e^m and t_2^n*

$\langle e^2(S'\Gamma')\,eS\Gamma|\}\,e^3S\Gamma\rangle$

$S\Gamma$ \ $S'\Gamma'$	$^1A_1$	$^3A_2$	^1E
^2E	$\frac{1}{\sqrt{6}}$	$\frac{1}{\sqrt{2}}$	$-\frac{1}{\sqrt{3}}$

$\langle t_2^2(S'\Gamma')\,t_2S\Gamma|\}\,t_2^3S\Gamma\rangle$

$S\Gamma$ \ $S'\Gamma'$	$^1A_1$	^1E	$^3T_1$	$^1T_2$
$^4A_2$	.	.	1	.
^2E	.	.	$-\frac{1}{\sqrt{2}}$	$\frac{1}{\sqrt{2}}$
$^2T_1$	.	$-\frac{1}{\sqrt{3}}$	$\frac{1}{\sqrt{2}}$	$\frac{1}{\sqrt{6}}$
$^2T_2$	$\frac{\sqrt{2}}{3}$	$\frac{1}{3}$	$-\frac{1}{\sqrt{2}}$	$-\frac{1}{\sqrt{6}}$

$\langle t_2^3(S'\Gamma')\,t_2S\Gamma|\}\,t_2^4S\Gamma\rangle$

$S\Gamma$ \ $S'\Gamma'$	$^4A_2$	^2E	$^2T_1$	$^2T_2$
$^1A_1$	.	.	.	1
^1E	.	.	$\frac{1}{2}\sqrt{3}$	$-\frac{1}{2}$
$^3T_1$	$\frac{1}{\sqrt{3}}$	$-\frac{1}{\sqrt{6}}$	$\frac{1}{2}$	$-\frac{1}{2}$
$^1T_2$	.	$\frac{1}{\sqrt{2}}$	$\frac{1}{2}$	$-\frac{1}{2}$

Table A 39 (*cont.*)

$\langle t_2^4(S'\Gamma')\,t_2 S\Gamma|\}\,t_2^5 S\Gamma\rangle$

$S\Gamma$ \ $S'\Gamma'$	$^1A_1$	^1E	$^3T_1$	$^1T_2$
$^2T_2$	$\frac{1}{\sqrt{15}}$	$-\frac{\sqrt{2}}{\sqrt{15}}$	$\frac{\sqrt{3}}{\sqrt{5}}$	$\frac{1}{\sqrt{5}}$

$\langle t_2^3(S_1\Gamma_1)\,t_2^2(S_2\Gamma_2)\,^2T_2|\}\,t_2^5\,^2T_2\rangle$

$S_1\Gamma_1$ \ $S_2\Gamma_2$	$^1A_1$	$^3T_1$	^1E	$^1T_2$
$^4A_2$	.	$-\frac{1}{\sqrt{5}}$	.	.
$^2T_1$	.	$-\frac{\sqrt{3}}{2\sqrt{5}}$	$\frac{1}{\sqrt{10}}$	$-\frac{1}{2\sqrt{5}}$
^2E	.	$\frac{1}{\sqrt{10}}$	.	$\frac{1}{\sqrt{10}}$
$^2T_2$	$\frac{1}{\sqrt{15}}$	$\frac{\sqrt{3}}{2\sqrt{5}}$	$\frac{1}{\sqrt{30}}$	$-\frac{1}{2\sqrt{5}}$

$\langle t_2^2(S_1\Gamma_1)\,t_2^2(S_2\Gamma_2)\,S\Gamma|\}\,t_2^4 S\Gamma\rangle$

Rows: $S_1\Gamma_1$; columns: $S_2\Gamma_2$

$S\Gamma = {}^1A_1$	$^1A_1$	$^3T_1$	^1E	$^1T_2$
$^1A_1$	$\frac{1}{3}\sqrt{2}$	.	.	.
$^3T_1$	.	$-\frac{1}{\sqrt{2}}$	.	.
^1E	.	.	$-\frac{1}{3}$	.
$^1T_2$	.	.	.	$-\frac{1}{\sqrt{6}}$

$S\Gamma = {}^3T_1$	$^1A_1$	$^3T_1$	^1E	$^1T_2$
$^1A_1$	.	$-\frac{1}{3\sqrt{2}}$	.	.
$^3T_1$	$-\frac{1}{3\sqrt{2}}$	$\frac{1}{\sqrt{3}}$	$\frac{1}{3}$	$-\frac{1}{\sqrt{6}}$
^1E	.	$\frac{1}{3}$	.	.
$^1T_2$	.	$-\frac{1}{\sqrt{6}}$	.	.

$S\Gamma = {}^1E$	$^1A_1$	$^3T_1$	^1E	$^1T_2$
$^1A_1$	.	.	$-\frac{1}{3\sqrt{2}}$	.
$^3T_1$	.	$-\frac{1}{\sqrt{2}}$	.	.
^1E	$-\frac{1}{3\sqrt{2}}$	.	$\frac{1}{3}\sqrt{2}$	.
$^1T_2$	.	.	.	$-\frac{1}{\sqrt{6}}$

$S\Gamma = {}^1T_2$	$^1A_1$	$^3T_1$	^1E	$^1T_2$
$^1A_1$	.	.	.	$-\frac{1}{3\sqrt{2}}$
$^3T_1$	.	$-\frac{1}{\sqrt{2}}$	.	.
^1E	.	.	.	$\frac{1}{3}$
$^1T_2$	$-\frac{1}{3\sqrt{2}}$	.	$\frac{1}{3}$	$\frac{1}{\sqrt{6}}$

(Data kindly supplied by C. K. Jørgensen and (for $CrCl_3$, $Cr(CN)_6^{3-}$ and Fe ox_3^{3-}) by C. E. Schäffer. I give their assignments for the bands; these should not be regarded as necessarily final although it is unlikely that many will need revision. Doubt about the ^1E assignments for $3d^8$ has been expressed by Liehr & Ballhausen (1959) and about the chemical identity of the rhodium species by Shukla & Lederer (1959). The ground term can usually be inferred from magnetic measurements (see ch. 10). The valency, in the sense of ch. 1, is given in brackets. ox = oxalate, en = ethylene diamine.)

Configuration	Metal	Ligands	Ground term	Positions of excited terms
$3d^2$	V (III)	H_2O	$^3T_1$	17,100 ($^3T_2$), 25,200 ($^3T_1$)
		ox^{--}	$^3T_1$	16,500 ($^3T_2$), 23,500 ($^3T_1$)
$3d^3$	V (II)	H_2O	$^4A_2$	12,400 ($^4T_2$), 18,500 ($^4T_1$), 28,000 ($^4T_1$)
	Cr (III)	H_2O	$^4A_2$	15,000 (^2E), 17,400 ($^4T_2$), 24,700 ($^4T_1$), 37,800 ($^4T_1$)
		ox^{--}	$^4A_2$	14,350 (^2E), 17,500 ($^4T_2$), 24,000 ($^4T_1$)
		Cl^- in $CrCl_3$ solid	$^4A_2$	13,550 ($^4T_2$), 18,800 ($^4T_1$), 27,600 ($^4T_1$?)
		NH_3	$^4A_2$	15,300 (^2E), 21,500 ($^4T_2$), 28,500 ($^4T_1$)
		en	$^4A_2$	15,100 and 15,600 (^2E), 21,900 ($^4T_2$), 28,600 ($^4T_1$)
		CN^-	$^4A_2$	26,600 ($^4T_2$), 32,400 ($^4T_1$)
	Mn (IV)	F^-	$^4A_2$	16,200 (^2E), 21,800 ($^4T_2$), 28,200 ($^4T_1$)
$3d^5$	Mn (II)	Cl^- in $MnCl_2$ solid	$^6A_1$	18,500 ($^4T_1$), 22,000 ($^4T_2$), 23,700 ($^4A_1$, ^4E), 26,900 ($^4T_2$), 28,300 (^4E)
		F^- in MnF_2 solid	$^6A_1$	25,300 ($^4A_1$, ^4E), 28,200 ($^4T_2$), 30,000 (^4E)
		H_2O	$^6A_1$	18,800 ($^4T_1$), 23,100 ($^4T_2$), 25,000 and 25,250 ($^4A_1$, ^4E), 28,100 ($^4T_2$), 29,700 (^4E), 32,800 ($^4T_1$)
		en	$^6A_1$	15,900 ($^4T_1$), 20,300 ($^4T_2$), 23,700 ($^4A_1$, ^4E)
	Fe (III)	F^-	$^6A_1$	14,300 ($^4T_1$), 19,700 ($^4T_2$), 25,350 ($^4A_1$, ^4E), 28,800 ($^4T_2$)
		H_2O	$^6A_1$	12,600 ($^4T_1$), 18,500 ($^4T_2$), 24,300 and 24,600 ($^4A_1$, ^4E)
		ox^{--}	$^6A_1$	10,700 ($^4T_1$), 15,150 ($^4T_2$), 22,100 ($^4A_1$, ^4E)
$3d^6$	Fe (II)	H_2O	$^5T_2$	10,400 (^5E)
		CN^-	$^1A_1$	31,000 ($^1T_1$), 37,000 ($^1T_2$)
	Co (III)	H_2O	$^1A_1$	16,600 ($^1T_1$), 24,900 ($^1T_2$)
		ox^{--}	$^1A_1$	16,500 ($^1T_1$), 23,800 ($^1T_2$)
		NH_3	$^1A_1$	13,000 ($^3T_1$), 21,000 ($^1T_1$), 29,500 ($^1T_2$)
		en	$^1A_1$	13,700 ($^3T_1$), 21,500 ($^1T_1$), 29,500 ($^1T_2$)
		CN^-	$^1A_1$	32,200 ($^1T_1$), 38,600 ($^1T_2$)

Table A 40 (*cont.*)

Con-figuration	Metal	Ligands	Ground term	Positions of excited terms
$3d^7$	Co (II)	H_2O	$^4T_1$	8,100 ($^4T_2$), 11,300 (^2E), 16,000 ($^4A_2$), 19,400 and 21,550 ($^4T_1$)
		NH_3	$^4T_1$	9,000 ($^4T_2$), 18,500 ($^4A_2$), 21,100 ($^4T_1$)
$3d^8$	Ni (II)	Br^- in $CsNiBr_3$	$^3A_2$	10,300 (^1E), 11,300 ($^3T_1$), 17,200 ($^1T_2$), 20,000 ($^3T_1$)
		Cl^- in $CsNiCl_3$	$^3A_2$	11,300 ($^3T_1$), 12,500 (^1E), 18,700 ($^1T_2$), 21,500 ($^3T_1$)
		F^- in $KNiF_3$	$^3A_2$	12,500 ($^3T_1$), 15,300 (^1E), 21,000 ($^1T_2$), 23,700 ($^3T_1$)
		H_2O	$^3A_2$	8,500 ($^3T_2$), 13,500 ($^3T_1$), 15,400 (^1E), 22,000 ($^1T_2$), 25,300 ($^3T_1$)
		NH_3	$^3A_2$	10,750 ($^3T_2$), 13,150 (^1E), 17,500 ($^3T_1$), 28,200 ($^3T_1$)
		en	$^3A_2$	11,200 ($^3T_2$), 12,400 (^1E), 18,350 ($^3T_1$), 24,000 ($^1T_2$), 29,000 ($^3T_1$)
$4d^3$	Mo (III)	Cl^-	$^4A_2$	9,600 (^2E, $^2T_1$), 14,800 ($^2T_2$), 19,200 ($^4T_2$), 23,900 ($^4T_1$)
	Tc (IV)	Br^-	$^4A_2$	13,500 ($^2T_2$)
		Cl^-	$^4A_2$	14,300 ($^2T_2$)
$4d^6$	Rh (III)	Br^-	$^1A_1$	18,100 ($^1T_1$), 22,200 ($^1T_2$)
		Cl^-	$^1A_1$	14,700 ($^3T_1$), 19,300 ($^1T_1$), 24,300 ($^1T_2$)
		H_2O	$^1A_1$	25,500 ($^1T_1$), 32,800 ($^1T_2$)
		ox^{--}	$^1A_1$	25,100 ($^1T_1$)
		NH_3	$^1A_1$	32,700 ($^1T_1$), 39,100 ($^1T_2$)
		en	$^1A_1$	33,200 ($^1T_1$), 39,600 ($^1T_2$)
$5d^3$	Re (IV)	Br^-	$^4A_2$	9,000 (^2E, $^2T_1$), 13,200 ($^2T_2\,E''$), 15,200 ($^2T_2\,U'$)
		Cl^-	$^4A_2$	9,100 (^2E, $^2T_1$), 14,100($^2T_2\,E''$), 15,800 ($^2T_2\,U'$), 27,500 ($^4T_1$)
$5d^4$	Os (IV)	Br^-	$^3T_{10}$	11,200 (^1E, $^1T_2$), 17,240 ($^1A_1$)
		Cl^-	$^3T_{10}$	10,800 and 11,700 (^1E, $^1T_2$), 16,100 ($^1A_1$)
$5d^6$	Ir (III)	Br^-	$^1A_1$	16,800 ($^3T_1$), 22,400 ($^1T_1$), 25,800 ($^1T_2$)
		Cl^-	$^1A_1$	16,300 and 17,900 ($^3T_1$), 24,100 ($^1T_1$), 28,100 ($^1T_2$)
		en	$^1A_1$	33,100 ($^3T_1$), 40,200 ($^1T_1$)
	Pt (IV)	Cl^-	$^1A_1$	22,100 ($^3T_1$), 28,300 ($^1T_1$)
		F^-	$^1A_1$	22,500 and 24,500 ($^3T_1$), 31,400 ($^1T_1$), 36,400 ($^1T_2$)

Table A 41. *Matrix elements of the quantities a_z and a^- of (12.22) for use in the calculation of hyperfine energies*

a_z	ϵ^+	ζ^+	-1^-
ϵ^+	-2	0	$\frac{3}{\sqrt{2}}$
ζ^+	0	-2	$-\frac{3}{\sqrt{2}}$
-1^-	$\frac{3}{\sqrt{2}}$	$-\frac{3}{\sqrt{2}}$	-1

a_z	ϵ^-	ζ^-	1^+
ϵ^-	2	0	$-\frac{3}{\sqrt{2}}$
ζ^-	0	2	$-\frac{3}{\sqrt{2}}$
1^+	$-\frac{3}{\sqrt{2}}$	$-\frac{3}{\sqrt{2}}$	1

a_z	1^-	0^+
1^-	-1	$-\frac{1}{2}\sqrt{6}$
0^+	$-\frac{1}{2}\sqrt{6}$	2

a_z	-1^+	0^-
-1^+	1	$\frac{1}{2}\sqrt{6}$
0^-	$\frac{1}{2}\sqrt{6}$	-2

a^-	ϵ^+	ϵ^-	ζ^+	ζ^-	1^+	1^-	-1^+	-1^-	0^+	0^-
ϵ^+	.	.	.	.	.	.	$\frac{3}{\sqrt{2}}$	.	.	$-2\sqrt{3}$
ϵ^-	2	.	.	.	.	.	.	$-\frac{3}{\sqrt{2}}$	.	.
ζ^+	.	.	.	.	.	.	$-\frac{3}{\sqrt{2}}$	.	.	$2\sqrt{3}$
ζ^-	.	.	2	.	.	.	.	$\frac{3}{\sqrt{2}}$	.	.
1^+	$-\frac{3}{\sqrt{2}}$	.	$-\frac{3}{\sqrt{2}}$	.	.	.	.	.	.	.
1^-	.	$\frac{3}{\sqrt{2}}$	.	$\frac{3}{\sqrt{2}}$	-1	.	.	.	.	.
-1^+	.	.	.	.	.	-6	.	.	$\frac{1}{2}\sqrt{6}$	.
-1^-	.	.	.	.	.	.	-1	.	.	$-\frac{1}{2}\sqrt{6}$
0^+	.	$-2\sqrt{3}$	.	$-2\sqrt{3}$	$-\frac{1}{2}\sqrt{6}$	.	.	.	.	.
0^-	.	.	.	.	.	$\frac{1}{2}\sqrt{6}$	.	.	-2	.

APPENDIX 3

SUMMARY OF RELEVANT PARTS OF PERTURBATION THEORY

(See Dirac, 1947; Pryce, 1950*b*)

The states $|0\rangle$, $|i\rangle$ are always normalized eigenstates of the unperturbed Hamiltonian $\mathscr{H}_0$ with energies E_0, E_i, respectively. $\mathscr{H}$ is the full Hamiltonian for the system.

(1) $\mathscr{H} = \mathscr{H}_0 + \mathscr{H}_1 + \mathscr{H}_2$, $\mathscr{H}_1$ presumed small, order of magnitude ϵ, $\mathscr{H}_2$ of order ϵ^2. Unperturbed state $|0\rangle$ of $\mathscr{H}_0$ non-degenerate.

Write
$$\mathscr{H}|X\rangle = E|X\rangle \quad \text{with} \quad E = E_0 + E_1 + E_2 + \ldots,$$
$|X\rangle = |0\rangle + |X_1\rangle + |X_2\rangle + \ldots$, where the suffix n denotes order of magnitude ϵ^n. Then

$$(\mathscr{H}_0 + \mathscr{H}_1 + \mathscr{H}_2)(|0\rangle + |X_1\rangle + |X_2\rangle + \ldots)$$
$$= (E_0 + E_1 + E_2 + \ldots)(|0\rangle + |X_1\rangle + |X_2\rangle + \ldots)$$

leading to equations of order 1, ϵ, ϵ^2:

$$\mathscr{H}_0|0\rangle = E_0|0\rangle, \tag{A 3.1}$$

$$\mathscr{H}_1|0\rangle + \mathscr{H}_0|X_1\rangle = E_1|0\rangle + E_0|X_1\rangle, \tag{A 3.2}$$

$$\mathscr{H}_2|0\rangle + \mathscr{H}_1|X_1\rangle + \mathscr{H}_0|X_2\rangle = E_2|0\rangle + E_1|X_1\rangle + E_0|X_2\rangle. \tag{A 3.3}$$

Write $|X_1\rangle = \sum_{i\neq 0} a_i|i\rangle$, $|X_2\rangle = \sum_{i\neq 0} b_i|i\rangle$. Then multiply (A 3.2) by $\langle 0|$, giving

$$E_1 = \langle 0|\mathscr{H}_1|0\rangle. \tag{A 3.4}$$

This is the first-order correction to the energy. Multiply by $\langle i|$, giving

$$\langle i|\mathscr{H}_1|0\rangle + a_i E_i = E_0 a_i$$

or
$$a_i = -\frac{\langle i|\mathscr{H}_1|0\rangle}{E_i - E_0}. \tag{A 3.5}$$

This is the first-order correction to the wave-function.

Now multiply (A 3.3) by $\langle 0|$, giving

$$\langle 0|\mathscr{H}_2|0\rangle + \sum_{i\neq 0} a_i\langle 0|\mathscr{H}_1|i\rangle = E_2$$

whence, by (5):
$$E_2 = \langle 0|\mathscr{H}_2|0\rangle - \sum_{i\neq 0}\frac{|\langle 0|\mathscr{H}_1|i\rangle|^2}{E_i - E_0}. \tag{A 3.6}$$

This is the second-order correction to the energy.

The process can be continued indefinitely. The function $|X\rangle$ is only correctly normalized to first order in ϵ, to second order it must be multiplied by

$$(1+\Sigma\,|a_i\,|^2)^{-\frac{1}{2}}.$$

(2) $\mathscr{H} = \mathscr{H}_0 + \mathscr{H}_1$, $\mathscr{H}_1$ of order ϵ. Unperturbed state $|0\rangle$ of $\mathscr{H}_0$ degenerate. $|0\rangle = \sum_{M=1}^{n} c_M\,|0M\rangle$, $\quad \mathscr{H}_0\,|0M\rangle = E_0\,|0M\rangle$.

(*a*) Under the full site symmetry group G, it may be that the representation spanned by the $|0M\rangle$ contains no repeated representation. In that case take linear combinations of the $|0M\rangle$ (and also of the $|i\rangle$) which are components of bases of irreducible representations. Use a standard choice of basis for these components. Then no matrix element of $\mathscr{H}_1$ can connect different components and the entire matrix of $\mathscr{H}_1$ breaks up into pieces, one for each component of each irreducible representation. For each of these pieces the relevant unperturbed state of $\mathscr{H}_0$ is non-degenerate so the previous theory still applies, the sum over i being now restricted to excited states forming the same component of an irreducible representation as the ground state.

Example. $G = O^*$, electronic configuration d^8. $\mathscr{H}_0$ electrostatic Hamiltonian, $\mathscr{H}_1$ spin-orbit coupling. $|0\rangle$ is the ${}^3A_{2g}$ term of d^8, spanning T_2 with components $|0T_2M\rangle$, $M = 1, 0, -1$. First-order correction to wave-function

$$|X_1 M\rangle = |0T_2M\rangle - \sum_i \frac{\langle iT_2M\,|\,\mathscr{H}_1\,|\,0T_2M\rangle}{E_i - E_0}\,|iT_2M\rangle.$$

The coefficients $\langle iT_2M\,|\,\mathscr{H}_1\,|\,0T_2M\rangle$ are independent of M.

(*b*) As (*a*) but with the unperturbed state degenerate for one or more components.

Equation (A 3.4) now becomes

$$\langle 0M'\,|\,\mathscr{H}_1\,|\,0\rangle = E_1\langle 0M'\,|\,0\rangle$$

or

$$\sum_M \langle 0M'\,|\,\mathscr{H}_1\,|\,0M\rangle c_M = E_1 c_{M'} \quad (M' = 1, 2, \ldots, n). \tag{A 3.7}$$

These are the secular equations associated with the matrix of $\mathscr{H}_1$ within the states $|0M\rangle$. The eigenvalues E_1 are n in number and are the solutions of the secular determinantal equation

$$|\langle 0M'\,|\,\mathscr{H}_1\,|\,0M\rangle - E_1\delta_{MM'}\,| = 0. \tag{A 3.8}$$

They are the first-order corrections to the energy. They determine sets of values of the c_M which then give the correct zero-order wave-functions.

(3) $\mathscr{H} = \mathscr{H}_0 + \mathscr{H}_1 + \mathscr{H}_2$, $\mathscr{H}_1^2$ and $\mathscr{H}_2$ of comparable order. Ground state $|0\rangle$ of $\mathscr{H}_0$ degenerate. $|0\rangle = \sum_{M=1}^{n} c_M\,|0M\rangle$ with $\langle 0M'\,|\,\mathscr{H}_1\,|\,0M\rangle$ all zero.

This case is important in electron resonance, where $\mathscr{H}_1$ is the spin-orbit coupling and $\mathscr{H}_2$ is the coupling with the external magnetic field and with nuclear moments.

Equation (A 3.8) reduces to $E_1 = 0$, so there is no first-order correction to the energy and the c_M are as yet undetermined. The first-order correction to the wave-function is still given by (A 3.5) but, because we do not know the c_M, is not completely determined yet. Multiply (A 3.3) by $\langle 0M|$

$$\langle 0M|\mathscr{H}_2|0\rangle+\sum_{i\neq 0}a_i\langle 0M|\mathscr{H}_1|i\rangle = E_2\langle 0M|0\rangle,$$

or $$\sum_{M'}\left\{\langle 0M|\mathscr{H}_2|0M'\rangle-\sum_{i\neq 0}\frac{\langle 0M|\mathscr{H}_1|i\rangle\langle i|\mathscr{H}_1|0M'\rangle}{E_i-E_0}\right\}c_{M'} = E_2c_M. \quad \text{(A 3.9)}$$

This equation determines the second-order corrections to the energies and the first-order corrections to the wave-functions. For future reference we abbreviate it to

$$\sum_{M'}V(MM')c_{M'} = E_2c_M. \quad \text{(A 3.10)}$$

(4) Hypotheses as under 3. Alternative method of calculation.

Equation (A 3.9) has exactly the same form as (A 3.7) and suggests that if we choose our zero-order functions as the perturbed ones

$$|\alpha M\rangle = |0M\rangle-\sum_i\frac{\langle i|\mathscr{H}_1|0M\rangle}{E_i-E_0}|i\rangle, \quad \text{(A 3.11)}$$

and use the full Hamiltonian $\mathscr{H}$ as our perturbing Hamiltonian we might obtain the same results. Our plan, then, is to solve the secular equations associated with the matrix

$$A(MM') = \langle\alpha M|(\mathscr{H}_0+\mathscr{H}_1+\mathscr{H}_2)|\alpha M'\rangle. \quad \text{(A 3.12)}$$

The $|\alpha M\rangle$ are not, in general, an orthonormal set. We have

$$S(MM') = \langle\alpha M|\alpha M'\rangle = \delta_{MM'}+\sum_i\frac{\langle 0M|\mathscr{H}_1|i\rangle\langle i|\mathscr{H}_1|0M'\rangle}{(E_i-E_0)^2}$$

and

$$\begin{aligned}A(MM') &= E_0\delta_{MM'}+\Sigma\frac{\langle 0M|\mathscr{H}_1|i\rangle\langle i|\mathscr{H}_1|0M'\rangle E_i}{(E_i-E_0)^2}\\ &\quad -2\Sigma\frac{\langle 0M|\mathscr{H}_1|i\rangle\langle i|\mathscr{H}_1|0M'\rangle}{E_i-E_0}+\langle 0M|\mathscr{H}_2|0M'\rangle\\ &= E_0S(MM')-\Sigma\frac{\langle 0M|\mathscr{H}_1|i\rangle\langle i|\mathscr{H}_1|0M'\rangle}{E_i-E_0}+\langle 0M|\mathscr{H}_2|0M'\rangle\\ &= E_0S(MM')+V(MM')\\ &= E_0S(MM')+\sum_{M''}S(MM'')V(M''M'),\end{aligned} \quad \text{(A 3.13)}$$

where the last equality is true to our order in perturbation theory.

It follows at once from (A 3.2) that instead of (A 3.7) for the first-order energies we have

$$\sum_{M'}A(MM')c_{M'} = E_1\sum_{M'}S(MM')c_{M'}$$

which, using (A 3.13), becomes

$$E_0\sum_{M'}S(MM')c_{M'}+\sum_{M'',M'}S(MM'')V(M''M')c_{M'} = E_1\sum_{M'}S(MM')c_{M'} \quad \text{(A 3.14)}$$

and in matrix notation $\quad E_0S\mathbf{c}+SV\mathbf{c} = E_1S\mathbf{c}.$

Multiply on the left by S^{-1} $\quad E_0\mathbf{c}+V\mathbf{c} = E_1\mathbf{c}. \quad$ (A 3.15)

This is identical with (A 3.10) if we put $E_1 = E_0 + E_2$, and justifies the use of the $|\alpha M\rangle$ to first order in $\mathscr{H}$.

In our applications we shall often need energies including the term linear in both $\mathscr{H}_1$ and $\mathscr{H}_2$. This is part of the third-order correction to the energy and is obtained from the equation

$$\mathscr{H}_2|X_1\rangle + \mathscr{H}_1|X_2\rangle + \mathscr{H}_0|X_3\rangle = E_3|0\rangle + E_2|X_1\rangle + E_1|X_2\rangle + E_0|X_3\rangle.$$

On multiplying by $\langle 0|$ we get

$$\sum_i (a_i\langle 0|\mathscr{H}_2|i\rangle + b_i\langle 0|\mathscr{H}_1|i\rangle) = E_3.$$

b_i is obtained by multiplying (A 3.3) by $\langle i|$ and we find our new correction to be

$$-\sum_i \frac{\langle 0|\mathscr{H}_2|i\rangle\langle i|\mathscr{H}_1|0\rangle + \langle 0|\mathscr{H}_1|i\rangle\langle i|\mathscr{H}_2|0\rangle}{E_i - E_0}$$

which would also follow from (A 3.11).

APPENDIX 4

THE HAMILTONIAN FOR A CHARGED PARTICLE IN AN ELECTROMAGNETIC FIELD

In this appendix we derive the laws of motion for a charged particle in an electromagnetic field in Hamiltonian form. We use the notation of § 3.1 and suppose the particle to have a charge minus e and mass m. We use a general gauge, not necessarily Coulomb gauge.

The laws of motion in classical mechanics are obtained by using the force on the particle given by (3.2)

$$\mathbf{F} = -e\left(\mathbf{E}+\frac{1}{c}\mathbf{v}\wedge\mathbf{H}\right). \tag{A 4.1}$$

The equations of motion are then given by $m\ddot{\mathbf{r}} = \mathbf{F}$. First, we must transform this equation into Lagrangian form. In detail the equation is

$$m\ddot{\mathbf{r}} = m\dot{\mathbf{v}} = \mathbf{F} = -e\left\{-\frac{1}{c}\frac{\partial\mathbf{A}}{\partial t}-\nabla\phi+\frac{1}{c}\mathbf{v}\wedge(\nabla\wedge\mathbf{A})\right\}$$

$$= \frac{e}{c}\frac{\partial\mathbf{A}}{\partial t}+e\nabla\phi-\frac{e}{c}\nabla(\mathbf{v}\,.\,\mathbf{A})+\frac{e}{c}(\mathbf{v}\,.\,\nabla)\,\mathbf{A}. \tag{A 4.2}$$

We use the fact that
$$\frac{d\mathbf{A}}{dt} = \frac{\partial\mathbf{A}}{\partial t}+\Sigma\frac{\partial\mathbf{A}}{\partial x_i}\dot{x}_i = \frac{\partial\mathbf{A}}{\partial t}+(\mathbf{v}\,.\,\nabla)\,\mathbf{A} \tag{A 4.3}$$

to simplify (A 4.2) and readily arrive at

$$m\dot{\mathbf{v}} = \frac{e}{c}\frac{d\mathbf{A}}{dt}+e\nabla\phi-\frac{e}{c}\nabla(\mathbf{v}\,.\,\mathbf{A}). \tag{A 4.4}$$

We re-arrange this and put the two time derivatives on the left-hand side and it then becomes

$$\frac{d}{dt}\left(m\mathbf{v}-\frac{e}{c}\mathbf{A}\right) = \nabla\left(e\phi-\frac{e}{c}\mathbf{v}\,.\,\mathbf{A}\right). \tag{A 4.5}$$

We must choose a Lagrangian function and it is evident that if we write

$$L = \tfrac{1}{2}mv^2+e\phi-e\frac{\mathbf{v}\,.\,\mathbf{A}}{c} \tag{A 4.6}$$

then (A 4.5) shows that Lagrange's equations

$$\frac{d}{dt}\left(\frac{\partial L}{\partial\dot{x}_i}\right) = \frac{\partial L}{\partial x_i} \tag{A 4.7}$$

are satisfied by L of (A 4.6). In classical mechanics the momentum conjugate to x_i is given by partially differentiating the Lagrangian L with respect to $\dot{x}_i$. Thus

$$p_i = \frac{\partial L}{\partial\dot{x}_i} \quad \text{so} \quad \mathbf{p} = m\mathbf{v}-\frac{e}{c}\mathbf{A}. \tag{A 4.8}$$

The Hamiltonian for the system is given in terms of the conjugate coordinates and momenta and the Lagrangian L by the well-known formula

$$\mathscr{H} = \sum_i p_i \dot{x}_i - L = \mathbf{p} \, . \, \mathbf{v} - L. \tag{A 4.9}$$

Then we substitute for L and $\mathbf{p}$ from (A 4.6) and (A 4.8) to give our final expression

$$\mathscr{H} = mv^2 - \frac{e}{c}\mathbf{v} \, . \, \mathbf{A} - L = \tfrac{1}{2}mv^2 - e\phi = \frac{1}{2m}\left(\mathbf{p} + \frac{e}{c}\mathbf{A}\right)^2 - e\phi. \tag{A 4.10}$$

Equation (A 4.10) is the desired expression for the Hamiltonian of a charged particle as modified by the presence of an electromagnetic field, and affords a justification for the procedure adopted in quantum theory of replacing $\mathbf{p}$ by $p + (e/c)\,\mathbf{A}$ and for the addition of the term in the scalar potential ϕ. The formula can easily be generalized to the case where there are n charged particles. If we suppose their mutual interactions to be already included in the unperturbed Hamiltonian then the result follows immediately by the same method as that used above by merely summing over the n particles.

APPENDIX 5

GAUGE INVARIANCE IN PERTURBATION THEORY

In this appendix I show in detail how gauge invariance is maintained in first-order and second-order perturbation theory. We showed in ch. 3 that the accurate predictions of quantum theory are gauge invariant. If, however, we use perturbation theory it will only be obvious that the predictions of nth-order perturbation theory are gauge invariant if we can be sure that the solutions of the wave-equation can be expanded as a convergent power series in the perturbation. This is often far from obvious and it seems, therefore, of interest to give an independent proof for the two cases which are of the greatest practical interest to us, namely, the first- and second-order theory for a system in a general static magnetic field. Also it is instructive to see the precise manner in which the gauge invariance is maintained in these two cases.

I shall give the demonstration for the problem of just one particle for purposes of simplicity because the extension to n particles is, in this problem, completely trivial. Let us take a Coulomb gauge, this slight restriction again being in the cause of simplicity. Then we have the scalar potential $\phi = 0$ and let us suppose the vector potential to be given by $\mathbf{A} - \operatorname{grad} \chi_1$ with $\operatorname{div} \mathbf{A} = \nabla^2 \chi_1 = 0$. As before the quantity χ_1 represents the arbitrariness of the gauge and we wish to compare the predictions for a general χ_1 with the predictions when $\chi_1 = 0$.

The Hamiltonian for our system is given by

$$\mathscr{H} = \frac{1}{2m}\left(\mathbf{p} + \frac{e}{c}\mathbf{A} - \frac{e}{c}\operatorname{grad}\chi_1\right)^2 + V \tag{A 5.1}$$

and we write it in a slightly more suitable form for our purposes by putting $\chi_1 = i\hbar\chi$ and can then rewrite (A 5.1) as

$$\begin{aligned}\mathscr{H} &= \frac{1}{2m}\left(\mathbf{p} + \frac{e}{c}\mathbf{A} + \frac{e}{c}(\mathbf{p}\chi)\right)^2 + V \\ &= \mathscr{H}_0 + \mathscr{H}_1 + \mathscr{H}_2.\end{aligned} \tag{A 5.2}$$

In the latter form of writing (A 5.2) has been broken up in parts which are respectively zero, first and second order in the perturbation. In detail this means that

$$\left.\begin{aligned}\mathscr{H}_0 &= \frac{1}{2m}\mathbf{p}^2 + V, \\ \mathscr{H}_1 &= \frac{e}{mc}[\mathbf{A} + (\mathbf{p}\chi)] \cdot \mathbf{p}, \\ \mathscr{H}_2 &= \frac{e^2}{2mc^2}[\mathbf{A} + (\mathbf{p}\chi)]^2.\end{aligned}\right\} \tag{A 5.3}$$

In (A 5.3) $\mathscr{H}_0$ is the Hamiltonian for the system in the absence of a field. Then the first-order energy E_1 is the first-order perturbation energy arising from $\mathscr{H}_1$ and is

$$E_1 = \frac{e}{mc}\langle 0|\,[\mathbf{A}+(\mathbf{p}\chi)]\,.\,\mathbf{p}\,|0\rangle. \qquad \text{(A 5.4)}$$

The second-order perturbation energy consists of two parts, one being the first-order energy due to $\mathscr{H}_2$ and the other the second-order energy due to $\mathscr{H}_1$ and is, therefore, given by (A 3.6) of Appendix 3 as

$$\begin{aligned} E_2 &= \langle 0\,|\,\mathscr{H}_2\,|\,0\rangle + \sum_n \frac{|\langle n\,|\,\mathscr{H}_1\,|\,0\rangle|^2}{E_0-E_n} \\ &= \frac{e^2}{2mc^2}\langle 0|\,[\mathbf{A}+(\mathbf{p}\chi)]^2\,|0\rangle \\ &\quad + \frac{e^2}{m^2c^2}\sum_n \frac{|\langle n|\,[\mathbf{A}+(\mathbf{p}\chi)]\,.\,\mathbf{p}\,|0\rangle|^2}{E_0-E_n} \\ &= E_2^1+E_2^2, \quad \text{say.} \end{aligned} \qquad \text{(A 5.5)}$$

We wish to show that E_1 and E_2 are the same for $\chi = 0$ as for general χ. Let us define quantities δE, corresponding to each E occurring above, as the difference between the E for general χ and for $\chi = 0$. Then δE_1 is simply

$$\delta E_1 = \frac{e}{mc}\langle 0|\,(\mathbf{p}\chi)\,.\,\mathbf{p}\,|0\rangle. \qquad \text{(A 5.6)}$$

In order to establish our results we need first to prove a simple theorem. We let ξ be any scalar function of space coordinates alone, and we establish a formula for its matrix elements. We do this by taking matrix elements with respect to the eigenstates of the unperturbed Hamiltonian $\mathscr{H}_0$ and simply use the known form of $\mathscr{H}_0$ given in (A 5.3) and the fact that the kets concerned are indeed eigenstates of $\mathscr{H}_0$. Then we have

$$\begin{aligned} (E_n-E_0)\langle n\,|\,\xi\,|\,0\rangle &= \langle n|\,(\,\mathscr{H}_0\xi-\xi\mathscr{H}_0)\,|0\rangle \\ &= \langle n|\,\frac{1}{2m}(\mathbf{p}^2\xi-\xi\mathbf{p}^2)\,|0\rangle \\ &= \frac{1}{2m}\langle n|\,(\mathbf{p}^2\xi)\,|\,0\rangle + \frac{1}{m}\langle n\,|\,(\mathbf{p}\xi)\,.\,\mathbf{p}\,|0\rangle. \end{aligned} \qquad \text{(A 5.7)}$$

We can interchange 0 and n throughout and then get a corresponding result for the matrix element $\langle 0\,|\,\xi\,|\,n\rangle$. Equation (A 5.7) together with the form obtained by exchanging n and 0 is our theorem. We now take two special functions for ξ in (A 5.7). First, we put $\xi = \chi$ and $n = 0$ which gives

$$0 = \langle 0|\,(\mathbf{p}\chi)\,.\,\mathbf{p}\,|0\rangle, \qquad \text{(A 5.8)}$$

and, therefore, $\delta E_1 = 0$. So we have demonstrated that the first-order predictions of the theory are gauge invariant. Let us now retain $\xi = \chi$ and take a general value for n. Then we obtain the relation

$$(E_n-E_0)\,\langle n\,|\,\chi\,|\,0\rangle = \frac{1}{m}\langle n|\,(\mathbf{p}\chi)\,.\,\mathbf{p}\,|0\rangle \qquad \text{(A 5.9)}$$

which is useful in discussing δE_2. As a second special case we take $\xi = \chi^2$ and $n = 0$. In this case (A 5.7) becomes

$$0 = \tfrac{1}{2}\langle 0|\,(\mathbf{p}^2\chi^2)\,|0\rangle + \langle 0|\,(\mathbf{p}\chi^2)\cdot\mathbf{p}\,|0\rangle$$
$$= \langle 0|\,(\mathbf{p}\chi)^2\,|0\rangle + 2\,\langle 0|\,\chi(\mathbf{p}\chi)\cdot\mathbf{p}\,|0\rangle. \qquad \text{(A 5.10)}$$

We now write the explicit form of the two parts of δE_2. They follow immediately from (A 5.5) and the definition of δE. The first part is

$$\delta E_2^1 = \frac{e^2}{2mc^2}\langle 0|\,[2\mathbf{A}\,.\,(\mathbf{p}\chi) + (\mathbf{p}\chi)^2]\,|0\rangle \qquad \text{(A 5.11)}$$

which is already in a formally simple form. The second part is more complicated, but we simplify it by using the two results (A 5.9) and (A 5.10) that we have just obtained. Then by straightforward manipulation, remembering that the $|n\rangle$ form a complete set we find

$$\delta E_2^2 = \frac{e^2}{m^2c^2}\sum_{n\neq 0}(E_0 - E_n)^{-1}\{\langle 0|\,\mathbf{p}\chi\cdot\mathbf{p}\,|n\rangle\langle n|\,\mathbf{p}\chi\cdot\mathbf{p}\,|0\rangle$$
$$+\langle 0|\,\mathbf{A}\,.\,\mathbf{p}\,|n\rangle\langle n|\,\mathbf{p}\chi\cdot\mathbf{p}\,|0\rangle + \langle 0|\,\mathbf{p}\chi\cdot\mathbf{p}\,|n\rangle\langle n|\,\mathbf{A}\,.\,\mathbf{p}\,|0\rangle\}$$
$$= \frac{e^2}{m^2c^2}\cdot m\sum_{n}\{\langle 0|\,\chi\,|\,n\rangle\langle n|\,\mathbf{p}\chi\cdot\mathbf{p}\,|0\rangle - \langle 0|\,\mathbf{A}\,.\,\mathbf{p}\,|n\rangle\langle n\,|\,\chi\,|\,0\rangle$$
$$+\langle 0\,|\,\chi\,|\,n\rangle\langle n|\,\mathbf{A}\,.\,\mathbf{p}\,|0\rangle\}$$
$$= \frac{e^2}{mc^2}\langle 0|\,[\chi(\mathbf{p}\chi)\cdot\mathbf{p} - \mathbf{A}\,.\,(\mathbf{p}\chi)]\,|0\rangle \qquad \text{(A 5.12)}$$

and then add (A 5.11) to (A 5.12) to obtain

$$\delta E_2^1 + \delta E_2^2 = \frac{e^2}{2mc^2}\langle 0|\,[2\chi(\mathbf{p}\chi)\cdot\mathbf{p} + (\mathbf{p}\chi)^2]\,|0\rangle = 0, \qquad \text{(A 5.13)}$$

where we have used (A 5.10) and the justification for extending the summation in (A 5.12) from $n \neq 0$ in the first line to all n in the second line follows from (A 5.8).

This completes our demonstration of gauge invariance for a system in a magnetic field. We see that it is clearly quite essential to consider the second-order term arising from $\mathscr{H}_1$ whenever we consider the first-order term arising from $\mathscr{H}_2$. It will very often be the case that we shall choose our gauge judiciously so that this second-order term is zero, but if then we should change to a gauge which does not make it zero, we must not then forget about it. As a particular illustration of this point we may refer to the fact that a change of origin in the expression $\mathbf{A} = \frac{1}{2}\mathbf{H}\wedge\mathbf{r}$ is a change of gauge, and that although in considering an atomic system we shall almost always take our origin at the nucleus, if we did not do so we would get a much larger term arising from $\mathscr{H}_2$ which would be compensated by an equally large term from $\mathscr{H}_1$ having the opposite sign.

APPENDIX 6

ATOMIC SPECTRAL PARAMETERS

The parameters B, C and ζ derived from experiment are given below. Many of them, especially for the second- and third-transition series, are calculated by procedures which are very far from a best fit by least squares. Literature references or calculation procedures (for parameters evaluated by me) are given below the tables. The parameters with literature references are usually more carefully fitted to the data than those without.

Quite apart from the fact that least squares fitting is not necessarily the best procedure, the concept of the best value for one of these empirically determined parameters is strictly meaningless. The value varies with the method of calculation and the latter should vary with the purpose for which the parameter is required. The reader ought to bear this in mind when he uses the parameters given here.

A6.1. The first transition series. MI is the neutral atom M

B	Ti	V	Cr	Mn	Fe	Co	Ni	Cu
I	560[a]	578[b]	790[c]	720[d]	806[d]	798[d]	1025[d]	—
II	682[e]	659[a]	710[b]	873[f]	869[d]	878[d]	1037[d]	1216[d]
III	718[g]	766[a]	830[c]	960[c]	1058[h]	1115[d]	1084[d]	1238[d]
IV	—	861[i]	1030[c]	1140[c]	—	—	—	—
V	—	—	1039[j]	—	1144[k]	—	—	—

C	Ti	V	Cr	Mn	Fe	Co	Ni	Cu
I	1840[a]	2273[b]	2520[c]	3087[d]	3506[d]	4167[d]	4226[d]	—
II	2481[e]	2417[a]	2790[b]	3130[f]	3638[d]	3828[d]	4314[d]	4745[d]
III	2629[g]	2855[a]	3430[c]	3325[c]	3901[h]	4366[d]	4831[d]	4659[d]
IV	—	4165[i]	3850[c]	3675[c]	—	—	—	—
V	—	—	4238[j]	—	4459[k]	—	—	—

ζ	Sc	Ti	V	Cr	Mn	Fe	Co	Ni	Cu
I	67	111	158	223	239[l]	391	517	603[y]	817
II	53	88	136	222[l]	254	356	456	603	828[y]
III	79	121	167	230	347[l]	410	533	649	829
IV	—	154	209	273	352	—	—	—	—
V	—	—	248	327	402	514	—	—	—

A6.2. The second transition series

B	Y	Zr	Nb	Mo	Tc	Ru	Rh	Pd
I	274[m]	254[j]	300[p]	455[n]	—	—	596[p]	—
II	349[j]	454[r]	260[q]	440[z]	—	474[p]	667[j]	—
III	—	539[j]	532[p]	—	—	—	620[p]	826[j]
IV	—	—	602[s]	—	—	—	—	—
V	—	—	—	682[s]	—	—	—	—

C	Y	Zr	Nb	Mo	Tc	Ru	Rh	Pd
I	1120[m]	1975[j]	2390[p]	1771[n]	—	—	3236[p]	—
II	1760[j]	1765[r]	1990[q]	1987[z]	—	1806[p]	2313[j]	—
III	—	1640[j]	2095[p]	—	—	—	4002[p]	2620[j]
IV	—	—	1367[s]	—	—	—	—	—

ζ	Y	Zr	Nb	Mo	Tc	Ru	Rh	Pd	Ag
I	212	335[aa]	475	552	647[l]	878[l]	968[l]	1412[y]	—
II	212	339	490	672[l]	656	887	1212[l]	1316[u]	1830[y]
III	290	403[aa]	554	—	—	990	1235	1615[aa]	1844[u]
IV	—	500	670	817	—	—	—	—	—
V	—	—	748	887[aa]	—	—	—	—	—

A6.3. The third transition series

B	Hf	Ta	W	Re	Os	Ir	Pt
I	281[s]	345[w]	371[w]	847[x]	—	—	—
II	435[v]	483[w]	—	—	—	—	—

C	Hf	Ta	W	Re	Os	Ir	Pt
I	—	1289[w]	1900[w]	1182[x]	—	—	—
II	1530[v]	1841[w]	—	—	—	—	—

ζ	Lu	Hf	Ta	W	Re	Os	Ir	Pt	Au
I	798	1307	1657[w]	2089[w]	2285[l]	—	—	4052[y]	4910
II	—	1336[v]	1776[w]	2561[l]	—	—	—	3368	5091[y]

A6.4. References and notes

Experimental data: Moore, Atomic energy levels, *National Bureau of Standards*, no. 467 (1949, 1952, 1958); Iglesias, *J. Opt. Soc. Amer.* (1955), **45**, 856 for Nb III; Klinkenberg, *Physica* (1954), **21**, 53 for Lu I. Fitting of parameters: All parameters ζ were calculated from the formula of ex. 2, p. 113, unless otherwise stated. An asterisk shows that the polarization correction $\alpha L(L+1)$ was included.

a. Many, *Phys. Rev.* (1946), **70**, 511.

b. Schweitzer, *Phys. Rev.* (1950), **80**, 1080.

c. Orgel, *J. Chem. Phys.* (1955), **23**, 1819.
d. Skinner & Sumner, *J. Inorg. Nucl. Chem.* (1957), **4**, 245.
e. Meshkov, *Phys. Rev.* (1953), **91**, 871.
f. Trees, *Phys. Rev.* (1951), **83**, 756. $\alpha = 0$. If $\alpha = 69$, $B = 879$ and $C = 3139$.
*g**. $\alpha = 35$. Trees, *Phys. Rev.* (1955), **97**, 686.
*h**. $\alpha = 81$. Trees, *Phys. Rev.* (1951), **84**, 1089.
i. Cady, *Phys. Rev.* (1933), **43**, 322.
j. From (4.77).
*k**. $\alpha = 85$. Garstang, *Mon. Not. R. Astr. Soc.* (1957), **117**, 393.
l. Weighted mean of values from ^6D, ^4D of the lowest $d^4(^5D)s$ or $d^6(^5D)s$ (^6D only for Re I and W II), from ^5F, ^3F of $4d^7(^4F)\,5s$ or from d^9 and $d^8s\,^4F$ (for Rh I).
m. Condon & Shortley, *Theory of Atomic Spectra*. Cambridge (1953).
*n**. $\alpha = 38$. Trees & Harvey, *J. Res. Nat. Bur. Stand.* (1952), **49**, 397.
p. By approximate fitting to the linear formulae of d^3 or d^7.
q. As under *p*, from d^4.
r. Ufford, *Phys. Rev.* (1933), **44**, 732.
s. From $15B = {}^3P_1 - {}^3F_3$. C for Nb IV from $5B + 2C + \frac{1}{2}\zeta = {}^1D_2 - {}^3F_3$.
t. Fitted to ^4F, ^4P of d^7.
u. Green, *Phys. Rev.* (1941), **60**, 117.
v. Gehatia, *Phys. Rev.* (1954), **94**, 618.
w. Ta I* (with $\alpha = 112$; if $\alpha = 0$, then $B = 278$, $C = 2100$, $\zeta = 1650$), Trees & Kamei; W I* ($\alpha = 46$), Racah; Ta II ($\alpha = 0$; α if included is small and negative), Trees, Cahill & Rabinowitz. Taken from *J. Res. Nat. Bur. Stand.* (1955), **55**, 335.
x. Least squares fit to ^6S, ^4P, ^4D, ^4F, ^4G. Weighted as in § 4.7.
y. Laporte & Inglis, *Phys. Rev.* (1930), **35**, 1337. Intermediate coupling for d^9s.
z. By approximate fitting to the linear formulae of d^5.
aa. From $^3F_3$–$^3F_4$ of d^2 or d^8.

A 6.5. Some recent work

Recently a very detailed analysis including interactions within and between the configurations d^n, $d^{n-1}s$ and $d^{n-2}s^2$ has been given for the ions M II of the first transition series by Racah and Shadmi, *Bull. Soc. Counc. Israel* (1959), 8 *F*, 15. Their values of B, C, ζ and α for use with the d^n configurations are reproduced below. The parameters used in the text of the book are always from A 6.1-4.

M II	Sc	Ti	V	Cr	Mn	Fe	Co	Ni	Cu
B	499	594	648	701	816	810	—	—	—
C	1574	1802	2152	2490	2812	3244	—	—	—
ζ	52	94	125	212	278	368	—	603	—
α	27	49	52	58	67	74	—	78	88

APPENDIX 7

DEDUCTIONS FROM WIGNER'S FORMULA

In this appendix we deduce from Wigner's formula expressions for certain matrix elements which are needed in the theory of magnetic properties. We wish to find the dependence on j and m in a j_1j_2jm scheme of matrix elements of a vector $\mathbf{p}$ which is of type T with respect to $\mathbf{j}_1$ and commutes with $\mathbf{j}_2$. We assume that $\mathbf{j}_1$ commutes with $\mathbf{j}_2$. The dependence on m is given by (2.80).

The possibility of finding further information stems from the fact that the dependence of $\mathbf{p}$ on m_1 and on m_2 in the $j_1j_2m_1m_2$ scheme is given immediately by the considerations of § 2.8 ((2.80) for m_1; $\mathbf{p}$ is diagonal with respect to j_2 and independent of m_2). We write

$$\langle j_1j_2m_1m_2\,|\,p_z\,|\,j_1j_2m_1m_2\rangle = m_1\langle j_1j_2\,\vdots\,p\,\vdots\,j_1j_2\rangle.$$

Of course $\langle j_1j_2jm\,|\,\mathbf{p}\,|\,j_1'j_2'j'm'\rangle = 0$ unless $j_2 = j_2'$ and either $j_1 = j_1'$ or $j_1 = j_1'\pm 1$. We consider only the first possibility and first let $j = j'$ also.

It follows from (2.80) that

$$\begin{aligned}
j\langle j_1j_2j\,\vdots\,p\,\vdots\,j_1j_2j\rangle &= \langle j_1j_2jj\,|\,p_z\,|\,j_1j_2jj\rangle \\
&= \sum_{m_1,m_1',m_2,m_2'} \langle j_1j_2jj\,|\,j_1j_2m_1m_2\rangle\langle j_1j_2m_1m_2\,|\,p_z\,|\,j_1j_2m_1'm_2'\rangle\langle j_1j_2m_1'm_2'\,|\,j_1j_2jj\rangle \\
&= \langle j_1j_2\,\vdots\,p\,\vdots\,j_1j_2\rangle \sum_{m_1,m_2} m_1\,|\langle j_1j_2jj\,|\,j_1j_2m_1m_2\rangle|^2 \\
&= \langle j_1j_2\,\vdots\,p\,\vdots\,j_1j_2\rangle\Big\{\sum_{m_1,m_2}(j_1+m_1+1)\,|\langle j_1j_2jj\,|\,j_1j_2m_1m_2\rangle|^2-(j_1+1)\Big\}.
\end{aligned}$$

We simplify this further by using Wigner's formula.

$$\begin{aligned}
&\sum_{m_1,m_2}(j_1+m_1+1)\,|\langle j_1j_2jj\,|\,j_1j_2m_1m_2\rangle|^2 \\
&\qquad= \sum_{m_1+m_2=j}\binom{j_2+m_2}{j_1-m_1}\binom{j_1+m_1}{j_2-m_2}\binom{j_1+j_2+j+1}{j_1+j_2-j}^{-1}(j_1+m_1+1) \\
&\qquad= (j_1-j_2+j+1)\binom{j_1+j_2+j+1}{j_1+j_2-j}^{-1}\sum_{m_1+m_2=j}\binom{j_2+m_2}{j_1-m_1}\binom{j_1+m_1+1}{j_2-m_2} \\
&\qquad= \frac{(j_1-j_2+j+1)(j_1+j_2+j+2)}{2(j+1)},
\end{aligned}$$

where we have used (2.44). It follows immediately that

$$\langle j_1j_2j\,\vdots\,p\,\vdots\,j_1j_2j\rangle = \langle j_1j_2\,\vdots\,p\,\vdots\,j_1j_2\rangle\frac{j_1(j_1+1)-j_2(j_2+1)+j(j+1)}{2j(j+1)}. \quad \text{(A 7.1)}$$

In the particular case $\mathbf{p} = \mathbf{j}_1$, $\langle j_1 j_2 \vdots p \vdots j_1 j_2 \rangle = \hbar$, so we have obtained the complete matrix element for $\mathbf{j}_1$.

The dependence on j for $j' = j \pm 1$ can be obtained by a precisely similar procedure. I shall do this in outline only; the reader will be able to fill in the details easily if he wishes. Write

$$P(j) = (j - j_1 + j_2)(j + j_1 + j_2 + 1),$$

$$Q(j) = (j_1 + j_2 - j)(j + j_1 - j_2 + 1),$$

$$R(j) = j(j+1) - j_1(j_1+1) + j_2(j_2+1),$$

$$R'(j) = j(j+1) + j_1(j_1+1) - j_2(j_2+1).$$

Consider $j' = j+1$. Then it is immediate that

$$\langle j_1 j_2 m_1 m_2 \mid j_1 j_2 j+1\, j \rangle = \langle j_1 j_2 m_1 m_2 \mid j_1 j_2 j j \rangle \left(\frac{2j+3}{P(j+1)\,Q(j)} \right)^{\frac{1}{2}} [R(j) - 2m_2(j+1)],$$

and so

$$\sqrt{(2j+1)}\,\langle j_1 j_2 j \vdots p \vdots j_1 j_2 j+1 \rangle = \langle j_1 j_2 j j \mid p_z \mid j_1 j_2 j+1\, j \rangle$$

$$= \langle j_1 j_2 \vdots p \vdots j_1 j_2 \rangle \left(\frac{2j+3}{P(j+1)\,Q(j)} \right)^{\frac{1}{2}} \sum_{m_1, m_2} (m_1 R(j) - 2m_1 m_2 (j+1)) \times |\langle j_1 j_2 j j \mid j_1 j_2 m_1 m_2 \rangle|^2.$$

From our discussion of the case $j = j'$ we have

$$\sum_{m_1, m_2} m_1 |\langle j_1 j_2 j j \mid j_1 j_2 m_1 m_2 \rangle|^2 = \frac{R'(j)}{2(j+1)},$$

and by considering $(j_1 + m_1 + 1)(j_2 + m_2 + 1)$ we get

$$-2(j+1) \sum_{m_1, m_2} m_1 m_2 |\langle j_1 j_2 j j \mid j_1 j_2 m_1 m_2 \rangle|^2 = R'(j)(j_2+1) - \frac{P(j+1)}{2j+3}(R'(j) + j - j_2).$$

With these results and a little straightforward manipulation we derive

$$\sum_{m_1, m_2} (m_1 R(j) - 2m_1 m_2 (j+1))\, |\langle j_1 j_2 j j \mid j_1 j_2 m_1 m_2 \rangle|^2 = \frac{P(j+1)\,Q(j)}{(2j+2)(2j+3)},$$

whence

$$\langle j_1 j_2 j \vdots p \vdots j_1 j_2 j+1 \rangle = \frac{\sqrt{[P(j+1)\,Q(j)]}}{2(j+1)\sqrt{[(2j+1)(2j+3)]}} \langle j_1 j_2 \vdots p \vdots j_1 j_2 \rangle. \quad \text{(A 7.2)}$$

If $\mathbf{p} = \mathbf{j}_1$ then $\langle j_1 j_2 j \vdots p \vdots j_1' j_2' j' \rangle$ is diagonal both with respect to j_1 and j_2 so (A 7.1) and (A 7.2) together with $\langle j_1 j_2 \vdots p \vdots j_1 j_2 \rangle = \hbar$ give us the complete matrix of $\mathbf{j}_1$ in the $j_1 j_2 j m$ scheme. The matrix of $\mathbf{j}_2$ is given by (A 7.1) with j_1 and j_2 interchanged and by (A 7.2) with a minus sign in front (I remark that the product $P(j+1)\,Q(j)$ is unaffected by interchanging j_1 and j_2). This is most easily seen from the relation $\mathbf{j}_1 + \mathbf{j}_2 = \mathbf{j}$.

In ch. 5 we need the matrix elements of L_z in the $SLJM$ scheme and then (A 7.2), together with (2.80), becomes

$$\langle SLJ-1M\,|\,L_z\,|\,SLJM\rangle$$
$$= \langle SLJ-1 \vdots L \vdots SLJ\rangle (J^2-M^2)^{\frac{1}{2}} = -\hbar f(SLJ)\,(J^2-M^2)^{\frac{1}{2}},$$

where $$f(SLJ) = \left\{\frac{[J^2-(L-S)^2]\,[(L+S+1)^2-J^2]}{4J^2(4J^2-1)}\right\}^{\frac{1}{2}}.$$

This is a mere rewriting of (A 7.2) with $j+1 = J$.

An alternative way of deriving these formulae together with the formulae for $j_1 = j_1' \pm 1$ may be found in Condon & Shortley (1953), p. 67. We do not need the case $j_1 = j_1' \pm 1$.

APPENDIX 8

L.C.A.O. THEORY

(See Eyring, Walter & Kimball, 1948)

A8.1. Introduction

From a formal point of view, an attractive method of calculating approximate energies and eigenfunctions for a complex ion is to use the molecular orbital scheme and to express the one-electron functions as linear combinations of the atomic orbitals of the constituent atoms. Then the partly-filled shells may be taken as combinations of the valence orbitals on the metal ion and the ligands. For example, in FeF_6^{3-} these valence orbitals would be $3d$, $4s$, $4p$ on the iron and $2s$, $2p$ on each of the fluorines. This method has not had much quantitative application yet in our field, but it is useful qualitatively as a guide to our physical intuition. I give a very brief summary of some parts of it here.

A8.2. Two identical atoms

Suppose we have two identical atoms at a distance R apart and that they consist of filled shells with just one electron outside. The hydrogen molecular ion H_2^+ would be the simplest example of this; there the filled shells are non-existent. Suppose further that the orbital for the single electron is a linear combination of two real atomic orbitals ϕ_1 and ϕ_2, one from each atom but otherwise identical.

The Hamiltonian for the single electron is

$$\mathscr{H} = -\frac{\hbar^2}{2m}\nabla^2 + f_1 + f_2 + g, \tag{A 8.1}$$

where f_1, f_2 are the potential energies due to filled shells and nucleus of atoms 1 and 2, respectively, and g represents the interaction between the filled shells. g is a function of R alone. Then

$$\left.\begin{aligned} \mathscr{H}\phi_1 &= (f_2+g+\epsilon)\,\phi_1, \\ \mathscr{H}\phi_2 &= (f_1+g+\epsilon)\,\phi_2, \end{aligned}\right\} \tag{A 8.2}$$

where ϵ is the energy of the system when $R = \infty$. Write

$$S = \langle\phi_1\,|\,\phi_2\rangle,$$

$$\alpha = \langle\phi_1\,|\,f_2\,|\,\phi_1\rangle = \langle\phi_2\,|\,f_1\,|\,\phi_2\rangle,$$

$$\beta + \alpha S = \langle\phi_1\,|\,f_1\,|\,\phi_2\rangle = \langle\phi_2\,|\,f_2\,|\,\phi_1\rangle,$$

and (Appendix 3) we have the secular equation

$$\begin{vmatrix} a+g+\epsilon-E, & \beta+S(a+g+\epsilon-E) \\ \beta+S(a+g+\epsilon-E), & a+g+\epsilon-E \end{vmatrix} = 0,$$

$$E = a+g+\epsilon \pm \frac{\beta}{1 \pm S}. \tag{A 8.3}$$

a largely cancels g out (why?) and β is probably always negative in actual compounds so the lowest state of the system has an energy approximately $\beta(1+S)^{-1}$ relative to the energy of the two atoms separated to infinity. We say there is a chemical bond between the two atoms. The one-electron ground-state orbital is $(2+2S)^{-\frac{1}{2}}(\phi_1+\phi_2)$. β is called the resonance integral and, in a fairly real sense, is responsible for the bond.

The above considerations are easily extended to n electrons outside the filled shells, but we must now add to the one-electron energies the energies of electrostatic interaction between the electrons. Formally these can be written down immediately; however, the problem is now considerably more complicated and involves a large number of integrals which must be guessed or calculated.

A 8.3. Polyatomic systems

For a polyatomic system with symmetry group G we classify ϕ_i according to the irreducible representations. Then both S and β are zero unless ϕ_1 and ϕ_2 are corresponding components of bases for the same irreducible representation (§ 8.7). So in case a particular irreducible representation Γ occurs n times among the valence orbitals there will be an $n \times n$ secular equation to determine the one-electron energies. In case $n = 2$ one expects the lower function $c_1\phi_1+c_2\phi_2$ to have c_1c_2 positive and the upper to have c_1c_2 negative.

If we have real atomic orbitals we may classify them by their l_z^2 values along a particular internuclear axis. Working in units of $\hbar$ we say they are $\sigma, \pi, \delta, \phi, \ldots$ orbitals according as l_z^2 is $0, 1, 4, 9, \ldots$. Resonance integrals generally decrease in magnitude in the order $\sigma\sigma$, $\pi\pi$, $\delta\delta$, ...

Let us now discuss the octahedral complex ion FeF_6^{3-}. The valence orbitals on the iron are $3d$, $4s$, $4p$ spanning e_g+t_{2g}, a_{1g}, t_{1u}, respectively, of O. The valence orbitals on the fluorine atoms are the 12 σ-orbitals $2s$ and $2p\sigma$ ($|2p0\rangle$ along an FeF axis) and the 12 π-orbitals $2p\pi$. The σ-orbitals have the same group-theoretic behaviour as two independent sets of six arrows (all identical within one set) outward along the FeF axes and span $a_{1g}+e_g+t_{1u}$. The π-orbitals behave as 6 pairs of arrows at right-angles to each FeF axis, each pair being mutually orthogonal (an arrow changes sign on reversing its direction). They span $t_{1g}+t_{1u}+t_{2g}+t_{2u}$. Comparing FeF_6^{3-} with the units Fe^{3+} and $6F^-$ we have[1] in the latter 5 electrons in e_g+t_{2g} and 48 filling $2a_{1g}+2e_g+t_{1g}+3t_{1u}+t_{2g}+t_{2u}$. In the complex ion we therefore expect (using the fundamental hypothesis, § 7.2) to have $2a_{1g}+2e_g+t_{1g}+3t_{1u}+t_{2g}+t_{2u}$ as the filled shells and e_g+t_{2g} as the partly-filled shells. The representations e_g and t_{2g} occur twice amongst the valence

[1] Omitting the inner cores.

orbitals so in the complex ion the one-electron orbitals of the partly-filled shells are each upper orbitals for the secular equation for their representation. The e_g orbitals are σ (and δ) orbitals along the FeF axes while the t_{2g} orbitals are π (and δ). Therefore e_g should lie above t_{2g} and we should have $\Delta = E(e_g) - E(t_{2g})$ positive. This is the simplest molecular orbital interpretation of the ligand field parameter Δ.

This type of interpretation may be easily extended to other kinds of complex ion. As given above it is, of course, only one part of the origin of Δ but it is a useful qualitative picture to bear in mind.

A8.4. Extension of treatment

Not much work has been done here for transition-metal ions although the general method is clear enough. I content myself with giving some references which may be helpful. In *Trans. Farad. Soc.*: Pople, J. A. (1953), **49**, 1375; Brickstock, A. & Pople, J. A. (1954), **50**, 901; Hush, N. S. & Pople, J. A. (1955), 600. In *J. Chem. Phys.*: Pariser, R. & Parr, R. G. (1953), **21**, 466, 767; Fumi, F. G. & Parr, R. G. (1953), **21**, 1864; Parr, R. G. & Pariser, R. (1955), **23**, 711; Pariser, R. (1956), **24**, 250. In *J. Phys. Soc. Japan*: Tanabe & Sugano (1956), **11**, 864. The last, alone, is concerned with transition-metal ions.

APPENDIX 9

SYMMETRY OF THE COUPLING COEFFICIENTS FOR THE OCTAHEDRAL GROUP

It is possible to redefine the phase choices in the coupling coefficients

$$\langle \Gamma_1 \Gamma_2 M_1 M_2 \mid \Gamma_1 \Gamma_2 \Gamma_3 M_3 \rangle$$

for the one-valued representations of the octahedral group so that the real quantities

$$V\begin{pmatrix} \Gamma_1 \Gamma_2 \Gamma_3 \\ M_1 M_2 M_3 \end{pmatrix} = \lambda_3^{-\frac{1}{2}} \langle \Gamma_1 \Gamma_2 M_1 M_2 \mid \Gamma_1 \Gamma_2 \Gamma_3 M_3 \rangle,$$

with λ_3 the degree of Γ_3, are invariant to all even permutations of their columns and multiplied by $\epsilon = \pm 1$ for odd permutations. $\epsilon = +1$ unless the (non-ordered) trio $\Gamma_1 \Gamma_2 \Gamma_3$ is $A_2 E^2$, T_1^3 or $T_1 T_2^2$, when $\epsilon = -1$. This exhibits an important inherent symmetry of the coupling coefficients and a general scheme of calculation can be founded upon it. However, the latter is at such an early stage that it is unsuitable to describe it here. The interested reader is referred to a series of papers by myself currently being published in *Molecular Physics* and also, for analogous properties of the Wigner coefficients, to Fano & Racah (1959).

BIBLIOGRAPHY

(*Other references are to be found in the footnotes and Appendices.*)

Abragam, A., Horowitz, J. & Pryce, M. H. L. (1955). *Proc. Roy. Soc.* A, **230**, 169.
Abragam, A. & Pryce, M. H. L. (1951*a*). *Proc. Roy. Soc.* A, **205**, 135.
Abragam, A. & Pryce, M. H. L. (1951*b*). *Proc. Roy. Soc.* A, **206**, 164.
Abragam, A. & Pryce, M. H. L. (1951*c*). *Proc. Roy. Soc.* A, **206**, 173.
Aitken, A. C. (1946). *Determinants and Matrices.* Oliver and Boyd.
Araki, G. (1948). *Prog. Theor. Phys.* **3**, 152, 262.
Bacher, R. F. (1933). *Phys. Rev.* **43**, 264.
Bacher, R. F. (1939). *Phys. Rev.* **56**, 385.
Bagguley, D. M. S. & Owen, J. (1957). *Rep. Prog. Phys.* **20**, 304.
Ballhausen, C. J. & Moffitt, W. (1956). *J. Inorg. Nucl. Chem.* **3**, 178.
Belford, R. L., Calvin, M. & Belford, G. (1957). *J. Chem. Phys.* **26**, 1165.
Bethe, H. A. & Salpeter, E. E. (1957). *Quantum Mechanics of One- and Two-electron Atoms.* Springer-Verlag.
Bjerrum, J. & Jørgensen, C. K. (1956). *Rec. Trav. chim. Pays-Bas*, **75**, 658.
Bleaney, B., Bowers, K. D. & Ingram, D. J. E. (1955). *Proc. Roy. Soc.* A, **228**, 147.
Bleaney, B., Bowers, K. D. & Pryce, M. H. L. (1955). *Proc. Roy. Soc.* A, **228**, 166.
Bleaney, B., Bowers, K. D. & Trenam, R. S. (1955). *Proc. Roy. Soc.* A, **228**, 157.
Bleaney, B. & Hayes, W. (1957). *Proc. Phys. Soc.* B, **70**, 626.
Bleaney, B. & Ingram, D. J. E. (1951). *Proc. Roy. Soc.* A, **205**, 336.
Bleaney, B. & O'Brien, M. C. M. (1956). *Proc. Phys. Soc.* B, **69**, 1216.
Bleaney, B. & Stevens, K. W. H. (1953). *Rep. Prog. Phys.* **16**, 108.
Bowers, K. D. & Owen, J. (1955). *Rep. Prog. Phys.* **18**, 304.
Brown, D. A. (1958). *J. Chem. Phys.* **28**, 67.
Calvin, M. & Melchior, N. (1948). *J. Amer. Chem. Soc.* **70**, 3270.
Condon, E. U. & Shortley, G. H. (1953). *The Theory of Atomic Spectra.* Cambridge.
Cooke, A. H. & Duffus, H. J. (1955). *Proc. Phys. Soc.* A, **68**, 33.
Dirac, P. A. M. (1947). *Quantum Mechanics.* Oxford.
Dunitz, J. D. & Orgel, L. E. (1957). *J. Phys. Chem. Solids*, **3**, 20.
Elliott, J. P., Judd, B. R. & Runciman, W. A. (1957). *Proc. Roy. Soc.* A, **240**, 509.
Eyring, H., Walter, J. & Kimball, G. E. (1948). *Quantum Chemistry.* John Wiley.
Fano, U. & Racah, G. (1959). *Irreducible Tensorial Sets.* Academic Press.
Figgis, B. N. (1958). *Nature, Lond.* **182**, 1568.
Figgis, B. N., Lewis, J., Nyholm, R. S. & Peacock, R. D. (1958). *Disc. Farad. Soc.* **26**, 103.
Finkelstein, R. & Van Vleck, J. H. (1940). *J. Chem. Phys.* **8**, 790.
Freeman, R., Murray, G. R. & Richards, R. E. (1957). *Proc. Roy. Soc.* A, **242**, 455.
Garstang, R. H. (1957). *Proc. Camb. Phil. Soc.* **53**, 214.
Gaunt, J. A. (1929). *Phil. Trans.* A, **228**, 151.
George, P. (1956). *Rec. Trav. chim. Pays-Bas*, **75**, 671.
George, P. & McClure, D. S. (1959). *Progress in Inorganic Chemistry*, **1**, 381.
Gibson, J. F., Ingram, D. J. E. & Schonland, D. (1958). *Disc. Farad. Soc.* **26**, 72.
Griffith, J. S. (1956*a*). *J. Inorg. Nucl. Chem.* **2**, 1, 229.
Griffith, J. S. (1956*b*). *Proc. Roy. Soc.* A, **235**, 23.
Griffith, J. S. (1956*c*). *Rec. Trav. chim. Pays-Bas*, **75**, 676.
Griffith, J. S. (1958*a*). *Trans. Farad. Soc.* **54**, 1109.
Griffith, J. S. (1958*b*). *Disc. Farad. Soc.* **26**, 81.
Griffith, J. S. & Orgel, L. E. (1957*a*). *Trans. Farad. Soc.* **53**, 601.
Griffith, J. S. & Orgel, L. E. (1957*b*). *J. Chem. Phys.* **26**, 988.

Griffith, J. S. & Orgel, L. E. (1957*c*). *Quart. Rev. Chem. Soc. Lond.* **11**, 381.
Griffiths, J. H. E., Owen, J. & Ward, I. M. (1953). *Proc. Roy. Soc.* A, **219**, 542.
Hartmann, H. & Schläfer, H. L. (1954). *Z. Angew. Chem.* **66**, 768.
Hartree, D. R. (1955). *Proc. Camb. Phil. Soc.* **51**, 126.
Hartree, D. R. (1956). *J. Opt. Soc. Amer.* **46**, 350.
Hartree, D. R. (1957). *The Calculation of Atomic Structures.* Wiley.
Hayes, W. (1958). *Disc. Farad. Soc.* **26**, 58.
Heine, V. (1957). *Phys. Rev.* **107**, 1002.
Heitler, W. (1954). *The Quantum Theory of Radiation.* Oxford.
Holmes, O. G. & McClure, D. S. (1957). *J. Chem. Phys.* **26**, 1686.
Howard, J. B. (1935). *J. Chem. Phys.* **3**, 813.
Hush, N. S. & Pryce, M. H. L. (1958). *J. Chem. Phys.* **28**, 244.
Ilse, F. E. & Hartmann, H. (1951). *Z. Phys. Chem.* **197**, 239.
Ingram, D. J. E. (1955). *Spectroscopy at Radio and Microwave Frequencies.* Butterworth.
Irving, H. & Williams, R. J. P. (1948). *Nature, Lond.* **162**, 746.
Irving, H. & Williams, R. J. P. (1953). *J. Chem. Soc.* p. 3192.
Jackson, L. C. (1938). *Proc. Phys. Soc.* **50**, 707.
Jackson, L. C. (1959). *Phil. Mag.* **4**, 269.
Jarrett, H. S. (1957). *J. Chem. Phys.* **27**, 1298.
Jordahl, O. M. (1934). *Phys. Rev.* **45**, 87.
Jørgensen, C. K. (1954*a*). *Acta Chem. scand.* **8**, 1495.
Jørgensen, C. K. (1954*b*). *Acta Chem. scand.* **8**, 1502.
Jørgensen, C. K. (1955). *Acta Chem. scand.* **9**, 1362.
Jørgensen, C. K. (1956). *Report to the Xth Solvay Congress, Brussels.*
Jørgensen, C. K. (1957). *Energy Levels.* København.
Jørgensen, C. K. (1958). *Disc. Farad. Soc.* **26**, 110.
Judd, B. R. (1957). *Proc. Roy. Soc.* A, **241**, 122.
Judd, B. R. & Wong, E. (1958). *J. Chem. Phys.* **28**, 1097.
Kamimura, H. (1956). *J. Phys. Soc. Japan,* **11**, 1171.
Kiess, C. C. (1956). *J. Res. Nat. Bur. Stand.* **56**, 167.
Kleiner, W. H. (1952). *J. Chem. Phys.* **20**, 1784.
Klemm, W. & Steinberg, H. (1936). *Z. anorg. Chem.* **227**, 193.
Koide, S. & Pryce, M. H. L. (1958). *Phil. Mag.* **3**, 607.
Koide, S. (1959). *Phil. Mag.* **4**, 243.
Koster, G. F. (1958). *Phys. Rev.* **109**, 227.
Kotani, M. (1949). *J. Phys. Soc. Japan,* **4**, 293.
Lamb, W. E. (1941). *Phys. Rev.* **60**, 817.
Laporte, O. (1942). *Phys. Rev.* **61**, 302.
Laporte, O. & Platt, J. R. (1942). *Phys. Rev.* **62**, 305.
Liehr, A. D. & Ballhausen, C. J. (1957). *Phys. Rev.* **106**, 1161.
Liehr, A. D. & Ballhausen, C. J. (1959). *Molecular Physics,* **2**, 123.
Longuet-Higgins, H. C., Öpik, U., Pryce, M. H. L. & Sack, R. A. (1958). *Proc. Roy. Soc.* A, **244**, 1.
Low, W. (1958). *Ann. N.Y. Acad. Sci.* **72**, 71.
Many, A. (1946). *Phys. Rev.* **70**, 511.
Mayer, M. G. & Jensen, J. H. D. (1955). *Elementary Theory of Nuclear Shell Structure.* Wiley.
Mellor, D. P. & Maley, L. (1947). *Nature, Lond.* **159**, 370; (1948), **161**, 436.
Moffitt, W. (1956). *J. Chem. Phys.* **25**, 1189.
Moffitt, W. & Liehr, A. D. (1957). *Phys. Rev.* **106**, 1195.
Moore, C. E. (1949, 1952, 1958). *Atomic Energy Levels.* National Bureau of Standards, no. 467.
Nyholm, R. S. (1956). *Report to the Xth Solvay Congress, Brussels.*

Ono, K., Koide, S., Sekiyama, H. & Abe, H. (1954). *Phys. Rev.* **96**, 38.
Öpik, U. & Pryce, M. H. L. (1956). *Proc. Roy. Soc.* A, **238**, 425.
Orgel, L. E. and Dunitz, J. D. (1957). *Nature, Lond.*, **179**, 462.
Orgel, L. E. (1952). *J. Chem. Soc.* p. 4756.
Orgel, L. E. (1955*a*). *J. Chem. Phys.* **23**, 1004.
Orgel, L. E. (1955*b*). *J. Chem. Phys.* **23**, 1819.
Orgel, L. E. (1956). *Report to the Xth Solvay Congress, Brussels.*
Orgel, L. E. (1957). *J. Phys. Chem. Solids*, **3**, 50.
Ostrofsky, M. (1934). *Phys. Rev.* **42**, 167.
Owen, J. (1955). *Proc. Roy. Soc.* A, **227**, 183.
Palumbo, D. (1958). *Il Nuovo Cimento*, **8**, 271.
Pauling, L. (1940). *The Nature of the Chemical Bond.* Oxford.
Phillips, M. (1933). *Phys. Rev.* **44**, 644.
Proctor, W. G. & Yu, F. C. (1951). *Phys. Rev.* **81**, 20.
Pryce, M. H. L. (1950*a*). *Phys. Rev.* **80**, 1107.
Pryce, M. H. L. (1950*b*). *Proc. Phys. Soc.* A, **63**, 25.
Pryce, M. H. L. (1959). *Phys. Rev. letters*, **3**, 375.
Pryce, M. H. L. & Runciman, W. A. (1958). *Disc. Farad. Soc.* **26**, 34.
Racah, G. (1942*a*). *Phys. Rev.* **61**, 186.
Racah, G. (1942*b*). *Phys. Rev.* **62**, 438.
Racah, G. (1943). *Phys. Rev.* **63**, 367.
Racah, G. (1949). *Phys. Rev.* **76**, 1352.
Racah, G. (1952). *Phys. Rev.* **85**, 381.
Racah, G. (1954). *Bull. Res. Counc. Israel*, **3**, 290.
Racah, G. (1955). *Acta Univ. Lund*, **50**, no. 21, p. 31.
Rohrlich, F. (1956). *Phys. Rev.* **101**, 69.
Runciman, W. A. (1958). *Rep. Prog. Phys.* **21**, 30.
Rutherford, D. E. (1954). *Vector Methods.* Oliver and Boyd.
Schlapp, R. & Penney, W. G. (1932). *Phys. Rev.* **42**, 666.
Schur, I. (1906I). *S.B. Akad. Berlin*, p. 164.
Selwood, P. W. (1943). *Magnetochemistry.* New York: Interscience.
Shortley, G. H. & Fried, B. (1938). *Phys. Rev.* **54**, 739.
Shukla, S. K. & Lederer, M. (1959). *J. Less-Common Metals*, **1**, 202.
Shull, C. A., Straussen, W. A. & Wollan, E. O. (1951). **83**, 336.
Stevens, K. W. H. (1952). *Proc. Phys. Soc.* A, **65**, 209.
Stevens, K. W. H. (1953). *Proc. Roy. Soc.* A, **219**, 542.
Sugano, S. & Tanabe, Y. (1958). *Disc. Farad. Soc.* **26**, 43.
Sugano, S., Tanabe, Y. & Tsujikawa, I. (1958). *J. Phys. Soc. Japan*, **13**, 880–910.
Tanabe, Y. & Kamimura, H. (1958). *J. Phys. Soc. Japan*, **13**, 394.
Tanabe, Y. & Sugano, S. (1954). *J. Phys. Soc. Japan*, **9**, 753–79.
Tanabe, Y. & Sugano, S. (1956). *J. Phys. Soc. Japan*, **11**, 864.
Tinkham, M. (1955). *Proc. Phys. Soc.* A, **68**, 258.
Tinkham, M. (1956). *Proc. Roy. Soc.* A, **236**, 535–63.
Trees, R. E. (1951*a*). *Phys. Rev.* **82**, 683.
Trees, R. E. (1951*b*). *Phys. Rev.* **83**, 756.
Tsuchida, R. (1938). *Bull. Chem. Soc. Japan*, **13**, 388, 436, 471.
Ufford, C. W. & Shortley, G. H. (1932). *Phys. Rev.* **42**, 167.
Van Santen, J. H. & Van Wieringen, J. S. (1952). *Rec. Trav. chim. Pays=Bas*, **71**, 420
Van Vleck, J. H. (1932*a*). *The Theory of Electric and Magnetic Susceptibilities.* Oxford.
Van Vleck, J. H. (1932*b*). *Phys. Rev.* **41**, 208.
Van Vleck, J. H. (1934). *Phys. Rev.* **45**, 416.
Van Vleck, J. H. (1937). *J. Phys. Chem.* **41**, 67.
Van Vleck, J. H. (1939). *J. Chem. Phys.* **7**, 72.

Van Vleck, J. H. (1958). *Disc. Farad. Soc.* **26**, 96.

Walsh, W. M. & Bloembergen, N. (1957). *Phys. Rev.* **107**, 904.

Wells. A. F. (1956). *The Third Dimension in Chemistry*. Oxford.

Weyl, H. (1946). *Classical Groups*. Princeton.

Weyl, H. (1952). *Symmetry*. Princeton.

White, H. E. & Ritschl, R. (1930). *Phys. Rev.* **35**, 1146.

Whittaker, E. T. & Watson, G. N. (1946). *Modern Analysis*. Cambridge.

Yamada, S., Nakahara, A., Shimura, Y. & Tsuchida, R. (1955). *Bull. Chem. Soc. Japan*, **28**, 222.

Yamatera, H. (1958). *Bull. Chem. Soc. Japan*, **31**, 95.

INDEX OF SYMBOLS

SUBJECT INDEX